Sedimentary Geology

AN INTRODUCTION TO SEDIMENTARY ROCKS AND STRATIGRAPHY

SECOND EDITION

Donald R. Prothero
Occidental College

Fred Schwab
Washington & Lee University

W. H. Freeman and Company
New York

TO OUR WIVES,
TERESA LEVELLE AND
CLAUDIA AARONS SCHWAB,
FOR THEIR AMAZING PATIENCE AND TOLERANCE

Acquisitions Editor: Valerie Raymond
Marketing Manager: Mark Santee
Senior Project Editor: Georgia Lee Hadler
Cover Designer: Blake Logan
Cover Photograph: Dick Dietrich. "Chimney Rock Reef in Captitol"
                          National Park, Utah.
Text Designer: Rae Grant
Illustrations: Fine Line
Illustration Coordinator: Shawn Churchman
Photo Editor: Vikii Wong
Production Coordinator: Paul Rohloff
Composition: Progressive Information Technologies
Printing and Binding: RR Donnelley & Sons Company

Library of Congress Cataloging-in-Publication Data

Prothero, Donald R.
        Sedimentary geology: an introduction to sedimentary rocks and stratigraphy/Donald R.
        Prothero, Fred Schwab.—2nd ed.
        p. cm.
        Includes bibliographical references and index.
            ISBN-13: 978-0-7167-3905-0 (ISBN-10: 0-7167-3905-4)
    I. Sedimentation and deposition. 2. Rocks, Sedimentary. 3. Geology,
Stratigraphic. I. Schwab, F. L. (Frederic L.)
II. Title.
QE571.P772004
552'.5—dc21   2003049091

Printed in the United States of America

Third printing

W. H. Freeman and Company
41 Madison Avenue
New York, NY 10010
Houndmills, Basingstoke RG21 6XS, England

www.whfreeman.com

Contents

Preface

To the Instructor

OVER THE PAST THREE DECADES, WE have introduced many talented undergraduate students to sedimentary geology: in the classroom, in the laboratory, and in the field. The first edition of this book was a direct outgrowth of our earlier experiences, and this second edition builds on the strong success of that earlier edition. This text is written especially for undergraduates and is designed specifically for use in a first course in both sedimentary rocks and stratigraphy. We emphasize general principles that students need to master. We intentionally avoid overwhelming students with details, exceptions, or overly specialized examples. Coverage is deliberately weighted in favor of the varieties of sedimentary rocks such as conglomerate, sandstone, mudrock, limestone, and dolostone that make up 99% of the sedimentary rock column. There is a general summary of aqueous geochemistry because a clear understanding of weathering and chemical sedimentation requires it. Similarly, principles of fluid mechanics are covered so that sedimentary structures, sediment entrainment, and sediment deposition can be adequately understood. Not every detail and nuance of the stratigraphic code is discussed, but a reading of the text will provide students with a good grasp of the relative strengths and weaknesses of various methods of dating and correlation.

We believe that this new edition is a significant improvement over the first edition. That first edition enjoyed remarkable success, perhaps because it so fortunately and correctly targeted the market. Why is this edition better? First of all, a number of users kindly sent us various suggestions about what needed improvement, culling, or expansion. The occasional imprecision was eliminated. We expanded coverage in some areas, such as weathering and soil formation. We tried to do a better job of understanding and interpreting the sedimentary rock record in the context of an Earth that has evolved through time. We added a Glossary partly in recognition of how prolific jargon, terminology, and nomenclature have become—this feature should help students stay on track as they proceed through the various subject areas.

Sedimentary Geology assumes only a single-course background in introductory geology. Additional exposure to historical geology, mineralogy, and petrology is helpful but not crucial. We review or introduce relevant concepts from these fields, as well as from physics, chemistry, and statistics. The level of detail reflects our experience with undergraduate readers and the preferences of instructors. For example, there is little detailed discussion of how rock and mineral components can be discriminated optically. This would require too much space and time, and is probably more adequately presented in published manuals selected by the individual instructor. We recognize that most faculty prefer to design their own laboratories and field trips in order to best capitalize on their own local geology and their personal passions and expertise. We also have not covered to any substantial degree topics like well-logging and subsurface analysis. Undergraduates can better acquire these specialized skills on the job, especially if their understanding of sedimentary geology rests on a strong solid base.

The nucleus for the book is Prothero's 1990 textbook *Interpreting the Stratigraphic Record*. Most of the chapters from that book were substantially modified, updated, and shortened. Schwab added new chapters that emphasized the sedimentary rock record expressly for a comprehensive volume that would cover both stratigraphy and sedimentary rocks. We have worked together harmoniously and diligently in order to blend our writing styles. Style, approach, and pedagogy are, we hope, cohesive and uniform. This second edition of *Sedimentary Geology* builds on the strengths of the first: it is intentionally balanced, yet current. Any success earned by this text deservedly belongs to the many bright, well-motivated students who over the years were never shy about letting us know what works and what doesn't. Finally, we hope this text conveys to the students who read, and we hope, enjoy it, just how fascinating the world of sedimentary rocks can be.

To the Student

We revised this textbook to help you understand the Earth's sedimentary rock record. The book's tone is intentionally conversational and, we hope, reader-friendly. This new edition incorporates a number of suggestions that readers and users of the first edition sent our way. A number of relatively minor errors that appeared in the first edition have been eliminated. We've expanded coverage in a few areas in response to readers' demands.

For example, there is far better coverage of soils and weathering, a bit more emphasis on timely topics such as glacial sedimentation, the role of meteorite impacts on sedimentation, and the long-term secular greenhouse and icehouse states of our evolving planet.

We've also put together a reasonably comprehensive glossary of key terms from the text. Nomenclature and jargon typically get out of hand in any scientific discipline, and a concise but comprehensive glossary seemed the best way to keep the complex terminology of our field in perspective.

In addition, we've put together a list of interesting web sites relevant to the study of sedimentary geology at www.whfreeman.com/sedimentary geology.

We would like to share with you the reasons we became "soft rock" geologists and that compelled us first to write, and then rewrite, this text.

1. *Sedimentary geology is probably the most practical and valuable course in the undergraduate geology curriculum.* We live on a planet whose surface is dominated by sediment and sedimentary rocks. Geologists, regardless of interest or objective, will invariably encounter the Earth's sedimentary shell. One of the ultimate goals of geology is to decipher the terrestrial rock record. While igneous rocks and metamorphic rocks are historical "snapshots," they record only brief, short-lived episodes in Earth's history. It is the sedimentary rock record that acts as an almost continuous movie film of that history. The stratified record provides a rational, almost complete documentary record of our planet's history.

2. *A background in sedimentary geology is essential for most jobs in geology.* Most jobs in geology require some familiarity with the Earth's sedimentary rock record. This was more obvious in the middle to later twentieth century, when the energy business traditionally employed two out of three geologists. That figure has now been reduced to only one out of three geologists,

but it is as true as ever that coal, oil, natural gas, and nuclear fuels are housed in stratified rocks. The newer, rapidly exploding areas of employment in environmental geology are primarily "soft-rock" based. A good third of all geologists today are environmental geologists. They seek water in sedimentary rocks, they're preoccupied with cleaning up air and water pollution, they fight to remediate damaged sites. What areas of specialty knowledge are important to the environmental sciences? Certainly aqueous geochemistry, fluid flow, and a knowledge of the temporal and spatial distribution of stratified rocks, precisely the areas with which sedimentary geologists are most familiar.

3. *Sedimentation and stratigraphy: conciseness, flexibility, and adaptability.* This book comprehensively covers two principal fields of sedimentary geology: sedimentary petrology and stratigraphy. Sedimentary petrology deals primarily with properties of sedimentary rocks (composition, texture, sedimentary structures), their classification, and nomenclature. Stratigraphy defines and describes natural bodies of rock (mainly, but not exclusively sedimentary rocks). Sedimentary petrologists focus particularly on how a rock forms, what it is derived from, and how the material was transported from the source and deposited in a particular setting (such as a delta, alluvial fan, submarine fan). Stratigraphers are obsessed by questions of rock age, fossil content, position in a succession, and correlation in time and space. Curricular modifications of the past decade or so have necessarily trimmed the undergraduate calendar markedly. A full-term separate course in sedimentation, followed by a second full-term course in stratigraphy, are no longer viable options in many cases. While this book can easily serve as a text for such a two-term classical approach, it has been intentionally designed as a solid base for a single course, multi-objective format.

4. *Sedimentary rocks: fascinating, intriguing, and fun!* We authors are the truly lucky ones. We've found a subject area that is both challenging and fun, and this book gives us a marvelous opportunity to share our excitement with you, to tempt you to come along with us and further explore this fascinating area of geology. We are rewarded monetarily for doing something we well might do for free—if we could afford it—because it's so entertaining to us. Untrained observers looking at a ledge of sandstone see simply an ordinary rock. A trained sedimentary geologist, on the other

hand, sees a fascinating glimpse of ancient history. A hungry, carnivorous dinosaur scrambling up the banks of a meandering river formed as periodic flash floods deposited and grains eroded from lofty, granitic mountain peaks 20 kilometers to the east. Likewise, a simple block of limestone in a slab of building stone comes to life in the mind of a carbonate sedimentary geologist, conjuring up the image of an ancient tropical lagoon filled with bizarre, extinct marine plants and animals. And from the bluffs bordering the Grand Canyon, where the casual tourist sees a photogenic stack of colored rock bands, the skilled stratigrapher sees a record of the ancient Earth that presents an intriguing challenge to decipher.

Acknowledgments

We thank Ray Ingersoll, Dewey Moore, Ray Siever, and Don Woodrow for reviewing substantial portions of the manuscript of the first edition. For reviewing the second edition, we thank K. Siân Davies-Vollum, Pomona College; David N. Lumsden, The University of Memphis; J. Fred Read, Virginia Polytechnic Institute and State University; Bruce M. Simonson, Oberlin College; Mark A. Wilson, The College of Wooster. We thank all the reviewers acknowledged in *Interpreting the Stratigraphic Record;* much of what we learned from them influenced the new parts of this book as well as the old. We also thank the many colleagues who are acknowledged in the captions for the generous use of their photographs. Clifford Prothero also helped by printing many of the photographs used in this book.

Fred Schwab's work on this volume honors the three sedimentary geologists who most influenced him professionally. Bob Reynolds of Dartmouth College first introduced him to sedimentary rocks. Bob Dott of the University of Wisconsin showed him how much fun it can be to study them in the field and the classroom. Ray Siever of Harvard University, by example, steered him to a career largely devoted to understanding these fascinating deposits. Schwab also thanks John D. Wilson, President Emeritus of Washington & Lee University, and Ed Spencer, his department chairman for the past three decades, the two colleagues most responsible for nurturing an academic setting in which teaching, research, and writing mutually flourish. He also thanks his four favorite field assistants (and kids), Kimberly, Bryan, Jeffrey, and Jonathan, for continued support and encouragement during these efforts.

Our editor, Valerie Raymond, was a constant inspiration in bringing this project to completion. Many other people at W. H. Freeman and Company have contributed greatly to this book: Georgia Lee Hadler, senior project editor; Diana Siemens, copy editor; Rae Grant, designer; Vikii Wong, photo editor; Shawn Churchman, illustration coordinator; and Paul Rohloff, production coordinator.

I

Sedimentary Processes and Products

Entrenched meanders cut through Permian sediments at Goosenecks of the San Juan River, Utah.
Road in upper left corner shows scale. (Courtesy of Donald L. Baars.)

1 Sedimentary Rocks: An Introduction

The mouth of the Russian River in northern California shows the process of sedimentation in a microcosm. Sediments are eroded from the weathered hills (at right) and are transported down the river into the sea (note the plume of muddy water at the mouth of the river). Once the sediments settle out of the water and are deposited, they can become sedimentary rock. (Shelton, 1966.)

WE SUBSTANTIALLY REVISED THE FIRST EDITION OF THIS book while retaining our original objectives: to help you better understand (1) the processes that erode, transport, and deposit sediments (**sedimentology**); (2) the characteristics and origins of sedimentary rocks (**sedimentary petrology**); and (3) the complex distribution of the sedimentary rock record in space and time (**stratigraphy**). The first two areas are the subjects of Chapters 1 through 14.

The field of stratigraphy is covered in Chapters 15 through 19.

Analysis of sedimentary rocks involves *description* and *interpretation*. Description is straightforward: "What can we see when we examine a sedimentary rock? What characteristics does it exhibit?" Interpretation is more subjective because it requires us to make inferences about the features described. The following case studies illustrate these contrasting approaches.

Sedimentary Rock Description: A Case Study

To describe any igneous, sedimentary, or metamorphic rock, it must be carefully examined in the field at outcrops, as a hand specimen, or by using thin sections and a petrographic microscope. Detailed description allows the distinguishing properties of any rock to be identified and characterized, and it is a necessary first step to understanding the rock's origin. Although the description of sedimentary rock properties is straightforward, it does require a sound understanding of the theoretical factors that control rock features.

Place a hand specimen of sedimentary rock in front of you and examine it as you read this chapter. What physical properties are visible and how can they be characterized?

Obviously, your response will depend on the sedimentary rock selected. Unfortunately, randomly choosing just any sedimentary rock specimen to illustrate the principles of sedimentary rock description might be a wasted exercise. For example, very fine grained, homogeneous rocks such as shale or rock salt reveal few distinguishing features. Describing them is a quick and easy task, but not a

Figure 1.1 Hand sample of a coarse, poorly sorted conglomerate with well-rounded cobble- and pebble-sized clasts. (D. R. Prothero.)

particularly enlightening one. The description of a coarser-grained sedimentary rock such as conglomerate (essentially lithified gravel) reveals much more about the rock's origin.

In the following discussion, we describe a specific conglomerate (Fig. 1.1) that may differ from the sedimentary rock that you have before you. Our reference conglomerate is composed mainly of pebbles of pre-existing rocks and minerals. The technical term for chunks or broken fragments is **clasts** (from the Greek *klastos,* meaning "broken"). Although the term *clast* does not imply a specific size (grain diameter), a standardized clast size scale is used. For example, clasts with maximum diameters of 4 to 64 mm are pebbles. Our conglomerate also contains subordinate amounts of finer clasts with diameters from 2 to 1/16 (or 0.0625) mm; we call these *sand.* By convention, coarser pebbles are collectively lumped as **framework** and the finer sand as **matrix.** A third component, chemical **cement,** glues the sand and pebbles together to form a cohesive rock.

A short list of physical properties can be used to characterize a rock specimen: color, composition, texture, sedimentary structures, fossil content, and geometry or architecture. Table 1.1 summarizes these properties for our conglomerate specimen. Although this table is simplified, it also intentionally includes a few examples of the technical terminology (jargon) that can complicate straightforward scientific description.

Color

Color is easy to describe and is one of the more striking properties of a sedimentary rock. Color usually reflects some aspect of the rock's composition. Bulk color can reflect the color of major mineralogical components. The net color of a conglomerate depends on the kinds of pebbles that compose it; for example, white quartz, pink feldspar, or speckled black and white volcanic rock fragments. The matrix might be a different color. Color can also be controlled by minor constituents such as the cement filling the spaces between pebbles and sand grains. Carbon-rich cements impart a black to dark gray color; iron-rich cements produce a reddish to orange color. Staining or weathering of a rock surface can also produce color changes. Despite these complications, color can be summarized straightforwardly. Color is not treated as an independent property, however, but as an aspect of sedimentary rock composition.

TABLE 1.1 Physical Properties of Sedimentary Rocks (Specifics of a Representative Example; see Fig. 1.1)

Color	>2 mm (pebble framework): White to gray
	$2-\frac{1}{16}$ mm (sand-sized matrix): White to brown to gray
Composition	>2 mm pebble- and cobblesize framework components: 95% quartz, 5% metaquartzite
	$2-\frac{1}{16}$ mm sand-sized matrix: 90% or more monocrystalline quartz
	Cement (trace): Siliceous (chert and chalcedony)
Texture	Type: Clastic (as opposed to crystalline)
	Grain sizes (two distinct groupings):
	A coarser-grained pebble (4–64 mm) framework
	A finer, coarse sand (1–2 mm) matrix
	(Note: The presence of trace amounts of a presumably crystalline cement, not visible in Fig. 1.1, is implied by the cohesiveness of the conglomerate.)
	Variation in clast diameter: Moderately sorted
	Shape: Pebble and sand grains are subequant (an elongation to pebbles)
	Roundness:
	Pebbles: Very well rounded (ultrasmooth corners)
	Sand: Well rounded
	Grain surface textures: 90% of grains are frosted
	Fabric: Weak subparallel alignment of pebble long axes
Sedimentary structures	Thickly bedded; top of bedding surfaces marked by 1-cm-high symmetrical ripple marks; internally cross-bedded (troughs, 6 cm high) and laminated; abundant worm burrows
Fossil content	Scattered, poorly sorted, broken fragments of heavily ribbed, thick-shelled marine brachiopods (Devonian)
Sedimentary rock geometry	Blanket-shaped conglomerate bodies with constant thickness and length-to-width ratios of roughly 1:1 interbedded with laminated and cross-laminated well-sorted quartz arenite

Composition

Although the composition of sedimentary rocks can be described in terms of chemistry or mineralogy, the more conventional method is mineralogical. Why?

First, determining the overall chemical composition of a sedimentary rock (routinely expressed in terms of major oxides) is a complex procedure requiring sophisticated technical equipment. Such procedures are impractical both in the field and for the rapid description of sedimentary rock samples in hand specimen.

More important, describing the composition of a sedimentary rock using bulk chemistry is misleading because it often obscures important genetic distinctions. For example, the chemical composition of a conglomerate composed of pebbles of quartz, a quartz-rich sandy matrix, and silica cement would closely resemble the chemical composition of a different type of sedimentary rock known as bedded chert. (Both would be approximately 99% SiO_2.) Bedded chert consists of interlocking crystals of chalcedony and microcrystalline quartz. Many cherts form when fine-grained siliceous oozes made up of the shells of floating pelagic plankton recrystallize after being buried on the abyssal ocean floor. But quartz-rich gravel and intermixed sand may be deposited by surf and longshore currents along shorelines.

As another example, the chemical composition of a deposit of quartz pebbles cemented with precipitated calcium carbonate might mimic that of a limestone in which quartz sand grains are embedded. Similar chemistries falsely imply identical rocks and similar modes of origin, when important differences exist. For practicality and accuracy, the composition of a sedimentary rock either at an outcrop or as a

hand specimen is described in terms of mineralogy, not chemistry.

Characterizing the composition of a sedimentary rock in terms of the mineralogy (or petrology) of its components is quick and straightforward and provides a clearer insight into the rock's origin. Crude estimates of the relative abundance of major mineralogical components (for example, quartz, feldspar, micas, and rock fragments) can be made visually, especially if individual grains are large and distinct. Pebbles in coarse-grained rocks such as conglomerate can be counted and categorized. All the pebbles in a chalked-off area on the surface of an outcrop may be counted, or all the grains that make contact with a string placed across an exposure may be tabulated. Analyzing the mineralogical composition of finer-grained rocks such as sandstone and limestone requires point-counting of thin-sectioned samples with a petrographic microscope.

Texture

Texture refers to the size, shape, and arrangement of the grains that make up a sedimentary rock.

Texture Types   There are two fundamentally different textural types: *clastic* and *crystalline*. Conglomerates exhibit mainly clastic texture. They contain individual fragments (clasts) of pre-existing rocks and minerals that were transported and deposited as discrete particles. In clastic textures, grain boundaries touch one another tangentially. When grains are interlocked or intergrown, the texture is referred to as crystalline. Crystalline textures result from the in situ precipitation of solid mineral crystals. Most igneous rocks have crystalline textures that formed when magmas cooled and solidified. A single sedimentary rock can exhibit both clastic and crystalline texture. For example, although the coarser framework and finer matrix of conglomerate are clastic, the cement that provides the rock's cohesiveness is a low-temperature, crystalline-textured precipitate.

Grain Size   Clasts or crystals are conventionally categorized by their maximum grain diameter. The diameter can be estimated visually, but accurate measurements require more sophisticated methods. It is often necessary to disaggregate (break apart) consolidated sedimentary rocks and separate grains on the basis of size by passing them through a nest of wire mesh sieves of different sizes. It is also practical to group grain diameters into categories called size classes; for example, boulders, pebbles, cobbles, sand, silt, or clay (see Table 5.1).

Variation in grain size in clastic sedimentary rocks is known as **sorting**. A well-sorted sedimentary rock shows little variation in grain diameter; a poorly sorted sedimentary rock exhibits large deviations from the mean grain size (Fig. 1.2).

Shape and roundness (angularity) are other aspects of texture that are particularly applied to clastic sedimentary rocks.

Shape   Are the clasts equidimensional (*equant*)? Are they disklike sheets or flakes? Are they needlelike (prismatic) or elongate? Shape is often described in terms of **sphericity**. Equant grains (whether they be cubes or spheres) have high sphericity; those with one or more dimensions of unequal length have lower sphericity.

Roundness (Angularity)   The roundness or angularity of grains refers to the sharpness or smoothness of their corners.

Clast shape and roundness can be categorized by using standardized grain silhouettes (Fig. 1.3). For conglomerates, this can be done visually in hand specimen, but the analysis of finer-grained clastic sedimentary rocks requires more complicated

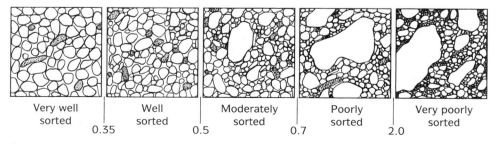

| Very well sorted | Well sorted | Moderately sorted | Poorly sorted | Very poorly sorted |

0.35   0.5   0.7   2.0

Figure 1.2 Standard images for visually estimating sorting. Numbers are sorting (standard deviation) values expressed in phi units that can be calculated using the standard formula shown in Table 5.3. (After Compton, 1962: 214; by permission of John Wiley, New York.)

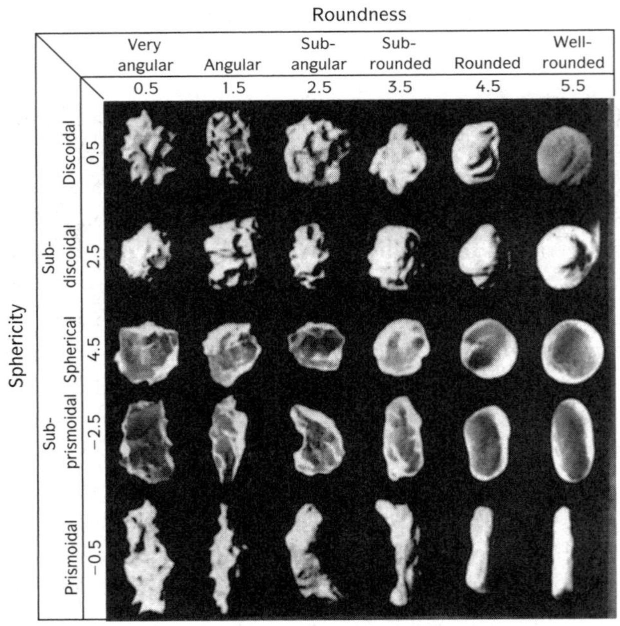

Figure 1.3 Standard images of roundness and sphericity for quantitative estimates of grain shape. (Powers, 1953; by permission of the American Geological Institute.)

analytical methods. The shape and angularity of crystals in crystalline sedimentary rocks are not usually analyzed (with some important exceptions), because they provide little information about rock genesis.

Even though shape and roundness are related (that is, an increase in roundness is usually accompanied by increased sphericity), they can be quite independent properties. An equant grain freshly broken out of a rock can be very angular (see Fig. 1.3), and many natural clasts that are disk-shaped or elongate can be very well rounded.

Other clastic textural features are discussed in more detail in later chapters. These features include (1) sand and coarser clast **grain surface features** such as crescent-shaped pits or a surface gloss or frosting; and (2) **textural fabrics,** specifically the common orientation of such alignable components as the long axes of pebbles or the short axes of flaky mineral fragments.

Fossil Content

Examination of a sedimentary rock often reveals organic remains, either hard parts (fossil shells and bones or their replacements) or such traces of organisms as tracks, trails, and burrows (**ichnofossils**). Fossil content can be characterized not only by the specific type of organic remains found but also by the characteristics of such remains (for example, whole or broken).

Sedimentary Structures

Some features of sedimentary rocks are best studied in outcrop; for example, sedimentary structures. These are large-scale, three-dimensional features of sedimentary rocks (discussed in detail in Chapter 4). The most common sedimentary structure is **stratification,** the bedding or layering exhibited by all sedimentary rocks. Primary stratification is a consequence of deposition of the clasts grain by grain over time and is originally horizontal. A single **sedimentation unit** (a band or layer of sediment of similar composition and texture deposited under consistent, almost identical conditions) can vary in thickness laterally and vertically.

Most other sedimentary structures can be classified in relation to stratification. Ripple marks, raindrop imprints, and mudcracks develop on the top of bedding planes; sole marks include a variety of structures developed on the base of stratification surfaces; cross-bedding and graded bedding occur within individual strata.

Sedimentary Rock Geometry

The geometry of a sedimentary unit is a large-scale feature of sedimentary rocks that may not be obvious in a single field exposure. The geometries or three-dimensional shapes of sedimentary rock bodies range from sheetlike blankets to elongate ribbons or shoestrings. Accurate description of sedimentary rock architecture requires excellent three-dimensional exposure to permit measurement of lateral variations in internal organization, thickness, and extent.

Summary

These properties, though not comprehensive, do include the essential descriptive aspects of sedimentary rocks. The properties are straightforward and can be recognized and described by a geologist with a trained eye using relatively unsophisticated analytical tools. With practice, the descriptive process becomes routine. No matter how unexciting, rock description is the essential first step in the analysis and interpretation of a sedimentary rock. Even by itself, description is useful because it sets out the criteria by

which one sedimentary rock can be distinguished from another, a prerequisite for accurate classification, interpretation, and mapping.

Finally, the challenging task of description is simply the means to a more ambitious end: using these descriptive characteristics to *understand* and *interpret* the origin and evolution of sedimentary rocks.

Sedimentary Rock Interpretation: A Case Study

What do we really want to know about any sedimentary rock? What information can be inferred from each of the physical characteristics just discussed? We seek the answers to rather simple questions.

1. When was the sedimentary rock unit deposited, and over how broad a region?

2. With what other rock units is the sedimentary rock contemporaneous?

3. From what kinds of source rocks were the sediments derived?

4. Where was that source located? Was it near or far from the depositional site, and in what direction?

5. Was the source a mountainous highland or an area of low relief?

6. How was the material transported to the depositional site from the area where it was weathered and eroded? Was it blown by the wind, bounced along the channel of a flowing river, moved by the surf and longshore currents, or carried by a sheet of slow-moving glacial ice?

7. In what kind of physical setting did the sedimentary rock form? Was it deposited by an ancient river delta system? Is it a lithified desert dune complex?

8. How have the color, composition, texture, and other physical properties of the sedimentary rock been modified in the time since deposition?

Answering these questions helps us understand the genesis of a sedimentary rock. Answers to such questions are formally embodied as *stratigraphy, provenance, dispersal, transporting agent, depositional setting, paleogeography, sedimentary tectonics,* and *diagenesis.* Figure 1.4 summarizes these interpretive aspects for a typical sedimentary rock.

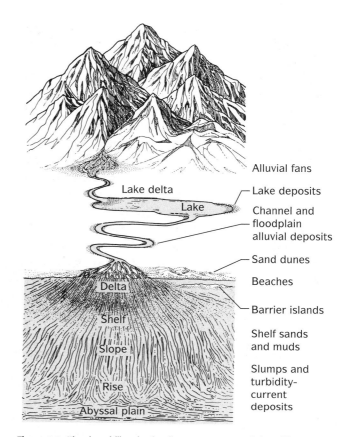

Figure 1.4 The downhill path of sediment transport and deposition. From weathered particles freshly eroded from the mountains to the depths of the ocean, various sedimentary environments are encountered. (After Press and Siever, 1986: 301; by permission of W. H. Freeman and Company, New York.)

Stratigraphy

Stratigraphy studies the distribution of stratified rocks through time and space. If an ultimate goal of sedimentary geology is to reconstruct the history of the Earth at various stages of its development, it is imperative to determine how widely scattered sections of the sedimentary rock record are temporally related (*correlation*). Very precise reconstructions of sediment source and depositional setting are of questionable value if they cannot be correlated correctly and tied to the standard geologic time scale. A precise description of the stratigraphy of a sedimentary unit includes its stratigraphic age (for example, Early Cambrian), and a detailed knowledge of the units with which it is correlated.

Provenance

The term **provenance** is derived from the French verb *provenir*, "to come forth," and it refers to all

aspects of the sources from which a particular sedimentary rock is derived. Of particular concern are the source area's composition, location, and topographic relief. What kinds of rocks made up the source area? Were they igneous, metamorphic, or sedimentary rocks—or perhaps a mixture of all three? Can the source areas be identified more specifically as granite or basalt, schist or marble, limestone or dolomite? Location includes both the distance to the source and the direction of the source. Source relief refers to whether the source was high (mountainous) or low. A precise description of provenance is the following statement: "This sedimentary rock was derived from a mountainous andesitic volcanic island arc located 100 km east of the depositional area."

Dispersal

Dispersal describes the pattern by which material is eroded from the source, transported away, and deposited elsewhere. The pattern of sediment transport and redistribution must be described precisely in terms of time, space, and mechanism. Such description requires a thorough understanding of both the transporting agent and the depositional setting. A paleocurrent dispersal system might be described as follows: "This sedimentary unit was transported from west to east down a series of subparallel submarine canyons etched into a broad, north-south trending continental shelf 200 km wide."

Transporting Agent and Depositional Setting

These two aspects of a sedimentary rock are interrelated. The **transporting agent** is the mechanism responsible for moving a sedimentary component from where it was produced by weathering to where it was finally deposited. Common transporting agents are river (fluvial) systems, submarine turbidity flows, slow-moving continental glacial ice sheets, and episodic windstorms. **Depositional setting** or environment can be described piecemeal in terms of such factors as salinity, water depth, water temperature, and current flow velocity. It is better summarized in terms of a geomorphic unit (that is, a three-dimensional landform) such as a delta, a tidal flat, or a terminal glacial moraine. This description provides us with a mental picture that conveys what

that portion of the Earth looked like at the time and place pinpointed by the particular sedimentary rock in question.

Paleogeography and Sedimentary Tectonics

It is possible to reconstruct the **paleogeography** of a region—the areal distribution through time of the physical geography—by integrating inferences about provenance, dispersal, transporting agent, and depositional setting. Paleogeographic maps can be broadly based global or regional reconstructions showing the gross distribution of continental blocks, ocean basins, mountain belts, and continental lowlands. With finer focus, paleogeographic maps show mountain fronts, alluvial fan complexes, glaciated mountain valleys, and so on. The historical evolution of a region, regardless of scale, can be summarized as a series of sequential paleogeographic maps. It is crucial to understand **sedimentary tectonics**—the dynamic context in which a sedimentary rock is deposited (for example, the area along a continental margin bordering an active convergent plate margin). Sedimentary rocks are arguably the most useful tools for reconstructing ancient plates and plate margins.

Diagenesis

Diagenesis is a comprehensive term for all changes (short of metamorphism) in texture, composition, and other physical properties that occur in a sedimentary rock after it is deposited as a sediment up until the time it is examined. Diagenetic processes of compaction, recrystallization, and cementation play a crucial role in converting sediment to sedimentary rock. They can also alter or obscure the original sedimentary rock texture, composition, color, and sedimentary structures, thus making it impossible to know what such properties were like originally. For example, a sandstone originally composed of roughly equivalent proportions of limestone and volcanic rock clasts might be altered long after deposition by acid groundwater percolating through the rock. Selective dissolution of all limestone clasts could produce a diagenetically altered end product containing no trace of a carbonate source rock.

Sedimentary Geology: Goals

Sedimentary geologists interpret provenance, dispersal, transporting agent, depositional setting, paleogeography, sedimentary tectonics, and diagenetic history. These interpretations are based on such properties as mineralogy and texture, which can be linked to these factors. It is crucial to understand which inferences can be based on texture, which on sedimentary structures, and so on. For example, fossil content helps fix the age of a sedimentary rock but implies nothing about source relief. Crossbedding and ripple marks provide information about dispersal, depositional setting, and paleogeography, but yield no information about rock age. Table 1.2 summarizes these various interpretive aspects of sedimentary rocks, linking each with the descriptive properties from which they are typically inferred. The discussions of major rock categories (Chapters 5, 6, 11, 13, and 14) include the physical characteristics of each and detail the linkage between their descriptive properties and their genesis.

Sediments and Sedimentary Rocks: Major Categories

Many sedimentary rocks are produced by a cycle of weathering, transport, deposition, and diagenesis of sediment (Fig. 1.5), each stage of which puts its own imprint on the sediment. (Particles of sediment can also originate from explosive volcanism or by organic precipitation.) Weathering is the destructive breakdown of pre-existing igneous, metamorphic, and sedimentary rocks by physical disintegration and chemical decomposition. It occurs at or near the

TABLE 1.2 Sedimentary Rock Genesis: The Relevant Database

Interpretive Property	Best Indicators
Stratigraphy: Unit age, distribution, and correlative rock units	Fossil content
Provenance:	
Source area composition	Sedimentary rock composition
Source area location	Primary directional structures
	Regional variations in texture and thickness
Source area relief	Sedimentary rock composition
	Texture
	Geometry
Dispersal	Primary directional structures
	Texture
	Geometry
Transporting agent and depositional setting	Texture
	Sedimentary structures
	Geometry
	Fossil content
Paleogeography and sedimentary tectonics	Stratigraphy
	Provenance
	Dispersal
	Transporting agent
	Depositional setting
Diagenesis	Composition
	Texture
	Sedimentary structures

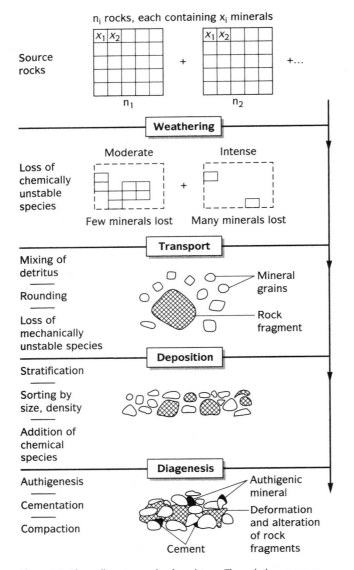

Figure 1.5 The sedimentary cycle of sandstone. Through the processes of weathering, transport, deposition, and diagenesis, the weathered fragments of pre-existing rocks are turned into sedimentary rock. (After Pettijohn, Potter, and Siever, 1987: 26; by permission of Springer-Verlag.)

Earth's surface, where pre-existing rocks at relatively low temperatures and pressures come into contact with water (the hydrosphere), living organisms (the biosphere), and the atmosphere. Weathering generates a variety of products: soil, disaggregated rock debris, and constituents dissolved in groundwater and runoff. The removal of weathering products from the weathering site constitutes **erosion. Transportation** is the movement of weathering products (either as discrete fragments of pre-existing material or as components dissolved in water) from the sites where they are produced to the sites where they

accumulate. When transportation ends, **deposition** of sediment begins. Sedimentary rocks are produced by burial, compaction, recrystallization, and cementation (collectively, **lithification,** the making of sedimentary rocks from sediment).

Because weathering products are transported, deposited, and ultimately transformed into sediment and sedimentary rocks in distinctive ways, three broad categories of sedimentary rock are recognized: detrital, biogenic, and chemical (Table 1.3). A fourth category, "other sedimentary rocks," accommodates sedimentary rock types generated by processes other than weathering.

Nature is seldom as neat and tidy as the classification schemes scientists devise to organize natural phenomena. Many sedimentary rock types straddle these boundaries and do not fall into a single pigeonhole. The biochemical category, for example, includes two varieties of rock called sedimentary ironstone and iron formation. Neither is exclusively produced by organisms; some varieties are produced by inorganic chemical processes. Bearing in mind that classification is artificial, what are the salient characteristics of each major grouping?

Detrital sedimentary rocks consist mainly of clasts of pre-existing rocks and minerals that have been physically transported and deposited as discrete fragments. Because they represent material eroded from the Earth, they are also termed **terrigenous.** They are produced when transporting agents such as running water or the blowing wind pick up (entrain) soil and loose rock debris and transport the clasts away from the weathering site, eventually depositing them as layers of gravel, sand, or mud. Terrigenous material can also be subdivided into (1) **extrabasinal** grains weathered and eroded from sources outside the depositional basin into which they were later deposited; and (2) **intrabasinal** grains eroded and reworked from essentially contemporaneous materials deposited within the same basin. Terrigenous material consisting mainly of bits and pieces of quartz and other silica-rich minerals is called **siliciclastic.** Consolidation (lithification) of detrital sediment (most of which is terrigenous siliciclastic material) by compaction and cementation transforms it into sedimentary rock. Individual subcategories are distinguished based on predominant clast size (particle diameter). In conglomerates, a large proportion of fragment diameters exceed 2 mm. In sandstone, most clasts range from 2 to 1/16 (or 0.0625) mm. Mudrock consists of clasts with diameters of less than 1/16 mm.

TABLE 1.3 Principal Sedimentary Rock Categories

1. Siliciclastic (epiclastic, detrital, or terrigenous) sedimentary rocks
Genesis: The physical disintegration and chemical decomposition of pre-existing rocks generates fragments of rocks and minerals (termed clasts, regardless of size). Clasts are picked up and transported as discrete particles by moving bodies of water, wind, and ice as well as by various kinds of gravity-induced movements. Deposition occurs on a particle-by-particle basis as transporting agents slow, stop, or melt.

Siliciclastic sedimentary rocks are further subdivided on the basis of principal clast size into conglomerate and breccia, sandstone, and various types of mudrock (claystone, siltstone, mudstone, and shale).

2. Biogenic, biochemical, or organic sedimentary rocks
Genesis: Physical disintegration and chemical decomposition of pre-existing rocks and minerals generate chemical components dissolved in runoff and groundwater. This dissolved material is transported into standing bodies of water (playas, lakes, the ocean), where it is extracted directly or indirectly by organisms and precipitated as solid, crystalline minerals.

Biogenic, biochemical, or organic sedimentary rocks are further subdivided on the basis of their principal chemical component. Major categories are carbonate rock (limestone and dolostone), chert (silicate rock), and phosphate rock.

3. Chemical sedimentary rocks
Genesis: Physical disintegration and chemical decomposition of pre-existing rocks generate chemical components dissolved in runoff and groundwater. This dissolved material is transported into standing bodies of water (playas, lakes, the ocean), where it is precipitated largely by purely chemical, inorganic processes.

The principal purely chemical sedimentary rock type is evaporite. Precambrian banded iron formations and Phanerozoic ironstones can be tentatively placed here, although they can probably just as easily be categorized as biogenic, biochemical, or organic sedimentary rocks.

4. Other sedimentary rocks
Genesis: This category includes all clastic sedimentary rocks that are produced by processes other than the physical and chemical weathering of pre-existing rocks.

The major varieties of this group of sedimentary rocks are subdivided on the basis of the mechanism by which the clasts are produced. Principal types include pyroclastics (generated by explosive igneous activity), meteoritics (produced by the impact of extraterrestrial bodies), and cataclastics (related to collapse or tectonism).

Biogenic, biochemical, or **organic sedimentary rocks,** as the names imply, are formed by organic activity. Although organisms play the predominant role in creating these rocks, they do so in a variety of ways. In some cases, organic metabolism simply modifies the chemical environment and sediment is precipitated directly from saturated fresh or ocean water. Alternatively, organisms precipitate carbonate, phosphate, and silicate minerals to manufacture their shelly and bony components. Plants concentrate carbonaceous material directly by photosynthesis as they manufacture plant tissue, which becomes the raw material of coal. Other types of carbonaceous sedimentary deposits such as oil shale (kerogen) and petroleum (mainly methane and paraffins) are generated by bacterial activity.

Complicating the premise that this category is completely distinct from clastic sedimentary rocks is the fact that after biochemically and organically precipitated minerals are formed, they typically experience a brief period as physically transported intrabasinal grains (allochems and intraclasts) before becoming consolidated sedimentary rocks.

Subdivisions of this category are distinguished on the basis of overall sedimentary rock chemistry. The three main varieties are (1) carbonate rocks (limestone and dolomite), (2) silicate rocks (chert), and (3) phosphate rocks.

Chemical sedimentary rocks result from the nonorganic, physicochemical precipitation of solid crystalline minerals as the dissolved constituents reach saturation. The best examples of chemical sedimentary

rocks are layered evaporite sequences composed of anhydrite, gypsum, and halite (rock salt) that grow from brines that develop as seawater evaporates. Precambrian iron formations and Phanerozoic ironstones also fall within this category, although such deposits may also be produced by organisms.

Other sedimentary rock types consist of various sedimentary rocks whose origin is unrelated to weathering. Nevertheless, most share the clastic (fragmental) textures and variations in clast size that characterize sediments generated by weathering. **Pyroclastic** sedimentary rocks, like volcaniclastic terrigenous sedimentary rocks weathered from volcanic source rocks, contain abundant volcanic glass shards and rock fragments but are produced by explosive volcanism rather than weathering. **Meteoritic** clastics are created when extraterrrestrial bodies impact the crust at high velocity. **Cataclastic** sedimentary rocks are coarse-grained, angular sedimentary deposits of restricted extent. Some are produced by gravity-driven movements such as landslides. Others form when the roofs of caverns (produced by groundwater solution) collapse, creating collapse (solution) breccias. Still others are generated along fault and fracture zones and where sedimentary rocks crumple and fragment as they are folded.

The Earth's Sedimentary Shell
Thickness, Volume, and Distribution

Most of the solid Earth consists of igneous and metamorphic rocks (typically 90% to 95% of the upper 16 km or 10 miles of the crust). Table 1.4 and Fig. 1.6 summarize the extent of exposure, the volume, and the mass of Earth's sediment and sedimentary rock shell.

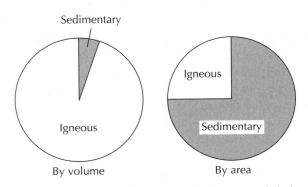

Figure 1.6 Relative abundance of igneous and sedimentary rocks in the crust of the Earth. (Data from Clarke, 1924: 34.)

Most of the surface of the solid Earth is either sediment or sedimentary rock! Ignoring surface soil cover (the raw material of sediment), almost the entire surface of the Earth (approximately 90%) is covered with sediments or sedimentary rocks. Roughly three-fourths of Earth's *land area* (30% of the surface area) is mantled with a relatively thin (thicker in mountain systems) veneer of sediments, sedimentary rocks, and metasedimentary rocks. Across the three-fourths of Earth's surface that is under water, sediments are almost everywhere. Most of the continental shelf, slope, rise, and deeper ocean basins are mantled with sediment. Igneous rocks are restricted to the crest of the mid-ocean ridge-rise system and marginal volcanic arcs. Any extraterrestrial field geologist landing on Earth for an initial survey would need this textbook, rather than texts dealing with igneous or metamorphic rocks.

This extensive surface exposure stands in sharp contrast to the small *volume* of sediments and sedimentary rocks compared with the volume of igneous and high-grade metamorphic rocks. Adjusting for topographic differences within the crust, only about 5% to 10% of the volume of the outermost, 16-km-thick, solid terrestrial shell is sediment, sedimentary rock, and metasedimentary rock. Although this value can be estimated in several ways, all suggest that if Earth's sediment and sedimentary rock shell were spread around the globe as a uniformly thick carpet, its thickness would total only 800 to 2000 m.

The overwhelming areal exposure of sedimentary rocks in contrast with their limited volume underscores the position of Earth's sediment and sedimentary rock shell as an outer envelope of solid material concentrated at or near the interface of the hydrosphere, biosphere, atmosphere, and lithosphere. This is the logical consequence of the rock cycle (or system), the process by which the different rock types are transformed into one another. Weathering—the combination of physical and chemical processes that produces the raw material of sediment—and erosion—the process that leads to sediment transportation and deposition—can occur only near Earth's surface. And any sediments or sedimentary rocks carried below the shallow, near-surface zone of weathering and erosion will become metamorphosed or melted by the increased temperature and pressure at depth, which effectively removes them from the sedimentary rock inventory.

How are these data on sedimentary shell area, thickness, and volume determined?

TABLE 1.4 Various Estimates of the Relative Distribution of Sedimentary Rocks

Relative abundance of sedimentary rocks (Pettijohn, 1975)
Based on surface area exposure
Areas above sea level 75%
Areas below sea level 25%
Based on volume (as a percentage of upper 16 km of the solid Earth)
Sedimentary rocks 5%
Igneous and metamorphic rocks 95%

Sediment–sedimentary rock shell compared to the overall crust and mantle
(Ronov and Yaroshevsky, 1969)

	Volume (millions of km³)	Mean Global Thickness (m)
Sedimentary rock shell, direct measurement	400–1,000	800–1,000
Sedimentary rock shell, chemical methods	1,000–2,000	1,000–2,000
Continental crust	62,100	40,000–70,000
Oceanic crust	26,600	6,000
Total crust	89,700	6,000–70,000
Total mantle	898,000	2,900,000

Thickness and volume of the sedimentary rock shell in various tectonic provinces

	Volume (millions of km³)	Mean Thickness (m)
Poldervaart (1955)		
Deep oceanic regions (trenches and abyssal plain)	80.4	300
Cratonic platforms	52.5	500
Young (post-Precambrian) fold belts	126.0	5,000
Suboceanic regions (continental shelves)	372.0	4,000
Total volume	630.9	
Ronov and Yaroshevsky (1969)		
Oceanic regions	120	400
Subcontinental crust (shelf and rise areas)	190	2,900
Cratonic platforms	110	1,800
Phanerozoic fold belts	390	1,800
Total volume	940[a]	10,000

[a]Includes 10% volcanics.

Direct Measurement Areal exposures of sediments and of igneous, metamorphic, and sedimentary rocks are directly measured from maps showing bedrock geology and surficial geology of the continental blocks and ocean basins. Subsurface maps compile thickness data derived directly from drill holes or indirectly from reflection and refraction seismology. Table 1.4 compares variations in overall thickness of the sediment–sedimentary rock package in different regions; for example, continental cratons versus continental shelves. The total sedimentary rock volume, determined by combining estimates of surface area with the thickness of the sediment–sedimentary rock column, yields a globe-girdling sedimentary rock shell between 800 and 1000 m thick!

Geochemical Measurement The total volume of the sedimentary rock package can also be determined by comparing the chemistry of seawater, sedimentary rocks, and Earth's crust (Clarke, 1924). The procedure

is crudely analogous to determining how much of a solid powdered dye must be dissolved in a beaker of originally clear water of known volume to bring it to a measurable tint.

Estimates based on chemistry require two pieces of data: (1) the overall abundance of a specific component (typically sodium, potassium, or calcium) dissolved in seawater (stated as a percentage); and (2) the percentage abundance of that constituent in the source rocks from which it is ultimately derived (via chemical weathering).

For example, if the present volume of seawater were spread about the Earth, it would generate a global sea roughly 1 km deep. Seawater contains 1.14% sodium. The average granite contains 2.83% sodium. If one assumes a hypothetical granitic continent 16 km (10 miles) thick as the source of sodium dissolved in seawater, the resulting ratio ($1 \text{ km} \times 1.14\%/2.83\% \times 16 \text{ km} = 1.14/45.28$) suggests that the total sodium contained in seawater requires the dissolution of only about 1/35 (45.28/1.14) of the 16-km-thick granitic continent, or 450 to 500 m—by inference equal in thickness to the total terrestrial sedimentary shell. This figure must be increased by a factor of one-third (to roughly 750 or 800 m), because sedimentary rocks typically contain 0.9% sodium (in other words, only two out of every three sodium ions remain in seawater).

Various independent studies, using a wide variety of dissolved components and different kinds of potential source rocks, yield other results. The thickness of a globe-girdling sedimentary shell, however, straddles a narrow range from 800 to 2000 m.

The fact that these contrasting methods yield totals that differ only by as much as a factor of roughly two suggests that the figure is accurate. The higher total from the chemical approach reflects the recent conclusion by some geologists that considerably more weathered sodium (or potassium, or calcium) exists and either is incorporated into undiscovered (or destroyed) evaporite deposits or is locked up in conventional sandstone and mudrock.

The Role of Weathering, Sedimentation, and Recycling in Continental Crustal Evolution

Weathering, erosion, and sedimentation may have played significant roles in the conversion of primitive Archean basic or ultrabasic volcanic arcs and protocontinents into the granitic continents that

exist today. For example, Taylor and McClennan (1985) argue that weathering and sedimentation promote chemical differentiation in a variety of ways. Sedimentary rock sequences rich in continental crustal components—such as silica (SiO_2) in quartz-rich sandstone and potassium (K_2O) in feldspar-rich alluvial fans filling continental rift valleys—are seldom returned to Earth's interior via subduction, because they rarely accumulate in trenches and on the abyssal seafloor. Instead, they tend to be permanently incorporated into continental blocks, enriching the crust in SiO_2 and K_2O.

How can chemical differentiation via weathering and selective fractionation of components during transport and deposition be effective in light of the almost trivial volume of the sedimentary rock shell compared to that of the overall crust and mantle? As Table 1.4 shows, a global sedimentary rock shell 1000 to 2000 m thick (a volume of 500 to 1000 million km^3) (Pettijohn, 1975) is only about 1/90 of the total crustal volume (90,000 km^3) and as little as 1/1000 of the volume of the mantle!

Surprisingly, the total volume (or mass) of sediment generated over time might greatly exceed the volume or mass of Earth's existing sedimentary rock shell! If, as many geologists contend, much of Earth's sedimentary rock shell has been recycled—that is, repeatedly weathered, eroded, retransported, and redeposited—the total volume of sediment might greatly exceed the present sedimentary rock volume. For example, 100 km^3 of sediment eroded from a block of granitic crust—deformed, uplifted, weathered, eroded, and redeposited as a second, entirely recycled 100 km^3 of sediment—represents a total volume of sedimentation of 200 km^3, double the actual volume of sediment or sedimentary rock that survives.

Considerable data exist to support this idea that the Earth's sedimentary shell is continually recycled.

1. Dissolved sodium presently enters the sea at a rate that is 40 to 50 times that necessary to account for all sodium now dissolved in seawater. This implies much faster rates of weathering and sedimentation than most geochemists concede (Gregor, 1968).

2. Detailed studies of the provenance of modern sediments in the coterminous United States show that most (79%) are weathered and eroded from sedimentary rocks rather than from igneous or metamorphic rocks (Gilluly, Reed, and Cady, 1970).

3. Detailed analysis of basement rocks in the coterminous United States reveals that large volumes of Phanerozoic crystalline rocks, perhaps as much as 60 million km³, are recrystallized from sedimentary rock sequences (Gilully, Reed, and Cady, 1970). If the Phanerozoic crystalline continental crust of the United States (1/20 of the global continental surface; the Phanerozoic is about 1/6 of geologic time) is typical, the total volume of global sedimentation must greatly exceed 1000 to 2000 million km³.

4. The uneven volume of preserved sedimentary rock sequences as a function of age further supports massive recycling. If Earth's sedimentary shell results from the progressive accumulation of sediment over time, the volume of sedimentary rocks should vary systematically by age. For example, the volume of rocks of Precambrian age—from 4000 Ma (Megennia, or million years before present) to roughly 543 Ma—should greatly exceed that of rocks of Phanerozoic age (543 Ma to the present). Precisely the opposite is true (Garrels and Mackenzie, 1971). Even taking into account the probability that older sequences are less well exposed because they are covered or masked by younger sequences, only a fraction of very old sedimentary rocks remains (Fig. 1.7). Estimates show that the volume of Mesozoic and Cenozoic sequences (a 250-million-year time interval) roughly matches the volume of Paleozoic sequences (a 250- to 300-million-year time interval). The entire volume of sedimentary rocks of Precambrian age (representing a time interval roughly 10 times that of the Paleozoic) approaches the volume of all Phanerozoic (Paleozoic, Mesozoic, and Cenozoic) sequences.

Based on all these factors, many geologists argue that the volume of sedimentary rocks produced over geologic time exceeds the volume of Earth's sedimentary rock shell by a factor ranging from at least 5 or 6 to as much as 40 or 50 (Gilully, Reed, and Cady, 1970; Garrels and Mackenzie, 1971). At the very least, we must accept sediment recycling as an important phenomenon and concede the possibly significant roles of weathering and sedimentation as mechanisms for chemical differentiation of the crust from the mantle.

What Rock Types are Most Abundant?

Now that we have an idea of the total volume of the Earth's sedimentary shell, it would be interesting to know what proportion of that shell is siliciclastic sedimentary rocks, what proportion is biogenic, what proportion is chemical, and so on. There are several ways to arrive at an answer.

A count of the sedimentary rock specimens found in the reference sample collection of a typical sedimentary petrology laboratory would reveal that a large number of sedimentary rock types exist. A cursory examination of sedimentary rock exposures closest to that laboratory, however, would demonstrate that only a few sedimentary rock types are common. In fact, careful surveys show that the three most abundant sedimentary rock types, collectively totaling 90% to 95% of Earth's sedimentary shell, are carbonate (limestone and dolostone), sandstone, and mudrock (Table 1.5). Where do these estimates come from? What is their significance?

Estimating the relative abundance of major sedimentary rock types is done in two ways: (1) by direct observation and measurement of rock types visible in exposed sections; and (2) by numerically juggling geochemical data comparing the composition of major sedimentary rock types with the composition of their likely ultimate source rocks.

Stratigraphers studying sedimentary rock sequences on a centimeter-by-centimeter basis traditionally assign exposed rocks to a single rock type. Local exceptions aside, such rock types as banded iron formations and bedded evaporites are uncommon. Conversely, limestone and the various kinds of

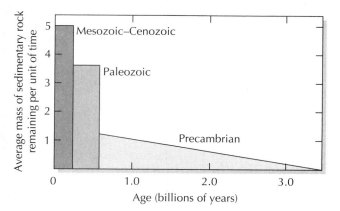

Figure 1.7 Relative existing volume of sedimentary rocks of differing ages plotted as blocks whose areas are proportional to rock volume. The comparatively small volume of Precambrian rocks (one-third of the total volume; 85% of geological time) supports the conclusion that the sedimentary rock column is repeatedly recycled. (After Garrels and Mackenzie, 1971: 258.)

TABLE 1.5 Relative Abundance of the Major Sedimentary Rock Types

	Percentage		
	Carbonate	Sandstone	Mudrock
Direct measurement methods[a]	22	31	47
Geochemical methods[b]	8	13	79
Average of both methods	15	22	63

[a]Pettijohn, 1975: 20.
[b]Pettijohn, 1975: 21.

mudrock (shale, claystone, mudstone, and siltstone) are found almost everywhere. Table 1.5 lists the overall relative abundance of the three major rock types based on field description. The figures shown are an average of four compilations listed by Pettijohn (1975).

Table 1.5 also shows the mean percentage of major sedimentary rock types based on geochemical calculations (again, an average of four calculations that appear in Pettijohn, 1975). With this procedure, the average chemical compositions of mudrock, sandstone, and carbonate are numerically weighted so that their proportions generate a chemical composition identical either to granite or to some other igneous rock thought to be a reasonable ultimate source of sedimentary rocks.

The results from these two methods are intriguing. Both pinpoint the same three rock types and rank them in the same order. Which procedure is closer to the truth, and how can significant differences in detail between the two be explained? The values obtained using weighting of chemical compositions are probably more accurate. Rock types differ drastically in terms of their potential for preservation in the stratigraphic record. Most stratigraphers measure stratigraphic sections within continental platform areas or in the less deformed, less metamorphosed portions of mobile belts, which biases these estimates in favor of shallow-water sedimentary sequences such as quartz-rich sandstone, limestone, and dolostone. Finer-grained detrital mud is typically winnowed from these areas and transported to deeper-water continental rise areas or onto the abyssal floor of the ocean basins. These areas are intensely deformed and metamorphosed and are consequently ignored by most stratigra-

phers. Traditionally, stratigraphers also ignore the significant amounts of mud in limestone, dolostone, and sandstone.

In any case, even though the final figures are not absolutely accurate, they serve a useful purpose. The amount of coverage by individual rock type in this book bears little relationship to the overall abundance of the rocks. For example, even though almost two-thirds of the sedimentary rock shell is mudrock, mudrocks are discussed in only one chapter. On the other hand, such rock types as evaporites and banded iron formations, which do not even appear in the inventory, together warrant a separate and distinct chapter. Why such discrepancies?

First, the coverage in this book is based not on relative abundance but on how successfully analyses of various rock types answer the questions posed earlier in this chapter. For example, questions about sedimentary provenance, dispersal, paleogeography, and tectonics can be answered with reasonable precision using sandstone data. Mudrocks provide little information about these characteristics. Second, certain types of sedimentary rock—for example, such organic deposits as coal and petroleum—are important economic resources. It is crucial to understand the origin and evolution of such deposits to improve our ability to find and exploit them. Finally, the formation of such rock types as bedded evaporites and banded iron formations requires unusual conditions that have existed rarely in geologic history. This fact explains the relative scarcity of these deposits compared to rocks such as mudstone, sandstone, and limestone, and it underscores the relevance of such lithologies as keys to our understanding of the Earth.

CONCLUSIONS

Despite its minor volume (as little as 1/20 of the upper-most 16 km of the solid Earth), we can learn much from the present sedimentary rock shell. It is important that we understand it for several reasons.

1. Sedimentary rocks are economically important. They contain the world's entire store of petroleum, natural gas, coal, and fertilizer. Sediments and sedimentary rocks constitute a principal reservoir for groundwater.

2. Sedimentary rocks are the primary repositories of the fossil record, on which rests our understanding of the evolution of life.

3. Because changes in the sedimentary rock record over time both reflect and control the character of the atmosphere, hydrosphere, and biosphere, understanding sedimentary rocks is crucial to deciphering the origin and evolution of these spheres.

4. The sedimentary rock record offers the clearest insights into the history and evolution of our

planet. When properly understood and interpreted, sedimentary rocks allow us to reconstruct paleogeography beginning at the local scale. (Where were the mountains? Which way did river systems flow? In which direction did the shoreline run?) On a global scale, accurate interpretations of paleogeography, paleoclimatology, and depositional settings based on sedimentary rocks allow us to draw conclusions about the distribution of continental blocks and ocean basins as well as the origin and evolution of mountain systems.

5. Earth's history is a continuum of events whose character changes through time and space. Folds and faults—the structural features produced by orogenic episodes—and igneous and metamorphic rock complexes provide but fleeting glimpses of Earth's history. They are snapshots of what the Earth was like in past moments, whereas sedimentary rocks are analogous to motion picture film. This book aims to show how that film can best be developed, analyzed, and understood.

FOR FURTHER READING

Garrels, R. M., and F. T. Mackenzie. 1971. *Evolution of Sedimentary Rocks.* New York: W. W. Norton.

Taylor, S. R., and S. M. McClennan. 1985. *The Continental Crust: Its Composition and Evolution.* Oxford: Blackwell Scientific Publications.

Tucker, M. E. 2001. *Sedimentary Petrology: An Introduction to the Origin of Sedimentary Rocks,* 3d ed. Oxford: Blackwell Science.

2 Weathering and Soils

Deeply weathered granitic rocks from the San Gabriel Mountains break down into coarse sedimentary particles, which accumulate below them. (D. R. Prothero.)

IF YOU STROLL AMONG THE OUTCROPS ON A TYPICAL granitic batholith, such as the San Gabriel Mountains, you will notice that most of the rocks are granodiorites and quartz monzonites. In any given sample, the dominant minerals are feldspars, with plagioclases constituting 30% to 50% of the rock and K-feldspars typically ranging from 5% to 35%. In almost any igneous rock, quartz is typically only 5% to 10%, with a maximum of about 25% to 30% in a quartz-rich granite and none whatsoever in diorites or gabbros (or their volcanic equivalents, andesite and basalt).

But if you scramble down into the washes and ravines that drain the granitic mountains and examine the sand washing off those same rocks, you will see that quartz is already a much higher percentage of the sand and that feldspars are much less common.

If you follow a mountain stream down to the ocean and look at the sand on the beach, you will find that nearly all the sand is quartz and that feldspars are relatively rare. In most beach sands, quartz may be 80% to 95% of the sand-sized fraction. Clearly, some process is at work to decrease feldspar abundance, because feldspars go from being the most abundant mineral in Earth's upper crust to being one of the scarcest in the detritus that was disintegrated and decomposed from that crust.

Where did all those feldspars (and the enormous quantities of aluminum, calcium, and sodium that make them up) go? Where did all that extra quartz come from? Because quartz is pure silica, we can deduce that some aluminum, calcium and magnesium must have been lost from the initial proportions found in igneous rocks. Since they can't leave the planet, these chemicals must go somewhere. Most of them end up as components of clay minerals.

The same conclusion could be drawn by examining ancient sandstones. Although the minerals in these rocks were derived largely from igneous and metamorphic parents, much as they are in modern sediments, many ancient sandstones have few feldspars. The average sandstone is 50% to 70% quartz, and many sandstones are 99% pure quartz. Just as happens today, some process in the geologic past must have altered the proportion of minerals from a parent rock to a final sedimentary end product. That process is **weathering.** Sediments and sedimentary rocks would not exist without weathering, the combination of processes by which pre-existing rocks physically disintegrate and chemically decompose into soil, loose clasts, and dissolved components. Weathering products constitute the raw materials from which sedimentary rocks are made.

Weathering is the simple consequence of exposing pre-existing rocks to the conditions at Earth's surface: low temperature and pressure, organic activity, and chemically active substances such as water and atmospheric gases. Physical and chemical weathering are the means by which pre-existing rocks and minerals change and come into equilibrium with this surface environment.

Soil consists of untransported products of physical weathering, typically loose, unconsolidated, mainly resistant, compositionally altered mineral residue (for example, grains of quartz, feldspar, mica, and rock fragments). Soil becomes sedimentary rock if it is eroded, transported (generally, by any of four agents: wind, ice, running water, and gravity), and deposited as **sediment.** Compaction and cementation convert sediment to sedimentary rock. Chapter 3 discusses how weathered clastic residues are actually entrained, carried away, and eventually deposited.

Chemical weathering generates various types of material. Some products are ions in solution (for example, potassium, sodium, and silica). These are transported as dissolved constituents in groundwater and surface runoff and are eventually precipitated as sediment. Other chemically weathered material is modified solid mineral residue (for example, clay minerals). Clay minerals are eroded and transported much as are coarser-grained clasts.

This chapter outlines the processes of physical and chemical weathering and summarizes how rocks disintegrate and decompose as a function of climate and topography.

Physical Weathering: Disintegrating Rock into Clasts

There are four major mechanisms of physical (mechanical) weathering: freeze-thaw, insolation, stress release (unloading), and organic activity. Each is a slow, unspectacular process that leads to the same result. Solid, unyielding, erosion-resistant rock is converted into smaller, movable, unconsolidated rock and mineral debris.

Freeze-Thaw

In *freeze-thaw,* the active agent is water; the active catalyst is hourly, daily, weekly, or longer term temperature changes. At temperatures hovering near 0°C, water freezes into ice and melts into water repeatedly. When water freezes, a 9% to 10% volume expansion occurs. Water freezing along cracks and fissures developed in solid masses of rock must expand. Forces as great as several kilograms per square centimeter gradually split the rock apart. The term *ice-wedging* is used interchangeably with freeze-thaw. It was coined to describe situations in which ponded films of water in fractures transform into solid masses of ice that wedge apart masses of rock.

Several factors promote freeze-thaw. The process works best where fractures are abundant. Fractures might be columnar joints produced when lava cools, joints formed during extension or bending of bedrock, or joints formed along bedding planes. A moist climate in which the daily temperature range roughly straddles the 0°C mark further increases the likelihood of freeze-thaw. A similar,

less common process occurs when salts such as halite and gypsum crystallize in cracks and crevices. Evaporation of trapped brines can lead to the growth of more voluminous crystals that force apart the solid rock mass.

Insolation

Insolation refers to stresses generated when minerals are exposed to changing temperatures and undergo differential thermal expansion and contraction. When the latticework of adjacent minerals enlarges and collapses as bedrock surface temperatures rise and fall, expansion and contraction cracks develop and cause the solid rock to disintegrate. This process is common in arid climates such as the Sahara and Mojave deserts, where daily temperature fluctuations of 20° to 30°C are common. In wetter climates, moisture facilitates insolation. Minerals such as clays hydrate and swell, then contract and desiccate, generating additional stress and strain. Insolation creates mechanical weathering products that are indistinguishable from those produced by freeze-thaw.

Stress Release

Stress release occurs when rocks buried beneath overlying material experience high confining pressures. As surface weathering and erosion proceed, overburden is removed, confining pressures drop, and the deeper-seated rock mass expands. A series of expansion cracks or joints develops roughly parallel to the ground surface; these joints evolve into a series of

Figure 2.1 Spheroidal weathering in a boulder of San Marcos Gabbro, Mesa Grande, San Diego County, California. (W. T. Schaller; by permission of the U.S. Geological Survey.)

onionlike sheets or slabs of rock separated by crudely curved, subparallel cracks. Because these cracks are likely passageways for water, they become sites of freeze-thaw. Exfoliation—the spalling of curved slabs of rock from bedrock surfaces—and spheroidal weathering (Fig. 2.1)—a process in which solid rock masses cut by intersecting, roughly cubical joint patterns weather into spheroidal cores—are attributable to both freeze-thaw and stress release from unloading.

Organic Activity

Organisms that live on or in weathering bedrock promote physical weathering. Plant roots seek out small pockets of soil formed above developing cracks. As plant growth continues, the root system lengthens and thickens, gradually prying apart the crack by generating stresses similar in magnitude and orientation to freeze-thaw stresses. Microscopic and megascopic organisms living within soil and altered bedrock can also fragment them further as they ingest and burrow through the material.

Chemical Weathering Reactions

Dissolving Constituents

Chemical weathering proceeds in two distinct ways. (1) Some constituents—for example, such minerals as calcite and halite—dissolve completely. The constituents dissolved from such minerals are carried away by groundwater and runoff and can be precipitated elsewhere with or without the assistance of organisms. (2) Other constituents, such as feldspars and micas, are altered into new minerals (especially clay minerals). These new minerals, altered or weathered residues, form when selected components are removed and carried away. Because they typically are finer-grained than the original material, they are more readily removed from the weathering site.

Chemical weathering of rocks and minerals involves several simultaneous chemical reactions: hydrolysis, hydration, simple solution, and oxidation-reduction. These reactions proceed most easily in the presence of both water and air.

Simple Solution (Solid Mineral + Acid or Water = Ions in Solution) Simple solution (dissolution) is the chemical reaction of solid rocks and minerals

with water or acid. Bonds between ions in rigid crystalline lattices are broken, and the freed ions are disseminated in solution. Solubility can be partial or complete. Let's examine some specific examples.

The mineral quartz is not very soluble. Less than 6 ppm (parts per million) is dissolved in normal fresh water. Crystals of quartz exposed in an outcrop of granite typically show little corrosion because of this minimal solubility. They appear fresh and unscathed by solution, standing out in relief above more easily decomposed minerals such as feldspar. The weathering reaction can be expressed as

$$SiO_2 + 2\,H_2O \rightarrow H_4SiO_4 \ (6\ ppm)$$
quartz   water   silica in solution as
                     hydrosilicic acid

Calcite is much more soluble than quartz. In the laboratory, dousing a block of limestone with weak acid will result in noticeable, instantaneous corrosion. In the natural world, exposures of limestone become pitted over periods of only days as they react with rainfall. Most natural rainfall becomes carbonic acid as raindrops fall through Earth's atmosphere and absorb small amounts of carbon dioxide gas:

$$H_2O + \quad CO_2 \quad \rightarrow \quad H_2CO_3$$
water   carbon dioxide   carbonic acid

Carbonic acid and acids in general contain abundant hydrogen ions. Thanks to their valency and small size, they have a strong affinity for anions and will displace other cations in mineral structures. Limestone dissolves as the hydrogen ions displace calcium ions, generating both dissolved calcium and biocarbonate:

$$CaCO_3 + H_2CO_3 \rightarrow \quad Ca^{2+} \quad + 2\,HCO_3^-$$
limestone   rainfall   dissolved   dissolved
                       calcium   bicarbonate

Halite (NaCl) is extremely soluble, even in distilled water. Grains of halite dropped into a beaker of water disappear completely. A salty taste develops as sodium and chlorine ions become dispersed in the water, producing a brine. The solubility of sodium and chlorine is measured in thousands of parts per million, enormous compared with the solubility of quartz:

$$NaCl + H_2O \rightarrow Na^+ + Cl^-$$
halite   water   ions dissolved
                 in water

Hydration and Dehydration (Solid Mineral + Water = New Hydrated Mineral; Dehyration Is the Reverse) Some weathering processes involve the chemical combination of pre-existing minerals with water (hydration) or the removal of water from some pre-existing mineral (dehydration). These processes produce new minerals in greater equilibrium with the environment. Two common reactions are the dehydration of gypsum to form anhydrite:

$$CaSO_4 \cdot 2\,H_2O \rightarrow \quad CaSO_4 \quad + 2\,H_2O$$
gypsum   anhydrite   water

and the hydration of iron oxide (hematite) to form "limonite" (this is essentially corrosive rusting):

$$Fe_2O_3 + 3\,H_2O \rightarrow \quad 2\,Fe(OH)_3$$
hematite   water   "limonite"
                   (oxidized iron)

A third hydration reaction occurs when the clay mineral kaolinite combines with water to form the clay mineral gibbsite; dissolved silica is released as a by-product.

Hydrolysis (Hydrogen Ion + Mineral with Mobile Cations = Entirely Dissolved Mineral or Partially Altered Mineral in Which Hydrogen Ions Replace Mobile Ions That Are Put into Solution) Hydrolysis is defined as the replacement of cations in a mineral structure by hydrogen ions derived either from water or, more likely, from acid. Hydrolysis releases to the solution the cations replaced in the mineral structure by hydrogen and either converts the original mineral into a different mineral or dissolves it completely. (Simple solution of such minerals as calcite is really hydrolysis where no solid relict of the original survives.)

Most silicate minerals weather primarily by a series of hydrolysis reactions, and silicate minerals such as pyroxenes, amphiboles, micas, and feldspars, along with quartz, make up the bulk of Earth's primary crust. The specifics of the hydrolysis process and the extent to which an original mineral is decomposed depend on the material.

Dark-colored (mafic) minerals such as olivine and pyroxene can dissolve completely:

$$Mg_2SiO_4 + 4\,H^+ \rightarrow 2Mg^{2+} + H_4SiO_4$$

olivine   from water   ions in   dissolved
      or acid   solution   silica

$$2\,CaMgSi_2O_6 + 16\,H^+ + 2\,O_2 \rightarrow$$

pyroxene   from water   atmospheric
      or acid   oxygen

$$2\,Ca^{2+} + 2\,Mg^{2+} + 4\,H_4SiO_4$$

ions in solution   dissolved
                 silica

Light-colored (felsic) minerals, especially feldspars such as orthoclase and plagioclase, dissolve partially, producing dissolved silica and cations and leaving fine-grained, easily transportable clay minerals:

$$KAlSi_3O_8 + H^+ \rightarrow Al_2Si_2O_5(OH)_4 + K^+ + H_4SiO_4$$

K or Na or   acid   clay mineral   ions in   dissolved
Ca feldspar   (kaolinite)   solution   silica

Oxidation-Reduction (Atmospheric Oxygen Gains Electrons and Is Reduced as Mineral Constituents Lose Electrons and Are Oxidized, Producing New "Rusted" Minerals) Oxidation and reduction are inexorably linked. Oxidation does not occur without reduction, and vice versa. Oxidation is the process by which an atom or ion loses electrons. Reduction is the process by which an atom or ion gains electrons. The best oxidizing agent is atmospheric oxygen, O_2; non-ionized atoms of oxygen (zero valency) in the atmosphere combine readily with other existing ions and gain electrons to become anions of oxygen (O^{2-}). As a result, oxygen is reduced, but the ion from which the oxygen atom gain electrons is oxidized. The most obvious examples of this process involve the oxidation of ferrous iron (Fe^{2+}) to ferric iron (Fe^{3+}). In the natural world, this is the process of rusting, in which dull, metallic, ferrous iron is changed to reddish-orange ferric iron. For example,

$$(Fe^{2+})SiO_3 + O_2 + 2\,H_2O \rightarrow$$

pyroxene   atmospheric   water
         oxygen

$$Fe^{3+}(OH^-)_3 + H_4SiO_4$$

"limonite"   dissolved
           silica

$$2\,FeS_2 + O_2 \rightarrow Fe_2O_3 + 2\,S$$

pyrite   atmospheric   hematite   dissolved
      oxygen   sulfur

The Controls

Several factors dictate which chemical weathering reactions will be at work. Climate is of paramount importance. Higher temperatures promote chemical weathering because kinetic thermal energy facilitates any reaction. Therefore, chemical weathering should be more extensive in warmer climates (lower latitudes) than in colder climates (higher latitudes and/or higher elevations). Moisture (rainfall) is also important because most chemical weathering requires water. Thus, warm and humid climates will be most favorable for chemical weathering. Even though the temperature is favorable in desert areas, not enough water is present for hydrolysis, solution, and hydration to occur. In arctic regions and high mountainous terrains, abundant precipitation exists, but in the wrong form, falling as snow and ice rather than rain.

Hydrolysis and simple solution depend on the ready availability of hydrogen ions. The abundance of hydrogen ions—more specifically, the activity of the hydrogen ions in a solution—is controlled by the acidity or alkalinity of a solution; that is, the pH. The pH is a way to designate the concentration of H^+ ions in solution, or $[H^+]$.

The pH of a solution expresses hydrogen ion concentration as the negative of the log to the base 10 of hydrogen ion concentration, or $-\log_{10}[H^+]$. The pH of solutions can range from 1 to 14 because the ratio of water to hydrogen and hydroxyl (the dissociation constant) is 1×10^{-14}. That is, the dissociation constant for water is

$$\frac{[H^+][OH^-]}{[H_2O]} = 10^{-14}$$

Solutions with precisely equal amounts of acid components $[H^+] = 10^{-7}$ and base components $[OH^-] = 10^{-7}$ have a pH of 7 and are termed neutral. Solutions with an excess of hydrogen ions have pH values from 1 to just below 7 and are called acids. Acids are the most effective agents of hydrolysis because they readily provide hydrogen ions to replace cations in the mineral lattice.

Most natural waters have pH values between 4 and 9. The graph in Fig. 2.2 displays the pH (and Eh, discussed shortly) of waters in various natural environments. This graph is important because it permits inferences about the likelihood of hydrolysis and oxidation-reduction in waters of particular settings. For example, rainwater, streams, and groundwater are

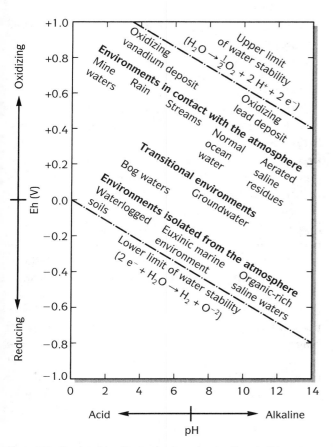

Figure 2.2 Geochemistry of natural waters, showing Eh and pH in various settings. These chemical conditions are classified into oxidizing and reducing conditions (vertical axis) and acidic versus alkaline (basic) conditions (horizontal axis). (After Blatt, Middleton, and Murray, 1980: 241; by permission of Prentice-Hall, Inc., Englewood Cliffs, N.J.)

The likelihood of oxidation depends largely on the availability of free atmospheric oxygen (the principal natural agent of oxidation). The term **Eh** expresses the potential for either oxidation or reduction. Eh, short for redox potential, is measured using an electrolytic cell. Its value includes magnitude and sign, either positive (oxidizing) or negative (reducing). The higher the magnitude (in volts), the more likely it is that a particular ion or atom will be either oxidized or reduced. Iron is the most important element affected by the oxidation-reduction process.

Figure 2.2 shows variations in Eh for waters in selected natural environments. Natural waters that are reducing characterize depositional settings physically isolated from the atmosphere. In such settings as the waterlogged soils of swamps and the oxygen-starved, stagnant marine waters at the bottom of the Black Sea, reduction rather than oxidation occurs. Any iron found will be in the form of the ferrous (Fe^{2+}) ion (for example, as the mineral pyrite). Any organic matter found therein will not decompose because organic decay is just natural oxidation; that is, the chemical "burning" of hydrocarbons as they slowly combine with free oxygen. Oxidation, iron present predominantly in the form of reddish ferric iron (Fe^{3+}), and organic tissue decay characterize settings in which there is free interaction among atmospheric oxygen, rainwater, stream water, groundwater, and ocean water.

Weathering in the Natural World

The process by which source rock breaks down is complex because it depends on several factors: source composition, climate, drainage, topographic relief, and the relative rates of chemical and physical weathering. Nevertheless, the sedimentary geologist must attempt to infer what is happening at the source from the composition and texture of the material weathered from it.

Source Composition

Source composition—specifically the mineralogy, texture, and rock structure—is of paramount importance. Consider some examples. Two rock types, a marble and a quartzite, might be identical in every respect but mineral composition. In identical physical settings, however, their styles of weathering will differ dramatically. In a moist, relatively warm climate,

slightly acidic (pH values from 4 to 6.5), which makes them active agents of hydrolysis and solution. Acidity is directly attributable to the absorption of atmospheric carbon dioxide by water as it falls through the atmosphere. (Recent concerns about acid rain reflect the fact that increased industrialization based on the accelerated combustion of fossil fuels adds ever-increasing amounts of such acid-producing gases as carbon dioxide and hydrogen sulfide to the atmosphere, making falling rain more acidic and corrosive.) Most soils have pH values from 4 to 5, so further chemical decomposition by hydrolysis continues in soil zones. Mine waters generated as groundwater percolates through sulfide-rich mineral deposits are notoriously acidic and corrosive. Conversely, the pH of seawater is slightly higher than 8, so little submarine hydrolysis occurs. (Many low-temperature chemical reactions do occur, however, as pre-existing minerals such as clays equilibrate with seawater.)

calcite grains corrode and dissolve as acidic rainfall washes over the marble. The dissolved components are carried away by groundwater and runoff and might eventually be precipitated as carbonate sediments in a setting such as the modern Bahama Banks. Conversely, the quartz grains composing the quartzite will dissolve slightly, if at all. In colder winter months, ice-wedging along bedding planes and joints may disintegrate the quartzite mechanically, producing an apron of quartz-rich rubble. Most of the quartz grains eventually will be transported away from the weathering site by mass wasting and runoff, perhaps ending up as alluvial or deltaic clastic deposits. Segregation by size might occur during transport; likewise, there might be some additional mechanical weathering by abrasion (quartz is hard and lacks cleavage).

Does physical weathering also affect the marble? Ice-wedging could produce an apron of calcite fragments around the exposed marble. Some of this detritus, like the quartz grains flanking the quartzite, might experience a brief period of clastic transport and deposition. Chemical weathering would continue to corrode this detrital calcite, however. After mechanical disintegration, individual calcite grains would have more exposed surface area, a condition that generally promotes chemical weathering. Corrosion would intensify, producing additional dissolved calcium and carbonate for precipitation elsewhere. Calcite fragments entrained as clasts would abrade rapidly to finer-grained detritus (calcite is soft and has excellent rhombic cleavage).

The details of how source rock lithology controls weathering differ, but there are general rules. Fine-grained rocks (for example, volcanic rock, slate, and mudrock) decompose chemically more readily than coarse-grained rocks (for example, plutonic rock, gneiss, and sandstone). This tendency reflects the amount of grain surface area per unit volume, which controls the availability of chemical reaction points (that is, incomplete bonds) in a mineral.

In this context, it is useful to think of particular minerals as being *in equilibrium* with certain conditions in the crust or on the surface. Igneous minerals—especially olivine, pyroxene, hornblende, and the plagioclase feldspars—form in magma chambers at high temperatures; this is their equilibrium state. Once they are exposed to Earth's surface conditions (lower temperatures and chemical weathering), they are out of equilibrium and essentially doomed to destruction (however slowly). To reach equilibrium, the silica, aluminum, iron, and other elements must seek a new mineral state that is stable at Earth's surface. These new minerals are mostly quartz, clay minerals, iron oxides, and aluminum oxides.

Decades ago, Goldich (1938) examined the distribution of major rock-forming silicate minerals in several soil profiles and proposed a mineral-stability series that lists minerals in descending order of their propensity to survive chemical decomposition. This listing is almost precisely the same as the listing known as Bowen's reaction series, used to describe the order in which major silicate minerals solidify from igneous magmas. Olivine, pyroxene, and calcic plagioclase feldspar—the higher-temperature, higher-pressure minerals that form first from igneous melts—are least resistant to weathering. This is largely because their simple crystal lattices (olivine is composed of simple tetrahedra; pyroxene is a single-chain silicate) are less tightly bonded together, and so the bonds are easily broken. These minerals also have abundant cations (especially iron, magnesium, and calcium) that are subject to chemical weathering and are eagerly grabbed by plant roots in biological weathering. Quartz, muscovite, and K-feldspar crystallize later from cooler melts in conditions more like those of the surface. These minerals are all complex sheet or framework silicates, with many Si–O bonds that are harder to break. Consequently, they are less susceptible to weathering. Rocks that are aggregates of unstable minerals such as olivine, pyroxene, and calcic plagioclase (basalt, peridotite, dunite, and gabbro) will decompose more easily than rocks composed of resistant minerals such as quartz, muscovite, and K-feldspar (rhyolite, granite).

Fissured and jointed rocks are more susceptible to weathering because cracks provide access to the fluids that promote chemical decomposition and ice-wedging. For example, foliated metamorphic rocks and well-bedded sedimentary rocks weather more rapidly than massive coarse-grained igneous rocks.

Climate

The climate at the weathering site is crucial. Daily temperature fluctuations and the frequency with which the temperature rises above and falls below the freezing point determine the importance of ice-wedging and insolation. Total precipitation, especially the amount in the form of rainfall, governs the extent of hydrolysis, hydration, and solution. Higher mean annual temperatures increase chemical weathering rates.

Drainage

Like all chemical reactions, weathering reactions are reversible. For reactions to continue, reaction products must be removed. This removal stops re-reaction of products with one another in the reverse direction. For example, when feldspar weathers by hydrolysis to produce clays and dissolved constituents such as silica, potassium, sodium, and calcium, the reaction products can easily re-react if they remain trapped or ponded at the soil site (LeChatelier's principle). Re-reaction is common in poorly drained soil zones developed above bedrock. Conversely, where drainage is good and re-action products are quickly swept away, re-reaction cannot occur.

Topographic Relief

Topographic relief (the difference in elevation between low and high points in a region) and slope steepness strongly influence weathering. Slope steepness controls the rate at which weathering products are eroded from the weathering site and transported elsewhere. Once material is in transit, little additional chemical weathering occurs.

Mass wasting and runoff occur at higher rates in steep-sloped mountainous areas than in areas of low-lying, monotonous plains. Consequently, in mountainous regions, weathered material is removed rapidly and decomposition is less extensive. Soils developed where slopes are gentle and relief is low remain in place as precipitation continues to filter through them, permitting extensive decomposition.

Relative Rates of Chemical and Physical Weathering

The precise mix of weathering products generated in a region reflects the relative rates of mechanical disintegration and chemical decomposition. Rocks exposed in polar regions and high latitides are typically mantled with a veneer of physically disintegrated, undecomposed rock rubble. Where topographic relief is high and steep slopes and cliffs are common, aprons of talus develop. Negligible chemical alteration accompanied by significant mechanical disintegration also typifies hot, dry desert climates. Weathering in desert and arctic regions produces large volumes of clastic sediment but few raw materials for chemical and biochemical sediments. Conversely, ice-wedging and insolation are unimportant in low-lying, low-latitude areas with tropical climates, but chemical decomposition is extensive. Tremendous volumes of dissolved silica, potassium, calcium, and sodium are produced and become ready candidates for chemical or biochemical sedimentation elsewhere. Clastic material generated in such tropical settings will be clay minerals, the future constituents of mudrock.

Soils and Paleosols

Soils

The weathering processes discussed in the previous sections affect not only fresh bedrock, transforming it to sediment, but also sediments that have been deposited but not yet lithified into rock. When any accumulation of weathered material builds up on Earth's surface, we call it a **soil.** Soils can be very thin, as in the case of the tiny amount of weathered material formed under lichens as they attack naked rock outcrops. Soils can also be many meters thick in tropical rain forests, where the rain and roots penetrate deeply and weather material far below the surface. Soils consist of the weathered bedrock material that produces much of the sediment, the organic material added by living organisms, and additional chemical elements that move through the soil in groundwater.

Several processes contribute to the formation of soils. First, animals and plants interact with the sediment, absorbing nutrients and leaving behind their wastes and remains. Second, burrowing organisms such as ants, worms, and gophers churn the soil, so its texture is very different from freshly weathered sediment. Finally, rainwater percolates through the sediment and moves chemical elements through the soil. Near the surface, there may be a **zone of leaching** where the water dissolves ions and carries away the fine clays, moving them down into the **zone of accumulation.** Here the ions precipitate as new minerals, and the clays settle out. In the tropical rain forests, the zone of leaching is very thick. Intense weathering from the heavy rains, dense plant roots, and high temperatures breaks down most of the minerals in the upper part of the soil and transfers their ions to a thick zone of accumulation several meters below the surface. This leaves aluminum-rich clays such as kaolinite (see Chapter 6) plus iron oxides, resulting in the characteristic reddish **lateritic** soils of the tropics.

Many soils have a vertical zonation of horizons known as a **soil profile** (Fig. 2.3). The idealized soil profile has an **O-horizon** (for "organic") at the top, composed mostly of **humus** (partially decayed

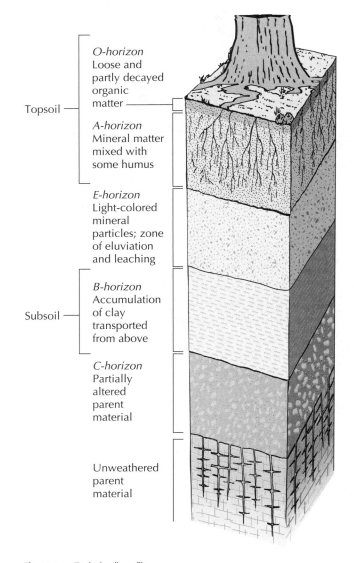

Topsoil

O-horizon
Loose and
partly decayed
organic
matter

A-horizon
Mineral matter
mixed with
some humus

E-horizon
Light-colored
mineral
particles; zone
of eluviation
and leaching

Subsoil

B-horizon
Accumulation
of clay
transported
from above

C-horizon
Partially
altered
parent
material

Unweathered
parent
material

Figure 2.3 Typical soil profile.

organic matter) from the activity of organisms. Below the O-horizon, there is an **A-horizon,** which consists of humus mixed with sediment. These two horizons together are the main components of topsoil. The A-horizon grades down into a transitional **E-horizon** (for **eluviation,** the washing down of fine components). This horizon is also known as the A2-horizon in older literature, and it tends to be lighter-colored with fewer organic components than the horizons above it. The next level down, the **B-horizon,** is where the zone of accumulation begins and ions that have leached downward from higher horizons are precipitated. Often, the B-horizon is slightly to richly reddish in color because iron oxides have precipitated and there is no dark organic matter. Finally, the base of the soil profile is known as the **C-horizon;** it consists of material derived

from the substrate that has been chemically weathered and broken apart. If the substrate is bedrock, the C-horizon grades downward into unweathered bedrock. If the substrate is loose sediment, the C-horizon grade down into unweathered sediment.

Of course, this description of a profile is an idealization. There are tremendous variations in this pattern depending upon climate, rainfall, slope steepness, and substrate. In many temperate climates, for example, we find **pedalfer** soils, which have well-developed O-, A-, B- and C-horizons, much like the idealized example just discussed. In desert climates, however, we find **pedocal** soils, which have much less organic matter (and no O-horizon) and a thin A-horizon with little organic matter. Because of cyclical leaching and drying, pedocal soils often have high concentrations of calcite in the B-horizon, which can accumulate and cement the soil to form a caliche. The lateritic soils of the tropics have very thick, deeply weathered A-horizons. Most of the minerals and organic materials are dissolved and weathered away, leaving only kaolinite clays and iron oxides. There is no zone of accumulation or B-horizon because all the leached materials move into the rivers as a result of the intense weathering and transport caused by the high rainfall.

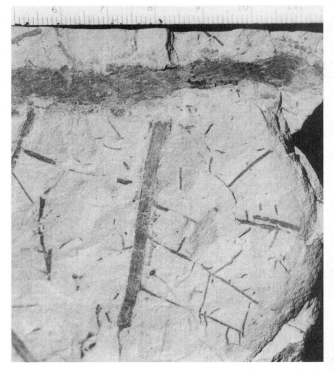

Figure 2.4 The organic matter in the root traces in this paleosol have been partially replaced by iron oxides. This example is from the lower Miocene Mollala Formation, western Oregon. (Courtesy of G. Retallack.)

Paleosols

Paleosols are fossil soils that have been buried and preserved in ancient rocks. In the past, geologists may have noticed distinctive color bands and even root casts in ancient sedimentary sequences, but they rarely realized that they were ancient weathered horizons and soils. Since the 1980s, however, there has been an explosion of research and interest in paleosols (a field called **paleopedology**). Now there are many books published on the subject (for example, Retallack, 2001; Reinhardt and Sigleo, 1988), and many studies of paleosol sequences are conducted on rocks from the Precambrian to the Pleistocene all over the world. Paleosols have been used to reconstruct ancient climate and vegetation patterns where there is no other evidence to work with. They have helped determine the level of certain gases in the ancient atmosphere and have even been used to infer the existence and activities of organisms for which there are no body fossils. For example, body fossils of land animals are not known until the Silurian, but burrows and fossil soils suggest that there may have been millipedes and simple plant communities on land in the Ordovician (Retallack and Feakes, 1987). Paleosols are surfaces of weathering and erosion, so they often mark unconformities (especially subtle disconformities that are hard to recognize by other field criteria). They may also be useful in understanding the completeness of the rock record and in deciphering how and when events that caused the unconformity, such as climatic or tectonic changes, might have occurred.

A number of features help geologists recognize paleosols in the field (Fenwick, 1985; Retallack, 1988, 2001). The best evidence comes from layers enriched in organic matter and layers enriched in reddish iron oxides that become more intense in color toward the top. Other characteristics include the noticeable decrease in weatherable minerals toward the top of the presumed soil profile and the disruption of the original bedding by organic activity, such as worm burrows or root traces (Fig. 2.4). Root traces usually taper and branch downward from the ancient soil surface, although some roots spread laterally, and a

Type	Platy	Prismatic	Columnar	Angular blocky	Subangular blocky	Granular	Crumb
Sketch							
Description	Tabular and horizontal to land surface	Elongate with flat top and vertical to land surface	Elongate with domed top and vertical to surface	Equant with sharp interlocking edges	Equant with dull interlocking edges	Spheroidal with slightly interlocking edges	Rounded and spheroidal but not interlocking
Usual horizon	E, Bs, K, C	Bt	Bn	Bt	Bt	A	A
Main likely causes	Initial disruption of relict bedding; accretion of cementing material	Swelling and shrinking on wetting and drying	Same as prismatic, out with greater erosion by percolating water and greater swelling of clay	Cracking around roots and burrows; swelling and shrinking on wetting and drying	Same as angular blocky, but with more erosion and deposition of material in cracks	Active bioturbation and coating of soil with films of clay, sesquioxides, and organic matter	Same as granular; including fecal pellets and relict soil clasts
Size class	Very thin < 1 mm	Very fine < 1 cm	Very fine < 1 cm	Very fine < 0.5 cm	Very fine < 0.5 cm	Very fine < 1 mm	Very fine < 1 mm
	Thin 1 – 2 mm	Fine 1 – 2 cm	Fine 1 – 2 cm	Fine 0.5 – 1 cm	Fine 0.5 – 1 cm	Fine 1 – 2 mm	Fine 1 – 2 mm
	Medium 2 – 5 mm	Medium 2 – 5 cm	Medium 2 – 5 cm	Medium 1 – 2 cm	Medium 1 – 2 cm	Medium 2 – 5 mm	Medium 2 – 5 mm
	Thick 5 – 10 mm	Coarse 5 – 10 cm	Coarse 5 – 10 cm	Coarse 2 – 5 cm	Coarse 2 – 5 cm	Coarse 5 – 10 mm	Not found
	Very thick > 10 mm	Very coarse > 10 cm	Very coarse > 10 cm	Very coarse > 5 cm	Very coarse > 5 cm	Very coarse > 10 mm	Not found

Figure 2.5 Characteristics of different kinds of soil peds. (After Retallack, 1988.)

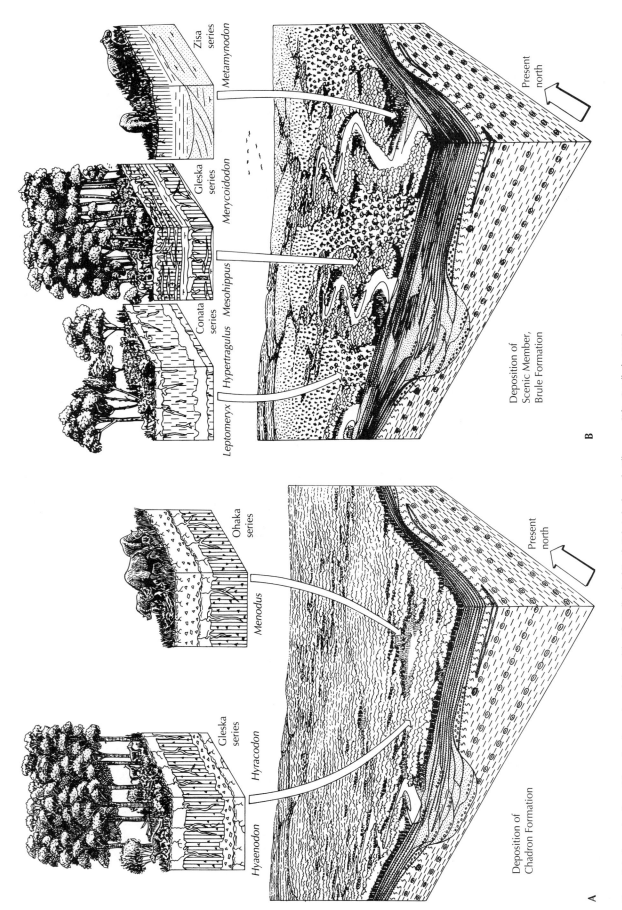

Figure 2.6 Reconstruction of the soil series and vegetation of the Big Badlands of South Dakota in the early Oligocene. (After Retallack, 1983.)

few kinds branch upward and out of the soil. Root traces with original organic matter are most often preserved in waterlogged anoxic environments. In more oxidized soils, roots leave tubular features that are filled with material from the surrounding paleosol matrix, and they may be highlighted in iron oxides. These features can be seen in outcrops but are especially evident in thin sections.

As a result of the activity of animals and plants, the wetting and drying cycles, and many other processes, paleosols develop characteristic soil structures that replace the original bedding and sedimentary fabric of the rock. These structures typically occur as an irregular network of planes (called **cutans**) surrounded by stable clumps of soil material known as **peds** (Fig. 2.5). Peds give the soil its typical hackly fracture when broken, and they have a wide variety of sizes and shapes. In the field, they can often be recognized by the cutans that surround them, which typically occur as clay skins around the peds. As Fig. 2.5 shows, ped shapes range from platy to prismatic to blocky to columnar to granular to crumbly. Each shape is typical of certain parts of the soil horizon and may also be specific to certain climatic conditions. In addition to cutans and peds, there may also be hard lumps called **glaebules** (including nodules and concretions; see Chapter 7) that may be cemented by calcite, iron oxides, or siderite (iron carbonate).

The intensity of recent research interest in paleosols is largely driven by the many important geological insights and discoveries that paleopedology has produced. Retallack (2001) reviewed a number of studies on paleosols from Precambrian to Pleistocene age that have given us important insights into changes in climate, atmospheric gases, and the nature of land life. For example, the Eocene-Oligocene rocks of the Big Badlands of South Dakota have long been famous for their striking color bands, but only a few limited inferences had been made about their paleosols. Retallack (1983) did a detailed study of these paleosols and found a fascinating story of changes in climate and vegetation that were not evident in any other source of data. The Badlands rocks were too oxidized to contain many plant body fossils, but analysis of the soil horizons and root traces nonetheless showed the nature of the missing, unfossilized plant community. Climate went from fairly wet (over a meter of annual rainfall) in the late Eocene, when dense forests covered the landscape, to a mixed landscape in the early Oligocene, when small forests separated open and scrubby areas receiving less than 500 mm of annual rainfall (Fig. 2.6).

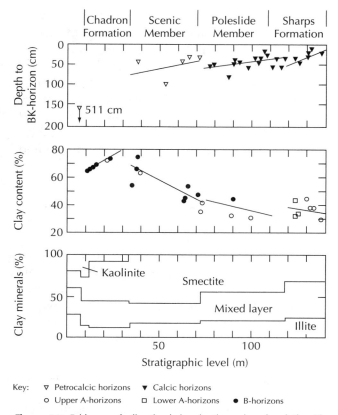

Figure 2.7 Evidence of climatic drying in the paleosols of the Big Badlands of South Dakota. (After Retallack, 1986.)

The late Oligocene paleosols are products of a much drier climate, with extensive dune formations and calcareous nodules produced by caliche processes in the groundwater. None of this climatic change was decipherable before the paleosols were analyzed.

Certain criteria are used to recognize such changes in climate and vegetation. For example, the depth to the Bk-horizon (the area within the B-horizon that produces calcareous nodules and concretions) is often considered a good proxy of annual rainfall. The deeper the Bk-horizon, the wetter the conditions; very shallow nodules are characteristic of dry conditions (Fig. 2.7). Clay content is also a good proxy of climatic change. Wetter conditions are associated with higher clay contents caused by soil weathering, and drier conditions produce much less clay. In addition, the nature of the clay minerals (see Chapter 6) can also be diagnostic. Kaolinitic clays (found in laterites) are associated with wet conditions. Unusual minerals—for example, gibbsite, diaspore, boehmite, and iron hydroxides such as goethite—are found in extremely wet climates (greater than 2 m of annual rainfall). Smectites and mixed-layer clays tend to be found in drier

environments. Illites are associated with the driest climates—but one must be careful with this interpretation, because illite is also the stable end product of the diagenesis of other clays, such as kaolinites and smectites. If we put all these characteristics together (see Fig. 2.7), we can see that the paleosols of the Big Badlands show a clear trend from wet, tropical conditions in the late Eocene (as indicated by paleosols with a deep Bk-horizon and abundant clays with significant kaolinite) to drier conditions in the Oligocene (as shown by the very shallow Bk-horizons and reduced clay content consisting largely of smectites and illite).

Many other types of soils and paleosols have been described and classified, but a book like this one cannot cover all of them in detail. See the "For Further Reading" section at the end of this chapter for more information on soils and paleosols.

CONCLUSIONS

Physical disintegration and chemical decomposition of pre-existing rocks generate the raw materials from which Earth's sedimentary rock record is built. How the dissolved constituents produced by chemical weathering travel to depositional sites and are precipitated as the various chemical and biochemical sedimentary rock types is discussed in the chapters describing those rocks. The processes by which the physical residues produced by mechanical weathering are entrained at their place of origin, are transported elsewhere, and are eventually deposited are addressed in the next chapter.

FOR FURTHER READING

Balasubramanian, D. S., et al., eds. 1989. *Weathering: Its Products and Deposits.* Vol. 1, *Processes;* Vol. 2, *Deposits.* Athens, Greece: Theophrastus Publications.

Berner, R. A. 1971. *Principles of Chemical Sedimentology.* New York: McGraw-Hill.

Bronger, A., and J. A. Catt, eds. 1989. *Paleopedology: Nature and Application of Paleosols.* Destedt, Germany: Catena Verlag.

Catt, J. A. 1986. *Soils and Quaternary Geology: A Handbook for Field Scientists.* Oxford: Clarendon Press.

Krauskopf, K. B. 1967. *Introduction to Geochemistry.* New York: McGraw-Hill.

Lerman, A., and M. Meybeck, eds. 1988. *Physical and Chemical Weathering in Geochemical Cycles.* Dordrecht, Germany: Kluwer Academic.

Mason, B. 1966. *Principles of Geochemistry.* New York: John Wiley.

Nahon, D. B. 1991. *Introduction to the Petrology of Soils and Chemical Weathering.* New York: John Wiley.

Reinhardt, J., and W. R. Sigleo, eds. 1988. *Paleosols and Weathering through Geologic Time: Principles and Applications.* Geological Society of America Special Paper 216.

Retallack, G. J. 2001. *Soils of the Past.* London: Blackwell Science.

Wedepohl, K. H. 1971. *Geochemistry.* New York: Holt, Rinehart, and Winston.

Wright, V. P. 1986. *Paleosols: Their Recognition and Interpretation.* Princeton, N.J.: Princeton University Press.

3 Clastic Transport and Fluid Flow

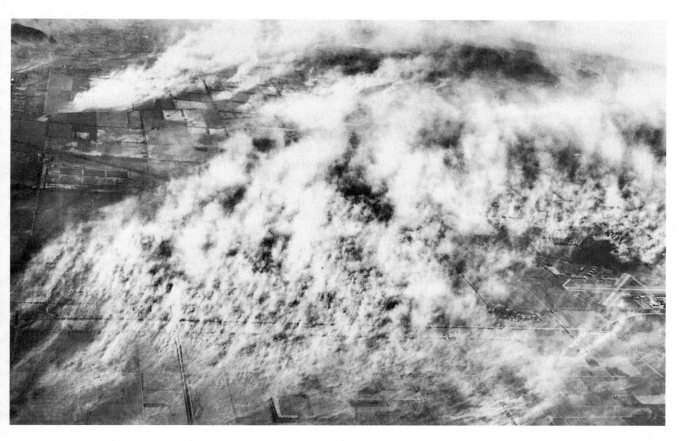

The process of sediment transport is vividly shown by this dust storm raised by northeast winds over the vineyard district of southern California. (Shelton, 1966.)

WEATHERED ROCK AND MINERAL FRAGMENTS ARE transported from source areas to depositional sites (where they are subject to additional transport and redeposition) by three kinds of processes: (1) dry (non-fluid-assisted), gravity-driven mass wasting processes such as rockfalls (talus falls) and rockslides (avalanches); (2) wet (fluid-assisted), gravity-driven mass wasting processes (sediment gravity flows) such as grain flows, mudflows, debris flows, and some slumps; and (3) processes that involve direct fluid flows of air, water, or ice.

Mass Wasting

Mass wasting processes are important mechanisms of sediment transport. Although they move soil

and rock debris only short distances (a few kilometers at most) downslope from the site at which they originated, these processes play a crucial role in sediment transport by getting the products of weathering into the longer-distance sediment transport system. They also disrupt drainage systems and modify groundwater paths.

In dry mass-wasting processes, fluid plays either a minor role or no role at all. In rock or talus falls, for example, clasts of any size simply fall freely; the presence of fluid is incidental. Fluid is not necessary for the downslope movement of bodies of rock or sediment in slumps or slides, either. They can slump or glide downslope en masse without significant internal folding or faulting, although fluid near the base of such masses provides

31

lubrication and promotes shear failure along the slippage surface.

A classic example of a dry mass movement took place in the Swiss village of Elm in 1881. A steep crag almost 600 m high was undercut by a slate quarry. Over about 18 months, a curving fissure grew slowly across the ridge about 350 m above the quarry. In late summer, runoff from heavy rains poured into the fissure and saturated it. One September afternoon, the entire mass started to slide, filling the quarry and falling freely into the valley. Once it reached the valley floor, the churning mass ran up the opposite slope to a height of 100 m, then swept back down into the valley in a debris avalanche that killed 115 people. Ten million cubic meters of rock fell about 450 m and spread into a carpet about 10 to 20 m deep covering 3 km^3.

Observations of the slide showed that the rocks traveled at 155 km/hr (about 100 mph). To move at such velocities, the rock mass must have been in free fall through most of its descent, buoyed up by a trapped carpet of air beneath it. This air cushion is analogous to the carpet of air that keeps the puck floating in a game of air hockey. Similar air cushions have been reported in snow avalanches, and the blasts of trapped air can knock down masonry buildings. Neither air nor water is essential for such movement, however. Gigantic mass movements have been described on Mars and the Moon.

Fluid Flow, in Theory and in Nature

Fluid plays an important role in all other models of sediment transport, both in such wet, gravity-driven mass movements as debris flows and mudflows and in mechanisms that move weathering products long distances, such as rivers, dust storms, and glaciers. Consequently, some knowledge of **hydraulics,** the science of fluid flow, is essential to understanding sediment transport. Hydraulics involves complex, abstract mathematics, a discipline with which many sedimentologists are uncomfortable. Sedimentologists are principally interested in understanding hydraulics well enough to make inferences about sediment transport and deposition from clastic sedimentary rock textures and sedimentary structures. Let us explore this intriguing field.

Matter can be a solid, a liquid, or a gas. Liquids (like water) and gases (like air) are fluids. A fluid is any substance that is capable of flowing. Although fluids resist forces that tend to change their *volume,* they readily alter their *shape* in response to external forces. Conversely, solids do not flow and they resist changes in *both* shape and volume.

The ability of a fluid to **entrain** (pick up), transport, and deposit sediment depends on many factors, principally fluid density, viscosity, and flow velocity. The *density* of a fluid is its mass per unit volume. The density of seawater is 1.03 g/cm^3 and that of fresh water is 1.0 g/cm^3. The density of glacial ice is 0.9 g/cm^3. The density of air is very low, less than 0.1% that of water. The *viscosity* of a fluid is a measure of its resistance to shearing. Air has a very low viscosity, the viscosity of ice is very high, and water has a viscosity intermediate between the two.

Many of the differences in clastic grain size (for example, the mean and maximum grain sizes) in glacial, alluvial, and eolian sediments reflect the different fluid densities and viscosities of ice (coarse, poorly sorted detritus), running water, and air (well-sorted, very fine grained sand and silt).

Flow velocity determines the type of fluid flow, of which there are two fundamentally different kinds: **laminar** and **turbulent** (Fig. 3.1). In laminar flow (characteristic of water flowing at low velocity), individual molecules of matter (masses of water or air) move uniformly as subparallel sheets or filaments of material. **Streamlines** (flow lines), visible when droplets of dye are injected into a slow-moving stream of water, do not cross one another. They persist as long, drawn-out coherent streaks. Parallel streams of smoke emanating from a burning cigarette in an absolutely still room exhibit laminar flow for several centimeters before breaking down into crisscrossing eddies and vortices of turbulence. Because particles of fluid move essentially parallel to the underlying boundary surface (for example, the ground surface or the floor of a laboratory flume), laminar fluid motion is basically *only downcurrent* or *downwind.*

In turbulent flow (characteristic of water flowing at high velocity), masses of material move in an apparently random, haphazard pattern. Eddies of upwelling and swirling develop. Particles of matter move both *downcurrent* and parallel with the lower bounding surface *and also up and down in the fluid.* As a result, dye streamlines are intertwined and deteriorate rapidly downstream.

Only very slowly moving (or very viscous) fluids exhibit laminar flow; most natural fluid flow is turbulent. This fact has important implications for the erosion, transport, and deposition of sediment. Fluid

Laminar flow, low Reynolds number

A

Turbulent flow, high Reynolds number

B

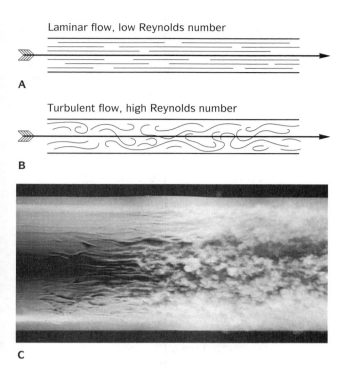

C

Figure 3.1 Contrasting flow streamlines for laminar and turbulent flow. (A) In laminar flow, discrete parcels of fluid (streamlines) move in a parallel, sheetlike fashion and propel any sedimentary clasts downstream. (B) In turbulent flow, streamlines become intertwined, and up-and-down eddies develop. Turbulent flow not only propels clasts downstream but also can lift particles into the flow. (C) The transition from laminar (*left*) to turbulent flow in water on a flat plate as seen by dye injection. Such a sharp transition is known as a hydraulic jump. (Siever, 1988: 42; by permission of W. H. Freeman and Company, New York.)

flows with upward turbulent eddies are more effective agents of erosion and transportation. The rising eddies in turbulent rivers and windstorms not only entrain detritus but also keep entrained material in transit because the turbulently rising streams of fluid counteract the tendency of grains to settle downward through them. Although laminar flow can help to transport material downcurrent, it moves material less effectively than turbulent flow because it lacks the ability to keep particles of sediment up in the moving current. Consequently, the only major nonturbulent agents of erosion and deposition are ice and mud-supported gravity flows.

Several equations are useful in understanding the basic mechanisms of hydraulics and sediment deposition. Two of these are the mathematical expressions used to compute the Reynolds number and the Froude number. These numbers allow inferences to be made about the relationships among fluid flow, the type of bedforms produced along the bounding surfaces of the moving fluid, and the mechanisms by which entrained particles move.

Reynolds Number

In 1883, the English physicist Sir Osborne Reynolds reported a classic series of experiments addressing the problem of how laminar flow changes to turbulent flow. He found that the transition from laminar to turbulent flow occurs as velocity increases, viscosity decreases, the roughness of the flow boundary increases, and/or the flow becomes less narrowly confined. In other words, the transition is controlled by the interaction of four variables, making it complicated to predict or understand. Reynolds combined these four parameters into a formula that relates velocity, geometry of flow (defined as pipe diameter by engineers or as depth of a stream by hydrologists), dynamic viscosity, and density. This combined expression is called the Reynolds number, R_e. In mathematical terms,

$$\text{Reynolds number} = \frac{\text{fluid inertial forces}}{\text{fluid viscous forces}}$$

$$R_e = \frac{2r\,V\rho}{\mu}$$

where V = velocity, ρ = density, μ = viscosity, and r = radius of the cylinder of moving fluid; in an open surface flow, the depth of the flow can be used for r.

As this equation indicates, the Reynolds number is a dimensionless number that expresses the ratio of the relative strength of the *inertial* and *viscous forces* in a moving fluid. The numerator of the equation approximates the inertial forces; that is, the tendency of discrete parcels of fluid to resist changes in velocity and to continue to move uniformly in the same direction. High inertial forces disrupt laminar flow, changing parallel stream-lines into turbulent eddies. Fluid inertial forces increase with higher flow velocity and/or a denser, more voluminous fluid mass. The denominator of the equation estimates the viscous forces. Viscous forces are directly related to fluid viscosity; they make a fluid resistant to shearing or deformation.

What are the practical consequences of fluid inertial forces and fluid viscous forces for sediment transport? Whether a flow is laminar or turbulent (with the greater potential of turbulent flow to entrain and transport particles) is related to its Reynolds number. Laminar flow occurs only where viscous forces greatly exceed inertial forces; that is, where Reynolds numbers are relatively smaller, typically falling below a critical range that lies between 500 and 2000. Such low values are characteristic of unconfined fluids that move across open surfaces, such as windstorms,

surface runoff sheet flows, slow-moving streams, highly concentrated mudflows, and continental ice sheets. Fluids with Reynolds numbers above the critical 500-to-2000 range, such as fast-moving streams and turbidity currents, have inertial forces that greatly exceed viscous forces. Their flow is turbulent.

The Reynolds number reflects several factors: fluid viscosity, current velocity, and the minimum volume or "thickness" of fluid. Increasing the viscous flow forces in a fluid suppresses turbulence. Viscous fluids such as maple syrup and the silicone gel known as Silly Putty®, and slow-moving natural geological agents such as ice and mudflows, exhibit laminar flow. They can move large volumes of sediment only because their high viscosity retards particle settling.

Because turbulent flow typically occurs when inertial forces greatly exceed viscous forces, it is characteristic of high-velocity windstorms and broad, deep, fast-moving rivers, both of which transport large volumes of sediment. Conversely, thin, watery, fast-moving films of surface sheet flow and shallow, slow-moving tidal channel currents exhibit laminar flow and transport only fine-grained materials short distances.

The exact Reynolds number at which the transition from laminar to turbulent flow occurs within the range from 500 to 2000 is variable. It depends on the fluid channel and the precise dimensions of the fluid. An additional factor particularly applicable to windblown transport is the **boundary layer effect,** produced when fluids move adjacent to a stationary boundary (for example, a stream channel developed in previously deposited sediment). The practical consequence of a boundary layer is that turbulent eddies develop within it. Many fluids that usually exhibit laminar flow, such as air (windstorms), contain a boundary layer within which flow is turbulent, which increases their capacity to erode and transport sediment. In windstorms blowing across deserts, the viscosity of air is low enough that laminar flow occurs high above the ground surface, but the moving upper air mass rides upon a basal boundary layer several hundred meters thick in which the flow is turbulent.

Froude Number

The Froude number is the ratio between *fluid inertial forces* and *fluid gravitational forces.* It compares the tendency of a moving fluid (and a particle borne by that fluid) to continue moving with the gravitational forces that act to stop that motion. (Again, the force of inertia expresses the distance traveled by a discrete portion of the fluid before it comes to rest.) Like Reynolds numbers, Froude numbers are dimensionless. The equation for the Froude number, F_r, is

Froude number

$$= \frac{\text{fluid inertial forces}}{\text{gravitational forces in flow}}$$

$$= \frac{\text{flow velocity}}{\sqrt{(\text{acceleration of gravity})(\text{force of inertia})}}$$

$$F_r = \frac{V}{\sqrt{gD}}$$

where V = velocity, D = depth of flow, and g is the gravitational constant.

The relationships among the bedforms or surface waves (ripples and dunes) produced beneath moving currents of wind or water, the flow streamlines within the current itself, and the surface waves developed on the upper surface of the fluid change with the Froude numbers; so too does the type of flow.

When the Froude number is less than 1, the velocity at which waves move is greater than the flow velocity, and waves can travel upstream. This kind of flow is called tranquil, streaming, or subcritical. But if the Froude number exceeds 1, waves do not flow upstream, and the flow is called rapid, shooting, or supercritical. So a Froude value of 1 represents the critical threshold between tranquil and rapid flows. Tranquil flow gives way to rapid flow (often where the channel becomes steeper) with a smooth transition, but when a rapid flow suddenly decreases to a tranquil flow, there is an abrupt change known as a **hydraulic jump**—a sudden increase in depth accompanied by much turbulence (see Fig. 3.1C). If you have ever watched a mountain stream or rapid runoff in storm drains, you have seen examples of hydraulic jumps. The stream is moving with shallow rapid flow and appears to be flowing quickly and smoothly. Then, without warning, it suddenly erupts into a turbulent upstream-breaking wave as the depth increases and the flow becomes subcritical. In most such cases, you are witnessing a flow that has just dropped below the threshold of Froude number 1.

Froude numbers are also important to understanding the ripples and other structures that form at the base of rapidly moving streams. We will discuss these concepts in Chapter 4.

Entrainment, Transport, and Deposition of Clasts

It is difficult to make the transition from laboratory flume experiments—in which the relationships among unidirectional currents of flowing water, bedforms, and sediment transport can be studied under controlled conditions—to the real world. The goal of sedimentologists specializing in hydraulics is to reconstruct all aspects of a flow (velocity, viscosity, and slope and their variations over time) using sediment grain size characteristics and the sedimentary structures produced during deposition. This objective has not yet been reached. It may not be achievable where such complex transporting agents as bidirectional tidal and continental shelf currents or density (turbidity) currents and sediment gravity flows are involved. Nevertheless, some notable relationships have been discovered.

Entrainment: How Are Sediments Lifted into the Flow?

First, we need to understand how particles get picked up, or entrained, into a flow. Two main forces (Fig. 3.2A) are usually involved: the *fluid drag force* (F_D)

exerts a horizontal force (that is, parallel to the flow) on the particle and tends to roll it along. In many cases, the torque produced by this rolling will lift the grain slightly as it rolls over other particles and bring it up off the bottom. But the *fluid lift force* (F_L) is primarily reponsible for raising the particle vertically into the current. The net fluid force (F_F) on the particle is thus the result of the horizontal fluid drag vector (F_D) and the fluid lift force vector (F_L), producing a net movement upward and downstream.

Lift force is an example of a well-known law of hydraulics called **Bernoulli's principle.** In simplest terms, Bernoulli's principle states that the sum of velocity and pressure on an object in a flow must be constant; if the velocity increases, then the pressure must decrease, and vice versa. Thus, wherever a flow speeds up, it exerts less pressure than slower-moving parts of the flow.

The most familiar example of Bernoulli's principle can be seen every time an airplane flies. The cross section of a wing, known as an airfoil (Fig. 3.2B), is designed so that the top surface is convex and the bottom surface is flat. As the wing moves through the air, the air deflected over the top of the wing must travel a longer distance over the curved top surface than the air moving straight along the bottom. If the two masses of air meet after the airfoil

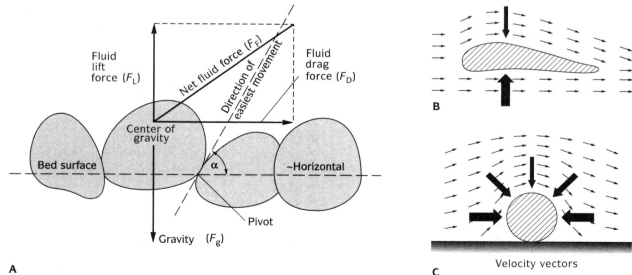

A

B

C Velocity vectors

Figure 3.2 (A) The forces that act upon a particle on a streambed. Although the force of gravity tends to hold the particle down, the fluid lift and drag forces tend to pull the particle up off the streambed and downstream. (After Siever, 1988: 46; by permission of W. H. Freeman and Company, New York.) (B) Streamlines over an airfoil. The flow moving over the top of the wing must move farther, and therefore faster, than the flow beneath the wing. According to Bernoulli's principle, the faster-moving flow exerts less pressure, so the pressure below the wing is greater than that above. This causes a net lift on the wing. (C) The same principle applies to a rounded sand grain on a streambed. The faster flow (and lower pressure) on the top of the grain results in net lift.

passes through them, the air deflected along the top must move *faster* to keep up with the air flowing along the bottom, and the two masses of air come together in the same place. From Bernoulli's principle, we know that the faster-moving air above the airfoil must also have less pressure than the slower-moving air along the bottom. The net difference in pressure between the top and bottom of the wing results in a net lift on the wing, and the airplane rises.

Although a spherical particle is not exactly the same as an airfoil, the application of Bernoulli's principle is similar (Fig. 3.2C). The fluid flowing over the top surface is deflected and must move farther and faster than the flows moving along the sides and bottom. This faster flow means that there is less pressure on the top of the grain than there is on other areas, and the grain is lifted up from the bottom. Once the grain is up in the flow, the pattern of streamlines around the particle becomes symmetrical and there is no further net lift. At this point, other forces must work to keep the particle in motion.

Transport: How Do Sediments Move Once They Have Been Lifted?

Regardless of the agent involved, sedimentary clasts are transported and deposited only in certain ways (Fig. 3.3). Some clasts are moved by **traction;** that is, they are rolled and dragged along the base of a moving fluid. Other materials are moved by **saltation;** that is, they abruptly leave the bottom and are temporarily suspended, essentially hopping, skipping, and jumping downcurrent in an irregular, discontinuous fashion. Many saltating grains strike others, causing them to ricochet and jump into the saltating layer. Traction load and saltation load taken together

constitute the **bedload. Suspension** constitutes a third mode of transport. Suspended load consists of those grains that float more or less continually within the moving fluid. Because sedimentary clasts are denser than the medium that is transporting them, they eventually settle out. However, particles of some materials, such as clays, are so tiny that they do not settle out until the flow has stopped moving entirely, and even then they may take hours to days or weeks to settle.

Clast size has an important effect on sediment entrainment, transport, and settling velocity, the factors that control deposition. The relationship among grain size, entrainment, transport, and deposition is summarized by a classic diagram initially developed by Shields (1936) and subsequently embellished as the Hjulstrom diagram (Fig. 3.4). This graph—based largely on empirical data from flume studies but supplemented with fluid inertial, viscous, and gravity force theory—shows the minimum (or critical) velocity necessary for erosion (entrainment), transportation, and deposition of clasts of varying size and cohesiveness.

The upper field of the graph shows the critical current velocity in centimeters per second (plotted on the logarithmic vertical scale) necessary for a particle of given diameter (plotted on the logarithmic horizontal scale) to be entrained. For most of the graph, the expected relationship holds: higher critical current velocities are necessary to entrain coarser clasts. Note, however, that this holds true only for medium-sized sand and coarser grains. For fine-grained sand, silt, and clay, the critical velocity does not continue to decrease with decreasing grain size. Why not?

The reason for this unexpected behavior is twofold. First, silt- and clay-sized particles do not always

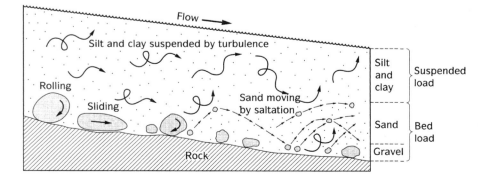

Figure 3.3 The types of movement of particles in a stream. The stream's bedload consists of sand and gravel moving on or near the bottom by traction and saltation. Finer silt and clay are carried in the suspended load and do not settle out until the flow slows down or stops. The dissolved load of soluble ions is not shown here.

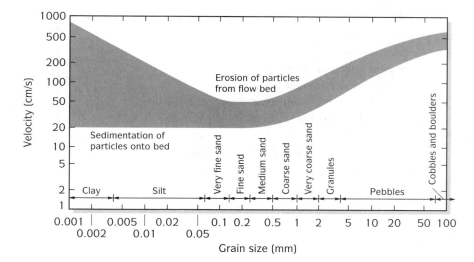

Figure 3.4 Hjulstrom's diagram, showing the critical velocity for movement of quartz grains on a plane bed, as modified by Sundborg (1956). For a given grain size, the lower line is the velocity at which all particles of that size fall to the bed; the upper line is the velocity at which all particles of that size are picked up from the bed or continue to be transported. The top of the transition zone corresponds to water depths of 10 m, the middle to depths of 1 m, and the bottom to depths of 0.1 m or less. The boundary between erosion and sedimentation is broad because of fluctuations in many of the properties of fluids and grains that are not included in measurements of grain size or velocity. Fine silts and clays typically flocculate, or clump together, to form sand- and pebble-sized clayballs that behave as larger particles rather than as discrete, noncohesive clay and silt grains. (After Press and Siever, 1986: 182; by permission of W. H. Freeman and Company, New York.)

behave as discrete grains but sometimes clump together (due to electrostatic attraction of oppositely charged material), or **flocculate,** to form clayballs and other larger lumps. These lumps have an effective diameter equivalent to that of sand- or pebble-sized particles and thus require the same amount of energy as sand or pebbles to be lifted. Second, fine-grained mud forms a more cohesive bed that is inherently difficult to entrain.

The central field in the graph shows the range of velocities necessary to transport clasts (irrespective of transport style) as a function of clast diameter. Once clasts are entrained, there is a straightforward relationship between the critical current velocity and the grain size. Again, there are important effects related to the cohesiveness of silt and clay. These finer clasts can move as aggregated clumps rather than individual fragments. The precise mode of overall transport varies. In general, coarser particles are less likely than finer-grained clasts to be carried high up into the moving current. Consequently grains moved by traction are coarser than grains moved by saltation. The suspended load is finest of all.

Why is there a boundary *zone* of transport, rather than a simple sharp boundary line between erosion and deposition? It turns out that although flow velocity is the main predictor of whether a grain of given size will be eroded or deposited, there are other factors as well. The most important variable is the depth of the flow. The lower boundary of the transport zone is valid for very shallow flows (about 0.1 m deep); the upper boundary applies to very deep flows (10 m or deeper). Apparently, in a very shallow flow, a greater percentage of the flow exists as a thick, turbulent layer that lifts particles; such a flow requires lower velocities to do the same amount of lifting that a deep, mostly laminar flow can produce.

The lower field of the graph shows the relationship between grain size diameter and the current velocity present as clasts are deposited. As a moving stream or a dust storm slows, the material in transit is deposited rapidly, and the size of clasts progressively decreases.

Deposition: What Forces Control the Settling of Particles?

As soon as a particle is lifted above the surface of a bed, it begins to sink back again. The distance it travels as it does so depends on the drag force of the current (just discussed) and the settling velocity of the particle. The velocity with which a clast settles through a fluid is calculated using **Stokes' law of**

settling. We can think of settling velocity in terms of the gravitational force pulling the particle down versus the drag force of the fluid resisting this sinking. Initially, the particle accelerates due to gravity, but soon the gravitational and drag forces reach an equilibrium, resulting in a constant **terminal fall velocity.**

The drag force exerted by a fluid on a falling grain is proportional to the fluid density (ρ_F), the diameter (d) of the grains (in centimeters), the drag coefficient (C_D) and the fall velocity (V) in the following relationship:

$$\text{Drag force} = C_D \pi \left(\frac{d^2}{4}\right) \left(\frac{\rho_F V^2}{2}\right)$$

The upward force due to the buoyancy of the fluid is given by

$$F_{\text{upward}} = \frac{4}{3} \pi \left(\frac{d}{2}\right)^3 \rho_F g$$

The downward force due to gravity can be expressed as

$$F_g = \frac{4}{3} \pi \left(\frac{d}{2}\right)^3 \rho_S g$$

where ρ_S is the particle density. As the particle stops accelerating and achieves fall velocity, the drag force of the fluid on the particle is equal to the downward gravitational force minus the upward force of buoyancy:

$$\text{Drag force} = F_g - F_{\text{upward}}$$

or

$$C_D \pi \left(\frac{d^2}{4}\right)\left(\frac{\rho_f V^2}{2}\right) = \frac{4}{3} \pi \left(\frac{d}{2}\right)^3 \rho_S g - \frac{4}{3} \pi \left(\frac{d}{2}\right)^3 \rho_F g$$

Rearranging terms, this formula can be expressed in terms of fall velocity as

$$V^2 = \frac{4 g d \left(\rho_S - \rho_F\right)}{3 \, C_D \rho_F}$$

For slow laminar flow at low concentrations of particles and low Reynolds numbers, C_D is equal to $24/R_e$. Substituting $24 d\mu / V\rho_F$ (see p. 33, and remember that $2r = d$) for C_D, we get

$$V = \frac{1}{18}\left[\frac{(\rho_S - \rho_F)g d^2}{\mu}\right]$$

This is Stokes' law of settling. The equation may look intimidating until you notice that most of the formula consists of constants. Even the difference in densities between particle and fluid ($\rho_S - \rho_F$) is constant for a given situation. If you combine all these constants into a single constant C, then

$$C = \frac{(\rho_S - \rho_F)g}{18 \, \mu}$$

and Stokes' law reduces to $V = CD^2$.

In other words, when density and viscosity are constant, settling velocity increases with the diameter of the particle, or *larger grains fall faster* (not really a surprising result). In addition, settling velocity decreases with higher viscosities and increases with denser particles.

What are the the implications of this law for sedimentation? For one thing, high-density minerals such as magnetite and olivine settle more rapidly than low-density minerals such as quartz and feldspar. Also, slow-moving, highly viscous fluids such as mudflows and density currents can transport coarser-grained materials than less viscous fluids such as rivers and the wind, despite the normally higher velocity of these less viscous fluids.

Technically, Stokes' law applies only to particles finer than 0.2 mm; because of turbulence, coarser particles fall somewhat more slowly than predicted by Stokes' law, but the effect is so small that Stokes' law is still a good estimate. The fall velocity also decreases at lower temperature, which increases viscosity. Stokes' law ideally describes the settling velocity of perfect spheres through a fluid at rest. Flake-shaped grains can settle more slowly than spheres of material of identical density. Likewise, angular grains generate small turbulent eddies that retard settling velocity. The term *hydraulic equivalency* refers to clasts that settle at identical velocities despite substantial differences in size, shape, angularity, and density. Hydraulic equivalency explains sediment mixes of fine-grained, silt-sized magnetite, fine-sand-sized biotite flakes, and medium-sand-sized quartz.

Putting a fluid in motion also changes the ground rules under which particles settle, especially if velocities are swift enough (beyond laminar flow) to

generate upward-moving turbulent eddies that can maintain clasts in transit.

Sediment Gravity Flows

So far, we have discussed two classes of flows that can move particles: simple dry mass wasting, in which particles move downhill without a fluid lubricating or suspending them (such as rockfalls and landslides); and conventional fluid flows, such as those encountered in a stream current or windstorm. A third flow class, wet or fluid-assisted mass wasting, has properties of both. For example, **density currents** (or fluid-assisted sediment gravity flows) are suspensions of grains in fluid (like a conventional fluid flow). Grains are principally *moved by gravity, rather than simply being carried by the motion of the fluid itself* (as in stream transport or windstorms). Four kinds of sediment gravity flows are generally recognized (Fig. 3.5): grain flows, fluidized sediment flows, debris flows, and turbidity currents.

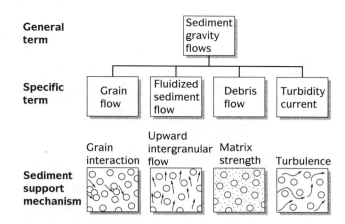

Figure 3.5 Classification of the four major types of sedimentary gravity flows, showing the interactions between fluids and grains that keep sediment moving during transport.

Grain Flows

Grain flows occur when cohesionless sediment (for example, dry sand) moves downward under the pull of gravity (Fig. 3.6A). Although there is air or water trapped between the grains, it merely acts as a

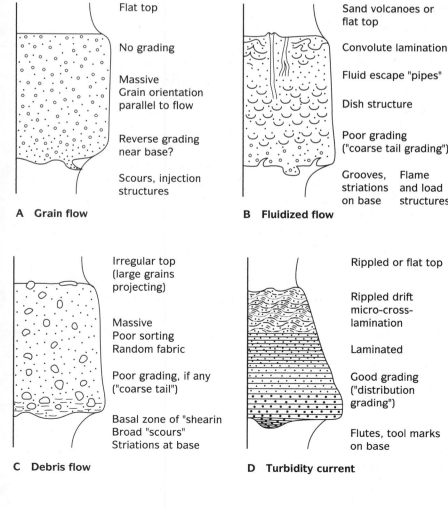

A Grain flow
- Flat top
- No grading
- Massive Grain orientation parallel to flow
- Reverse grading near base?
- Scours, injection structures

B Fluidized flow
- Sand volcanoes or flat top
- Convolute lamination
- Fluid escape "pipes"
- Dish structure
- Poor grading ("coarse tail grading")
- Grooves, striations on base   Flame and load structures

C Debris flow
- Irregular top (large grains projecting)
- Massive Poor sorting Random fabric
- Poor grading, if any ("coarse tail")
- Basal zone of "shearin Broad "scours" Striations at base

D Turbidity current
- Rippled or flat top
- Rippled drift micro-cross-lamination
- Laminated
- Good grading ("distribution grading")
- Flutes, tool marks on base

Figure 3.6 Sedimentary structures of the four major types of sedimentary gravity flow deposits. (After Middleton and Hampton, 1976.)

Figure 3.7 Avalanching sand grains on the lee side of a dune demonstrate grain flow, a sedimentary gravity flow in cohesionless sand. (D. R. Prothero.)

lubricant and does not actually propel the grains. Instead, the flow is sustained by the dispersive pressures caused by grain-to-grain collisions as the particles ricochet off one another. The most familiar examples of grain flows are the sand avalanches that occur whenever a sand dune becomes steeper than the angle of repose (Fig. 3.7). The next time you are walking in loose, dry sand (or you can pile up the sand in a sandbox to simulate this), watch the steep side of the dunes. At some critical angle, the sand grains begin to avalanche downward spontaneously until they reach a shallower, stable angle. Gravity is moving the grain flow, and the fluid (air) between the grains is simply acting as a medium through which the grains pass. Grain flows also occur in the deep sea, where sand cascades down the sides of steep submarine canyons.

Fluidized Sediment Flows

Fluidized sediment flows are concentrated dispersions of grains supported by pore water between the grains (Fig. 3.6B). They begin to flow laterally when something increases the pressure on the interstitial pore water, turning the once-firm sand into a soupy liquid. The most familiar example is *quicksand*. Contrary to what we see in the movies, quicksand is not an inexorable monster that sucks in unsuspecting victims. It is simply a bed of sand with water saturating the pores. When something (for example, a person's foot) puts additional pressure on the sand-water mixture, the increased pore pressure causes the water to move and to try to escape upward. As the water moves, the grains shift and settle past one another, a process called **liquefaction.** Once liquefied, the sediment will continue to flow around the source of pressure until that source stops exerting

any further force, producing downslope transport and redeposition of mud and sand. At that point, the water will sink back into the pores and the whole mass will become solid sand again.

Once we understand this process, we can see that the Hollywood stereotypes of quicksand are silly. As long as you don't thrash around, thereby keeping up the pressure to fluidize the mixture, the sand remains solid. If you should happen to step in real quicksand, the thing to remember is that it's just a very dense body of water. If you continue to stand vertically, you will continue to sink (as you do in a swimming pool) until you reach your point of neutral buoyancy. Even if you thrash around, you will not sink above your head, since you are less dense than water (after all, you float in a swimming pool), and you are even less dense than a thick slurry of sand and water (so you should float higher than you do in water). To get out of quicksand, swim just as you would in a pool. If you lie flat, you will distribute the weight of your body across a much wider area than just the soles of your feet, so the pressure at any given point is less. Eventually, you will float (as you do in a pool), and if you can find some way to make forward progress (by grabbing a stationary object such as a rope or a pole), you will get out.

One interesting aspect of liquefaction is that it develops slowly and requires continuous pressure to be maintained. The initial pressure applied starts the pore fluids moving, but if the pressure is removed immediately, the sand may not move. Continued pressure, however, sustains the flow, and it can trap whatever is on top. For example, it is not unusual for a four-wheel-drive truck to cross an apparently dry streambed without sinking. A second vehicle in the caravan, however, sinks down to the axles. The first vehicle got the fluid moving, and by the time the second vehicle reached the sand, it was liquefied.

Liquefaction occurs in many other instances besides quicksand. Shock waves sent through saturated soil by an earthquake cause liquefaction. In some places, the liquefied sand in a riverbed may actually boil up to form sand volcanoes. Any object (such as a building foundation) that puts pressure on this suddenly soupy substrate will sink. This was vividly demonstrated by the sagging buildings in the Marina district of San Francisco, which began to sink and lean and eventually broke up and burned during the 1989 Loma Prieta earthquake. Unlike some other parts of the Bay Area, the Marina district was built on old bay fill, which was fully saturated and thus liquefied during the earthquake.

Mudflows and Debris Flows

Mudflows are composed of a slurrylike mass of liquefied mud that moves downhill under the force of gravity. If there are larger particles (ranging up to boulder size), then they are known as **debris flows** (Figs. 3.6C, 3.8). Such slurries are dense enough to support very large particles but not solid enough to resist flowing downhill.

Mudflows and debris flows are most common in steep mountain canyons when a mass of mud and debris becomes saturated during a heavy rainstorm and suddenly begins to flow down the canyons.

These flows have the consistency of wet cement and can move as fast as water in a flash flood. In some mountainous regions, they can be deadly, picking up houses, cars, trees, and other large objects; carrying them long distances; hurling them at other houses and trees; and crushing them in their wake (Fig. 3.9). Some debris flows have been truly catastrophic. The 1985 eruption of the Nevado del Ruiz volcano in Colombia caused a sudden melting of its mountain glacier, releasing a thick slurry of liquid volcanic ash moving at about 100 km/hr (60 mph). Downstream, the city of Armero was wiped out by a thick carpet of mud and debris that struck in the middle of the night, killing 25,000 people almost instantly.

A

B

C

Figure 3.8 Sequence of aerial photos of a debris flow moving along a canyon bottom near Farmington, Utah, in June 1983. (A, B) The steep (2 m high) bouldery front advances from left to right at about 1.3 m/s and acts as a moving dam, holding back the muddy slurry behind it. (C) The main slurry is traveling at about 3 m/s and is viscous enough to carry cobbles and boulders in suspension. (Courtesy of the U.S. Geological Survey.)

Figure 3.9 Workmen stand on the roof of a home in La Crescenta, California, that was buried to the eaves by a debris flow in February 1978. This mudflow carried rocks, boulders, and even cars. It was triggered by a combination of heavy rains, steepened slopes, and devegetation by wildfires. (Courtesy of the U.S. Geological Survey.)

Figure 3.10 Inversely graded debris-flow deposit, Miocene Violin Breccia, Ridge Basin, California. Note the large rocks in the top of the debris flow that projected upward and were buried by fine material at the base of the next flow. (D. R. Prothero.)

The terrifying effect of debris flows was described vividly by John McPhee in his book, *The Control of Nature:*

The water was now spreading over the street. It descended in heavy sheets. As the young Genofiles and their mother glimpsed it in the all but total darkness, the scene was suddenly illuminated by a blue electrical flash. In the blue light they saw a massive blackness, moving. It was not a landslide, not a mudslide, not a rock avalanche; nor by any means was it the front of a conventional flood. In Jackie's words, "It was just one big black thing coming at us, rolling, rolling with a lot of water in front of it, pushing the water, this big black thing. It was just one big black hill coming toward us."

In geology, it would be known as a debris flow. Debris flows amass in stream valleys and more or less resemble fresh concrete. They consist of water mixed with solid material, most of which is sand size. Some of it is Chevrolet size. Boulders bigger than cars can ride long distances in debris flows. The dark material coming toward the Genofiles was not only full of boulders; it was so full of automobiles it was like bread dough mixed with raisins. On its way down Pine Cone Road, it plucked up cars from driveways and the street. When it crashed into the Genofiles' house, the shattering of safety glass made terrific explosive sounds. A door burst open. Mud and boulders poured into the hall. We're going to go, Jackie thought. Oh, my God, what a hell of a way for the four of us to die together.

The house became buried to the eaves. Boulders sat on the roof. Thirteen automobiles were packed around the building, including five in the pool. A stuck horn of a buried car was blaring. The family in the darkness in the fixed tableau watched one another by the light of a directional signal, endlessly blinking. The house had filled up in six minutes, and the mud stopped rising near the children's chins. (McPhee, 1989: 184–186)

Some beds exhibit **reverse** (or **inverse**) **grading,** with the coarsest material at the top and the finer material at the bottom (Fig. 3.10). The mechanism that produces inverse grading is not well understood, but such grading is commonly found in debris flows, probably because the dispersive pressures of a grain flow tend to push the larger particles to the top of the flow where they encounter less friction. The finer grain sizes, on the other hand, can move more easily in the base of the flow, where the shear stress against the bottom is greater. The intermediate sizes will be arrayed through the reversely graded bed so that they equalize the shear pressure gradient between the top and the base.

Turbidity Currents

Turbidity currents are gravity flows in which the sediment is supported by upward turbulence of the fluid within the flow (Fig. 3.6D). They occur when an earthquake or other shock stirs up a mixture of sand

A

B

Figure 3.11 (A) Experimental turbidity current of denser, more turbulent muddy water flowing (from right to left) beneath less dense, clear water. The current flows down a sloping laboratory flume in water depths of about 50 cm. Note how the turbidity current is completely discrete and separate from the water through which it moves and has a turbulent, eddying front and top where it meets the water. (Courtesy of G. Middleton and R. G. Walker.) (B) Graded bedding in a turbidite, ranging from coarse sand and gravel at the base (bottom of hammer) to fine silt at the top. This example is from the Pliocene Pico Formation, Santa Paula Creek, Ventura County, California. (D. R. Prothero.)

and mud on the bottom of a lake or the ocean. This liquefied mass of suspended sediment is denser than the lake water or seawater that surrounds it, so it can move downhill under the force of gravity (Fig. 3.11A). The mass can move downhill even on very gradual slopes. Unlike other bottom currents, however, *a turbidity current is not propelled by the water within it, but by gravity;* the water simply suspends the particles.

Turbidity currents typically produce normal (as opposed to inverse) **graded bedding,** where the grains in a single bed grade gradually from coarse grain sizes at the base to fine grain sizes at the top (Fig. 3.11B). Graded bedding usually results from the settling of a single poorly sorted suspension of sand, silt, and clay in which the coarser fractions settle out more rapidly than the finer fractions, following Stokes' law. Because of the difference between the density of the turbidity current and that of the surrounding clear water, the turbidity current flows as a discrete unit; very little mixing with the surrounding, less dense water occurs. The poorly sorted material remains suspended until the flow begins to slow down. Then the turbulence ceases, and the material settles out according to size. The depositional product of a turbidity current is called a **turbidite.**

Thick sequences of normally graded beds were long known in the geologic record, but for decades they were a puzzle. The coarse sand in such beds seemed to indicate nearshore deposition, but they were interbedded with deep-water shales. The foraminifera in the sandstones indicated shallow water, but those found in the shales were from the ocean depths. How could the two become so intimately and repeatedly interbedded?

Several scientists had observed density currents in shallow lakes, but it was the Dutch geologist Philip Kuenen who put all the pieces together in the 1940s and 1950s. Combining earlier observations of density currents with his own experiments in a laboratory tank, he showed that density currents could move enormous distances over relatively gradual slopes at velocities high enough (typically 0.4 to 3.0 m/s) to transport sand to the deep seafloor.

In the 1950s, oceanographer Maurice Ewing recognized that a natural experiment on turbidity currents had already taken place. During the November 18, 1929, earthquake off the Grand Banks of Newfoundland, the transatlantic telegraph cables on the continental slope had been broken. They had snapped first at the top of the slope and then at progressively later times down the slope. Back in 1929, no one could explain this peculiar pattern, but Ewing realized that the culprit was a turbidity current that had been triggered at the top of the continental slope and had broken cables on its way down as if they had been trip wires. From the precise times that the telegraph service had been interrupted, Ewing could calculate how fast the turbidity current was moving. On the steepest slope, it was moving at 100 km/hr (or about 65 mph)! As it reached the flatter continental rise and abyssal plain, it slowed to 25 km/hr (about 15 mph). The turbidity current covered more than 280,000 km^2 of the North Atlantic seafloor with well over 100 km^2 of sediment.

CONCLUSIONS

In this brief chapter, we could not present a detailed, comprehensive discussion of all the physical processes that transport sediments. The student interested in further information (with much more physics and math than we have used) should consult some of the references listed at the end of the chapter. The basic concepts we have discussed are crucial to understanding modern sediments and ancient sedimentary rocks.

In the natural world of sediment transport and sediment deposition, a fundamental battle is waged between the fluid dynamic forces that tend to entrain clasts and maintain them in transit and the forces encompassed by Stokes' law that control their settling. In the next chapter, we will see how an understanding of the physics of fluid transport can be applied to understanding sedimentary structures.

FOR FURTHER READING

Allen, J. R. L. 1985. *Principles of Physical Sedimentology*. London: George Allen and Unwin.

Chamley, H. 1990. *Sedimentology*. New York: Springer-Verlag.

Hsü, K. J. 1989. *Physical Principles of Sedimentology, A Readable Textbook for Beginners and Experts*. Berlin: Springer-Verlag.

Leeder, M. R. 1982. *Sedimentology, Process and Product*. London: George Allen and Unwin.

Pye, K., ed. 1994. *Sedimentary Transport and Depositional Processes*. Cambridge, Mass.: Blackwell Scientific Publications.

Selley, R. C. 1982. *An Introduction to Sedimentology*. London: Academic Press.

Selley, R. C. 2000. *Applied Sedimentology*, 2d ed. San Diego: Academic Press.

Shapiro, A. H. 1961. *Shape and Flow: The Fluid Dynamics of Drag*. New York: Doubleday.

Yang, C. T. 1996. *Sediment Transport: Theory and Practice*. New York: McGraw-Hill.

4 Sedimentary Structures

Close-up of aggrading ripples (climbing ripple drift) form the Colorado River, Arizona. These bedforms are indicative of a rapid unidirectional flow with excess sediment supply. (Courtesy of Paul E. Potter.)

SEDIMENTARY STRUCTURES ARE MACROSCOPIC, THREE-dimensional features best seen in outcrop. There are two major classifications: primary (physical) and secondary. **Primary sedimentary structures** occur in clastic sedimentary rocks and are produced by the same currents that deposit the sediment in which they are present. Their presence, scale, and orienta-tion reflect the conditions of transport and deposi-tion. Consequently, they are reliable indicators of origin. **Secondary sedimentary structures** form after sediment is deposited. They are produced by a va-riety of organic, physical, and chemical processes. This chapter discusses biogenic and mechanically produced secondary structures. Chapters 7 and 11

describe the various chemically produced sedimentary structures such as concretions, nodules, geodes, and stylolites, all of which result from diagenetic dissolution and recrystallization.

Primary Sedimentary Structures

Primary structures are features of sedimentary rocks generally formed without the influence of organisms. They are produced mechanically under the influence of the same hydrodynamic and/or aerodynamic conditions that control the entrainment, transport, and deposition of particles, as discussed in Chapter 3.

Identification and analysis of primary sedimentary structures can help resolve a number of questions about a stratigraphic sequence. Such questions include

1. Which way is "up"? (Top and bottom indicators, known as **geopetal structures,** include cross-bedding and graded bedding.)

2. How was the current system that dispersed the sediments oriented? **Directional structures** indicate current direction.

3. What agent or agents transported and deposited a particular sedimentary layer?

Plane Bedding

The simplest sedimentary structure is plane bedding. Simple horizontal beds form in practically all sedimentary environments and under a variety of conditions, so further descriptive detail is needed to interpret them. Bedding is so common in sedimentary rocks that its origin is often overlooked because it is assumed to be inevitable. Three basic mechanisms can form plane bedding: sedimentation from suspension, horizontal accretion from a moving bedload due to a change in the competence of a flow, and encroachment into the lee of an obstacle. Plane beds often represent rapid deposition, usually by a single hydrodynamic event. Most deposited beds have been reworked, however, so *preservation potential* is also important. For example, submarine landslide deposits are buried rapidly, so they have a high preservation probability, but beach deposits are almost always reworked after they are deposited.

Finer-scale plane bedding (less than 1 cm thick) is usually called **lamination.** A number of mechanisms

Figure 4.1 Seven years of Pleistocene varves near Seattle, Washington. Total thickness shown is about 12 cm. (Courtesy of J. Hoover Mackin.)

can form laminae. The classic example is the alternation of light and dark layers, such as glacial varves (Fig. 4.1). Alternation of mineral composition also occurs, as is seen in heavy mineral lags among the normal quartz sands on some beaches. Lamination can also result from the alternation of grain sizes caused by changes in the strength of the current during deposition. In some cases, apparent lamination is not a primary feature at all but is a secondary color banding due to diagenetic effects.

Lamination in muds is usually the result of slow, steady deposition. An absence of lamination in mudstones is probably due either to flocculation (clumping of clays before they settle) or to secondary **bioturbation** (disturbance by organisms). Usually, laminated sands have been deposited rapidly, often by a single hydrodynamic event such as the swash-backwash of surf; traction by steady flow; the avalanching of sand down a dune face; or the migration of ripples, which leaves a heavy mineral lag. Truly massive deposits that show no bedding are anomalous. Such deposits often show cryptic bedding when studied by X-ray techniques or when stained (Fig. 4.2). The lack of plane bedding is usually a result of bioturbation, deposition from highly concentrated sediment dispersions, or rapid deposition from suspension.

Bedforms Generated by Unidirectional Currents

As soon as flow attains a force sufficient to erode particles from the bed, sediments are transported in a set of structures on the surface of the bed called

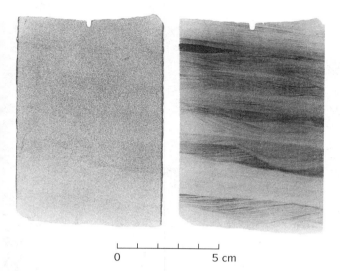

Figure 4.2 A polished slice of core (*left*) and a positive print of an X-radiograph (*right*) of the Mississippian Berea Sandstone, Illinois. Only vague banding is visible on the polished slab, but X-radiation reveals an apparent dip of 10°, scour and fill, and cross-bedding. (After Hamblin, 1965.)

bedforms. If these bedforms are later buried and preserved, they can form sedimentary structures. Flume studies have shown that there is a predictable sequence of bedforms that depends on velocity, grain size, and depth of flow (Fig. 4.3A). In sand that is finer than 0.7 mm (coarse sand or finer), the first features to form are *ripples.* Typically, their spacing is 10 to 20 mm or less, and their height is less than a few centimeters. As the flow velocity increases, the ripples enlarge until they form sand waves and finally dunes, which have spacings from 0.5 to 10 m or more and heights of tens of centimeters to a meter or more (Fig. 4.3B).

In deeper currents, greater flow velocity is required to produce the larger bedforms. Changes in water temperature or clay content alter the viscosity of the flow and thus affect the settling velocity. These factors alter the bedforms regardless of the other variables. Ripples, sand waves, and dunes all form in

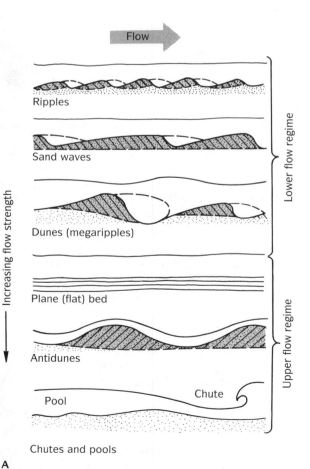

A

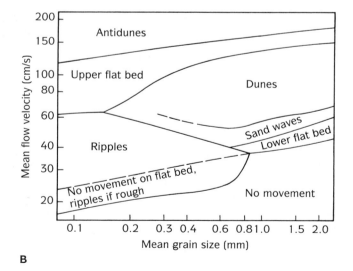

B

Figure 4.3 (A) Sequence of bedforms produced under conditions of increasing flow strength. (After Blatt, Middleton, and Murray, 1980: 137; by permission of Prentice-Hall, Inc., Englewood Cliffs, N.J.) (B) Changes in bedforms resulting from different flow velocities (vertical axis) and grain sizes (horizontal axis). (After Lewis, 1984: 42.)

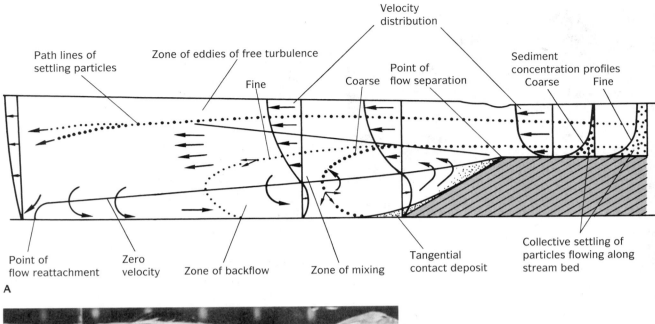

Figure 4.4 (A) Flow pattern and sediment movement over migrating ripples or dunes. Velocity profiles are shown by the vertical lines. (After Jopling, 1967: 298; © 1967, by permission of the University of Chicago Press.) (B) In a laboratory flume, the trajectories of sand grains on the lee side of a ripple (migrating from right to left) can be seen. Layers of dark sand are also included to show the development of cross-bedding. (Siever, 1988: 65; courtesy of A. V. Jopling.)

much the same manner. Small irregularities on the surface of the streambed cause a slight turbulence as the flow is diverted up and around them. Eventually, the flow over an obstacle no longer hugs the bottom but separates from it at the **point of flow separation** (Fig. 4.4), which is at the crest of the ripple or dune. The flow meets the bottom again at the **point of flow reattachment.** Beneath this zone of laminar flow is the zone of turbulence and backflow on the lee side of the ripple. This is the **zone of reverse circulation.** Sediment migrating up the ripple or dune avalanches down into this zone and is deposited by the weaker currents. This process generates the inclined foreset beds that produce cross-bedding. Because the ripple or dune is eroded on the upstream side and accreted on the downstream side, these bedforms migrate downstream. Meanwhile, most of the fine-grained

suspended load of silt and clay is carried downstream, resulting in segregation of grain sizes.

The shape of the ripples depends primarily on a balance between the bedload and the material that is settling from suspension. If there is little suspended load, the ripples are steep, with a sharp angle between the foreset and bottomset beds. If there is a large suspended load, the lee slope builds steadily, forming curved cross-strata and a tangential contact between foreset and bottomset beds. Ripples and dunes are dynamic features that change constantly. The downstream end of the zone of backflow (the point of reattachment) fluctuates continuously, so only its approximate position can be identified. Beyond the point of reattachment, turbulent eddies scour downstream and form troughs with their long axes parallel to the flow. As the ripples or dunes mi-

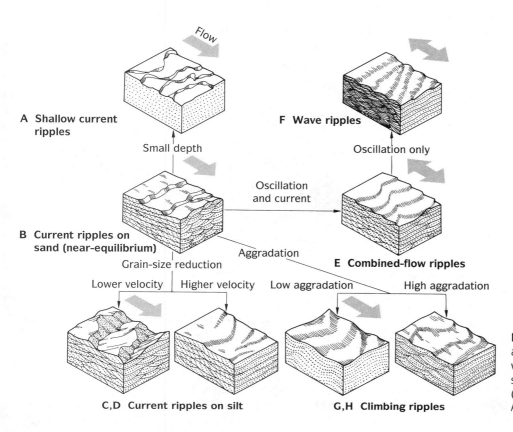

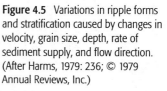

A Shallow current ripples

Small depth

F Wave ripples

Oscillation only

Oscillation and current

B Current ripples on sand (near-equilibrium)

Grain-size reduction

Aggradation

E Combined-flow ripples

Lower velocity   Higher velocity   Low aggradation   High aggradation

C,D Current ripples on silt   G,H Climbing ripples

Figure 4.5 Variations in ripple forms and stratification caused by changes in velocity, grain size, depth, rate of sediment supply, and flow direction. (After Harms, 1979: 236; © 1979 Annual Reviews, Inc.)

grate downstream, they fill the troughs in front of them. This natural association of troughs and ripples produces normal trough cross-stratification.

Dunes form by the same processes as ripples, only on a much larger scale (centimeters in the case of ripples, meters in the case of dunes). Whereas ripples are unaffected by changes in depth and are strongly affected by changes in grain size, dunes are more strongly affected by depth and less affected by grain size. Dune height is limited only by depth of flow, but ripples can reach only a certain maximum height. Ripples tend to migrate in one plane (except in the case of climbing ripple drift, discussed later). Dunes, on the other hand, often migrate up the backs of other dunes.

With increased flow velocity, dunes are destroyed, and the turbulent flow, which was out of phase with the bedforms, changes to a sheetlike flow, which is in phase with the bedforms. This point is also marked by Froude numbers greater than 1, indicating that the flow has become rapid, shooting, or supercritical. Intense sediment transport takes place along **plane beds** (see Fig. 4.3A) which are produced by sand deposition on a planar surface. At even higher velocities, plane beds are replaced by **antidunes,** which produce low, undulating bedforms that can reach 5 m in

spacing. Their fundamental feature is that their crests are *in phase* with the surface waves, so they migrate by accretion on the upstream side. In ancient deposits, antidunes are characterized by faint, poorly defined laminae. Antidunes generally show low dip angles (less than 10°) and are associated with other indicators of a high flow velocity. Because they migrate upstream, antidunes should leave evidence of a flow contrary to the flow direction shown by other current-direction indicators (see Box 4.1). It seems that antidunes are rare in the rock record, probably because they are re-worked where the current slows before final burial. Finally, at the highest flow velocities, the antidunes wash out and are replaced by chutes and pools (see Fig. 4.3A).

The three-dimensional geometry of cross-stratification is a useful indicator of flow and sediment load. Starting with stationary current ripples (Fig. 4.5A), simple trough cross-stratification develops from migrating ripples and dunes (Fig. 4.5B). Tabular cross-stratification (Fig. 4.5C, D), on the other hand, is produced by migrating sand waves. Horizontal stratification can be produced by plane-bed conditions at high flow velocities. Often, the migration of a ripple is interrupted; the ripple is eroded back and then buried by a new advancing bedform. Such an interruption

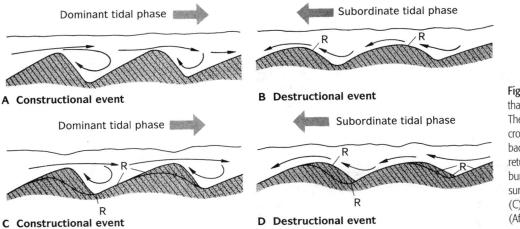

Dominant tidal phase →

← Subordinate tidal phase

A **Constructional event**

B **Destructional event**

Dominant tidal phase →

← Subordinate tidal phase

C **Constructional event**

D **Destructional event**

Figure 4.6 The sequence of events that forms reactivation structures. The dominant tidal phase builds cross-beds (A), which are eroded back during tidal retreat (B). The return of the constructional tide buries this erosional reactivation surface, R, with new cross-beds (C), and the process repeats (D). (After Klein, 1970: 1118.)

produces a tiny erosional surface between cross-strata, known as a **reactivation surface** (Fig. 4.6).

Figure 4.5 shows the natural sequence of ripple features resulting from changes in flow conditions, grain size, and sediment supply. As flow increases, incipient ripples develop into full-scale trough cross-beds at equilibrium. If the grain size then decreases, the shape of the current ripples changes, depending on flow velocity (see Fig. 4.5C, D). If the current becomes less unidirectional, sinuous combined-flow ripples result (Fig. 4.5E). A fully oscillatory current (such as in waves) produces straight, symmetrical ripple marks with a distinctive lenticular cross section (Fig. 4.5F; see also Fig. 10.8). If the sediment supply increases, then the ripples build upward, or **aggrade.** Low aggradation produces **climbing ripples** (Fig. 4.5C; see also p. 45). High aggradation produces sinuous ripples that are in phase (Fig. 4.5H).

Bedforms Generated by Multidirectional Flow

Although they form in a different manner, wave ripples on beaches are similar to current ripples. A rotating eddy precedes a wave as it moves onshore, precipitating the sand load into troughs and ripples. As the wave crest passes, the eddy rises with the crest and disperses into the backwash. The coarser grain sizes are left on the beach, and the finer sand is washed offshore, so beach sands are very well sorted. Wave ripples are not easy to distinguish from current ripples, but there are some differences. Wave ripples are usually symmetrical (or only slightly asymmetrical) with peaked crests and rounded troughs. If they are asymmetrical at all, they indicate a current direction toward the shore. Their cross-laminae also dip shoreward.

Other waveforms are confined to tidal regions. Unlike on the beach, fine sediment in the tidal zone is moved onshore because incoming tides flow in slowly, allowing the sediment to settle. Retreating tides move out too slowly to scour away much of this deposition. As a result, tidal ripples are generally unidirectional, with weak backflow structures. Cross-beds are oriented in two directions, often with reactivation surfaces caused by the reversal of current direction during a tidal cycle. This is known as **herringbone cross-bedding** (Fig. 4.7). The bidirectionality

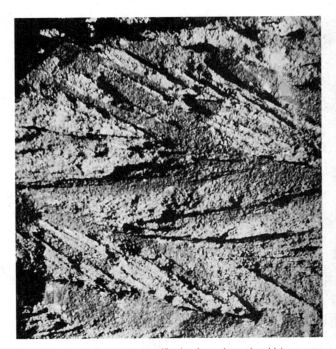

Figure 4.7 Herringbone cross-stratification from alternating tidal currents, Cambrian Cadiz Formation, Marble Mountains, California. (D. R. Prothero.)

BOX 4.1 PALEOCURRENT ANALYSIS

Sedimentary structures can be used to interpret depositional environments and ancient hydraulics in many ways. One of the most valuable pieces of data is the flow direction indicated by unidirectional or bidirectional currents. For example, the flow direction and source of ancient river systems can often be determined from ancient cross-bedding orientations; the downslope direction of a turbidity current can be determined from the orientation of flute casts and other directional sole marks. Paleocurrents may be crucial to testing certain hypotheses. For example, if the flow is unidirectional, flowing away from ancient source areas, and perpendicular to the ancient shoreline, it is probably fluvial or deltaic in origin. If the cross-beds are bidirectional, perpendicular to the shoreline, and 180° apart, they were probably caused by onshore-offshore tidal currents or waves. Unidirectional marine paleocurrents oriented parallel to the shoreline might be the result of longshore currents. Such information could be used to determine whether a cross-bedded sandstone in the marine-nonmarine transition is fluvial-deltaic, tidal, or longshore current in origin.

A number of paleocurrent features can be measured, including tabular and trough cross-bedding, the trends of channel axes, the alignment or imbrication of fossils or clasts, grain alignment in sandstones, sole marks (especially flute casts, drag marks, and groove casts), current and oscillation ripples, and even overturned soft-sediment folds (they indicate downslope). If these structures are well exposed in flat-lying strata, their trend or azimuth can be measured directly with a Brunton compass. In deformed strata, however, this trend must be corrected for the dip of the bedding. This is done using a stereonet.

First, the dipping plane of the bedding is represented as a great circle on a piece of tracing paper (Fig. 4.1.1). Then the paper is rotated to place the strike of the great circle along the north-south axis. The angle, or **rake,** between the current structure and the strike line (as measured in the field) is then plotted along the great circle (Fig. 4.1.2A). This gives an apparent azimuth of the paleocurrent direction (260° in this example). Finally, the bedding plane is rotated back to horizontal (Fig. 4.1.2B). During this rotation, the intersection between the paleocurrent and the plane of the bedding will also rotate along one of the small circles to the edge of the stereonet (horizontal). This gives the true trend of this current in the horizontal plane. (For bedding dips of less than 25°, the difference between

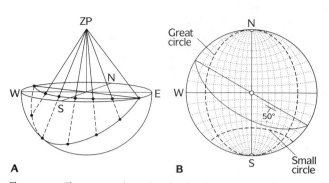

Figure 4.1.1 The stereonet is used to visualize three dimensions on a two-dimensional plot. (A) Projections of a plane with a dip of 50° and a dip direction of 210° (strike N60°W, dip 50°SW). ZP, zenith point. (B) Stereographic projection of the plane shown in (A). Also shown are projections of great circles (the intersection of a sphere with any plane passing through the center of the sphere) and small circles (the intersection of a sphere with any plane not passing through the center of the sphere). (After Lindholm, 1987: 44; by permission of Allen and Unwin, London.)

corrected and uncorrected paleocurrents is so slight that it is not necessary to correct at all.)

In other cases, we have only side views of the structure in three dimensions and cannot see the trend of the flow in outcrop clearly. For example, a rock may protrude and give two different views of the cross-bedding (as exposed by random joint faces), but there are no faces that are exactly perpendicular to the flow direction to allow measurement of the true trend. In these instances, we can measure the apparent dip of

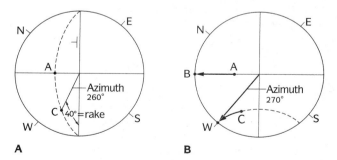

Figure 4.1.2 The correction of a linear structure for tectonic tilt using the stereonet. (A) Plot the plane of bedding as a great circle and the linear structure as a line. In this example, the bedding has a dip of 50° and a dip direction of 320° (strike N50°E, dip 50°W). The rake of the linear structure is 40°; the azimuth of a vertical plane, which passes through the linear structure, is 250°. (B) Restore the bedding to horizontal (point A to point B). Move the intersection point of the linear structure with the great circle projection of the bedding point (point C) along the nearest small circle (dotted line) to the edge of the stereonet. Read the azimuth of the linear structure. In this example, it is 270° (due west). (After Lindholm, 1987: 44; by permission of Allen and Unwin, London.)

(continued)

(Box 4.1 continued)

the cross-bedding on each of two faces in a single cross-bed set. We also measure the strike and dip of each of the two rock faces. On the stereonet, these are shown as great circles, and the two apparent dips occur as points on each great circle. Rotating the stereonet so that these two points align along a common great circle produces the great circle of the plane of the cross-bedding dune or ripple face. The dip direction of this plane is the true current direction.

If there are more than two or three paleocurrents, a summary of the vectors is needed. The most common of these is known as a **rose diagram** (Fig. 4.1.3). Rose diagrams are circular graphs that summarize data on current vectors (the row data appear as the table in Fig. 4.1.3). The compass is divided into convenient sectors (like the segments of an orange), typically of 20° to 30° of arc. All the corrected paleocurrent vectors that fall within a given sector are then summarized as "pie wedges," with the length of the pie wedge indicating the total number of vectors in that segment. The rose diagram shows the degree of scatter within unidirectional currents and often reveals that there are bimodal or polymodal vectors in the data set, indicating highly variable or multidirectional currents. Through visual inspection and comparison with other data, the significance of each mode should be apparent.

Although the rose diagram gives a good visual representation of the vector trend and the scatter of the data, a more rigorous statistical analysis is needed (especially if we want to compare rose diagrams from two or more places).

Two common methods trigonometric and graphical are shown in Fig. 4.1.4. Once the vector mean is known, we also need to know the scatter of

the vectors, or vector dispersion, known as the **consistency ratio** (analogous to the standard deviation in univariate statistics). These ratios allow a more rigorous comparison, such as determining whether two vector distributions are statistically the same or clearly come from different directions.

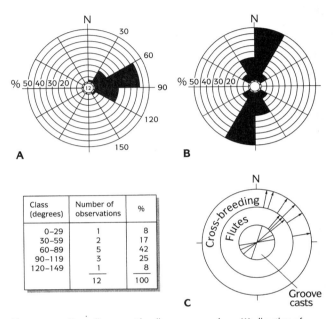

Class (degrees)	Number of observations	%
0–29	1	8
30–59	2	17
60–89	5	42
90–119	3	25
120–149	1	8
	12	100

Figure 4.1.3 Rose diagrams. The diagrams may show (A) direction of movement data (12 cross-bed dip azimuths in degrees); (B) line of movement data (compass bearing of 8 groove casts in degrees); or (C) data from several different structures (compass bearing of 4 groove casts, 3 flute casts, and 6 cross-bed azimuths). C shows the raw data on which the rose diagrams are based. (After Lindholm, 1987: 46; by permission of Allen and Unwin, London.)

of tidal outflow currents often superimposes a weaker ripple system on the dominant sinuous ripples produced by rising tides. These two systems produce **interference ripples,** or "tadpole nests" (Fig. 4.8). The most distinctive features of tidal regions are caused by the mixing of sand- and mud-sized fractions from the asymmetrical currents. Small lenses of sand in muddy beds, called **lenticular bedding** (Fig. 4.9A, B), occur when sand is trapped in troughs in the mud as sand waves migrate across a muddy substrate. If mixing produces minor mud layers in a sandy substrate, the pattern is called **flaser bedding** (Fig. 4.9A, C). An equal mixture of sand and mud (Fig. 4.9A) characterizes **wavy bedding.**

Wind-transported sand behaves differently from water-transported sand, although wind-generated ripples look superficially like water-generated ripples. Sand particles in wind move mostly by saltation (jumping and bouncing) and to a lesser extent by surface creep. Particles that are too large to move by saltation and creep accumulate as a lag, forming a desert pavement in areas of wind deflation. Because saltation is more effective than scouring in moving sand, erosion is heaviest on the exposed upwind side of a sand dune, where the impact of windblown particles is greatest. Deposition occurs on the protected lee side; because there is no zone of backflow, the lee sides do not scour. This is

$$\tan x = \frac{\Sigma n \sin x}{\Sigma n \cos x} = \frac{11.3541}{-3.2085} = -3.539$$

arctan $-3.539 = -74°$ or $106° =$ vector mean

$$R = \sqrt{[(\Sigma n \sin x)^2 + (\Sigma n \cos x)^2]}$$
$$= \sqrt{(128.91 + 10.29)} = 11.8$$

$$L = \frac{R}{n} \times 100 = \frac{11.8}{15} \times 100 = 79 = \text{vector magnitude}$$

A   Trigonometric method

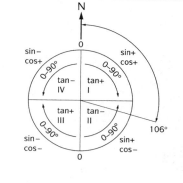

Length of resultant vector = 12 units

Vector magnitude $= \dfrac{12}{15} = 80\%$

B   Graphical method

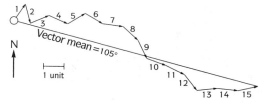

Azimuth		sin x	cos x
1	27°	+0.4540	+0.8910
2	172°	+0.1392	−0.9903
3	68°	+0.9272	+0.3746
4	112°	+0.9272	−0.3746
5	50°	+0.7660	+0.6428
6	123°	+0.8387	−0.5446
7	100°	+0.9480	−0.1736
8	137°	+0.6820	−0.7314
9	160°	+0.3420	−0.9397
10	111°	+0.9336	−0.3584
11	118°	+0.8829	−0.4695
12	146°	+0.5592	−0.8290
13	80°	+0.9848	+0.1736
14	96°	+0.9945	−0.1045
15	77°	+0.9748	+0.2250
Σn		+11.3541	−3.2085

Figure 4.1.4 Methods for calculating vector mean and vector magnitude. (A) Trigonometric method. The tangent of the mean vector is calculated by dividing the sum of the sines by the sum of the cosines. The vector mean is the arctan of this value. The *signs* of the trigonometric functions must be recorded accurately. In this example, the negative tangent (positive sine and negative cosine) lies in the second quadrant, and the resultant aziumuth ($-74°$) is plotted counterclockwise from zero at the bottom of the circle. According to standard geologic usage, this equals 106° (measured clockwise from zero, or due north) or S74°E in the quadrant scheme of some compasses. The vector magnitude (L) is determined by dividing R (11.8) by the number of measurements (15) multiplied by 100. (B) Graphical method. Each measured azimuth is plotted as a unit vector. One unit of length can be 1 cm, 1 inch, or whatever is convenient. In this illustration, the unit vectors are labeled to 1 to 15 (azimuths given in A above). The resultant vector, or the line that connects the origin to the end of the last unit vector, is the vector mean. The vector magnitude is obtained by dividing the length of the resultant vector (12 units) by the total length of the unit vectors (15 units) and multiplying by 100. (After Lindholm, 1987: 48; by permission of Allen and Unwin, London.)

the opposite of water ripples, which erode on the lee side.

Wind ripples migrate by eroding on their upwind side and building on their downwind side until they reach an equilibrium size for the wind strength and sand supply. They are usually composed of sand that is coarser than the substrate over which they migrate, and their crests are made of coarser particles than their troughs. Water ripples show the opposite condition in both these features. Wind ripples form by the winnowing of their crests, which leaves the coarser material behind, whereas water ripples accumulate coarser sediments in the troughs where the zone of backflow results in weaker currents and

reduced competence. Another major difference is that wind ripples are not limited by the shallow flow depths that restrict water ripples, so eolian dunes can be enormous (meters to tens of meters in height). Indeed, gigantic cross-strata are virtually always found only in eolian environments (see examples in Chapter 8).

Bedding Plane Structures

The sedimentary structures just discussed are formed during the deposition of the bed and are generally three-dimensional. Another class of sedimentary structures forms on the interface between beds, usually on the exposed surface of a recently

A

B

Figure 4.8 (A) Interference pattern formed in symmetrical ripples from two coexisting wave sets in a modern tidal flat. (Courtesy of J. D. Collinson.) (B) Ancient interference ripples from the Cambrian Cadiz Formation, Marble Mountains, California. (D. R. Prothero.)

deposited bed before it is finally buried. Such structures can be extremely useful because they indicate current directions and postdepositional deformation of the sediment.

Sole marks, found on the bottom surfaces of beds, are usually casts or molds of depressions that were formed in the underlying beds by currents. The filling, or sole mark, tends to have a higher preservation potential because it is buried immediately as the depression is filled. The most common form of sole mark is a **flute cast** (Fig. 4.10), which is shaped like an elongated teardrop that tapers upcurrent. It is formed by a slight irregularity on a mud substrate that causes flow separation and a spiral eddy. The eddy spirals around a horizontal axis parallel to the flow and scours out the rounded, deep end of the flute cast. As the spiral eddy diminishes, the scouring becomes shallower and wider until it no longer indents the substrate. Another class of sole mark is the **tool mark,** which is an indentation of the cohesive mud bottom made by any object, or "tool" (Fig. 4.11). Tool marks include groove casts, brush marks, skip marks, chevron molds, prod marks, and bounce marks. These names describe the types of indentations that are left by the various objects (for example, twigs, branches, pebbles, shell fragments, and fish vertebrae) that produce them.

Subaerially exposed mud also produces sedimentary structures that can be useful in identifying sedimentary environments. The most familiar of these are mudcracks and raindrop impressions, which nearly always indicate drying of a subaerial mudflat (see p. 99). Because curling mudcracks always curl upward, they are also good indicators of the top side of a bed. In undeformed strata, such indicators may not be very important, but when beds have been structurally deformed, the top is not necessarily obvious. In such cases, it is crucial to find geopetal structures, which indicate the top of the bed. Cross-beds usually have truncated tops (because the next cross-bed set scours down into the previous one) and tangential contacts between foresets and bottomsets, so they can often be used to determine the top (see p. 127 and Fig. 8.24). Ripple crests are usually sharp, whereas ripple troughs are always rounded and scooped. Normally graded beds are clear indicators of the top because the coarsest material settles out first and is concentrated at the bottom (see Fig. 3.11B). Sole marks are found only on the base of the bed; the depressions that molded them are therefore on top of the underlying bed.

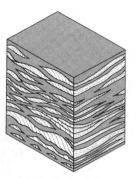
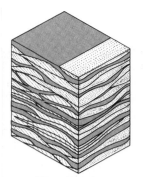
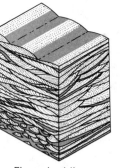

Lenticular bedding   Wavy bedding   Flaser bedding

A

B   **C**

Figure 4.9 (A) Diagrams showing lenticular, wavy, and flaser bedding. (B) Core showing lenticular bedding of sand lenses within a predominantly muddy sequence from the Mississippian Tar Spring Sandstone, Illinois. (C) Core showing flaser beds of mud stringers within a sandy sequence from the Pennsylvanian Carbondale Formation, Illinois. (B and C, Pettijohn and Potter, 1964: plates 18B and 17B; reprinted by permission of Springer-Verlag, New York.)

Figure 4.10 Flute casts from the Ordovician Normanskill Formation of New York. Flute casts are typically teardrop-shaped, with their tapered ends pointing downstream. The casts were produced when turbulent currents scoured the bottom and excavated tapered depressions. These flutes occur on the bottom surface of a turbidite bed, showing sole marks produced when the sediments forming this bed filled depressions in the layer that once underlaid it. The currents in this example flowed from lower right to upper left. (Courtesy of E. F. McBride.)

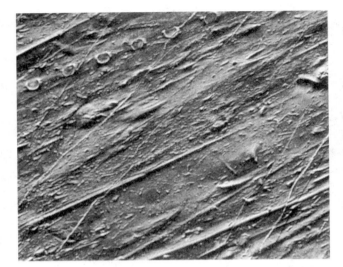

Figure 4.11 Tool marks from the base of the Carpathian flysch, Poland. The marks include circular skip casts from spool-shaped fish vertebrae, shallow brush marks, and deeper drag marks. (Courtesy of J. E. Sanders.)

Secondary Sedimentary Structures

Mechanically Produced Structures

Soft-sediment deformation structures form when sediment is deposited so rapidly that the beds are stable. Various sedimentary structures form via physical processes, but they are secondary (postdepositional), rather than primary.

In cases where denser material is deposited on top of less dense material, gravity plays an important role. If there is enough pore water, the whole mass becomes liquefied like quicksand. Strong forces applied before deformation deform still-soft sediment. If a mass of sediment slumps (a common occurrence on marine slopes), the sediment can be internally deformed. The most common deformations are load structures, irregular bulbous features formed when denser material sinks into less dense sediment (Fig. 4.12). Sometimes, droplet-shaped balls of sand sink into underlying mud, eventually breaking off to form **ball and pillow structures** (pseudonodules), which can be sizable (Fig. 4.13). Tonguelike protuberances of mud extending from the margin of these balls and pillows are known as **flame structures** (Fig. 4.14).

Deformation of soft sediment can produce **convolute bedding** as well as other features completely unrelated to intense deformation on a regional scale (Fig. 4.15). These features can fool the unwary geologist into postulating spurious structural events. The best way to distinguish convolute bedding from

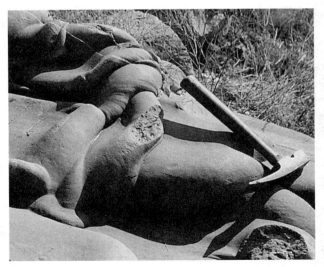

A

B

Figure 4.12 (A) Load casts from the Pennsylvanian Smithwick Formation, Burnett County, Texas. (Courtesy of E. F. McBride.) (B) Scaly or squamiform load casts (plus complex flute and groove casts) on the sole of an Ordovician turbidite that has been tilted vertically so that the bottom is exposed. (Siever, 1988: 123; by permission of W. H. Freeman and Company, New York.)

Figure 4.13 (A) Ball and pillow structure seen from below; from the Oligocene Annot Sandstone, Peira-Cava, Maritime Alps, France. (B) Cross section of ball and pillow structure showing internal lamination conforming to the boundary of the pillow; from the Ordovician Cynthiana Group, Pendleton County, Kentucky. (Courtesy of P. E. Potter.)

Figure 4.14 Flame structures and graded bedding in upper Pleistocene lacustrine sediments, Fraser River Valley, British Columbia. Field of view is 35 cm wide. (Courtesy of A. Rodman.)

true structural deformation is to see whether it is widespread and penetrative or restricted to a single bed (see Fig. 10.12). Also, convolute bedding (or lamination) is almost invariably closely associated with other soft-sediment deformation features.

Biogenic Structures

Sedimentary structures formed by the burrowing, boring, feeding, locomotion and resting of organisms are known as **trace fossils,** *Lebensspuren* (German for "living traces"), or **ichnofossils** (Greek *ichnos,* "trace"). Besides their importance as indica-

tors of stratigraphic age, trace fossils are useful clues to depositional conditions. In late Precambrian and early Paleozoic carbonates, for example, stromatolites are one of the most common trace fossils. **Stromatolites** are centimeter-sized hemispherical domal structures. They possess internal laminations that mimic the pattern seen when a knife cuts vertically or horizontally through a head of cabbage. Stromatolite structures form as a by-product of the metabolism of colonial blue-green algae (cyanobacteria), which generally thrive in the very shallow intertidal zone of marine estuaries and lagoons. They extend back in the geologic record to 3.5 billion years ago.

Figure 4.15 Convolute lamination in polished slabs of siltstone from the Ordovician Martinsburg Formation, Pennsylvania. (McBride, 1962: 53.)

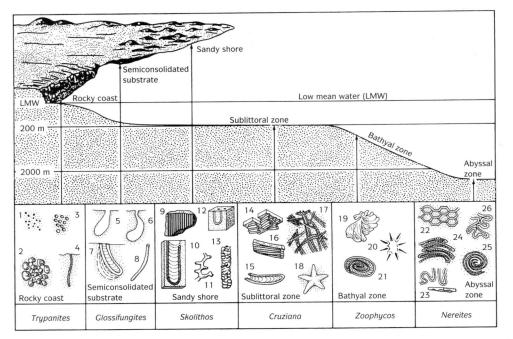

Figure 4.16 Summary diagram of the most common trace fossils and ichnofacies. Traces numbered as follows: 1 = *Caulostrepsis*; 2 = *Entobia*; 3 = unnamed echinoid borings; 4 = *Trypanites*; 5, 6 = *Gastrochaenolites* or related ichnogenera; 7 = *Diplocraterion*; 8 = *Psilonichnus*; 9 = *Skolithos*; 10 = *Diplocraterion*; 11 = *Thalassinoides*; 12 = *Arenicholites*; 13 = *Ophiomorpha*; 14 = *Phycodes*; 15 = *Rhizocorallium*; 16 = *Teichichnus*; 17 = *Crossopodia*; 18 = *Asteriacites*; 19 = *Zoophycos*; 20 = *Lorenzinia*; 21 = *Zoophycos*; 22 = *Paleodictyon*; 23 = *Taphrhelminthopsis*; 24 = *Helminthoida*; 25 = *Spirohaphe*; 26 = *Cosmoraphe*. (After Frey and Pemberton, 1984: 172; by permission of the Geological Association of Canada.)

Trace fossils are given taxonomic names as if they were valid Linnaean genera and species, but this is not really proper. Trace fossils are fossilized behavior, not body fossils. Few "ichnogenera" can be definitely associated with a known body fossil. It is likely that one type of trace was produced by several types of organisms or that one organism produced several types of races. This taxonomy is analogous to giving a different species name to footprints produced by the same individual wearing different shoes. Nevertheless, the practice of giving Linnaean names to trace fossils is so well established that it persists for lack of a better system.

Certain characteristic trace fossils have been clearly associated with specific depth and bottom conditions (Fig. 4.16). These associations are known as **ichnofacies.** A working knowledge of the more common ichnogenera and ichnofacies is very important because these trace fossils are almost as diagnostic as index fossils for certain purposes. In the following paragraphs, we will review only the most commonly encountered ichnofossils and ichnofacies. For further details consult Pemberton, Mac Eachern, and Frey (1992), Ekdale, Bromley, and Pemberton (1984), Bromley (1990), and Frey and Pemberton (1985).

Skolithos **Ichnofacies** Vertical tubelike burrows ("piperock") are commonly known as *Skolithos* and are believed to have been formed by tube-dwelling organisms that lived in rapidly moving water and shifting sands (Fig. 4.17A, B). Most of the tubes are 1 to 5 mm in diameter and can be as long as 30 cm. In some cases, they are densely clustered together and form thick layers of sandstone that resemble organ pipes (hence the name *piperock*). *Skolithos* piperock is particularly common in shallow marine Cambrian sandstones. The organism that made *Skolithos* is unknown, although some geologists have suggested phoronids (a burrowing wormlike lophophorate related to brachiopods) or tube worms. It is also possible that the trace-maker is extinct, since *Skolithos* is unknown after the Cretaceous.

Another common burrow in this ichnofacies is known as *Ophiomorpha* (Fig. 4.17A, C). These vertical cylindrical burrows are similar to *Skolithos*, except that they are slightly larger in diameter (0.5 to 3 cm) and have a bumpy outer surface caused by fecal pellets that lined the burrow. Typically, they are also less densely clustered than *Skolithos* and may have short horizontal connecting burrows between the vertical tubes. In cross section, they appear as circular or oval structures, often with a dark

ring of organic matter from the fecal pellet lining. Unlike *Skolithos,* however, we know what produces *Ophiomorpha* today (they are known back to the Permian.) The trace-maker is the burrowing ghost shrimp known as *Calianassa* (Fig. 4.17D).

A third common shallow marine ichnofossil is *Diplocraterion* (Fig. 4.17A, E, F). *Diplocraterion yoyo* tells a very specific story about the sea bottom. It is a burrow trace found between the arms of a vertical, U-shaped tube that presumably housed a burrowing, tubelike organism. When the openings were buried by sediment, the organism moved up in its burrow; when the upper part of the burrow was eroded away, the trace-maker dug in deeper. The sequence of U-shaped burrow traces thus responds like a yo-yo to the rise and fall of the sediment-water interface.

The characteristics of all these burrows suggest a rapidly shifting substrate that requires organisms to dig deep vertical burrows that must be rebuilt often when waves wash them away. Most of the burrowing organisms appear to be filter feeders that use the sediment strictly for shelter, not as a source of food. Sedimentological evidence also places this ichnofacies in shallow marine environments, and the known environmental preferences of living calianassid crustaceans further reinforces this interpretation. Thus, the *Skolithos* ichnofacies clearly indicates clean, well-sorted nearshore sands with high levels of wave and current energy.

Cruziana **Ichnofacies** Horizontal U-shaped troughs with many intermediate, riblike feeding traces are known as *Cruziana* and occur in moderate- to low-energy sands and silts of the shallow shelf (Figs. 4.18, 4.19). *Cruziana* is often preserved as the cast of the trough-shaped burrow, forming a convex sole mark, rather than as the original concave burrow itself. Many *Cruziana* are believed to represent the feeding traces of trilobites (Fig. 4.19B), since they are long troughs that appear to bear the scratch marks of trilobite legs as they burrowed through the shallow sediment. Their occurrence in rocks of Cambrian through Permian age (the same stratigraphic range as the trilobites) further reinforces this interpretation.

Another common trace fossil in this ichnofacies is *Thalassinoides* (Fig. 4.20). This is a general name for a complex three-dimensional network of cylindrical burrows that form an irregular web of crisscrossing tubes 1 to 7 cm in diameter. Apparently, this burrower was mining the shallow marine sands for their nutrients as well as seeking protection in its complex web of burrows. The organism or organisms that produced

A

B

C

D

E

F

Figure 4.17 (A) Common trace fossils of the *Skolithos* ichnofacies. 1 = *Ophiomorpha;* 2 = *Diplocraterion;* 3 = *Skolithos;* 4 = *Monocraterion.* (After Frey and Pemberton, 1984: 199; by permission of the Geological Association of Canada.) (B) Side view of piperock showing numerous parallel *Skolithos* burrows, from the Cambrian Zabriskie Quartzite, Death Valley, California. (Courtesy of M. L. Droser.) (C) Side view of the pellet-lined burrow known as *Ophiomorpha,* from the Cretaceous Fox Hills Sandstone of the Denver Basin. (Courtesy of R. J. Weimer.) (D) The living ghost shrimp *Calianassa,* exposed in its burrow; it produces *Ophiomorpha* burrows today. (Courtesy of R. J. Weimer.) (E) Side view of *Diplocraterion* burrows from Lower Cambrian Prospect Mountain Quartzite, Cricket Mountains, Millard County, Utah. (Courtesy of A. A. Ekdale.) (F) Top views of *Diplocraterion* burrows (note the paired set of holes) from the Lower Cambrian, Vik, Sweden. (Courtesy of A. A. Ekdale.)

Thalassinoides are unknown, although some modern calianassid burrows resemble them.

In addition to these two typical ichnogenera, there are a number of other less common trace fossils that are characteristic of this ichnofacies. They include (see Fig. 4.18) the star-shaped *Asteriacites,* the U-shaped *Rhizocorallium* (like a horizontal *Diplocraterion*), the C-shaped *Arenicolites,* the conical *Rossella,* and the deeper

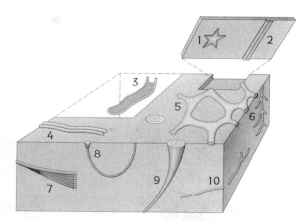

Figure 4.18 Common trace fossils of the Cruziana facies.
1 = *Asteriacites;* 2 = *Cruziana;* 3 = *Rhizocorallium;* 4 = *Aulichnites;*
5 = *Thalassinoides;* 6 = *Chondrites;* 7 = *Teichichnus;* 8 = *Arenicolites;*
9 = *Rossella;* 9 = *Planolites.* (After Frey and Pemberton, 1984: 200; by
permission of the Geological Association of Canada.)

Figure 4.20 *Thalassinoides* burrows are complex, three-dimensional
networks of traces at multiple levels, which usually collapse into a
jackstrawlike web of burrows when viewed in a two-dimensional bedding
plane. (D. R. Prothero.)

A

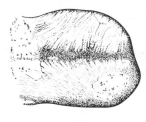

B

Figure 4.19 *Cruziana* traces (A) appear as bilobate convex structures with
parallel scratch marks from the legs of the burrowing trilobite, as shown in
(B). (Courtesy of T. P. Crimes.)

horizontal burrows known as *Planolites.* Most are
traces of organisms that used the substrate both as a
shelter and to mine the sediment for food particles.
Cruziana is also the most diverse of all ichnofossil communities, and it is commonly associated with finer
sediments than those associated with the *Skolithos*
ichnofacies. Based on all these lines of evidence, most
specialists consider the *Cruziana* ichnofacies to be indicative of shallow marine waters below normal wave
base but above storm wave base, typical of the middle
and outer shelf. Indeed, the top surfaces of storm deposits are often overprinted by *Cruziana* ichnofacies
activity that occurred on the fresh sea bottom right after a major storm.

***Zoophycos* Ichnofacies** Broad, looping infaunal
feeding traces known as *Zoophycos* occur in low-
energy muds and muddy sands (Fig. 4.21). Traditionally, they were considered indicators of deep waters along the continental slope below storm wave

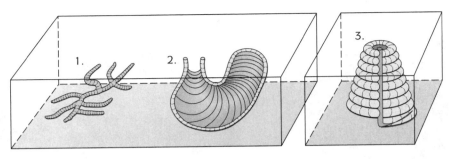

A

B

C

Figure 4.21 (A) Typical trace fossils of the *Zoophycos* facies.
1 = *Phycosiphon;* 2 = *Zoophycos;* 3 = *Spirophyton.* (After Frey and
Pemberton, 1984: 201; by permission of the Geological Association of
Canada.) (B, C) Typical *Zoophycos* traces, complex arcuate feeding traces

in three dimensions, from the Oligocene Amuri Limestone, Vulcan Gorge,
Canterbury, New Zealand, and the Eocene Saraceno Formation, Satanasso
Valley, Italy, respectively. (Courtesy of A. A. Ekdale.)

base but above the continental rise where turbidites accumulate. In the standard ichnofacies scheme, this placed *Zoophycos* between the *Cruziana* and *Nereites* ichnofacies (see Fig. 4.16). However, further study has shown that *Zoophycos* can be found in a great variety of depths (Frey and Seilacher, 1980). Indeed, they appear to represent a highly versatile, opportunistic trace-maker, because they occasionally occur in the *Cruziana* and *Nereites* ichnofacies. Instead of being good depth indicators, they are more closely associated with lowered oxygen levels and abundant organic material in the sediment in quiet-water settings. These conditions are indeed common on the outer shelf and continental slope, but they also occur in shallower waters of epeiric seas wherever the water is quiet enough but low in oxygen content.

Besides *Zoophycos,* relatively few other trace fossils are known from this community. The horizontal branched feeding trace known as *Phycosiphon* and the helically spiraling burrow known as *Spirophyton* are among the few commonly found with *Zoophycos.* The lack of diversity in the *Zoophycos* ichnofacies also suggests that it must represent a relatively hostile, oxygen-stressed environment where only a few low-oxygen-tolerant burrowers can thrive.

Nereites **Ichnofacies** The interpretation of the *Nereites* ichnofacies is relatively straightforward, in contrast to that of the *Zoophycos* ichnofacies. Meandering feeding traces on bedding planes are called *Nereites* and are usually found in the abyssal plains, often associated with turbidites and deep pelagic muds

(Fig. 4.22). Almost all the ichnogenera in this facies are superficial horizontal burrows in the top few centimeters of the muddy bottom. They all display a regular pattern of meandering or zigzagging across the bottom, reflecting the systematic mining of the organic-rich muds of the deep seafloor for detritus.

Other Ichnofacies Organisms can also bore their way into hard substrates. The presence of rock borings can indicate ancient shorelines and beach rock or an unconformity in which sediment was subaerially exposed. This is known as the *Trypanites* ichnofacies (see Fig. 4.16). In semiconsolidated substrates such as dewatered muds, the *Glossifungites* ichnofacies occurs. In addition to a mixture of *Diplocraterion, Thalassinoides, Arenicolites,* and *Rhizocorallium,* it may also include sacklike burrows known as *Gastrochaenolites.*

The absence of trace fossils can also be informative. If there are no trace fossils in a sequence that should be heavily burrowed, there might be reason to suspect that the water was anoxic and inhospitable to organisms. In sequences that are bioturbated, individual unburrowed beds were probably deposited very rapidly, so that the organisms could rework only the uppermost part.

In summary, a working knowledge of the common ichnogenera is extremely valuable. For environmental interpretation, and especially for determining

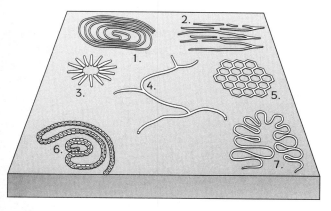

A

B

C

Figure 4.22 (A) Typical deep-water trace fossils of the *Nereites* facies. 1 = *Spirorhaphe;* 2 = *Urohelominthoida;* 3 = *Lorenzinia;* 4 = *Megagrapton;* 5 = *Paleodictyon;* 6 = *Nereites;* 7 = *Cosmorhaphe.* (After Frey and Pemberton, 1984: 203; by permission of the Geological Association of Canada.) (B) Two different meandering traces, *Spirophycus* (larger burrows) and *Phycosiphon* (smaller burrows), Permian Oquirrh Formation, Wasatch Mountains, Utah. (Courtesy of A. A. Ekdale.) (C) *Paleodictyon,* a netlike trace from the Middle Jurassic of the Ziz Valley, Morocco. (Courtesy of A. A. Ekdale.)

paleobathymetry and oxygen levels, ichnofossils are often the most diagnostic structures in the rock (far more definitive than the sediments themselves). Rocks with ichnofossils are much more common than those with diagnostic body fossils, so a good geologist must be ready to read the trace fossils wherever they occur.

CONCLUSIONS

When beginning geology students first examine a sandstone outcrop, all they see is rocks. The trained geologist, however, sees sedimentary structures and trace fossils and can immediately visualize the flow of the currents, the activities of organisms, and ultimately the entire environmental mosaic. As we will see in Chapters 8, 9, and 10, sedimentary structures are the most important-evidence for depositional interpretations. Sedimentary structures and trace fossils are the "alphabet" that geologists use to "read" sedimentary sequences. Without them, the stones are mute.

FOR FURTHER READING

Blatt, H., G. V. Middleton, and R. C. Murray. 1980. *Origin of Sedimentary Rocks.* Prentice-Hall: Englewood Cliffs, N.J.

Bromley, R. G. 1990. *Trace Fossils, Biology and Taphonomy.* Special Topics in Palaeontology. London: Unwin and Hyman.

Collinson, J. D., and D. B. Thompson. 1982. *Sedimentary Structures.* London: Allen and Unwin.

Donovan, S. K. 1994. *The Paleobiology of Trace Fossils.* Baltimore: Johns Hopkins University Press.

Ekdale, A. A., R. G. Bromley, and S. G. Pemberton, eds. 1984. *Ichnology: The Use of Trace Fossils in Sedimentology and Stratigraphy.* SEPM Short Course Notes 15.

Frey, R. W., and S. G. Pemberton. 1985. Biogenic structures in outcrops and cores. I. Approaches to ichnology. *Bulletin of Canadian Petroleum Geology* 33: 72–115.

Leeder, M. R. 1982. *Sedimentology, Process and Product.* London: Allen and Unwin.

Lindholm, R. C. 1987. *A Practical Approach to Sedimentology.* London: Allen and Unwin.

Maples, C. G., and R. P. West, eds. 1992. *Trace Fossils.* Knoxville, Tenn.: *Paleontological Society.*

Pemberton, S. G., J. A. MacEachern, and R. W. Frey. 1992. Trace fossil facies models: Environmental and allostratigraphic significance. In R. G. Walker and N. P. James, eds. *Facies Models: Response to Sea Level Change.* Toronto: Geological Association of Canada.

Pettijohn, F. J., and P. E. Potter. 1964. *Atlas and Glossary of Primary Sedimentary Structures.* New York: Springer-Verlag.

Ricci Lucchi, F., 1995. *Sedimentographica: A Photographic Atlas of Sedimentary Structures,* 2d ed. New York: Columbia University Press.

Rubin, D. S. 1987. Cross-bedding, bedforms, and paleocurrents. *SEPM Concepts in Sedimentology and Paleontology* 1: 1–187.

Selley, R. C. 1982. *An Introduction to Sedimentology.* London: Academic Press.

Selley, R. C. 1988. *Applied Sedimentology.* San Diego: Academic Press.

Tucker, M. E. 1982. *The Field Description of Sedimentary Rocks.* Milton Keynes, England: Open University Press.

Walter, M., ed. 1976. *Stromatolites.* New York: Elsevier.

II

Siliciclastic Sediments and Environments

Sand dunes near Stovepipe Wells, Death Valley, California. (Photo by George Grant, courtesy of U.S. Department of the Interior.)

5 Sandstones and Conglomerates

Giant dropstone released by melting iceberg drifting away from the Gondwana ice sheet into deep marine sands. The layered sandstones were bent downward from the impact and then smoothly buried the dropstone once normal quiet marine deposition had resumed. Outcrop from the Permian of the Paraná Basin, Brazil. (Courtesy of J. C. Crowell.)

THE TERM *SILICICLASTIC SEDIMENTS OR SEDIMENTARY rocks* refers to deposits composed of clasts of pre-existing rocks and minerals, most of which consist of quartz, feldspar, common rock fragments, and other silicate minerals. Because these deposits are derived from the erosion of detritus weathered from pre-existing rocks, they are also commonly and correctly described as *detrital* ("detached from"), *epiclastic* ("derived from the surface"), and *terrigenous* ("from the Earth"). The individual clasts in such deposits form by both physical and chemical weathering. They are transported and deposited as discrete bits and pieces by a variety of erosional agents: mass wasting, wind, water, and ice. After final deposition as discrete,

TABLE 5.1 Major Categories of Terrigenous Sediments and Sedimentary Rocks

Clast		Unconsolidated Sediment		Sedimentary Rock		
Diameter (mm)	Name	Rounded, Subrounded, Subangular Clasts	Angular Clasts	Rounded, Subrounded, Subangular Clasts		Angular Clasts
>256						
	Boulder					
256						
	Cobble	Gravel[a]	Rubble[a]	Conglomerate	I.	Breccia
64						
	Pebble					
4						
	Granule					
2						
	Sand		Sand	II. Sandstone (clast roundness variable)		
$\frac{1}{16}$						
	Silt		Silt[b]	III. Mudrock (clast roundness variable)		Siltstone[b] Mudstone Claystone[b]
$\frac{1}{256}$		Mud				
	Clay		Clay[b]			

[a]A descriptive prefix derived from the most common coarse clast type (by size and/or composition) can be used to specify very coarse clastic sediment or sedimentary rock; for example, granite boulder rubble; rhyolite cobble conglomerate.
[b]Mud is an unconsolidated mixture of silt and clay. Mudrock is lithified mud. Most terrigenous sedimentary rocks finer than sand are intermixtures of silt and clay. Siltstone, claystone, and mudstone are collectively grouped as mudrock. Shale is fissile mudrock; that is, it breaks into thin slabs along planar surfaces.

unconsolidated fragments, they eventually become lithified into the major siliciclastic sedimentary rocks, which collectively constitute at least two-thirds—perhaps as much as three-fourths—of Earth's sedimentary shell.

Table 5.1 shows the categories of siliciclastic sediments and sedimentary rocks defined on the basis of clast diameter. Three distinct groups are recognized: (1) conglomerate and breccia, (2) sandstone, and (3) mudrock. When clasts of various sizes—clay, silt, sand, granules, and coarser clasts—are intermixed, which is common, opinions differ about how best to categorize such mixtures.

The characteristics, origin, and geological significance of each group are summarized in this chapter and the next. Differences in the detail of coverage reflect differences in our ability to describe and understand these three rock assemblages. Because of their fundamental similarities, sandstones and conglomerates are covered together in this chapter; mudrocks are discussed in the next chapter.

Conglomerate and Breccia

Conglomerate (also called roundstone or puddingstone) is lithified gravel made up of rounded to subangular clasts whose diameters exceed 2 mm. **Breccia** (sharpstone) is lithified rubble made up of angular clasts coarser than 2 mm. The roundness, or angularity, of the grains is measured using standard grain silhouettes (see Fig. 1.3). Very coarse clastic rocks are collectively referred to as *rudites* or rudaceous sedimentary rocks (Latin) or *psephites* (Greek). More precise descriptive names incorporate the most obvious or predominant clast size or composition; for example, quartz-pebble conglomerate, granite-cobble breccia.

The literature on conglomerate and breccia is less extensive than that dealing with sandstone and carbonate because the former constitute no more than 1% to 2% of the sedimentary rock shell and are of limited regional extent. This restricted distribution and lack of fossils make stratigraphic correlation difficult. Conglomerate and breccia are best studied in the

field. In many cases, detailed counts of individual grains, either exposed in a limited area of an outcrop or in contact with a rope draped across the exposure, are invoked to characterize texture and composition. No other sedimentary rock group provides more insights about provenance, depositional environment, paleogeography, and tectonic setting.

Composition

Most clasts in conglomerate and breccia are fragments of rocks and minerals produced by the disintegration of bedrock. These occur both as coarser-grained framework and finer-grained matrix (filling the space between framework grains). Composition is analyzed in two ways. Framework grains are identified by pebble counts done in the field, and matrix (if sand or finer) is studied in thin section. Clasts are typically glued together by a small amount of siliceous, calcareous, or ferruginous cement. Three principal categories of coarser than sand-sized clasts are distinguished: (1) mineral fragments that occur as major components, (2) mineral fragments that occur as accessory constituents, and (3) fragments of rock.

Mineral Fragments Occurring as Major Components (5% or More)   Clasts of a single mineral such as quartz or feldspar tend to be less abundant in conglomerate and breccia than in sandstone because few igneous, metamorphic, or sedimentary rocks have original grains coarse enough to disintegrate into pebbles and coarser detritus. Source rocks with mineral grain diameters coarser than 8 mm (fine pebbles) include quartz veins, pegmatites, deep-seated plutons, high-grade metamorphic rocks, breccia, and conglomerate.

Quartz is the most abundant major mineral in conglomerate and breccia. It is harder than other rock-forming minerals, has no cleavage, and is practically insoluble. Large clasts of K-feldspar, plagioclase feldspar, and mica can also be abundant but seldom last as long as quartz because they corrode, disaggregate, and abrade with transport. The sand matrix is similar in composition to sandstones interbedded with the conglomerate or breccia.

Mineral Clasts Occurring as Accessory Constituents (Less Than 5%)   Other fragments composed of single minerals occur as accessories in conglomerate and breccia. Their presence is incidental to the sedimentary rock type, much as garnet crystals are scattered through a granite. Minerals occur in accessory amounts either because their original abundance in source rocks is low or because they are easily destroyed by weathering. Included in this category are micas such as muscovite and biotite and such heavy minerals (specific gravity > 2.9) as olivine, pyroxene, amphibole, zircon, magnetite, and hematite.

Rock Fragments   Rock fragments are typically the most abundant component in very coarse-grained terrigenous rocks and are invariably the most interesting. Careful analysis of their composition provides direct information on provenance. Rock fragments can consist of almost any variety of igneous, metamorphic, or sedimentary rock, although smaller clast diameters are correlated with finer-grained varieties. Clasts of harder, less easily decomposed lithologies are more likely to survive weathering at the source and breakdown during transport. Thus, fragments of durable, fine-grained rocks such as rhyolite, slate, and quartzite are more abundant than less resistant, coarse-grained rocks such as marble, limestone, and gabbro, even if these lithologies were originally present in equal amounts at the source. Less stable clasts survive under conditions of high source area relief and/or an arid or arctic climate; these conditions permit the rate of physical disintegration to surpass that of chemical decomposition.

Texture

Conglomerate and breccia textures are studied at the outcrop using methods of quantitative grain size analysis that differ from those used for sandstone. Grain diameters of particles coarser than sand are visually assigned to individual size classes. Large clast size also permits fabric, grain surface features, grain shape, and grain roundness to be studied in the field. More specific data on grain size and sorting can be obtained by using a caliper to measure the long, short, and intermediate axes of individual grains.

By definition, the framework fraction consists of clasts whose grain diameters exceed sand size (>2 mm). The interstitial space between framework grains can be empty (pore spaces); filled with finer-grained detrital matrix; or occupied by cement, fluid (water or oil), or natural gas.

Two distinct varieties of conglomerates (and breccias) are defined on the basis of texture: orthoconglomerates and paraconglomerates (Pettijohn, 1957).

Orthoconglomerates (literally, "true" conglomerates) consist mainly of gravel-sized framework grains. The proportion of matrix (sand and finer ma-

Figure 5.1 (A) An orthoconglomerate with closely packed cobbles and pebbles that contact one another and thus are self-supporting. This is the underside of a vertically tilted bedding surface from the Cretaceous debris flows in Wheeler Gorge, Ventura County, California. (B) A paraconglomerate contains clasts supported by a matrix of sandstone and mudstone. In this example from the Miocene Topanga Formation, Sunland, California, the clasts range from 10 to 70 cm in diameter. (D. R. Prothero.)

terial) is 15% or less. As a result, orthoconglomerates have an *intact, grain-supported framework*; that is, individual framework grains are in tangential contact and support one another. Framework grains would remain essentially in place if the matrix component were somehow removed (Fig. 5.1A).

Paraconglomerates have a matrix of sand and finer clasts. The proportion of matrix is at least 15%; most have more than 50% matrix and are actually sandstone or mudrock in which pebbles, cobbles, and boulders are scattered. Paraconglomerates can

have a grain-supported fabric, but those with high proportions of matrix have an unstable, *nonintact, matrix-supported framework* (Fig. 5.1B). If the matrix were removed, framework grains "floating" in it would collapse. The terms **diamictite** and **diamixtite** are also used for poorly sorted detrital rocks in which pebbles and larger grains float in a sandy or muddy matrix. The distinctive textural characteristics of orthoconglomerates and paraconglomerates are used for classification.

General Textural Characteristics

Sorting and Modality Because a broad range of clast diameters occurs in conglomerate and breccia, these rocks are almost invariably less well sorted (see Fig. 1.2) than finer-grained terrigenous rocks. Some are *unimodal;* that is, they contain a single modal size class more prominent than the adjacent classes, which uniformly drop off in abundance. Many are *bimodal* or *polymodal;* that is, they have two or more prominent size classes in addition to the modal class. Orthoconglomerates deposited by rivers tend to be bimodal (a framework modal class and a sandy matrix modal class) because deposition mixes coarser bedload with finer suspended load. Paraconglomerates are less well sorted than orthoconglomerates and are almost always at least bimodal; most are polymodal. These characteristics reflect the deposition of paraconglomerates by transport agents that rarely separate clast sizes: glaciers, mass wasting, and turbidity currents.

Shape, Roundness, and Grain Surface These textural characteristics correlate with transporting agent and depositional setting. For the most part, *clast shape* reflects the inherent physical properties of a particular rock type rather than transport history. Foliated metamorphic rocks such as schist and slate tend to disintegrate into elongate, flattened clasts. Massive rocks such as granite and marble generate equidimensional (equant) pebbles, cobbles, and boulders. In a few cases, clast shape might reflect the transporting agent. Wind-faceted cobbles exhibit distinctive einkanter and dreikanter shapes; glacial transport produces cobbles with a flatiron form (Fig. 5.2).

The *roundness* of clasts that are coarser than sand is controlled by both rock type and abrasion history. The intensity of abrasion varies with transport distance and agent. Laboratory tumbling mill experiments (Daubrée, 1879) and field studies of modern gravels (Plumley, 1948) show that pebbles and coarser clasts—especially soft, corrodible limestone and shale—become well rounded with only a few tens of kilometers

A

B

Figure 5.2 (A) Ventifacts are rocks that have been polished and faceted by wind abrasion. (M. R. Campbell, courtesy of U.S. Geological Survey.) (B) Glacial till stones from the Pleistocene of Illinois show parallel striations, faceting, and snubbed edges and corners. The larger cobble is about 13 cm in diameter. (Pettijohn, 1975: 173; by permission of Harper Collins, Inc. New York.)

by etching and differential solution and do not indicate a specific transporting agent or depositional setting. *Surface polish gloss* or *frosting* refers to the ability of a clast surface to scatter or diffuse light, giving the grain the appearance of frosted glass. Transport by wind is principally responsible for this feature because the high-velocity grain-to-grain impacts generated during dust storms produce numerous microfractures on the grain surface (see Fig. 5.2A). Some pebbles and cobbles with shiny surface gloss, however, are interpreted as gastroliths or stomach stones, so called because it is thought that they were produced by grain-to-grain collisions of stones ingested by dinosaurs to assist digestion.

Fabric or Internal Organization   Individual clasts—usually nonequant, elongate rock and mineral fragments—are fabric elements. Some exhibit no pre-

Figure 5.3 Rounding takes place very rapidly after clasts break away from the bedrock. The clasts at the top were found in a talus pile immediately below their source at the crest of the San Gabriel Mountains, California. The clasts on the bottom are much better rounded, yet they traveled only 5 km down Aliso Creek on the north flank of the range. Scale in inches. (D. R. Prothero.)

of river transport (Fig. 5.3). Even cobbles and boulders of more resistant lithologies, such as quartzite, are well rounded when transported as little as 100 km (Kraus, 1984; Lindsey, 1972).

Grain surface features are easily visible on pebbles, cobbles, and boulders. Such features are also called microrelief. They include striations (typically narrow, straight scratches), crescent-shaped percussion marks, indentations or pits, and surface polish or frosting. Striations are usually produced by glacial ice transport (see Fig. 5.2B), although they can also be seen on stream cobbles. Crescentric percussion marks are produced by the high-velocity impact of clasts transported by streams with steep gradients. Surface indentations or pits on grain surfaces originate mainly

A

B

Figure 5.4 (A) Imbrication of flat limestone slabs in Whiterock Creek, Brown County, Ohio. Stream flowed from right to left. (From Pettijohn and Potter, 1964: plate 43A; by permission of Springer-Verlag, New York; courtesy of Paul E. Potter.) (B) Well-developed imbrication in Pleistocene glacial gravels, north end of Wind River Canyon, Wyoming. Current flowed from right to left. (Courtesy of R. H. Dott, Jr.)

ferred alignment; others show a systematic orientation termed **imbrication** (Fig. 5.4).

In some modern stream gravels, the long axes of cobbles and pebbles are aligned subparallel with one another and dip upstream. Others have subparallel alignment of long axes with downstream dips. Still others have subparallel long axes transverse rather than parallel to the current flow. Coarse marine gravels and ice-deposited Pleistocene tills have pebble and cobble long axes aligned parallel with the transport direction. Conglomerates and breccias deposited by sediment gravity flows such as turbidity currents and landslide debris flows exhibit no internally organized fabric.

Classification, Origin, and Occurrence

Although there are more than 50 sandstone classification schemes, the few conglomerate and breccia classifications that exist differ in terms of the defining characteristics used to subdivide and name distinctive varieties. Factors considered useful for classification include framework-to-matrix ratio, stability of the framework, clast lithology, clast size, and overall fabric.

Table 5.2 shows the scheme best suited for classifying epiclastic conglomerates and breccias. This table is based on an earlier classification proposed by Pettijohn (1975) and modified by Boggs (1992). The flow diagram in Fig. 5.5 permits the classification to be used easily in the field or with hand specimens. The classification uses visible textural and compositional features. To the extent that these features reflect differences in origin, this classification satisfies our desire for a system that is both descriptive and genetic.

The classification scheme first separates framework and matrix, solely on the basis of size. Framework includes only those clasts whose diameters exceed 2 mm (coarser than sand).

Step 1. Extraformational and intraformational conglomerate and breccia are separated by comparing the composition of framework and matrix grains.

Intraformational conglomerate and breccia have an interior (intrabasinal) source; that is, they are eroded from the same sedimentary rock unit of which they are a part, rather than being derived from rocks located outside the depositional basin. Consequently, intraformational conglomerate and breccia have framework grains identical in composition to those of the matrix. Most are produced by brief interludes of physical weathering that disrupt normal sedimentation. The deposition of carbonate mud might be disrupted temporarily by a violent storm that generates waves and stirs up newly deposited mud. Penecontemporaneous fragmentation, entrainment with minimum transport, and abrasion followed by almost immediate redeposition produce a rock in which limestone mud clasts float in an identical limestone mud matrix. Similarly, weak, bottom-hugging turbidity currents that erode and redeposit chips of mud generate shale pebble conglomerate or breccia.

Only two principal types of intraformational conglomerate and breccia are common: shale pebble (or cobble, or boulder) (Fig. 5.6A) and limestone pebble (or cobble, or boulder). Framework clasts in both types are flat, tabular disks with long axes aligned

TABLE 5.2 Descriptive Classification of Epiclastic Conglomerates and Breccias

Provenance	Framework Grain-to-Matrix Ratio[a]	Fabric	Framework Clast Composition
Extraformational	Orthobreccia or orthoconglomerate: 4:1 or greater (matrix <20%, tends to be sand-sized)	Intact or grain-supported; framework grains in tangential contact; removal of matrix does not collapse framework	Oligomict: Most (>90%) of framework clasts composed of hard, resistant rocks and minerals (vein quartz, quartzite, and chert)
	Parabreccia[b] or paraconglomerate: <4:1	Typically unstable or matrix-supported; framework grains float in the matrix; removal of matrix collapses the framework	Petromict: (polymict) More than 10% of framework grains are clasts of various types of unstable rocks and minerals (granite, limestone, feldspar, volcanic rocks)
Intraformational	Various limestone and shale pebble, cobble, and boulder "edgewise" conglomerates and breccias		

[a]Matrix is that portion of a conglomerate or breccia (or their unconsolidated equivalents) that has clast diameters of 2 mm and less (sand and finer). Framework is that portion of clasts in a conglomerate or breccia (or their unconsolidated equivalents) whose grain diameters exceed 2 mm (sand-sized). Framework grains in breccias are angular; framework grains in conglomerates are subangular, subrounded, or rounded.
[b]The terms *diamictite* or *diamixtite* are used for paraconglomerates and parabreccias in which a small proportion (typically <10%) of framework clasts float in a matrix of mudrock.

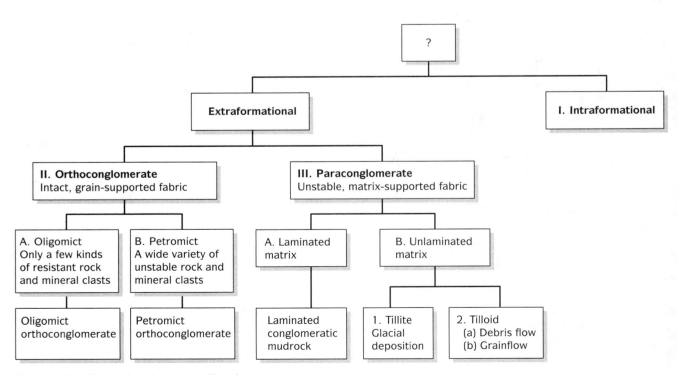

Figure 5.5 Classification of conglomerates and breccias.

A

B

Figure 5.6 Semiconsolidated mud chips form when dried-up and curled mudcracks are ripped up and transported. (A) These mud chips from a tidal flat in Germany show imbrication in the downstream direction (flow toward lower left.) (Pettijohn and Potter, 1964: plate 44A; by permission of Springer-Verlag, New York; courtesy of Paul E. Potter.) (B) When lithified into rock, mud chips form a flat-pebble, or "edgewise," conglomerate. (Courtesy of R. Siever.)

parallel or subparallel to stratification. Weathering etches out the less resistant matrix surrounding the framework clasts. As a result, the corners of the biscuitlike framework clasts protrude prominently, leading to the nickname "edgewise conglomerate" or "edgewise breccia" (Fig. 5.6B).

Limestone edgewise conglomerate is common in littoral or neritic limestone deposits where shallow water permits periodic emergence and desiccation of recently deposited lime mud or where wave energy disrupts unconsolidated sediment. Typically, shale pebble conglomerate and breccia form in settings where sudden surges of current velocity occur, such as in river systems and submarine canyons.

Extraformational conglomerate and breccia are derived from source areas outside the depositional basin. Detritus weathered from external sources is carried away and deposited elsewhere. As a result, framework clasts differ markedly from matrix in composition. Framework material is exotic; that is, not derived by the erosion and redeposition of matrix material.

Step 2. Orthoconglomerates (orthobreccias) and paraconglomerates (parabreccias) are separated by examining the proportion of matrix.

Orthoconglomerates are matrix-poor (80% or more framework grains) and have an intact, stable, grain-supported fabric. Paraconglomerates are matrix-rich.

Their fabric may be grain-supported, but most have an unstable, nonintact fabric that is supported by matrix grains. Removing the matrix would cause the framework to collapse.

Orthoconglomerates are transported and deposited on a grain-by-grain basis by conventional (normal, Newtonian) fluids, specifically water or air. These agents winnow, wash, and sort sediment from which most matrix (especially silt and finer material) has been removed. Fast-moving rivers and shallow surf deposit orthoconglomerates. Paraconglomerates are transported and deposited by glacial ice, turbidity currents, and mass movements such as landslides and debris flows.

Step 3A. Conglomerates are further divided into oligomict and petromict varieties on the basis of framework grain composition.

In **oligomict** (orthoquartzose) conglomerate or breccia, more than 90% of the framework clasts (granule and coarser grains) consist of fragments of only a few varieties of resistant rocks and minerals such as metaquartzite, vein quartz, and chert. The term **petromict** (or polymict) is used if clasts of many different kinds of metastable and unstable rocks are abundant; for example, basalt, slate, and limestone. A more precise classification is possible by specifying predominant clast size and lithology (for example, quartz

pebble oligomictic orthoconglomerate; slate cobble petromictic parabreccia).

Oligomict orthoconglomerates imply wholesale decomposition and disintegration of immense volumes of rock, reflecting climate and topography that promote chemical decomposition and physical disintegration of all but the most resistant components. Multiple recycling may be important. Many of these conglomerates may be streamflow deposits that fill stream channels as sheetlike gravel blankets or as bars developed transverse to the flow direction. Some occur as thin layers or lens-shaped bodies in fluvial sequences; these layers are generated after storms, when sheetlike surges of sediment-laden floodwater spread out on the perimeter of alluvial fans and glacial outwash fans. Still others are well-stratified, well-sorted deposits produced in near shore marine settings where wave energy washes and reworks gravel supplied by streams or coastal erosion. Such wave-worked (beach-face) deposits grade seaward into the shoreface, which extends from low tide to the depth where waves begin to break (roughly 10 m). Other shoreface deposits are paraconglomerates with a predominantly sandy matrix; these deposits are produced where longshore currents and rip currents fail to winnow matrix from gravel.

Petromict orthoconglomerates are much more abundant than oligomict orthoconglomerates and occur in sequences thousands of meters thick. Coarse clasts of volcanic, metamorphic, and sedimentary rocks predominate. Petromict orthoconglomerates are mainly alluvium eroded from high-relief sources. They make up large portions of modern and ancient alluvial fans and are a common component in orogenic clastic wedges. Some are deposited in desert or glacial-periglacial regions, where even low-lying sources disintegrate physically rather than decompose chemically.

Step 3B. Paraconglomerates (and parabreccias) are further subdivided on the basis of their inferred origin as well as the size and internal organization of their matrix.

The following questions must be asked to identify specific types. Is the matrix sand or mud? Is the matrix internally laminated or chaotic? Is the framework imbricated, sorted, and vertically graded? Is the deposit sheetlike or lenticular? With what other types of sediment is the deposit associated?

Paraconglomerates containing a matrix of delicately laminated mudrock in which coarser framework grains float are called **laminated pebbly** (or **cobbly,** or **bouldery**) **mudrock.** These rocks consist of widely scattered angular to subangular cobbles, pebbles, and boulders floating within laminated mudrock (see p. 66). Laminae are distorted above and below the larger clasts, bend down abruptly beneath clasts, and may be broken. These clasts are called **dropstones** because they have been sporadically dropped from above into the soft, muddy substrate (Fig. 5.7A). Post-depositional compaction of the plastic matrix around the more resistant clasts produces the drapelike pattern of overlying laminations. Various transport mechanisms can carry dropstones out into offshore positions from which they plunge into muddy sediments. Most dropstone-bearing laminated mudrocks owe their origin to ice rafting. When floating bodies of

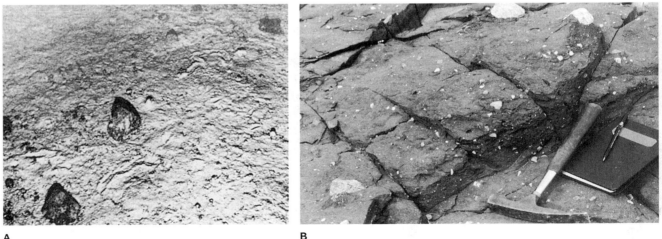

A   **B**

Figure 5.7 (A) Iceberg-rafted cobbles as big as 10 cm in diameter can be dropped to the seafloor far from their original source. These examples were photographed on the sea bottom between South America and Antarctica at a depth of 4000 m. In some places, dropstones have been carried as far as 3000 km from Antarctica. (Official NSF photo, USNS

Eltanin, Cruise 10; courtesy of Smithsonian Oceanographic Sorting Center.) (B) Lower Gondwana (Dwyka) tillite, Port St. Johns, South Africa. This famous Permian tillite clearly shows large glacial till stones randomly sprinkled through a fine matrix. (Courtesy of J. C. Crowell.)

gravel-laden pack ice or icebergs melt, they release clasts. Other laminated pebbly mudrocks contain dropstones generated by volcanic explosion. Pyroclastic ejecta propelled into the air plummet into mud and accumulate in subjacent lakes, lagoons, and estuaries. Dropstones also might have been rafted as soil detritus affixed to the root systems of floating trees or might even have been released from clumps of drifting kelp.

Paraconglomerates in which the matrix is disorganized and nonlaminated are either tillite (if a glacial origin can be inferred), or tilloid (deposited by mass movement). Sediment deposited by melting glaciers is **till;** lithified till is **tillite.** Because tillite implies special climatic conditions, it is important to be able to recognize these deposits (Eyles, Eyles, and Miall 1985).

Glaciers erode and deposit sediment in a distinctive manner. Glacial sediment load is entrained in various ways. On alpine (valley) glaciers, large blocks of rock slump or free-fall onto the ice surface from steep valley walls; finer dust is swept there by the wind. Glacial plucking entrains material of various sizes along the base of continental glaciers and along the base and sides of alpine glaciers. All material incorporated on, within, and near the base of glacial ice settles slowly. The high viscosity and low velocity of moving ice permit no sorting and little abrasion, although grinding of the bedload against valley walls and floor can generate additional fine-grained clasts. Glacial loads are plastered onto the substrate as ground moraine (lodgment tillite) or accumulate wherever there is a balance between the rates of ice flow and melting (this accumulation produces terminal and recessional moraines).

As a result of these entrainment mechanisms, clasts transported and deposited by ice are notoriously poorly sorted, angular, and lithologically heterogeneous (Fig. 5.7B). Tillites often have bimodal or polymodal distributions and contain clasts that range from very fine clay-sized material to house-sized boulders. Grains are angular to subangular, surface striations are abundant (see Fig. 5.2B), and pebbles and cobbles exhibit a distinctive triangular flatiron shape (Von Engelen, 1930). Stratification or any kind of internal organization is absent, although some modern tills possess a weak, difficult-to-detect preferred fabric with clast long axes aligned parallel to the ice flow. Clasts of a wide variety of rocks, derived from whatever lithologies lie upstream from the deposit, float in the predominantly clay-sized "rock paste" matrix. There must be supporting field evidence—such as an intimate association with varved clay sequences, striated bedrock surfaces, or

unambiguous lithified glacial outwash gravels—to identify a paraconglomerate definitively as tillite.

Ancient till deposits, especially those generated by widespread continental glaciers, are distributed sporadically in time and space, reflecting the unevenness of global glacial episodes and the low preservation potential of tillite. Valley (alpine) and valley mouth (piedmont) glaciers are more common, but they leave tills of limited areal extent. Schermerhorn (1974) and Hambrey and Harland (1981) provide complete surveys of pre-Pleistocene glacial deposits.

In contrast to tillites, most **tilloids** are deposited by dry and wet gravity-driven mass wasting processes such as rockslides, debris flows, and turbidity currents. Gravity acts directly on the sediment, and any fluid that is present facilitates transport by reducing internal friction and providing grain support. Deposition is rapid, and there is little reworking or sorting (Lowe, 1982; Walker, 1984).

In Chapter 3, we described debris flows as muddy mixtures of water, silt, and clay that move downslope and are capable of supporting coarser clasts (see Figs. 3.8 and 3.9). This slurry of material stabilizes ("freezes") as thick, internally disorganized beds of poorly sorted sediment (Leeder, 1982; Selley, 1982).

Subaerial debris flows occur on the surface of alluvial fans and outwash fans developed in front of glaciers. They are produced when surges of water from heavy rainfall or the melting of snow and ice mobilize gravel-sized detritus and a cohesive matrix of finer material. Slurries move down even gentle slopes. When they eventually stop, they deposit very poorly sorted, usually ungraded, internally disorganized material that is often mud-rich and matrix-supported. Typically, the bottoms of individual flows are sharp and show little evidence of channeling. A scoured, striated basal zone of shearing can occur. Coarser units are generally thicker; finer units thinner.

Subaqueous debris-flow deposits are associated with turbidite sequences rather than with alluvial fan and glacial outwash sediments. Such deposits are somewhat better sorted than subaerial debris-flow deposits, exhibit a weak preferred fabric (subparallel long axes; a vague imbrication), and are either ungraded or show poor to fair size grading that is inverse, normal, or inverse to normal. Unlike subaerial debris-flow deposits, no correlation exists between the maximum clast size in a unit and the unit thickness.

Subaqueous grain-flow deposits are quite unlike subaqueous and subaereal debris-flow deposits. Beds produced by subaqueous grain flows are seldom thicker than a few centimeters and are interbedded

with plane-laminated and cross-laminated sandstones. Clast-supported, they consist of poorly to moderately sorted gravel-sized clasts and a sandy, silty, or clay-sized matrix. Reverse (inverse) grading, arguably their most obvious characteristic, is often (but not invariably) present at the base of each unit. Pebble long axes are aligned parallel with flow direction, and imbrication is common.

Tilloids deposited by turbidity currents may be either clast-supported or grain-supported. These flows suspend high concentrations of sediment in upward-moving turbulent eddies. Pebble and cobble long axes are aligned subparallel with transport direction; an obvious upflow imbrication is typical. Normal grading is common, as is a general fining-upward in grain size and thinning-upward in scale of cross-bedding and stratification thickness. These deposits can also show horizontal stratification and cross-stratification.

Impact-generated breccias have attracted great interest since the 1980 hypothesis relating the Cretaceous-Tertiary mass extinction event to the collision of a large (>10 km in diameter) hypervelocity extraterrestrial body or bolide with Earth (Alvarez et al. 1980). Bolide impact potentially generates two distinct varieties of very coarse-grained deposits.

1. Breccias are produced by the enormous shock energy of collision. The shock energy fragments the target rock and disintegrates the impactor. Most debris falls into the impact crater or in a narrow band around it. Examples include the roughly 35-million-year-old Toms Canyon crater fill in Chesapeake Bay (Poag, 1999) and the extensive breccia deposits in and around the Chicxulub site straddling the shoreline of the Yucatán Peninsula (Frankel, 1999).

2. Breccias can also be generated by the potentially gigantic (up to 300 or 400 m high) super-tsunamis produced when a bolide splashes down into the sea. Impact-related tsunami deposits have been associated with both the Tertiary Chesapeake invader (the Exmore beds) and the Chicxulub impactor (Cretaceous-Tertiary sandy breccias exposed near the town of Beloc, Haiti, and near the mouth of the Brazos River on the Gulf Coast of Texas).

Sandstone

Sandstone is the indurated equivalent of unconsolidated sand. Sand includes clasts with diameters from 2 mm to 1/16 (0.0625) mm. Sandstones are also referred to as *psammites* (Greek) and *arenites* (Latin).

Sand grains constitute the framework of sandstone, and the pore spaces between framework grains may be empty or partly or entirely filled. Pore filling can be any combination of (1) finer-grained primary or secondary clastic matrix, (2) cement (typically calcite, quartz, chert, or hematite), and (3) fluids such as gas, air, oil, and groundwater.

Sandstones are arguably the best-known sedimentary rock type. They constitute between 10% and 20% of Earth's sedimentary rock record. Geologists find them particularly intriguing because they provide unique and fascinating insights into Earth's origin and evolution. Along with conglomerate and breccia, they are the most reliable indicators of sedimentary provenance. Sandstone textures and sedimentary structures reveal depositional setting, dispersal, and transporting mechanism. Sandstones are major reservoirs of groundwater and petroleum, are useful as building stones, and can be valuable sources of metallic ores. Because they resist weathering and erosion, they are well exposed and are the dominant control on topography.

Composition

Sandstone composition is analyzed using a petrographic microscope and thin sections. Only crude description and preliminary identification can be done in the field or from hand specimens. Discriminating individual mineral species and rock fragments using standard optical petrographic microscopy can be tedious and difficult, but it is far superior to traditional hand specimen methods. The added time, effort, and money is justified, and the procedure is straightforward and efficient.

Standard references allow sedimentary petrography to be self-taught (for example, Scholle, 1978, 1979; Adams, Mackenzie, and Guildford, 1984). The following discussion highlights the characteristics of each major mineral group.

Quartz   Sand grains can consist of any mineral, but *monocrystalline* (single-crystal) quartz grains are by far the most abundant type of sandstone grain (Fig. 5.8A). Although monocrystalline quartz grains typically constitute 60% to 70% of sandstone, some sandstones are almost 100% quartz; others contain none. When composite grains of multiple interlocking quartz crystals are defined as rock fragments, they are termed *polycrystalline quartz* (Fig. 5.8A, B).

The high overall abundance of quartz in sandstone is unsurprising. Quartz is a common constituent in rocks such as granite, gneiss, and schist, which make

A

B

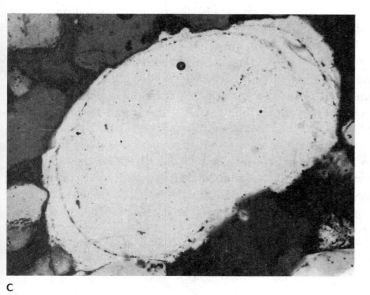

C

Figure 5.8 (A) This sandstone includes clasts of monocrystalline quartz (labeled M), polycrystalline quartz (composite grains labeled P), and quartz with undulose, or wavy, extinction (labeled U). Most grains are 0.5 to 1.0 mm in diameter. (D. R. Prothero.) (B) Polycrystalline quartz grain of metamorphic origin (a quartzite rock fragment). Metamorphic origin is indicated by the polygonal crystal boundaries at 120° and by the large number of quartz crystals that form the grain. Granitic polycrystalline quartz has fewer crystals and little or no intercrystal suturing. (Walker and Pettijohn, 1971: 2119; by permission of R. G. Walker and the Geological Society of America.) (C) Monocrystalline, well-rounded quartz sand grains (about 0.5 mm in diameter) with multiple rounded quartz overgrowths, Pennsylvanian-Permian, Weber Sandstone, Utah. Border of detrital part of grain is marked by the inner oval ring of water-filled vacuoles. (Odom, Doe, and Dott, 1976: 867; courtesy of I. E. Odom.)

up much of Earth's crust. Because quartz resists disintegration and decomposition, there is more quartz than other rock-forming minerals in sediments.

Given the prevalence of monocrystalline quartz grains in sandstone, many efforts have been made to discriminate varieties of quartz derived from a particular provenance. The results have been disappointing, but some important differences in quartz grain types have been documented (Basu et al., 1975). Monocrystalline quartz grains differ in (1) the character and composition of gas and mineral inclusions, and (2) the extent to which single grains darken and blacken in thin section under crossed polarizers (uniform or nonundulatory versus wavelike, undulatory extinction) (see Fig. 5.8A). Monocrystalline quartz grains derived from hydrothermal veins often contain fluid-filled vacuoles. Undulatory extinction characterizes quartz derived from plutonic and high-grade metamorphic source rocks; nonundulatory extinction indicates volcanic rock sources or grains recycled from older sandstones.

Many quartz grains recycled from sedimentary rocks show a thin rim of silica cement (quartz, chert, or opal). These **authigenic,** "grown in place" overgrowths are interstitial cement precipitated during an

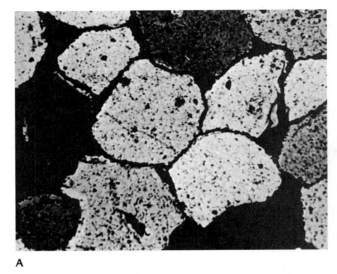

A

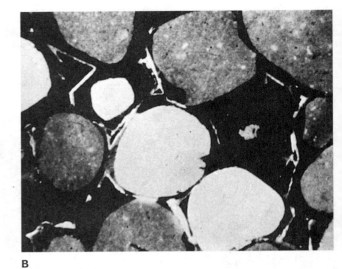

B

Figure 5.9 Photomicrographs of medium-sand-sized quartz grains in the Devonian Hoing Sandstone, Illinois. (A) Under crossed nicols. (B) Under cathodoluminescence. What appear to be pressure-solution contacts under crossed nicols are actually secondary quartz overgrowths abutting in original pore space between undeformed, well-rounded sand grains. (Courtesy of R. F. Sippel.)

earlier episode of lithification (Fig. 5.8C). The rounded borders of original grain nuclei are outlined by small specks of hematite, clay minerals, organic matter, and other materials. These borders are circumscribed by abraded rims of overgrowth material. Cathodoluminescence petrography can distinguish between seemingly identical monocystalline quartz grains that have a different provenance (Fig. 5.9). This process irradiates thin sections (lacking a cover slip) with a beam of electrons to produce **luminescence,** the emission of light from a solid material caused by the excitation or disruption of orbiting shell electrons. Small differences in the chemical composition of minerals such as quartz produce variations in the color and brightness of luminscence (Miller, 1988).

Feldspars Feldspar is generally less abundant than quartz in sandstone, averaging between 10% and 15% of sandstone composition. This is surprising in light of the larger overall amount of feldspar in such crystalline rocks as granite and gneiss. However, feldspars are more easily decomposed than quartz; they are not as hard and they cleave. (These properties often permit feldspar and quartz to be distinguished optically.)

High feldspar content in a sandstone carries specific implications about source area climate and topography. It means that chemical weathering is not extensive, probably because of climate and/or high source relief. Low precipitation in an arid setting, or an arctic climate in which precipitation occurs as snow and ice rather than as rain, limits hydrolysis and pro-

duces feldspar-rich debris. Even in climates that usually promote decomposition to clays (for example, a warm, humid climate), feldspars can survive if relief is high because fast-moving streams will erode feldspar before it can decompose.

The two principal feldspar families, potassium (potash or K-feldspar—orthoclase, sanidine, and microcline) and plagioclase (Na-Ca) feldspar, differ in abundance in sedimentary rocks (Fig. 5.10). Potassium feldspars are more prevalent because they are more common in continental crust and resist decomposition better. Numerous attempts to associate feldspar types with specific source rocks have had mixed success.

Rock (Lithic) Fragments The overall abundance of rock fragments in sandstone varies greatly. Lithic fragments provide the most specific information about sandstone provenance. Whether sand consists of rock fragments or clasts of mineral grains is partly a function of source rock grain size. Because coarse rocks such as granite, sandstone, and gneiss physically disintegrate into mineral grains and do not generally survive as composite grains, clasts of volcanic rocks, slate, phyllite, shale, and chert are more abundant than those of plutonic igneous rocks, gneiss, and schist. As is the case with boulders, cobbles, and pebbles, the survivability of a rock fragment is a function of mineralogy. Physically and chemically resistant minerals such as quartz (for example, a quartz siltstone or a bedded chert) survive weathering and erosion better than more easily decomposed materials such as limestone or basalt.

A

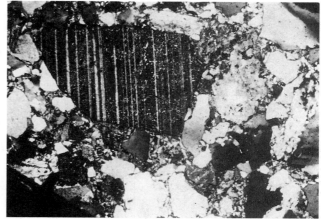

B

C

Figure 5.10 Three types of feldspar grains are common in arkosic sandstones and are easily recognized in thin section. (A) Microcline (a potassium feldspar) shows the characteristic grid, plaid, or tartan twinning. (B) Plagioclase (a sodium-calcium feldspar) exhibits parallel twinning that gives it a striped appearance. (C) Orthoclase (a potassium feldspar) is usually untwinned. It can be distinguished from quartz by its cleavage (note the straight sides) and by its tendency for diagenetic alteration. This grain has been partially altered to clays and fine muscovite mica ("sericite"), giving it a dirty, mottled appearance. By contrast, the quartz grains in the same slide are clear and unaltered. (D. R. Prothero.)

Differentiating between rock types in thin section is challenging. Clasts of volcanic rocks (referred to as VRF or Lv) are recognizable if they are porphyritic because lathlike feldspar phenocrysts stand out from the dark, finer-grained groundmass (Fig. 5.11A). However, it is difficult to differentiate nonporphyritic volcanic rock fragments from sedimentary rock fragments (SRF or Ls) such as chert (Fig. 5.11B), from some plutonic rock fragments (Fig. 5.11C), from metamorphic rock fragments (MRF or Lm) such as slate and schist (Fig. 5.11D), and from other types of sedimentary rock fragments (Fig. 5.11E).

Several varieties of polycrystalline quartz grains are recognizable. Those derived from high-grade metamorphic rocks (metaquartzite) are a mosaic of large numbers of elongate crystals (see Fig. 5.8B). Chert fragments appear as speckled composite quartz grains, with no elongation of individual crystals (see Fig. 5.11B).

Accessory Minerals Accessory minerals typically have densities that exceed those of the common rock-forming minerals quartz and feldspar (which have a specific gravity of about 2.6). Examples include garnet, rutile, zircon, corundum, kyanite, olivine, and pyroxene; these rarely occur in more than accessory abundances (1% to 2%). Because of their density, such minerals are finer than the less dense minerals with which they are deposited. The denser minerals are often concentrated in laminae or as placers.

Dense (sometimes misnamed "heavy") minerals are easy to identify in thin section and imply specific sediment sources and diagenetic history. Because they are scarce, they are not ordinarily studied in conventional thin sections. Instead, sandstone is disaggregated and the denser minerals are then separated from the less dense minerals. This is done using solutions of water and bromoform mixed to produce liquids of varying density. The denser minerals sink and the less dense

A

Figure 5.11 Textures of common lithic fragments found in sandstones and conglomerates. (A) Volcanic rock fragments show the typical fine-grained crystalline texture, with larger phenocrysts floating in the groundmass. (Courtesy of R. H. Dott, Jr.) (B) Chert rock fragments (cryptocrystalline quartz) show a finely speckled texture, with interference colors of first-order grays and blacks. (D. R. Prothero.) (C) This plutonic igneous rock fragment (probably a granite or granodiorite) is composed of coarse crystals of microcline (showing tartan twinning) and quartz (untwinned). (D. R. Prothero.) (D) This metamorphic rock fragment (a schist) shows a foliated texture, with stretched quartz grains and micas in alignment. (Courtesy of R. H. Dott, Jr.) (E) This sedimentary rock fragment (a shale) is clearly discrete from the surrounding sand grains and cement, and its own internal grain size is much finer than that of the surrounding matrix. (Scholle, 1979: 29; by permission of the American Association of Petroleum Geologists.)

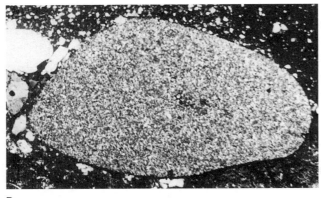

B

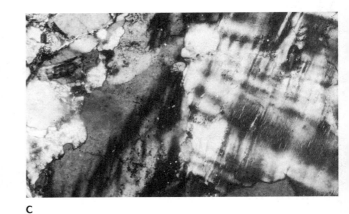

C

D

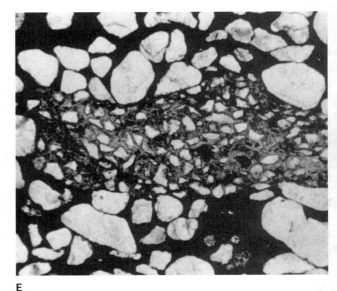

E

ones float. Grain mounts of the concentrate are analyzed with a binocular microscope.

During the first half of the twentieth century, sedimentary geologists hoped to infer provenance based on the dense mineral fraction. But this approach never proved as fruitful as expected. Dense minerals provide some insights into provenance and diagenetic history, but our improved ability to discriminate and understand the less dense minerals in sandstone makes studies of "heavy" minerals unnecessary.

Micas and Clay Minerals   The major micas—biotite, muscovite, and chlorite—occur in sandstone as silt and sand. Because their flakelike or discoidal shape slows their settling, micas tend to be slightly coarser than the more equant quartz and feldspar grains with which they are associated.

Micas are detrital components—much like quartz, feldspar, and rock fragments—weathered from preexisting rocks. They are helpful in pinpointing specific sources. For example, chlorite suggests a low-grade metamorphic rock source.

Clay minerals such as kaolinite, gibbsite, and illite (discussed in detail in Chapter 6) can also be abundant in sandstone. Because of their fine grain size, these minerals are concentrated as matrix. Fine grain size means that techniques such as X-ray diffraction or differential thermal analysis must be used to identify clay minerals. Even when clays are accurately identified, however, it is difficult to know whether they are detrital or the product of diagenesis.

Texture

The texture of a sandstone includes grain size, size variation, roundness, shape, surface features, and overall fabric (arrangement of the clasts in space). Texture is analyzed for many reasons, in addition to simple description. Often, stratigraphic units can be differentiated on the basis of mean grain size alone. Sandstone porosity (the ratio of the volume of empty space to that of solid material) and permeability (the degree to which pores are interconnected) are of practical importance in petroleum geology, hydrology, and waste disposal. Regional variations in texture allow inferences to be made about sediment dispersal. For example, the systematic downstream decrease in diameter of fluvial sand is well documented.

These reasons aside, the principal motivation behind most studies of sandstone texture is the belief that such studies allow the unequivocal identification of transporting agent and depositional setting. Underlying this belief is the controversial assumption that the physical processes prevailing during sediment transport and at the depositional site impart a distinctive fingerprint to sand texture. In the following discussion, we will define each textural characteristic and the methods used to measure it qualitatively and quantitatively. Then we will evaluate the contribution of each characteristic to our understanding of sandstone genesis.

Grain Size and Grain Size Variation: Methodology
When we use the term *size,* we usually mean diameter, but various methods measure diameter differently.

Sieve size analysis measures grain diameter by allowing sand to settle through a nest of sieves. Each sieve is a cylinder floored with a wire-mesh screen with square apertures of fixed dimensions. Aperture holes correspond to sand class boundaries (2 mm, 1 mm, 0.5 mm, 0.25 mm, 0.125 mm, and 0.0625 mm). In sieve analysis, therefore, clast diameter is the width of the smallest aperture through which grains pass. Sieve analysis can be used for unconsolidated sand or for ancient sandstones that can be disaggregated.

Sandstone texture can also be measured with a *petrographic microscope,* a method preferred for ancient, well-indurated sandstones that cannot be disaggregated easily. (Features other than texture are also best examined in thin section.) A petrographic microscope is fitted with an ocular micrometer, an eyepiece bearing a scale subdividing the field of view into discrete metric units. Grain diameter measured in thin section, therefore, corresponds to the maximum diameter visible. Because the plane of a thin section cuts randomly through grains (Fig. 5.12), data obtained from thin sections must be corrected arithmetically (Friedman, 1962).

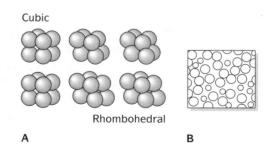

Cubic

Rhombohedral

A   B

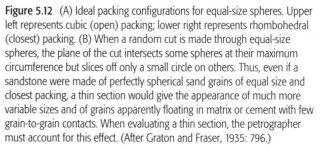

Figure 5.12 (A) Ideal packing configurations for equal-size spheres. Upper left represents cubic (open) packing; lower right represents rhombohedral (closest) packing. (B) When a random cut is made through equal-size spheres, the plane of the cut intersects some spheres at their maximum circumference but slices off only a small circle on others. Thus, even if a sandstone were made of perfectly spherical sand grains of equal size and closest packing, a thin section would give the appearance of much more variable sizes and of grains apparently floating in matrix or cement with few grain-to-grain contacts. When evaluating a thin section, the petrographer must account for this effect. (After Graton and Fraser, 1935: 796.)

More sophisticated methods of measuring grain size use expensive equipment and require considerable expertise. Image analysis, for example, measures grain size electronically using a TV camera attached to a modified petrographic microscope (Ehrlich et al., 1984).

Several methods measure grain size indirectly by recording the rate at which sediment falls through a column of water and settles on the bottom of a tube (using Stokes' law, discussed in Chapter 3, to calculate the effective diameter). Conventional pipette analysis and sophisticated automatic rapid sediment analysis are the only suitable procedures for unconsolidated materials that are finer than sand.

For pebbles and coarser clasts, the sizes of individual grains can also be measured using a caliper. Grain size is expressed in terms of the short, long, and intermediate axes. Data can be restated as nominal diameter by calculating the diameter of a hypothetical perfectly round sphere whose volume duplicates that of the actual clast.

Regardless of methodology, size is summarized in similar ways. All procedures list the proportion of grains in each major size class (Fig. 5.13) as a percentage (for example, weight percent of sediment retained in each sieve, or the number of grains falling within a particular class as a percentage of total grains counted using a petrographic microscope). Grain size distribution is presented in two ways: *graphs of grain size distribution* are plotted and compared; and *statistical measures* such as mean grain size and sorting are calculated arithmetically by taking values from the graphs and plugging them into standard formulas.

Both methods express grain diameter in millimeters or phi (ϕ) units. The **Udden-Wentworth scale** (see Fig. 5.13), which specifies grain diameters in millimeters, is a geometric scale. There is a fixed ratio between each successive size class boundary, each size being half as large as the preceding size (for example, 2, 1, 1/2, 1/4, 1/8, and 1/16, or 2, 1, 0.5, 0.25, 0.125, 0.0625, and so on). This scale is graphically and mathematically cumbersome.

The phi (ϕ) scale uses a logarithmic-based unit of measurement and expresses millimeter grain diameters in terms of exponents of the base 2. For example,

$$8 \quad 4 \quad 2 \quad 1 \quad \tfrac{1}{2} \quad \tfrac{1}{4} \quad \tfrac{1}{8} \quad \tfrac{1}{16} \quad \tfrac{1}{32} \quad \text{mm} =$$
$$2^3 \quad 2^2 \quad 2^1 \quad 2^0 \quad 2^{-1} \quad 2^{-2} \quad 2^{-3} \quad 2^{-4} \quad 2^{-5}$$

The phi unit grain size is the negative logarithm of grain diameter in millimeters to the base 2:

Grain diameter in phi units =
$$-\log_2 \text{ of grain diameter in mm}$$

For example, the grain size in phi units of a clast with a diameter of 2 mm is -1ϕ; the grain size in phi units of a clast with a diameter of 1/8 mm is 3ϕ. It is much easier to compare whole numbers rather than the fractions. In addition, the change of sign allows most of the common grain sizes (1 mm and finer) to be expressed as positive values; the less common pebbles have the negative phi values. This makes the mathematical and statistical manipulation of grain sizes much easier. All the statistical formulas described here express grain size in phi units rather than millimeters.

Fig. 5.14 shows the standard graphs used to present sandstone size data. **Simple histograms** are bar diagrams plotted on graph paper using two arithmetic scales, one horizontal and the other vertical (Fig. 5.14A). The height or length of the bars plotted on the vertical axis (ordinate or *y*-axis) corresponds to the proportion of grains in each size class (weight percentage from sieve size analysis; number of grains out of the total counted for thin section analysis). Each bar coincides with a single size class. By convention, grain diameter decreases to the right on the horizontal axis (abscissa or *x*-axis). A **frequency curve** (Fig. 5.14B) is a smooth curve that can be fitted to the bar diagram. It can be superimposed directly onto a histogram simply by joining the midpoint marks of each size class bar.

A **cumulative histogram** or **bar diagram** (not shown) is drawn by plotting cumulative weight percentage on the vertical axis, beginning with the coarsest size class. Again, both the horizontal and vertical axes have arithmetic scales. A **cumulative frequency curve** (Fig. 5.14C) can be fitted to the cumulative histogram by connecting the right-hand side (finer size limit) of each size class, which produces a smooth curve.

The most useful graph is a **probability graph.** Cumulative percentages are plotted on probability paper (5.14D). This paper has a vertical axis (ordinate) that is a log probability scale and a standard horizontal arithmetic abscissa scale (on which size class limits are marked). Normal bell-shaped distributions appear as straight lines when plotted on probability paper.

A quick inspection of any of these graphs allows us to draw general conclusions about some aspects of sandstone grain size distribution. For example, the most abundant (modal) class is identifiable from the

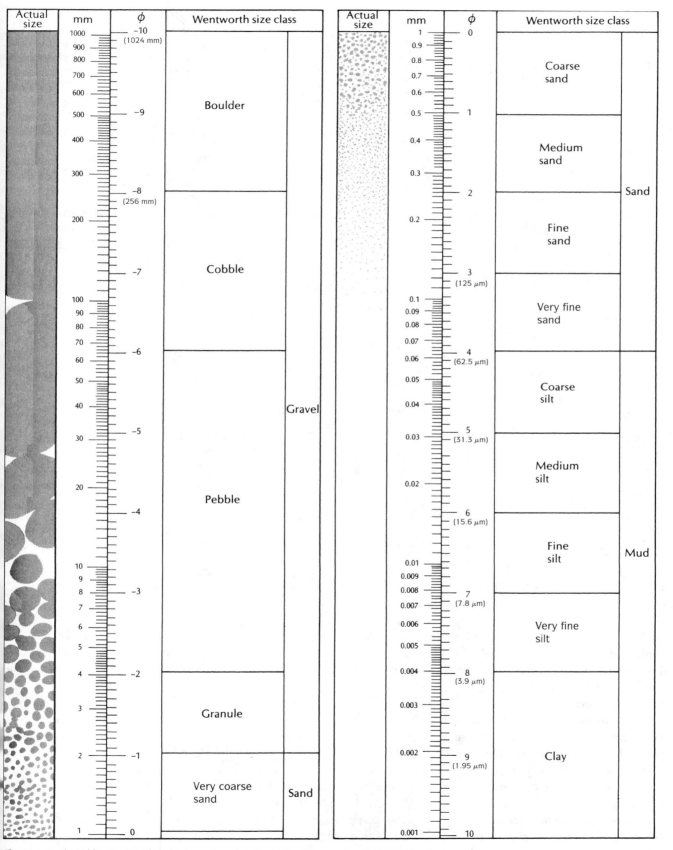

Figure 5.13 The Udden Wentworth grain size scale and millimeter-to-phi conversion chart. (After Lewis, 1984; 59; by permission of Chapman and Hall, London.)

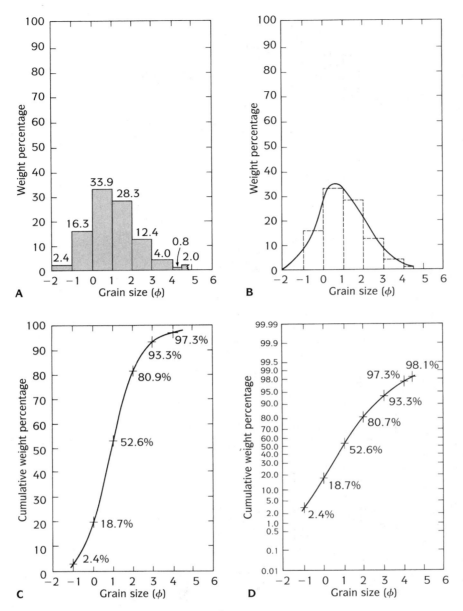

Figure 5.14 Results of a single sieve analysis displayed in various ways. (A) Histogram. (B) Noncumulative size-frequency curve. (C) Cumulative curve on normal graph paper with a standard arithmetic ordinate scale. (D) Cumulative curve on probability graph paper. The ordinate scale collapses the values in the center (the 10%–80% range) and stretches out the values on the tails (less than 10% and greater than 90%), making the curve in (C) into a straight line. On probability paper, therefore, each normal distribution plots as a straight line segment. The phi values can be read directly from these plots and plugged into the formulas shown in Table 5.3 to calculate statistics quickly and easily. (After Blatt, 1992: 474; by permission of W. H. Freeman and Company, New York.)

histogram; sorting, a measure of variation in grain size, is expressed by the breadth of the base (x-axis) of both simple and cumulative histograms. The slope of grain size distribution on a probability curve steepens with improved sorting.

More precise estimates of sandstone grain size can be determined arithmetically. Grain diameters expressed in phi units are read from the probability plot. (Data from sieve analyses are uncorrected, but thin section raw data are converted to sieve size analysis equivalents.) Table 5.3 lists the formulas used to calculate standard statistical measures: (1) various aspects of central tendency or averageness such as *mean size;* (2) *size sorting* (standard deviation); (3) *skewness*, or symmetry, which compares sorting in coarser and

finer halves of the population; and (4) *kurtosis*, or peakedness, a comparison of sorting in the central portion of the size distribution with sorting in the ends or "tails."

These statistical measures are used because the fundamental purpose of sandstone grain size studies is to identify the transporting agent and depositional setting. Classic uniformitarianism is the foundation for this approach. Size data from an ancient sandstone are compared with data from modern sands of differing known origins. When textural fingerprints match, it is argued that transporting agent and depositional setting can be identified.

What might these statistical measures tell us about the agents that transport and deposit sediment?

TABLE 5.3 Formulas for Calculating Statistical Measures Using Phi Unit Values from Probability Plots

I. Measures of central tendency

A. Mean $= \dfrac{\phi16 + \phi50 + \phi84}{3}$

B. Median $= \phi50$

C. Mode = Midpoint of most abundant class interval on histogram

D. Modal class = Most abundant class interval on histogram

II. Sorting (inclusive graphic standard deviation)

$\dfrac{\phi84 - \phi16}{4} + \dfrac{\phi95 - \phi5}{6.6}$

$<0.35\phi$	Very well sorted
$0.35\phi - 0.50\phi$	Well sorted
$0.50\phi - 0.71\phi$	Moderately well sorted
$0.71\phi - 1.00\phi$	Moderately sorted
$1.00\phi - 2.00\phi$	Poorly sorted
$>2.00\phi$	Very poorly sorted

III. Skewness (symmetry) (inclusive graphic skewness)

$\dfrac{\phi84 + \phi16 - 2\phi50}{2(\phi84 - \phi16)} + \dfrac{\phi95 + \phi5 - 2\phi50}{2(\phi95 - \phi5)}$

$> +0.30$	Strongly fine-skewed
$+0.30$ to $+0.10$	Fine-skewed
$+0.10$ to -0.10	Near-symmetrical (unskewed)
-0.10 to -0.30	Coarse-skewed
<-0.30	Strongly coarse-skewed
<1.0	Excessively peaked (leptokurtic)
1.0	Normally peaked (mesokurtic)
>1.00	Deficiently peaked (platykurtic)

IV. Kurtosis

$\dfrac{\phi95 - \phi5}{2.44(\phi75 - \phi25)}$

The various measures of **central tendency** are **mean** (average) **size, modal** (most frequently occurring) **size and class,** and **median size** (the diameter that splits the distribution into precisely equal halves; that is, half of all the data are below the median, and half are above). In graphical plots such as histograms, the mean straddles the central portion of the distribution. The median, the diameter of the 50th percentile, is located precisely halfway between the two ends of size-frequency, cumulative-frequency, and probability plots.

These measures of averageness are supposed to record potentially distinctive characteristics of the depositional agent, presumably its kinetic energy and competence (which vary with velocity and viscosity). Some transporting agents do produce distinctive textural imprints. Mean, median, and modal sizes of sediment carried by the wind (high velocity, low viscosity) are much finer than those of grains transported by ice (low velocity, high viscosity). Sand deposited by fast-moving (steep-gradient) rivers typically ex-

hibits a coarser mean than sand deposited by rivers flowing down lower gradients. Beach sandstones are repeatedly winnowed by waves and longshore currents; they are typically unimodal with a single modal size. Conversely, sandstones deposited by density currents are bimodal. A primary peak coincides with the framework modal class, and a secondary peak marks the matrix modal class.

Variation in grain size or sorting expresses the number of significant size classes in a population. The implied significance of sorting is that transporting agents differ in their ability to entrain, transport, and deposit grains of different sizes. Sorting may reflect variations in velocity and the ability of a particular process to transport and deposit certain sizes preferentially. Wave-related currents in the surf zone and blowing wind sort sand better than do turbidity currents and rivers.

Skewness is a statistical measure of the symmetry of a distribution. In a normal bell-shaped distribution, mean, median, and mode coincide; the two

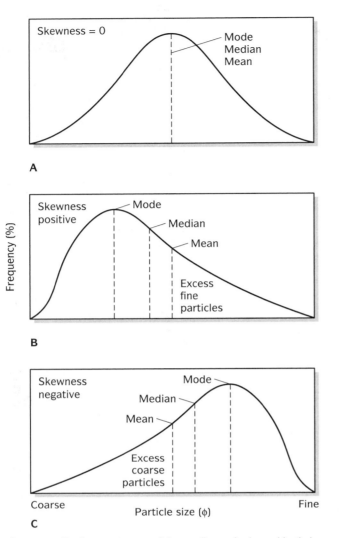

Figure 5.15 Size-frequency curves of three sediments having an identical sorting but different skewness. (A) Sediment is unskewed and shows a symmetrical distribution. Mean, mode, and median coincide. (B) Sediment is positively skewed; the coarser half of the population shows better sorting than the finer half (that is, both the median and the mode are shifted toward finer grain sizes). (C) Sediment is negatively skewed; the finer half of the population is better sorted than the coarser tail (both the median and the mean are shifted away from the mode toward the coarser grain sizes). (After Friedman and Sanders, 1978: 75; by permission of John Wiley, New York.)

halves of the distribution are mirror images (Fig. 5.15A). In asymmetrical or skewed distributions, the median and mean shift from the mode (central peak) toward coarser or finer sizes. With negative skewness (Fig. 5.15C), coarser grains are less well sorted than finer grains. This produces a long, more gently sloped coarse tail. With positive skewness (Fig. 5.15B), finer grains are more poorly sorted than coarser grains, producing a long, more gently sloped fine tail. Skewness is genetically significant because

transporting agents differ in their ability to entrain, transport, and deposit coarse versus fine material.

Kurtosis or peakedness compares sorting in the central portion of a population with that in the two tails. Normally peaked distributions (for example, a bell-shaped distribution) are described as **mesokurtic.** Excessively peaked distributions (better sorting in the central portion of the population than in the tails) are **leptokurtic;** deficiently peaked (flattened) distributions are **platykurtic.** Although some sedimentologists have argued that distinct depositional agents differ in their ability to sort coarse and fine tails relative to more abundant grain sizes, no analysis using only kurtosis has successfully discriminated transport agents.

Grain Size Distribution as a Key to Transporting Agent and Depositional Setting Many attempts have been made to link sand grain size distribution data with depositional history. Friedman (1967) pioneered the use of two-component (**bivariate**) diagrams that plot one statistical parameter against another (Fig. 5.16). Very few sandstone types of differing

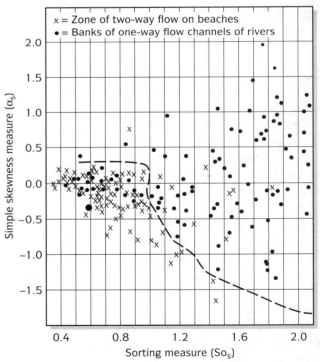

Figure 5.16 Various statistics have been used to try to identify sands from different sedimentary environments. In this example, the values for skewness (y-axis) and sorting (x-axis) are plotted for beach (×) and river (•) sands. They appear to plot in two partially overlapping fields separated by the dotted line. Beach sands tend to be slightly better sorted than river sands and are not as positively or negatively skewed. (After Friedman and Sanders, 1978; 78; by permission of John Wiley, New York.)

origin can be definitively recognized on this basis. For example, although Fig. 5.16 shows that beach sands are better sorted (lower standard deviation) than river sands, the sorting value is not unique. Likewise, although beach sands tend to show negative skewness (a less well sorted, coarse tail), their skewness values overlap those of river sands.

Visher (1969) developed an approach that also has had limited success. He correlated **changes in the slope of probability plots** as a function of grain size, arguing that these changes are distinctive for specific transporting agents (Fig. 5.17). For example, he suggested that grain size probability plots for river sands are uniquely segmented into three differently sloped components because rivers move material in three ways: by suspension, by saltation, and as bedload. Other transporting agents generate apparently different distributions. Beach sands seem to have distinctive slope breaks among four populations: traction load, two saltation loads (swash and backwash), and suspension load (Fig. 5.17A, B). Dune sands are extremely well sorted, with almost all the sand in a single large saltation population (Fig. 5.17C). Turbidite sands are extremely poorly sorted, with a single population consisting of everything from gravel to clay, generating a gentler, smooth slope (Fig. 5.17D). But few other meaningful conclusions can be based on allegedly unique probability plots.

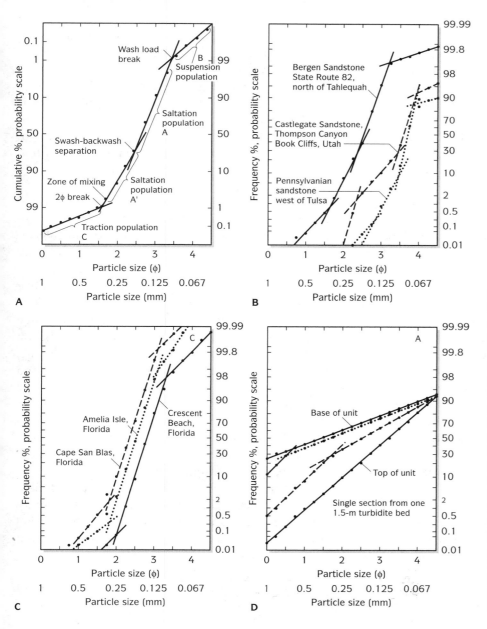

Figure 5.17 A classic paper by Glenn Visher (1969) attempted to extract the maximum information from probability plots. (A) Modern beach sands tend to have four overlapping normal populations (indicated by line segments): a traction population, two saltation populations (swash and backwash), and a suspension population. (B) Several ancient beach sands appear to have the same four-segment distribution. (C) Dune sands tend to be better sorted (steeper slope), with a single large saltation population and only minor traction and suspension populations. (D) Turbidite sands are very poorly sorted (shallow slope), with a single population ranging from very coarse to very fine grains. In the years since this paper was published, sedimentologists have become less confident that environment can be read so easily from grain size distributions. (After Visher, 1969; 1079; 1100; 1084; 1101; by permission of the SEPM.)

Passega (1964) developed yet another graphical approach, known as C-M and L-M diagrams. Instead of using such standard statistical measures as mean, sorting, and skewness, he plotted all grain size data as a cumulative probability curve. The diameter of the first (1%) percentile (C, or coarsest), the median diameter (M, or median), and the percentile of grains finer than 0.031 mm (L) are then determined. The results have been inconsistent. Values initially plotted from known modern settings defined separate fields, but subsequent applications resulted in the overlapping of fields.

Sedimentary petrologists continue to search for the ideal textural key that they hope will link transport agent, depositional setting, and grain size distribution. Several factors make this goal elusive. Why are such absolute textural fingerprints so difficult to find?

Factors other than size control how clasts are entrained, transported, and deposited by wind, water, and ice. Fine-grained clasts of dense minerals such as magnetite and gold are **hydraulically equivalent** to coarser but less dense clasts of quartz and feldspar. Differently sized clasts of such a mineral mix will settle through a fluid at identical velocities. Beach sands composed entirely of quartz will be better sorted than beach sands consisting of magnetite and quartz. Differences in particle shape (flakes versus spheres) and particle roundness or angularity (sharp-edged angular clasts versus smooth rounded clasts of identical diameter) will also affect settling velocity.

Inheritance also affects grain size distributions. Textural analysis is based on the incorrect assumption that depositional agents operate on grain size distributions that were originally infinitely varied and that therefore can be totally reshaped into diagnostic distributions. Yet well-sorted beach sand moved downslope by submarine slumping and density currents will remain well-sorted. In addition, such transporting mechanisms as streams and longshore currents might behave identically in terms of hydrodynamics.

Finally, diagenesis may modify original textural characteristics significantly. For example, modern turbidite sands typically contain only 7% to 8% matrix, whereas ancient turbidite sandstones contain 15% or more. The increase in matrix results from disaggregation of rock fragments (Cummins, 1962).

In summary, sandstone texture, depositional agent, and environment are not distinctively linked. Grain size distribution clearly separates some deposits—for example, submarine fan turbidites and eolian dunes—but inland dunes, onshore dunes, and beach sandstones cannot be discriminated. A study of grain size should be one of several approaches used to tie a sandstone to the agent that transported and deposited it, but it is not an independent, self-sufficient procedure. It should be used in concert with other genetically diagnostic features such as sedimentary structures and fossil content.

Shape and Roundness Sand grain shape (form) and roundness (angularity) are useful properties for describing and differentiating sandstone units. They are of limited value in identifying provenance, dispersal, and depositional mechanism.

A variety of methods are used to describe the shape or form of a sand-sized clast. Zingg (1935) used a caliper to measure the long, intermediate, and short axes of pebbles and defined four shape categories: equant, tabular or oblate, bladed, and prolate. These categories are more useful for loose sand grains than they are for a cemented sandstone. Visual comparison of sand grains with standard reference silhouettes is quicker and simpler (see Fig. 1.3). Both shape (evaluated in terms of sphericity) and roundness are categorized by counting as few as 50 grains either in thin section or using a binocular microscope for unconsolidated sand grains.

Fourier analysis is a more recent and sophisticated technique (Ehrlich and Weinberg, 1970). It uses complex microscopy, computers, and advanced mathematics to digitize grain shapes in an attempt to identify provenance and characterize depositional setting.

Sand particle shape does not identify transport history or depositional environment definitively. Transport agents do not shape sand grains; rather, shape is inherited, and it is controlled by composition (mineralogy). For example, sand grains of biotite, slate, and phyllite are almost always flake-shaped because they are sheet silicates. Nevertheless, shape does influence how a sand grain is transported and dispersed. Flakes are more easily transported than similarly sized spherical (equant) particles.

Sand grain roundness or angularity differs from grain shape. It is a function of the curvature of the corners of a clast rather than overall grain dimensions (see Fig. 1.3). Abrasion history determines sand grain roundness. The rate at which abrasion occurs is a function of clast size, mineralogy, and transport agent.

1. Size is important because the intensity of grain-to-grain impacts increases with momentum.

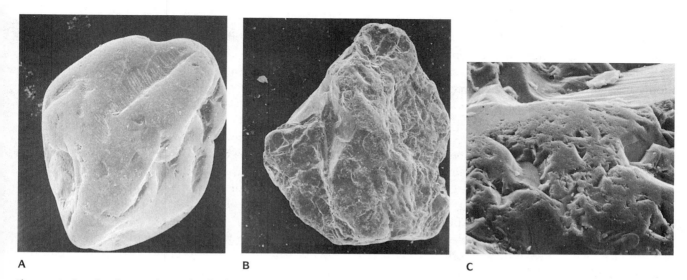

A   B   C

Figure 5.18 Scanning electron micrographs of grain surfaces can be useful in interpreting sedimentary environments. (A) Rounded grain from upper part of beach, well worn from abrasion (magnified 65 times). (B) Angular grain from surf zone, broken and pitted from wave action and dissolution (magnified 50 times). (C) Enlargement of grain in (B) showing etching and pitting from dissolution (*lower center*) and striations from glacial action (*upper right*). (Courtesy of M. Seidl.)

Coarser sand rounds faster than fine sand, cobbles round much faster than pebbles, and so on.

2. The hardness and cleavage of a clast are controlled by mineralogy. Soft, cleavable clasts of limestone or feldspar are more easily abraded than grains of quartz.

3. The velocity and viscosity of the transporting agent dictates the frequency and intensity of grain-to-grain collisions that chip off and smooth sharp corners. Dust storms and shoreline surf abrade sand intensely. Conversely, little or no abrasion of sand grains occurs as they ride passively on glaciers or settle slowly through the slow-moving, viscous ice.

The rounding of sand grains is a slow process, usually requiring long-distance transport. River transport over tens of thousands of kilometers is necessary just to increase the roundness of grains from angular to subangular, and there are no river systems that long. Most well-rounded sand grains have probably experienced periods of intense abrasion that occurred during wind transport or during episodes in which they were repeatedly washed back and forth along shorelines. Multiple cycles of uplift, erosion, and deposition, each accompanied by a modest increase in grain roundness, are contributing factors.

Roundness, like shape, affects transport. Rounded grains are more easily transported by rolling; angular corners on clasts generate fluid eddies that retard settling. Although roundness and shape are distinct properties, an increase in roundness—for example, by the chipping off of sharp edges and protruding corners of an equant grain—also improves its sphericity, although more modestly.

Grain Surface Features   The surfaces of sand grains display a variety of small-scale features ranging from pits, scratches, and ridges to polish or frosting (Fig. 5.18). These features are produced by abrasion during transport or by chemical corrosion or etching during diagenesis. Understanding the origin of surface features is far from easy. First, grains can display surface features inherited from a previous episode of transport and diagenesis. Second, the mechanism by which surface nicks and chips are produced is not unique to a particular transporting agent. Finally, studying sand grain surface features is technically difficult because of their small scale. Transmission or scanning electron microscopes are required. Instead of thin sections, grain mounts are prepared. More than 20 surface features can be tentatively linked with specific ancient transport mechanisms and depositional environments (Krinsley and Doornkamp, 1973).

Fabric   The term *fabric* refers to the orientation or arrangement of grains in a sandstone, how they are packed together, and the type of grain contacts. Sandstone fabric controls porosity and permeability. Describing sandstone fabric is difficult because the fabric elements are sand-sized. Sandstone fabric is either **random (isotropic)** or **oriented.** An oriented

fabric reflects the parallel alignment of elongate or disk-shaped grains. It is produced when sand is deposited by strong, directionally constant currents. The long axes of grains of fluvial sands are commonly aligned parallel with the direction of current flow. Grains rolled along as bedload line up transverse to current flow. Elongate and disk-shaped sand grains can be imbricated.

Sandstone packing refers to how densely individual grains are stacked together. Packing determines porosity, for it is a measure of how much empty space remains between grains. Packing ranges between two extremes. With **cubic packing,** each sphere of an overlying layer rests directly above a sphere of the layer beneath (see Fig. 5.12). With **rhombohedral packing,** each sphere of an overlying layer rests in the depression between spheres of the layer below. Cubic packing is the looser arrangement, generating about 50%

porosity. Rhombohedral packing is tighter, generating about 25% porosity. Packing in sandstones is complex because natural sediment never consists only of perfectly spherical sand grains of identical diameter.

Textural Maturity Folk (1951, 1966) defined four stages of sandstone textural maturity (Fig. 5.19A). An evaluation of sandstone textural maturity is based on three criteria: (1) the proportion of "clay" clasts (Folk classified as clay any grains with a diameter of 30 μm (micrometers) or less; that is, fine silt and below); (2) the sorting of the sand framework; and (3) the roundness of the sand grains.

A sandstone is texturally **immature** if the proportion of clay-sized material exceeds 5%, regardless of the degree of sand grain sorting and rounding. Immature sandstones also tend to be high in feldspars and rock fragments. According to Folk, immature sandstones

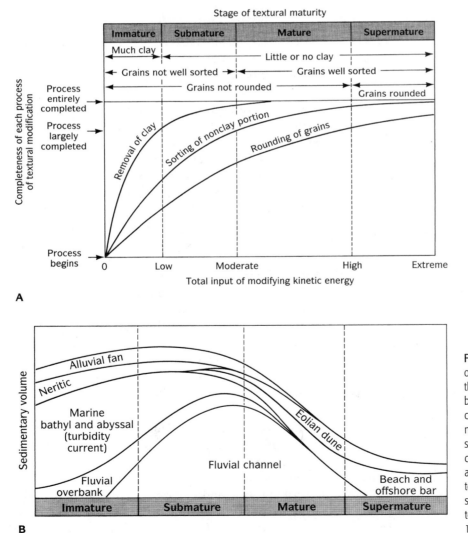

Figure 5.19 Folk (1951) pioneered the concept of textural maturity. (A) As sands are transported, they lose clay and become better sorted and better rounded. Using these three criteria, a sand can be characterized as immature, submature, mature, or supermature. (B) A particular sedimentary environment often has sands of characteristic maturity. Fluvial overbank, alluvial, and marine turbidite sands tend to be immature to submature; fluvial channel sands tend to be submature to mature; eolian and beach sands tend to be mature to supermature. (After Folk, 1951; 128; by permission of the SEPM.)

are particularly characteristic of alluvial fans, turbidite sands, and fluvial overbank sands (Fig. 5.19B).

Texturally **submature** sandstones have a clay component of less than 5%. Sand grains are no more than moderately sorted and are only slightly more rounded than in immature sandstones. Feldspars and lithic fragments are still common, but less so than in immature sandstones. According to Folk, these sandstones are found in a variety of environments but are particularly common in fluvial channels and in turbidites (see Fig. 5.19B).

Mature and **supermature** sandstones also have less than 5% clay, but the sand framework is well sorted or very well sorted. Supermature sandstones have rounded or subrounded grains; mature sandstones have grains that are subangular to subrounded. Virtually no feldspar or rock fragments (except possibly cherts) occur in mature or supermature sandstone. According to Folk, mature sandstones are particularly common in fluvial channels. Mature and supermature sandstones are also common on beaches and in eolian dunes, because of the high energies and abrasive potential of these environments (see Fig. 5.19B).

Classification

More than 50 sandstone classification systems have been proposed since the middle of the twentieth century. Those that have stood the test of time, judged by the frequency with which they are used today, employ the same aspects of texture and composition

as the scheme we will discuss shortly. But first we will present some cautionary notes.

Field analysis, the crucial first step in unraveling the origin of a particular deposit, can discriminate only broad classes of sandstone based largely on sedimentary structures, gross textural and mineralogical characteristics of the deposit, and the other rocks with which the sandstone is laterally and vertically associated. The classification scheme used here, though descriptive, requires point counting (typically, 500 points) of petrographic thin sections. Some field geologists argue that schemes that pigeonhole sandstones using a procedure requiring such laboratory analyses not only are impractical and essentially useless but also generate cumbersome nomenclature and promote largely misleading generalities. We believe that detailed petrographic classification of sandstone is simply a means to better understanding. To classify is to define. Definition forces sedimentary petrologists to identify those properties of sandstone that allow genetically different varieties to be recognized. Nomenclature generated during classification can clarify rather than obscure, and communication improves. In this spirit, we will now discuss petrographic sandstone classification, followed by a summary of what we believe to be the genetically significant characteristics of the major sandstone categories identified by that classification.

Figure 5.20 shows the petrographic sandstone classification most commonly used. It is based on

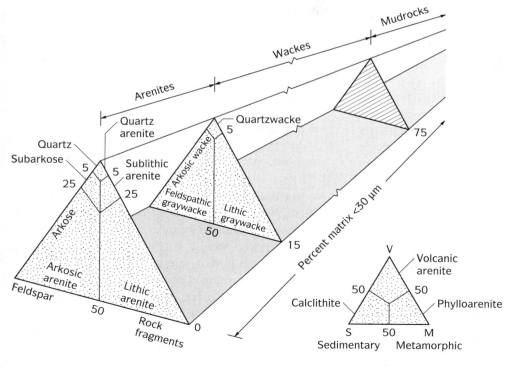

Figure 5.20 Classification scheme for terrigenous sandstones. The front panel shows arenites, the central panel wackes, and the rear panel mudrocks. The small panel at lower right allows more specific naming of lithic wackes and arenites on the basis of their rock fragments. (After Dott, 1964; 629; by permission of the SEPM.)

earlier classifications first proposed by P. D. Krynine in the 1940s and F. J. Pettijohn in the 1950s and 1960s and later modified by a number of workers, especially R. H. Dott, Jr. (1964). This classification can be used for both unconsolidated modern sand and ancient sandstone.

Two defining parameters are used to subdivide terrigenous sandstones:

1. The **percentage of matrix** (regardless of origin) is defined as any clastic material finer than 30 μm (coarse silt). For classification purposes, the amount and kind of other interstitial material such as cement is irrelevant.

2. The **composition of sand framework grains,** specifically the percentages of quartz (Q), rock fragments (R or L), and feldspar (F). Other framework clasts are not considered.

This classification scheme eliminates subjective judgment. The petrologist counts points in a thin section and determines the amount of matrix and the percentage of sand grains of quartz, feldspar, and rock fragments. Once the sample is categorized and named, an extensive body of sandstone literature is available as a valuable database.

Let us examine the classification in more detail. Using a 15% matrix content, two distinct groups of sandstone are distinguished: **arenites** and **wackes** (or *graywackes*). Specific varieties of each group occupy positions within the separate triangular panels for arenites and wackes that appear in Fig. 5.20. A third triangular panel represents mudrock, to portray the natural continuum that exists between sandstone and finer-grained clastic rocks.

Arenites are texturally "clean," matrix-free (or at least matrix-poor) sandstones. They owe their cohesiveness to cement precipitated in what were originally empty intergranular pores. Wackes are argillaceous, matrix-rich, texturally immature, or "dirty" sandstones. Matrix originates in two principal ways. *Primary* detrital matrix is transported and deposited with the coarser sand-sized framework grains. *Secondary* matrix is produced by diagenesis: rock fragments are squashed and disaggregated, and feldspar decomposes to form clay. Although size alone identifies a matrix, the origin of that matrix must be inferred to understand the sandstone containing it.

The percentage of quartz, feldspar, and rock fragments that occur as sand-sized grains allows arenite and wacke to be separated into subtypes. Other framework constituents such as the micas and heavy minerals are ignored. Therefore, raw percentages of Q, F, and L must be restated on the basis of 100%. Counting procedures vary, particularly with respect to rock fragments. The Gazzi-Dickinson technique restricts rock fragments to aphanitic materials containing individual crystals no larger than 0.0625 mm. Using this procedure, coarser composite grains would be identified by the particular mineral component centered in the field of view. Other point-counting methods classify such coarse-grained clasts as rock fragments.

Three principal varieties of arenite and wacke are recognized based on the percentage of quartz, feldspar, and rock fragments. **Quartz arenites** and **quartz wackes** are sandstones with framework grains composed mainly of quartz (95% or more). Sandstones containing a lower percentage of quartz can be further categorized by comparing the relative proportion of rock fragments and feldspar. The prefix **lithic** is used for arenites and wackes in which rock fragments exceed feldspar. Those in which feldspar exceeds rock fragments are called **feldspathic** or **arkosic**. Names such as arkose, subarkose, sublitharenite, and arkosic wacke were coined to refer to variations of these more general categories. The smaller panel at the lower right in Fig. 5.20 is used to generate specific names for rock-fragmentrich sandstones.

This classification incorporates the essential characteristics of any good scheme: it is fundamentally descriptive but also correlates with sandstone genesis. The correlation with sandstone origin is no accident. The physical properties were selected to discriminate among sandstones, although the linkage between physical properties and rock genesis is much more clear-cut in the case of igneous rock classification (where the texture reflects cooling rate and the mineral composition is related to magma chemistry and temperature of crystallization).

What relationships connect sandstone texture and composition to origin?

Matrix-Rich Wacke Versus Matrix-Poor Arenite When Pettijohn (1957) selected the percentage of matrix finer than 0.03 mm as the criterion to separate wacke and arenite, he proposed the **fluidity index.** He pointed out that most matrix-poor (typically 1% to 2% matrix) sandstones (arenites) exhibit sedimentary structures, fossils, and other features consistent with transport and deposition by such fluids as water and wind. Arenites contain cement simply because they originally had empty pore

space. Matrix-rich sandstones (wackes), on the other hand, tend to exhibit size grading, sole markings, and other features produced when transport and deposition are by quasi-liquid flows such as density currents and mass flows (although later examination of turbidite sands sampled from abyssal plains and submarine fans complicated this simple picture). Many matrix-rich modern sands are remarkably well sorted but contain less than 6% to 8% matrix—less than the 15% matrix required for a wacke sand or sandstone. They do, however, often contain rock fragments and feldspar that can disaggregate and weather into matrix.

Ratio of Rock Fragments to Feldspar Pettijohn (1957) argued that the proportion of feldspar to rock fragments is actually a reliable **index of sandstone provenance.** He pointed out that sand sources are either supracrustal or subcrustal rocks. Supracrustal sources form at or very near Earth's surface; for the most part, they are fine-grained (aphanitic) volcanic rocks, low-grade slate and phyllite, and such sedimentary rocks as chert and mudrock. These fine-grained rocks disintegrate into sand grains that are composite—that is, rock fragments—rather than clasts of single minerals. Subcrustal rocks form at depth and include such igneous rocks as granite and diorite and such higher-grade metamorphic rocks as schist and gneiss. Disintegration of these rocks generates sand grains of single minerals such as quartz and feldspar, not composite grains. Therefore, the ratio of feldspar to rock fragments ostensibly separates sandstone derived from two distinct provenances.

Percentage of Quartz or Ratio of Quartz to Feldspar + Rock Fragments This ratio is an **index of compositional maturity,** reflecting the differences between sand with lots of soft, unstable, decomposable rock fragments and feldspar and sand composed of only the most physically resistant and chemically stable materials, mainly monocrystalline and polycrystalline quartz. This index is based on the premise that physical disintegration and chemical decomposition operate on soil and sediment over extremely long spans of time. Weathering and recycling should ultimately generate sand residues composed of only the most resistant materials.

Like any scientific scheme designed to categorize and describe natural phenomena, this classification has its shortcomings. Nevertheless, it works to the extent that it separates sandstones into **four major sandstone suites** or **families** that differ not only in physical properties but also in depositional environment, provenance, and tectonic setting.

Characteristics, Significance, and Occurrence

Table 5.4 lists four major sandstone families and their estimated abundance (Pettijohn, 1975). In the following summaries, arenites are separated into *quartz arenites, feldspathic (arkosic) arenites,* and *lithic arenites; wackes* are lumped together as a single family. These summaries characterize the mineralogy, chemistry, texture, sedimentary structures, associated fossil content, sandstone body geometry, lateral and vertical lithological associations, depositional

TABLE 5.4 The Four Major Sandstone Families

Interstitial Space	Principal Category	Framework Mineralogy[a]	Sandstone Family	Abundance[b] (%)
Pore space empty or filled with cement; <15% matrix	Arenite	Qfl	Quartz arenite	33.9
		QFl	Feldspathic (arkosic) arenite	20.1
		QfL	Lithic arenite	25.5
Pores filled with matrix (15% or more)	Wacke	Qfl, QFl, QfL, etc.	Wacke	19.3
Total				99.2

[a]Q = quartz as a major constituent (>5%); F = feldspar as a major constituent (>5%); f = feldspar in accessory abundance (<5%); L = rock fragments as a major constituent (>5%); l = lithic fragments in accessory abundances (<5%).
[b]Abundance percentages are an average of four different estimates published by Pettijohn (1975). Three of these estimates disregarded hybrid sandstones: the fourth estimated their abundance at 5%.

Figure 5.21 Photomicrograph of a supermature quartz arenite, the Ordovician St. Peter Sandstone of Minnesota. The grains of nonundulose monocrystalline quartz are extremely well rounded and well sorted and make up 99% of the rock. There are almost no other minerals, no clay matrix, and relatively little cement. Supermature quartz arenites cannot be produced by intense weathering alone; they must be recycled from older sandstones. The extreme rounding suggests that they were part of a dune sand before they were deposited in a marine setting. (Pettijohn, Potter, and Siever, 1972: 181; © Springer-Verlag, New York.)

environment, origin, geological occurrence, examples, and historical significance.

Quartz Arenites (Orthoquartzites) Typically, quartz arenites (Fig. 5.21) are white to light gray sandstones, although they are often stained pink, brown, or red by iron oxide cement. They consist almost entirely of sand-sized monocrystalline quartz grains (many with abraded authigenic overgrowths). Resistant grains of chert, metaquartzite, and such "heavy" minerals as zircon, tourmaline, and rutile also can be present. Chemical composition reflects this restricted mineralogy: 95% to 97% SiO_2, 0.5% to 1.0% Al_2O_3 (traces of matrix and detrital feldspar), and 1.0% CaO and CO_2 (carbonate cement).

Quartz arenites typically have a supermature texture and composition. They are usually well-bedded and can exhibit ripple marks; lamination; cross-lamination; and, in some cases, large-scale cross-bedding. Body fossils are rare; when present, they are fossils of very shallow neritic or hypersaline organisms. Such trace fossils as worm burrows (*Skolithos* facies) are locally abundant.

Most quartz arenites occur as regionally extensive blanket-shaped bodies; for example, as basal transgressive sandstone sheets found within the major Paleozoic cratonic sequences. The thickness of individual sheets varies from a few meters to several hundred

meters. They are commonly interbedded laterally and vertically with sedimentary rocks that form along stable cratonic margins: shallow marine mudrock, limestone, dolomite, and shoreline conglomerate (oligomict orthoconglomerate). These bodies are generally deposited on unconformable surfaces.

Many quartz arenites are shallow marine (but above storm wave base) sands that accumulated along or near the shoreline as beach, shoreline dune, tidal flat, spit, barrier island, or longshore bar deposits. Repeated recycling of detritus weathered from stable, low-lying cratonic continental block sources probably played an important role in their genesis. Other quartz arenite bodies are subaerial windblown dune deposits.

Although quartz arenites constitute only one-third of all sandstones, they are unevenly distributed. Slightly more than half are of late Precambrian (1000 Ma or younger) and early Paleozoic age. This distribution in space and time is puzzling. It is unusual for deposits to be so narrowly restricted in time yet so widespread in extent when present. Their supermature texture and composition tightly constrain provenance, weathering, and depositional setting. Their absence from sequences of Archean and early Proterozoic age probably reflects the lack of abundant quartz-rich source rocks and/or the absence of broad, tectonically stable continental cratons. The predominance of late Precambrian–earliest Paleozoic quartz arenites suggests a lengthy interval of tectonic stability that promoted intense weathering. Still later episodes of tectonic quiescence that coincided with slow regional marine transgressions or arid global climate must have occurred repeatedly in the Paleozoic and Mesozoic, because quartz arenites that can be correlated with those of North American deposits are found on other cratons.

Feldspathic (Arkosic) Arenites The major framework grains found in this sandstone type are monocrystalline quartz and feldspar. Feldspar content typically reaches 40% to 50% (Fig. 5.22A). Orthoclase and microcline exceed plagioclase when continental crust is the dominant source; where plagioclase predominates, a volcanic arc source is indicated. The white, gray, or pink color of feldspar imparts a similar tint to feldspathic arenites, which is further enhanced by ferruginous cement. Other abundant framework grains are micas (muscovite and biotite) and rock fragments. High quartz and feldspar abundance produce an SiO_2 content ranging from 60% to 80% and a high percentage of Al_2O_3. The amount of K_2O (2% to 4%) ex-

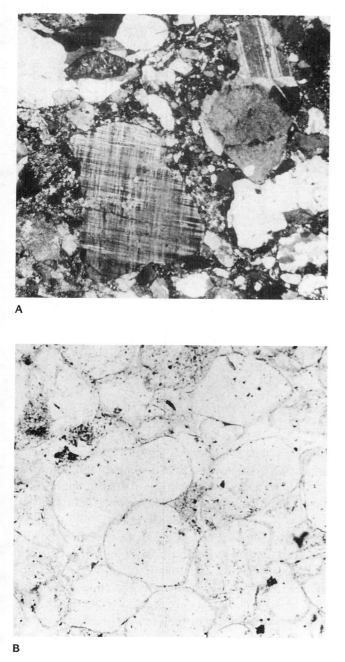

A

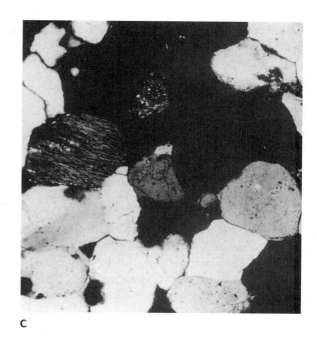

Figure 5.22 Arkoses are rich in feldspars and indicate relatively little transport or weathering. (A) A typical immature arkose from the Triassic rift valleys of the Newark Supergroup; this example is from Hot Springs, North Carolina. Large grains (about 1 to 2 mm in diameter) of microcline (tartan twinning) and plagioclase (striped twinning) are visible. (D. R. Prothero.) (B, C) A desert arkose from the Precambrian Torridonian Sandstone, northwest Scotland, under plane light (B) and crossed nicols (C). The feldspars are fresh and unaltered (chemically immature), but the grains are well rounded and sorted (texturally mature). These features suggest extensive transport and weathering in a dry climate where chemical weathering is minimal, such as in a desert sand dune. (Courtesy of John F. Hubert; by permission of W. A. Benjamin and Co., Menlo Park, Calif.)

B

C

ceeds that of Na_2O when K-feldspar is more abundant than plagioclase.

Feldspathic arenites are not as mature texturally or compositionally as quartz arenites. They are typically coarser; grains are less well sorted and less well rounded. "Desert arkose" is an exception (Fig. 5.22B, C). It is compositionally immature but texturally supermature, consisting of well-sorted, well-rounded grains of monocrystalline quartz and feldspar. Desert arkose is produced in arid climates when wind transports physically disintegrated grains of feldspar and quartz.

Bedding and internal organization are ordinarily less well developed than in quartz arenite, although cross-bedding and horizontal lamination can occur. Body fossils of shallow marine or terrestrial organisms may be rare or common; trace fossils are rare.

Some feldspathic arenites are residual or "sedentary" arkoses, formed as local lenses and layers found at or near the base of transgressive quartz arenite sequences. Many arkoses are found in regionally restricted wedge-shaped nonmarine alluvial fan and fan delta deposits. They form as sheet-flow and debris-flow sediments and accumulate on the surface

of alluvial fans, within river channels and point bars, and on and along shorelines. Their high feldspar content indicates that coarse, feldspar-rich rocks such as granite and gneiss are being eroded. Survival of feldspar, with little decomposition to clay minerals, signals a dry or arctic climate and/or a steep mountainous topography.

Arkose-bearing sequences ranging in age from Precambrian through the Neogene are known. Ignoring the relict arkosic deposits that constitute the basal portion of quartz arenite–rich transgressive sequences, individual units are restricted in extent. Arkose accumulates locally where there is vigorous uplift (faulting?) of feldspar-rich source rocks. The lack of arkoses of early Precambrian (Archean) age probably reflects a scarcity of granitic and other K-feldspar-bearing source rocks early in Earth's history.

Many classic feldspathic arenites coincide temporally and spatially with episodes of continental distension and the development of fault-bounded, rift-related grabens and half-grabens bordering steep basement rock. Feldspathic arenites (and wackes) are also deposited adjacent to active and/or dissected magmatic arcs.

Lithic Arenites (and Sublitharenites) Clasts of monocrystalline quartz (30% to 80%) and rock fragments (5% to 50%) are the most important constituents in this sandstone family (Fig. 5.23). The mix of lighter-colored quartz and feldspar clasts with darker-colored rock fragments gives these sandstones a speckled, salt-and-pepper appearance. Sand flakes of detrital mica are common; feldspar content is low (a few percent). Broad compositional diversity due to wide variations in rock fragment content and type is a distinguishing characteristic. Further categorization based on specific rock fragment types is sometimes helpful.

The variation in mineral constituents generates a wide range in chemical composition. The SiO_2 content can approach the high percentage of quartz arenites (90% or more) or match the 50% to 60% of wackes. The amount of Al_2O_3 and K_2O depends on whether such rock fragments as mudrock are abundant. Grains of limestone and dolomite increase CaO, MgO, and CO_2 content. Lithic arenites have submature to mature textures.

Lithic arenites that accumulate as alluvial deposits are well bedded and exhibit tabular and trough cross-bedding, ripple marks, internal lamination, current lineation, scour-and-fill structures and fining-upward cycles. Fossils are uncommon, although deposits laid

Figure 5.23 Lithic sandstones may contain a wide variety of rock fragments. This example, from the Eocene Coaledo Formation of Oregon, contains abundant volcanic rock fragments (rounded grains with randomly oriented fine plagioclase crystals), mica grains from schists (long fibrous minerals), and feldspars (twinned grains). Most grains are 1 to 2 mm in diameter. (Courtesy of R. H. Dott, Jr.)

down in deltaic sequences that are interbedded with shallow marine shelf mudrock contain abundant fossils. These deposits exhibit internal lamination, oscillation ripple marks, and well-developed bedding.

Many lenticular sandbar deposits that interfinger with channel conglomerate and floodplain mudrock are lithic arenite, as are some sheetlike shallow marine shelf and deeper-water abyssal plain deposits. Sandstones found on many alluvial fans and river basins developed adjacent and in front of recently uplifted mountainous sources often belong in this category. Lithifying river sands of the Colorado, Mississippi, and Amazon systems would produce lithic arenite. And most modern turbidite sands accumulating in submarine fan systems are lithic arenites.

Many orogenic clastic wedges consist largely of lithic arenite. This is not surprising, because physical disintegration of mountainous supracrustal rocks invariably generates detritus rich in rock fragments. Therefore, lithic arenites typically coincide temporally and spatially with subduction-related active magmatic arcs and collisional orogeny.

Wacke (Graywacke) Wackes are physically hard, dark, enigmatic rocks. Clasts of monocrystalline quartz are often the most abundant framework com-

Figure 5.24 Graywackes are poorly sorted sandstones with abundant clay matrix suspending angular grains of many types. This example, from the Ordovician Normanskill turbidites near Kingston, New York, includes many angular grains of quartz (untwinned) and plagioclase (parallel twins), as well as rock fragments in a fine clay matrix. Most larger grains are 0.5 to 1 mm in diameter. (D. R. Prothero.)

ponent (25% to 50%), although the proportion fluctuates. There are varieties of wacke that are equivalent in all respects except ratio of matrix to quartz arenite, feldspathic (arkosic) arenite, and lithic arenite (Fig. 5.24). Feldspar clasts are most often angular to subangular and can be twinned (dominantly sodic) plagioclase as well as K-feldspar. Grains of chert, mudrock, limestone, polycrystalline quartz, and volcanic rocks are also quite common. Clasts of detrital muscovite, biotite, and chert occur in accessory amounts.

The SiO_2 content of wackes ranges from 50% to 70%, reflecting the moderate amount of quartz and feldspar. Because they contain abundant matrix rich in clay minerals and chlorite, they are also rich in Al_2O_3, MgO, and FeO + Fe_2O_3. The ratio of Na_2O to K_2O is controlled by the relative proportion of albitic plagioclase to potassium feldspar.

The fact that wackes are identified by 15% or more matrix means that they are immature by definition. Although some have a sand framework that is also poorly sorted and angular or subangular, many consist of well-rounded, well-sorted sand. Grain size distribution is bimodal with a primary peak for framework grains and a secondary peak for matrix. The fact that modern turbidite sands have 6% to 8% matrix, rather than the 15% or more in ancient deposits, suggests that roughly half of this component is of secondary origin.

Many wackes were deposited by waning turbidity currents. They routinely display graded bedding, sole markings, and the systematic upward changes in sedimentary structures and grain size characteristic of turbidites (see Chapter 10). Deep-water abyssal and bathyal body fossils, pelagic fauna and flora, and retransported shallow-water organic remains are all found within wacke sandstone sequences. Some wackes were deposited within submarine fan complexes, but there are also examples of more distal deposits settled out of more diffuse density flows that spread out and dispersed across broad, flat abyssal plain surfaces. These latter deposits are interbedded with abyssal plain mudrock, bedded (pelagic) chert and limestone, and submarine basaltic flows.

Wackes are the dominant sandstone of the Archean, because the only emergent areas early in Earth's history were narrow, nongranitic volcanic arcs bordered by troughs. Archean wackes form much of the sedimentary rock carapace of Precambrian greenstone belts. The subsequent growth of granitic continental blocks during the Proterozoic made possible the formation of better-sorted, feldspar-rich and quartz-rich "platformal" arenites, but many classic wacke-bearing Phanerozoic continental slope, continental rise, and ocean trench successions have been described.

CONCLUSIONS

The substantial efforts made to improve the means by which the coarse and very coarse terrigenous clastic sedimentary rocks are analyzed, classified, and interpreted are easily explained. For example, even though conglomerate and breccia are not particularly abundant in the stratigraphic record, they are unmatched in their potential to identify provenance and provide a clear insight into transport agent and depositional setting. Sandstones—which are almost equal in reliability and precision as indicators of source composition, relief, location, and depositional environment—happen to be much more widely distributed in space and time. Taken together, no other class of sedimentary rocks has more to tell us about the Earth.

FOR FURTHER READING

Alvarez, L. W., et al. 1980. Extraterrestrial cause for the Cretaceous-Tertiary extinction. *Science* 208:1095–1108.

Frankel, C. 1999. *The End of the Dinosaurs.* Cambridge: Cambridge University Press.

Hambrey, M. J., and W. B. Harland, eds. 1981. *Earth's Pre-Pleistocene Glacial Record.* Cambridge: Cambridge University Press.

Koster, E. H., and R. J. Steel, eds. 1984. *Sedimentology of Gravels and Conglomerates.* Memoir 10. Calgary, Alberta: Canadian Society of Petroleum Geology.

Pettijohn, F. J., P. E. Potter, and R. Siever. 1987. *Sand and Sandstone.* 2d ed. New York: Springer-Verlag.

Poag, C. W. 1999. *Chesapeake Invader.* Princeton, N. J.: Princeton University Press.

Schermerhorn, L. J. G. 1974. Late Precambrian mictites: Glacial and/or nonglacial. *American Journal of Science* 274:673–824.

Scholle, P. A. 1979. *A Color Illustrated Guide to Constituents, Textures, Cements, and Porosities of Sandstones and Associated Rocks.* Memoir 28. Tulsa, Okla.: American Association of Petroleum Geologists.

Siever, R. 1988. *Sand.* New York: W. H. Freeman and Company.

6 Mudrocks

Terrestrial mudrocks can often be distinguished from marine shales by evidence of subaerial exposure. This example from a modern mudflat near Fort Dodge, Iowa, shows both mudcracks and raindrop-impact craters. (Pettijohn and Potter, 1964: plate 94; by permission of Springer-Verlag, New York.)

CLAYS—WHICH ARE SHEET SILICATES THE SIZE OF colloidal particles, viruses, or the particles of smoke from a cigarette—are the most abundant minerals on the surface of the Earth. Clays cover about 75% of the land surface and blanket most of the deep seafloor in pelagic oozes. As we saw in Chapter 1, mudrocks are by far the most abundant class of sedimentary rock, making up between 50% and 80% of Earth's total sedimentary rock shell. By any measure, clays and mudrocks are among the most important materials on the surface, and that should be reason enough to study them.

The term **mudrock** refers to all siliciclastic sedimentary rocks composed predominantly of silt-sized (1/16 to 1/256, or 0.0625 to 0.0039, mm) and clay-sized (< 1/256, or < 0.0039, mm) particles. Mudrock includes two lithologies in which one of these size ranges predominates—**siltstone,** with 50% or more silt-sized material, and **claystone,** with 50% or more clay-sized material—as well as lithologies that are a mix of the two. **Mudstone** is indurated mud, which is a mixture of silt with between one-third and two-thirds clay. **Shale** (Fig. 6.1) is any mudrock that exhibits lamination or fissility or both. **Argillite** is

99

Figure 6.1 Well-developed fissility in the Mississippian Antrim Shale, Alpena County, Michigan. Note hand lens for scale. (Dietrich and Skinner, 1979: 201; by permission of John Wiley, New York.)

mudrock that has been subjected to low-grade metamorphism. Mudrocks are also referred to as **pelites,** pelitic sedimentary rocks (from the Greek for mud), **lutites** (from the Latin for mud), and argillaceous sedimentary rocks.

Considering the great importance and abundance of these rocks, it is surprising that there are so few published mudrock studies and so few specialists who work on mudrocks. This disparity reflects the difficulties of studying these rocks and the ambiguities of the resulting data. What are the obstacles to understanding this rock type?

Even though mudrocks are the dominant rocks by volume, they are typically not well exposed. In most climates, mudrocks weather deeply to form valleys or lowlands with few or no outcrops. In addition, mineral composition and texture cannot be studied easily in the field or laboratory. The mineralogy and texture of silt-sized clasts can be determined only in a rudimentary fashion with a petrographic microscope, and this method is of no use for clay-sized materials. Even X-ray diffraction methods able to identify individual clay minerals provide only crude estimates of clay mineral abundance. Grain size data obtained using sedimentation tubes have produced few meaningful conclusions. And the sedimentary structures in mudrocks are difficult to see and are of limited use in interpretation.

In addition to the practical problems, there are other, more subtle reasons for the comparative neglect of mudrocks. Sedimentary research has long been driven by economic motives (especially the needs of the petroleum industry), and for a long time mudrocks were considered the uninteresting "source rock" from which the organics had migrated away, not the profitable "reservoir rock" where hydrocarbons accumulated (these are usually sandstones or limestones). When the possibility of extracting hydrocarbons from oil shale emerged in the 1970s, there was a sudden expansion of mudrock research. Then the price of oil collapsed, and with it, oil shale research, exploration, and investment also declined.

Another historical factor is the fact that most marine mudrocks are deposited in the deep ocean, which makes them extremely difficult for land-based geologists to study. Because very little was known about them, the understanding of deep-marine shales exposed on land was also very limited. When various oceanographic vessels began deep-sea coring in the 1950s and 1960s, and when the Deep Sea Drilling Project began routinely drilling cores in the ocean floor around the world in the 1970s and 1980s, the interest in the clays on the ocean floor suddenly increased.

Despite these problems, recent improvements in analytical and interpretive techniques permit sedimentary geologists to cope better with mudrocks (see, for example, Potter, Maynard, and Pryor, 1980; Hardy and Tucker, 1988; McManus, 1988).

Texture

There is no general agreement about the best way to categorize mudrock types based on the proportion of sand, silt, and clay. Boggs (1992) suggested that the typical mudrock ranges from 80% silt, 17% clay, and 3% sand to 2 parts silt and 1 part clay. In any case, there is less textural variation among mudrocks than among coarser-grained siliciclastics. The various mechanisms that entrain, transport, and deposit silt and fragments finer than silt evidently do not discriminate among them well. No diagnostic textural fingerprints can be used to match grain size distribution with transporting agent.

Clasts in mudrocks tend to be more angular and less spherical than the clasts in other siliciclastic rocks. This is probably a consequence of both mineralogy and size. Silt, and especially clay-sized material, consists mainly of flaky fine micas and clay minerals, rather than quartz, feldspar, and rock fragments. Micas and

clay minerals are carried by weak currents that cause little abrasion.

Many mudrocks exhibit a preferred fabric caused by the parallel alignment of flat, flake-shaped micas and clays. This fabric is expressed as **fissility:** the tendency of mudrock to part or split along thin, closely spaced parallel surfaces. **Shale** is a general term used to refer to any mudrock possessing fissility (see Fig. 6.1). The thickness of the peel-like fissile sheets ranges from less than 0.5 mm (*papery* parting) to more than 10 mm (*slabby*), with various terms for intermediate thicknesses (*fissile,* 0.5 to 1 mm; *platy,* 1 to 5 mm; and *flaggy,* 5 to 10 mm). Although such terms can be used to describe shale in the field, they have dubious genetic implications (Potter, Maynard, and Pryor, 1980).

Composition

The bulk mineralogy of mudrocks resembles that of sandstone, except that few if any rock fragments are present. Quartz and feldspar are as important in mudrock as they are in sandstone. Fine-grained micas and clay minerals replace the rock fragment component. Boggs (1992) speculated that the typical shale contains 30% quartz, 10% feldspar, and 50% clay minerals (or fine micas), with the remaining 10% being cement (mainly carbonate or iron oxide).

Courses in mineralogy rarely discuss the clay minerals in much detail, so we will review the important ones found in most modern and ancient muds. Because clays are very complex chemically and mineralogically, however, the simplified discussion below ignores much of the variability and most of the less common clay types found in the real world.

Clays are **phyllosilicates,** minerals that consist of tightly bonded sheets of *silicate tetrahedra* (t sheets) internally linked together with shared corner oxygen anions (Fig. 6.2). Attached to these tetrahedral sheets and mutually sharing certain components with them are *octahedral* sheets (o sheets). They are composed of hydroxyl (OH) anions arranged around cations such as calcium, aluminum, and magnesium. As a separate mineral made of aluminum hydroxide, this combination is known as *gibbsite,* so the octahedral layer is sometimes called the gibbsite-like sheet.

From these two basic building blocks (plus additional cations and anions), a great variety of clay minerals can be constructed. Most are composed of sandwiches of tetrahedral or octahedral sheets, which repeat over and over again. Various cations and an-

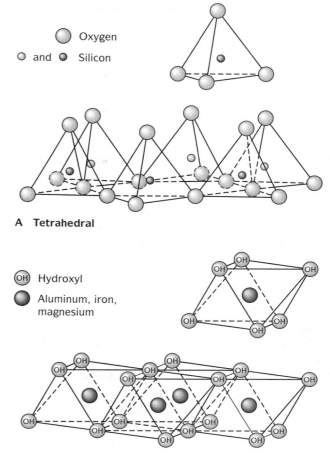

Figure 6.2 Clay minerals are built of two basic components: (A) silicon-oxygen tetrahedral layers and (B) aluminum-oxygen octahedral layers. (After Grim, 1968: 52; by permission of McGraw-Hill, Inc., New York.)

ions are found between the sandwiches, resulting in much chemical variability.

The simplest arrangement characterizes the **kandite** group of clays, most commonly represented by the mineral **kaolinite,** whose name is often used to refer to the kandite group (Fig. 6.3). Kaolinite is composed of an open-faced sandwich of one tetrahedral sheet

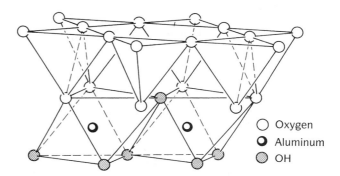

Figure 6.3 Structure of kaolinite, the most common mineral in the kandite group. (After Grim, 1968: 58; by permission of McGraw-Hill, Inc., New York.)

and one octahedral sheet in each repeating layer, or a 1:1 arrangement. The layers are tightly bonded together, with only hydrogen cations between them, so their structure is $-t-o-H^+-t-o-H^+-t-o-$. . . . Unlike most other clays, there is no room for water, hydroxyls, or larger cations between the layers. Consequently, kaolinite clays are chemically and mineralogically simple compared to other clay minerals.

The most common clays, however, have a three-sheet sandwich structure (Fig. 6.4). In this arrangement, the middle layer is an aluminum-hydroxyl octahedral sheet sandwiched between two silicon-oxygen tetrahedral sheets, or a 2:1 arrangement of sheets in each layer. This three-component layer is the basic unit, and it is repeated over and over again, with a variety of chemicals in the interlayer spaces. In the notation just used, the structure is $-t-o-t-$cations$-t-o-t-$cations$-t-o-t-$. . . . The differences between 2:1 clays are caused by the variations in these interlayer cations and anions.

The simplest 2:1 clays have only potassium cations in the interlayers; we know them as the familiar white mica, **muscovite.** Muscovite is common in high-grade metamorphic rocks, in granitic and pegmatitic plutonic igneous rocks, and as a diagenetic replacement of feldspars. It is relatively rare, however, in the near-surface conditions where most clays and muds are

found. Instead, the stable K-rich 2:1 clay at the surface is **illite,** the most abundant clay mineral of all. It has about 70% to 80% as much potassium as muscovite in the interlayers and has slightly more silicon and less aluminum in other parts of the structure than muscovite. The potassium cations give the structure strong ionic bonding, so it does not expand readily and water and hydroxyls cannot easily percolate through; they are found only on the edges of the sheets unless the structure has been degraded.

The next class of 2:1 clays is known as **smectites;** the most familiar of these are called **montmorillonites.** A small amount of Mg^{2+} substitutes for Al^{3+} in the gibbsite-like octahedral sheets, creating a small net negative charge (smaller than the charge in illite). To balance this charge, the interlayers are filled mostly with Na^+, K^+, and Ca^{2+} cations. Because the charge is smaller than that of illite, water is readily absorbed in the interlayers, making the montmorillonite structure very expandable in one direction (but the structural integrity of the sheets is maintained). These clays almost double in volume when they absorb water, and they are notorious for expanding so much during saturation that hillslopes made of montmorillonite shales (and the houses built on them) can slide during heavy rains. Some sodium smectites, such as those found in

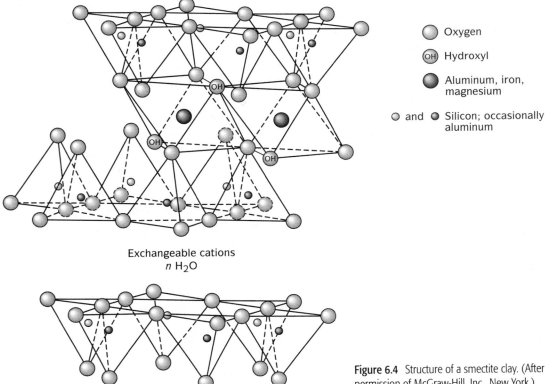

Exchangeable cations
n H$_2$O

○ Oxygen

(OH) Hydroxyl

● Aluminum, iron, magnesium

○ and ● Silicon; occasionally aluminum

Figure 6.4 Structure of a smectite clay. (After Grim, 1968: 79; by permission of McGraw-Hill, Inc., New York.)

Wyoming bentonites, expand to 10 *times* their original volume.

A third type of 2:1 clay mineral, known as **glauconite,** is much like montmorillonite except that most of the cations are iron, not magnesium. We will discuss glauconite later in this chapter.

After kaolinites, illites, and smectites, the fourth common class of clay minerals is the **chlorites** (Fig. 6.5). Chlorite is familiar as the green mica that gives greenschist-grade metamorphic rocks their color, but they are also common in the clay size range. Like smectites, chlorites have a 2:1 sandwich of tetrahedral-octahedral-tetrahedral components. The interlayer space of these minerals, however, contains a new component: a layer of Mg^{2+} and Fe^{2+} surrounded in octahedral, sixfold coordination by hydroxyls, or $(Mg, Fe)(OH)_6$. Because the composition and lattice structure correspond to the mineral **brucite,** the layer is called the brucite-like layer. In the notation used earlier, the chlorite structure is $-t-o-t-brucite-t-o-t-brucite-t-o-t-. \ldots$ Fe^{2+} is usually much less common than Mg^{2+} in the brucite layer, so chlorites are usually found in areas of magnesium-rich parent rocks, where weathering is not so intense that the iron will be oxidized.

Typical clay minerals are rarely composed of a single structural type. Instead, most natural clays are **mixed-layered,** with either regular or random mixtures of illite-montmorillonite being most common (Fig. 6.6).

Another highly expandable clay is known as **vermiculite.** These clays are familiar from nurseries and greenhouses, where you can buy them by the bag to put in planters. Vermiculites have a complex arrangement of water molecules bonded together in the interlayers (quite similar to the interlayer structure in chlorite, but instead of hydroxyls around a cation, there are water molecules alone). This makes them extremely expandable and able to absorb immense amounts of water (Fig. 6.7). Hence, they are commonly used wherever water needs to be absorbed or especially where absorbed water needs to be retained until it is needed. For example, when plant roots run out of easily accessible water in the

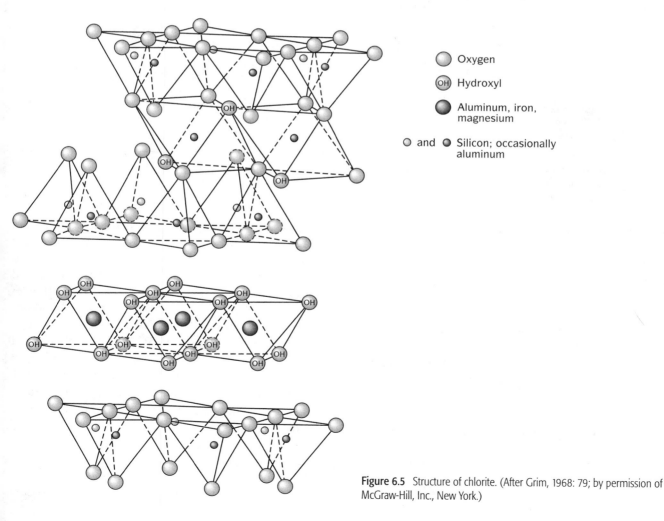

○ Oxygen

(OH) Hydroxyl

● Aluminum, iron, magnesium

○ and ● Silicon; occasionally aluminum

Figure 6.5 Structure of chlorite. (After Grim, 1968: 79; by permission of McGraw-Hill, Inc., New York.)

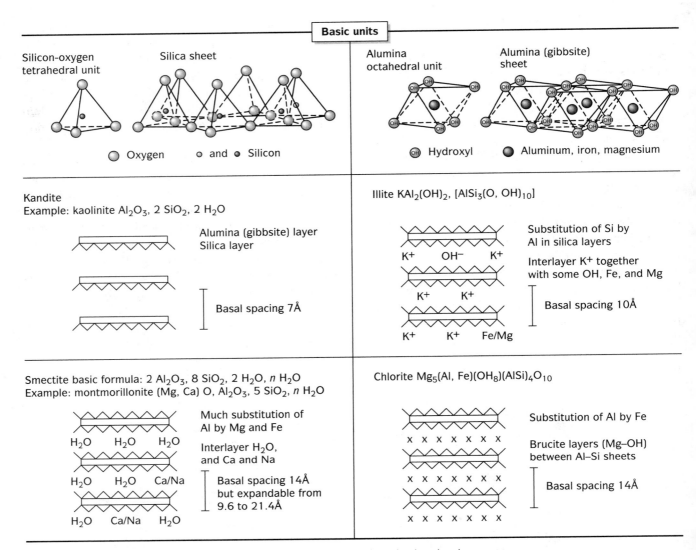

Figure 6.6 Summary of the common structural and chemical differences among the major clay minerals. (After Tucker, 1981:82; by permission of Blackwell Scientific Publications, Oxford.)

soil, the vermiculite in the soil traps additional water and helps keep the plant alive.

The clay mineralogy of a mudrock reflects several factors: provenance (both source rock type and weathering), diagenetic history and rock age, and depositional setting. We can use the five most important minerals in mudrocks (gibbsite, kaolinite, illite, montmorillonite, and chlorite) to illustrate the strengths and limitations of clay minerals as instruments for understanding the sedimentary rocks that contain them.

Clay Mineralogy and Provenance

Clay minerals can be produced by the weathering of igneous, metamorphic, or sedimentary rocks. Although igneous rocks initially contain no clay miner-

als, the feldspars that largely constitute them decompose to clay minerals, especially when the topography is low and the climate warm and humid. Metamorphic rocks also contain feldspars and micas that decompose easily to clay. Weathering of pre-existing sedimentary rocks, especially other mudrocks, generates detritus rich in clay.

The particular clay mineral produced is dictated by the extent of decomposition. Extensive alteration generates such clay minerals as kaolinite and gibbsite because interlayer cations such as Na, K, and Ca are completely leached away. Unfortunately, no trace of the original mica or feldspar type survives, so these degraded clay minerals provide no hint of source rock type.

In the modern environment, certain clay minerals are associated with specific conditions. As we saw

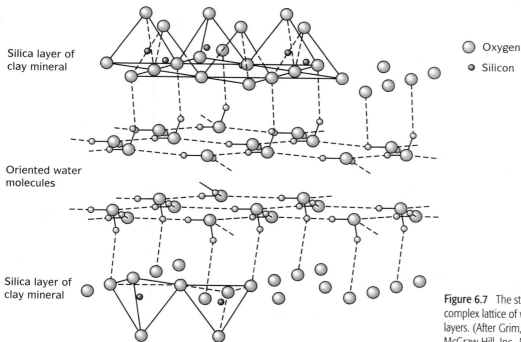

Silica layer of clay mineral

Oriented water molecules

Silica layer of clay mineral

○ Oxygen

● Silicon

Figure 6.7 The structure of vermiculite has a complex lattice of water molecules between the layers. (After Grim, 1968: 107; by permission of McGraw-Hill, Inc., New York.)

earlier, kaolinite is rich in aluminum but not in other cations. Consequently, it is abundant in areas of prolonged leaching, which concentrates aluminum, particularly if the parent material is also high in aluminum (for example, granites and rhyolites). Such leaching occurs most often in tropical soils with acid conditions and good drainage. The soils of the tropics are often so enriched in aluminum that they become a hard, crusty material known as **laterite**. Under some conditions, the aluminum concentration is so great that the soil can be mined for aluminum; this is known as **bauxite** ore.

Among all 2:1 clay minerals, the K-rich clay illite is by far the most abundant. Today, it is most often found in temperate environments, where there are alternate wet and dry conditions and the soil is neutral or slightly alkaline. The high potassium (an alkali metal) is favored under such conditions. Illites are the most common weathering products of feldspars and micas. Illites are also often produced by the weathering of pre-existing shales, and they are common in deeply buried muds and shales, where they are formed by the transformation of smectite.

Montmorillonitic smectites are higher in Fe and Mg, so they are commonly found as weathering products of ferromagnesian rocks, such as basalts and gabbros. Although they can form under a variety of conditions, they are particularly abundant in arid areas with alkaline soils. Stagnant water and poor leaching allow the retention of Mg, Ca, and Na cations in their structure,

and with their abundant interlayer water, they are favored where extreme wetting and drying occur.

Chlorites, which have magnesium and iron in the hydroxyl sheet, are produced by the weathering of ferromagnesian minerals in such rocks as basalt and gabbro. Like smectites, they are favored by alkalinity, especially where there is standing water and slightly reducing conditions. The ease of oxidation of the Fe^{2+} in the hydroxyl sheet means that chlorites can occur only in areas where there is little or no chemical weathering. Today, they are found most often in high-latitude temperate regions.

Depositional Setting

On the deep seafloor, where clays from the land eventually settle out, a predictable pattern can be seen (Fig. 6.8). Kaolinites are most abundant in the tropical belts, especially adjacent to the mouths of major jungle rivers. Chlorites are most common in the high latitudes, especially in the North Pacific, North Atlantic, and Antarctic oceans. Montmorillonites are found throughout the sea bottom but especially in the areas along mid-ocean ridges and near island arcs, where they are produced by the weathering of basalts. The rest of the seafloor, especially in temperate regions, is predominantly illite.

Biscaye (1965) examined the clay minerals found immediately beneath surface muds off the coast of eastern North America. They are entirely illite. He

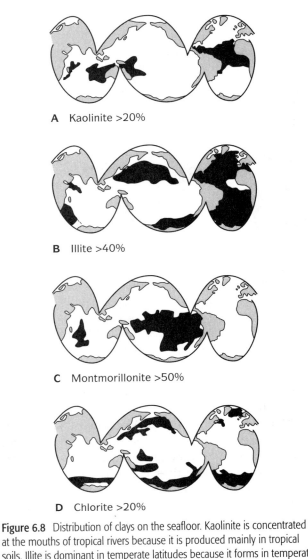

A Kaolinite >20%

B Illite >40%

C Montmorillonite >50%

D Chlorite >20%

Figure 6.8 Distribution of clays on the seafloor. Kaolinite is concentrated at the mouths of tropical rivers because it is produced mainly in tropical soils. Illite is dominant in temperate latitudes because it forms in temperate soils. Chlorite is more common in higher latitudes where chemical weathering is reduced. Montmorillonitic smectites are especially common along mid-ocean ridges, where they form from weathered basalts. (After Griffin, Windom, and Goldberg, 1968: 433.)

concluded that diagenesis had altered all clays to illite, erasing the record of climate and latitude. Secular changes in clay mineralogy verify that long-term diagenesis dictates clay mineralogy more than do provenance and climate. Kaolinite and gibbsite are the major clay minerals in Mesozoic and Cenozoic rocks; illite and chlorite dominate Paleozoic rocks. *Clay minerals now present in a mudrock might be quite different from those present when it was deposited.*

Illite and chlorite are more abundant in marine rocks than in nonmarine rocks. Unfortunately, clay mineralogy cannot be used as a reliable index of overall depositional setting because many clays reequilibrate chemically, reacting with both seawater and the fluids contained in the rocks in which they occur. *Diagenesis may alter the chemical signature of depositional setting originally registered in clay minerals.*

Glauconite

There is one important exception that deserves mention: the iron-rich 2:1 clay known as **glauconite.** Some authors (for example, Odin and Matter, 1981) prefer the term *glaucony,* although this is not widely used in the United States. The Greek word *glaukos* means "blue-green," and the blue-green color of this mineral is one of its distinctive characteristics. The iron-rich chlorite known as chamosite and a variety of other iron-rich clays are often associated with glauconite. Glauconites are often concentrated as spherules and pellets (Fig. 6.9) in shallow marine sandstones, giving these rocks a distinctive color; such rocks are known as **greensands.** Because glauconites are more often a component of sandstones than of mudrocks, they are typically studied by sandstone petrologists.

Unlike just about any other siliciclastic mineral, glauconites are a product (and thus a good indicator) of a specific range of environmental conditions. They form exclusively in agitated, oxidized, normal shallow marine waters. They are known from water depths of 50 to 200 m, but mostly at the shallow end of that range. However, it appears that the glauconite itself is typically produced under locally reducing conditions, such as when it replaces a decaying fecal pellet or the dead tissue inside a foraminifer shell. Most important, large concentrations of glauconite occur only in shallow shelf environments that are starved of clastic sediment or have very slow sedimentation rates. Odin and Matter (1981) estimated that, to form, glauconites must reside at the sediment-water interface for 1000 to 10,000 years, and it takes much longer for highly complex glauconites to develop. Consequently, thick greensand deposits are typically found at the top of shallowing-upward marine sequences and just below unconformities. Apparently, they represent shallow marine surfaces that were exposed to extensive burrowing and diagenesis when the clastic supply ceased.

Once formed, glauconites can be redeposited in energetic, well-oxygenated settings, further concentrating some greensands; they can also survive burial in reduced subsurface environments. Most petrologists, however, regard glauconite as a powerful tool for paleoenvironmental reconstruction, because it is practically the only siliciclastic mineral that indicates a specific depositional environment.

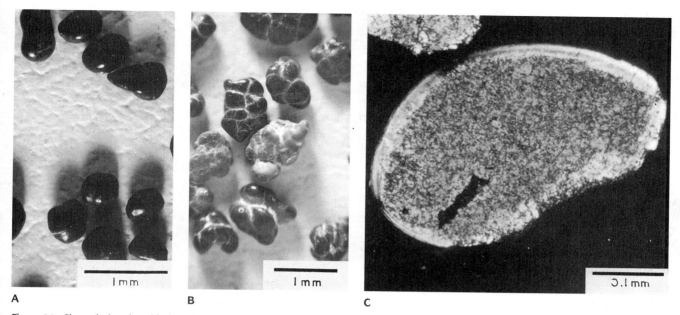

Figure 6.9 Glauconite is an iron-rich clay mineral that commonly forms as pellets on the seafloor. (A) Typical glauconite pellets, probably of fecal origin. (B) Glauconite formed after foraminiferal tests. (C) Photomicrograph of glauconite pellets, showing characteristic speckled texture. (Courtesy of D. L. Triplehorn.)

Bentonite

Another distinctive mudrock that has many important uses is **bentonite**. Bentonite is a mixture of 2:1 smectites formed from the weathering of volcanic ash that has settled onto the ocean floor. If the clays are more kaolinitic and were formed when volcanic ash degraded in the acidic waters of coal swamps, they are known as **tonsteins**. In addition to clays, bentonites and tonsteins contain unaltered igneous

materials, such as euhedral biotite flakes, euhedral volcanic quartz, magnetite, euhedral zircon, apatite, zeolites, and a variety of feldspars (especially the volcanic K-feldspar, sanidine). Although most of the volcanic glass shards have long since degraded to clays, occasionally the relict texture of volcanic glass can be seen (see Fig. 14.13A).

Typically, bentonites are found interbedded with shallow marine limestones or shales (Fig. 6.10). Although bentonite beds up to 15 m thick have been re-

Figure 6.10 Bentonites are altered volcanic ashes that settle to the sea bottom. (A) Outcrop of Middle Ordovician Deike bentonite (crumbly gray layer at chest level) in the middle of a thick sequence of resistant limestones, South Carthage, Tennessee. This immense ashfall, from a volcanic arc source just off the present Atlantic coast, covered the entire eastern United States in the Ordovician. The ash layer was more than a meter thick in the Appalachians and extended as far as Wisconsin and Minnesota. (Courtesy of W. D. Huff.) (B) The Cretaceous shales of the Western Interior are punctuated by many ashfalls and bentonites (two white layers in lower center.) (Courtesy of D. L. Eicher.)

ported, they are seldom more than 0.3 m thick because there are limits on how much glass and volcanic matter can be ejected during a single eruption. The thickness of the bentonite bed and the size of the unaltered fragments decrease logarithmically with distance from the volcanic vent (Fisher and Schmincke, 1984). Volcanic fragments also tend to be graded in size within the bed, showing that they result from a single settling event. Widespread bentonites have been used to determine paleowind directions (Elder, 1988).

Bentonites have their greatest stratigraphic value, however, as a tool for correlation. Because they represent a single, rapid geologic event, they can be used as a time horizon as well as a unique stratigraphic marker in a given stratigraphic correlation setting. Certain bentonites can be traced for hundreds of kilometers and matched up with precise chemical or mineralogical signatures. Unfortunately, most bentonites are too weathered to provide good radioisotopic dates, but occasionally there have been successes.

The other importance of bentonite is economic. Thick deposits of bentonite are widely mined commercially for a variety of industrial uses: fuller's earth for decolorizing oils, drilling mud for oil wells, bonding agents to hold together foundry molding sands, and many others.

Classification

Other than the consensus regarding the distinctions among mudrock, siltstone, claystone, shale, and argillite, no single system of mudrock classification and nomenclature has won widespread acceptance. Table 6.1 lists various criteria used to classify mudrocks and the categories generated by each scheme. We prefer the straightforward use of mudrock color for classification. It is practical in the field and for hand specimens, although it is not without problems. In some cases, mudrock color reflects nonessential aspects of the lithology. Small percentages of such

---

TABLE 6.1 Schemes of Classifying Mudrock

I. Grain size

Percentage of clay-sized constituents	To the touch	Name[a]
0–32	Gritty	Siltstone
33–65	Loamy	Mudstone
66–100	Slick	Claystone

II. Color
1. A primary color is assigned (white, black, gray, red, brown, etc.)
2. It is important to differentiate between surface staining and color as seen on fresh surfaces.

III. Chemical composition
1. Individual mudrock samples are chemically analyzed.
2. The composition of samples is compared to a standard reference average mudrock composition.
3. The names of particular oxides that are more abundant than the average mudrock are applied to the unknown mudrock; for example, ferruginous, high alumina, potassic, calcaerous, phosphatic, and siliceous.

IV. Detrital mineralogy of the silt-sized fraction
This method requires microscope analysis. The major categories are quartzose, feldspathic, micaceous, and chloritic mudrocks.

V. Lithology of associated sedimentary rocks
This method is useful when the results of field work can be combined with microscope examination of conglomerate, breccia, sandstone, limestone, and dolomite. A mudrock might be categorized as "associated with quartz arenite and oligomict conglomerate" or "interbedded with wacke and tilloid."

[a]The term *shale* is used to refer to mudrock (whether siltstone, claystone, or mudstone) that exhibits fissility, the ability to break into thin sheets. Adjectives that supplement these textural terms are based on the thickness of sheets (papery to slabby) and the presence or absence of internal lamination.

constituents as iron oxide and carbonaceous material can color a mudrock. The color of specimens can also be greatly altered by surface staining that penetrates the interior.

Chemical composition is also used to distinguish mudrock types. First, the relative percentage of chemical constituents in the average mudrock must be known. Mudrocks whose geochemistry differs significantly from this standard can then be designated, for example, as ferruginous or calcareous. This procedure has its drawbacks. It cannot be used for field identification, nor is it practical for analysis of hand specimens. Samples must first be analyzed. Also, serious disagreement exists over precisely what chemical composition constitutes the normal or average mudrock. There is a crude correlation between mudrock color and chemistry. Red to reddish brown mudrocks are typically ferruginous; black mudrocks are carbonaceous.

The mineralogy of the silt-sized component in mudrock is also used for classification. In this method, specimens must be studied in thin section. Mudrock types are characterized as quartzose, feldspathic, chloritic, or micaceous. Mudrocks are also differentiated on the basis of the coarser siliciclastic materials with which they are associated: sandstone, conglomerate, and breccia. Neither of these methods has been very popular.

Origin and Occurrence

Relatively weak transporting currents deposit mudrock. Isolating the site of mudrock accumulation temporally and/or spatially from stronger currents is crucial to the deposition of thick, monotonous mudrock sequences. The depositional setting of a mudrock is not inferred from its texture or composition. Greater reliance is placed on fossil content, sedimentary structures, and the type and origin of sedimentary rocks interbedded with the mudrock.

Mudrocks of marine origin include those deposited as abyssal plain sequences far from land. They consist of pelagic materials that slowly settled out of suspension, terrigenous materials transported far out into the ocean basins by weak density currents, and windblown materials from the continents. Some marine mudrocks are transported and deposited by **contour currents** (see Chapter 10). These are weak, bottom-hugging currents that flow parallel to the depth contours of the continental slope and rise.

Other marine mudrocks are deposited nearer shore in the deeper, more protected parts of continental shelf areas and on the floor of shallow epeiric seas (for example, the Mancos and Pierre shales of the Cretaceous inland sea of western North America; the middle Paleozoic Chattanooga and New Albany shales of the Appalachians). Bands of mudrock commonly occur oceanward from the shore zone, where flakes of fluvially transported materials become attached to one another by flocculation as they enter the more saline ocean. Mud also accumulates on tidal flats (see Chapter 9).

Mudrocks of continental origin include lacustrine deposits and the thick, fine-grained siliciclastics found within meandering river systems (see Chapter 8). Rivers that flood periodically cause channelized flow to escape the confines of natural levees. The flow spreads quickly and then slows, depositing predominantly finer-grained material across the floodplains. Many streams and rivers draining glaciated regions are choked with rock flour, fine silt produced by glacial grinding. Sediments deposited in the distal or deltaic portion of such systems can be predominantly silty or clayey. Finally, windstorms constantly entrain, transport, abrade, and redeposit fine-grained material as loess across wide portions of the globe. Material from the Sahara Desert is found in the deep Atlantic thousands of kilometers from shore.

Proterozoic Mudrock Facies

Compared to studies of sandstones and limestones, there are relatively few studies of ancient mudrock facies. Most geological literature simply describes such rocks as red shale or black shale and often leaves it at that. In some specific cases, however, there is more information to be gleaned than one would suspect. For example, Schieber (1989) studied the mid-Proterozoic Newland Formation from the Belt Supergroup of Montana. He recognized six different shale facies based on such features as mechanical strength, carbonate content, silt content, and sedimentary structures (Figs. 6.11 and 6.12). For example, he described *silty shales* having silty lenses and laminae, silt-filled desiccation cracks, and lenticular to flaser bedding, all indicators of intertidal conditions; thus, they appear to be the most nearshore of all the shale facies. Schieber also described *carbonaceous swirl shales* having rolled up and folded flakes of microbial mats, as well as alternations of carbonaceous stromatolitic microbial layers and silty storm layers. These are both

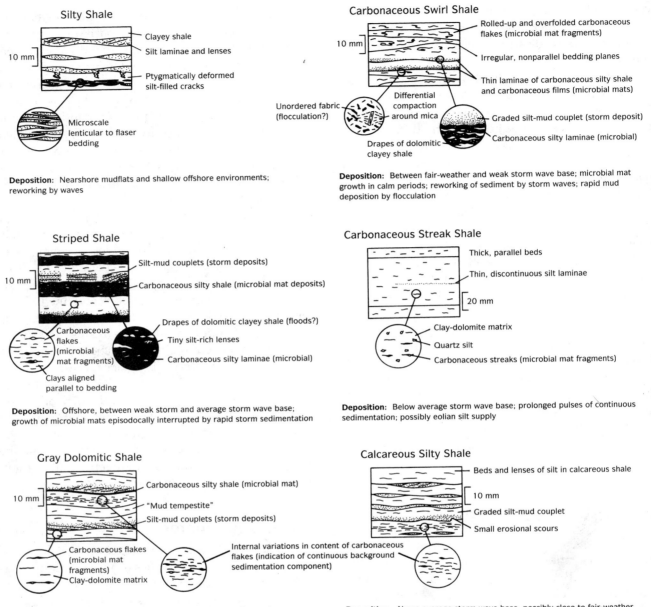

Figure 6.11 Summary of shale facies and suggested depositional environments. "Weak storm wave base" is a water depth below fair-weather wave base and above average storm wave base. (After Schieber, 1989: 207.)

indicators of nearshore deposition as well, but just below fair-weather wave base where weak storms rip up microbial mats and roll them out to sea. Other shale facies (see Figs. 6.11 and 6.12) indicate deposition below the wave base of weak storms but still above average storm wave base, or deposition in quiet waters below storm wave base, where the main influences might be gravity flows and silt from occasional storms that is washed farther out to sea.

Schieber's (1989) shale facies models suggest that mudrocks are much more informative than most geologists think. There is an important caveat, however: these rocks are Proterozoic in age, deposited before the appearance of multicellular animals (especially burrowing animals). In the Proterozoic, only stromatolitic microbial mats exerted biological influences on the seafloor. Many of these shale facies may not occur in Cambrian or younger rocks, because the advent of

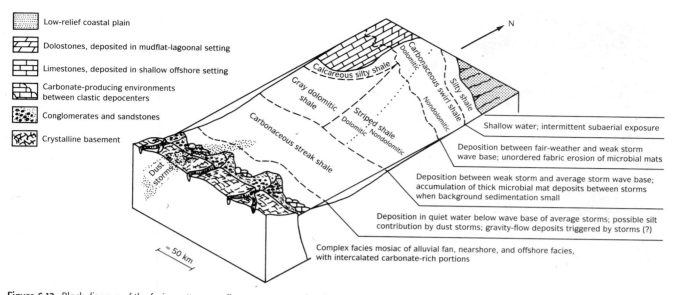

Low-relief coastal plain

Dolostones, deposited in mudflat-lagoonal setting

Limestones, deposited in shallow offshore setting

Carbonate-producing environments between clastic depocenters

Conglomerates and sandstones

Crystalline basement

Shallow water; intermittent subaerial exposure

Deposition between fair-weather and weak storm wave base; unordered fabric erosion of microbial mats

Deposition between weak storm and average storm wave base; accumulation of thick microbial mat deposits between storms when background sedimentation small

Deposition in quiet water below wave base of average storms; possible silt contribution by dust storms; gravity-flow deposits triggered by storms (?)

Complex facies mosiac of alluvial fan, nearshore, and offshore facies, with intercalated carbonate-rich portions

Figure 6.12 Block diagram of the facies patterns, sediment sources, and sedimentary processes in the Precambrian Newland Formation. (After Schieber, 1989: 217.)

bioturbation and the great variety of marine organisms in and on the sea bottom have changed all the rules. Nevertheless, Schieber's studies suggest that Phanerozoic mud-rocks may yield more information if geologists begin to do the necessary detailed work.

Black Shales

One of the most distinctive and controversial of all sedimentary rock types is the *black shales* (Fig. 6.13). Their black color is due to a very high content of un-oxidized organic matter. They also contain reduced iron in the form of pyrite and are usually finely laminated, indicating little bioturbation or storm disturbance. Typically, they contain no bottom-dwelling marine organisms, but they may contain planktonic (floating) organisms, such as microfossils or graptolites, or nektonic (swimming) organisms, such as fish and marine reptiles, which sank to the bottom after they died. Occasionally, they contain extraordinary fossils in which even the organic films of the soft tissues are preserved (Fig. 6.14). All these features show that black shales were deposited under reduced, low-oxygen conditions. If they are **anaerobic** (less than 0.1 ml of oxygen per liter of seawater), only bacteria tolerant of low oxygen levels can live, and no bottom scavengers can break up the carcasses or stir up the mud. Under **dysaerobic** conditions (between 0.1 and 1.0 ml oxygen per liter of seawater), organic

Figure 6.13 Hundreds of feet of black shale from the Devonian Genesee Group, Taughannock Falls, Lake Ithaca, New York. (D. R. Prothero.)

A

B

Figure 6.14 Black shales are famous for the extraordinary preservation of their fossils. Because they are typically anoxic and have no scavengers, even the delicate soft tissues can leave impressions. (A) An unusually well preserved trilobite (complete with gills and appendages) appears in this X ray of the Devonian Hünsruck Shale of Germany. (Raup and Stanley, 1976: 23.) (B) The fishlike reptiles known as ichthyosaurs from the Posidonia Shales of Holzmaden, Germany, are often preserved with the body outlines, including the dorsal fin, tail fin, and even stomach contents and newly born young in the birth canal. (Stanley, 1989: 442; by permission of W. H. Freeman and Company.)

material accumulates as well, and only a few types of burrowing worms can live. In older literature, you might encounter the term *euxinic* for these conditions, but it largely has been abandoned in favor of the more precise terms just mentioned.

These aspects of black shale are clear. As Pettijohn (1957, p. 625) put it: "That the pyritic black shales were deposited under anaerobic conditions is unquestioned. Whether, however, the basin of accumulation was shallow or deep and whether it was landlocked or freely connected with the sea or even a stagnant area of open sea has been much debated." Influencing the debate has the been the choice of which modern analog is used to understand black shale deposition. The dominant model has been the modern Black Sea, which is a relatively deep basin

with a restricted mouth connecting it to the Mediterranean (the Bosporus and Dardanelles straits). (In Roman times, the Black Sea was known as the *Pontus Euxinus,* from which the term *euxinic* was derived. In Greek, the name means "the sea that is hospitable to strangers," a facetious name for a very stormy and dangerous body of water).

In the Black Sea, relatively little oxygenated marine or fresh water flows into the basin, and it tends to stay near the surface, forming a freshwater "lid." Beneath about 150 m, the waters are completely anoxic, contain dissolved H_2S, and are completely devoid of animal life. These pyritic black muds of the well-named Black Sea have strongly influenced most research into ancient black shales.

More recent research, however, suggests that this deep, restricted basin may not be such a good model for most ancient black shale deposits. Such deposits tend to be found in much more open oceans with no evidence of great depth or restricted water flow. In addition, many lines of evidence suggest that not all black shale deposits were necessarily that deep. Many black shales are found interstratified with fossiliferous and burrowed muds, suggesting rapid fluctuations in oxygen content that could not occur in a permanently deep, restricted basin. This is particularly true of the black shales found throughout the Western Interior Seaway of North America in the Cretaceous (see Chapter 10), but it might even apply to the black shales found in the Cretaceous strata below the modern ocean basins. Other possible modern analogs of oceanic conditions with high organic productivity and little exchange of oxygen with the surface waters have been proposed (see Arthur and Sageman, 1994, for a review). It appears that there may be *no* good modern analog for the great black shale deposits of the past.

Indeed, the distribution of black shales in the Phanerozoic further reinforces this point. Most of the famous black shales, such as the Devonian-Mississippian Chattanooga and New Albany shales or the many black shales of the Jurassic and Cretaceous, are known from times when there were unusually high stands of sea level. During these times, the seas flooded the cratons, and black shale accumulated even on the continents. These high sea-level stands are correlated with greenhouse atmospheric conditions and a complete lack of polar ice caps.

A variety of models have been suggested to account for the production of black shales under these conditions. Most hinge on the fact that under greenhouse conditions, there would be less vigorous

deep-water circulation than there is today. In the modern oceans, for example, the deep but relatively oxygenated bottom currents are formed by cold, dense waters around the icy poles (particularly the Antarctic Bottom Water and the North Atlantic Deep Water). These waters sink and then flow along the ocean bottoms from the poles to the equator. On a greenhouse Earth, there would have been no cold polar waters to sink, and the less vigorous circulation would have allowed the oceans to become stratified with poorly oxygenated waters forming at depth. This is especially true in the equatorial belt, where surface waters are warmer and less dense than those lower in the water column, so that vertical overturn and oxygenation of bottom waters cannot occur.

In addition, warmer oceans during greenhouse conditions mean that gases would be less soluble, so that there would be less oxygen uptake by the water. In short, the well-oxygenated oceans of today are probably not a good analog for the warmer, stratified, greenhouse oceans that may have prevailed during the mid-Paleozoic or Cretaceous. Instead, these conditions require special climate modeling methods that use modern examples only as a general guide to the chemical and physical parameters of ancient anoxic oceans.

CONCLUSIONS

Mudrocks are the most abundant type of sedimentary rock on Earth, yet they have been neglected for more than a century because they are difficult to study. Recent improvements in our ability to analyze clay minerals and renewed interest in the importance of shales for solving many stratigraphic problems have greatly increased the impetus for research. With enough detailed analysis and new techniques of chemical and mineralogical analysis, much more information can be extracted from mudrocks than previous generations of geologists have appreciated.

FOR FURTHER READING

Arthur, M. A., and B. B. Sageman. 1994. Marine black shales: Depositional mechanisms and environments of ancient deposits. *Annual Review of Earth and Planetary Sciences* 22: 499–551.

Chamley, H. 1989. *Clay Sedimentology.* Berlin: Springer-Verlag.

Millot, G. 1987. *Geology of Clays: Weathering, Sedimentology, Geochemistry.* New York: Springer-Verlag.

Moore, D. M., and R. C. Reynolds, Jr. 1989. *X-Ray Diffraction and the Identification and Analysis of Clay Minerals.* Oxford: Oxford University Press.

Potter, P. E., J. B. Maynard, and W. A. Pryor. 1980. *Sedimentology of Shale.* New York: Springer-Verlag.

Schieber, J. 1989. Facies and origin of shales from the mid-Proterozoic Newland Formation, Belt Basin, Montana, USA. *Sedimentology* 36:203–219.

Schieber, J. 1993. Evidence for high-energy events and shallow-water deposition in the Chattanooga Shale, Devonian, central Tennessee, USA. *Sedimentary Geology* 93:193–208.

Scott, E. D., A. H. Bouma, and W. R. Bryant. 2003. *Siltstones, Mudstones and Shales: Depositional Processes and Characteristics.* Tulsa, Oklahoma: SEPM.

Weaver, C. E. 1989. *Clays, Muds, and Shales.* Developments in Sedimentology 44. Amsterdam: Elsevier.

7 Siliciclastic Diagenesis

These unusual pipy concretions look like petrified plumbing, but they formed by calcite precipitated in the groundwater as it percolated horizontally. Note also the calcified root traces and other concretions in this outcrop of the upper Oligocene Arikaree Group, near the top of Scottsbluff National Monument, Nebraska. (D. R. Prothero.)

THE LOOSELY PACKED SEDIMENTS DESCRIBED IN THE previous chapters are not yet sedimentary rocks. That transformation involves an additional set of processes, most of which require that the sediment leave the influence of the surface and undergo burial. The processes that lithify loose sediment are known as **diagenesis.** Some changes occur very near the surface at low pressures and temperatures, but many require deeper burial, several kilobars of pressure, and temperatures up to 300°C. The upper limits of diagenesis are hard to distinguish from low-grade metamorphism. Conventionally, processes that take place at

temperatures lower than 300°C and pressures less than 1 to 2 kbar are considered diagenetic, and those that occur beyond that range are metamorphic. Many of the diagenetic processes discussed in this chapter, however, continue outside these limits, and some metamorphic changes occur within them.

Some diagenetic processes begin very early. When groundwater is full of dissolved minerals, such as calcite, it is not unusual for surficial materials to become cemented. Indeed, calcite-cemented soil horizons, known as **caliches,** are common in the world's deserts. They result when the evaporation of carbonate-laden

groundwater causes the calcite to precipitate in the pore spaces, cementing the soil horizon. The silica-cemented equivalent is known as **silcrete**. In limestones, cementation is even easier, and it often occurs at near-surface conditions. It is not unusual for carbonate beach sands to become cemented with modern bottles and other garbage in them, showing that this **beach rock** formed in very recent times. We will discuss carbonate diagenesis in Chapter 11. In this chapter, we will focus on the diagenesis of sandstones, conglomerates, and shales.

Why do we care how a sediment becomes rock? Diagenesis is important for a number of economic reasons, as well as for purely academic interest. A loose, well-sorted sand typically has a very high porosity and permeability, but after diagenesis it can become impermeable. Permeability strongly affects groundwater flow and determines whether the rock will allow hydrocarbons to migrate through it or accumulate. In fact, the major impetus in diagenesis research in the past three decades has come from the petroleum industry, which is vitally interested in predicting whether or not a certain rock has sufficient porosity and permeability to become a good oil conduit or reservoir.

Several processes are part of the spectrum of diagenesis. This chapter will focus on the most important ones for sandstones and shales. Because this is an introductory undergraduate textbook, we will not discuss in depth the petrological or geochemical details that characterize most research in diagenesis today. To learn more, the interested student should refer to some of the further readings listed at the end of this chapter.

Compaction

One of the earliest and most important diagenetic processes is **compaction**. Loosely packed sand has a porosity of about 25%; theoretically, porosity can reach as high as 47.6% with perfectly packed spheres of equal size. Most sandstones, however, have much less porosity or none at all, mostly because of compaction. Wet muds are 60% to 80% water and compact readily when put under any pressure (such as when you step in them). The clay minerals themselves are ductile and platy, and their layers compact easily when water is driven out of the interlayers.

The fact that these two sediment types compact differently can be an important clue to the conditions under which the sediment is lithified. This can be seen in cases of **differential compaction**. When both

Figure 7.1 Shales are much more prone to compaction than sandstones. Here, a sandstone dike that intruded into shale did not compact and was crumpled into folds; the shales simply became more compact and dense. (Truswell, 1972:582; by permission of the SEPM.)

sands and muds are subjected to burial pressure, the muds compact almost twice as much as the sands. A striking example is a cross-cutting sandstone dike (Fig. 7.1). The dike forms when an overpressured liquefied sand layer squeezes upward into cracks in the overlying sediments. If the dike intruded before compaction, it will be deformed where it cuts through compacted shales by all the shortening that took place as compaction occurred. A similar dike cutting through compaction-resistant sandstones would be relatively undeformed.

The most important practical consequence of differential compaction occurs when clastic rocks are buried in deep sedimentary basins. The shales typically undergo so much compaction that their present thickness is approximately half their original thickness. This may be an important consideration if one wishes to correlate the thicknesses of beds across distances; the shale thicknesses are underestimates, and a mathematical formula is needed to compensate for this lost thickness. Similarly, studies of basin subsidence must take into account the effect of compaction on the shales to get an accurate estimate of the original thickness of the sedimentary pile, the rate of subsidence, and many other important parameters.

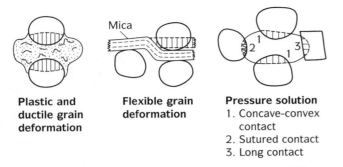

Plastic and ductile grain deformation

Flexible grain deformation

Pressure solution
1. Concave-convex contact
2. Sutured contact
3. Long contact

Figure 7.2 Grains of different composition respond to pressure in different ways. These are some of the common textures that result from pressure. (After Wilson and McBride, 1978:679, by permission of the American Association of Petroleum Geologists, Tulsa, Oklahoma.)

Compaction can result in fabrics that are visible in thin section. The most obvious are the deformation, distortion, or flattening of pebbles, fossils, and any other object whose original shape is known (Fig. 7.2). On a microscopic scale, flat objects such as mica flakes can be bent or deformed between sand grains. Rock fragments (especially fine-grained ones such as shales, slates, phyllites, and volcanics) are often deformed by the pressure applied to them and may break down to clays, giving the false impression that a clean sandstone once had a clay matrix; this is known as **pseudomatrix.** Under greater pressures, grains begin to dissolve at the point where they contact other grains, a process called **pressure solution.** As a result, the two grains may have a wavy, sutured contact, or one grain may press a concavity entirely into the other. Finally, the sand grains may show clear evidence of fractures running through them, especially if the compaction was near the surface where the temperatures were lower and the grain did not dissolve or heal its fractures easily.

These and other petrographic clues have been used widely by petrographers to predict the depth and temperature of burial of a given rock. For example, Smosna (1989) generated a formula (which includes variables for percentage of matrix and phyllitic rock fragments) that predicts the depth of burial. Recent work, however, suggests that in many sandstones, pressure solution increases only in the first 1 to 2 km of burial; below that depth, the system appears to remain stable. Other methods use the number and type of the grain-to-grain contacts to predict burial depth. Although the widespread applicability of these formulas to other cases has not yet been tested, these formulas do give a valuable indication of the likely depth of burial and therefore the likelihood that a sandstone reservoir might still contain hydrocarbons.

Cementation

Although compacted shales can become completely lithified, compaction alone is not enough to lithify sands or gravels that have little muddy matrix. For example, in experiments where quartz sand with only 20% schist fragments or muddy matrix and water are placed in a pressure cylinder, pressure alone can produce a lithified rock. But when clean, pure quartz sand and water are placed in the same cylinder, no amount of pressure seems to lithify the mixture into a rock. Normal grain-to-grain contacts are not abundant enough to generate much suturing or welding of the grains, and the sand fails to become sandstone. Instead, clean porous sands are much more likely to become rock if they are **cemented** by new minerals that precipitate into the pore spaces as groundwater flows through.

One of the common cements in pure quartz sandstones is (not surprisingly) quartz. Dissolved silica precipitates on the surfaces of the framework quartz sand grains, sometimes in multiple layers (Fig. 7.3). In many instances, these **silica overgrowths** crystallize in the same optical crystallographic orientation as the

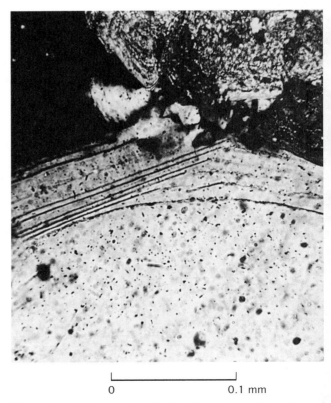

0   0.1 mm

Figure 7.3 Detrital quartz grain with nine distinct overgrowths, each bounded by well-defined crystal faces, from the Ordovician Shakopee Formation of Minnesota. (Austin, 1974:360; by permission of the SEPM.)

quartz in the core, so that they extinguish at the same angle under the petrographic microscope. Indeed, the overgrowth cementation can completely obscure the shape of the original grain and may give the false impression that a well-rounded grain is much more angular than it really is (see Fig. 5.9). But in some cases, each overgrowth traps a fine film of dark opaque matter (organic material or iron oxides) that shows the outline of the original grain and allows the petrographer to distinguish the core grain from the overgrowth. In other cases, quartz grains are interlocked, without obvious crystal growth in the visible pore spaces. Such grains were probably welded together and cemented by silica from pressure solution.

One of the great puzzles of diagenesis is the source of silica cement, especially in pure quartz sandstones. Shallow marine waters have only about 1 ppm silica and fluvial waters about 13 ppm silica, neither of which is enough to yield even a trace amount of silica cement. Studies of pressure solution show that this process yields only about one-third of the silica needed. The transformation of clay minerals releases much silica, which may help to explain silica in muddy sandstones. But for pure quartz sandstones, clay-derived silica would have to be imported, and there is no evidence that it travels very far from the shales to adjacent sands. Similarly, the breakdown of volcanic rock fragments and glass yields silica, but pure quartz sandstones rarely occur in volcaniclastic settings. Siliceous microfossils (diatoms and radiolarians) and sponge spicules might conceivably release opaline silica, but such microfossils are most abundant in deep marine shales and are seldom found near pure quartz sandstones.

Eliminating these other sources compels us to reconsider silica in groundwater. Calculations of silica solubility chemistry imply that enormous amounts of groundwater must move through a rock to produce silica cement, because groundwater flows very slowly and silica takes a long time to precipitate. Sibley and Blatt (1976) argued that the only practical way to pump this much groundwater through the sediment is by vertical circulation. This implies that cementation took place at shallow depths, because groundwater flows even more slowly at greater depths.

Experimental work on the solubility of silica (Fig. 7.4) yields some important conclusions about the conditions needed to produce silica cements. The solubility of both quartz and amorphous (opaline) silica increases with increasing temperature. In Fig. 7.4, we can see that amorphous silica is always more soluble than quartz, so if it is present, it is far more likely to

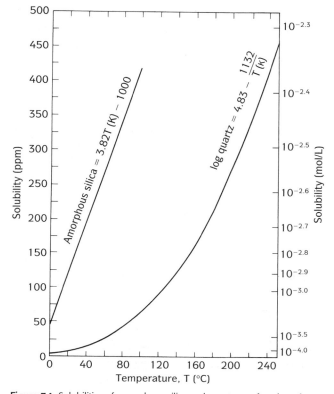

Figure 7.4 Solubilities of amorphous silica and quartz as a function of temperature. At temperatures above 100°C, amorphous silica is so unstable that it crystallizes rapidly to quartz, forming chert. (After Blatt, 1992:134; by permission of W. H. Freeman and Company, New York.)

contribute to the cement. However, it is so easily dissolved at low temperatures (much less than 100°C) that it cannot precipitate at higher temperatures. This means that cementation by opaline silica must occur near the surface and cannot occur at great depths.

The solubility of quartz, on the other hand, rises very slowly until it reaches threshold temperatures in the range of 100°C to 150°C (see Fig. 7.4). Thus, the presence of quartz cements implies deeper burial and higher temperatures than does the presence of opaline cements. Quartz nucleates and precipitates so slowly that deep burial and long time spans are virtually required.

Calcite is an even more common cement in sandstones, because it is much more soluble than silica. It is particularly common in near-surface conditions, where the fluctuation of groundwater pH between acid and alkaline conditions rapidly dissolves and reprecipitates calcite. As with silica, however, an enormous flow of water is needed to cement a rock with calcite. Calculations of the typical calcium and carbonate ion concentrations in seawater suggest that the pore water volume in a mass of loose sand must

be flushed and replaced approximately 2700 times to fill it with calcite! Such calculations imply that either high permeability or a steep hydraulic gradient, plus many millions of years, is required to produce a calcite cement.

Another common cement in sandstones is iron oxide, mostly red-brown hematite (Fe_2O_3) and yellowish iron hydroxide, such as "limonite" (made mostly of goethite, $FeO \cdot OH$). As little as 1% to 2% of these minerals in a rock is usually sufficient to give it a distinctive rusty red color. Originally, most of the iron in these minerals comes from the weathering of iron-rich silicate minerals such as biotites, pyroxenes, and amphiboles and from the leaching of iron-rich clays. The chemistry of iron oxides is determined largely by the oxidation state of the atmosphere and water in which the oxides occur. Dissolution out of these source minerals occurs mostly in reducing conditions (those with little or no free oxygen). Once dissolved, iron cations are highly mobile as long as oxygen content remains low. When iron is exposed to oxygen, it precipitates readily and is very hard to redissolve. Under most naturally occurring values of pH and Eh, hematite is completely insoluble (see Fig. 2.2). Only extraordinary conditions can liberate its iron again.

Many iron oxides have magnetic properties that can be used to date the time of cementation. When the iron oxide precipitated, it locked in the position of Earth's magnetic pole at the time it formed. Once that pole position has been determined, it can be compared to the apparent polar wander curve for that continent and used to estimate the age of cementation. Of course, magnetic poles wander very slowly, so this method gives results that are good only to the nearest 20 million years or so.

Diagenetic Structures

In Chapter 4, we discussed primary sedimentary structures produced by normal depositional processes. A wide variety of the colors and shapes found in sedimentary rocks are due to later diagenetic effects. Some of these are bizarre or spectacular enough to attract the attention of amateur rockhounds, but geologists generally have dismissed them as curiosities. When such diagenetic structures obscure or overprint the original depositional features, geologists try to develop a mental screen or filter to see through the overprint to the original structures. As a result, there has been much less research into diagenetic structures than one would expect.

Most of these diagenetic structures result from a groundwater mineral that has collected and precipitated to form an unusual object or pattern. The simplest example is the diagenetic color bands often found in porous sandstones. When these color bands were caused by water flowing along the original bedding, the bands highlight the bedforms. Often, however, these *Liesegangen* bands crosscut the primary structures, giving false impressions about the bedding (Fig. 7.5). This is particularly obvious when the color bands form irregular, curving patterns that no sedimentary bed could ever produce. Under these circumstances, we must look very hard at the outcrop to be sure that the conspicuous color banding has not misled us.

Many diagenetic structures are three-dimensional. The simplest are **concretions** (regular rounded objects that form around a nucleus) and **nodules** (irregular objects that usually don't have an obvious nucleus). Most of these structures form as concentrations of some minor mineral in a host rock (for example, calcite in sandstone or shale, chert in limestone, pyrite in black shale). Some form along bedding and appear to be influenced by it; examples include the calcareous nodular layers found in many shales and mudstones and the nodular flints found in the chalks of the Cretaceous (Figs. 7.6 and 7.7). In many cases, groundwater has seeped along bedding planes and concentrated in these beds because less permeable layers have restricted its flow. In other cases (such as the nodules in mudstones), these structures were produced in ancient soil horizons and were concentrated when soil-forming processes precipitated calcite.

Many concretions and nodules form around some object in the sediment. Often, fossils serve as a site of nucleation for the calcite or silica in the groundwater, and many of the concretions have fossils in the middle (Fig. 7.8). In some famous localities, paleontologists have struck it rich by collecting and splitting open nodules. Apparently, ammonia and other chemicals released by the decaying organic matter in the organism changes the pH so that calcium carbonate or silica can precipitate.

Other diagenetic objects clearly have grown with no influence of the original bedding. The most spectacular are sand crystals (Fig. 7.9), which are simply loose sand trapped by the growth of a crystal of the cementing mineral. They include sand calcite crystals (sand grains trapped by large euhedral calcite crystals), sand gypsum crystals, and the barite rosettes so commonly found in rock shops. Many concretions have shapes that suggest all sorts of objects

A

B

Figure 7.5 *Liesegangen* bands form when groundwater percolates through porous rocks and precipitates colorful bands of iron oxides. Occasionally, the color bands may parallel the bedding, but in most cases they crosscut bedding and deceive the geologist about the true direction of bedding. (A) In this upper Proterozoic sandstone from the eastern Grand Canyon, the bedding descends from upper right to lower left; the colorful vertical *Liesegangen* bands are nearly perpendicular to the bedding. (B) In this sandstone, the bedding is nearly horizontal, but the colorful swirls clearly do not follow the bedding. (D. R. Prothero.)

A

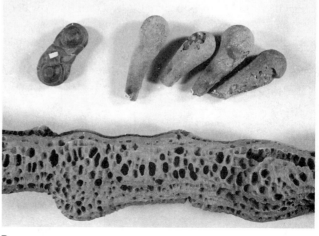

B

Figure 7.6 Concretions formed by precipitated groundwater can produce an amazing variety of shapes. (A) These nodular concretions are common in many types of shales and mudstones; they are usually produced by precipitation of soil carbonate. From the Pliocene, Val di Zena, Italy. (Courtesy of E. F. McBride.) (B) A variety of odd-shaped concretions, including teardrop shapes; long, snaky pipy concretions; and the disklike or dumbbell-shaped concretions known as "imatra stones" (*upper left*). (D. R. Prothero.)

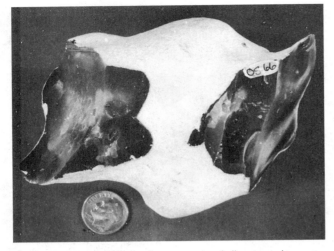

Figure 7.7 Flint and chert often form concretions in limestones, because they precipitate out of silica-rich groundwater percolating through the cracks (see Fig. 13.2.) This flint nodule from the Cretaceous Dover Chalk is composed of black chert inside, but it still has a coating of the white chalk on the outside. (D. R. Prothero.)

Figure 7.8 Concretions tend to nucleate around fossils. In some cases, the decaying organisms change the oxidation state and pH of the sediment around them, facilitating the precipitation of groundwater minerals. Paleontologists have long known that splitting concretions in certain collecting areas is a highly profitable way to find good fossils. At upper left are the two halves of a siderite (iron carbonate) concretion from the famous Pennsylvanian Mazon Creek beds of northeastern Illinois; this concretion precipitated around a fern frond. At upper right is a cannonball concretion formed around the ammonite *Sphenodiscus;* these are common in the Jurassic of Europe. At the bottom is a concretion formed around several straight-shelled ammonites of the genus *Baculites* from the Cretaceous Pierre Shale of South Dakota. (D. R. Prothero.)

(see Fig. 7.6B). In some cases, they seem to resemble human objects or an animal, such as a worm, a brain-cast, a face, or a wide spectrum of phallus-shaped objects. Paleontologists often receive such objects from amateurs who are convinced that these pseudofossils are real. Some concretions are truly spectacular in size (up to 9 m in diameter), forming large "toadstools" of cemented sandstone perched on a shale pedestal (Fig. 7.10) and "stone cities."

Among the most popular diagenetic objects are the concretions known as **geodes** or "thunder eggs" (Fig. 7.11). These spherical bodies are typically formed of concentric layers of chalcedony on the inside, and the internal cavity may be partially or completely filled with euhedral crystals of quartz or calcite that grew into the void. Although many ideas have been proposed to explain geodes, recent research suggests that

Figure 7.9 These sand calcite crystals from the upper Oligocene Arikaree Group in South Dakota formed when carbonate in the groundwater within a highly porous, unconsolidated sand crystallized into large, euhedral calcite crystals and trapped the host sand grains in their lattice. Note the crystal faces and terminations. (D. R. Prothero.)

Figure 7.10 Large concretions (up to 9 m in diameter) are common in certain formations. River channel sandstones are particularly prone to becoming cemented by groundwater into large "toadstools," "rock cities," or "*Kugelsandstein.*" These examples are from the lower Oligocene Brule Formation, Toadstool Park, western Nebraska. (D. R. Prothero.)

Figure 7.11 Geodes are particularly beautiful examples of diagenetic features. Most form when a cavity in the host rock (usually left by a decaying organism) expands and then gradually fills with concentric layers of silica or calcite, forming the delicate color bands that make them so popular with collectors. (D. R. Prothero.)

some sort of void must first exist in the host rock; usually, this is the hollow cavity of a fossil. Apparently, a mass of gelatinous silica collects in the void and begins to crystallize in concentric layers around the rim. As the geode grows, it expands outward and eventually bursts the fossil cavity, so that many geodes no longer have visible remains of the original fossil. Many geodes continue to expand, distorting the laminations around them; others even show an "exploded bomb" structure in their fractured outer rind.

Authigenesis, Recrystallization, and Replacement

In addition to compaction and cementation, a third common diagenetic process is **authigenesis,** in which *new minerals grow from old recycled chemicals.* For example, the growth of hematite or pyrite in sediments is authigenic; these minerals derive their iron from other

kinds of minerals. Similarly, the growth of clay minerals in most sedimentary environments is authigenic, because clay minerals derive their chemicals from feldspars rather than from preexisting clays of the same type. One clay can grow authigenically from the ingredients of another type. As we saw in Chapter 6, most ancient shales are made of illite, but apparently that illite grew authigenically from other, less stable clays such as montmorillonite. Under higher temperatures and pressures, even feldspars can grow authigenically. Indeed, authigenic feldspar overgrowths on quartz are not unusual in sandstones.

In **recrystallization,** *an existing mineral retains its original chemistry, but the crystal sizes grow larger.* For example, amorphous opaline silica often recrystallizes to form coarsely crystalline quartz. This process, however, is much more typical of limestones; fine lime mud readily recrystallizes into coarser "sparry" calcite.

Replacement is fairly common in sandstones and shales. Here, *a completely different mineral takes over the space once occupied by another.* Replacement requires the simultaneous dissolution of the first mineral and the precipitation of the second. The hallmark of replacement is evidence of a mineralogy that is clearly not original. For example, most marine invertebrate fossils are made of calcite. A mollusk shell made of silica or a brachiopod shell made of pyrite is clearly a result of replacement (Fig. 7.12). The reverse also occurs; siliciclastic grains that have been replaced by calcite are especially common.

In thin section, feldspar grains that are beginning to weather show a "decayed", mottled appearance that is a result of replacement by fine-grained muscovite, known as **sericite.** In sandstones that are buried at depths of 2 km and subjected to temperatures of about 100°C, both potassium and calcium feldspars alter to sodium plagioclase in a process called **albitization** (albite is the high-sodium end member of the plagioclase solid-solution series). Feldspars are often replaced by calcite, also.

Volcanic rock fragments (especially glassy ones) are particularly unstable. In thin section, we can see that they have been replaced by clays. This replacement is particularly noticeable when the distinctive angular shape of a volcanic glass shard has been retained but the glass has all been replaced by other minerals. Other common alteration products of volcanic rocks (as long as carbonates are absent) are the hydrous aluminosilicates known as **zeolites.** Certain zeolites (such as the sodium zeolite, analcime, and the calcium zeolite, heulandite) are

Figure 7.12 Replacement is easy to recognize when the fossil or other object is clearly composed of an unusual mineral that could not have been its original composition. For example, all articulate brachiopod shells are made of low-magnesium calcite. Yet some of these brachiopod shells have been replaced either by pyrite (*top row*) because they were fossilized in a reduced, pyritiferous black shale or by silica (*bottom row*), which has replaced all the fossils from this Permian limestone from West Texas. By etching these silicified fossils in acid, the limestone is dissolved away, leaving these delicately preserved specimens complete with fragile spines. (D. R. Prothero.)

stable at low temperatures (Fig. 7.13). Laumontite (another calcium zeolite) is stable at intermediate temperatures. Prehnite and pumpellyite are high-temperature zeolites also known from low-grade metamorphic rocks.

We have already seen that glauconite usually forms on the seafloor by slow replacement of fecal pellets or decaying organisms, and it can retain the shape of the object it replaced. The clay matrix of many rocks is highly susceptible to replacement, especially by calcite or chert. In these cases, however, there will be tiny specks of clay trapped in the replacement mineral; these inclusions show that the calcite or chert is secondary replacement, not primary cement.

Diagenetic Histories

We have not discussed all the known diagenetic processes. Clearly, many different changes can occur in a sediment during burial. We have seen that some changes are favored near the surface and others take place at higher temperatures and pressures. A major research question in recent years has been "When did these diagenetic effects occur?" Related questions include: "Based on these diagenetic clues, how long was this sediment buried? How deep was it buried?

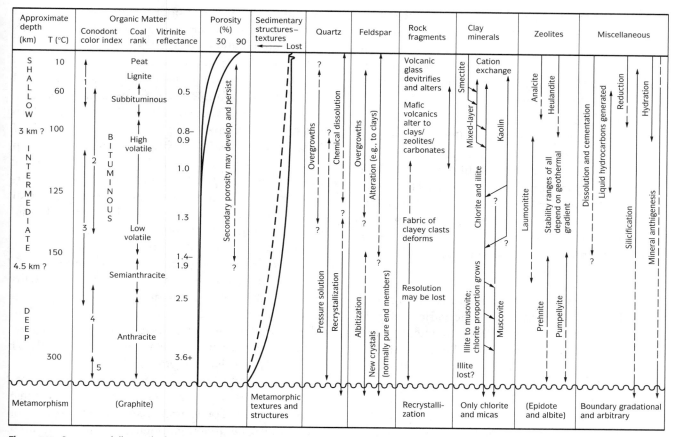

Figure 7.13 Summary of diagenetic changes occurring at various depths and temperatures. (After Lewis, 1984:34; by permission of Chapman and Hall, New York.)

How hot did it get, and to what pressures was it subjected?" If we can reconstruct the diagenetic history of a rock unit, we might get a better idea of the kinds of changes it has undergone. Not only is this information interesting in itself, but it might tell us whether the rock was part of a great collisional tectonic event, how deep the basin subsided, or whether the rock has undergone so much heating and compression that no porosity remains or no hydrocarbons survive.

Over the years, a number of indicators have been discovered, some of which give very precise estimates of the maximum temperature and/or depth of burial of a rock (see Fig. 7.13). Such indicators include the following:

1. Conodont color alteration. The microscopic phosphatic toothlike fossils known as conodonts undergo color changes with increased temperature (Harris, 1979; Epstein, Epstein, and Harris, 1977). Unaltered conodonts are pale yellow, at higher temperatures, the carbon in the organic material is fixed and changes their color from yellow to brown to black. A color alteration scale for conodonts with values from 1 (pale yellow, <80°C) to 5 (black, >300°C) gives a fairly precise estimate of the maximum temperature experienced by a rock. Unfortunately, conodonts are found only in Upper Cambrian through Triassic shallow marine rocks, so they cannot be used everywhere.

2. Vitrinite reflectance. The woody organic matter in lignites and coal changes and become shinier as it is subjected to higher and higher temperatures. By shining light on the surface of a coal under a microscope and measuring the light intensity reflected back, the maximum diagenetic temperature can be estimated between 100° and 240°C. This method is widely used by coal geologists to assess the metamorphic rank of coal (whether it is merely lignite, or bituminous, or even anthracite). Coals formed at higher temperatures, such as anthracite, tend to burn hotter and produce more energy per ton.

3. Transformations of clay minerals. As we discussed in Chapter 6, various clay minerals have different ranges of stability. Above 100°C, smectites begin to break down and form mixed-layer clays; above 200°C, mixed-layer clays change to

illite. At about 150°C, kaolinite also begins to change to illite or chlorite. As temperatures increase, chlorites begin to replace illite as well; other illites turn into muscovite. By 300°C, all clay minerals are gone, and only micas (mostly muscovite or chlorite) remain.

4. Zeolite facies. As noted earlier, the hydrous aluminosilicates known as zeolites are common alteration products of volcanic rocks in the absence of carbonates. Some are highly specific with regard to their temperature stability. Heulandite and analcime form at temperatures below 100°C; laumontite forms at about 100°C and is stable up to 150°C; prehnite and pumpellyite are stable above these temperatures. Thus, zeolites serve as excellent geothermometers.

Many other methods are summarized in Fig. 7.13. Geologists have developed a wide variety of paleothermometers to decipher the diagenetic history of a rock if any of the key ingredients (coal, conodonts, clays, zeolites, or fluid inclusions) is present. The Fig. 7.13 illustration can be viewed in several ways. For example, one can read *down* each column and trace a specific mineral through higher temperatures. Or one can read *across* the columns and gain a general impression of the characteristics of shallow, intermediate, and deep diagenesis. In short form, these transformations are summarized in Table 7.1.

Once a geologist knows how to read these paleothermometers, it is possible to reconstruct many aspects of the diagenetic history of a rock. By determining which cement has replaced earlier minerals, the relative sequence of minerals can be determined. A number of clever studies have reconstructed the history of specific rock units in great detail. For example, Loucks, Bebout, and Galloway (1977) deciphered the history of cementation in the Oligocene Frio Formation from the subsurface of the Texas Gulf coastal plain (Fig. 7.14). After initial weathering, the first diagenetic changes in the Frio Formation were compaction, calcite cementation, feldspar dissolution, and then authigenic feldspar overgrowths and clay coatings on the quartz grains. At later stages of burial (and depths below 1 km), the major event was extensive quartz cementation, which decreased the porosity from its initial 40% to almost zero. Slightly later, calcite cement further decreased porosity. When the Frio Formation reached depths of about 2.5 km, the feldspar framework grains and volcanic rock fragments dissolved away completely, producing **secondary porosity.** With further burial (greater than 3 km and 150°C), these secondary pores were filled by Fe-rich calcite and dolomite cement.

An even more sophisticated example of reconstructing multiple episodes of diagenesis using SEM (scanning electron microscopy) and thin-section petrography was demonstrated by Flesch and Wilson (1974) for the Upper Jurassic Morrison Formation of New Mexico (Fig. 7.15). Various textural criteria can be used to tell the relative age of two diagenetic events. For example, the minerals coating the outside rim of the pore formed before the minerals that crystallized in the center. If a grain is partially dissolved or replaced by another mineral, the replacement is younger than the mineral it replaces.

A wide variety of textures can be seen in the Morrison Formation. Both quartz and feldspar overgrowths occur on the sand grains, and many grains are cemented by a coating of clays and chalcedony. Pore fillings are typically calcite, chalcedony, or kaolinite. Many of the detrital feldspars have been partly or completely dissolved or replaced.

Such a varied assemblage of minerals suggests that many changes in the groundwater chemistry must have occurred. For example, secondary growths of feldspar, calcite, montmorillonite, and illite require a basic (high pH) solution rich in dissolved cations such as sodium or potassium. But the formation of kaolinite and gibbsite, and the dissolution of detrital feldspars, requires acidic (low pH) waters low in cations. The presence of hematite requires oxidizing waters, but the pyrite was produced under reducing conditions. As we have seen, the chalcedony cement must have precipitated from opaline silica, which requires a high concentration of silica in near-surface conditions; the quartz overgrowths, on the other hand, formed at depth with no opaline predecessor.

From these and many other clues, and from a knowledge of the regional geologic history, Flesch and Wilson (1974) worked out the diagenetic sequence of events. First, the Morrison Formation underwent shallow burial in the Late Jurassic and Early Cretaceous, shortly after it was deposited. The initial diagenetic changes were feldspar overgrowths, followed by the growth of smectite, illite, and chlorite clays in place of pre-existing rock fragments. At a later stage, quartz overgrowths and fibrous chalcedony took over the remaining pore space.

Sometime in the Early Cretaceous, however, the subsidence ceased, and the next generation of cement is calcite formed at shallower depths. Then the unit underwent exposure and weathering, so that the

TABLE 7.1 Diagenetic Changes in Sedimentary Rocks at Increasing Depths and Temperatures

Depth and Temperature	Quartz	Feldspars	Rock Fragments	Other Changes	Organic Materials
Shallow 0–3 km 0°–100°C	Formation of quartz overgrowths	K-feldspars replaced by sericite and kaolinite; plagioclase replaced with montmorillonite	Volcanic glass replaced by clays and zeolites such as analcime and heulandite	Compaction and cementation until porosity vanishes; smectites replaced by mixed-layer clays	Vitrinite reflectance <0.9 Conodont color index 1–2
Intermediate 3–5 km 100°–150°C	Chert becomes more coarsely crystalline; pressure solution of quartz	Continued alteration of K-feldspars to sericite	Mafic volcanics alter to clays and zeolites such as laumontite; lithic fragments deform and break up	Mixed-layer and kaolinite clays go to illite and increasingly to chlorite	Lignite becomes subbituminous coal Vitrinite reflectance 0.9–1.4 Conodont color index 2–3
Deep 5–7 km 150°–300°C	Pressure solution of quartz	Feldspars become Na-plagioclase (albitization)	Mafic volcanics replaced by zeolites such as prehnite and pumpellyite	Clays mostly chlorite with minor illite	Subbituminous coal becomes bituminous Vitrinite reflectance 1.4–1.9 Conodont color index 3
Low-grade metamorphic >7 km >300°C	Recrystallization of polycrystalline quartz	K-feldspars gone; all feldspars become Na-plagioclase	Growth of micas and epidote	Only chlorite and muscovite	Bituminous coals become anthracite Vitrinite reflectance >2.0 Conodont color index 3

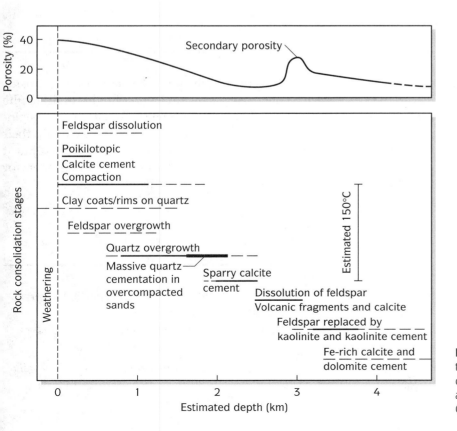

Figure 7.14 Diagenetic history of cements in the Oligocene Frio Sandstone of Texas. The changes in porosity are shown in the curve along the top. (After Loucks, Bebout, and Galloway, 1977:109–117.)

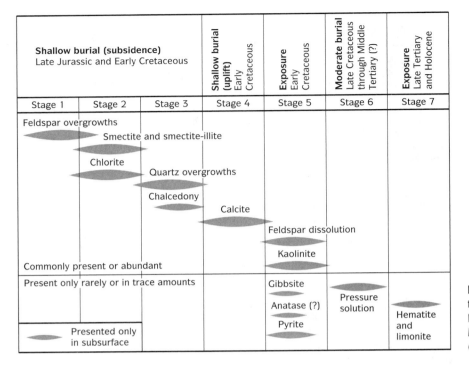

Figure 7.15 Types of chemical alterations and their timing in the Upper Jurassic Morrison Formation sandstones, northwestern New Mexico, based on thin section and SEM studies. (After Flesch and Wilson, 1974:187; Fig. 7.)

feldspars were dissolved and kaolinite formed (along with minor gibbsite, anatase, and pyrite). In the Late Cretaceous and the Tertiary, the Morrison Formation again underwent burial, producing pressure solution in the quartz. Finally, the unit was uplifted and exposed to its present position; the last diagenetic effect is a young overprint of hematite and iron hydroxide cement.

CONCLUSIONS

It is apparent that diagenetic analyses can be very complicated. In many cases, petrologists can unravel a complex sequence of overprinted events and estimate depths and temperatures of burial precisely. Analyses using thin sections, SEM, chemical analyses, conodont color index, and vitrinite reflectance can decipher much valuable information that may reveal whether a rock has been heated too much for oil to survive or whether it has any useful porosity. In other cases, such evidence might indicate the severity of continental collisions. Of course, beginning students will not pick up all these clues in their first experiences with sandstone petrography, but it is worthwhile to keep the eyes and mind open for any evidence of diagenesis in a thin section.

FOR FURTHER READING

McDonald, D. A., and R. C. Surdam, eds. 1984. *Clastic Diagenesis*. Memoir 37. Tulsa, Okla.: American Association of Petroleum Geologists.

Scholle, P. A., and P. R. Schluger, eds. 1979. *Aspects of Diagenesis*. SEPM Special Publication 26.

Stonecipher S. A. 2000. *Applied Sandstone Diagenesis: Practical Petrographic Solutions for a Variety of Common Exploration Development, and Production Problems*. SEPM Short Course Notes 50.

Taylor, J. C. M. 1978. Sandstone diagenesis. *Journal of the Geological Society of London* 135:1–133.

8 Terrestrial Sedimentary Environments

Gigantic cross-beds from the eolian Navajo Sandstone, Zion National Park, Utah. (Courtesy of W. K. Hamblin.)

CLASSIC STRATIGRAPHY, AS TREATED IN THE TEXTBOOKS OF the 1950s and 1960s, emphasized the *description* of sedimentary rocks and stratigraphic units. The interpretation of these rocks received less coverage and was often organized along descriptive lines. For example, distinctive rock types, such as redbeds, were discussed as if they had all formed under the same conditions. An approach that is more *genetic* and less descriptive has emerged in the past three decades as

geologists have begun to study modern depositional environments in detail to see what types of sediments and stratigraphic sequences are formed. These data are used to analyze ancient stratigraphic sequences for analogs of modern environments.

The genetic approach has become so fruitful that it is now used in virtually all interpretations of stratigraphic sequences. Take, for example, a sandstone exposed in an outcrop. A good depositional model not

only allows the geologist to make a reasonable reconstruction of the ancient environmental conditions but also has economic importance. A geologist who understands the depositional circumstances of a sandstone can make predictions about the subsurface geometry and the extent of that sandstone. If it happens to be an important reservoir rock for water or hydrocarbons, this knowledge may be crucial in determining where to drill.

This depositional-systems approach has two distinct parts. By studying the modern depositional analog, we are concentrating on the *process*, or cause, of deposition. There are many depositional processes. These include physical processes, such as wave and current activity, gravity flows, sea-level changes, and tectonism; biological processes, such as biochemical precipitation, bioturbation, and photosynthesis; and chemical processes, such as solution, precipitation, and authigenesis. Other static elements of the environment, such as water depth and chemistry, sediment supply, and local climate and topography, are also part of the depositional process.

The stratigraphic record, however, consists of sedimentary rocks that are the *products*, or effects, of these depositional processes. We must use the properties of these sedimentary bodies to infer the ancient processes that produced them. These properties include their overall geometry, physical and biological sedimentary structures, porosity and permeability, acoustical features, resistivity, and radioactivity. After looking at a number of ancient sequences that have similar sedimentary features, we distill out the local variability and construct a *facies model* that generalizes these features. The facies model is an environmental summary that will suggest which features in the modern depositional system would explain the features found in the ancient example. This model can be used not only to find a modern depositional analog but also as a framework for future observations and a predictor of what features should be found in other examples. The facies model, in turn, can be tested by examining modern analogs or ancient sequences with new features not seen before.

The important thing to keep in mind, however, is that a facies model is a deliberate idealization. Most of the variability of real examples has been distilled out. As a consequence, each time a geologist attempts to analyze real depositional sequences, there will be variations that are not part of the model. Some of these variations are insignificant; others are crucial to distinguishing between two very similar environments. Some differences result from the fact that the case

under study may be intermediate between two idealized models and have properties of both. This is often the case in environments that are typically adjacent in the real world. For example, many deltaic sequences interfinger with meandering fluvial floodplain sequences, and many fluvial sequences are hybrids between braided and meandering systems. A geologist interpreting these sequences should not be too rigid about fitting the interpretation into idealized pigeonholes. Real examples are much more complex than the idealized models we like to work with.

Nevertheless, there is value in learning about these simplified models. They greatly reduce the complexity of the real world into a useful framework that can guide our thinking, and they suggest what features to look for. In many cases, the models suggest which sedimentary features would discriminate between two likely alternatives. Without these models to focus our observations, we would be back in the days of purely descriptive stratigraphy and would not know where to begin looking for features that might be significant for interpretation.

In this chapter and the next two, we will examine some of the major depositional systems that are responsible for the bulk of the stratigraphic record of siliciclastic rocks. Along with the discussion of each major system, an ancient example is discussed in an associated box. Many environments produce extremely variable outcrop patterns, however, and not every outcrop formed in a given environment will closely resemble the example. For this reason, the reader should not take the examples too literally. They are meant to demonstrate some of the major themes seen in each environment, and variability is the norm.

In these three chapters, there is only enough space to give a brief introduction to the more important depositional systems found in the stratigraphic record. For more detail, the student is urged to consult some of the many excellent books listed at the end of each chapter.

All the depositional systems discussed in this chapter are nonmarine. These systems tend to be more familiar because of our own experience with streams, rivers, lakes, sand dunes, and other subaerial features. Most terrestrial systems, however, have a low preservation potential because they are above the *base level* of erosion, the level on Earth's surface above which sediments must eventually erode and below which they can accumulate. Most of the time, the base level is near or below sea level. Nonmarine deposits are usually preserved in the rock record only when they fill a basin that is sinking below base level, such as

a graben or a subsiding downwarp. In addition, the late Cenozoic is an era of unusually low sea level relative to the rest of the Phanerozoic, so a much larger area of the globe is nonmarine at this moment. During most of the Paleozoic and Mesozoic, however, large, shallow epicontinental seas dominated the continents, and areas of nonmarine deposition were fewer.

Alluvial Fans

The sedimentary cycle begins as bedrock is weathered away from uplifts and picked up by mountain streams. The first major sedimentary body to be formed of these newly eroded sediments is usually an alluvial fan. An **alluvial fan** is a cone-shaped deposit of coarse stream sediments, sheet-flood deposits, and debris flows that forms where a narrow canyon stream suddenly disgorges into a flat valley. The sudden change of a stream from a narrow, confined channel with a steep gradient to the broad flats of the valley causes a sudden drop in the hydraulic power of the stream. The decreased competence of the stream allows the coarser material to drop and accumulate.

Alluvial fans are best known from desert environments (Fig. 8.1), although they also occur in humid environments. They are usually triangular in map view and wedge-shaped in cross section (Fig. 8.2), radiating from the mouths of mountain canyons. A large number of fans spreading out from canyons along a mountain front can coalesce to form a pediment along the base of the mountain. In arid regions, this **pediment** is known as a **bajada.** The slopes of alluvial fans range from 1° to 25° but average around 5° to 10°. The larger the particle size, the steeper the slope. For an alluvial fan to be active, there must be continued elevation and erosion of the highland to supply the coarse debris. This process occurs most often along rising fault scarps, which are frequently seen in tectonically active desert areas today. Where the headland stops rising, the supply of coarse debris eventually stops, and the fan degrades and begins to merge with the surrounding valley deposits.

Sediment in alluvial fans is transported in three ways: streamflow, debris flow, and mudflow. In arid regions, the most significant streamflow occurs in flash floods, which take place when sudden desert thundershowers dump large amounts of rain in a limited area. This flow is channeled down the mountain canyons as rapidly moving waves of water with

Figure 8.1 Single alluvial fan at the mouth of a canyon in the steep east wall of Death Valley, California. Badwater at left; highway for scale. Note the braided streams coursing along the surface of the fan. (Shelton, 1966: 155.)

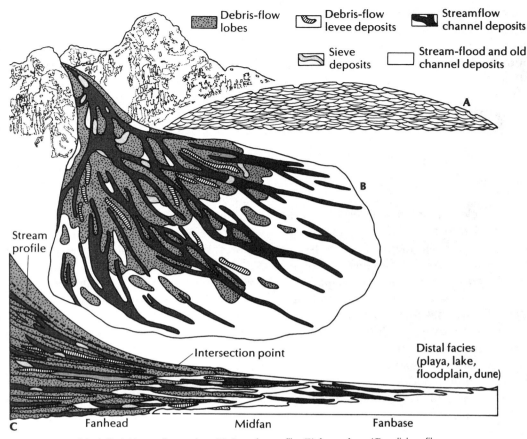

Figure 8.2 Model of alluvial fan sedimentation. (A) Cross-fan profile; (B) fan surface; (C) radial profile. Vertical profiles are greatly exaggerated. (From Spearing, 1974: Fig. 1C; by permission of the Geological Society of America.)

tremendous erosive power. There are many stories of enormous walls of water that suddenly sweep down on surprised hikers in canyons where no rain has been falling, and of the enormous boulders, vehicles, and even houses that these flash floods can carry. In *Basin and Range*, John McPhee relates the story of a bartender in Nevada who boarded up the doors and windows of his saloon when he heard the rumble of an approaching flash flood. So large were the boulders brought by the flood that this was the only protection he had against broken doors and windows.

As the flow of a flash flood decreases, it drops pebbles, cobbles, and boulders wherever the strength of the flow falls below the critical competence needed to transport them. This results in a flow that is choked with more detritus than it can carry, which forms a **braided stream.** The surfaces of alluvial fans in deserts are typically covered with radiating systems of braided stream channels, most of them dry except during the rare flood (see Fig. 8.1). Each flood cuts a new channel, which causes older channels to be filled

with gravel. During the high-water point of the flood, the excess water tops the channel banks and spreads out across the fan, forming a shallow sheet of sand or gravel that contains no clay or other fine material. These midfan sheet-flood deposits are typically well sorted, well stratified, and cross-bedded; they usually form lobes that emerge from the channel at the **intersection point** of the channel profile and the fan surface (Fig. 8.3). Because there is little silt or clay, the water can pass through the porous gravels on the fan without blocking the pores. Thus the lobe-shaped deposit, which is called a **sieve deposit** (Fig. 8.4), becomes progressively coarser toward the front of the lobe, where the porous gravel accumulates. Sieve deposits usually form in the proximal fan or upper midfan (see Fig. 8.2).

Freely flowing water is not the only mode of transport in an alluvial fan. When sediment becomes saturated with water, it can flow as a viscous plastic mass that behaves more like quicksand than like flowing water because the grains are supported by the pressure of the water-soaked matrix trapped

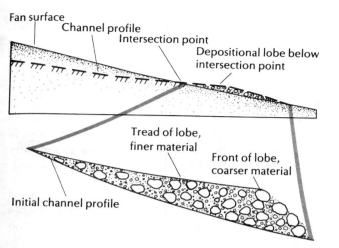

Figure 8.3 Diagram of sieve lobe deposit. (Hooke, 1967: 450; © 1967, by permission of the University of Chicago Press.)

between them. As we saw in Chapter 3, debris flows can carry enormous boulders, as well as every smaller particle, including clay, resulting in an extremely poorly sorted, chaotic jumble of boulders, cobbles, pebbles, sand, silt, and clay (see Figs. 3.8 and 3.9). There is no stratification unless a series of flows accumulate on top of one another. Sometimes the base of a debris flow shows reverse (or inverse) grading, in which the grain size increases upward (see Fig. 3.10). Debris flows typically form lobate, tabular bodies of uniform thickness and are usually found in the upper reaches of the fan. If the viscous

fluid flow has little coarse material and is composed mostly of sand, silt, and mud, it is called a mudflow. Like debris flows, viscous mudflows form restricted, narrow lobes, but mudflows that are more fluid can form enormous sheet-flood deposits and move at velocities of 10 km/hr (6 mph). There are many examples in the geological literature of catastrophic mudflows that bury villages, overrun fleeing people, and float houses and debris for kilometers.

Many examples of ancient alluvial fan deposits are found in the stratigraphic record (see Box 8.1). Because alluvial fans require rapid uplift, they are usually formed and buried in rapidly downdropping grabens, foreland basins, and strike-slip basins. A typical profile through an alluvial fan (see Fig. 8.2) shows a mixture of unsorted debris flows, stream-channel conglomerates (often called **fanglomerates** if they occur in a fan), cross-bedded sandstones, and sieve deposits. Grain sizes range from coarse gravels and cobbles at the top of the fan to fine sand near the base. These features are seen in the typical stratigraphic column (Fig. 8.5), which shows a coarsening-upward sequence of cross-bedded sandstones and fanglomerates. The basal part of the sequence consists of cross-bedded sandstones from the distal fan. As the uplift continues and the fan builds outward and upward, the coarser deposits of the proximal fan accumulate thicker and thicker packages. Many of these are individual debris flows that are meters in thickness. At the very end of the fan cycle, the uplift ceases, and a

A

B

Figure 8.4 Sieve lobe deposits in a modern alluvial fan, Eureka Valley, California. (A) Hummocky topography formed by sieve lobes. The slope is 16° along the radial profile through the fan valley and sieve lobes, 20° along older and smoother parts of the fan to the right. (B) Sieve lobe deposit from the same area. (Courtesy of W. B. Bull.)

BOX 8.1 DEVONIAN FANGLOMERATES OF NORWAY

One of the best-exposed examples of ancient alluvial fan deposits occurs in the Hornelen Basin in western Norway (Steel et al., 1977; Gloppen and Steel, 1981). This is one of several fault-bounded basins formed by extension of the region during the Devonian Caledonian orogeny (Steel and Gloppen, 1980; Steel, 1988). Although the basin is about 25 km wide and 70 km long, it has 25 km of sedimentary fill composed of a succession of stacked coarsening-upward sequences about 100 to 200 m thick (see Fig. 8.5). These form spectacular sandstone and conglomerate cliffs visible even from the air (Fig. 8.1.1). Within each cycle are coarsening-upward subcycles, ranging in thickness from 10 to 25 m. These cycles, as well as the thicker ones, are attributed to varying rates of tectonic movement along the extensional fault system. There are also cycles caused by major floods or by the switching of deposition from one fan lobe to another. These units are usually less than a few meters thick and fine upward.

The alluvial fan deposits are concentrated close to the fault-bounded margins of the basin (Fig. 8.1.2A) and interfinger with finer basin-fill deposits. These include sheet-flood and stream sandstones and flood-basin sandstones and shales. The fan deposits themselves consist of nearly all the typical facies. There are debris-flow deposits (Fig. 8.1.2B) that contain boulders several meters in diameter. These conglomerates are poorly sorted, lack any pervasive stratification, and are typically clast-supported. Many of the beds show inverse grading. Often, clasts project above the top of

the bed and are buried by the finer, inversely graded base of the next bed. Some of the elongate clasts are oriented vertically, indicating high matrix strength that prevented any further settling or alignment.

Some of the conglomerates have been interpreted as sieve deposits (Fig. 8.1.2C). These deposits are less than a meter thick and are composed of a well-sorted, open framework of angular pebbles and cobbles. They are lenticular and are sometimes banked against larger boulders or blocks. The open framework between clasts has been filled in at a later time by fine red siltstone.

The bulk of the deposits are stream-deposited sandstones and fanglomerates, which show much better sorting and rounding. They occur as laminated sandstones (Fig. 8.1.2D) or show planar and trough cross-stratification (Fig. 8.1.2E). Individual beds vary from 15 to 100 cm in thickness. These are found in lenticular channels in finely laminated sands, interpreted as sheet-flood deposits. Some of the debris-flow deposits apparently continued below the water level, forming subaqueous debris flows (Fig. 8.1.2F). These debris flows are characterized by inverse-to-normal grading within the bed.

Although some of these features can be found individually in other environments, the coarse conglomerates with the distinctive debris-flow and sieve deposits, combined with the geometry of the deposit along the margins of the basin, clearly indicate that much of the Hornelen Basin conglomerate and sandstone formed in Devonian alluvial fans.

Figure 8.1.1 Eastward view along part of the axis of the Hornelen Basin, western Norway, showing basinwide cyclicity of alluvial fan coarsening-upward sequences. Scarps are approximately 100 m high. In addition to the cyclicity, there is a general trend from fine-grained (foreground) to coarse-grained alluvial (background) fans. (Courtesy of R. J. Steel.)

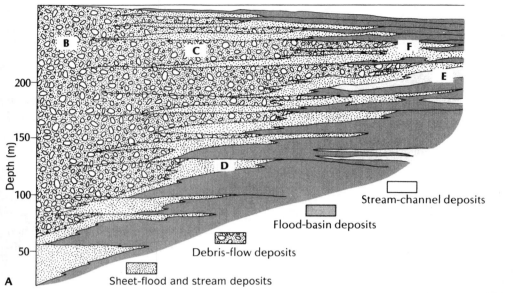

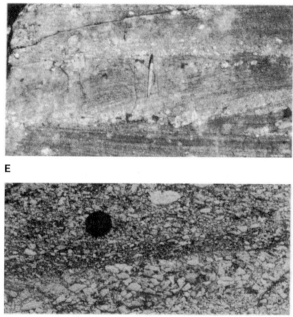

Figure 8.1.2 (A) Profile of the Hjortestegvatnet fan showing the characteristic interfingering of deposits. (B) Boundary between two debris flows. The upper flow is inversely graded and is scoured into the lower flow; one large clast from the lower flow protrudes into the upper flow. (C) Possible sieve deposit made of sorted angular quartzite pebbles with fine-grained secondary matrix filling the cavities. (D) Sheet-flood deposits composed of thin graded beds with pebble lags. (E) Stream flood fanglomerates showing well-developed cross-bedding. (F) Inversely graded subaqueous debris flows. (Photos courtesy of R. J. Steel.)

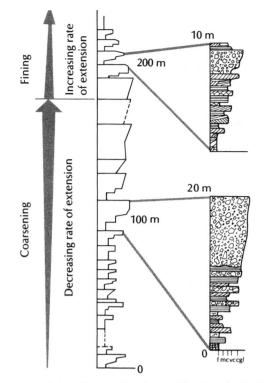

Figure 8.5 Vertical profile through cyclic alluvial fan deposits showing the characteristic coarsening-upward profile. Based on the Devonian Hjortestegvatnet fan of Norway. (Steel and Gloppen, 1980: 79.)

thin, fining-upward sequence forms during the decay of the fan.

Most alluvial fans form in mountain fronts far from the ocean, and their sediments have a long way to go to reach the sea. Consequently, conglomerates are seldom found in shallow marine sequences and are found even more rarely in deep marine deposits. In regions with rapid tectonic uplift, however, especially along steep faults, it is not unusual for alluvial fans to form right next to the shoreline. These **fan deltas** pour coarse conglomerate and debris flows directly into the deep marine system. It seems startling to imagine sequences with debris-flow conglomerates mixed with pelagic oozes (the most extreme opposites of grain size and transport energy), but such mixtures are not unusual along active faulted oceanic margins. Occasionally, there are instances of conglomerates mixing with deep marine turbidites, as in the famous Cretaceous sequence at Wheeler Gorge, Ventura County, California.

Alluvial fans are typically limited in their lateral extent and merge abruptly with the deposits of the basin. Their thicknesses can be enormous if the source uplift continues for a long time. For example, the fault graben fill of the Miocene-Pliocene Ridge Basin near the San Andreas fault in California has more than

9000 m of sandstones, shales, and fanglomerates (see Box 19.1). The Triassic graben fills of the Newark Supergroup of the Appalachians exceed 7000 m in places (Bull, 1972).

Diagnostic Features of Alluvial Fans

Tectonic Setting   Alluvial fans are typically found in rifting continental grabens, foreland basins, collisional overthrust mountain belts, and other highlands undergoing rapid uplift. They are associated with meandering fluvial valleys and playa lakes.

Geometry   Wedge-shaped and limited in lateral extent, alluvial fans extend from only a few tens of meters to kilometers from the source highland. Their thickness can be tremendous (7000 m or more) if subsidence is persistent.

Typical Sequence   Alluvial fans are composed of coarsening-upward sequences of cross-bedded sandstone, channel-lag conglomerates, and unsorted debris-flow deposits. Sometimes a fining-upward sequence forms during the decay of the fan.

Sedimentology   There is an extreme range in grain size, from boulders to clay; particle size decreases downfan. Conglomerate (fanglomerate) and cross-bedded sandstone are most common. Debris-flow deposits are unsorted, can show reverse grading, and can contain boulders. Sieve deposits form conglomerates with no matrix of finer material, unless it has infiltrated at a later time. Lenticular bodies composed of cross-bedded channel sand and pebble conglomerates formed by channel cut-and-fill and debris flows can occur; they are most common near the apex of the fan and decrease downslope. Paleocurrents radiate from the apex as well. Ripple marks and convolute lamination can occur in the finer-grained sheet-flood deposits. Sediments can be very immature and angular, with abundant coarse rock fragments and feldspars. Sheet-flood deposits are typically oxidized, so redbeds are common.

Fossils   Because alluvial fans are usually highly oxidized, fossils and organic matter are rare. High-energy floodwaters, with accompanying deposition of coarse conglomerates, also destroy fossils.

Braided Fluvial Systems

Any flowing body of water that has insufficient discharge to carry its sediment load, or has easily erodi-

Figure 8.6 Braided river channels, Muddy River, Alaska. (Courtesy of Bradford Washburn.)

ble banks, forms the classic braided pattern (Fig. 8.6). We have already seen how most of the flow in an alluvial fan is braided. Braiding is typically found in the upper reaches of a fluvial system as it comes out of the source area. Thus, braided streams usually have steep gradients, abundant coarser sediments (chiefly sand and gravel), and rapid discharge fluctuations, all of which lead to the rapid shifting of channels over their easily eroded banks. Braided systems are also very wide and shallow during their slack-water stages. "A mile wide and a foot deep" was the way the Oregon Trail pioneers described the Platte River, which was so different from the deep, constantly flowing rivers they had known in the East and Midwest.

Large, rapid discharge fluctuations are responsible for the overloading of sediment. At flood stage, the river can carry virtually all of its load. Most of the time, however, little water is flowing, and the competence of the flow is insufficient to move most of the bedload. Some reworking occurs, but most of the changes in the morphology of the braided fluvial complex occur during flood stage. Typically, an obstacle that is difficult to erode causes a flow separation and accumulates sediment in its lee. These deposits continue to build up until they form **longitudinal bars,** one of the most distinctive features of braided systems (Fig. 8.7). Longitudinal bars continue to aggrade downstream and sometimes become islands, which can be stabilized by vegetation. Longitudinal bars are more common in the upper reaches of a braided complex and are usually full of gravel. They show high-flow-velocity features, including plane bedding. The abundant pebbles are usually well imbricated. On their downstream end, cross-bedding may develop; sandier, protected channels between the

longitudinal bars usually develop ripples and dunes. **Transverse** (or **linguoid**) **bars** (see Fig. 8.7), which are broader than longitudinal bars and more lobate in shape, are more abundant in the sandier, downstream reaches of a braided fluvial system. Such bars are in fact megaripples, formed during flood conditions and modified by slack-water conditions. Transverse bars are composed of low-flow-velocity features, particularly ripples, dunes, and trough cross-bedding.

There are also wedge-shaped sandbars that form at oblique angles to the flow. These **cross-channel bars** form where a small channel discharges into a larger one or where the flow is forced to spread laterally or flow obliquely. These bars start with a nucleus that is high during slack-water stages. Sand, flowing around the nucleus, forms "horns" on either side that continue to grow. Eventually, as it grows and accretes other sandbars, the bar can expand into a sand flat. The cross-beds in these bars often show flow directions perpendicular to the main current.

The bars, which tend to merge and divide with the changes in flow, are eroded back along their sides and downstream ends. During flood stage, the old channels, constructed by sandbars, are swamped and obliterated. As the water recedes and the stream restabilizes, new channels form that often cut the old ones. The resulting channel cross section is thus a complex of lenticular channel sequences that crosscut one another in great profusion. In longitudinal section (see Fig. 8.7), some bars appear lenticular, but many form long, continuous cross-bedded sand bodies. A typical stratigraphic column for a braided river (see Fig. 8.7) shows abundant trough cross-stratification in channels that are scoured into older channels. The base of the channel sequence contains the coarsest particles, commonly

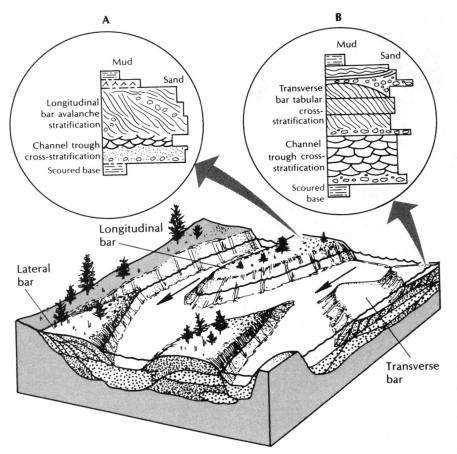

Figure 8.7 Depositional model for a low-sinuousity braided channel. Sequence A is dominated by the migration of a gravelly longitudinal bar. Sequence B records deposition of successive transverse bar cross-beds across a channel. (After Galloway and Hobday, 1983: 70.)

a lag of pebbles or cobbles. Often, cross-channel bar migration produces cross-beds that are perpendicular to the flow direction of the trough cross-beds. As the channel fills and becomes a sand flat, there is both lateral accretion of cross-channel bars and vertical accretion of laminated sands and muds that fill the abandoned channel. These bar-top muds are almost the only fine sediment in such systems because most of the clays and muds remain in suspension in the fast-moving water. These sand flats are often stabilized by vegetation, causing bar-top deposits to be full of burrows and root casts. With the gravel at the base and the mud and silt at the top, the typical braided system forms a fining-upward sequence (see Fig. 8.7). Unlike the meandering system described shortly, it fines upward abruptly at the top of a cycle; the meandering sequence fines upward gradually. An example of an ancient braided river is discussed in Box 8.2.

Diagnostic Features of Braided Fluvial Systems

Tectonic Setting Braided systems occur in the upper reaches of alluvial plains, relatively near the upland source. Like alluvial fans, they can be associated

with rapidly downdropping basins because they require an upland to provide the coarse material and high stream gradient.

Geometry Elongate, fairly straight lenticular or sheetlike sand bodies grade laterally into finer deposits of an alluvial plain.

Typical Sequence There is a fining-upward sequence of channel-lag gravels, abundant sandy trough cross-beds filling channels, and occasional tabular cross-beds migrating across channels, topped by vertically accreted, laminated sand and mud with burrows and root casts. Unlike meandering rivers, braided systems are ephemeral and rapidly shifting, so the sequence may cross the channel and repeat several times.

Sedimentology Gravel is more common in longitudinal bars of the upper reaches of the system, but sand is dominant throughout. Unlike meandering systems, there is very little silt and mud. There is abundant tabular and trough cross-stratification; vertically accreting plane beds are less common. Longitudinal

BOX 8.2 TRIASSIC FLUVIAL SANDSTONES OF SPAIN

In the central Iberian Range of Spain, just north of Madrid, is a sequence of Triassic sediments that demonstrates many of the features of sandy braided systems (Ramos, Sopeña, and Perez-Arlucca, 1986). The Rillo de Gallo Sandstone (Fig. 8.2.1) is composed of medium-grained red sandstone with some conglomerate beds, and it forms the caprock in much of the Iberian Range. This unit is interbedded with conglomerates and thin-bedded sandstones and mudstones, which together represent the classic basal clastic sequence that is found throughout the Lower Triassic in Europe. In the type area in Germany, it is known as the Buntsandstein. As in the rest of Europe, the Spanish Buntsandstein is capped by the other two facies of the trinity that gave the Triassic its name. Above the redbeds are carbonate tidal sediments equivalent to the German Muschelkalk, followed by mudstones and evaporites of the Keuper facies.

The Spanish Buntsandstein sequence is believed to represent the filling of grabens that formed when the Atlantic began to rift open in the Triassic (Alvaro, Capote, and Vegas, 1979). At the base of the sequence are pebble conglomerates formed in alluvial fans (Ramos and Sopeña, 1983). Overlying and interbedded with these conglomerates are the Rillo de Gallo sandstones, which form cliffs and ledges that are several hundred meters thick and can be traced along the range for more than 100 km (see Fig. 8.2.1). They pinch out rapidly, however, as they approach the boundary faults of the ancient graben. The sandstones are medium- to coarse-grained, subrounded to subangular, with less than 15% matrix. In most places, the sandstones are predominantly quartz (56% to 98%) and moderately mature, but near the border faults, the feldspar and rock fragment content can reach 30%.

Sketches of the outcrop pattern (Fig. 8.2.2A) clearly show a cross section that would be expected from sandbar migration in a braided river channel (Fig. 8.2.2B). Several common facies are shown by the sedimentary structures in the sandstones. Facies TB consists of tabular cross-strata about 4 m thick, with flat bases and tops. The foresets dip at angles of 12° to 19°, with the angle increasing downstream. This facies is attributed to the growth and migration of transverse bars (Fig. 8.2.2B). Some of the tabular cross-stratification (facies TBv in Fig. 8.2.2C) shows clear reactivation surfaces, indicating that minor bedforms migrated over the lee side of the major bedform. This facies represents both lateral growth of the foresets and, vertical growth of the topsets. Other tabular cross-stratified layers pass downstream into trough cross-stratification (facies TBt). Tabular cross-stratified conglomerates (facies TC) are interbedded with the sandstones (Fig. 8.2.2C). Finally, there are trough cross-stratified sandstone units (facies T) with scoured channel (ch) bottoms. The troughs can be up to 2 m thick and 20 m wide. This facies is attributed to migrating channel-fill deposits formed by the migration of megaripples during flood stage.

As the sketch of the cross section (see Fig. 8.2.2A) clearly shows, the Rillo de Gallo Sandstone is the product of a braided system. The abundance of gravel, the predominance of tabular cross-stratification, and the almost total lack of fine-grained deposits all strongly support this interpretation. A meandering system would show almost no gravel and would consist of sandstone channels in floodplain mudstones. Meandering systems seldom show the repeated stacking of tabular cross-stratified sandstones and the frequent channeling and trough cross-bedding of braided systems.

Figure 8.2.1 Panoramic view of Rillo de Gallo Sandstone, Guadalajara, Spain. Facies T and TB are labeled. (Courtesy of A. Ramos, A. Sopeña, and M. Perez-Arlucea.)

(continued)

(Box 8.2 continued)

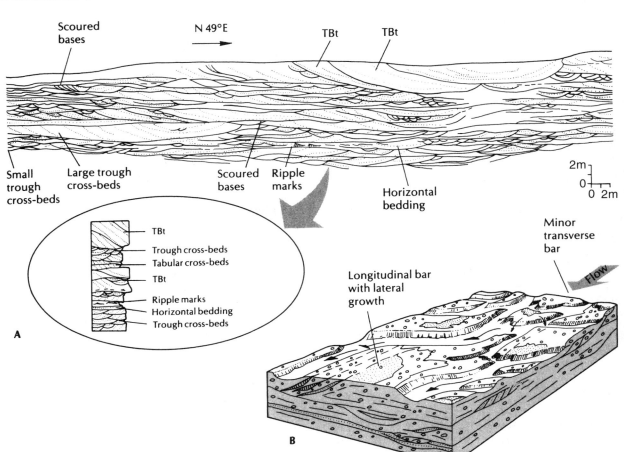

Figure 8.2.2 (A) Field sketch of main sequence from Rillo de Gallo Sandstone. Facies labels are explained in the text. (B) Block diagram of depositional model suggested for the Rillo de Gallo Sandstone. (After Ramos, Sopeña, and Perez-Arlucea, 1986: 863.) (C) View of outcrop. Facies TBv (*lower right*) shows tabular cross-stratification, reactivation surfaces, and vertical accretion simultaneous with forward progradation of bedforms. Paleocurrent direction is oblique to the picture. The small channel (ch) on top of the sandbar was filled with cross-stratified sand. Sandstone facies TB and tabular cross-stratified conglomerates (TC) are seen in the middle of the picture. Sandstone facies T is composed of large trough cross-strata. (Courtesy of A. Ramos, A. Sopeña, and M. Perez-Arlucea.)

bars can show high-flow-velocity plane beds, and ripples and dunes are common at lower flow velocities.

Fossils Braided systems are usually unfossiliferous except for root casts and burrows on vegetated sand flats.

Meandering Fluvial Systems

In the lower reaches of a fluvial system, the gradient is much less steep than in the upper reaches, and most of the coarser material has been left behind. The relatively straight braided channels become more sinuous as they get farther from the source uplands until the fully sinuous **meandering** system is established. In braided systems and steep mountain streams, most of the energy of the river is expended incising channels, since these systems are well above the base level of erosion. However, as rivers approach base level, they cannot cut their channels much deeper. Hence, the energy of their large volume of moving water goes into cutting the channel *sideways* instead.

As the streamflow becomes slower, deeper, and steadier, secondary currents can develop. The dominant pattern is a spiral secondary flow (Fig. 8.8A) that results as water moves around a bend and is deflected

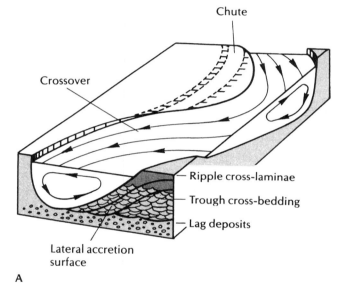

A

B

C

Figure 8.8 (A) Model of meandering river showing secondary flow patterns and the internal structure of point-bar deposits. (After Blatt, Middleton, and Murray, 1980: 637; © 1980, by permission of Prentice-Hall, Inc., Englewood Cliffs, N.J.) (B) Detail of the outcrop in (C), showing sedimentary structures characteristic of decreasing flow regimes through the point bar. The trough cross-beds at the base are nearly 1 m thick; the rippled cross-beds near the top are only a few centimeters thick. (D. R. Prothero.) (C) Cross section of an ancient point-bar deposit showing well-developed lateral accretion surfaces (epsilon cross-beds) accentuated by thin mudstone layers. From the Triassic Moenkopi Formation, north of Winslow, Arizona. (D. R. Prothero.)

toward the outer bank of the bend by centrifugal forces. The deflection is stronger near the surface, where the flow velocity is high, and weaker near the base, where the flow velocity is retarded by friction with the bed. This spiral current tends to erode the outside of each bend and then to transport the material laterally and downstream to the inside bend, where the flow is slower and less erosive. The flow spiral in each bend is reversed from that in the previous bend, so there must be a nonspiraling flow, or **crossover** (see Fig. 8.8A), in the straight stretches between bends.

Because the natural tendency of meanders is to erode their outside banks and deposit sediment along their inside banks, they are continuously migrating laterally and becoming more sinuous. When a river becomes extremely sinuous, the broad meanders have only narrow necks separating them (Fig. 8.9). During floods, the erosive power of the water increases and eventually breaches these necks, forming **cutoffs.** Cutoffs are straight stretches of rapid flow that divert most of the water from the meander. Eventually, the abandoned meander is isolated from the river and forms an **oxbow lake,** which gradually fills up with silt, mud, and vegetation. Sometimes a major flood

can cause the flow to breach the side of a channel and start a new sinuous course across some other part of the floodplain. This process is called **avulsion.**

The most characteristic product of a meandering river is the **point-bar sequence** (see Figs. 8.8 and 8.9), which forms at the inner bank of a meander. As the meander becomes more sinuous, the point bar accretes laterally. At the edge of the bank where the bed is just high enough to begin to slow the current, gravel and coarse sand are deposited, forming a thin, discontinuous **channel lag** (see Fig. 8.8A). As this base builds up, it diverts the strong gravel- and sand-carrying current farther out into the stream, where a new packet of channel lag begins to accrete. Over the old channel-lag deposit, a high-flow-velocity plane lamination and large-scale trough cross-beds of migrating dunes are deposited (see Fig. 8.8B). As these features are built, the current that is strong enough to deposit them is in turn diverted outward (as the gravel deposition again moves farther from shore), and low-flow-velocity sand ripples and a climbing ripple lamination are deposited by the remaining shallow current. If this point bar is now abandoned by the stream so that it

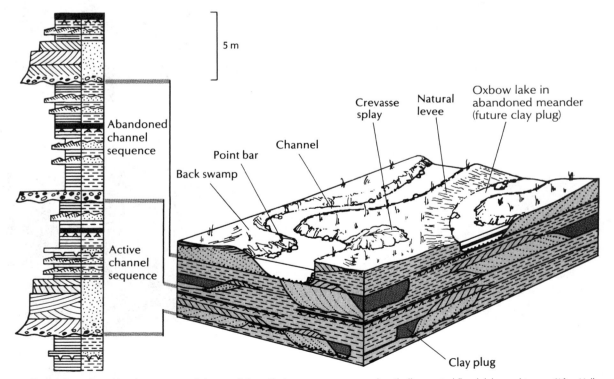

Figure 8.9 Typical three-dimensional geometry and characteristic vertical sequence of a meandering floodplain deposit, showing point bars, crevasse splays, and clay plugs filling oxbow lakes, all interrupting thick sequences of vertically accreted floodplain mudstones. (After Hallam, 1981, and Selley, 1978.)

is not cut by subsequent channels, it will finally be capped by fine muds on its highest surface by the lowest-flow-velocity currents that manage to wash across the top of the bar.

The product of this process is the lens-shaped point-bar sequence, a series of sedimentary structures that reflect, upward through the sequence, the decrease in flow velocities (see Figs. 8.8A, C). If these processes continue long enough for the outer edge of the bar to migrate some distance laterally out into the streambed, they produce a tabular, well-stratified body of sand that is progressively younger away from the axis of the meander. The top surface of this sand body is covered by arcuate swells and swales known as **scroll bars,** each of which is a product of the lateral accretion of point-bar sands (Fig. 8.10). It is important to remember that although each level of

Figure 8.10 Aerial photograph of spectacular scroll bars of an Ohio River point bar near Henderson, Kentucky, exhibiting more than a kilometer of lateral accretion. (Courtesy of U.S. Geological Survey.)

the sequence looks continuous and almost undifferentiated from its oldest to its newest deposition, all the strata of a point-bar sequence are formed simultaneously. The channel lag is farthest out from the axis of the meander, the plane lamination and trough cross-beds are next closest to the bank, and so on to the lowest-flow-velocity deposits at water's edge. The time planes are not the horizontal boundaries between depositional features but the sloping surfaces that parallel the inside slope of the meander (see Fig. 8.8A). The deposit builds *laterally* from the axis of the meander into the stream. Occasionally, this sloping surface is visible in cross sections of the point bar. These are known as **lateral accretion surfaces** or "epsilon cross-beds" (see Fig. 8.8B). There is a tendency, during mapping and correlation, to treat similar-looking rock units as being of the same age, but the point-bar sequence is a classic example of a *time-transgressive but laterally continuous deposit.* This must be kept in mind when examining ancient deposits (see Box 8.3).

Sedimentary features other than point bars also form in the meandering system. During flood stage, the point bar is inundated and scoured by flows in shallow flood channels, or **chutes,** which may or may not cut off the meander (see Fig. 8.8A). These **chute bars** form small channels in the point-bar sequence that fill with gravel and large-scale tabular cross-beds. The oxbow lakes fill with laminated clay, silt, and organic matter that settle out of suspension after the meander is abandoned. These form isolated **clay plugs** (see Fig. 8.9) in the sandy meander belt, which may block later erosion when the meander attempts to cut across them again.

The most striking difference between the meandering and braided fluvial systems is that the sandy deposits of the meandering river are narrow belts lying in a sequence of muddy floodplain deposits (see Fig. 8.9), whereas the braided system is much broader with relatively little mud or silt. The meandering channel is bordered by a **natural levee,** a low, wedge-shaped ridge that usually confines the flow within the channel (see Fig. 8.9). When floods top the banks, the levees are bypassed, only to be rebuilt as the water level drops. The rebuilt levees form long, sinuous, ribbonlike bodies of laminated muds and small ripple cross-beds. Because they are the only high features on the floodplain, they often become overgrown and may preserve root casts, soil horizons, and organic debris, as well as mudcracks and raindrop impressions.

BOX 8.3   PALEOCENE AND EOCENE FLOODPLAIN DEPOSITS OF WYOMING

One of the best exposed examples of ancient mean-dering fluvial deposits can be seen in the Powder River Basin of Wyoming (Ethridge, Jackson, and Youngberg, 1981). Here, the coal and uranium that accumulated in these deposits has prompted massive strip-mining operations. The mine walls provide clean, vertical cuts through the floodplain muds and channel sandstones, giving cross sections that look almost like idealized block diagrams. The Powder River Basin lies between the Black Hills of South Dakota on the east, the Bighorn Mountains of Wyoming on the west, and the Laramie Range on the south. These ranges were uplifted by the Laramide orogeny in the Paleocene and early Eocene, so the basin may have had much the same configuration then as it does now. The ancient drainages converged toward the center of the basin and then drained north into Montana, as they do today.

Most of the basin was apparently covered with low, swampy floodplains, on which both mudflats and coal-producing swamps were abundant. These environments produced the thick sequences of silt-stones and claystones of the Paleocene Fort Union Group and the Eocene Wasatch Formation. When the floodplain was covered by thick vegetation, low-grade lignite coal seams were produced. In some places, coal seams reach 10 m in thickness, and a few are almost 30 m thick. The great thickness, low sul-fur content, and proximity to the surface of these coal seams make this region one of the primary coal-mining areas in North America.

Punctuating the thick coal-bearing mudstone se-quence are the sandstones of the fluvial channels (Fig. 8.3.1A). Many of these channel sands occur as isolated lenses in the mudstones. Surrounding these channels are thin stringers of fine sandstone pro-duced by **crevasse splays** (Fig. 8.3.1B). These consist of fine-grained to very fine-grained sandstone with ripples, small-scale trough cross-bedding, parallel lamination, and scour-and-fill structures. The upper parts of the crevasse splays are often full of roots and burrows, and their upper portion merges into the swamp deposits.

In addition to the lenticular sandstones of the fluvial channels, there are thick tabular sand bodies that ap-pear to represent migrating point-bar sequences (Fig. 8.3.1C). These fine upward from medium-grained to very coarse-grained sandstone at the base to very fine-grained sandstones and siltstones at the top. Horizon-tal lamination and large-scale cross-beds are found

Figure 8.3.1 (A) Block diagram of Paleogene floodplain deposits, Powder River Basin, Wyoming. (B) Cross-bedded crevasse-splay sand sheets alternating with floodplain muds and coals. (C) Lateral accretion surfaces in a 5-m-thick point-bar channel sand, Bear Creek Uranium Mine. (D) Clay plug filling abandoned meander channel, approximately 8 m thick. (E) Freshwater mollusks in pond deposit. (B, C, and D courtesy of F. G. Ethridge. E by D. R. Prothero.)

near the bases; the cross-beds become progressively smaller in scale toward the top. The most characteris-tic diagnostic features of the point bar, however, are the large-scale lateral accretion surfaces, or epsilon cross-beds. In the Powder River Basin, these can be as much as 5 m thick.

As the point bars and meanders migrated, they became more sinuous until the meander was cut off, leaving abandoned channels (Fig. 8.3.1D) that became oxbow lakes. These eventually filled with fine-grained, parallel-laminated, organic-rich clay-stones and siltstones and eventually were buried and capped by laminated floodplain mudstones. Many of these lake deposits contain freshwater mollusks (Fig. 8.3.1E).

In summary, the early Tertiary rocks of the Powder River Basin show all the classic features expected of a meandering fluvial system: a predominance of flood-plain muds and silts punctuated by coal seams, lentic-ular sand channels with crevasse splays, tabular point-bar sand bodies with lateral accretion surfaces, and abandoned meander channels filled with laminated, organic-rich clays and silts.

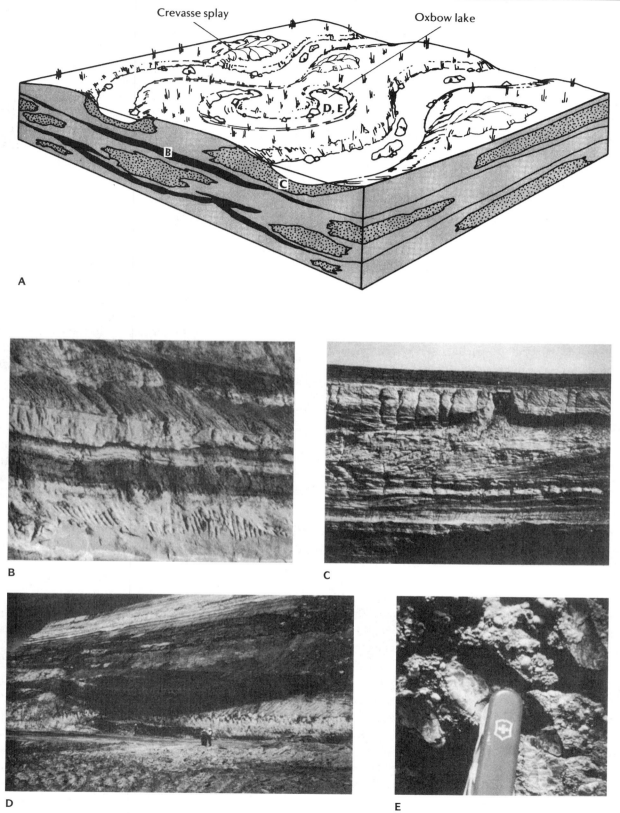

Crevasse splay

Oxbow lake

A

B

C

D

E

During rising water conditions, the levee is often breached, and the water spills out onto the floodplain, forming crevasse splay deposits (see Fig. 8.9). These are broad tongues of sandy and muddy sediment that radiate across the floodplain away from the channel. They show small-scale cross-beds and climbing ripple drift. As discussed in Chapter 4, climbing ripple drift indicates a rapidly aggrading condition with excess sediment load relative to the water flow. This is to be expected because the water velocity is greatly reduced after the water leaves the channel and spreads as a shallow sheet across the floodplain, finally coming to a stop. With the largest floods, the entire floodplain is covered with a few centimeters of fine silt and mud. These floodplain muds are uniformly laminated as they settle out of suspension. The floodplain is often marshy and can build up humus, soil horizons, and other features of vegetation. The natural levee, crevasse splay, and floodplain deposits are all examples of vertical accretion, in which each lamination represents a separate event and is synchronous. By contrast, the lateral accretion of the channel and point-bar sequence is time-transgressive.

Diagnostic Features of Meandering Fluvial Systems

Tectonic Setting Meandering systems are most often found in low parts of the craton. They can be preserved in downdropping basins or in aggrading coastal sequences. They are associated with floodplain muds and lake deposits. They grade downstream into the deltaic system and upstream into a braided system.

Geometry Channel sequences typically form long, ribbonlike bodies of sand ("shoestring sands") within a thick sequence of shales. Channel sands may be scattered randomly through the sequence, depending on where the channel shifts after avulsion (see Fig. 8.9).

Typical Sequence As in the braided system, there is fining upward from a basal channel-lag gravel to the sandy point-bar sequence of plane beds, trough crossbeds, and ripple drift (see Fig. 8.8A). Unlike the braided system, however, meandering systems have a much larger fine-grained component of laminated muds formed in the oxbow lakes, natural levees, crevasse splays, and floodplain.

Sedimentology Grain sizes range from channel-lag gravels to floodplain muds. Laterally accreted point-bar sands show decreasing-flow-velocity sedimentary structures: plane beds, trough cross-beds, and ripple cross-lamination. Floodplain muds are finely laminated and vertically accreted and may show climbing ripple drift, mudcracks, raindrop impressions, soil horizons, organic matter, and fossils.

Fossils Organic matter and fossil wood are common, particularly in the floodplain. Land vertebrates and invertebrates can occur in the floodplain muds or the channel sands. Freshwater mollusks are particularly diagnostic.

Lacustrine Deposits

A lake is a landlocked body of standing, nonmarine water, and **lacustrine** deposits are formed by ancient lakes. In terms of geologic time, most lakes are temporary features. They form in a basin or impoundment and become sediment traps that eventually fill up. Lakes can vary enormously in their dimensions, from small, ephemeral ponds to bodies of water as large as the Caspian Sea (372,000 km^2) and the Great Lakes (245,240 km^2). Some lakes are deep (Lake Baikal in Siberia is 1742 m deep), but some are so shallow that their levels fluctuate dramatically with season and climate. For example, Lake Eyre in Australia (9300 km^2 in area) is covered with water only a few times in a century and is a dry lake bed the rest of the time. About 60% of lakes are freshwater, but many are saltier than the ocean.

Ancient lacustrine deposits received relatively little attention in the past, but recently they have become important source rocks for oil shale, uranium, and coal. Many ancient lakes were significantly larger than all but the largest lakes today. For example, the Triassic lakes that produced the Popo Agie Formation of Wyoming, Utah, and Colorado had a surface area of more 130,000 km^2, and Pleistocene Lake Dieri in Australia covered more than 110,000 km^2. Lake deposits can also vary tremendously in thickness. Lake Eyre and its Pleistocene predecessors have accumulated only 20 m of sediment, but Devonian lacustrine sediments in Scotland are reported to be up to 4000 m thick. Lakes are most common in regions of internal drainage, where a closed basin accumulates water. Lakes are usually found in regions of tectonic depression such as rift grabens. They also form in volcanic calderas (for example, Crater Lake), glacial depressions (for example, the Great Lakes and most of the 10,000 lakes of Minnesota), karst sinkholes, and meteorite craters, or are impounded in river val-

leys behind glacial moraines, lava flows, alluvium, or landslide debris. For a lacustrine deposit to be preserved, however, it must accumulate a thick sequence that is subsequently buried. This happens most often in grabens and broad regional downwarps. Consequently, these areas contain most of the ancient lake deposits.

Because both lakes and shallow seas are large bodies of standing water, they produce fairly similar sedimentary sequences. In both cases, the dominant sediment is low-energy silt and mud with occasional carbonates. In most other respects, however, lake shales differ from marine shales. On average, lake basins are much smaller than epicontinental seas, so their deposits tend to be much less laterally continuous than marine shales and limestones. Along the shore of a lake, there is a rapid change in facies, interfingering with a narrow belt of fluvial deposits and even alluvial fans, which are less likely to occur along a marine coastline. Typically, lakes form a series of these facies belts arranged concentrically from the mudstones or marls in the center to the coarsest sandstones on the margins. Because most lakes fill with sediment over time, they tend to show a sequence that is shallowing and coarsening upward (Fig. 8.11).

Many ancient lakes, however, are large enough to produce facies patterns similar to those of marine sequences. Deep lakes that are affected by pulses of sedimentation during runoff peaks can have turbid-ity currents, which produce graded beds. The marginal fluvial sequences can prograde and fill in the lake, forming lacustrine deltaic deposits. Usually, however, lacustrine deltas and turbidites are of a much smaller scale than their marine analogs.

Lakes also have some properties that produce sediments seldom found in marine sequences. Most lakes have negligible tides; their waves are generally small, the shoreline belt and its associated sedimentary structures are limited, and the peculiar patterns of lake circulation have a much greater effect on the sediments than the wave-dominated flow seen in the marine environment. Typically, a lake has periods of overturn (such as spring and fall), when the entire lake circulates, and periods when the water is density stratified (such as summer and winter). This regular alternation of stagnation and overturn produces fine lamination that can be very rhythmic and laterally extensive. Because this lamination can reflect both seasonal cycles and larger-scale climatic cycles, it is of great importance as a paleoclimatic tool. In proglacial lakes, the cycles of stagnant frozen sediments in winter and oxygenated sediments in summer overturn produce striking glacial varves (see Fig. 4.1). During periods of stagnation in any density-stratified lake, organic matter becomes concentrated in the deeper water. The lack of oxygen from reduced circulation and from the excess nutrients leads to stagnant, reducing conditions. Under these circumstances, there will be a concentration of organic matter, producing black shales that are high in low-grade hydrocarbons, or **kerogens.** These are the oil shales that have received so much attention in recent years. Stagnant lake bottoms also discourage scavengers and the processes of decay and so can produce unusually complete fossils, such as the famous fossil fish of the Eocene Green River Formation (see Box 8.4).

Where the clastic input into a lake is limited, chemical sedimentation can predominate. Chemical precipitates are usually either saline or carbonate. Saline lakes are better known because they are abundant in the desert basins of the world (see Chapter 14). Wherever evaporation exceeds inflow, the salinity can exceed 35,000 ppm dissolved solids, producing a saline lake. (Normal marine salinity is about 33,000 to 37,000 ppm). The most abundant chemicals are SiO_2 and ions such as Ca^{2+}, Mg^{2+}, Na^+, K^+, HCO_3^-, CO_3^{2-}, SO_4^{2-}, and Cl^-. As evaporation proceeds and increases the ionic concentration, dense brines sink to the bottom and precipitate evaporite minerals. Carbonates are the first to be produced, followed by gypsum. If the process continues, halite is precipitated, followed by

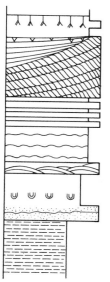

Rootlets

Desiccation cracks

Climbing ripples

Parallel-laminated sandstone and mudstone

Wave ripples

Hummocky cross-stratification

Burrows

Graded beds

Homogeneous mudstone

Figure 8.11 Ideal vertical sequence developed by lake margin regression due to fluvial influx. (After Galloway and Hobday, 1983.)

BOX 8.4   THE EOCENE GREEN RIVER FORMATION OF THE ROCKY MOUNTAINS

Figure 8.4.1 Classic exposures of extensive, finely laminated Green River lacustrine shales. View northward across Hells Hole Canyon, Uintah County, Utah. (W. Cashion, courtesy of U.S. Geological Survey.)

One of the largest and best-studied ancient lacustrine deposits in the world is the Eocene Green River Formation of Wyoming, Colorado, and Utah. It is one of the world's richest deposits of oil shale, as well as having economic amounts of trona and other evaporite minerals. It extends over 100,000 km^2 and reaches a maximum thickness of 3 km in Utah (Ryder, Fouch, and Elison, 1976). The Green River Formation was deposited in four separate basins: the Green River and Washakie basins in Wyoming, the Piceance Basin in Colorado, and the Uinta Basin of Utah. The Wyoming basins were filled by Eocene Lake Goshiute, and the southern basins contained Eocene Lake Uinta. Lake Goshiute was apparently a shallow playa during most of its history, whereas Lake Uinta was a deeper, perennial lake. Although their detailed histories are different and each basin has its own formal stratigraphic names, their overall histories are the same. Together, they spanned almost 10 million years of deposition in the region.

The predominant lithology of the Green River Formation is miles and miles of finely laminated shale (Fig. 8.4.1). The shales include marlstones, silty organic-poor dolomicrites, dolomitic claystones, volcanic tuffs, and oil shales. The oil shales contain much organic matter in the form of kerogen, making them valuable sources of hydrocarbons. The kerogen is associated with calcite layers and is believed to have accumulated during algal blooms. This would indicate quiet waters and depths of 5 to 30 m (Ryder, Fouch, and Elison, 1976). The shales show cyclic patterns of deposition that were probably caused by climatic fluctuations in lake level. These cycles reach 5 m in thickness and can be traced for 20 km without a change in thickness. These climatic fluctuations are believed to have operated on time scales of 20,000 to 50,000 years per cycle.

Lake Uinta was apparently a stratified, shallow carbonate- and organic-rich perennial lake (Desborough, 1978). Along its margins (Fig. 8.4.2A), the lake shales pass into the sandstones and siltstones of the streams

Figure 8.4.2 (A) Block diagram showing relationships of the facies of the Green River Formation. (B) Extraordinary preservation of freshwater fossils, including this frog, is common in the Green River Formation. (C) Mudcracked shales from the desiccated lake margins. (D) Freshwater algal stromatolites from the lake basin. (E) Intertonguing between lacustrine shales and fluvial sands of the Uinta Formation. (F) Finely varved oil shale, the most common rock type in the open lake basin. (G) Authigenic trona crystals in dolomitic marlstone of the saline mudflat. (B courtesy of L. Grande, E by D. R. Prothero, C, D, F, and G by W. H. Bradley, courtesy of U.S. Geological Survey.)

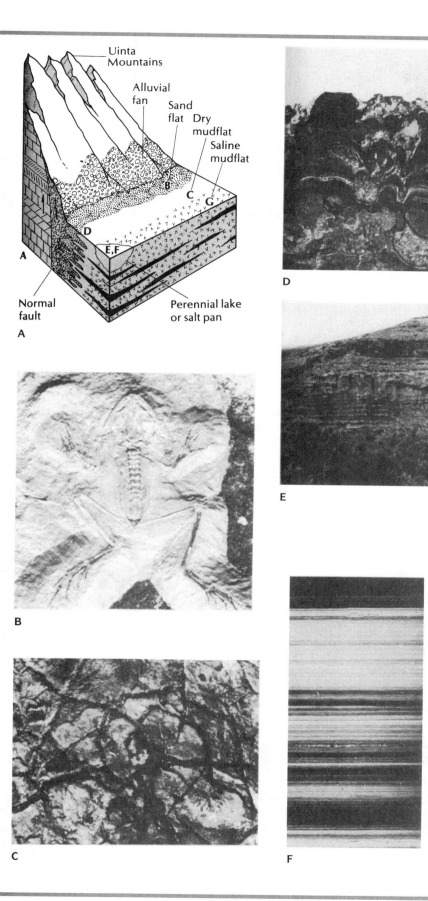

A

(*continued*)

(*Box 8.4 continued*)

and deltas along the lakeshore. In some places, the lake margin is dominated by carbonate tidal flats with stromatolites, oolitic and pisolitic grainstones, and green pyritic marlstone. Seasonal stratification was so strong that the deeper parts of the lake were reducing and anoxic, allowing the deposition of oil shale. By contrast, the surface waters must have been well oxygenated because the Green River Formation is famous for its well-preserved fossils of freshwater fish. These organisms and many birds, frogs, and other unlucky victims (as well as much plant debris) sank to the stagnant bottom, where there were no scavengers to disaggregate them. They were slowly buried by the fine clay, resulting in world-famous preservation (Fig. 8.4.2B).

Lake Goshiute, on the other hand, appears to have been considerably shallower and more ephemeral. During the deposition of the Wilkins Peak Member of the Green River Formation, this playa lake often dried up (Eugster and Hardie, 1975). The best indicators of this drying are thin-bedded dolomitic mudstones with abundant mudcracks (Fig. 8.4.2C), which show complex histories of opening and filling. These suggest an exposed playa covered by occasional sheet floods that deposited the fine laminae of silt and mud and filled older mudcracks. In some places, there are flat-pebble conglomerates made of ripped-up dolomitic mudchips. The mudstones are interbedded with wave-rippled calcarenites and oolites, suggesting shallow, agitated waters during wet episodes. Algal stromatolites occur along the lake margin

(Fig. 8.4.2D). At the edges of Lake Goshiute, the mudflat deposits are interbedded with massive and cross-bedded sandstones of the marginal alluvial fans and streams that encroached on the lake basin (Fig. 8.4.2E). Some of these sandstones are 10 m thick and can be traced for distances of 100 km. They probably represent braided streams that traversed the old lake floor during dry phases.

The center of Lake Goshiute was occupied by a lacustrine sequence that changed with lake level. During wet periods, the lake bottom was filled with stagnant, organic- and carbonate-rich water that produced oil shales and dolomitic mudstones similar to those of Lake Uinta (Fig. 8.4.2F). During periods of drying, the lake bottom became hypersaline, and thick (1 to 11 m) beds of trona and halite were precipitated (Fig. 8.4.2G). These conditions were similar to those found in modern soda lakes, such as Lake Magadi in Kenya (Eugster, 1970). Abundant volcanic ash contributed to the unusual alkali chemistry of this system.

Although the thick, extensive sequence of Green River shales might at first suggest a marine environment, there are clear indications that the shale was lacustrine. The sequence was formed in a landlocked basin far from the ocean and contains freshwater fish and ostracods, land plants, and many other diagnostic organisms. The abundance of mudcracks, flat-pebble conglomerates, and evaporites in Lake Goshiute also suggests an ephemeral lake. The often stagnant water, producing organic-rich oil shale, is unique to lacustrine environments.

a series of potassium and magnesium salts, many of which are unique to dry lakes. These can include a number of unusual hydrated sodium and calcium carbonates and sulfates (such as gaylussite, trona, glauberite, mirabilite, and thenardite), halides (halite and sylvite), and even borate minerals. Desiccation features, such as mudcracks, are commonly associated with the minerals found in the late stages of the drying of a lake.

Lacustrine carbonates, on the other hand, are produced where there is neither excess evaporation nor much clastic input. Unlike marine carbonates, freshwater limestones are produced mostly by inorganic precipitation. Fresh water typically contains abundant carbonate from dissolved atmospheric

CO_2 and from dissolved bedrock carbonate. The carbonate ion concentration is strongly controlled by changes in pH, which fluctuates continuously in freshwater lakes. The precipitation of calcite is facilitated by two factors: plants use carbon dioxide and raise the pH, and warmer temperature lowers the solubility of calcite. Lacustrine carbonates are usually low-magnesium calcite, precipitated in finely laminated beds with mudstone and marl. Occasionally, biogenic activity can produce freshwater limestone. Freshwater ostracods are the most common, but freshwater gastropods and bivalves are also important. Freshwater calcareous algae form a kind of stromatolite called an **oncolite**, which is a subspherical or tabular body formed when encrust-

ing algae trap sediment (see Fig. 11.3D). Some reach tens of centimeters in diameter and are known as **algal biscuits.**

These freshwater fossils, though rare, are important as environmental indicators. In many cases, lacustrine and marine deposits are difficult to tell apart, but the absence of normally abundant marine invertebrate fossils or the presence of certain freshwater invertebrates is often the best criterion. In addition, the scale and geometry, the rapid facies changes, and the tectonic setting can all yield clues to distinguish marine from lacustrine deposits.

Diagnostic Features of Lacustrine Deposits

Tectonic Setting Lacustrine deposits are typically found in fault grabens or broadly downwarped basins with internal drainage or limited outflow. They are associated with other nonmarine environments, particularly fluvial sands and alluvial fans.

Geometry Lacustrine deposits are circular or elongate in plan view and lenticular in cross section. Their area ranges from a few meters to 100,000 km^2; such deposits are usually thin (less than 200 m thick), but sequences can exceed 1000 m in thickness.

Typical Sequence As the lake dries up, there is a coarsening upward from laminated shales, marls, and limestones to rippled and cross-bedded sandstones and, possibly, conglomerates. The sequence often shows cyclic alternation of laminae.

Sedimentology There are finely laminated mudstones, commonly rich in kerogen, along with marls and freshwater limestones. The margins of the lake have fluvial muds and sands, with cross-beds, wave ripples, and climbing ripple drift. Thin turbidite sequences can form in the lake basin, along with black shales. Hypersaline lakes form a regular sequence of carbonate, gypsum, halite, and other evaporites, often associated with desiccation features such as mudcracks.

Fossils Fossils are important for distinguishing lacustrine from marine sequences. The complete absence of normal marine invertebrates is significant, as is the presence of nonmarine ostracods, diatoms, mollusks, and freshwater fish and insects. Preservation is excellent in some anoxic lake basins.

Eolian Deposits

Eolian (wind-formed) sand dunes are among the best-known depositional environments because of their spectacular cross-beds (see Box 8.5) and their use as paleoclimatic indicators. Sand dunes are best known from the desert environment, although they do occur in coastal settings (discussed in Chapter 9). Most sand deserts occur in a narrow belt between 20° and 30° latitude in areas of persistent high pressure or behind mountain ranges that have a rain shadow. Thus, they have been used as indicators of paleolatitude, as well as being clear signs of arid climate and strong prevailing winds.

Because air is only about one-thousandth as dense as water, eolian processes are very different from subaqueous processes. Wind has much less competence than water to pick up particles, so the coarsest material is left behind as a **deflation lag,** or **desert pavement,** which protects underlying fine material from being eroded. Most of the sand-sized and finer particles are carried along in a traction carpet moving just above the surface, and the individual grains move by saltation and intergranular collision. As a result, eolian sands tend to be very well sorted and well rounded, with microscopic surface pitting and frosting. The silt and clay are carried away by the wind and deposited where the wind diminishes. Most dune sands are nearly pure quartz, although there are unusual dunes, such as the gypsum sands of White Sands, New Mexico.

The most characteristic feature of eolian dunes is their enormous cross-beds. Unlike water, there is no depth limitation on the "sea of air," so dune height is limited only by the strength of the wind and the supply of sand. Individual cross-bed sets can reach 30 to 35 m in thickness, with foresets that dip 20° to 35° (Fig. 8.12). This thickness of individual sets is found throughout the entire unit; it is not just an unusual maximum. Eolian cross-beds have long, sweeping sigmoid shapes with longer, more asymptotic bottomsets than those of marine dunes. Geometrically, most cross-beds are planar-tabular or trough- and wedge-shaped. Detailed study of the faces of eolian dunes (Fig. 8.13) shows that the ripples move laterally across the face of the dune. This suggests that much of the cross-bed is formed not by sand avalanching down the face of the dune but by sand blown horizontally across its lee face.

In addition to the abundance of large-scale cross-beds, low-amplitude wind ripples typically form a

BOX 8.5 JURASSIC DUNES OF THE NAVAJO SANDSTONE, UTAH AND ARIZONA

One of the most photogenic examples of ancient cross-bedded dune sands is the famous Navajo Sandstone of the Utah-Arizona border region. This thick sandstone makes up the spectacular white cliffs of Zion National Park in Utah, and the combination of fractures and cross-beds gives "Checkerboard Mesa" its distinctive surface pattern. In northeastern Arizona, the Navajo Sandstone is about 300 m thick, and it reaches 600 m thick in southwestern Utah (Harshbarger, Repenning, and Irwin, 1957). Its correlative, the Aztec Sandstone in the Mojave Desert of California and Nevada, is more than 900 m thick. This sandstone also extends to southwestern Wyoming, where it is known as the Nugget Sandstone and is about 150 m thick.

Along its eastern and southeastern margins in Colorado and northern Arizona, the Navajo Sandstone thins to a wedge and then vanishes. This enormous sand sea must have extended 1000 km north-south and 400 km east-west, comparable to a small Sahara (Fig. 8.5.1).

The Navajo Sandstone consists of homogeneous, well-rounded, well-sorted, frosted quartz sand with an average grain diameter of 0.2 mm. Its most striking feature is the well-developed cross-stratification. Both wedge-planar (see Fig. 8.12) and tabular-planar (Fig. 8.5.2) cross-beds are well developed. Individual cross-bed foresets tend to be long (15 m) and sweeping, with concave-upward surfaces. Dips are 20° or greater. Individual cross-bed sets are typically 5 to

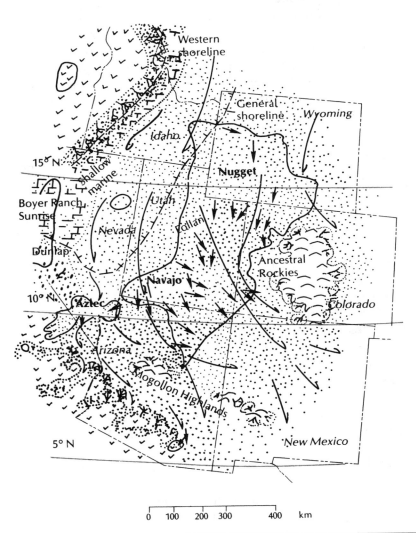

Figure 8.5.1 Paleogeographic reconstruction of environmental conditions during the deposition of the Navajo-Nugget-Aztec Sandstones. (After Kocurek and Dott, 1983:109.)

Figure 8.5.2 Tabular-planar cross-beds in the Navajo Sandstone, north of Kanab, Utah. (E. Tad Nichols, courtesy of U.S. Geological Survey.)

10 m thick and in some places reach almost 35 m in thickness.

Ripple marks are relatively rare in the Navajo Sandstone. Where they occur, their crests are parallel to the crests of the dunes. Only ripple marks that form from crosswinds on the lee sides of the dunes have a good chance of burial and preservation. Fossils also are rare in the Navajo dune sands. Ostracods and other freshwater crustaceans have been reported from interdune areas, and a few dinosaur and pterosaur tracks (Fig. 8.5.3A) are known, as well as the skeletons of bipedal dinosaurs and early mammals (Fig. 8.5.3B). More common are contorted beds and other slump structures. These kinds of folds are often seen in modern dunes when the sands have been wetted; they are usually caused by oversteepening on the lee side of the dune.

Compared with the spectacular cross-bedded sand dunes, the interdune deposits have received little attention. The Navajo Sandstone and the underlying Kayenta Formation, which interfingers with the Navajo and contains eolian sands throughout, contain many facies that are interpreted as interdune deposits (Middleton and Blakey, 1983). In many places,

the thick sets of eolian cross-bedded sandstones are separated by thin (less than 1 m) horizontal planar-laminated sandstones. These appear to represent thin sheets of interdune sand deposited on an erosional surface between episodes of dune migration.

In the intertonguing interval are thicker sequences of planar-laminated, fine-grained sandstone and siltstone. These occur as beds 2 to 6 m in thickness, sandwiched between eolian cross-bedded sandstones. Thin laminae of very coarse-grained sandstone and granules also occur in this facies. These beds are believed to be the product of interdune ponds and mudflats. In a few places, sand-filled V-shaped desiccation cracks can be found, which indicate alternate wetting and drying of these mudflats. Occasionally, the beds are capped by a calcareous horizon that probably represents a caliche. They also include cryptalgal laminated carbonates with ostracods and mudcracks that were probably formed in small calcareous ponds.

In summary, many lines of evidence in the Navajo Sandstone clearly point to an eolian origin and rule out formation by tidal sand ridges. The enormous thickness, abundance, and steepness of the cross-bedded sands, which are tabular-planar and wedge-

(continued)

(Box 8.5 continued)

Figure 8.5.3 Pterodactyl trackway (*top*) up the dune face in the Navajo Sandstone. (Courtesy of J. Madsen.) Skeleton of a synapsid (formerly called a "mammal-like reptile") (*bottom*) from the Navajo Sandstone, Navajo National Monument. (Courtesy of Shuler Museum of Paleontology, Southern Methodist University.)

planar cross-stratified, clearly point to a dune origin. The high angle and the size of the dune foresets are particularly characteristic of desert sand dunes. In addition, the rapid directional fluctuations of the cross-bed sets would only be expected with rapidly fluctuating wind directions. The mudcracked inter-dune deposits indicate subaerial exposure. Finally, the presence of dinosaurs and their tracks and of freshwater ostracods rules out a marine origin.

small-scale cross-lamination within the larger cross-bed sets. These usually migrate up and along the lee face of a dune; this migration does not occur in marine sands. Wind ripples are typically lower in amplitude and more asymmetrical than water ripples. Other types of deposits can be found between the dune fields. Thin mudstones with mudcracks and raindrop impressions are the remains of ephemeral lakes between dune fields. Occasionally, coarse defla-tion lags from blowout areas are also found interbed-ded with dune deposits. A typical eolian sequence (Fig. 8.14) is dominated by thick dune cross-beds and

A

B

Figure 8.12 (A) Large-scale wedge-planar cross-beds (middle) between tabular cross-beds in the Navajo Sandstone, Zion National Park, Utah. (E. Tad Nichols, courtesy of U.S. Geological Survey.) (B) Steep, sharply truncated cross-bedding is typical of modern sand dunes; this example is from Great Sand Dunes National Monument, Colorado. (Pettijohn and Potter, 1964: plate 32A; by permission of Springer-Verlag, New York, courtesy of G. P. Merk.)

lesser amounts of deflation gravels, mudcracked pond shales, and small-scale cross-lamination. Unlike most other depositional sequences, however, eolian deposits have no regular progression of sedimentary structures. Dune migration and ephemeral lake formation are much more random processes than the orderly sequence of facies belts found in most sedimentary environments.

Recently, some geologists (reviewed by Walker, 1984) have suggested that many examples of classic eolian cross-bedded sandstone were actually formed by large-scale marine sand waves and tidal ridges (discussed in Chapter 10). Walker and others have shown convincingly, however, that tidal sand waves have dips of no more than 6°, in contrast to the much steeper dips found in eolian dunes.

In addition, the other evidence—mudcracked layers; the absence of marine faunas and the presence of terrestrial trackways; tens to hundreds of meters of monotonous, thickly cross-bedded sands; and association with other terrestrial facies—argues strongly against a marine origin for these extensive, steeply cross-bedded sandstones.

Diagnostic Features of Eolian Deposits

Tectonic Setting   Eolian deposits are found in inland basins at latitudes between 10° and 30° or behind mountains in rain shadows. They are commonly associated with alluvial fans, playa lakes, and other desert deposits.

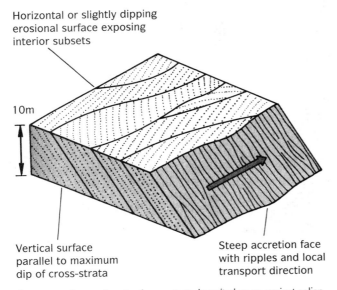

Horizontal or slightly dipping
erosional surface exposing
interior subsets

10m

Vertical surface
parallel to maximum
dip of cross-strata

Steep accretion face
with ripples and local
transport direction

Figure 8.13 Composite sets of cross-strata deposited on an ancient eolian dune complex. Sand drifted by the wind along the dune slopes deposited the low crests of eolian ripples on their surfaces. (After Harms, 1979: 242 © 1979 by Annual Reviews, Inc.)

Geometry   Dune fields can cover hundreds of square kilometers and form thick tabular bodies with individual sets up to 35 m thick.

Typical Sequence   The sequence is dominated by enormous cross-bed sets that are meters in thickness with foreset dips of 25° to 30°; lesser amounts of mud-cracked shales and deflation-lag gravels are present. The sequence of sedimentary structures is random.

Sedimentology   Extremely well sorted, well rounded, quartz-rich sand with no fine matrix and few coarse gravel lags is found. Under the SEM, grain surfaces are typically frosted or pitted and are highly oxidized and coated with iron oxides. Large-scale cross-beds composed of smaller-scale low-amplitude wind ripples form cross-lamination that moves along or up the lee face of the dune.

Fossils   Rare vertebrate footprints, root casts, and insect burrows are all typical.

Glacial Deposits

Although they occur only in limited regions of the Earth (primarily at high altitudes or high latitudes), and only during times when the Earth has cold, ice-house climates (as it has for the past 33 million years), glacial deposits are nonetheless very important. Today only 10 percent of Earth's surface is covered by glaciers and glacial deposits, but during the great glacial episodes, as much as 30% of the surface was glaciated. In areas that were once overrun by the great Pleistocene ice sheets (such as New England, New York, the northern Midwest and Great Plains, and

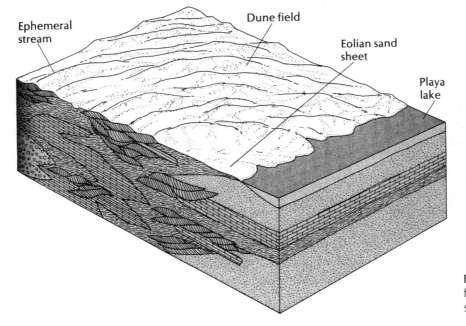

Ephemeral
stream

Dune field

Eolian sand
sheet

Playa
lake

Dune — Interdune surface
Alluvial fan
Nearshore
Fluvial
Playa and sabkha

Figure 8.14   This block diagram shows lateral facies relationships of dune and interdune systems. (After Lupe and Alhbrandt, 1979: 246.)

most of Canada and Alaska), most of the bedrock is overlain by thick glacial deposits or scoured by the grinding action of glaciers. The same is true of the northern parts of Eurasia and high alpine regions around the world. Before the icehouse conditions that began during the Oligocene, there were other important episodes of glaciation: the great Gondwana glaciers of the Pennsylvanian and Permian and the Gondwana ice sheets of the Ordovician. Other important episodes include the great "Snowball Earth" glaciations of the early Proterozoic (Huronian, around 2 billion years ago) and late Proterozoic (Varangian, about 700 million years ago). Glacial deposits interbedded with marine carbonates prove that there were sea-level glaciers at the equator during the Snowball Earth glaciations and that the Earth nearly froze over (Prothero and Dott, 2003). All these great glacial episodes have similar facies associations and characteristics that are consistent over more than 2 billion years, helping us to recognize them as important indicators of climate.

Although glacial environments have been studied primarily by geomorphologists and Quaternary geologists, rather than by sedimentary and economic geologists, they deserve just as much attention as other sedimentary environments. Glacial deposits have many practical implications for environmental and economic geologists. In many places in the world, the major aquifers of a region are glacial sands, so hydrogeologists must understand glacial processes to predict the behavior of groundwater. Modern glacial deposits may not be important oil reservoirs, but many of the late Paleozoic basins of Australia, Argentina, Brazil, Bolivia, Saudi Arabia, Jordan, and Oman contain important oil reserves in ancient glacial sediments (Levell et al., 1988; Franca and Potter, 1991). In addition, glacial sediments are important placer deposits of many metallic ores, such as gold and copper (Eyles and Eyles, 1992). Some geologists use the presence of distinctive clasts of precious metal in glacial tills to track down their sources and uncover previously unsuspected ore bodies. For example, many of the till deposits in the central Midwest (Ohio, Indiana, Illinois, Iowa) have large pebbles of native copper that can be traced back to their source area, the rich copper deposits of the Upper Peninsula of Michigan. In northern Canada, the presence of uranium ore in many glacial deposits led to the discovery of important uranium deposits beneath the tundra. Thus, glacial deposits can be of great economic significance as well as important indicators of ancient climate.

Glacial sedimentary environments have many of the characteristics of some of the other terrestrial facies (for example, glacial outwash streams are choked with excess sediment, so they are heavily braided), but they also have distinctive deposits found only in the glacial realm. The primary reason for this distinctiveness is the nature of the transporting agent, the glacier itself. A glacier is a huge, slowly flowing river of ice and is capable of feats of erosion and transport that no body of liquid water can match. Glaciers scour deep into the bedrock over which they slide, carving steep-sided U-shaped valleys in mountainous areas and sculpting vertical concave surfaces known as **cirques** and **arêtes** on the mountainside. As they erode, they carry everything from huge boulders down to fine, powdery, ground-up rock known as **glacial flour,** all indiscriminately mixed together within the ice. The large rocks at the base of the glacier are like teeth on a rasp, grinding down the bedrock and carving long, straight glacial striations in the direction of ice flow. Thus, the erosive power and the sediment transport capacity of a glacier is unlike that of any other erosional agent on Earth's surface.

These same characteristics mark the deposits that glaciers leave behind. Glaciers can carry everything from boulders to clay, and they tend to drop it all in a mass of unsorted debris known as glacial till (see Fig. 5.7B). As described in Chapter 5, a till (or tillite) is primarily defined by its poor sorting, with a huge range of grain sizes. Tills and tillites are also recognized by their lack of internal stratification. The clasts are dropped by melting ice and are usually not sorted by any weaker current of liquid water that would separate the clays from coarser material. However, not every body of poorly sorted, unstratified paraconglomerate is automatically a glacial till. Gravity-driven processes, such as debris flows, can deposit similar types of sediment (see Chapter 5). The word **diamictite** or **diamict** (Greek for "thoroughly mixed") is a descriptive nongenetic term used for rocks that have the poor sorting and lack of stratification often found in tills. The term *tilloid*, on the other hand, describes a rock that has till-like characteristics but is now thought to be of debris-flow origin.

So how does one decide whether a diamictite is a true till or a debris-flow tilloid? The best clues come not from the diamictite itself but from the related facies interfingered with the diamictite and overlying and underlying it. If it is associated with other glacial facies (described below), it is probably a true glacial till. If it is associated with other gravity deposits such as turbidites, then it is probably a tilloid. Another

Figure 8.15 The Great Nisqually Glacier forms a long valley on the south flank of Mount Rainier, Washington, although it has been retreating in the past century due to global warming. (A) The present glacier, with large morainal tills filling most of the valley it once occupied. (B) Farther down the valley are outwash boulders and gravels and a braided outwash stream with chalky white water. The color of the water is a result of glacial flour. (D. R. Prothero.)

diagnostic clue is the presence of linear striations on the larger clasts, which only a glacier can produce (see Fig. 5.2B).

Glacial tills are dropped by the snout of a melting glacier as it retreats, often piling into huge ridges and mounds known as **moraines. Terminal moraines** pile up at the farthest advance of a glacier. As it pauses on its retreat, the glacier may deposit additional **recessional moraines** in the glacial valley (Fig. 8.15A). Glacial debris also accumulates on the flank of the ice to form **lateral moraines.** Where two or more glaciers merge, the lateral moraines between them combine to form a **medial moraine** running down the length of the glacier.

In a typical glacial facies sequence, these moraines occur at the base of the sedimentary column and are known as a **lodgment till** (Fig. 8.16). In the area nearest the glacier (the **lodgment till zone**), the only deposits will be the tills from the many moraines. However, the retreating snout of the glacier produces many other facies, in association with the moraines, that are sorted by wind or water rather than dumped by ice. The melting ice flows down a series of proglacial **outwash streams** that carry the sand and glacial flour but leave behind gravel bars of glacial cobbles and boulders (see Fig. 8.15A, B). Typically, the water in these outwash streams is milky white from all the fine glacial flour that is being carried along (see Fig. 8.15B). Because outwash streams are choked with debris and are carrying a much greater sediment load than their flow can handle, they are always braided (see Figs. 8.6 and 8.15B). They leave behind braided gravel bars and highly cross-bedded glacial outwash sands, often deposited directly on top of the lodgment till (see Fig. 8.16).

The moraines also make excellent dams, which trap meltwater pouring out of the glacier and cause it to form ponds and **proglacial lakes.** Such lakes have very unusual characteristics: their primary sediment supply is fine clays and glacial flour settling from the

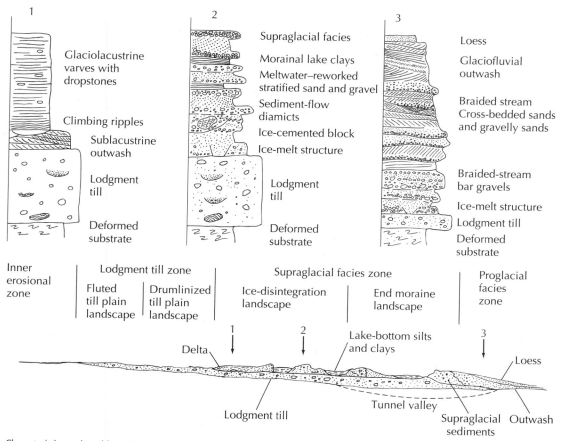

Figure 8.16 Characteristic stratigraphic sections and cross-sectional profile through a terrestrial glacial deposit from the glacier itself (*left*) to the proglacial zone beyond the terminal moraine (*right*). (After Sugden and John, 1976.)

glacial outwash, and they are prone to freezing over each winter, then thawing in summer. Consequently, they may go through cycles of oxidation (producing light-colored silts and clays) during the summer. When they freeze over in the winter, they trap reduced organic matter, generating dark organic-rich laminae. Thus, proglacial lake sediments are typically varved (see Fig. 4.1), with fine centimeter- or millimeter-scale lamination; these laminae have often been used to study climatic cycles. Another characteristic of proglacial lake varves is that the quiet deposition of silts and glacial flours can be interrupted by larger dropstones (see Fig. 5.7) that melt out of floating chunks of ice.

As one progresses farther down from the glacier, the lodgment till sheet becomes thinner and thinner and is covered mostly by glacial outwash sands and gravels (see Fig. 8.16). Finally, at some distance from the active ice sheet, there may be windblown deposits of glacial silt and clay known as glacial **loess** (pronounced "lurss"). In many regions—such as the Great Plains of North America, the Polish and Ukrainian plains of Europe, and the plains of central Asia and China—thick deposits of loess cover huge areas of land. These fine silts become rich soils and are responsible for the ability of these "breadbaskets" to produce most of the world's major grain crops. Without the legacy of glacial loess, the world would have starved to death a long time ago.

Diagnostic Features of Glacial Deposits

Tectonic Setting Glacial deposits can occur in any tectonic setting that produces high mountains and alpine glaciers. During periods of global glaciation, they can cover large areas of continents (or entire continents, such as Antarctica), regardless of tectonic setting.

Geometry Mountain glacial deposits occur in narrow valley-fill sequences, but continental glacial deposits are sheetlike and can cover hundreds of square kilometers.

Typical Sequence Most glacial deposits have a lodgment till at the base, which may be overlain by braided gravel bars or cross-bedded sands and gravels from the outwash stream, or possibly by varved lacustrine mudstones, which may contain dropstones.

Sedimentology Glacial tills, with their lack of sorting and stratification, are the most diagnostic deposit of this facies. Glacial varves are also unique to the glacial sedimentary environment. However, gravity-driven debris flows may closely mimic tills, and nonglacial lakes can produce sediments that resemble

varves, so care must be taken to be sure that such deposits are truly of glacial origin.

Fossils The bones of animals that lived near glaciers are commonly buried in the till or in the varved lake deposits. If the animals die and are buried in quiet lake sediments, their skeletons may be complete and articulated; if the bones are ground along in the glacier or rolled around by the outwash gravels, they tend to be broken and abraded. In many glacial lakes, leaves and freshwater fossils such as diatoms and ostracodes can be found in the fine sediments.

CONCLUSIONS

Nonmarine sedimentary deposits such as those formed by ancient alluvial fans, rivers, lakes, deserts, and glaciers are important indicators of ancient climate and paleogeography and are the source of much of the world's coal, oil, and uranium. Although they are the most familiar environments from personal ex-

perience, they are also relatively rare in the stratigraphic record, because they are also above base level and subject to erosion. Only when they occur in rapidly subsiding basins do they have much chance of preservation.

FOR FURTHER READING

Bridge, J. S. 2003. *Rivers and Floodplains: Forms, Processes, and Sedimentary Record.* New York: Blackwell Science.

Davis, R. A., Jr. 1992. *Depositional Systems,* 2d ed. New York: Prentice-Hall.

Ethridge, F. G., and R. M. Flores, eds. 1981. *Recent and Ancient Nonmarine Depositional Environments: Models for Exploration.* SEPM Special Publication 31.

Galloway, W. A., and D. K. Hobday. 1996. *Terrigenous Clastic Depositional Systems,* 2d ed. New York: Springer-Verlag.

Reading, H. G., ed. 1996. *Sedimentary Environments and Facies,* 3d ed. Oxford: Blackwell Scientific Publications.

Reineck, H. E., and I. B. Singh. 1973. *Depositional Sedimentary Environments.* New York: Springer-Verlag.

Rigby, J. K., and W. K. Hamblin, eds. 1972. *Recognition of Ancient Sedimentary Environments.* SEPM Special Publication 16.

Scholle, P. A., and D. Spearing, eds. 1982. *Sandstone Depositional Environments.* Amer. Assoc. Petrol. Geol. Memoir 31.

Selley, R. C. 1978. *Ancient Sedimentary Environments,* 2d ed. Ithaca, N.Y.: Cornell University Press.

Spearing, D. 1974. *Summary Sheets of Sedimentary Deposits.* Geological Society of America Maps and Charts Series MC-8.

Walker, R. G., and N. P. James, eds. 1992. *Facies Models: Response to Sea Level Change,* 3d ed. Geoscience Canada Reprint Series 1. Toronto: Geological Association of Canada.

9 Coastal Environments

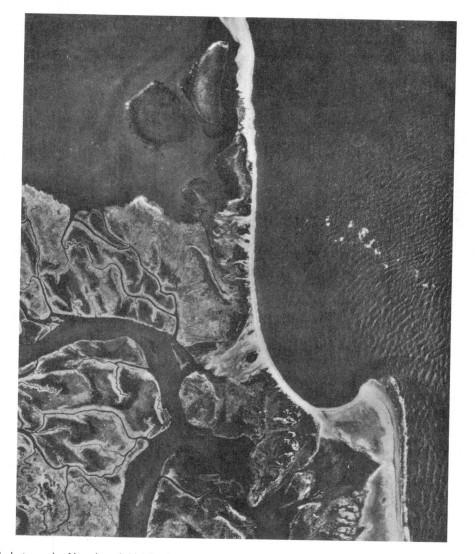

Aerial photograph of beach and tidal flat deposits, Little Egg Inlet, near Atlantic City, New Jersey. (Courtesy of U.S. Geological Survey.)

THE BOUNDARY BETWEEN MARINE AND NONMARINE ROCKS is usually a gradual transition. Meandering rivers begin to show the influence of the ocean in their deltas or tidal estuaries. Barrier islands, though fully marine, are formed from terrestrial sediments and can have subaerial deposits on them, such as sand dunes. In fact, there are a number of transitional environments that are neither fully marine nor fully nonmarine: the delta, peritidal flats, lagoon, marshes, and barrier island complex together form a group of closely associated facies, which we will discuss in this chapter.

Deltas

Deltas form where rivers carry more sediment into the sea than marine erosion can carry away. The Greek historian Herodotus (c. 490 B.C.) first used this name for the deposits at the mouth of the Nile, which

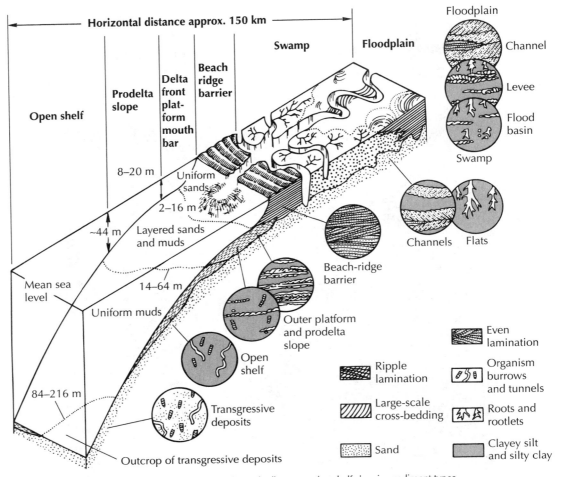

Figure 9.1 Typical cross section across the Niger Delta and adjacent marine shelf showing sediment types, morphology, and distribution. (After Allen, 1970: 149.)

resemble the Greek letter *delta* (Δ) on a map. Deltas seldom form on active, subducting continental margins because there is no stable shallow shelf on which sediments can accumulate. In addition, active margins usually have coastal mountain ranges that limit the size of the river basin that provides the deltaic sediment.

Geologists have devoted much research and study to deltas over the past few decades because they are important host rocks for coal and petroleum. Also, because deltas **prograde** (build out) from the margins of basins and often fill them over geologic time, they have great stratigraphic significance. Deltas make up a large part of many basin-fill sequences (see Box 9.1).

Deltas are influenced by a complex combination of fluvial and marine processes; each delta has more than a dozen distinct environments of deposition. These environments can be grouped into three broad divisions: the **delta plain,** with its meandering floodplains, swamps, and beach complex; the steeper **delta front;** and the broadly sloping **prodelta,** which grades

into the open shelf (Fig. 9.1). Because deltas are constantly supplied with sediment from their river basins and so prograde outward, the time planes run parallel to the sloping depositional surface (Fig. 9.2). Deltaic

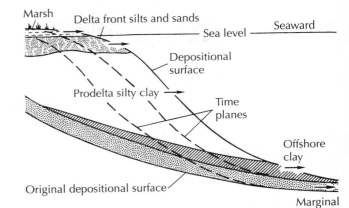

Figure 9.2 The constructional phase in a delta cycle; time planes parallel the depositional surface. (After Shepard, Phleger, and van Andel, 25; 1960: by permission of the American Association of Petroleum Geologists, Tulsa, Okla.)

sands and muds, like point-bar sequences, are sedimentary bodies that build by lateral accretion and are therefore time-transgressive.

The shape of a delta is not always the triangle that suggested the name to Herodotus. Delta shape is influenced by sediment input, wave energy, and tidal energy. A *river-dominated delta* has a large volume of sediment and tends to be lobate when there is a moderate sediment supply (Fig. 9.3A) and elongate when the sediment supply is large (Fig. 9.3B). If the sediment supply cannot keep up with the erosive powers of tides, the delta tends to be very small. A *tide-dominated delta* (Fig. 9.3C) has many linear channels parallel to the tidal flow and perpendicular to the shore. A *wave-dominated delta* (Fig. 9.3D) is smoothly arcuate; the wave action reworks the sediment and make such deltas much sandier than other types of deltas.

River-dominated deltas are lobate because of a unique hydrodynamic interaction between river water and seawater. There is a sharp contrast in water density where the less dense fresh water of the river enters the denser saline seawater. As a consequence, the river water forms a plane jet that spreads out and forms a layer over the seawater (Fig. 9.4). Be-

cause there is only limited mixing at the margins of the jet, there is relatively little frictional inertia. The jet often goes a long way seaward before it gradually begins to break up and mix with the seawater. The sand carried in the stream is deposited along the sides of the jet in subaqueous levees, where friction and mixing slow the flow. Farther offshore, where friction and spreading begin to slow the jet, sediment is dropped in **distributary mouth bars** (Figs. 9.4 and 9.5). The finest suspended matter is carried even farther, eventually settling out along the prodelta slope.

The hydrodynamics of the river mouth not only influence the map view of the delta shape but also produce variations in its cross-sectional geometry. Most of the delta complex is composed of fine silts and clays that settle out of suspension in the prodelta, delta front, or lagoon complexes. Scattered through these fine sediments are sandy wedges and tongues that are formed by the various distributaries of the river. These distributary mouth bars tend to build narrow lenses of sand and then to abandon them as the river switches to a new distributary. A typical prograding deltaic sequence is a complex of discontinuous sandy lenses scattered through the silty-

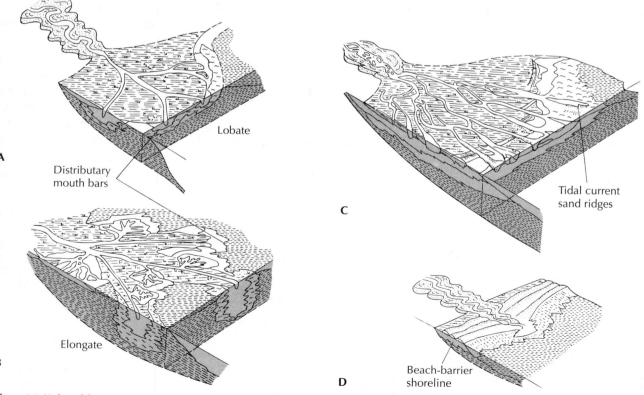

Figure 9.3 Various delta types: (A) Lobate and (B) elongate are river-dominated; (C) is tide-dominated; (D) is wave-dominated. (After Reading, 1986: 117: by permission of Blackwell Scientific Publishers.)

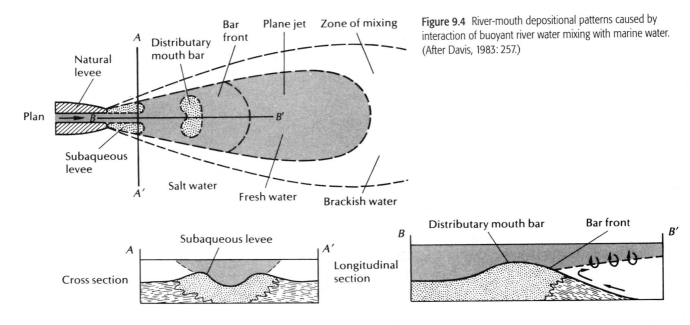

Figure 9.4 River-mouth depositional patterns caused by interaction of buoyant river water mixing with marine water. (After Davis, 1983: 257.)

muddy wedge (Fig. 9.6). In cross section, examples of ancient prograding deltas have complex geometries of levee muds and silts, lenticular distributary mouth bar sands, and frequent cross-cutting channels (Fig. 9.7). The distributary channel sands are abundantly cross-bedded, with plenty of ripple cross-lamination, scour-and-fill structures, and discontinuous clay lenses. The distributary mouth bar sands are even more complexly cross-stratified because of the complex current systems that pass over them; they may show the effects of both wave and current ripples. Wood, debris, and other organic matter carried down the river during floods end up in the distributary mouth bar. Sometimes decaying material produces gases that deform the overlying beds in gas-heave structures.

Between the distributaries on the delta plain are wide, shallow **interdistributary bays and marshes.** Like the floodplains of the meandering river, they are low, marshy mudflats cut by small, narrow channels of slow-moving water (Fig. 9.8). Because there are no fast-moving currents, most of the sediment is finely laminated silt and clay that settles out of suspension, usually after a flood. The abundant vegetation on these mudflats is the source of organic matter that can become peat or coal. The muds are also full of root casts and burrows, and there may be shell debris in the tidal channels. Much of the interdistributary sequence is built of sand sheets from crevasse splays deposited when floods breach the levees that protect the marsh.

At the outer limits of the distributary mouth bar is the distal bar, which occupies much of the upper foreset part of the delta front. Because it is at the fringes of the plane jet of river water, it shows interbedded layers of fine sand and mud (see Fig. 9.1). There are also cross-beds, cut-and-fill structures, and ripple marks from both the fluvial and the ma-

Figure 9.5 The Pass-a-Loutre distributary of the Mississippi, looking west, on January 2, 1984. Strong offshore winds pushed water away to expose the crest of the distributary mouth bar, which is normally submerged. (Courtesy of D. Nummedal.)

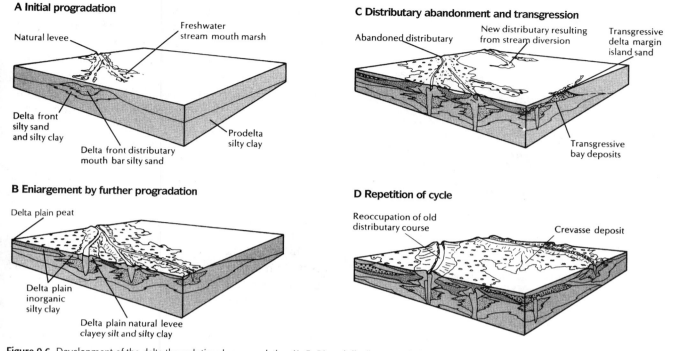

A Initial progradation

Natural levee

Freshwater
stream mouth marsh

Delta front
silty sand
and silty clay

Delta front distributary
mouth bar silty sand

Prodelta
silty clay

B Enlargement by further progradation

Delta plain peat

Delta plain
inorganic
silty clay

Delta plain natural levee
clayey silt and silty clay

C Distributary abandonment and transgression

Abandoned distributary

New distributary resulting
from stream diversion

Transgressive
delta margin
island sand

Transgressive
bay deposits

D Repetition of cycle

Reoccupation of old
distributary course

Crevasse deposit

Figure 9.6 Development of the delta through time by progradation (A, B, D) and distributary switching (C). (After Davis, 1983: 276.)

rine currents. The distal bar is within normal marine salinity but is very shallow; this makes it rich in life, with shell beds, burrows, and bioturbation. The distal bar often has a "stairstep" topography caused by rotated slump blocks of sand and mud; these **growth faults** are caused by the instability of dense sand lying over less dense mud. The delta front can be unstable because of oversteepening as well as over-

loading, so slumps, slides, and convolute bedding are very common.

The remaining slope and bottomset beds of the delta are composed of prodelta clays and silts. These sediments become progressively finer away from the delta and are usually finely laminated unless bioturbation is extensive. When floods enlarge the plane jet, they deposit an occasional sandy layer on the

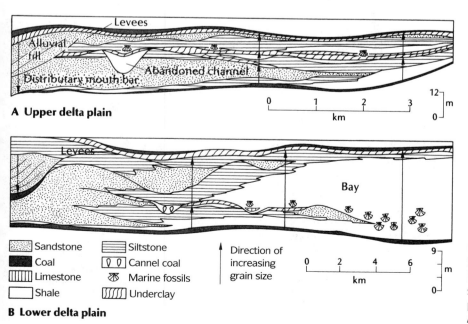

Levees

Alluvial
fill

Abandoned channel

Distributary mouth bar

A Upper delta plain

0   1   2   3
km

12
m
0

Levees

Bay

B Lower delta plain

0   2   4   6
km

9
m
0

▨ Sandstone   ≣ Siltstone   ↑ Direction of
■ Coal   ▣ Cannel coal   increasing
▥ Limestone   ✿ Marine fossils grain size
☐ Shale   ▨ Underclay

Figure 9.7 Cross section of (A) upper and (B) lower delta plain deposits, Allegheny Formation. Note the coarsening-upward grain size trends. (After Ferm, 1974: 87; by permission of the Geological Society of America.)

Figure 9.8 Interdistributary bays and marshes of the Mississippi River delta. A large crevasse splay (*center*) fans out to the right. The splay had breached the main distributary channel on the left during a 1976 flood and nearly filled the interdistributary bay by 1984. (Courtesy of J. M. Coleman.)

The classic stratigraphic profile of deltaic deposits (Fig. 9.9) shows a coarsening-upward sequence from the delta slope muds and silts to the distributary mouth bar sands. This is the opposite of the fining-upward sequence found in most meandering fluvial systems. The deposits on top of the distributary bar finger sands are abruptly finer, consisting of interdistributary and levee muds and silts; these deposits often include coals. In large prograding delta complexes, there is tremendous lateral and vertical variation due to the influence of several coarsening-upward cycles from sequential distributaries.

Diagnostic Features of Deltas

Tectonic Setting   Deltas occur along coastal plains of passive margins or in broadly downwarping cratonic basins. They are associated with meandering fluvial deposits and with shallow marine shelf deposits.

Geometry   Deltas are roughly triangular in plan view and wedge-shaped in cross section. They are tens to thousands of square kilometers in area and tens to thousands of meters thick.

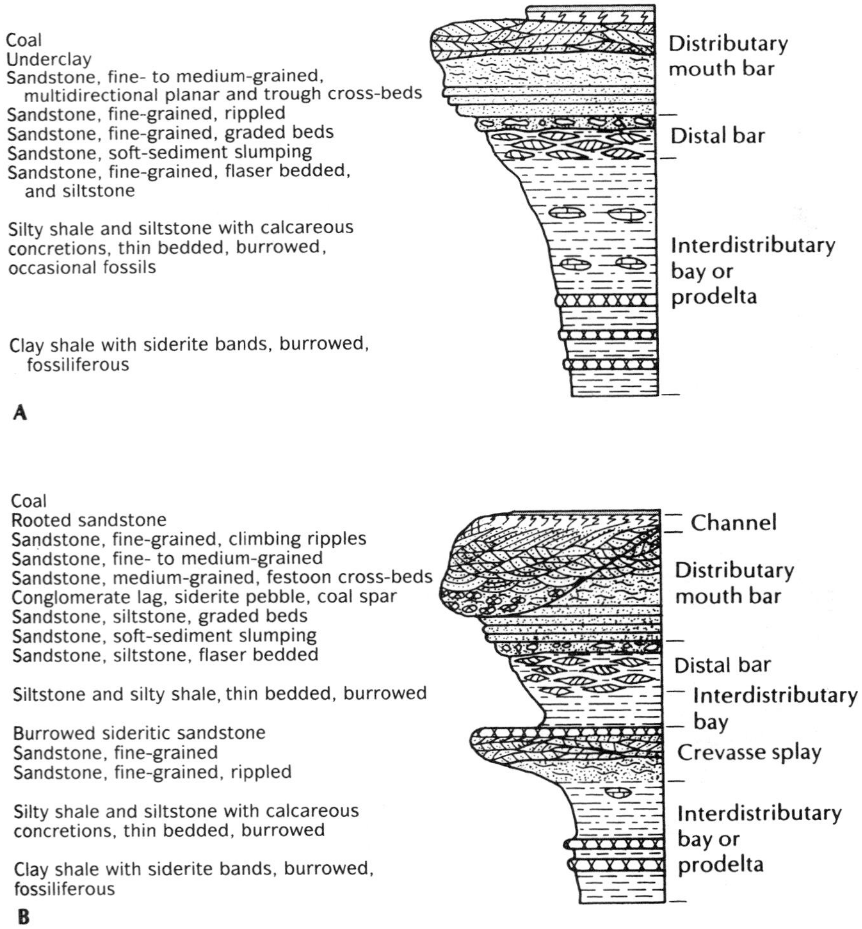

prodelta slope. Normal marine invertebrates are common in the prodelta muds, and these muds and shell beds grade outward onto the marine shelf (discussed in Chapter 10).

Coal
Underclay
Sandstone, fine- to medium-grained,
    multidirectional planar and trough cross-beds
Sandstone, fine-grained, rippled
Sandstone, fine-grained, graded beds
Sandstone, soft-sediment slumping
Sandstone, fine-grained, flaser bedded,
    and siltstone

Silty shale and siltstone with calcareous
concretions, thin bedded, burrowed,
occasional fossils

Clay shale with siderite bands, burrowed,
    fossiliferous

A

Distributary
mouth bar

Distal bar

Interdistributary
bay or
prodelta

Coal
Rooted sandstone
Sandstone, fine-grained, climbing ripples
Sandstone, fine- to medium-grained
Sandstone, medium-grained, festoon cross-beds
Conglomerate lag, siderite pebble, coal spar
Sandstone, siltstone, graded beds
Sandstone, soft-sediment slumping
Sandstone, siltstone, flaser bedded

Siltstone and silty shale, thin bedded, burrowed

Burrowed sideritic sandstone
Sandstone, fine-grained
Sandstone, fine-grained, rippled

Silty shale and siltstone with calcareous
concretions, thin bedded, burrowed

Clay shale with siderite bands, burrowed,
    fossiliferous

B

Channel

Distributary
mouth bar

Distal bar
Interdistributary
bay
Crevasse splay

Interdistributary
bay or
prodelta

Figure 9.9 Typical vertical sequences through lower delta plain deposits in the Pennsylvanian of eastern Kentucky. (A) Typical coarsening-upward sequence. (B) Same sequence interrupted by splay deposit. (After Baganz, Horne, and Ferm, 1975: 185.)

BOX 9.1 PENNSYLVANIAN DELTAS OF THE APPALACHIANS

Coal-bearing deltaic deposits formed in great abundance during the Pennsylvanian (or Late Carboniferous) in many parts of the world. Some of the best exposed and most studied of these deposits occur in the Allegheny-Cumberland Plateau region of the Appalachians, primarily in the states of Pennsylvania, Ohio, West Virginia, and Kentucky. They were formed by a series of rivers that drained northwestward out of the eroding Appalachian highlands in Maryland and Virginia (Fig. 9.1.1) toward the shallow seaways in western Pennsylvania and Ohio. Many of these ancient deltaic sequences are exposed in roadcuts and coal-mine walls, as described by Baganz, Horne, and Ferm (1975), Horne and Ferm (1976), Horne et al. (1978), and Howell and Ferm (1980).

The Appalachian sequences show a full spectrum of environments, from upper delta plain fluvial meander belts to barrier island complexes (discussed later). Between these environments are well-developed deltaic plains and distributary sequences, which in cross section (see Fig. 9.7) show a series of coarsening-upward sequences of floodplain and interdistributary bay silts and muds, punctuated by channel and distributary mouth bar sands and coal swamps. In vertical profile, these coarsening-upward sequences usually begin with interdistributary bay or prodeltaic muds at the base, which may be interrupted by sandy

crevasse splays (see Fig. 9.9). Eventually, the sequence becomes sandier, with abundant ripple bedding that represents deposits produced in the distal bar. This sequence is capped by fine-grained sandstones with some small-scale festoons and ripples that reflect deposition on a distributary mouth bar. In many places, this upper sandstone is cut by sandstone channels with multidirectional planar and festoon cross-beds and with climbing ripples from the main distributary channel. Where progradation is well advanced, the distributary channel evolves into a fluvial channel and bar deposits are eroded by meandering rivers, which leave laterally accreted point bars.

Fig. 9.1.2A shows the idealized geometry of this environment in detail. Some roadcuts near Pikeville, Kentucky, show a cross-sectional geometry that is almost identical to that of the model (Fig. 9.1.2B). Amid bay-fill mudstones are lenticular sandstone bodies 1.5 to 5 km wide and 15 to 25 m thick, which represent the distributary mouth bar sands. These sandstone lenses are widest at their bases and grade into mudstones along their sides and bases. Grain size increases upward in the sequence and toward the center of the bar. Laterally persistent graded beds (Fig. 9.1.2C) are found on the flanks of the bars, as are oscillation- and current-rippled surfaces. The central part of the bar exhibits multidirectional festoon cross-beds (Fig. 9.1.2D). The flanks and the front of the bar show much slumping and soft-sediment deformation, which is associated with over-steepening along the delta front of dense sands overlying less dense saturated muds (Fig. 9.1.2E).

Some distributaries are rapidly abandoned, so few continuous sheet sands are formed by the lateral accretion of point bars typical of the fluvial plain. Instead, the sand bodies are discontinuous and pinch out laterally. They are capped by channels with thin levees along the edges (see Fig. 9.1.2A). These channels are rapidly abandoned and are filled with a clay plug (Fig. 9.1.2G) composed of shales, siltstones, and organic debris that settles from suspension in the ponded water of an abandoned distributary. In a few cases, the abandoned channels became small estuaries, which were inhabited by marine and brackish-water invertebrates. Some are filled with thick coals and other plant debris (Fig. 9.1.2F), representing swamps that grew in the abandoned channel (Hook and Ferm, 1985). The clay plug is often burrowed or penetrated with root casts. If coarse-grained sediments are present, they are thin-rippled, cross-bedded sands and

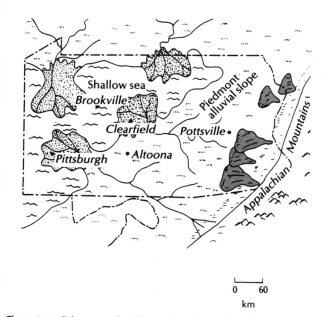

Figure 9.1.1 Paleogeography of Pennsylvania during deposition of the Lower Kittanning (Middle Allegheny Group). (After Williams et al., 1964: 12.)

(continued)

(*Box 9.1 continued*)

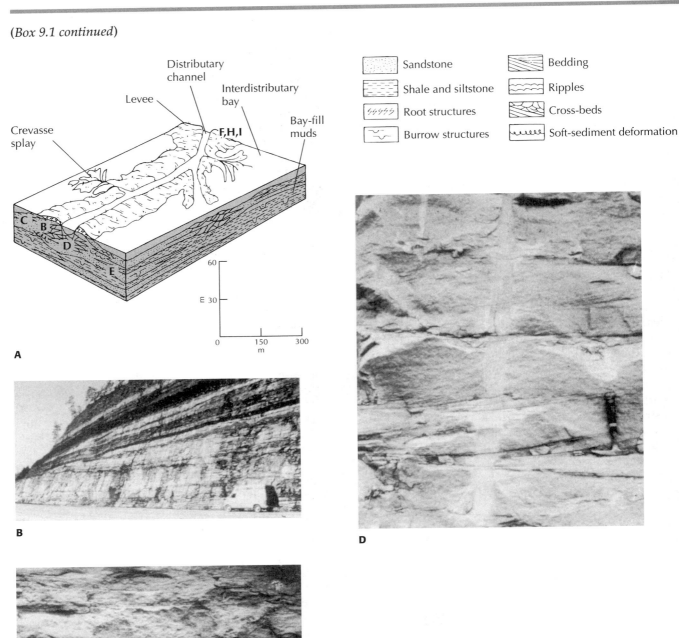

Figure 9.1.2 (A) Block diagram of distributary showing various subfacies. (B) Lateral and vertical gradational coarsening from bay-fill shales (*lower right*) to distributary mouth bar sandstones. (C) Individual sets of graded beds on the flanks of the distributary mouth bars. (D) Multidirectional festoon cross-beds in the central portion of the bar, due to the dispersal of river currents as well as wave and tidal currents over the bar crest. (E) Slump structures and flow rolls on the front and flanks of the distributary mouth bar. (F) Abundant plant debris on bedding surfaces at the distal margin of the splay deposit. (G) Abandoned channel fill with thin levees, near Ivel, Kentucky; levees dip away from the channel. (H) Levee deposits consisting of irregularly bedded, root-penetrated clay shales and siltstones with thin local coals. (I) Tree stump surrounded by shale and siltstone, indicating very rapid deposition of the fine-grained sediment. (B–I courtesy of J. C. Ferm.)

(continued)

(Box 9.1 continued)

silts deposited during floods or at sites near the distributary cutoff.

The largest levees (Fig. 9.1.2H) are 1.5 m thick and about 150 m wide, but most levees are smaller. They are composed of poorly sorted, irregularly bedded, partially rooted siltstones and sandstones. Thin coal beds may parallel the distributary.

Crevasse splays are another distinctive component of the Appalachian deltas. They are miniature versions of the delta, fanning out from a breach in a levee and forming their own thin sets of channels within the interdistributary bay mudstones. They coarsen upward but on a thinner, smaller scale; they also become finer-grained away from the breached levee until they grade laterally into interdistributary bay-fill muds. The largest crevasse splays reach 12 m in thickness and 30 m in width. In outcrop (see Fig. 9.1.2G), they appear as thin sand lenses that thicken in areas where splay channels developed. The splay channels are usually filled with ripple drift cross-lamination, indicating very rapid sedimentation that occurs in the waning flood currents when the water released by the breached levee has nearly drained away. In some splay deposits, the sedimentation was so rapid that tree stumps were buried upright and fossilized in place (Fig. 9.1.2I).

The Pennsylvanian sequences of the Appalachians are among the best-developed and best-exposed examples of ancient deltas. Laterally discontinuous distributary bar sandstones punctuate a sequence of interdistributary muds and prodeltaic clays; within these bars are well-developed channel sequences, levees, and crevasse splays. This association, along with the coals and other plant fossils, is clearly deltaic. Fluvial sequences and barrier island sequences would exhibit sandstone bodies of much greater lateral continuity and would not have the distinctive association of sedimentary structures that form in deltas.

Typical Sequence A coarsening-upward sequence of prodelta muds and clays is followed by distributary bar finger sands and then muds and coals of interdistributary marshes and levees.

Sedimentology There is a wide range of grain sizes from coarse sand to fine mud, generally becoming finer away from land. Coal and other organic matter can be important locally. There are a wide variety of sedimentary structures, depending on current strength and grain size. Distributary sands tend to show small cross-beds and ripple marks. Levee muds are mudcracked. Burrows, shells, and other biogenic evidence are common in the laminated muds of the interdistributary. Prodelta muds are finely laminated (unless bioturbated), with occasional sandy layers; growth faults, slumping, and soft-sediment deformation are common.

Fossils Organic material is very common in the interdistributary, where thick layers of peat or coal can form. Most of the world's great coal deposits come from ancient deltas. Shells, bioturbation, and root casts are also common in the marshes. Organic material can be trapped in the distributary sands, although this is less common. Prodelta muds have lesser amounts of organic debris but can have abundant shells and bioturbation.

Peritidal Environments

Coastal regions with rivers that do not carry large amounts of sediment do not have deltas. Instead, a system of coastal lagoons, estuaries, and tidal flats develops. The dominant influence on this system is the rise and fall of tides; these regions are typically very low and marshy at high tides and emergent during low tides. Like deltas, peritidal environments are common on the subsiding coastlines of passive tectonic margins, where the relief is minimal and the shoreline is often drowned by the postglacial rise in sea level. Peritidal environments have many other similarities to deltaic environments. The primary difference is the excess of river sediment that is supplied to a constructive delta, which causes it to prograde, whereas the peritidal environment shows the reworking effects of tides.

Many coastal areas have extensive wetlands or salt marshes, particularly on the borders of estuaries, delta channels, and lagoons. Unlike the deltaic floodplain marshes or other freshwater swamps, these wetlands are regions where fresh water and salt water mix; thus, the range of water salinity can be extreme. Salt marshes are highly stressed environments dominated by a few organisms that can tolerate the extreme ranges of salinity. They are usually clogged with vegetation, mostly with marsh grass, or with mangroves in

A

B

Figure 9.10 (A) Aerial view of tidal flats near Waddensee, the Netherlands. (Courtesy of KLM-Aerocarto.) (B) Patterns of ripples developed upon megaripples on tidal flats exposed at low tide. (Courtesy of R. A. Davis, Jr.)

some tropical settings. As one would expect, muds rich in organic material accumulate in the marshes, and in many parts of the geologic record, these have become peats or coals. Because of the limited water circulation and the excess of decaying plant matter, the water chemistry is often very reducing, so the muds are black with unoxidized organic matter. Typically, these black muds alternate with fine silt and sand layers from tidal channel flooding. Sometimes the bedding is slightly undulatory due to disturbance by plants; in extreme cases, the bedding is completely destroyed by bioturbation. Interspersed with the muds are small, sandy channel-fill deposits from the tidal channels. These are typically cross-stratified and include organisms (such as oysters) that can tolerate the brackish water.

In many protected coastal regions, the tidal flow is strong enough to prevent vegetation from becoming established. In these regions, broad areas of unvegetated sediments form **tidal flats** (Fig. 9.10). In general, both a wide tidal range and the absence of strong wave action are necessary to form tidal flats. For example, the 2.5- to 4-m tidal range of the North Sea coast of Europe produces tidal flats up to 7 km wide (Fig. 9.10A), and the 5-m tides of the Yellow Sea produce flats up to 25 km wide. These flats have extremely low relief, except for the tidal channels that are carved into them. During low tide, they are completely exposed, and all the effects of the receding tidewaters can be seen in the exposed sedimentary structures (Fig. 9.10B).

Tidal flats are unique among sedimentary environments in that they experience a daily cycle of revers-

ing flow direction and exposure. This produces **tidal bedding** (Fig. 9.11), a regular cycle of mud and sand deposition from the fluctuation in current. At low water and high water, there is relatively little flow, and the finer muds settle out. During advancing and receding tides, the currents move sand across the mudflats, producing cross-bedded sands between the muds. In many instances, the reversing directions of cross-bedding produce herringbone crossbeds (see Fig. 4.7), which resemble the pattern of ribs on a herring or other fish. These are unique to the tidal environment.

Another characteristic sedimentary feature of the tidal flats is flaser and lenticular bedding (see Fig. 4.9). The difference in current strength between the tidal cycles separates the muddy from the sandy fraction. Ripples formed by higher-velocity currents accumulate mud in their troughs; the mud is not scoured out by receding tides, and flaser beds form. As the abundance of mud increases, wavy bedding and eventually lenticular bedding results. Tidal cross-beds often show reactivation surfaces (see Fig. 4.6) because the retreating tides carve back the cross-beds that are built during the advance, only to have the next tidal surge rebuild the cross-bed over the reactivation surface. The asymmetry of the incoming and outgoing tides can have other effects; if the tides advance and retreat at right angles, they can produce features such as interference ripples (see Fig. 4.8). The difference in tidal flow results in superimposed bedforms from two flow velocities. For example, stronger incoming tides may build megaripples, on which smaller ripples are superimposed by outgoing tides (see Fig. 9.10B). The

A

Figure 9.11 (A) Development of tidal bedding according to Reineck and Wunderlich (1968a). Marker beds emplaced at high and low tide enclose bedload deposition of sand and suspension deposition of mud. Depositional phases are controlled by changes in velocity during a tidal cycle, which is also shown, as is depth change. During two tidal cycles, four couplets of sand and mud constituting a tidal bed are deposited. (B) Tidal bedding, Holocene tidal flats, northwestern Germany. Scale in centimeters. (Courtesy of F. Wunderlich). (C) Tidal bedding, Jurassic, Bornholm, Denmark. (Courtesy of B. W. Sellwood.)

B

C

upper reaches of the tidal flat are subaerially exposed except during unusually high tides. They have many features of subaerial exposure, such as mudcracks or algal mats. These mudflats can also harbor oysters and other organisms that cannot tolerate the strong flow of the lower tidal flat. Consequently, upper tidal flat muds are often heavily bioturbated.

The typical tidal flat sequence thus shows a fining-upward column (Fig. 9.12). The lower tidal flat is coarser (mostly sand) and is dominated by herring-bone cross-stratification. The midflat region shows the rapid changes between bedload sands and suspended muds in its abundant tidal bedding, flaser bedding, and related features. The upper tidal flat is formed entirely of suspended muds, which may be mudcracked, bioturbated, and colonized by salinity-tolerant organisms and algal mats. In three dimensions, the tidal sequence may move laterally and is punctuated by many tidal channel deposits. The tidal sequence can also be seen to intergrade with the salt-marsh muds and the foreshore sands.

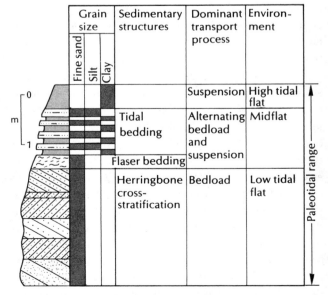

Figure 9.12 Typical paleotidal range sequence based on the middle member of the upper Proterozoic–Lower Cambrian Wood Canyon Formation, Nevada. (After Klein, 1972b: 540; by permission of the Geological Society of America.)

Diagnostic Features of Peritidal Environments

Tectonic Setting   Peritidal environments are common on stable, passive tectonic margins with broad, shallow coastal regions. They are an important component of the thick sequence that is built on passive margins.

Geometry   Tabular in shape, peritidal environments usually form local sand bodies a few meters in thickness that parallel the shoreline, because they are very sensitive to small changes in sea level and so do not accumulate very long in one place.

Typical Sequence   A fining-upward sequence of herringbone cross-bedded sands; flaser- and tidal-bedded sands and muds; and mudcracked, bioturbated, upper tidal flat muds is typical. Black, organic-rich salt-marsh muds accumulate near and on top of the tidal flat sequence.

Sedimentology   Grain size is finest in the upper tidal flat and coarsest in the tidal channels, with abrupt changes between mud and sand in the mid-flat. Many unique sedimentary structures—such as herringbone cross-bedding, flaser and lenticular bedding, interference ripples, and reactivation surfaces—occur. There may also be superimposed bedforms from two flow velocities. Abundant bioturbation, organic matter, and coals and peat are found in the upper tidal flat and salt marsh.

Fossils   Invertebrates that are tolerant of extreme salinity changes (such as oysters and certain crustaceans) can be very abundant. Plant remains are common in the salt marsh, and algal mats can form in the upper tidal flat. Upper tidal flat muds also show extensive bioturbation and other evidence of burrowing.

Barrier Complexes

Barriers are elongate sandy islands or peninsulas that parallel the shoreline and are separated from it by lagoons or marshes. They form on coastlines where there is an abundant supply of sediment and where the tidal range is small enough that the longshore currents and wave action are more important than the onshore-offshore flow of tidal currents (Fig. 9.13). Long, narrow barrier islands are found on *microtidal* coasts, which have a tide range of 2 m or less (Fig. 9.13A). Short barrier islands, broken by numerous tidal channels, are found on *mesotidal* coasts, which have a tide range of 2 to 4 m (Fig. 9.13B). If the sediment supply is too low, no barriers are formed; the coastline is unprotected and usually erodes back. Some barrier-free coasts are influenced only by the tides, producing a *macrotidal* coast (Fig. 9.13C).

Barrier islands, like deltas and tidal flats, are found primarily on passive continental margins adjacent to coastal plains. Particularly good examples have been studied along the Atlantic and Gulf coastal plains of North America. Barrier islands migrate landward and seaward with changes in sea level and often constitute a significant portion of transgressive or regressive shoreline deposits. The channels and spits of barrier islands also tend to migrate up and down the length of the barrier. Because barrier islands typically produce long shoestring sands of high porosity and permeability within impermeable shale sequences, they become excellent reservoir rocks for petroleum. Barrier sands have also proved to be host rocks for uranium, and beach sands can be placer mined for gold, diamonds, and heavy minerals. Marshes and lagoons behind barriers are important areas of coal accumulation.

The barrier island complex is subdivided into several distinct subenvironments.

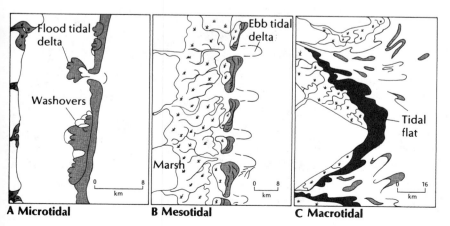

Figure 9.13 Variation in coastal sand bodies due to differences in tidal range. (A) Long, narrow microtidal barriers. (B) Short mesotidal barriers. (C) Linear tidal-current ridges of macrotidal estuaries, perpendicular to shore. (After Barwis and Hayes, 1979: 115.)

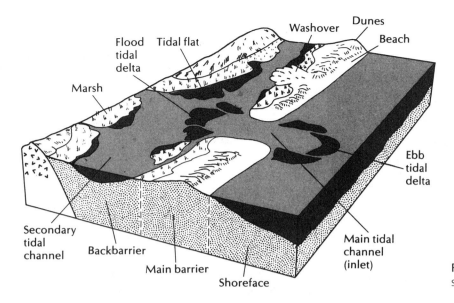

Figure 9.14 Subenvironments in a barrier island system. (After Walker, 1984.)

Shoreface

The main barrier separates the shoreface region from the backbarrier lagoons and tidal flats (Figs. 9.14 and 9.15). Along the shoreface is a series of features that are caused by wave action and the variation in tide levels (Fig. 9.16). The shoreface begins where the wave base first begins to meet the bottom, around depths of 10 to 20 m, depending on the size of the waves. On the lower shoreface, the wave effect is very weak, so offshore shelf processes are also important. As a consequence, the typical deposit (see Fig. 9.16) is fine sand intercalated with layers of mud. There may be planar lamination, although bioturbation and trace fossils often obliterate primary structures. The middle shoreface is strongly influenced by the shoaling and breaking of waves. Where the shoreward drag of the wave base is balanced by the backwash of breakers, the sand accumulates in longshore bars. Here the medium-fine-grained sand is very well sorted, with abundant broken shell layers. Low-angle, wedge-shaped planar cross-beds are most common on the seaward side of these bars, but ripple laminae and trough cross-beds occur on the landward sides. The middle shoreface is less strongly bioturbated than the lower shoreface, although extensive biogenic debris and activity are common. The upper shoreface, or surf zone, is affected by the plunging waves in the onshore-offshore direction and also by the wave-driven longshore current. Thus the sedimentary structures of the shoreface reflect varying flow directions, with multidirectional trough cross-beds interspersed with low-angle planar cross-beds and subhorizontal plane beds.

The shoreface is particularly susceptible to erosion during the unusually strong waves generated by a major storm. Most longshore bars are wiped out, and

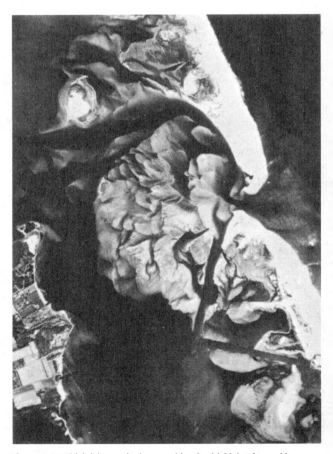

Figure 9.15 Tidal delta on the lagoon side of a tidal inlet, formed by a hurricane in September 1947. (Courtesy of M. M. Nichols, U.S. Department of Agriculture.)

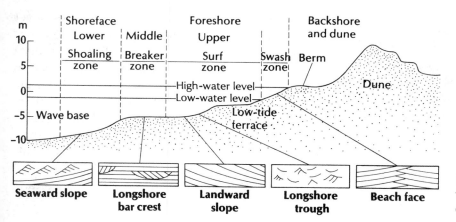

Figure 9.16 Characteristic sedimentary structures for the beach, shoreface, and nearshore bar environments. (After Walker, 1984: 148.)

the entire sequence is often scoured away and replaced by sand dropped during the receding storm. For this reason, shoreface sequences are not common in the stratigraphic record. Often, the entire shoreface sequence is represented by storm deposits, capped by the foreshore, backshore, and dune deposits that are emplaced after the storm recedes.

Foreshore

The foreshore, or swash zone, is the region between the high and low water levels of the beach. It is influenced primarily by the swash and backwash of the breakers, so the directions of cross-bedding are mostly perpendicular to the shoreline. The most characteristic feature of the swash zone is gently laminated planar cross-beds that dip shallowly seaward (about 2° to 10°) and are broadly lenticular (Fig. 9.17). Although symmetrical ripples are often formed from the regular

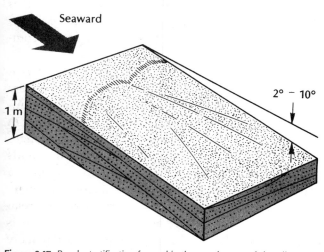

Figure 9.17 Beach stratification formed in the swash zone of shoreline waves. Fine, even laminae are deposited as thin flows run up the beach and then return as waves break. The laminae dip seaward at the angle of the beach slope, which for sand beaches ranges from about 2° to 10°. Slight discordances between sets of laminae reflect changes in wave size or direction or in tide level. (After Harms, 1979: 239; © 1979 by Annual Reviews, Inc.)

ebb and flow of the water on the beach, they are usually destroyed by the incessant reworking and so are seldom found in the stratigraphic record. Foreshore sands tend to be coarser than those found farther offshore because the wave energy is greater; they are also very well sorted due to the extensive reworking by swash and backwash. Extensive reworking often concentrates minerals by density, so beach sands may have well-developed dark laminae of heavy minerals segregated from the quartz sand.

Backshore

Above the high-water level is the backshore and dune area of the barrier complex (see Fig. 9.16). The top of the foreshore is marked by a sandy terrace, called the **berm,** whose crest is just above high-water level. The berm is formed during unusually high storm waves and remains there until the next big storm. Here the wind is more important than the waves, except when storm surges cause unusually high water levels. The sand brought in by these high water levels is redistributed by the wind and by storm washovers. The sand is formed into subhorizontal or landward-dipping plane beds with small cross-beds. At the crest of the barrier complex there is often a dune field, which is affected only by wind. As one would expect, there are extensive eolian trough cross-beds up to 2 m thick; these cross-beds are multidirectional due to fluctuating wind directions. They often have the curved bedding surfaces characteristic of eolian cross-beds. Barrier dune cross-beds differ from desert dunes chiefly in their smaller scale and limited, linear distribution. They also tend to be abundantly burrowed by marine crustaceans and may have well-developed roots, unlike desert dunes.

During hurricane-force storms, however, the dune crest is breached in many places, and waves that overtop the barrier deposit lobes or sheets of sand into the lagoon. These are called **washover deposits** (see

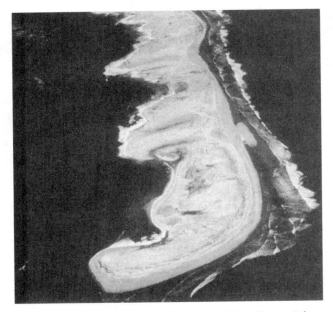

Figure 9.18 Washover fans created on the lagoonal side of Nauset Spit, Massachusetts, during a 1978 blizzard. (Courtesy of S. Leatherman.)

Fig. 9.14). They are dominated by landward-dipping horizontal strata from storm washover episodes (Fig. 9.18). Each layer is usually thin (a few centimeters to 2 m in thickness) and limited in lateral extent, forming a lobate sheet. Between washover episodes, the sand is reworked by the wind into dune trough cross-beds. Where the washover sand enters the lagoon, small-scale delta foresets form. Washover sands are generally fine- to medium-grained and are not as well sorted as beach or dune sands.

Lagoons and Tidal Flats

Behind the barrier is a region of lagoons and tidal flats, in which the quiet waters allow fine silt and clays to settle out of suspension (as in the tidal flats described previously) and form a sequence of mudstone and shale. Usually, there is an abrupt shift from the coarse, clean, well-sorted sands of the backshore dunes to the finely laminated clays of the lagoon. The lagoonal waters can become stagnant and anaerobic, forming black organic-rich muds. In other instances, the lagoons are overgrown with vegetation and form salt marshes, coal and peat swamps, or algal mats. In some parts of the world, the lagoons become hypersaline, and evaporite minerals form. Large parts of the area around the lagoon are within the tidal range, and a tidal flat sequence of herringbone cross-beds, flaser beds, and reactivation surfaces accumulates. Unlike ordinary tidal flats and salt marshes, however, the area around the backbarrier lagoon is punctuated by sands that show the effects of the nearby

barrier and the storms that breach it. Storm washover layers with their delta foresets are found frequently. Another characteristic feature is the flood tidal delta (see Figs. 9.14 and 9.15) formed by sand washed in through the tidal inlet. The sands of such deltas are typically well sorted, with planar cross-beds that are laterally discontinuous because they form a lobate sheet at the mouth of the tidal inlet. Lagoonal shale sequences are easily distinguished from other shale sequences by their proximity to backshore dune deposits and by frequent intrusions of washover and flood tidal inlet sands.

Barrier Island Dynamics

Barrier island complexes are dynamic systems, capable of great lateral migration and variability. Studies of modern tidal inlet channels have documented rapid migration by lateral growth and accretion (Figs. 9.19 and 9.20B). As the channel and spit sequence migrates laterally, it leaves behind a series of deposits that are very different from the classic barrier bar sands (Fig. 9.20A). The sequence begins with a channel floor lag gravel of shells and pebbles. It is then covered by a thick sequence of planar and trough cross-beds of the deep channel, which show many reactivation surfaces. The shallow channel deposits can be plane-parallel laminae or shallow, bimodal trough cross-beds. Eventually, the channel sequence is capped by beach spit deposits made mostly of steeply seaward-dipping planar cross-beds or landward-dipping cross-beds behind the spit. The lateral migration of the inlet (Fig. 9.20B) forms a sequence of channel deposits with time planes that cut obliquely through the sequence of sedimentary structures. The inlet sequence in the barrier complex is inherently time-transgressive.

Even more important than lateral longshore migration is the migration of the barrier complex onshore and offshore. Barrier island sands that have accumulated vertically in a single place are seldom found in the stratigraphic record. Instead, because of changes in relative sea level and sediment supply, the sands migrate onshore and offshore rapidly, forming a large part of the stratigraphic sequence in many basins. When there is excess sediment supply relative to sea-level change, a prograding sequence is formed (Fig. 9.20C). Although progradation usually takes place during falling sea level, it can also occur during rising sea level if the sediment supply is sufficient. In either case, the coarse sands of the beach, the finer sands of the upper shoreface, and the silts and muds of the lower shoreface build outward across the offshore muds and in turn are followed by

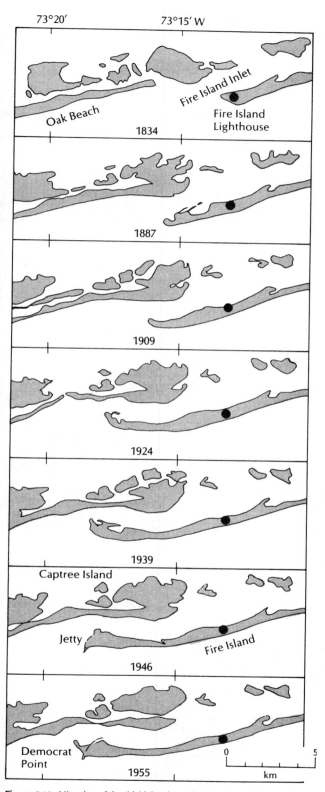

73°20'   73°15' W

Oak Beach   Fire Island Inlet

Fire Island Lighthouse

1834

1887

1909

1924

1939

Captree Island

Jetty

Fire Island

1946

Democrat Point

0   5
km

1955

Figure 9.19 Migration of the tidal inlet along Fire Island, south shore of Long Island, New York, from 1834 to 1955. (After Friedman and Sanders, 1978.)

the lagoonal muds of the backbarrier. Although horizontal layers of differing grain sizes are formed, it is clear that the time planes cut through the lithologic

units and that the entire sequence of sedimentary structures is time-transgressive.

A prograding (regressive) sequence thus produces a coarsening-upward profile (Fig. 9.21A), from offshore muds to shoreface silts and sands to the coarser sands of the beach and dune complex. As shown in Box 9.2, many examples of these sequences exist in the rock record. Because prograding barriers migrate laterally, they can produce extensive sand sheets and shoestring sands that interfinger with the underlying and overlying shales.

Transgressive barrier sequences are much less well known in the stratigraphic record, for reasons discussed in Chapter 15. The rise in sea level must be very rapid relative to the supply of sediment, or shoreface processes will erode the sediments back rapidly. If the sediment supply and sea-level rise are ideally balanced, the barrier island sands migrate shoreward over the lagoonal muds, producing a different kind of coarsening-upward sequence (Fig. 9.21B). The lagoonal muds are capped by flood tidal delta sands and washover sands, interspersed with tidal flat and marsh muds. Eventually, the washover sands are topped by the cross-bedded eolian dunes. Adding the lateral migration of the barrier inlet sequence (see Fig. 9.20C) to either the prograding or the transgressive pattern results in extremely complex barrier island sequences.

Diagnostic Features of Barrier Complexes

Tectonic Setting Barrier complexes occur in stable, sloping coastal plain settings, particularly along passive tectonic margin sequences or in broad cratonic seaways. They are closely associated with peritidal and deltaic sequences and with shallow marine shelf sands and muds.

Geometry A barrier complex typically forms elongate, shoestring sands within marine shale sequences that are tens of meters in thickness, a few kilometers in width, and tens to hundreds of kilometers long. If the barrier complex progrades, it can produce a tabular sandstone body extending for tens to hundreds of kilometers.

Typical Sequence A coarsening-upward sequence is typical. For prograding barriers, offshore muds are capped first by shoreface silts and sands and then by medium- and fine-grained beach and dune sands. For transgressive barriers, the lagoonal muds interfinger with washover and flood tidal inlet sands and are capped by dune sands.

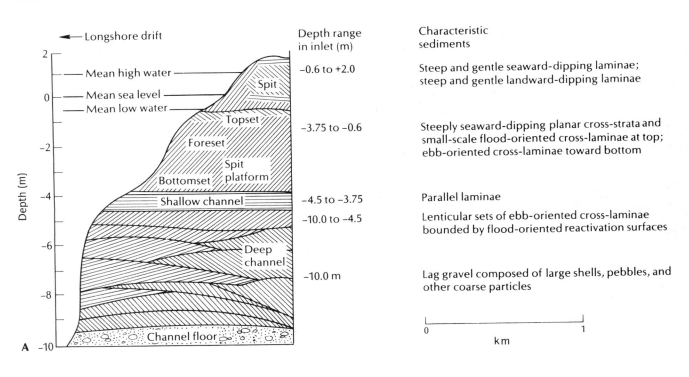

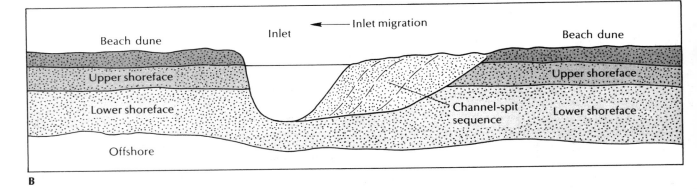

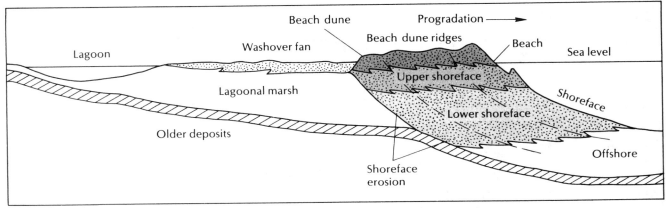

Figure 9.20 (A) Vertical sequence of sedimentary structures formed by the migration of Fire Island Inlet, New York. (After Walker, 1984.) (B) Section parallel to shore showing lateral migration of tidal inlet sequence. (C) Section perpendicular to shore showing onshore-offshore migration of barrier island sequence, here responding to regression. (B and C after Scholle and Spearing, 1982: 316; by permission of the American Association of Petroleum Geologists, Tulsa, Oklahoma.)

BOX 9.2 ORDOVICIAN SHORELINE SEQUENCES OF SOUTH AFRICA

One of the world's best exposures of ancient shoreline sequences occurs in the cliffs of Table Mountain, which towers over Cape Town, South Africa (Fig. 9.2.1). As described by Tankard and Hobday (1977), Rust (1977), and Hobday and Tankard (1978), the lower Paleozoic Table Mountain Group includes the Ordovician Graafwater Formation (Rust, 1973), a tidal flat deposit, capped by the Peninsula Formation, which is a barrier and shelf deposit with some fluvial deposits. To the north, the Graafwater intergrades at its base with the Piekenierskloof Formation, a braided alluvial deposit. To the south, the Peninsula Formation is capped by a series of younger units representing offshore sands and muds and also glacial deposits. The Cedarberg Formation contains Late Ordovician brachiopods, which suggests an age for the underlying units. The Table Mountain Group is one of the few good examples in the stratigraphic record of a nearly continuous transgressive barrier sequence. As discussed in Chapter 15, most transgressive sequences are erased unless the rise in sea level is rapid and is balanced by continuous sediment influx. Most coastal sequences in the stratigraphic record appear to be regressive, caused by prograding shorelines and/or a relative drop in sea level.

The 70-m-thick Graafwater Formation demonstrates many features that are characteristic of a tidal flat backbarrier deposit. Three basic facies are recognized by Tankard and Hobday (1977). The quartz arenite facies (Fig. 9.2.2A) is composed of supermature fine- to medium-grained quartz sandstones, with a few pebbles and cobbles up to 12 cm in diameter in basal channel lags. Bed thickness and grain size both decrease upward within each unit, which ranges in thickness from 5 to 120 cm. The individual sandstone units are separated by erosional surfaces draped with green siltstone (Fig. 9.2.2B). Trough cross-stratification is common in the sandstones, with minor planar cross-bedding. Even more characteristic are well-developed herringbone cross-beds (Fig. 9.2.2B) and reactivation surfaces (Fig. 9.2.2C). Straight-crested, slightly asymmetrical ripple marks about 4 cm in wavelength are common on the sandstones that are draped by mud layers. Wave-generated symmetrical ripples, double-crested ripples, and interference ripples also occur. These ripples indicate both unidirectional and oscillatory flow, both of which occur in tidal flats. The ripple marks show a distinctive interference pattern known as **ladderbacks** (Fig. 9.2.2D), and some have mudcracks, indicating subaerial exposure (Fig. 9.2.2E). The

Figure 9.2.1 Aerial view of Cape Town, South Africa, showing the 750-m-thick nearshore deposits of the Peninsula Formation making up Table Mountain in the background. (Courtesy of M. P. A. Jackson, University of Texas Bureau of Economic Geology.)

(continued)

(Box 9.2 continued)

combination of sandstone channels, herringbone cross-beds, reactivation structures, and ladderback ripple marks led Tankard and Hobday (1977) to attribute this facies to shallow subtidal and low-tide terrace environments.

Covering the quartz arenite facies is a heterolithic facies with rapidly alternating mudstones and quartz sandstones (hence the name). The sandstone beds range from 2 to 15 cm in thickness and are cross-bedded in solitary sets. The intercalated mudstones exhibit both flaser and lenticular bedding (Fig. 9.2.2F) and abundant mudcracks (Fig. 9.2.2E). These features indicate a midtidal flat origin for the heterolithic facies. It grades upward into the mudstone facies (see Fig. 9.2.2A), which consists of horizontally and wavy laminated siltstones and red-brown or maroon-brown mudstones. Trace fossils, such as *Skolithos*, indicate extremely shallow conditions. Tankard and Hobday (1977) attribute this facies to high-tide mudflats and supratidal flats.

The three facies of the idealized sequence (see Fig. 9.2.2A) bear a striking resemblance to the tidal flat model sequence (see Fig. 9.12). The facies alternate through the entire section to produce a semicyclic, fining-upward sequence (Fig. 9.2.3). This cyclicity is attributed to frequent shore zone regressions. The sedimentary structures indicate that the flow was asymmetrical, with dominant ebb currents and a strong longshore movement of water. From the thickness of the units, Tankard and Hobday (1977) postulate a tidal range of 2 to 3 m, comparable to that of mesotidal flats now found in the Netherlands. The alternation of the quartz arenite facies with the heterolithic and mudstone facies produces a repetitive, laterally continuous sequence of sands and thin mudstones that clearly shows their cyclicity (Fig. 9.2.4).

The tidal flat mudstones pass upward into and intergrade locally with landward-dipping sandstone plane beds of the Peninsula Formation (Fig. 9.2.5B). These plane beds dip at about 4° toward the ancient land direction and have small-scale cross-bedding in them. Hobday and Tankard (1978) attribute these beds to washover fans behind the barrier (see Fig. 9.18). This is indicated not only by their similarity in structure but also by their interfingering with tidal flat muds. In many places, the plane beds of the washover fans are interrupted by small-scale (30- to 180-cm-thick) channel sandstones, which may represent small tidal channels cut through the backbarrier flats. The presumed tidal channel sandstones have unidirectional foresets in individual channels, but adjacent or superimposed channels can have directions that are nearly opposite. Herringbone cross-beds and reactivation structures are also present; both features indicate periodic current reversals commonly produced by tides.

Another plane-bedded sandstone facies occurs in the Peninsula Formation, but these dip 3° to 8° in the ancient *seaward* direction (Fig. 9.2.5D). The thinly laminated plane beds are very evenly spaced and include heavy-mineral layers characteristic of modern beaches. Hobday and Tankard (1978) interpreted this facies as produced by the beach face of the barrier island sand spits.

Much of the Peninsula Formation is composed of a large-scale channel facies (see Fig. 9.2.4A, *top*; Fig. 9.2.5C). These channels are up to 40 m thick and are composed of trough cross-stratification with highly variable directions. At the base are thick (80 to 250 cm) trough cross-beds composed of coarse sand and pebbles with a seaward inclination. In some places, they include reactivation surfaces and herringbone cross-stratification. The cross-bed directions suggest a very high energy ebb flow, probably due to storm-surge ebbs and rip currents. The thick basal cross-beds are overlain by progressively smaller scale (15-cm-thick) sets that become thinner upward. The entire sequence of the large-scale channels is reminiscent of the sequence formed by the lateral migration of tidal inlets in a barrier island complex. These produce the bulk of the barrier island sand body (see Fig. 9.22B). The Peninsula Formation shows all the classic facies of the barrier island: washover fans, small backbarrier tidal channels, large-scale tidal inlet deposits, and laminated beach-face sandstones. Combined with the obvious tidal flat deposits of the Graafwater Formation, they constitute one of the most clear-cut and best-exposed ancient examples of tidal flat and barrier island sedimentation. A block diagram shows how the facies geometry of the Graafwater and Peninsula formations was produced (see Fig. 9.2.5A).

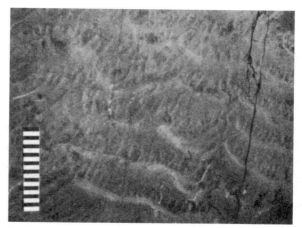

D

B

E

A

Facies	Sedimentary structures	Interpretation
Red-brown mudstone	Concretions	Supratidal flat
Maroon-brown mudstone	Horizontal and wavy siltstone laminae	High-tide mudflat
Alternating mudstone and quartz arenite	Desiccation cracks, cross-lamination, flasers; lenticular and wavy bedding	Mid-tide flat
Quartz arenite	Plane beds, ripples and cross-lamination, herringbone structures, reactivation surfaces	Low-tide terrace
	Large-scale solitary cross-beds, massive sandstone, channels, herringbone structures, reactivation surfaces	Shallow subtidal

C

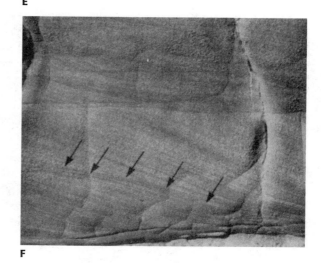

F

Figure 9.2.2 Characteristic sedimentary structures in the intertidal deposits of the Graafwater Formation. (A) Idealized model of sequence in the Graafwater, with depositional interpretation. (B) Flaser and lenticular bedding, showing some soft-sediment deformation due to dewatering. (C) Mudcrack polygons enhanced by sand filling. (D) Interference ripples (ladderback ripples). (E) Herringbone cross-bedding in the Loop Sandstone near Graafwater. (F) Reactivation structures dipping to right (*arrows*). (A and E courtesy of I. C. Rust; B, C, D, and F courtesy of A. B. Tankard.)

(*continued*)

(Box 9.2 continued)

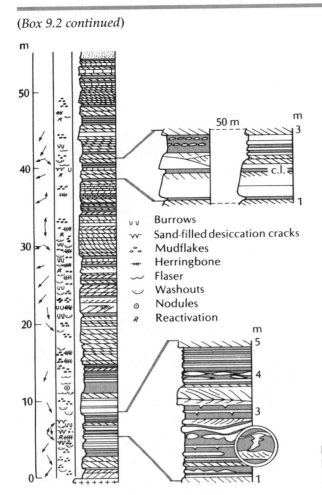

Burrows
Sand-filled desiccation cracks
Mudflakes
Herringbone
Flaser
Washouts
Nodules
Reactivation

Figure 9.2.3 Vertical section through the Graafwater Formation, Chapman's Peak Drive, South Africa, showing repetitive intertidal shales and sandstones. (After Rust, 1977: 149; by permission of Elsevier Science Publishers.)

Figure 9.2.4 Exposure of the vertically repetitive, cyclical intertidal deposits at Chapman's Peak Drive, shown in Fig. 9.2.3. (Courtesy of A. B. Tankard.)

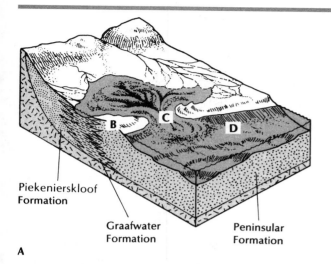

Figure 9.2.5 (A) Generalized model for the deposition of the lower Table Mountain Group, including braided alluvial plain, fan delta, tidal flat, tidal estuary, barrier island, washover fan, tidal channel, inlet, and offshore tidal sand bar. (B) Landward-dipping plane beds of barrier overwash (*lower sequence*) overlain by tidal channel deposits (*upper sequence*). (C) Rapidly alternating cross-bed sets from the tidal channel. (D) Seaward-dipping spit beach-face plane beds, overlying ebb-dominated tidal inlet sequence. (B – D courtesy of A. B. Tankard.)

Sedimentology Because of extensive reworking, beach sands can have mineralogically mature quartz mixed with occasional heavy-mineral lags. Lagoonal muds can be organically rich and can accumulate peat and coal; in a few cases, evaporites are known. A variety of sedimentary structures occur, but shallow-dipping tabular cross-beds and plane beds are most common on the shoreface and washover fan. Beach sands have a unique type of cross-stratification formed by swash and backwash. Eolian dunes produce larger-scale tabular and trough cross-beds but are abundantly burrowed and impregnated by root casts (unlike desert dunes).

Fossils The shoreface has a diverse and abundant shelly invertebrate fauna and many burrowing in-

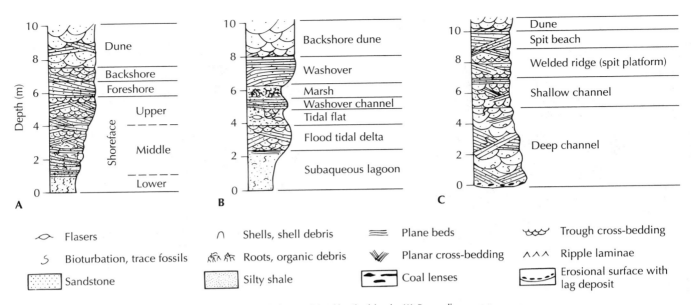

Flasers
Bioturbation, trace fossils
Sandstone

Shells, shell debris
Roots, organic debris
Silty shale

Plane beds
Planar cross-bedding
Coal lenses

Trough cross-bedding
Ripple laminae
Erosional surface with lag deposit

Figure 9.21 Stratigraphic sequences of three end-member facies models of barrier islands. (A) Prograding model. (B) Transgressive model. (C) Barrier-inlet model. (After Walker, 1984.)

vertebrates, with many types of burrows and trace fossils. The high-energy wave environment, however, typically breaks up and abrades most of the shell debris. Dunes have both burrows and root casts. Lagoons are dominated by plants and mud-flat-dwelling animals, depending on the salinity. If there is a wide fluctuation in salinity, only organisms that can tolerate those conditions are found there.

CONCLUSIONS

Coastal sedimentary environments such as deltas, tidal flats, and barrier islands are highly dynamic systems. They migrate both offshore and onshore in response to sea-level changes, migrate up and down the coast in response to currents, and prograde due to sediment supply. They are excellent indicators of ancient climates and paleogeography, as well as recording the migration of shorelines and changes in sea level. They are also important for their deposits of coal, oil, and other mineral resources.

FOR FURTHER READING

The works listed at the end of Chapter 8 by Davis; Galloway and Hobday; Reading; Rigby and Hamblin; Scholle and Spearing; Selley; and Walker and James are also the primary sources of information in this chapter. In addition, the following sources may prove useful.

Coleman, J. M. 1976. *Deltas: Processes of Deposition and Models for Exploration.* Champaign, Ill.: Continuing Education Publishing.

Davis, R. A., Jr., ed. 1978. *Coastal Sedimentary Environments.* New York: Springer-Verlag.

Davis, R. A., Jr., and R. L. Ethington, eds. 1976. *Beach and Nearshore Sedimentation.* SEPM Special Publication 24.

Davis, R. A., Jr., and D. Fitzgerald. 2003. *Beaches and Coasts.* New York: Blackwell Science.

Ginsburg, R. N., ed. 1975. *Tidal Deposits.* New York: Springer-Verlag.

Klein, G. D. 1977. *Clastic Tidal Facies.* Champaign, Ill.: Continuing Education Publishing.

Morgan, J. P., and R. H. Shaver, eds. 1970. *Deltaic Sedimentation: Modern and Ancient.* SEPM Special Publication 15.

10 Clastic Marine and Pelagic Environments

Thick sequence of Eocene turbidites and hemipelagic mudstones of the basin-plain facies near Zumaya, northern Spain. (Courtesy of G. Shanmugam.)

ALTHOUGH WE ARE MORE FAMILIAR WITH NONMARINE and coastal environments, marine sedimentary environments are far more important in the stratigraphic record. Not only is 75% of the Earth's surface covered with oceans, but an even more important fact is that marine environments are below the base level of erosion. Unless sea level drops and exposes them to subaerial erosion, marine sediments have a high preservation potential. In a rapidly subsiding basin, tens of thousands of meters of marine sediments can accumulate. For these reasons, marine sedimentary rocks make up most of the stratigraphic record and are studied far more than other deposits.

Clastic Shelf Deposits

About 5% of Earth's surface is covered by seawater that is less than 200 m deep. Those areas, known as

continental shelves, are continuous with the coastal plain sequences of the continents and have slopes of only 0.1°. In tropical regions, the shelves are the sites of carbonate sedimentation (which we will discuss in the next two chapters). Where the water temperatures are too cold for carbonate sedimentation or where siliciclastic sediment overwhelms carbonate sediment, the shelves are covered by detrital particles from the continents, mostly fine sands, silts, and muds.

In the geologic past, shallow marine siliciclastic sedimentation was much more widespread. During episodes of high sea level, all the continents were partially covered by vast areas of epicontinental, or **epeiric,** seas. Modern examples of epeiric seaways, such as Hudson Bay, the Yellow Sea, and the Caspian and Arafura seas, do not begin to approach the scale of ancient epicontinental seas. Epeiric marine sandstones and shales make up a major portion of the geologic record on the continents. These sequences can be very thick: the Proterozoic Transvaal Group of South Africa is 12,000 m in thickness, and marine sandstones reach thicknesses of 2000 m in the Cambrian-Ordovician rocks of southern Africa (see Box 9.2.).

Despite their importance in the stratigraphic record, modern continental shelf sediments are still not well understood. Naturally, it is difficult to study sediments under tens to hundreds of meters of water, and it is only recently that suitable techniques (diving craft and underwater photography, box coring, high-resolution seismic profiling, and side-scanning sonar) have been developed to undertake these studies. More important, present-day shelf sediments may not be the best analogs for ancient epeiric seas. The modern continental shelves were completely emergent during the last glacial maximum 20,000 years ago, and their sedimentary cover was undoubtedly affected by shoreline and fluvial processes. Between 10,000 and 7000 years ago, sea level rose rapidly (nearly a centimeter per year) due to melting of the glaciers, and seawater covered the outer shelf very rapidly. Emery (1968) suggested that most of the sand and gravel on the outer continental shelves is a relict from the last low-sea-level episode and has been little reworked since then. Swift, Stanley, and Curray (1971) have since argued that most shelf sediments seem to be in dynamic equilibrium with modern marine processes. In many instances, the sediment may have reached the shelf originally by fluvial or deltaic processes but then been reworked by marine processes. They called this the **palimpsest effect** after the ancient parchments, reused by medieval authors, that

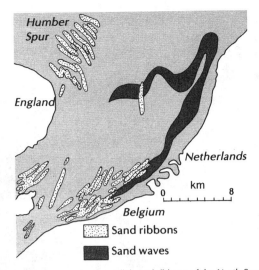

Figure 10.1 Numerous shore-parallel sand ribbons of the North Sea and a belt where sand waves are found. (After Galloway and Hobday, 1983: 173.)

bear traces of older writing under the new. Thus, modern continental shelf sediments may be only partial analogs of ancient epeiric sediments.

Continental shelf sedimentation is important for economic reasons. Shelf sands have high porosity and permeability and usually occur as isolated lenses or sheets in predominantly impermeable shales, so they are good stratigraphic traps for hydrocarbons. Some of the surrounding marine shales are also believed to be good source rocks for oil, and some shallow marine sands have produced enormous amounts of oil and gas.

The depositional processes in the continental shelf system are affected most strongly by where they occur on the shelf. The inner shelf is shallow enough to experience wind-driven, storm-driven, and tidal currents. The outer shelf is well below wave base of even the strongest storms or tidal currents, so most current flow is due to oceanic circulation, upwelling, and the effects of density-stratified water columns. It is believed that about two-thirds of the outer shelf sediments are relict or palimpsest because this region was covered during the most rapid phase of sea-level rise and has had relatively little time to reach equilibrium. The inner shelf, on the other hand, has accumulated a considerable blanket of sediment over the past 7000 years. It is probably the best analog for ancient epeiric seaways.

On shelves that experience mesotidal and macrotidal ranges (tides greater than 2 m), there are tidal currents of 50 to 100 cm/s. These currents flow back and forth across shallow shelves, such as those of the

Figure 10.2 A cross section of sand ribbons shows superposed smaller-scale bedforms that migrate obliquely up the gentle slope and at right angles across the steeper side. Complex internal stratification is inferred between the major sand-ribbon foresets. (After Galloway and Hobday, 1983: 172.)

North Sea, producing a strong reversing flow that is intensified as it passes through narrow straits. Tidal currents strongly affect the **sand ribbons,** or **tidal ridges,** that run parallel to the main direction of tidal flow; these features are long stringers of sand up to 40 m high, 200 m wide, and 15 km long (Fig. 10.1). Sand ribbons are asymmetrical in cross section (Fig. 10.2), with erratic cross-stratification on the gentle slope and well-developed foreset stratification with a slope of about 5° on the steeper slope. Apparently, sediment is moved *along* the steep face of the ridge by the stronger tidal currents. Tide reversals generate currents over the gentle slope (indicated by megaripples), and eventually the sand migrates to the top of the crest and slides down the steep side. At very high tidal current velocities (greater than 125 cm/s) there is active scour, which produces gravel lags (Fig. 10.3). If the tidal current is less than 100 cm/s, tidal sand waves form. At tidal currents slower than about 50 cm/s, sheets of smooth sand develop, and at even lower velocities, sand forms in patches on the muddy bottom.

Tidal sand waves have crests of 3 to 15 m and wavelengths of 150 to 500 m. In cross section, they are composed of a set of low-angle surfaces (dipping at 5° to 6°) with smaller cross-bed sets dipping down these low-angle master bedding surfaces (Fig. 10.4A). Because the length and the height of the cross-bed sets resemble eolian cross-stratification in some ways, various authors have argued that classic examples of eolian strata might be formed in tidal sand waves (discussed in Chapter 8). A closer look at the marine sand wave cross-beds, however, clearly shows some significant differences. Tidal sand wave cross-beds seldom dip more than 6°, whereas eolian foresets typically dip 20° to 30°. Tidal sand wave cross-bed sets are seldom more than a few meters thick, but eolian cross-beds are usually 10 to 35 m thick, and there can be hundreds of meters of these cross-beds in succession. The tidal sand wave foresets are actually master bedding surfaces covered by dipping sets of small cross-beds

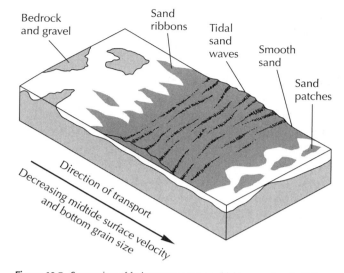

Figure 10.3 Succession of facies zones across a tidal transport path. (After Blatt, Middleton, and Murray, 1980: 690)

(Fig. 10.4B, *upper right*), whereas eolian crossbeds are remarkably uniform in their thickness and continuity (see Fig. 8.5.2).

Figure 10.4 (A) Walker's (1984) model of the internal geometry of marine sand waves. (B) Cross section of a large sand wave in the Lower Greensand (Lower Cretaceous) near Leighton Buzzard, England. The main cross-bed set is about 5 m thick but is truncated by thinner, horizontally bounded sets near the top. The compound internal cross-bedding within the larger sets distinguishes these sand waves from eolian deposits. (Courtesy of R. G. Walker.)

Most shallow inner shelves, which have weak tidal currents (velocities less than 25 cm/s), are quiet under normal conditions, but waves with very deep wave bases form during storms. These waves produce strong currents that move obliquely onto the shelf and shoreline and accentuate any bottom irregularities that are oriented obliquely to the shoreline. During the Holocene transgression, shallow shoreface ridges were apparently exposed to deeper water and stronger storm currents, which formed **linear sand ridges** up to 10 m high, 1 to 2 km wide, and tens of kilometers long. Periods of intense storm activity produced variable cross-bedding within these bars, greatly modifying them. During quiet periods, the bars were covered with wave ripples, and fine sediment accumulated in their troughs as their tops were reworked. These processes produced a coarsening-upward sequence. A gravel lag at the bottom seems to be a relict of an older erosional surface that was produced during lower sea levels in the Pleistocene.

Linear sand ridges are best known from storm-dominated coasts such as the northwest Atlantic shelf off North America. It appears that they began as abandoned barrier complexes built over the shallow marine mud because many other relict features are still visible on the shelf (for example, shoal retreat massifs produced by submergence of ancient peninsulas). On shelves where there is no evidence of relict barrier islands, the sand supply for these ridges may come from turbidity currents that are set in motion by major storms.

Another feature of storm-dominated sandy shelves is **hummocky cross-stratification** (Fig. 10.5). It is well known from ancient examples but has never been produced in the laboratory. This bedform is composed of low mounds and hollows of very fine sand and silt that have sharp bounding surfaces and no apparent directionality. Hummocky cross-stratification is formed at water depths of 5 to 15 m, where strong storm waves produce water displacements of several meters and velocities of more than 1 m/s. Apparently, a storm surge picks up and suspends the finer material but immediately drops it as the flow reverses. This sediment is deposited in irregular, hummocky sheets that are partially eroded by the next surge, producing the cross-bedding. Hummocky cross-stratification seems to form in the zone below the fair-weather wave base (since it is not reworked between storms) but above the storm wave base (Fig. 10.6).

Wave-ripple cross-bedding, another feature known from ancient marine sands, displays storm and wave

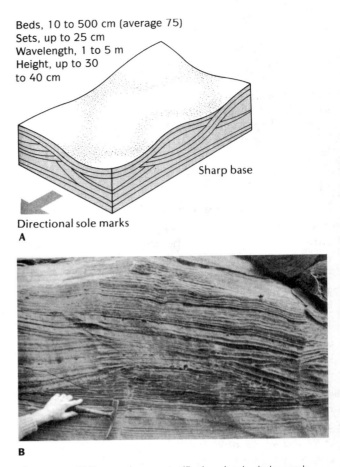

Beds, 10 to 500 cm (average 75)
Sets, up to 25 cm
Wavelength, 1 to 5 m
Height, up to 30 to 40 cm

Sharp base

Directional sole marks

A

B

Figure 10.5 (A) Hummocky cross-stratification, showing its low-angle, curved intersections and upward-domed laminae. (After Walker, 1984: 149.) (B) Hummocky cross-stratification from the Cape Sebastian Sandstone, Oregon. (Courtesy of J. Bourgeois.)

influence but no tidal effects. Wave ripples differ from current ripples not only in their symmetry but also in having an irregular, undulatory lower bounding surface, a less troughlike shape, bundled upbuilding of foreset laminae, swollen lenticular sets, and offshooting and draping foresets (Fig. 10.7). These bedforms are unusual in that they clearly display the effects of rapidly reversing wave flows. Yet they are made of poorly sorted sand and mud and often show lenticular and flaser bedding. If they are formed by high-energy storm waves, however, the finer material can be carried in suspension and deposited where the waves shoal and lose their energy. In such cases, rapid reworking by the oscillatory currents might explain not only the wavy, truncated ripples but also the drapes of mud that fill the hollows during shoaling.

The typical sequence of a shallow siliciclastic shelf is obviously affected by storm and tidal processes, but *changes in relative sea level* are the primary source of the sediments and structures on which these forces act.

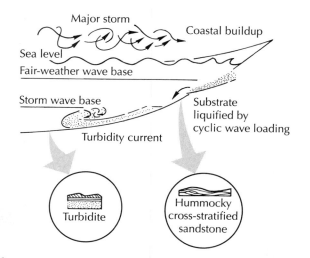

Figure 10.6 Storm winds create coastal buildup. Cyclic loading of the substrate by storm waves may liquefy the substrate. The liquefied sediment may flow and accelerate basinward, becoming a turbidity current with all the sediment in suspension. Deposition from this flow below the storm wave base would result in turbidites with Bouma sequences. Above the storm wave base, waves influenced by the bottom would rework the turbidity current deposits into hummocky cross-stratification. Hummocky cross-stratification could also form above the fair-weather wave base but would probably be reworked into other structures by fair-weather processes. (After Walker, 1984: 154.)

Many of the structures we have discussed began as shallow-water features during episodes of lower sea level and were modified during later rises in sea level. Because sea level has fluctuated many times in the geologic past, it may also be responsible for the form of many ancient shelf deposits. Storm-dominated shelf sequences, for example, would be expected to vary upward from burrowed shelf muds (well below storm wave base) to storm-graded beds (just below storm wave base) to hummocky cross-stratification (between normal and storm wave base) and finally to trough cross-bedding (reworking by helical storm currents). This coarsening-upward sequence would be expected on a prograding or regressive shelf, but the reverse would be found on a transgressive shelf (Fig. 10.8A, B). A transgressive tide-dominated shelf (Fig. 10.8C) would proceed from the high-energy gravel lags to the large-scale cross-beds of sand waves and sand ribbons to the bioturbated silts and muds found between the sand bodies; a regressive tidal shelf would produce the reverse sequence, coarsening upward. If both storm and tidal processes occur, the sedimentary structures would be mixed. In a shelf on which accumulation is balanced by relative sea level change, there would be little net change in grain size (Fig. 10.8D), and the combined storm and tidal influence would produce a mixture of gravel lags, large-scale cross-beds, hummocky cross-stratification, trough cross-beds, and occasional burrowed mud and silt layers.

Diagnostic Features of Clastic Shelf Deposits

Tectonic Setting Clastic shelf deposits are extensive in passive continental margins but are more restricted in convergent margins. They are also found in epicontinental seaways and gently downwarping cratonic basins. These deposits are associated with deeper marine shale sequences or with shallow marine limestones and with deltas and coastline deposits.

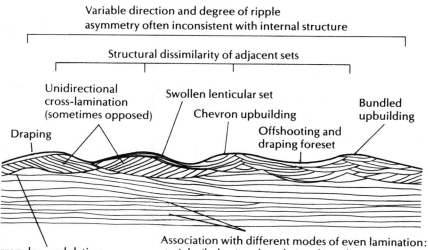

Figure 10.7 Diagnostic features of wave-ripple cross-bedding. (After de Raaf, Boersma, and van Gelder, 1977: 460; by permission of Blackwell Scientific Publishers.)

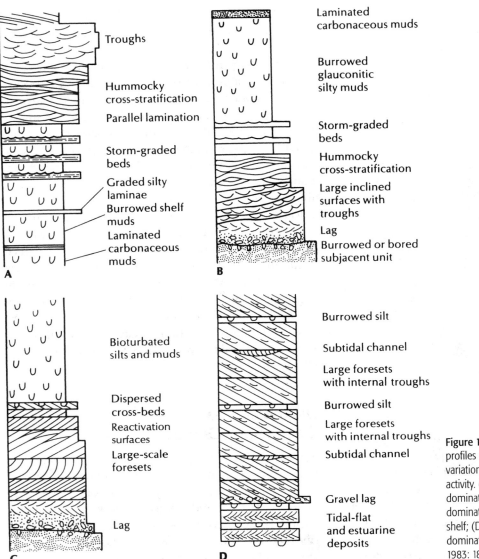

Figure 10.8 Four characteristic stratigraphic profiles of marine shelf sequences resulting from variations in relative sea level and storm or tide activity. (A) Prograding (regressive) storm-dominated shelf; (B) transgressive storm-dominated shelf; (C) transgressive tide-dominated shelf; (D) balanced accumulation, storm- and tide-dominated shelf. (After Galloway and Hobday, 1983: 180.)

Geometry Tabular sequences of shale with long sheetlike or lenticular sandstone bodies are found. The total shelf package may cover thousands of square kilometers and be hundreds of meters in thickness.

Typical Sequence Fining- or coarsening-upward sequences occur, depending on whether the sequence is transgressive or prograding. Burrowed glauconitic muds are typically followed by storm-graded beds, hummocky cross-stratification, and trough cross-beds (storm-dominated) or by large-scale cross-beds and lag gravels (tide-dominated).

Sedimentology Quartz and clay minerals are predominant, although carbonate minerals and shell frag-

ments can also occur. Glauconite is particularly characteristic of shallow marine muds and sands. Hummocky cross-stratification, wave-generated cross-beds, large-scale cross-stratification, and storm-graded beds are all characteristic of this environment. Burrowed silt and mud with abundant shell debris and trace fossils form the bulk of most shallow marine sequences.

Fossils The most diagnostic feature of a shallow marine deposit is its fossils. Because most marine invertebrates require a shallow, well-aerated bottom with a narrow range of salinities, they are found in greatest abundance on the shallow shelf. If such restricted organisms are found in a sedimentary sequence, it is almost certainly shallow marine.

Continental Slope and Rise Sediments

Between the nearly horizontal shelf and the deeper ocean floor is an abrupt boundary called the **shelf-slope break;** below the shelf-slope break is the steepest part of the ocean floor, the **continental slope** (Fig. 10.9), which is narrow (10 to 100 km) and slopes downward at an average angle of 4° to 6°. Sediments deposited on the slope are moved downslope by gravity and seldom remain on the slope. Much of the sediment on the continental slope has slumped or washed off the shelf-slope break during storms or earthquakes.

The characteristic sedimentary features of the slope are gravity-transported deposits: large exotic slide blocks (**olistholiths**), slumped and deformed shales, debris flows full of a chaotic assemblage of exotic brecciated blocks (**olisthostromes**), and turbidites (Fig. 10.10). Some of these features can be striking—and

easily misinterpreted. For example, spectacular syndepositional folding can be developed by slumping blocks of mud (Fig. 10.11A, B). To some geologists, these features might suggest that the sequence has been tectonically deformed, but the folding is entirely due to soft-sediment deformation. Likewise, olistostromes (Fig. 10.11C) are not easily distinguished from other types of sedimentary breccias or tectonic mélanges. Tectonic mélanges tend to be more highly sheared and deformed than olisthostromes (Hsü, 1974), although some olisthostromes are secondarily sheared. Olisthostromes tend to be much more limited in extent and are bounded by other deep-marine sediments because they are sedimentary bodies and are not formed by tectonic accretion. Olisthostrome clasts are often sedimentary boulders and may show some signs of rounding and transport; mélange blocks are formed by fracture and may be highly deformed. Mélanges are discussed further in Chapter 19.

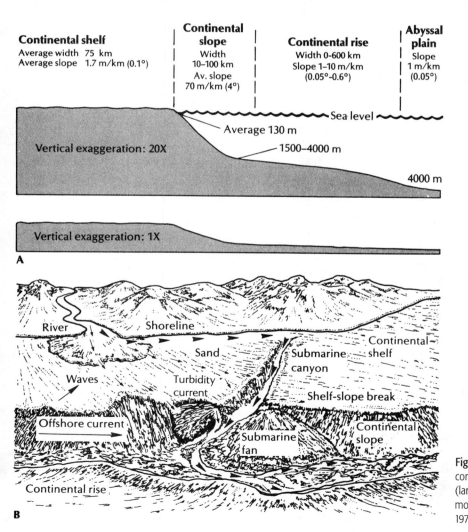

Continental shelf
Average width 75 km
Average slope 1.7 m/km (0.1°)

Continental slope
Width 10–100 km
Av. slope 70 m/km (4°)

Continental rise
Width 0–600 km
Slope 1–10 m/km (0.05°-0.6°)

Abyssal plain
Slope 1 m/km (0.05°)

Sea level

Average 130 m

1500–4000 m

4000 m

Vertical exaggeration: 20X

Vertical exaggeration: 1X

A

River
Shoreline
Sand
Submarine canyon
Continental shelf
Waves
Turbidity current
Shelf-slope break
Offshore current
Continental slope
Submarine fan
Continental rise

B

Figure 10.9 (A) Typical dimensions of a modern continental shelf. (B) Route of transport of sand (large arrow) and mud (thin arrow) from river mouth to deep seafloor. (After Reineck and Singh, 1973: 379.)

Processes   **Deposits**

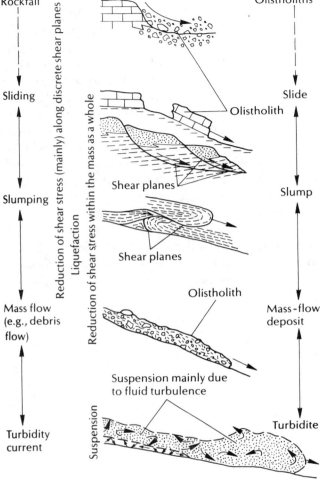

Figure 10.10 Mass-gravity transport processes and their deposits. (After Reading, 1986: 402; by permission of Blackwell Scientific Publishers.)

In addition to gravity-displaced deposits, the continental slope is also overlain by deposits of hemipelagic mud that settle out of suspension from the oceanic water column. These muds are typically reworked by **contour currents** that flow parallel to the slope (Fig. 10.12A). Contour currents are the product of normal oceanic circulation as water masses of different densities move relative to one another. They usually have speeds of 5 to 30 cm/s, although speeds of 70 cm/s have been recorded. Such speeds are sufficient to keep most of the clay in suspension and to transport the fine sand and silt. The resulting deposit, called a **contourite** (Fig. 10.12B, C), is finely laminated or faintly cross-bedded silt and sand interbedded with mud and forms thin, imbricated lenses that fill bottom scours. Contourites show neither the lateral continuity nor the graded bedding of turbidites.

Figure 10.11 (A) Large-scale gravity folds in the Miocene Castaic Formation, Ridge Basin, California. (B) Giant submarine slump fold within the Castaic Formation. (C) Submarine landslide conglomerate showing normal grading with a rippled Bouma C division capping the bed. The clasts are imbricated in an upslope direction at the top of the 1.5-m-thick channel sequence. (A and B, D. R. Prothero; C courtesy of H. E. Cook.)

At the base of the continental slope is the **continental rise,** a broad, gently sloping region that grades into the seafloor (see Fig. 10.9A). The slope-rise break

A

B

C

Figure 10.12 (A) Deep marine ripple marks on a fine sandy bottom at a depth of 3500 m in the South Pacific Ocean. Before the development of submarine photography around 1950, it was assumed that the deep sea was static and without currents. Thus, ripples were thought to be restricted to shallow water. (Official NSF photo, USNS *Eltanin* Cruise 15, courtesy of Smithsonian Oceanographic Sorting Center.) (B) Core of contourite sediment from the North American Atlantic continental rise off New England in 4745 m of water. Centimeter-wide beds of hemipelagic claystones alternate with coarser cross-laminated contourite siltstone. (Courtesy of C. D. Hollister.) (C) Possible Ordovician contourite showing ripple forms with wavelengths of about 9 cm and amplitudes of about 1 cm. Both base and top have a sharp contact with enclosing hemipelagic mudstone. Outcrop is from the Hales Limestone, Nevada. (Courtesy of H. E. Cook.)

occurs at a depth of about 1500 m, and the slope can be up to 600 km wide. The gradient of the continental rise is very gentle, from 0.05° to 1.0°, or about 1 to 10 m/km. Continental rises occur primarily on passive margins; on active margins, the slope usually drops directly into the subducting trench. The bulk of the continental rise sequence is made of material that has slid down the continental slope or down through the submarine canyons cut into the continental shelf (see Fig. 10.9B). Between episodes of sliding, hemipelagic muds settle out of suspension and eventually form thick sequences of shale.

The most important process of sediment transport on the continental rise is the turbidity current, a turbulent suspension of particles that is denser than the surrounding water. Like other density currents, a turbidity current flows as a discrete surge, separate from the medium through which it travels (see Fig. 3.7). Its high velocity and turbulent motion enable it to carry much larger particles than would normally be found far out on the rise. The turbidity current is the key to a mystery that long puzzled geologists: Why were fairly coarse sandstone beds of great thickness and lateral extent found in sequences that had deep marine shales and all the other hallmarks of very deep water? As discussed in Chapter 3, Kuenen and Migliorini (1950) first proposed the turbidity current as a mechanism for bringing sand and fine gravel to the deep seafloor.

The lithologic product of a turbidity current is called a turbidite. Every turbidite represents a single

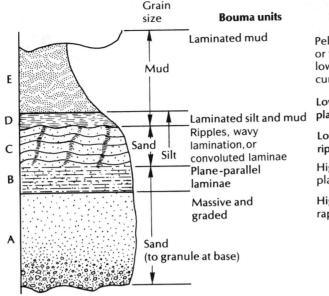

Figure 10.13 The classic Bouma turbidite sequence showing trends of sedimentary structures, grain sizes, and depositional conditions. (After Bouma, 1962: 49; by permission of Elsevier Science Publishers.)

catastrophic flow; most deposits have a regular and predictable sequence of sedimentary features known as the **Bouma sequence,** after the man who first analyzed turbidite sequences in detail. The classic Bouma turbidite sequence (Fig. 10.13) has a scoured base overlain by a massive graded bed (unit A), which represents the coarsest material to settle out of suspension as the turbidity current slowed. Above this is plane lamination (unit B), which is believed to represent high-flow-regime plane beds. Unit C displays ripples and wavy lamination, indicative of lower flow regime. Unit D is laminated silt, and unit E is laminated mud that settled out of suspension during the waning of the turbidity current. In some cases, unit E is topped by laminated hemipelagic mud that settled from suspension in the episodes between turbidity currents. In practice, these muds are often very difficult to distinguish from the laminated muds of unit E.

Not all turbidity currents produce complete Bouma sequences. Typically, turbidite sequences accumulate as broad sheets and lobes of sediment that spread out from submarine canyons and build extensive **submarine fan** complexes (see Fig. 10.9B). Deposits near the top (**proximal** part) of the rise or within the submarine canyons are the coarsest, with debris flows and coarsely graded beds predominating (Fig. 10.14). "Classic" turbidites are found in the middle part of the fan. At the lower (**distal**) end of the fan, only the finer material is left, producing thin-bedded turbidites dominated by the fine sand, silt, and clay. The fan channels are braided, much like the channels on alluvial fans, and they have distinctive levee and over-

bank deposits, like those seen in fluvial systems. As the fan sequences accumulate, they build the characteristic wedge shape of the continental rise. This pile of turbidites and hemipelagic muds can accumulate into enormous thicknesses of rhythmically alternating graded sandstones and shales. Many examples of rapidly sinking deep-water tectonic basins are known that have accumulated thousands of meters of turbidites and shales (see Box 10.1).

Other sedimentary structures are also characteristic of the slope and rise. The strong currents that scour the cohesive, muddy bottom cause many types of sole marks, such as flute casts and groove casts, and many of the tool marks described in Chapter 4. Because sand and mud layers of different densities are often interbedded, load casts and other types of soft-sediment deformation are common. In addition, there is a characteristic suite of trace fossils, particularly those bottom-dwelling organisms that systematically feed through the muddy bottom (see Fig. 4.22, *Nereites* ichnofacies).

With an understanding of the slope and rise processes, one can predict the characteristic stratigraphic sequence (see Fig. 10.14A). The slope is dominated by fine hemipelagic muds, interrupted by submarine canyon sandstones, and by chaotic olistostrome breccias and slump deposits. Slope sequences often have slump scars or soft-sediment deformation from slumping. The upper part of the rise is composed of turbidite fans, which prograde as they build outward. This implies that the sequence *coarsens upward* from the thin, outer-fan sandstones to the complete turbidites and channel sandstones of

Facies A Thick to massive, channeled and amalgamated, poorly sorted coarse Ss and Cgl, with thin or no mud intervals; all gradations to facies E

Facies B Thick to massive, lenticular sorted Ss with parallel to undulating laminae, common mud clasts, and erosional bases; thin mud intervals

Facies C Couplets of even, parallel-bedded M-F Ss and minor homogeneous muds; Ss may show complete Bouma succession, some broad, shallow channels; common sole marks

 Facies D Couplets of parallel-bedded, laterally continuous F-VF Ss/Siltst. and thicker muds. Ss with regular to convolute to ripple-drift laminations. Bouma base cutout sequences common

Facies E Thinner, irregular and discontinuous beds of slightly coarser Ss and Silst. than D; also thinner muds. Ss with basal graded and structureless intervals; sharp upper contacts with mud

 Facies F Thick intervals of mildly deformed chaotic deposits derived from sliding or mass flow

Facies G Thick muds with often obscure continuous parallel bedding

A

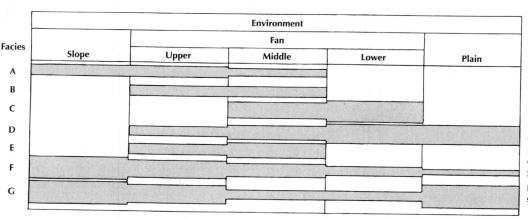

Shelf

Slope

Inner fan

Middle fan

Outer fan

Basin plain

Massive or stratified mudstone (facies G); discordant surfaces and facies F deposits from mass movements; lag deposits of sediment gravity flow; facies A in channels, locally succeeded by facies E

Facies G mudstones enclosing thick, broad channel-filling facies A, B, and F; facies E may be present

Essentially facies D and E; subordinate facies C; local intercalations of lenticular facies A, B, and F; typically showing thinning-and-fining-upward sandy cycles; paleocurrents show dispersion of about 90°

Facies D sediments with lenticular facies C; typically showing thickening-and-coarsening-upward sandy cycles; coarser deposits usually not channelized; paleocurrents spread over 90°, transverse or longitudinal to basin axis

Facies D, with thin intervals of facies G mudstone sometimes detectable; paleocurrents parallel to basin axis

Facies	Environment					Depositional processes
	Slope	Fan			Plain	
		Upper	Middle	Lower		
A						Debris flows, liquified flows
B						Debris flows, liquified flows, turbidity currents (high energy)
C						Turbidity currents
D						Turbidity currents (low energy)
E						Liquified flows, turbidity currents, traction currents (?)
F						Slumps, debris flows
G						Pelagic and hemipelagic sedimentation

B

Figure 10.14 (A) Geometry of submarine fans and the associations of facies, according to Mutti and Ricci Lucchi (1972.) (B) Distribution of submarine fan facies. (After Shanmugam and Moiola, 1985: 234; by permission of the Geological Society of America.)

the midfan to the large inner-fan channels with local olistostromes. If the lobes changed course, many such sequences may have accumulated vertically. The lower part of the rise and some of the abyssal plain are covered by pelagic muds, with local thin, very-fine-grained sandstones from the distal edge of each turbidite.

Because submarine fans represent pulses of sedimentation filling a deep marine trough, it has long been assumed that the accumulation of thick turbidite sequences was a reflection of tectonism that caused the trough to sink rapidly. Recently, however, Shanmugam and Moiola (1982), Mutti (1985), and Shanmugam, Moiola, and Damuth (1985) have shown that most episodes of submarine fan deposition are highly correlated with low sea levels. For example, during the Quaternary, turbidite deposition took place only during episodes of glaciation and low sea level. During interglacial periods, there was only pelagic deposition on the fan surface. Because

BOX 10.1 TERTIARY TURBIDITES OF THE NORTHERN APENNINES

The most famous and best exposed of the many well-studied ancient turbidite sequences around the world are those of the northern Apennine Mountains of Italy. The classic paper on these deposits by Mutti and Ricci Lucchi (1972; English translation, 1978) became the standard on which most later studies were based, and their facies terminology has influenced many sedimentologists. These thick flysch sequences were deposited in deep basins formed by the collision of Africa with the southern Mediterranean and the consequent rotation of major crustal blocks that make up the Italian Peninsula.

Mutti and Ricci Lucchi (1972) described the Apennine turbidites in terms of seven lithofacies, which they labeled A through G. *Facies A* (see Fig. 10.14) is represented by poorly sorted coarse-grained sandstones and conglomerates. These occur in massive beds (1 to 2 m thick) and are found in ancient channels. Thin lenses of silt or fine sand may be present, but they are rare. These conglomerates, believed to have been moved by debris flows, are apparently confined to the feeder channels in the slope and upper fan. *Facies B* (Fig. 10.1.1A) consists of medium- to coarse-grained sandstones in thick beds that are lenticular but more laterally continuous than facies A. Their most diagnostic structure is thick, parallel or broadly undulating current laminae. Facies B is closely associated with facies A and intergrades with it; it probably represents debris flows with some current reworking. The sedimentary structures suggest that they may be high-flow-velocity features, such as antidunes and plane beds. Both facies A and B appear to be produced by rapidly moving turbidity currents that fill the feeder channels along the slope and the inner fan.

Facies C (Fig. 10.1.1B) consists of the classic turbidites and is the dominant facies on the middle and outer fan (see Fig. 10.14). This facies is mostly medium- to fine-grained sandstones with well-developed Bouma sequences. The thick sandstone beds are generally planar and laterally continuous, although there may be broad, low-relief channels in places. The beds are typically 50 to 150 cm thick and may be separated by thin shale partings. These sandstones were deposited by the classical turbidity current mechanism described previously.

Facies D (Fig. 10.1.1C) consists of fine to very fine laminae with parallel or undulating surfaces or climbing ripple drift. The most diagnostic feature is the planar nature of the beds, which are typically 3 to 40 cm thick and show tremendous lateral continuity. Some beds can be traced for tens of kilometers. Facies D

sands typically alternate with facies G hemipelagic muds. Facies D appears to represent the lower-velocity portion of the Bouma cycle (Bouma intervals B–E) and results from low-energy turbidity currents that have left their coarse material behind. Facies D is found on the inner and middle fan but is more common on the outer fan; it is one of the predominant facies on the deep-sea plain.

Facies E (Fig 10.1.1D) sandstones differ from those of facies D in having higher sand-to-shale ratios (typically 1:1) and thinner, less regular, less continuous sandstone beds with frequent wedging and lensing. Typically, facies E sandstone beds are coarser grained and more cross-bedded than facies D and have ripples or dunes on the top. This facies was probably produced by liquefied grain flows and small currents and appears to have originated as overbank deposits from the channels in the upper and middle fan (see Fig. 10.14).

Facies A through E are products of turbidity currents in various guises; facies F and G are both associated with turbidites. *Facies F* is Mutti and Ricci Lucchi's (1972) label for any type of slump, slide, olistostrome, mudflow, or other type of chaotic deposit. These are found primarily on the slope and upper fan, where gravity sliding predominates. *Facies G* is pelagic and hemipelagic mudstones and shales. These are usually massive but can be vaguely stratified, especially when they are interbedded with one of the sandy facies. Of course, these muds occur in all parts of the deep sea, but they are dominant on the slope, inner fan, and abyssal plain. The occurrence of these facies and their depositional interpretation are summarized in Fig. 10.14.

The cliffs of the Apennines show spectacular examples of associations of facies that represent various parts of the marine slope-rise complex. The slope and inner fan are usually composed of thick sequences of facies G pelagic muds that have been cut by slumps, olistostromes, and thick fan channels (Fig. 10.1.2A) filled with facies A and B conglomerates and sandstones. Some of these channels reach 1 to 2 km in width. The middle fan is characterized by smaller, lenticular channels filled with facies A and B conglomerates and sandstones that represent more distal (outer) fan channels. These channels are carved into planar laminated turbidites of facies C, D, and E. In the outer fan (Fig. 10.1.2B), the fan channels have become very thin or have disappeared, so the sequence is planar, composed mostly of facies C and D turbidites with tremendous lateral continuity. This con-

A

B

C

D

Figure 10.1.1 Outcrops of submarine fan deposits from the classic examples of Mutti and Ricci Lucchi (1972.) (A) Facies B, fine- to medium-grained sandstone with almost no mud intervals. The dominant structures are thin laminae that gently undulate, like those formed by antidunes. The units are punctuated by numerous erosional surfaces. (B) Facies C, alternating thick, graded sandstones and thin mudstone interbeds. (C) Facies D, couplets of fine-grained, finely laminated sandstones with convolute bedding and climbing ripple drift, alternating with thicker mudstones. (D) Facies E, thinly stratified, irregular, discontinuous sandstones, slightly coarser than facies D, interbedded with thinner muds. The sandstones are not laminated but may be graded, and they often have dunelike top surfaces, which are in sharp contact with the overlying mudstone. Outcrops shown are from the Marnoso Arenacea Formation, Senio Vally, Italy. (Courtesy of E. Mutti.)

tinuity is even more striking in the deep-sea plain association (Fig. 10.1.2C), but the sequence is much thinner and finer grained than the outer fan. Typically, the deep-sea plain association is dominated by thin, fine-grained facies D turbidites alternating with facies G hemipelagites.

Although there are some superficial similarities between deep marine sandstones and those from other environments, their many unique or characteristic features make them easy to recognize. Well-developed graded bedding in Bouma sequences, coupled with distinctive sedimentary structures, is the

(continued)

(Box 10.1 continued)

A

B

C

Figure 10.1.2 Large-scale outcrops display various parts of the submarine fan. (A) Proximal (inner) fan association showing a well-developed submarine channel about 2 km wide, filled with facies A and B beds. (B) Distal (outer) fan association composed of thick, continuous, repetitive facies C and D turbidites. (C) Deep-sea plain association composed of facies D turbidites and facies G hemipelagic shales, showing extreme parallelism and lateral continuity of the beds. (Courtesy of E. Mutti.)

most characteristic feature. Only deep-sea fans and plains produce such thick and laterally continuous sequences of alternating sandstones and mudstones. Finally, there are many diagnostic trace fossils, and often foraminifers and other marine fossils as well, to distinguish these deposits from those of shallower water.

the lower sea level exposed and eroded the continental shelf, more sediment accumulated at the heads of submarine canyons. Eventually, this sediment became destabilized and avalanched down the canyons in turbidity currents. As in the case of shelf sediments, sea-level changes appear to be significant controlling influence of deposition.

Deep marine sandstones are important because they clearly indicate paleoceanographic conditions. Also, they sometimes are the main sedimentary fill of deep, rapidly dropping tectonic basins. Because turbidite sands are fairly coarse and are interbedded with impermeable shales, they have been important hydrocarbon reservoir rocks.

Diagnostic Features of Continental Slope and Rise Sediments

Tectonic Setting   Slope and rise sediments are found along all continental margins, but thick continental-rise wedges occur primarily in rifted passive margins. Tectonically downdropped or downwarped basins may also reach sufficient depths to accumulate thick sequences of shales and turbidites. Slope and rise deposits are typically associated with deep marine pelagic sediments.

Geometry   A wedge or thick lens builds against the continental margin. It may be thousands of meters

thick and hundreds of kilometers wide and extend thousands of kilometers along the base of the slope.

Typical Sequence Slope sequences are mainly hemipelagic muds, with contourites that are interrupted by submarine channel sands and by olistostromes and slump deposits. The main part of the fan is built of prograding turbidites, which coarsen upward from finer, thin, distal sands at the base to coarser, complete Bouma sequences to channel sands at the top. The lower fan and abyssal plain are covered with pelagic muds and episodic fine sand layers from distal turbidites.

Sedimentology The coarser fraction of a graded bed can contain many other minerals besides quartz, but the bulk of slope and rise sediment is fine sand, silt, and clay. Bouma sequences display a series of sedimentary structures reflecting decreasing flow velocities, from graded beds to higher-flow-velocity plane beds to lower-flow-velocity ripples and to finely laminated silts and muds. Olisthostromes and slump deposits are also characteristic of this system, as are sole marks, soft-sediment deformation, and certain types of ichnofossils.

Fossils Pelagic organisms, particularly benthic foraminifers, are the most diagnostic, but they are not very abundant. Clastic particles in turbidites can be made of broken shell debris from the continental shelf.

Pelagic Sediments

Deep ocean basins cover more than 50% of Earth's surface. This vast area of oceanic crust, which lies at depths of 4 to 6 km or more, is mostly a flat, featureless abyssal plain with occasional seamounts and mid-ocean ridges. In the past, geologists paid little attention to deep-sea sediments, partly because they were not likely to contain oil but mostly because so little was known about the deep sea until the advent of modern marine geology in the 1950s. According to Davis (1983), "It is safe to say that more has been learned about the geology of the deep ocean since 1965 than in all previous time."

Marine geology essentially began with the voyages of the British naval vessel HMS *Challenger* from 1872 to 1876. Before these voyages, it was widely believed that the waters of the deep sea were stratified by age, with trilobites still living at abyssal depths. The work of the *Challenger* scientists forever demolished this notion. John Murray (Murray and Renard, 1891) de-

scribed the deep-sea sediments that were recovered by the *Challenger*, and his general outline of deep-sea sedimentation is still valid. The modern era of marine geology began in 1947 with the Swedish Deep Sea Expedition, which used the first piston-coring device for taking a continuous sample of deep-sea sediments. In the 1950s and 1960s, marine geology continued to develop at institutions such as the Lamont-Doherty Geological Observatory in New York, the Woods Hole Oceanographic Institution in Massachusetts, and the Scripps Institution of Oceanography in California. In 1968, the maiden voyage of the *Glomar Challenger* (Fig. 10.15), the primary vessel of the Deep Sea Drilling Project (DSDP), began a revolution in the Earth sciences. The DSDP cores soon provided confirmation of seafloor spreading and the best and most detailed data ever gathered about the evolution of Earth's oceans and climates. Many other important discoveries resulted from the study of deep-sea cores, such as confirmation of the orbital variation theory of the ice ages and proof that the Mediterranean Sea dried up 6 million years ago (see Chapter 14). One of the most surprising discoveries was that the oldest oceanic crust is Jurassic in age and that all older seafloor has been subducted. The cycling of crust is so rapid that only the oceanic crust of the last 150 million years still survives, even though the Earth has had oceans for almost 4 billion years.

The major conclusion reached by Murray, confirmed by later work, is that *most of the terrigenous sediment from the continents (except for suspended clays) is trapped in the continental shelves and rises and never reaches the deep seafloor.* Instead, deep-sea sediments are **pelagic,** settling out of the overlying water column. The main constituents of deep-sea sediments are *terrigenous clays* from the continents; *biogenic skeletal material* of calcareous, siliceous, or phosphatic marine organisms; and minor amounts of *authigenic* components that develop diagenetically in deep-sea muds and oozes. In some places, deep-sea sediments also include occasional volcanic ash layers, eolian dust, or tektites from extraterrestrial impacts.

These constituents are not randomly distributed across the seafloor but show a definite pattern (Fig. 10.16). *Siliceous oozes* are predominant in equatorial and polar regions where oceanic upwelling occurs due to boundary currents between large water masses. *Calcareous oozes,* on the other hand, are predominant in tropical, subtropical, and temperate waters where the conditions are warm enough for calcite secretion and where the ocean floor is not so deep that they are dissolved. Terrigenous clays are found in the centers of

A   **B**

Figure 10.15 (A) The *Glomar Challenger,* the vessel that was the primary workhorse of the Deep Sea Drilling Project in the 1970s. It is now retired and was replaced by the *Glomar Explorer;* both are managed by Texas A&M University. The vessel is 122 m long and handles 7320 m of drill pipe. (Courtesy of Scripps Institution of Oceanography.) (B) Scale of deep-sea drilling operation at a depth of 5500 m. If the drawing were accurate, drill string width could not be detected at this scale, yet by using sonar beacons and steering devices on the drill, the vessel can relocate the same hole many times and recover continuous core. (From Kennett, 1982: 98; © 1982; by permission of Prentice-Hall, Inc., Englewood Cliffs, N.J.)

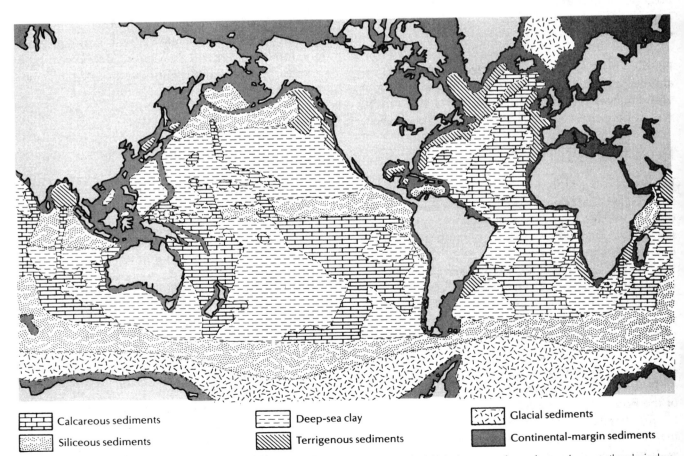

▦ Calcareous sediments	▤ Deep-sea clay	⬚ Glacial sediments
⠂ Siliceous sediments	▨ Terrigenous sediments	▰ Continental-margin sediments

Figure 10.16 The global pattern of deep-sea sediments. Calcareous oozes are restricted to low latitudes. Most siliceous oozes lie close to the poles, although some occur in the equatorial Pacific and Indian oceans. The areas of low productivity in the center of oceanic gyres have mostly pelagic clays. (After Riley and Chester, 1976: 180.)

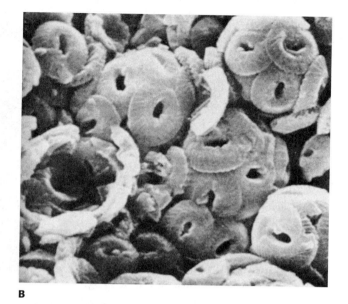

A

B

Figure 10.17 (A) Foraminifers (large objects) and radiolarians (smaller, coarse-meshed objects) from the Pacific Ocean floor. Diameter of radiolarians is about 100 to 200 μm. (Courtesy of Scripps Institution of Oceanography.) (B) Button-shaped calcareous plates (coccoliths) of microscopic algae known as coccolithosphorids. Most are less than a few tens of micrometers in size. (Courtesy of D. Noël.)

oceanic water masses where there is limited productivity and the seafloor is so deep that calcite dissolves. Glacial sediments are common in the polar regions, particularly around the Antarctic ice cap.

More than 47% of the deep seafloor is covered by calcareous oozes, which are made of the carbonate tests of planktonic microfossils. The dominant carbonate microfossils are foraminifers, particularly the family Globigerinidae, which are single-celled protozoans with a porous calcitic skeleton (Fig. 10.17A). Coccolithophorids, submicroscopic photosynthetic organisms covered with tiny porous calcareous plates, are also very important in many regions (Fig. 10.17B). Pteropods, tiny pelagic snails with cone-shaped or coiled shells made of aragonite, are common in certain shallow tropical waters, such as the Mediterranean Sea and the Gulf of California.

All these organisms live in great abundance in surface waters wherever the temperature is high enough, primarily in temperate and subtropical latitudes. In shallow waters, their skeletons can form great thicknesses of calcareous ooze. Near the surface, seawater is supersaturated with calcium carbonate ($CaCO_3$), so these organisms have plenty of material with which to build their skeletons. At a depth of about 500 m, however, the high pressure and cold temperature change the water chemistry so that calcite becomes undersaturated; this chemical change is related to the production of CO_2 by the respiration of organisms and also to the decay of dead organisms. When these organisms die, their calcareous skeletons

sink toward the seafloor, dissolving as they get deeper. At a certain depth, called the **lysocline,** the rate of carbonate dissolution reaches a maximum (Fig. 10.18). Dissolution then continues until a depth

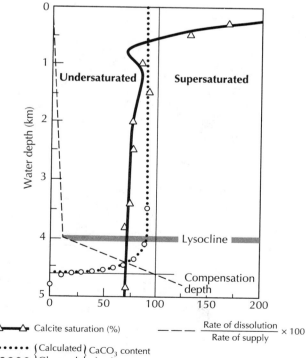

Figure 10.18 Variations in carbonate saturation with increasing depth in the equatorial Pacific Ocean. (After van Andel, Heath, and Moore, 1975: 40; by permission of the Geological Society of America.)

is reached at which the rate of calcite supply is balanced, or compensated, by the rate of dissolution; little or no calcareous material persists below this depth. This critical level is called the **carbonate compensation depth (CCD).** The depths of the lysocline and CCD vary, depending on the depth of the ocean basin and the kind of oceanic circulation patterns. Most calcium carbonate dissolution occurs at depths of 4 to 5 km, although in some places calcite survives to water depths of 6 km or more. Because the depth of most ocean basins is 4 to 5 km, only the shallower ocean basins, the floors of which are above the CCD, can retain the calcareous ooze.

It seems surprising that these tiny calcareous skeletons can survive dissolution from the first zone of undersaturation at 500 m to the CCD at 4000 m or more. This delay of dissolution is apparently caused by the protective organic coating on some shells (particularly coccoliths) or by the relatively large sizes of some foraminifers. In some regions, the sedimentation rate is so high that calcareous tests are buried rapidly enough to protect them from corrosive seawater. This is particularly true of calcareous oozes deposited on the mid-ocean ridges, which have been called the "snow-covered mountains of the deep

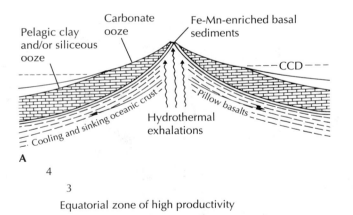

A

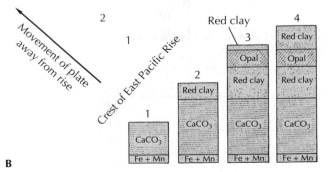

B

Figure 10.19 (A) Sedimentation across a mid-ocean ridge. (B) Modification of sediments as a plate migrates across belts of different biotic productivity in the ocean. (After Broecker, 1972: 53.)

ocean." The crest of the ridge (Fig. 10.19A) is typically covered by a thick layer of calcareous ooze. As spreading continues and the oceanic crust begins to sink away from the ridge, a bathymetric zonation begins to develop. First, at about 1.5 to 2 km, the pteropods dissolve because they are made of highly soluble aragonite. At the CCD, the foraminifers and coccoliths dissolve, unless they are buried. Below the CCD, only a carbonate-poor residual clay remains. Because oceanic crust is mobile, it can migrate under water masses and areas of varying sedimentation. The result might be a stratigraphic sequence in the crust that begins with ridge-crest calcareous oozes, followed by ridge-flank red clays, and then (if the crust migrates under a belt of equatorial upwelling) by a siliceous ooze (Fig. 10.19B).

Calcareous oozes form only in the deep sea because calcareous materials are swamped by terrigenous clastics on the continental margins. For example, the sediments of the Gulf of California are rich with microfossils, but they are so diluted by terrigenous clays that an ooze cannot form. In the deep sea, calcareous oozes predominate where the waters are warm and relatively shallow. They are not greatly restricted by currents, oceanic upwelling, fertility, or nutrients because the surface of the ocean is supersaturated with calcium carbonate (the most important mineral in calcareous plankton). In some cases, high fertility due to upwelling raises the CCD because fertility increases biogenic activity and therefore increases CO_2 concentration, which pushes the equilibrium toward dissolution. At the equator, however, high fertility lowers the CCD because calcareous shell building outstrips the supply of organic matter. In general, though, the primary limiting factors of calcareous oozes are depth and temperature.

Siliceous oozes are composed of the tests of planktonic organisms that build their skeletons of opaline silica, $SiO_2 \cdot nH_2O$. The most important of these are **radiolarians,** protozoans with radially symmetrical internal tests (see Fig. 10.17A), and **diatoms,** microscopic single-celled algae with porous, platelike or rodlike tests. Radiolarians dominate siliceous oozes in tropical waters, and diatoms are enormously abundant in polar siliceous oozes. Other minor components of siliceous oozes include sponge spicules (skeletal elements of siliceous sponges) and silicoflagellates (microfossils with simple tubular or ringlike tests).

Although radiolarians and diatoms are abundant in the same surface waters that teem with calcareous microplankton, their distribution on the seafloor is very different (see Fig. 10.16). Unlike calcite, *silica is under-*

saturated in all seawater. Silica is transported from the land by rivers, but in very small amounts. Because it is a rare nutrient, recycling is important. Siliceous oozes can form only where oceanic upwelling brings silica and other nutrients up from the bottom. This happens primarily in areas where oceanic water masses meet in zones of convergence or where cold bottom currents flow upward along a continental rise. Undersaturation is most severe in surface waters, where the abundant biotic activity depletes silica as rapidly as it becomes available. Silica has no compensation depth; its chance of dissolution is highest as the tests begin to sink and decreases with depth, the opposite of calcite (Fig. 10.20). For this reason, siliceous oozes can be found on any ocean bottom including the deepest, but they are scarce except where upwelling allows siliceous microplankton to flourish in the surface waters.

Murray and Renard (1891) described the sediments found over much of the ocean floor as "the red clay" and thought that this clay was a product of submarine eruption and weathering. When deep-sea clays were first studied by X-ray diffraction in the 1930s, however, it turned out that Murray was at least partly wrong. Most deep-sea clays are derived from the land and settle out of suspension in the deep sea. Their distribution on the ocean floor (Fig. 10.21; see also Fig. 6.8) clearly indicates their origin. Kaolinite, the predominant clay in tropical soils, is found offshore from the mouths of tropical rivers. Illite, the stable clay found in most temperate soils, is most abundant in the middle latitudes. Chlorite, found mainly in polar soils, is restricted to high latitudes. Montmorillonite, the dominant clay produced by weathering of subma-

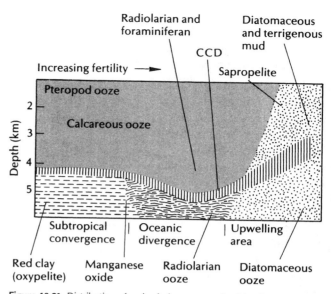

Figure 10.21 Distribution of major facies on a graph of depth versus fertility. (After Berger, 1974: 240.)

rine volcanics, is most common around mid-ocean ridges (especially the East Pacific Rise and the Mid-Atlantic Ridge) and near island arcs.

The ocean floor sediments contain not only clay minerals but also authigenic minerals (zeolites, manganese oxides and hydroxides); volcanogenic debris (mostly plagioclases and pyroxenes); eolian quartz and feldspar; manganese nodules; phosphatic fish teeth and earbones (otoliths); foraminifers, which build skeletons from sand; and a few sponge spicules and radiolarians. In short, "the red clay" is a term for everything that reaches the overlying water column and does not dissolve. As was shown in Fig. 10.16, this happens primarily in the centers of oceanic gyres, where there is only a single water mass and therefore little chance for mixing or upwelling.

In summary, deep-sea sediments are entirely pelagic and are derived mainly from terrigenous clays that settle out of suspension or from biogenic skeletal material. Their distribution follows a predictable facies pattern (Fig. 10.21). Calcareous oozes (which become limestones and chalks) are found wherever the water is sufficiently warm and shallow for them to form. Siliceous oozes (which become cherts, diatomites, and radiolarites) are found in abundance only in regions of upwelling. Where the ocean is not shallow enough, warm enough, or rich enough in nutrients from upwelling for calcareous or siliceous oozes to form, terrigenous clays dominate. An ancient example of this pattern is discussed in Box 10.2.

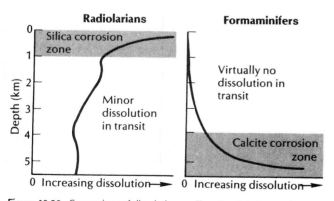

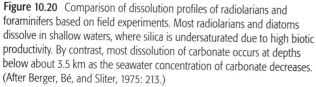

Figure 10.20 Comparison of dissolution profiles of radiolarians and foraminifers based on field experiments. Most radiolarians and diatoms dissolve in shallow waters, where silica is undersaturated due to high biotic productivity. By contrast, most dissolution of carbonate occurs at depths below about 3.5 km as the seawater concentration of carbonate decreases. (After Berger, Bé, and Sliter, 1975: 213.)

BOX 10.2 CRETACEOUS SHELF AND PELAGIC DEPOSITS OF THE WESTERN INTERIOR OF NORTH AMERICA

Cretaceous exposures in the Western Interior of North America show most of the environments discussed in Chapters 9 and 10. In the Cretaceous, the central part of North America was covered by a vast epicontinental seaway that stretched 5000 km from the Arctic to the Gulf of Mexico and 1400 km from what is now Utah to Iowa during its maximum transgression. In some places, more than 2200 m of marine sediments accumulated. The seaway was deepest over the modern-day Great Plains and shoaled westward over a region that would become the Rocky Mountains (Fig. 10.2.1). Consequently, a distinct set of facies was de-

posited around the margin of this great Cretaceous ocean. The center of the seaway was shallow (not much more than 300 m deep, according to Hancock, 1975) but had open marine conditions, producing pelagic shales and chalks. A series of shales and sandstones were deposited on the shallower shelf that rimmed the seaway; in some places, the transition to nonmarine environments is also preserved.

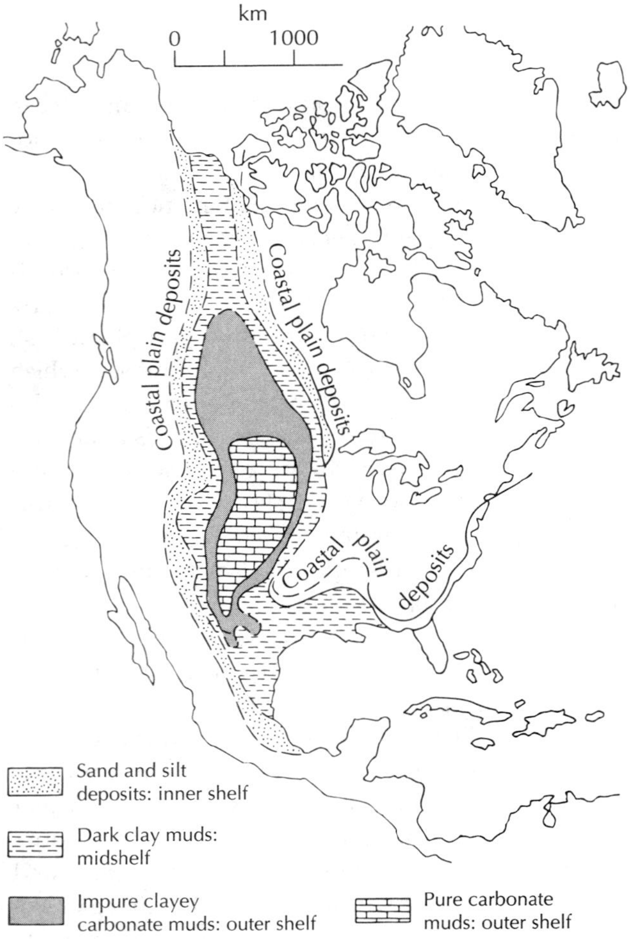

Sea level fluctuated continuously, so these facies belts did not accumulate in one place. Instead, they moved onshore and offshore with each change in sea level. In addition, extensive mountain building occurred in what is now the western United States throughout the Cretaceous, so orogenic episodes contributed much clastic sediment and caused local regressions. This produced a very distinctive facies pattern for the Western Interior Cretaceous (Fig. 10.2.2). In the center of the seaway to the east (the Plains region), deep-water shales and chalks prevailed. In the west (mostly Utah, Idaho, and Montana), orogenic activity produced nonmarine shales, sandstones, and conglomerates. In between, the marine and nonmarine shales were separated by a zigzagging series of sandstones representing various coastal and shallow shelf sands that were responding to the fluctuating shoreline. This shoreline area is concentrated in what is now New Mexico, Colorado, Wyoming, eastern Montana, Saskatchewan, and Alberta. When these areas were uplifted and deformed in the latest Cretaceous and the Cenozoic, many of these coastal and shallow marine deposits were exposed and are now found in the cliffs of the Rocky Mountains and Colorado Plateau.

The shallow marine shales of the Cretaceous seaway are punctuated by a number of long, narrow, shallow marine sand bodies (see Fig. 10.2.2). Because they are porous reservoir rocks that are enveloped by impermeable shales, they have been intensively studied and exploited for oil. Walker (1984) has reviewed some of the many studies of these enigmatic sand bodies. Dimensions vary greatly, but typically they are about 15 m thick, about 10 km wide, and tens of kilometers long (Fig. 10.2.3). A particularly well exposed example is the Semilla Sandstone Member of the Mancos Shale, located in the San Juan Basin of New Mexico (Fig. 10.2.4A). As described by La Fon (1981), this example shows a sequence (Fig. 10.2.4C) typical of most of these linear sand bars (Fig. 10.2.4B). The sequence coarsens upward from marine shales to mudstones and siltstones with thin cross-

Sand and silt deposits: inner shelf

Dark clay muds: midshelf

Impure clayey carbonate muds: outer shelf

Pure carbonate muds: outer shelf

Figure 10.2.1 General distribution of sediments in the Rocky Mountain seaway during peak transgression of the Greenhorn marine cycle (early Turonian.) (After Kauffman, 1969:229.)

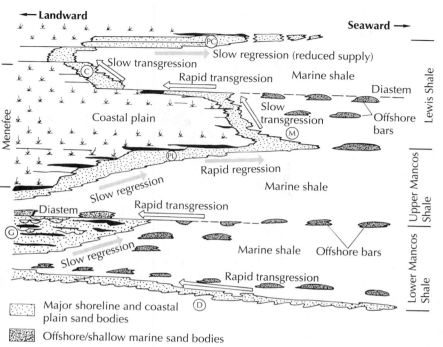

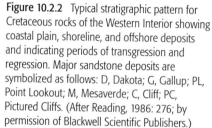

Figure 10.2.2 Typical stratigraphic pattern for Cretaceous rocks of the Western Interior showing coastal plain, shoreline, and offshore deposits and indicating periods of transgression and regression. Major sandstone deposits are symbolized as follows: D, Dakota; G, Gallup; PL, Point Lookout; M, Mesaverde; C, Cliff; PC, Pictured Cliffs. (After Reading, 1986: 276; by permission of Blackwell Scientific Publishers.)

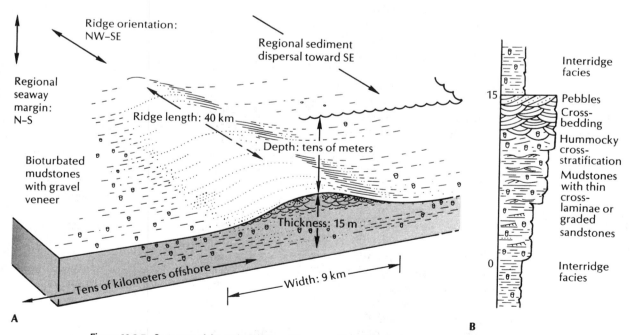

Figure 10.2.3 Summary of the main features of linear sand bars from the Western Interior Cretaceous. (A) Block diagram. (B) Typical stratigraphic sequence. (After Walker, 1984: 82.)

(continued)

(*Box 10.2 continued*)

laminated or graded sandstones. This silty sandstone facies contains small clay chips (1 mm thick and 2 cm across) and is extensively bioturbated and homogenized (Fig. 10.2.4D). As a consequence, the bedding is very poor, although there are shaly zones within this facies that are laminated. The silty sandstone facies reaches a maximum thickness of about 20 m but thins laterally until it pinches out beyond the margins of the bar (Fig. 10.2.4B). This facies contains a fauna of oysters, clams, gastropods, ammonites, crabs, and callianassid shrimp burrows (*Ophiomorpha*) that suggests well-agitated, shallow, normal marine waters.

The silty sandstone facies grades upward into a cross-bedded sandstone facies that caps the deposit (see Figs. 10.2.3 and 10.2.4). The cross-bedded sandstone facies is composed of thin-bedded (less than 15 cm thick), coarsening-upward, fine- to medium-grained sandstone. It is only slightly burrowed and contains abundant small-scale trough cross-beds (Fig. 10.2.4E). In some places, the beds show hummocky cross-stratification. The tops of beds are ripple-marked and are separated from overlying sandstone beds by clay layers about 1 cm thick. The base of each bed is usually paved with rounded, flat, oval clay chips up to 9 cm across. The base may also contain broken oyster shells, snail-shell fragments, shark teeth, and bleached phosphate pebbles. The top of the sand bar is capped with shale, phosphatic bone debris, and shark teeth, which suggests slow deposition; there is no evidence of emergence or of beach deposition.

The origin of Cretaceous offshore sand bars such as the Semilla Sandstone is still under discussion. They do not fit any of the modern analogs discussed earlier in this chapter. Most of the authors who have described these Cretaceous sand bars emphasized storms as the major transporting agent. Storms are one of the few forces that can transport sand this far offshore into the muddy shelf. In addition, the presence of hummocky cross-stratification and storm lags of mudchips and broken oyster shells indicates high transport energy. Tidal currents are thought to be of lesser importance. Modern tidal sand waves and ridges rest unconformably on transgressive surfaces with gravel lags, and these do not occur in most of the Cretaceous examples. Both tidal currents and storms can produce the cross-bedding that is found at the top of the sequence. A number of models have been proposed to describe the flow patterns of storm waves necessary to produce these offshore sand ridges, but no one model is clearly correct.

A

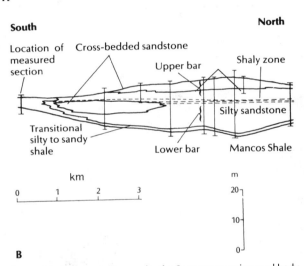

B

Figure 10.2.4 An ancient example of a Cretaceous marine sand body, the Semilla Sandstone of the San Juan Basin, New Mexico. (A) Outcrop showing silty sandstone facies (lower two-thirds of cliff) capped by cross-bedded sandstone facies (top of cliff.) (B) Cross section through Holy Ghost bar of the Semilla Sandstone. Note that the bar gradually thins and pinches out. (C) Stratigraphic section through the bar. (D) Lower silty sandstone facies, showing extensive bioturbation. (E) Cross-bedded sandstone facies showing erosional contacts and pockets created from erosion of clay pebble clasts. (Drawings after La Fon, 1981: 367; by permission of the American Association of Petroleum Geologists, Tulsa, Oklahoma; photos courtesy of N. La Fon.)

Bergman and Walker (1987) have shown that many of these offshore sand ridges have erosional surfaces capped by gravels, suggesting that the sand ridges were subaerially exposed during periods of low sea

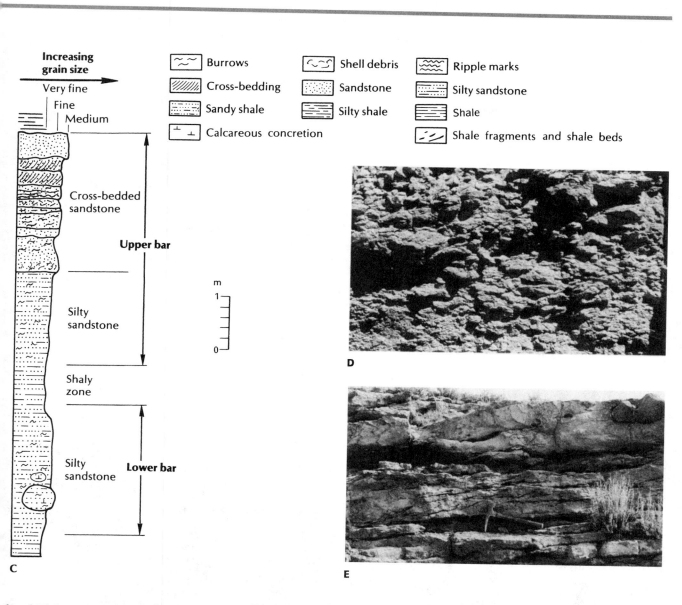

Increasing grain size →

Very fine
Fine
Medium

~~ Burrows

◠ Shell debris

≈≈≈ Ripple marks

/// Cross-bedding

Sandstone

Silty sandstone

Sandy shale

Silty shale

Shale

⊥ Calcareous concretion

Shale fragments and shale beds

Cross-bedded sandstone

Upper bar

Silty sandstone

Shaly zone

Silty sandstone

Lower bar

m
1

0

C

D

E

level. If this is so, then the presence of sand and gravel ridges on the Cretaceous shelf might be related to episodes of canyon cutting during periods of low sea level. When sea level rose again, the gravel source was cut off by the transgression. The sands and gravels were reworked and spread out, concentrating in the erosional channels and hollows. Long sand ridges surrounded by marine shales were the eventual result. As we saw with turbidites, sea-level changes were apparently the premier controlling factor in the Cretaceous as well.

Farther offshore from the coastal sands and offshore sand bars, marine conditions existed and pelagic sedimentation prevailed. The major sedimentary products of these conditions are the thick sequences of finely laminated shale and limestones that underlie most of the Great Plains: the Pierre Shale, the Mancos Shale, the Greenhorn and Niobrara limestones, and many other well-known units. Although the shales may appear monotonous at first, there is actually a distinct sequence of shale and chalk facies that formed in progressively deeper water. The facies that originated nearer to shore are predominantly pelagic shales, but the facies from the center of the seaway, which was deeper and relatively starved of mud, are clean limestones and chalks (Fig. 10.2.5A). Unlike some of the modern examples of pelagic sedimentation discussed earlier, the Western Interior Cretaceous seaway was never as deep as the modern ocean. Because water depths were seldom more than 300 m, there was no

(*continued*)

(Box 10.2 continued)

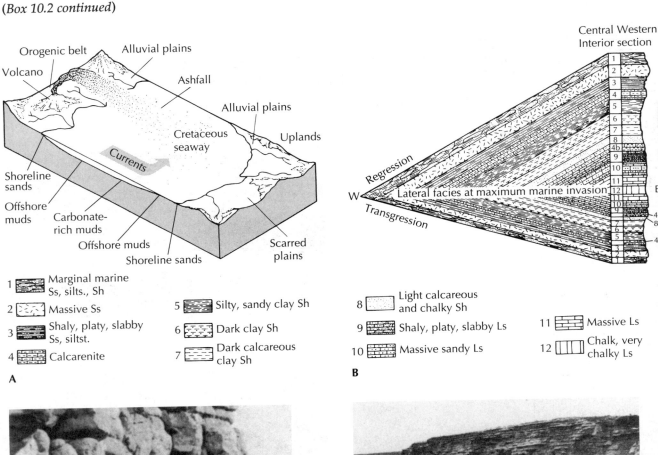

C

D

Figure 10.2.5 (A) The Western Interior seaway during the peak of the Greenhorn cycle. (B) Kauffman's (1969) model of Cretaceous marine sedimentation in the Western Interior, patterned after the Greenhorn cycle with complex interfingering removed. The western edge of the diagram would be in east-central Utah, and the eastern margin represents what is now western Kansas and Oklahoma. The numbers in the key correspond to Kauffman's 12 facies. (C) Contact between silty shales (facies 5) of the Blue Hill Shale and the overlying shoreface sandstones (facies 4) of the Codell Sandstone, exposed at Liberty Point, west of Pueblo, Colorado. (D) Contact between finely laminated black shales (facies 6) of the Graneros Shale and the overlying marly limestone (facies 7) of the Greenhorn Limestone in southwestern South Dakota. (E) Shaly, platy, slabby limestone with fragments of *Inoceramus* shells (facies 9) of the Lincoln Member, Greenhorn Limestone, Pueblo County, Colorado. (F) Exposure of massive, rhythmically bedded limestone (facies 11) of the Bridge Creek Member, Greenhorn Limestone, Russell County, Kansas. (G) Finely bedded chalk (facies 12) from open marine pelagic deposits, Niobrara Limestone, Gove County, Kansas. (Photos C and F from Pratt, Kauffman, and Zelt, 1985; photo D by N. H. Darton and photo E by G. R. Scott, courtesy of U.S. Geological Survey; photo G courtesy of Kansas Geological Survey.)

E

F

G

(continued)

(Box 10.2 continued)

significant carbonate dissolution. Nevertheless, most features of the Cretaceous seaway deposits clearly indicate open, pelagic conditions with relatively little influence from shoreline clastics upon the center of the seaway.

Kauffman (1969) recognizes at least 12 distinctive facies in succession, each resulting from part of the overall transgressive-regressive cycle (Fig. 10.2.5B). No more than 10 of these can be found in a single local section, however. Facies 1 and 2 are coastal-fluvial and shallow marine sandstones, respectively. Facies 3 is slabby, platy, shaly sandstone and siltstone, and facies 4 is a slightly more calcareous version of the same (Fig. 10.2.5C). These beds have thin sandstone laminae and starved ripples and are low in organic carbon. They are moderately bioturbated and contain scattered small oysters, fish teeth, fragments of inoceramid clams, and abundant benthic foraminifers. According to Kauffman (1969, 1985) and Hancock (1975), this facies represents a proximal offshore, or very shallow shelf, environment that was current-swept, warm, well aerated, and generally normal marine.

Facies 5, on the other hand, is a dark organic-rich clay shale that is very finely laminated and contains only a few silt laminae, probably reworked from distal storm deposits (Fig. 10.2.5C). It is very poorly bioturbated, and well-preserved and undisturbed volcanic ash layers are present. There are calcareous and sideritic nodules and many other geochemical features that suggest calcite saturation but reducing (low-oxygen) conditions. Only rare oysters and a few benthic foraminifers that are able to survive with little oxygen are found in this facies. According to Kauffman (1969, 1985), this facies probably represents middle offshore, quiet water, and highly reducing and oxygen depleted conditions. The stagnant bottom waters were hostile to benthic organisms and burrowing infauna, thus explaining the lack of fossils and bioturbation.

Facies 6 is a dark clay shale that is noncalcareous and completely lacking in silt or sand (Fig. 10.2.5D). There is very little organic carbon and only moderate bioturbation. The sparse fauna include a few low-oxygen-tolerant oysters, and most of the foraminifers are low-oxygen-tolerant agglutinated forms. This facies probably represents middle offshore (125 to 200 m deep), quiet water conditions that were occasionally slightly reducing but usually oxygenated. The lack of silt suggests that the facies was deep enough and far enough offshore that it was beyond the reach of distal deposits.

Facies 7 and 8 are calcareous clay shales that grade into Facies 9, a shaly, platy, slabby limestone (Fig.

10.2.5D, E). All these units are thin bedded; facies 8 and 9 contain abundant fecal pellets and a diverse warm-water fauna of snails, clams, ammonites, both planktonic and benthic foraminifers, and even corals. These factors suggest quiet, outer-shelf conditions, with warm, well-oxygenated, and well-circulated normal marine water.

Facies 10 is sandy, calcarenitic massive limestone. It contains some fine-grained quartz sand and is often mottled and bioturbated. Some of these beds are made almost entirely of inoceramid bivalve shell fragments. The quartz sand and the shell fragments suggest that this deposit formed in the open sea on a high-wave-energy bottom. Facies 11 is a massive, pure limestone that is very fine grained and well cemented. It often forms ledges. Its most striking feature is rhythmic cycles of marl passing up into limestone, in units 0.6 to 1 m thick (Fig. 10.2.5F). The alternation of marl and limestone, along with details of the petrography, suggest that these deposits reflect rapid changes in water depth (20 to 100 m or more) in response to changes in sea level. There appear to be many minor breaks or near breaks in sedimentation.

Facies 12 is the open-sea, pure-chalk facies found only in the deepest (150 to 300 m) water (Fig. 10.2.5G). It is finely bedded, with thin chalk beds that can be traced for hundreds of kilometers (Frey, 1972; Hattin, 1982). This great lateral persistence would be found only in open pelagic environments. Most of this chalk is composed of calcareous microfossils, primarily coccoliths and foraminifers. It is comparable to the thick Cretaceous chalk deposits found all over Europe, best known from the White Cliffs of Dover. In the Western Interior of North America, however, it is seldom white. It tends to be olive-gray to olive-black because it is rich in fecal pellets and organic matter. Common fossils include oysters, inoceramids, and fish bones and scales. Ammonites, belemnites, and rudistid clams are rare. The bottom was apparently very soupy because many types of oysters, bryozoans, barnacles, serpulid worms, and sponges attached themselves to any abandoned shell that could serve as a substrate. This facies is typical of the classic exposures of the Niobrara Chalk in Kansas (see Fig. 10.2.5G) and elsewhere in the Plains states.

The Western Interior Cretaceous seaway provides excellent examples of the offshore sand bar (barrier) environment and the open pelagic environment. While these examples are not typical of every comparable deposit in the geologic record, most of the salient features are present.

Diagnostic Features of Pelagic Sediments

Tectonic Setting Pelagic sediments form only in deep ocean basins. These sediments are uplifted and become part of subaerially exposed sequences in only a few places, such as subduction zones, where deep marine pelagic sequences (chalks and cherts) are scraped up into the accretionary wedge over a trench, along with ophiolites and other ocean-floor rocks. Some epicratonic basins are deep enough, or are so starved of clastic debris, that they have accumulated thick sequences of chalk, cherts, or black shales.

Geometry Deep-sea deposits form vast, thin, tabular sheets that are limited in area only by the size of the ocean basin. However, in fault-bounded pelagic basins, these deposits can reach thousands of feet in thickness.

Typical Sequence Homogeneous, thinly bedded, finely laminated chalks, cherts, or shales with little variation in grain size are typical. These grade laterally into coarsening-upward sandy turbidite sequences in many places.

Sedimentology Calcite (chalks), opaline silica (chert), and clay minerals (illite, kaolinite, chlorite, and montmorillonite) are the predominant components. Minor authigenic minerals, volcanic and eolian dust, phosphatic fish fragments, and manganese nodules also occur. Few sedimentary structures are known other than fine lamination and extensive bioturbation and burrowing.

Fossils Biogenic oozes are composed almost entirely of planktonic microfossils of types that clearly indicate their deep-sea origin. Deep-sea clays have minor amounts of planktonic microfossils (mostly siliceous or phosphatic).

CONCLUSIONS

Marine sedimentary deposits make up most of the stratigraphic record, since they are formed below the base level of erosion and have a high preservation potential. In fact, the sediments of the deep sea are formed by a steady "rain" of clays and microfossils from the surface, and they are the most complete record of geologic history preserved anywhere. These sediments also entomb most of the fossil record of marine invertebrates and vertebrates, and they can be faithful recorders of ancient climatic changes and oceanographic conditions. Many important economic deposits—including most of the world's oil, as well as clays, diatomites, and other economically valuable sedimentary rocks—are produced in the marine realm.

FOR FURTHER READING

In addition to the references listed at the end of Chapter 9, the following sources are worth consulting:

Bouma, A. H. 1962. *Sedimentology of Some Flysch Deposits: A Graphic Approach to Facies Interpretation.* New York: Elsevier.

Bouma, A. H., W. R. Normark, and N. E. Barnes, eds. 1985. *Submarine Fans and Related Turbidite Systems.* New York: Springer-Verlag.

Burk, C. A., and C. L. Drake, eds. 1974. *The Geology of Continental Margins.* New York: Springer-Verlag.

Doyle, L. J., and O. H. Pilkey, eds. 1979. *Geology of Continental Slopes.* SEPM Special Publication 27.

Dzulynski, S., and E. K. Walton, 1965. *Sedimentary Features of Flysch and Graywackes.* New York: Elsevier.

Heezen, B. C., and C. D. Hollister. 1971. *The Face of the Deep.* New York: Oxford University Press.

Kennett, J. P. 1982. *Marine Geology.* Englewood Cliffs, N.J.: Prentice-Hall.

Lisitzin, A. P. 1972. *Sedimentation in the World Ocean.* SEPM Special Publication 17.

Middleton, G. V., and A. H. Bouma, eds. 1973. *Turbidites and Deep Water Sedimentation.* SEPM Pacific Section Short Course, Anaheim, Calif.

Mutti, E., and F. Ricci Lucchi. 1972. Turbidites of the
    Northern Apennines: Introduction to facies
    analysis. *International Geology Review* 20(2):
    125–166. Trans. T. H. Nilsen, 1978. Reprinted
    by the American Geological Institute,
    Falls Church, Va.

Stanley, D. J., and D. J. P. Swift, eds. 1976. *Marine
    Sediment Transport and Environmental Management.*
    New York: John Wiley.

Swift, D. J. P., D. B. Duane, and O. H. Pilkey, eds. 1972.
    Shelf Sediment Transport: Process and Product.
    Stroudsburg, Pa.: Dowden, Hutchinson, and Ross.

Tillman, R. W., and C. W. Siemers, eds. 1984.
    Siliciclastic Shelf Sediments. SEPM Special
    Publication 34.

Tillman, R. W., D. J. P. Swift, and R. G. Walker. 1985.
    Shelf Sands and Sandstone Reservoirs. SEPM Short
    Course 13.

Warme, J. E., R. G. Douglas, and E. L. Winterer, eds.
    1981. *The Deep Sea Drilling Project: A Decade of
    Progress.* SEPM Special Publication 32.

III

Biogenic, Chemical, and Other Nonsiliciclastic Sedimentary Rocks

The Guadalupe Mountains of west Texas, viewed from the south. The prominent peak, El Capitan, is composed of massive limestones that constituted the core of the reef. To the right are the steeply dipping beds on the flank of the reef. To the left are lagoonal deposits of limestone and evaporites. (Shelton, 1966.)

11 Carbonate Rocks

Many limestones are composed entirely of fossils and fossil fragments. This richly fossiliferous limestone from the Upper Ordovician Platteville Group, near Dickeyville, Wisconsin, includes abundant D-shaped strophomenids and other brachiopods, numerous branching bryozoans, massive bryozoans (*bottom center*), kidney bean–shaped ostracodes, and even a fragment of a trilobite (*upper left corner*). (D. R. Prothero.)

THE EARLIER CHAPTERS DESCRIBED TERRIGENOUS sedimentary rocks, those composed mainly of siliciclastic material—fragments of pre-existing rocks and minerals transported and deposited as *grains*. This chapter covers limestone and dolostone, called carbonate rocks because they contain large amounts of carbonate (CO_3).

Carbonates are the most abundant nonterrigenous sedimentary rocks, constituting roughly one-tenth of the Earth's sedimentary shell (Table 11.1). In contrast to terrigenous rocks, they form chemically and biochemically. Dissolved *ions* carried from source to depositional site in solution eventually precipitate and form solid minerals. The distinction between

TABLE 11.1 Distribution Through Time and Space of Types of Modern Carbonate Sediment and Ancient Limestone

Type	Modern Carbonate	Ancient Limestone
1. Precipitation		
Direct inorganic precipitation	Rare travertine in caverns and tufa around springs	Both extremely rare; uncertain temporal extent.
Direct organic precipitation	Modern reefs, abundant in a band from 40°N to 40°S latitude in shallow marine areas where siliciclastic supply is low	Generally abundant throughout the Phanerozoic and back into late Proterozoic. Organic reef-makers differ as a consequence of evolution. Peak developments controlled by tectonism and organic evolution.
2. Lithification of unconsolidated sediment		
Nonmarine	Rare scattered shallow carbonate muds deposited on lake bottoms	Rare lacustrine marls of Phanerozoic and Proterozoic age.
Shallow marine	Abundant carbonate sediments of the neritic zones (intertidal, inner shelf, and outer shelf) circumscribing modern continents where siliciclastic sediment supply is low (the perimeter of the Florida Peninsula) or on isolated shallow marine plateaus (such as the Bahama Banks)	Abundant neritic carbonates of epicontinental (epeiric) sea deposits. Peak developments coincide with episodes of continental subsidence and/or ocean high stands. Abundant throughout the Phanerozoic; progressively less common back into the Proterozoic. No Archean examples.
Deep marine	Calcareous oozes abundant on abyssal plain areas that lie above the carbonate compensation depth, beneath surface waters where carbonate-secreting plankton thrive, and where sediment masking is minimal	Rare examples exist. Sedimentary rock sequences of deep-sea origin are rarely incorporated into continental blocks. None are older than Jurassic because carbonate-secreting plankton were not abundant until the Jurassic.
3. Miscellaneous	Scattered and rare unconsolidated limestone rubble, collapse rubble, and caliche	Very rarely produced or preserved. Unknown extent through time.

biochemical and chemical sedimentary rocks is not always clear-cut. Organisms have a role in the formation of many sedimentary rocks. For example, they can extract dissolved components from seawater to manufacture shells or skeletons that later are incorporated into sedimentary rocks. Organisms can also be indirectly involved, such as when their me-

tabolism modifies the geochemical setting enough to cause mineral precipitation.

The distinction between clastic sedimentary rocks and "crystalline" chemical and biochemical rocks such as carbonates can also be ambiguous. Besides broken or whole skeletal matter, many limestones contain composite grains formed of an interlocking mosaic of

calcite or aragonite. These composite grains include **ooids** and intrabasinal detrital carbonate called **intraclasts;** such grains often experience a history of transport and deposition as *clasts.* Thus, some limestones can be considered clastic sedimentary rocks, but they are not siliciclastic. Finally, because most dolostone is the product of diagenetic alteration of limestone, dolostone is secondarily a chemical sedimentary rock.

The Importance of Limestone

Limestone tells us much about the origin and evolution of the Earth. The texture and composition of limestone reveal as much about depositional setting as do the texture and composition of sandstone, conglomerate, and breccia. Limestones are the most important evidence remaining of shallow marine seas that covered much of the world during most of the Phanerozoic. Because they are almost invariably fossiliferous (many are made of nothing but fossils), they are probably the best documentation of organic evolution.

Limestones have an economic importance out of proportion with their abundance. Because they dissolve easily, they are often porous in the subsurface and become reservoir rocks for petroleum and natural gas. About 80% of the hydrocarbons in North America occur in carbonates, and about 50% of the world's petroleum is recovered from carbonate reservoir rocks (half of the Persian Gulf's immense pools of oil is trapped in dolostones). Because carbonate rocks are important aquifers, there is much research into how water flows through limestone caverns and fissures. High porosity also makes limestones excellent host rocks for ore-bearing solutions (such as the Mississippi Valley lead-zinc deposits). Cement is crushed limestone mixed with various clays and other components, and limestone is one of the most popular building materials in the world. For example, the Pyramids of Egypt and the Empire State Building are made of limestone. Limestone is widely used in agriculture to neutralize soil acid, in metallurgy as a flux for smelting iron and steel, and in other industrial processes that require an inexpensive base to neutralize acid.

Limestone deposition in the oceans is partially responsible for Earth's unique atmosphere. Without this process and organic evolution, Earth's atmosphere would resemble those of Venus and Mars (mostly carbon dioxide). Mercury, Venus, Earth, and Mars are small, dense bodies that lie close to the Sun. Planetary astronomers and historical geologists believe that these four terrestrial planets originally possessed identical atmospheres of carbon dioxide and water vapor degassed by volcanic activity from their interiors. Mercury lost its atmosphere completely because it lies so close to the Sun and because of its low gravity. The carbon dioxide–rich atmospheres that mantle Venus and Mars are simply residues that remained after water vapor was condensed. On Earth, by contrast, deposition of marine limestone and photosynthesis removed most of the carbon dioxide from the atmosphere, locking it up in sedimentary rock. These processes left a residual atmosphere composed of minor gases such as ammonia and nitrous oxide. Further alteration of these gases (by reaction with free oxygen produced from the photochemical dissociation of water vapor or as a by-product of photosynthesis) resulted in the present atmosphere of 80% nitrogen and 20% oxygen.

At present, limestone deposition is occurring at a much slower rate than it has in the geologic past. Two contrasting states of the Earth, **icehouse** and **greenhouse,** produce dramatic differences in seawater temperature and chemistry. During much of the Phanerozoic (for example, from the Ordovician through the Devonian and during the Jurassic and Cretaceous), warm greenhouse conditions prevailed. There were no polar ice caps, sea level was higher, the oceans were less stratified, and enormous volumes of limestone could be deposited. Under cold icehouse conditions (the Permo-Triassic, the late Cenozoic, and presently), sea level is low, there are extensive polar ice caps, ocean water is stratified, and little carbonate is formed. Modern shallow marine carbonates are now restricted to a few pathetic, ghostlike remnants of a once extensive distribution. The mineralogy of carbonates also reflects ocean water conditions, with high-Mg calcite precipitated under greenhouse conditions and aragonite and low-magnesium calcite formed under icehouse conditions. Carbonate environments are discussed further in Chapter 12.

Carbonate Mineral Chemistry

All carbonate minerals are formed by combining divalent cations (2+)—particularly calcium and magnesium, as well as minor amounts of iron, strontium, manganese, and barium—with carbonate anions $(CO_3)^{2-}$. Although there are about 60 natural carbonate minerals, only 3 are abundant in the Earth's crust (or at the surface): calcite, aragonite, and dolomite. The first two are polymorphs of $CaCO_3$. Calcite is a

soft mineral that readily fizzes in acid. Aragonite (the "mother of pearl" lining mollusk shells and pearls) is less stable than calcite under most conditions and tends to alter into calcite during diagenesis.

Calcite is the rhombohedral form of calcium carbonate; aragonite has an orthorhombic lattice. This difference is important. The orthorhombic lattice of aragonite is much more open, so cations can be substituted for one another more easily than is possible in the tighter rhombohedral lattice of calcite. Seawater is supersaturated with calcium carbonate, so organisms have no trouble removing it to make their skeletons. They often use cations other than calcium, none of which fits the lattice perfectly. Calcium ions (Ca^{2+}) have a radius of 0.99 Å; ions of Mg^{2+} (0.66 Å), Fe^{2+} (0.74 Å), and Mn^{2+} (0.80 Å) can fit in the calcium sites. But Ba^{2+} (1.32 Å) and Sr^{2+} (1.12 Å) are too large to fit in the calcite lattice. They can substitute for calcium in the less constricted aragonite lattice, however. Because magnesium is too small to fit in the large spaces available in the aragonite lattice, magnesium rarely replaces calcium in aragonite.

The free substitution of magnesium for calcium in calcite is common. Indeed, high-Mg (greater than 5% Mg) calcite is often treated as a distinct mineral, because many organisms (especially calcareous sponges, imperforate foraminiferans, octocorals, corallinacean algae, many crustaceans, and echinoderms) secrete high-Mg calcite (with as much as 15% to 25% magnesium replacing calcium) (Table 11.2). Other organisms (most corals, pteropod mollusks, *Halimeda* calcareous algae, and the pearly layer on mollusk shells) make their skeletons out of aragonite. Because aragonite is unstable, however, fossils of these organisms now consist of calcite that was transformed from aragonite. In some cases, the instability of aragonite actually decreases the chances of fossilization. This is particularly true of pteropods, tiny planktonic snails with cap- or cone-shaped shells. Although pteropods are abundant in some surface waters, they are seldom preserved in deep-sea sediment. Their aragonitic shells dissolve too quickly.

Other organisms effectively resist any substitutions and from the outset form skeletons of only stable, low-Mg calcite. These animals include articulate brachiopods, mollusks (except for the pearly nacreous layer), most foraminifers, and barnacles. Not surprisingly, these groups are well represented in the fossil record.

Dolomite, $CaMg(CO_3)_2$, is the third important carbonate mineral. The structure of dolomite consists of alternating layers of Ca and Mg ions separated by layers of CO_3. Ideally, it should have equal amounts of Ca and Mg, but the average dolomite is about 56% Ca and 44% Mg. A few large calcium ions are squeezed into smaller magnesium lattice spaces. Ferrous (Fe^{2+}) iron also substitutes for the magnesium. If enough ferrous iron is present, it forms ferroan dolomite or **ankerite,** with the composition $Ca(MgFe)(CO_3)_2$.

Carbonate Geochemistry

To better understand calcium carbonate precipitation, a brief review of chemical reactions, especially chemical reaction equilibrium constants, is appropriate.

Equilibrium Constants: A Case Study of the System $H_2O = H^+ + OH^-$

Consider the chemical reaction A + B = C + D. A and B are referred to as the *reactants*; C and D are the *products*. An equal sign is used to indicate that the reaction proceeds with equal speed in both directions. This type of chemical reaction is the exception in nature. Most chemical reactions are displayed with unequal-sized arrows pointing in opposite directions to convey symbolically the effect of the reaction's equilibrium constant.

The equilibrium constant for a reaction (k) equals the product of the reaction product concentrations divided by the product of the reactant concentrations. For example, the equilibrium constant (k_1) for the reaction of A + B (the reactants) to produce C + D (the products) is

$$k_1 = \frac{(C)(D)}{(A)(B)}$$

Like all chemical reactions, however, this reaction can occur in the reverse direction (C + D reacting to form A + B), for which a second equilibrium constant (k_2) is needed:

$$k_2 = \frac{(A)(B)}{(C)(D)}$$

Although these two reactions occur simultaneously, they normally proceed at different rates, reflecting the different values of the two equilibrium constants. For example, if k_1 is larger than k_2, the reaction A + B → C + D will proceed from left to right (as indicated by the arrow). Conversely, if k_2 is larger than k_1, the reaction A + B ← C + D proceeds from right to left.

TABLE 11.2   Variation in Skeletal Carbonate Mineralogy among Organic Groups

Taxon	Aragonite	Calcite mol % $MgCO_3$ 0 10 20 30	Both Aragonite and Calcite
Calcareous algae: Red		x—x	
Green	x		
Coccoliths		x	
Foraminifers: Benthonic	0	x–x	
Planktonic		x—x	
Sponges	0	x—x	
Coelenterates: Stromatoporoids[a]	x	x?	
Milleporoids	x		
Rugose[a]		x	
Tabulate[a]		x	
Scleractinian	x		
Alcyonarian	0	x—x	
Bryozoans	0	x—x	0
Brachiopods		xx	
Mollusks: Chitons	x		
Pelecypods	x	x–x	x
Gastropods	x	x–x	x
Pteropods	x		
Cephalopods (most)	x		
Belemnoids[a]		x	
Annelids (serpulids)	x	x—x	x
Arthropods: Decapods		x–x	
Ostracods		x–x	
Barnacles		x–x	
Trilobites[a]		x	
Echinoderms		x—x	

Note: x. common; 0, rare
[a] Not based on modern forms
Source: Scholle, 1978: by permission of AAPG, Tulsa, Okla.

An overall equilibrium constant, k_3, describes the reaction A + B → C + D as well as A + B ← C + D. It is calculated by dividing k_1 by k_2. The relative speed by which the two reactions proceed simultaneously is shown by using arrows of different sizes. For example, if $k_1 > k_2$, then reaction A + B ⇌ C + D proceeds from left to right.

An example of this is the reaction by which water (H_2O) dissociates into cations of hydrogen (H^+) and anions of hydroxyl (OH^-). The overall reaction is made up of two distinct reactions that occur simultaneously:

$$H_2O \rightarrow H^+ + OH^- \quad k_1 \text{ (reaction 1)}$$

$$H^+ + OH^- \rightarrow H_2O \quad k_2 \text{ (reaction 2)}$$

Equilibrium constant k_1 describes reaction 1, the dissociation of water into ions of hydrogen and hydroxyl. Equilibrium constant k_2 describes reaction 2, the com-

bining of hydrogen and hydroxyl anions to form water. The overall reactions should be shown correctly as

$$H_2O \xrightleftharpoons{} H^+ + OH^-$$

The force driving the reaction from left to right, k_1, is proportional to the concentration of H_2O. The force driving the reaction from right to left, k_2, is proportional to the concentration of hydrogen and hydroxyl. Again, the two reactions occur simultaneously but at differing rates. The overall equilibrium constant for the two simultaneous reactions, k_3, is calculated by dividing k_1 by k_2. The value of k_3 is 1×10^{-14}. This small value means that very little water dissociates into hydrogen and hydroxyl ions.

A practical consequence of this value of k_3 is the standard scale of pH that describes acidic or basic (alkaline) solutions. The pH of a solution is the negative log to the base 10 of the hydrogen cation concentration. When the concentration of hydrogen cations equals the concentration of hydroxyl anions, the solution is described as neutral. The pH is 7, and the hydrogen cations and hydroxyl anions each have a concentration of 1×10^{-7}. Acidic solutions (pH from 1 up to 7) have more hydrogen cations than hydroxyl anions. Basic solutions (pH from more than 7 up to a maximum of 14) have more hydroxyl anions than hydrogen cations.

Carbonate Precipitation: Reactions and their Equilibrium Constants

Table 11.3 lists the four reactions that occur simultaneously in seawater to form limestone.

In *reaction 1*, atmospheric carbon dioxide combines with water to form carbonic acid. This reaction's small equilibrium constant means that very little carbonic acid is produced when rain falls through the atmosphere or when atmospheric carbon dioxide is absorbed into seawater.

In *reaction 2*, the carbonic acid produced by reaction 1 spontaneously dissociates into hydrogen cations (H^+) and bicarbonate anions (HCO_3^-). The equilibrium constant for reaction 2 is much smaller than that for reaction 1. This means that very few hydrogen cations are produced.

In *reaction 3*, the bicarbonate produced by reaction 2 dissociates into more hydrogen cations and carbonate anions ($CO_3)^{2-}$. The equilibrium constant for reaction 3 is even smaller than that for reaction 2.

In *reaction 4*, dissolved calcium and dissolved carbonate combine (with or without the assistance of organisms) to form solid calcium carbonate (aragonite or calcite).

Ignoring for the moment the equilibrium constants for reactions 1, 2, and 3, **LeChatelier's principle** (adding components to the left side of a reaction will cause the reaction to proceed from left to right, and vice versa) implies that calcium carbonate formation is favored by conditions that promote absorption of atmospheric carbon dioxide into water. But *if the equilibrium constants for these reactions are taken into account, precisely the opposite is true.* Because reaction 2 generates hydrogen cations at a far greater rate than reaction 3 generates carbonate anions, any carbonate anions produced by reaction 3 will be consumed by the far larger number of hydrogen cations produced by reaction 2. In other words, *adding carbon dioxide to water actually decreases the concentration of carbonate; conversely, removing carbon dioxide from water increases the concentration of dissolved carbonate.* This conclusion has far-ranging implications for limestone formation. *The precipitation of limestone is promoted by any process that removes carbon dioxide from water.*

Controls on Carbonate Deposition

We can now predict which physical conditions favor formation of limestone simply by considering which conditions remove carbon dioxide gas from seawater. Simple kitchen chemistry helps. Consider a can of capped soda or beer, for example. What promotes the escape of gas? Three obvious choices come to mind: uncap the can to remove the pressure, heat it, or shake it. Carbon dioxide dissolved in seawater is analogous to the fizz in soda or beer. Three physical conditions control how much CO_2 can be dissolved: temperature, pressure (water depth), and degree of agitation.

TABLE 11.3 Carbonate Geochemistry: Reactions and Equilibrium Constants

1. CO_2 (gas) + H_2O (water) $\xrightleftharpoons{}$ H_2CO_3 $k = 10^{-1.43}$
2. $H_2CO_3 \xrightleftharpoons{} H^+ + HCO_3^-$ $k = 10^{-6.40}$
3. $HCO_3 \xrightleftharpoons{} H^+ + CO_3^{2-}$ $k = 10^{-10.33}$
4. $Ca^{2+} + (CO_3)^{2-} \xrightleftharpoons{} CaCO_3$ (limestone) $k = 10^{-8.33}$ (aragonite)
    $k = 10^{-8.48}$ (calcite)

Note: By convention, relative arrow lengths in each set of reactions indicate in which direction each normally proceeds.

1. *Temperature.* Raising the temperature of sea-water promotes limestone deposition. Modern carbonate sediment and ancient limestone form more readily in tropical seas than in polar waters.

2. *Pressure.* Reducing the pressure (or depth) of seawater promotes limestone deposition. Modern carbonate sediment and ancient limestone form more readily in shallow water than in deep water. In settings such as the modern Bahama Banks, the effects of temperature and pressure combine to promote carbonate deposition. The upwelling of cold, dense bottom water onto shallow shelf areas warms the water and reduces pressure. Carbon dioxide is removed and limestone is formed.

3. *Degree of agitation.* Breaking waves in the surf zone mix seawater with air. This agitation promotes limestone formation because additional carbon dioxide is absorbed by the atmosphere. For example, modern fringe reefs grow faster in directions that face breaking waves.

Warm temperature, shallow depth, and agitation greatly enhance the likelihood that limestone will form, but four other factors are also important: organic activity, sediment masking and clogging, light, and carbonate compensation depth.

4. *Organic activity.* Plants and animals either precipitate calcium carbonate directly or modify the geochemical environment enough for precipitation to occur. Animals such as clams, snails, brachiopods, and zooplankton extract their calcareous skeletons from seawater directly. Plants such as phytoplankton and algae promote calcium carbonate precipitation because they remove carbon dioxide from seawater by photosynthesis (combining it with water to produce organic tissue and energy).

5. *Sediment masking and clogging.* As we will discuss further in Chapter 14, the kind of sediment accumulating at any point in time and space reflects what isn't happening as much as what is; this is the **sedimentary masking effect.** Even where carbonate sediment is forming, if clay and silt are being supplied more rapidly, (calcareous) mudrock rather than limestone accumulates. Evidently, large accumulations of limestone can occur only when other kinds of sediment are being deposited at exceedingly slow rates. In addition, an influx of mud can clog the filter-feeding apparatus and gills of many marine organisms. When this happens, many invertebrates that promote carbonate deposition cannot survive.

6. *Light.* Because photosynthetic organisms (especially calcareous algae and hermatypic corals) require light for photosynthesis, most large carbonate accumulations form in water shallow enough (less than 20 m deep) for adequate light to penetrate. Muddy water further inhibits the growth of corals and algae that depend upon clear water for sufficient light to grow.

7. *Carbonate compensation depth (CCD).* As discussed in Chapter 10, the temperature and pressure of very deep seawater control the areal distribution of calcareous ooze on the modern abyssal plains. Calcareous ooze consists of the unconsolidated shells of floating pelagic organisms that thrive in the photic zone that extends from the water surface to about 200 m. After organisms die, their shells settle to the seafloor, much as snowflakes fall through the atmosphere and accumulate on the ground. Several factors control the rate at which calcareous ooze accumulates. The rate at which shells are supplied from above is very important and reflects organic productivity. This rate varies with water temperature and is largely a function of latitude, water depth, and the supply of organic nutrients. Equally important is the rate at which shells are destroyed by dissolution (see Fig. 10.18). This is controlled by the carbonate compensation depth, as discussed in Chapter 10. The CCD is analogous to the permanent snow line in mountains, which is simply the elevation above which there is year-round snow cover. The snow line coincides with the precise elevation at which the total rate of snowfall and the total rate of snow melting are in balance. Above the snow line, more snow accumulates annually than melts; below it, more snow melts annually than accumulates. The CCD marks the water depth at which slowly falling calcium carbonate sediment dissolves at precisely the same rate as it is supplied from above (see Fig. 10.19). Calcareous oozes accumulate only above the CCD; below it, calcium carbonate dissolves at a faster rate than it is supplied (Fig. 10.20). The depth of the CCD varies with water temperature. Colder temperatures increase the rate of solution. In modern oceans, the depth of the CCD ranges from 4500 or even 5000 m in warm equatorial waters to 3000 m in polar waters.

Limestone Components and Classification

The most useful schemes for classifying limestone recognize the diverse origin of carbonate rocks and the clastic aspect of their texture. Components are normally recognizable using thin sections, and a number of standard reference guides facilitate the procedure (For example, see studies published by Scholle, 1978; Milliman, 1974; and Adams, Mackenzie, and Guildford, 1984.) Where distinctive clastic components are absent, a general descriptive scale of crystalline grain size can be used (Table 11.4).

One of the most popular classifications was devised by R. L. Folk (1959, 1962). It separates **allochemical** and **orthochemical** components. Allochemical (from the Greek *allos*, "elsewhere" or "from outside") components (or **allochems**) are any grains of calcium carbonate that, after formation, are transported and deposited as clasts. They are analogous to rock and mineral fragments in the framework fraction of terrigenous sandstone. There are different types of allochems, including coated grains (such as ooids) and skeletal fragments (bioclasts). We will discuss their origin and distinguishing characteristics shortly.

Orthochemical (from the Greek *orthos*, "straight" or "true") components (or **orthochems**) are not transported and deposited as clasts. Orthochems are found precisely where they formed or have been moved only a short distance. There are two kinds of orthochems: (1) microcrystalline calcite matrix, or **micrite,** is fine-grained (finer than 4 microns in diameter) carbonate mud, analogous to matrix in wacke sandstone; (2) microcrystalline sparry cement, or **spar,** is relatively clear interlocking crystals of calcium carbonate, analogous to cement in arenite sandstone.

Figure 11.1A shows the four principal limestone families: (1) sparry allochemical rocks, (2) microcrystalline allochemical rocks; (3) micrite; and (4) biolithite. Sparry allochemical rocks resemble arenites and consist chiefly of such allochems as skeletal fragments or ooids glued together by interstitial sparry cement. Microcrystalline allochemical rocks are similar to wackes. They consist of such allochems as skeletal fragments or ooids floating in or intermixed with fine-grained microcrystalline matrix (micrite). Micrite is analogous to mudrock. It consists principally of microcrystalline calcium carbonate matrix in which are scattered less than 10% allochems. Biolithites are rocks that were cemented together into limestone while the organisms that constitute them were still alive and growing. These include reef limestone and stromatolites.

Allochems: Characteristics and Genesis

Allochems are subdivided into (1) skeletal (biogenic) grains and (2) nonskeletal grains.

Skeletal Allochems There are many types of skeletal grains. The various kinds present in a limestone reflect age and depositional setting. Geologic age is important because a variety of calcium carbonate–secreting organisms have evolved over the geologic past. Depositional setting dictates paleoecology. The specific biological community that exists in any given setting is determined by such factors as water depth, water temperature, salinity, and turbidity. Fossil components in limestone (and in sedimentary rocks generally) reflect both age and depositional environment.

Identifying organic remains is a complex task that is accomplished by using thin sections (Fig. 11.2). Thin sections randomly cut three-dimensional fossils in two dimensions. With experience and the use of published guides, a carbonate petrographer can identify specific skeletal remains by size, external shape, and internal organization (see, for example, Horowitz and Potter, 1971; Scholle, 1978).

TABLE 11.4 Standard Scale for Describing Crystalline Grain Sizes	
Terminology	Crystal Grain Size (mm)[a]
Extremely coarsely crystalline	>4
Very coarsely crystalline	1–4
Coarsely crystalline	0.25–1.0
Medium crystalline	0.062–0.25
Finely crystalline	0.016–0.062
Very finely crystalline	0.004–0.016
Aphanocrystalline or cryptocrystalline	<0.004

Source: Folk (1962).
[a] Carbonate rocks in which readily distinguishable allochemical components occur are broadly subdivided into major categories on the basis of the grain diameter of such components as follows: calcirudites (carbonate conglomerates and breccias) have clast diameters in excess of 1 mm; calcarenites (carbonate sandstones) have clast diameters from 0.062 to 1.0 mm: and calcilutites (carbonate mud) have clast diameters less than 0.062 mm.

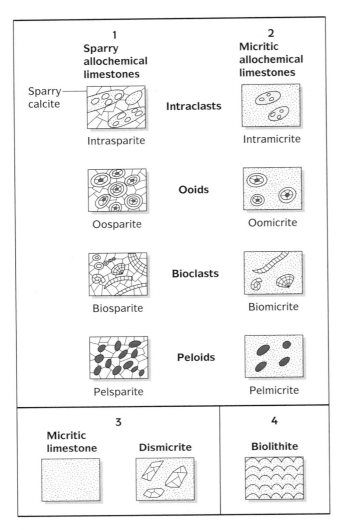

A

Figure 11.1 (A) R. L. Folk's (1959, 1962) classification of limestones, which uses prefixes to indicate the framework grains present (*bio-* for fossils, *pel-* for peloids, *oo-* for ooids, and *intra-* for intraclasts) and stems to indicate whether the interstitial calcite is micritic or sparry. If the rock is originally bound together (as in a reef rock), it is a *biolithite*. (B) Textural maturity classification of limestones proposed by Folk (1962). Textural maturity classes are based on the percentage of allochems present, their degree of sorting, and the extent of rounding (a function of abrasion history). (After Folk, 1962: 71, 76; by permission of the American Association of Petroleum Geologists, Tulsa, Okla.)

	Over $\frac{2}{3}$ lime mud matrix				Subequal spar and lime mud	Over $\frac{2}{3}$ spar cement		
Percent allochems	0–1	0–10	10–50	Over 50		Sorting poor	Sorting good	Rounded and abraded
Representative rock terms	Micrite and dismicrite	Fossiliferous micrite	Sparse biomicrite	Packed biomicrite	Poorly washed biosparite	Unsorted biosparite	Sorted biosparite	Rounded biosparite
Terminology	Micrite and dismicrite	Fossiliferous micrite	Biomicrite			Biosparite		
Terrigenous analogs	Claystone		Sandy claystone	Clayey or immature sandstone		Submature sandstone	Mature sandstone	Supermature sandstone

▮ Lime mud matrix   ▨ Sparry calcite cement

B

Common minerals (rare minerals in parenthesis)	Appearance of thin section in ordinary transmitted light	Appearance of thin section under crossed polars	Examples
Homogeneous prismatic Calcite (Aragonite)	No visible structure	Extinction in one direction; optic axes parallel and usually normal to surface of skeleton	Trilobites, ostracods
Granular Calcite Aragonite	Irregular grains (if fine and uniform in size, sometimes referred to as sugary or sucrosic)	Random orientation of optic axes	Foraminifers
Normal prismatic Calcite (Aragonite)	Polygonal prisms normal to outer surface Long Transverse	Each prism extinguishes as a unit Long Transverse	Punctate brachiopods *Inoceramus*
Foliated Calcite	Thin parallel leaves of calcite often having a wavy banded appearance	Variable orientation of optic axes of leaves	Bryozoans, pseudopunctate brachiopods, worm tubes, oysters
Nacreous Aragonite	Regular thin parallel leaves of aragonite (separated by organic films)	Parallel extinction	Mollusks
Single crystal Calcite	Coarse single calcite grain showing cleavage	Grain extinguishes as a unit	Echinoderms, sponge spicules
Crossed lamellar Aragonite (Calcite)	Layer of large lamellae, each lamella composed of small flat crystals, uniformly inclined in plane of larger lamella, giving herringbone pattern	Uniform orientation of optic axes in small crystals sometimes causes large lamellae to extinguish as a unit	Mollusks
Spherulitic fascicle Aragonite (Calcite)	Fibers radiating fanlike outward from a point center that is dark due to the concentration of very fine crystals	Each fiber extinguishes as a unit	Coelenterates

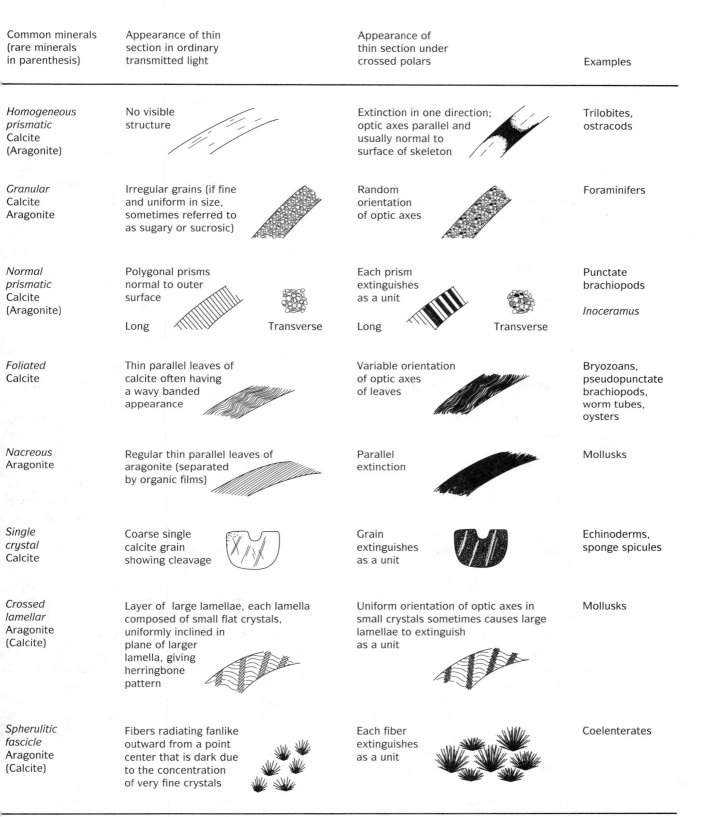

Figure 11.2 Common types of skeletal microstructure in thin section as seen in plane-polarized light and crossed nicols. (After Scoffin, 1987: 17; by permission of Blackie, Ltd., Glasgow.)

Nonskeletal Allochems Four major types of non-skeletal grains are recognized: (1) various coated grains (ooids, pisoids, and oncoids); (2) peloids; (3) clumped or aggregated grains (lumps, grapestones); and (4) limestone clasts (limeclasts)—that is, fragments of pre-existing limestone derived from intra-basinal and extrabasinal sources. These clasts are ripped up, transported, redeposited, and confined within limestone strata.

1. *Coated grains.* There are several types of coated calcium carbonate grains. All share a common spherical to subspherical shape, but they vary in size and degree of internal organization (Fig. 11.3). Ooids (also called ooliths) have a nucleus (often a skeletal fragment or a small clast of detrital quartz) around which concentric layers of calcium carbonate are wrapped (Fig. 11.3A, B). Ooids have diameters of less than 2 mm; most are the size of very fine sand (from 0.2 to 0.5 mm in diameter). Coarser coated grains that are identical to ooids in shape and internal organization are called **pisoids** (or pisolites) (Fig. 11.3C). Oncoids (or oncolites) are the same size as ooids or may be much larger, but they are irregularly shaped (Fig. 11.3D). They contain no obvious nucleus; individual coated laminae vary in thickness and show irregular overlap.

Modern ooids form in marine settings at shallow water depths. Although ooids can form in water as deep as 15 m, most form at less than 5 m. Maximum development occurs at depths of less than 2 m. Agitation by waves, tides, and storm currents promotes the growth of ooids in shoal areas around the perimeter of the Bahama Platform and in the tidal channels and deltas of the Persian Gulf. These modern ooids consist of acicular needles of aragonite. Some show tangential orientation with long axes subparallel with the ooid laminae. Radial microfabrics can also form, with aragonite needles fanning out in all directions from the nucleus, perpendicular to the laminae. Ancient ooids are composed of calcite rather than aragonite, and fibrous crystals typically fan out radially from the grain center. Recrystallization and dolomitization often obscure the original texture and mineralogy.

Ooids and pisoids can be biogenic, inorganic, or both. They apparently grow by simple accretion as wave, tidal, and storm currents sweep grain nuclei back and forth in shallow marine seawater supersaturated with dissolved calcium and carbonate. This is comparable to the growth of a snowball as it rolls across a slope blanketed with freshly deposited snow. The presence of radial microfabrics in modern ooids and pisoids, and the lack of any apparent mechanism for accreting successive discrete layers, however, makes many carbonate specialists skeptical about this mode of origin. Also, most modern ooids contain bits and pieces of organic matter that coat and permeate the carbonate laminae. This suggests that organisms play a role, but geologists disagree about whether the calcium carbonate is precipitated directly by organisms such as colonial algae, is a by-product of bacterial activity in organic matter, or occurs inorganically in a chemical milieu altered by photosynthesis.

The irregular thickness of individual laminae and their tendency to overlap suggest that oncoids form under less uniform conditions than ooids. Most are produced biogenically; modern examples have living encrustations of algae, foraminifers, and corals and form in various settings, including freshwater and brackish water reef complexes, shallow marine tidal flats and platforms, and (rarely) even deeper water carbonate settings. Many large oncoids appear to be made of pieces of ripped-up algal mat that have been rolled around into an irregular ball.

2. *Peloids.* Most peloids are fine-grained (0.1 to 0.5 mm) sand- to silt-sized clasts of microcrystalline carbonate that lack a coherent internal structure (Fig. 11.4A,B). The term implies neither size nor mode of formation. Most peloids are rounded to subrounded, but they can be spherical, subspherical, ellipsoidal, or irregular in shape. Peloids form in shallow marine low-energy platform carbonate settings such as the lagoonal areas of the Bahama Banks. They originate in various ways. Many peloids are fecal pellets of waste matter generated by such organisms as fish and shrimp. Others are produced by the micritization of other kinds of allochems: ooids, oncoids, intraclasts (limeclasts), and abraded skeletal grains.

3. *Grain aggregates.* Grain aggregates form when such carbonate particles as ooids and peloids adhere to one another (Fig. 11.4C). Most are fine- to medium-grained sand-sized masses that are given specific descriptive names: grapestones, lumps, and botryoidal lumps (Illing, 1954). Though distinctive in appearance, these masses are not genetically different. Interstitial pores are filled with micrite, but most aggregates are bound together by encrusting organisms. Grain aggregates form-

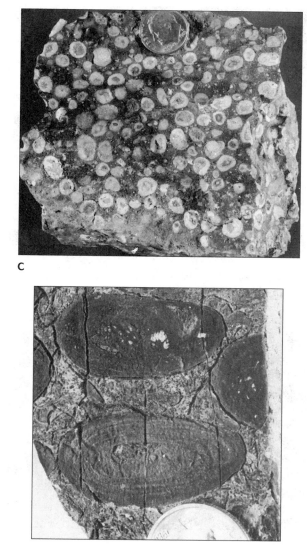

Figure 11.3 Several types of coated grains and concentrically banded particles occur in carbonate rocks. (A) Modern oolitic carbonate sand, Cat Cay, Bahamas. The dime shows the scale; most ooids are much smaller than 2 mm in diameter. (D. R. Prothero.) (B) Oolitic limestone with clear sparry cement showing the internal structure of ooids in thin section. This example is from the Cambrian Warrior Formation of Pennsylvania. 1, Radial-concentric ooids with peloidal core and purely radial center; diameter is 1.7 mm. 2, Purely radial ooids. 3, Radial ooids with two concentric coatings. Note that the diameters of the inner radial parts of the ooids is similar. (Heller et al., 1980: 944; by permission of the SEPM.) (C) Concentrically banded pisolites are much larger than ooids and usually form in calcrete soil horizons. Pisolites are also known to form in caves, in lakes, and even in normal marine settings by the same process as ooids. (D. R. Prothero.) (D) These irregular, ovoid, concentrically laminated structures are oncolites (from the Lower Cambrian Chambless Limestone, Marble Mountains, California; typical diameter is 1 to 2 cm.) They form when algal mats are ripped up and rolled about or encrusted around a nucleus. (D. R. Prothero.)

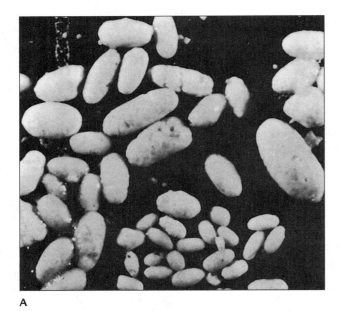

A

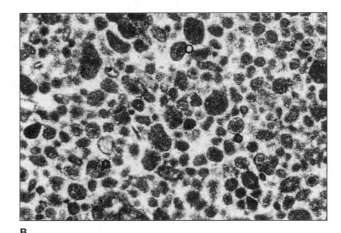

B

C

Figure 11.4 Pellets and peloids are a third type of carbonate framework grain. (A) Fecal pellets produced by various species of marine animals (10 × magnification.) (P. E. Cloud, courtesy of U. S. Geological Survey.) (B) Peloidal limestone with clear sparry calcite cement, from the Jurassic of the Venetian Alps, Italy. Length of elongate peloid in the right center is 0.9 mm. (Courtesy of A. Bosellini.) (C) Cemented clusters of peloids or ooids are called grapestone. (Siever, 1988: 18; by permission of W. H. Freeman and Company, New York.)

ing today on the Bahama Banks occur in a band that separates shallow-water, well-washed ooid shoals from deep-water lagoonal areas in which micrite and peloidal micrites are accumulating. Water depths are very shallow (typically 3 m or less), and only weak bottom currents are present. These conditions permit the growth of thriving surface mats of living microbes that help to bind grains together.

4. *Limestone clasts.* Limestone clasts (or limeclasts) are ripped-up and transported fragments of pre-existing carbonate derived from extrabasinal and intrabasinal sources (see Fig. 5.6B). Extrabasinal limeclasts differ in composition, texture, and overall character from the limey material in which they occur. Intrabasinal clasts are eroded from the same stratigraphic unit, or even the same stratum, in which they are found now. Partially consolidated lime muds can be ripped up and transported short distances by currents generated by storm waves or tidal activity.

Orthochems: Characteristics and Genesis

Micrite is calcium carbonate mud that occurs as matrix in allochem-bearing limestone or by itself as limey mudrock (calcilutite). It consists of silt and clay fragments (0.004 mm or finer) of aragonite (in modern sediment) or calcite (in ancient limestone). Micrite is typically translucent under the microscope with a dull brown cast. It is dull and opaque in hand specimen.

Spar is carbonate cement. Crystals of spar are generally coarser than micrite. Crystal diameters typically range from 0.02 to 0.1 mm, but crystals of spar as fine as 0.001 mm (microspar) occur, so spar and micrite sizes overlap. Under the microscope, spar is crystal clear without the hazy brownish cast of micrite. **Neomorphism**—the various diagenetic processes of recrystallization and replacement, including changes in mineralogy—is very common in carbonate rocks. Yesterday's micrite can become today's spar, and vice versa.

Micrite is polygenetic and is most commonly formed by the physical disintegration of calcareous green algal masses within which biogenically produced needles of aragonitic are disseminated. Micrite also forms when allochems such as peloids and limeclasts disaggregate; it can also be the by-product of mechanical abrasion and bioerosion (organic burrowing and ingestion) of carbonate grains. It is also possible that some micrite is produced by direct chemical precipitation, although this is still disputed. Favorable conditions for this process are found in seawater across broad stretches of the Bahama Banks.

Spar forms as a simple, primary, pore-filling cement or is generated secondarily by recrystallizating micrite.

Limestone Classifications of Folk and Dunham Compared

R. J. Dunham (1962) proposed a second classification scheme that is used as much as Folk's (1959, 1962) scheme. Both classifications distinguish allochems, matrix or micrite, and sparry calcite cement, and both emphasize texture. Both are so widely used, to the virtual exclusion of other limestone classification systems, that each warrants further discussion. Many geologists use the two schemes interchangeably, or side by side, so it is worthwhile to know them well.

The Folk Classification   Figure 11.1A shows Folk's four major limestone types. Sparry allochemical limestones and micritic allochemical limestones are subdivided on the basis of the kind and proportion of allochems and given composite names. The name stem indicates whether the interstitial material is spar or micrite. A prefix identifies the predominant allochem, and a suffix indicates allochem size as either finer (*micrite*) or coarser (*rudite*) than 1.0 mm. For example, a limestone with a sparry cement and

ooids and shell fragments would be an oobiosparite; one with peloids and intraclasts surrounded by lime mud would be an intrapelmicrite. Micritic limestone contains less than 10% allochems; specific varieties are named based on the predominant allochem. The recrystallized, bioturbated micrite is called **dismicrite.** Biolithites are limestones that were crystallized directly from the activity of reef-building corals or algae.

Folk (1962) introduced the concept of limestone textural maturity, which is determined by measuring the grain-to-matrix ratio (GMR) (see Fig. 11.1B). Textural maturity adds precision to limestone description and allows energy conditions at the depositional site to be implied. Stronger or more frequent currents (contingent in most instances on shallower depth) abrade away micrite; allochems become better sorted and, with continued abrasion, better rounded.

The Dunham Classification   The one major aspect of limestones not reflected in Folk's classification is whether the sparry calcite is primary cement or a secondary recrystallization of micrite. In other words, any limestone with a sparry calcite cement is a *sparite* even if it started out as a *micrite.* Since this determination is often hard to make, Folk's classification is much more descriptive and objective. Some geologists felt that the classification should also separate primary from secondary sparry limestones. The easiest way to do this is to incorporate grain support into the classification. Limestones in which all the grains touch others (grain-supported) were originally porous and were later cemented. This is **primary sparry cement.** Limestones with grains floating in spar (not grain-supported) were probably originally composed of grains floating in micrite that has been recrystallized. This is **secondary spar,** a replacement after micrite. These features are important in the Dunham classification scheme (Fig. 11.5A).

The Dunham classification emphasizes limestone texture, especially grain (allochem) packing and the ratio of grains to matrix. Allochem type is ignored. Five types of limestones are identified: **mudstone, wackestone, packstone, grainstone,** and **boundstone.** All except boundstone accumulate as clastic carbonates; individual components are not bound together during deposition.

Mudstone, wackestone, and packstone contain mud, which Dunham defined as any silt- or clay-sized grains, regardless of composition. Mudstone and wackestone are mud-supported. They are limestones

Depositional Texture Recognizable					Depositional Texture Not Recognizable
Original components not bound together during deposition				Original components bound together during deposition, as shown by intergrown skeletal matter, lamination contrary to gravity, or sediment-floored cavities roofed over by organic or questionably organic matter and too large to be interstices	Crystalline Carbonate
Contains mud (particles of clay and fine silt size)			Lacks mud and is grain-supported		
Mud-supported		Grain-supported			
Less than 10% grains	More than 10% grains				(Subdivide according to classifications designed to bear on physical texture or diagenesis)
Mudstone	Wackestone	Packstone	Grainstone	Boundstone	

A

Allochthonous Limestone — Original components not organically bound during deposition						Autochthonous Limestone — Original components organically bound during deposition		
Less than 10% >2 mm components				Greater than 10% >2 mm components		By organisms that build a rigid framework	By organisms that encrust and bind	By organisms that act as baffles
Contains lime mud (<0.03 mm)			No lime mud	Matrix-supported	Supported by grain components coarser than 2 mm			
Mud-supported		Grain-supported						
Less than 10% grains (>0.03 mm <2 mm)	Greater than 10% grains						Boundstone	
Mudstone	Wackestone	Packstone	Grainstone	Floatstone	Rudstone	Framestone	Bindstone	Bafflestone

B

Figure 11.5 Classification scheme for limestones proposed by Dunham (1962.) (A) An outline of the six original limestone varieties in Dunham's 1962 classification. (After Dunham, 1962: 117; by permission of the American Association of Petroleum Geologists, Tulsa, Okla.) (B) Modification of the Dunham classification by Embry and Klovan (1972), adding five additional categories for reef rocks (three of these were boundstones in Dunham's original scheme.) (After Embry and Klovan, 1972: 676; by permission.)

in which allochems are scattered through a rock that is basically micrite. Packstone contains less mud and is grain-supported. Grain-supported limestones typically have their allochems in tangential contact. Grainstone contains no mud, and allochem grains support one another. Limestones in which the components have been bound together from origin (such as reef rocks) are called boundstones (equivalent to Folk's biolithites). A sixth category, crystalline carbonate, refers to any limestone in which the original depositional texture is unrecognizable.

Embry and Klovan (1972) further modified the original Dunham classification to provide niches for limestones that contain allochems coarser than 2 mm (see Fig. 11.5B). Those with a matrix-supported texture are called **floatstone.** Those with a grain-supported texture coarser than 2 mm are called **rudstone.** Bound-

stones are further subdivided into **framestone, bindstone,** and **bafflestone.**

Limestone Diagenesis

Because calcium carbonate is so soluble, diagenesis occurs more readily in limestones than in terrigenous siliciclastic rocks. In fact, diagenesis can take place at surface conditions through changes in the pore water chemistry. This is how beachrock is formed, as mentioned in Chapter 7. Almost any carbonate material that has been subjected to even moderate burial has undergone significant diagenesis, and in many cases the limestone is so recrystallized that its original texture is unrecognizable. Thus, nearly every study of ancient limestones involves the interpretation of both the diagenetic history and the original depositional texture. When

metamorphism recrystallizes a limestone completely, it become a marble.

Diagenetic Textures

Diagenetic processes such as cementation occur as commonly in sandstones as they do in limestones, but other processes, such as dissolution and replacement, are much more common in carbonate rocks.

Dissolution produces pore space by dissolving pre-existing minerals. This process is particularly important in carbonate rocks because it often creates additional porosity that might serve as a hydrocarbon trap. Consequently, the petroleum industry has spent much time and effort on carbonate diagenesis research in recent years. Although high-Mg calcite and aragonite are stable in seawater, they are easily dissolved in fresh water and groundwater; high-Mg calcite is even more soluble than aragonite. For this reason, very little high-Mg calcite or aragonite survives in the pre-Cenozoic record, except in cases of unusual preservation.

Dissolution can leave a variety of distinctive and interesting textures in a limestone. Often it produces voids where aragonitic or high-Mg calcitic fossils were selectively dissolved but the low-Mg calcitic matrix persisted. This produces **secondary porosity** (in contrast to **primary porosity,** the original void space between grains in a clean, well-sorted sand). Selective dissolution may etch out the centers of ooids, for example, and leave concentrically banded rinds (Fig. 11.6A). Selective dissolution is particularly common in dolomitic rocks. If the matrix has been dolomitized but the framework skeletal grains remain as calcite, then dissolution leaves a mass of dolomite with voids shaped like fossils.

Particularly distinctive are the dissolution features known as **stylolites** (Fig. 11.6B). These structures are jagged, irregular seams dividing the limestone into two parts that interpenetrate. Each side of the seam has toothlike projections that fit into the cavities of the opposite side. These interpenetrating tooth-and-socket structures are highlighted by a residue of insoluble opaque minerals (such as hematite) or organic matter, which shows that much limestone has been dissolved to leave so much residue behind.

Stylolites are formed by pressure solution. When limestones undergo pressure during burial, they first dissolve rather than deform. The direction of pressure is usually perpendicular to the plane of the stylolites. In some cases, the amount of dissolution can also be estimated. For example, if a carbonate grain of known shape is crossed and partially dissolved by a stylolite

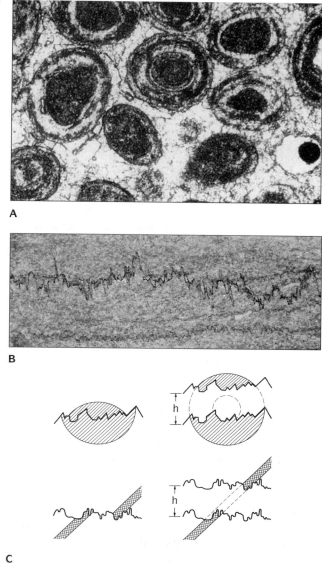

Figure 11.6 Diagenesis can be detected by its effects on the textures of limestone components. (A) The cores of these Pleistocene ooids from the Miami Limestone, Coral Gables, Florida, have been dissolved away, leaving secondary porosity. In some, the laminae have been selectively dissolved; in others, most of the ooid interior has been completely dissolved. These ooids are about 0.2 mm in diameter. (Courtesy of H. S. Chafetz.) (B) A stylolitic seam in a limestone showing the jagged boundary of dissolution highlighted by a residue of dark minerals. (D. R. Prothero.) (C) The volume of limestone dissolved along a stylolite can be estimated by the amount of material missing in an offset carbonate grain or seam or by the missing cross section of a circular crinoid stem.

(Fig. 11.6C), the original volume of that particle before dissolution can be estimated. Studies of stylolites show that, typically, about 25% of the original rock is missing; in some cases, as much as 90% of the original limestone has been dissolved away.

Cementation occurs when carbonate is precipitated into a pre-existing void space. It is one of the most

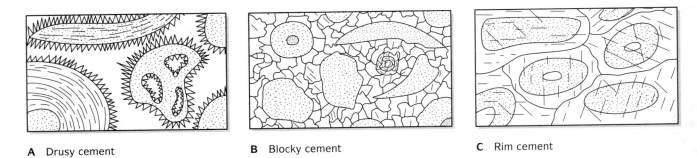

A Drusy cement   **B** Blocky cement   **C** Rim cement

Figure 11.7 Cement types in carbonate rocks. (After Blatt, Middleton, and Murray, 1980: 498; by permission of Prentice-Hall, Inc., Englewood Cliffs, N.J.)

common diagenetic processes, because limestones are easily dissolved and reprecipitated. In carbonate rocks, the cements often have distinctive habits and may reveal a sequence of cementing events. For example, the first generation of cement lining the voids often forms needlelike crystals that radiate away from the rim of the void and toward the center; these are known as **drusy cement** (Fig. 11.7). Other cements show larger, irregular, equant patches and mosaics of sparry calcite known as **blocky cement.** This cement often fills the remaining void space as a second generation of cement, following a first generation of drusy cement. In other limestones, cements form overgrowths around framework grains, resulting in **rim cement. A syntaxial overgrowth** consists of a framework grain composed of a single crystal of calcite (such as an echinoderm fragment) that has been overgrown by calcite cement precipitated in the same optical orientation. When thin sections of such grains are rotated on the microscope stage under crossed polars, the core grain and its overgrowth become extinct (go black) at the same time.

Replacement involves the simultaneous dissolution of original material and precipitation of a new mineral while preserving the original form. Many structures, such as ooids or fossils, can be replaced with all the fine details still intact, so there must have been slow, step-by-step dissolution and immediate cementation. Such a process suggests that the pore waters were in a delicate balance between the dissolution of one mineral and the precipitation of its successor.

Replacement is recognized when a structure of known composition, such as a fossil or ooid, is clearly made of a mineral that is not original. Aragonitic fossils or ooids that are now calcite, and calcium carbonate fossils that are now dolomite, are two common examples. As discussed earlier, micrite is originally aragonite but is often replaced by coarsely sparry calcite (secondary or neomorphic spar). In addition to

carbonates, other minerals can replace the components of limestone. The most spectacular examples occur when silica in the form of chert or chalcedony replaces carbonate. Many cherts preserve beautifully silicified fossils, ooids, peloids, or even the original calcite cementation texture of drusy and blocky cement (Fig. 11.8).

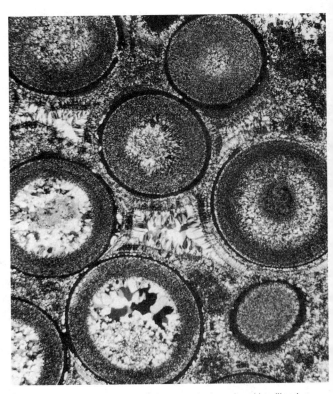

Figure 11.8 Some limestones can be so perfectly replaced by silica that the original texture is almost completely preserved. This is the famous State College Oolite from the Cambrian Mines Formation, State College, Pennsylvania. Except for some recrystallization of the cores of the ooids, the concentric structure of the ooids (1.0 to 1.5 mm in diameter), the drusy primary cement around the rims of the pores, and the blocky secondary cement in the centers of pores are well preserved, yet this limestone is completely silicified—no carbonate remains. (D. R. Prothero.)

Diagenetic Environments

Because seawater is supersaturated with calcium carbonate, it takes just a slight change in water chemistry, or impurities and sites of nucleation, for marine cementation to occur. A wide variety of environmental settings (Fig. 11.9) can produce diagenesis in carbonate rocks. In the *intertidal zone*, for example, there is rapid fluctuation of the water chemistry during tides. The result is beach rock, which can form in a matter of years under the right conditions. The cements in beach rock are composed of drusy needles of aragonite and high-Mg calcite. Apparently, beach rock is precipitated when seawater evaporates and loses CO_2, a reaction that favors carbonate precipitation, although some beach rocks occur in regions of frequent mixing of fresh and marine waters, which also produce rapid changes in water chemistry.

Cementation also takes place in the *shallow subtidal* region, filling the hollows in and between carbonate skeletal grains and within reef cavities. Again, most of the cements are aragonite and high-Mg calcite, often with a high strontium content; they typically have a crusty, bladed, or fibrous habit rather than the needle-like appearance of beach-rock cement. In some cases, cements precipitate as coatings of micrite lining the cavities of either reef framework organisms or of detached but untransported shells. Obviously, this type of cementation must occur very soon after the organism has died, but the exact mechanism by which such cementation occurs is not well understood. The availability of abundant sites for nucleation of calcite must be important. In the Bahamas, studies have shown that dissolved CO_2 is driven off by heating and turbu-

lence, resulting in carbonate precipitation in areas that are quiet enough to allow it.

The result of shallow subtidal cementation is a crusty surface known as a **hardground.** These surfaces are usually cemented most at the sediment-water interface, and cementation decreases within centimeters of the surface. They occur predominantly in shallow water, although they can occur as deep as 300 m. Hardgrounds are often found in ancient limestone and are considered evidence of interruptions in sedimentation, or hiatuses, that may span several thousand years (based on faunal discontinuities).

The most extensive areas of early carbonate sedimentation, however, occur above sea level, in zones where fresh meteoric (atmospheric) waters—that is, rain or runoff groundwater—flow through the pores of carbonate sediments. This area is known as the **vadose zone,** or zone of aeration above the water table. The **phreatic zone,** or zone of saturation below the water table, can also produce cementation. The most noticeable difference between marine and freshwater cements is that marine cements precipitate high-Mg calcite, whereas meteoric waters can only precipitate low-Mg calcite.

Some of the most detailed studies of vadose zone cementation focused on the Pleistocene sediments of the Caribbean, which were exposed to meteoric waters during the drop in sea level that occurred during glacial episodes. For example, Land, Mackenzie, and Gould (1967) documented five stages of diagenesis in the limestones of Bermuda. After the primary sediment is deposited (stage I), a thick fringe of drusy low-Mg calcite needles forms on the rims of the voids, and

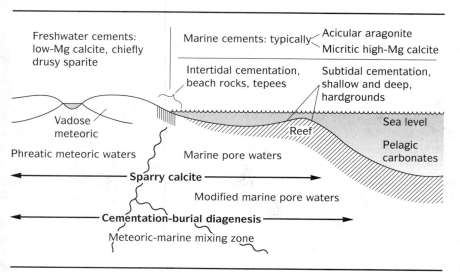

Figure 11.9 Principal environments of cementation in carbonate rocks. (After Tucker, 1981: 127; by permission of Blackwell Scientific Publishers, Oxford.)

syntaxial overgrowths occur (stage II). In stage III, the high-Mg calcite is replaced by low-Mg calcite and aragonite. In stage IV, aragonitic skeletal grains are dissolved and their voids are filled with low-Mg calcite cement (although the internal structure of the fossils is lost). Stage V involves final cementation to fill the remaining voids, although these Pleistocene limestones of Bermuda may still have up to 20% porosity. Rarely do these limestones become cemented to the point that they have only 5% porosity or less, like a typical ancient limestone.

Other studies of meteoric cementation (James and Choquette, 1984) showed that vadose and phreatic cementation produce recognizably different textures. For example, vadose-zone cements commonly include micritic envelopes around grains, thin layers of cement in the contact zone between grains (**meniscus cement**), and stalagtitic cements that hang down into the pores. Phreatic-zone cements, on the other hand, are composed largely of drusy overgrowths of equal thickness around the rim of the void and secondary blocky cement. The fact that some of the cements clearly indicate which way was up may be valuable in beds that have been deformed and overturned.

All these cases are examples of early diagenesis, typically occurring at near-surface conditions. *Later diagenesis* occurs when carbonate sediments have undergone burial and compaction. In addition to stylolites, compaction can be demonstrated by fossils, ooids, or other grains that have been visibly crushed or distorted in clean skeletal limestones. Micritic limestones, however, seldom show crushing of skeletal grains, because the micrite can absorb about a 50% decrease in volume without causing distortion of the framework grains.

As burial proceeds, however, the porosity of the limestone decreases to nearly zero, which inhibits the flow of diagenetic solutions. Yet the diagenesis of limestones becomes more intense at greater depths, even though there is almost no pore space left to transport fluids. Somehow, other processes, such as pressure solution, must operate at depth to allow diagenesis without the extensive flux of pore waters. Most deep carbonate cements are identified because they encase younger features—such as stylolites, crushed and healed coatings on grains, or dolomite crystals—that must have formed at depth. Later diagenetic cements also have distinctive chemical signatures (enriched in Fe and Mn and depleted in Sr compared to early diagenetic cements), and they are distinctive under cathodoluminescence.

Dolomite and Dolomitization

Many ancient carbonate rocks consist in whole or in part of the mineral **dolomite,** $CaMg(CO_3)_2$. Technically, the name **dolostone** should be applied to carbonate rock made of the mineral dolomite, but sedimentary petrologists persist in using the mineral name for the rock out of force of habit. The context typically indicates whether the mineral or rock is meant. *Dolomitization* is the process by which calcium carbonate rock alters to dolostone. This process obscures the original texture, which makes classification and genetic interpretation difficult. Where original textures and components survive, limestone classification is used and the degree of dolomitization is noted.

Dolomite is very similar to other carbonate minerals, so it is not always easy to identify in the field. It is less soluble than calcite and does not readily fizz in dilute acid. The standard field test is to powder a bit of the sample, which increases the surface area, and apply the acid; this causes a weak effervescence. In the laboratory, chemical stains readily separate dolomite from other minerals on a polished and etched surface or in a thin section. The stain known as Alizarin Red S colors calcite and aragonite deep red but does not affect dolomite.

Frequently, dolomite changes the texture of a rock in a noticeable way. For example, dolomitized rocks look "sugary," and in many instances dolostones have a golden-brown or tan color (in contrast to limestones, which are typically gray). In addition, the different solubilities of dolomite and calcite mean that dolostones weather differently. Often, the fossils in a limestone will be undolomitized, so that under chemical weathering and dissolution, the dolomitized matrix will remain and there will be voids where calcitic fossils have dissolved away. In other cases, dolomite selectively replaces the burrows in a micritic limestone, resulting in a mottled texture.

In thin section, it is difficult to tell rhombohedral calcite crystals from dolomite rhombs, since both have a similar crystal habit. The most reliable test is to stain the rock with Alizarin Red S before gluing the cover slip on the thin section, and this is done routinely in most petrographic research that involves dolomite. Most carbonate rocks that show extensive development of rhombohedral crystals, which often overgrow the original limestone textures, can be assumed to be dolomite (Fig. 11.10A). Such crystals grow at temperatures between 50° and 100°C. When large mosaics of blocky anhedral dolomite are found (Fig. 11.10B), they

A

B

Figure 11.10 Dolomite can have a distinctive texture of euhedral rhombohedral crystals in thin section. (A) Dolomite rhombs have replaced ooids, preserving a fabric of ooids as "ghosts." The largest rhombs are 0.3 mm in diameter. (Courtesy of R. C. Murray.) (B) At higher temperatures, dolomite crystals become much coarser and more intergrown, losing their euhedral rhombic shape. This example is from the Devonian Lost Burro Formation, California. (Courtesy of D. H. Zenger.)

are apparently replacements of earlier dolostone that formed at temperatures greater than 100°C (Gregg and Sibley, 1984).

The process of dolomitization has been the subject of vigorous debate. The most recent and extensive review of "the dolomite problem" is that of Warren (2000). Virtually all dolostones are composed of **secondary** (or replacement) **dolomite,** meaning that calcite and aragonite form first but are later converted to dolomite. This can be proved easily if relict textures of limestone are still visible despite the overprinting of dolomite. In some dolostones, there are no relict fossils, ooids, or other evidence of limestone parent material; there are just masses of rhombohedral dolomite crystals. Are these dolostones just completely replaced limestones, or did the crystals of dolomite grow directly from solution with no intervening stage as a calcium carbonate mineral? In other words, are they truly **primary dolomite?** Modern settings in which recently formed dolomite occurs are not necessarily sites of primary dolomite. Most sites reported as primary are actually locales in which the time interval between the formation of calcium carbonate minerals and their

conversion to dolomite is exceedingly short (that is, penecontemporaneous).

Experimental Studies and Dolomitization Reactions

Our inability to produce primary dolomite in the laboratory under normal marine conditions suggests that little primary dolomite forms naturally. It precipitates in the laboratory only under unrealistically high temperatures or only with bacterial mediation under more realistic terrestrial conditions. This difficulty in producing dolomite experimentally under conditions that mimic those of modern seawater apparently reflects the kinetics of dolomite crystallization; specifically, the time required to crystallize the ordered lattice structure of dolomite.

In dolomite, cations and anions occupy the eight corners of a rhombohedral unit cell. Every other cation site must be systematically (but alternately) occupied by ions of calcium, first, and then magnesium. Aragonite and calcite (including Mg-rich varieties), on the other hand, have a much less ordered

TABLE 11.5 Chemical Reactions That Produce Dolomite
1. Direct (primary) precipitation of dolomite. This reaction produces crystals of dolomite directly from seawater. $$Ca^{2+} + Mg^{2+} \, 2(CO_3)^{2-} \rightarrow CaMg(CO_3)_2$$
2. Dolomitization involving replacement of individual calcium cations by magnesium cations on a one-for-one basis. This reaction requires the addition of magnesium to the system and the removal of calcium from the system. $$Mg^{2+} + 2(CaCO_3) \rightarrow CaMg(CO_3)_2 + Ca^{2+}$$
3. Dolomitization involving dolomitizing fluids that provide magnesium cations and carbonate anions. This reaction eliminates the requirement of reaction 2 that calcium ions be removed from original calcite or aragonite. $$CaCO_3 + Mg^{2+} + (CO_3)^{2-} \rightarrow CaMg(CO_3)_2$$

structure. Seawater is supersaturated with respect to all dolomite as well as with respect to calcium carbonate. The routine precipitation of calcite and aragonite rather than dolomite evidently occurs for two reasons. First, metabolizing organisms prefer calcium and cannot extract the required precisely equal amounts of magnesium and calcium. Second, the requirements that calcium and magnesium be systematically ordered in dolomite slows growth sufficiently that calcium carbonate minerals (including randomly ordered Mg-rich varieties) form instead.

Table 11.5 lists three reactions that produce dolomite. Reaction 1 is the direct precipitation of dolomite from seawater. Reactions 2 and 3 are two reactions that convert calcium carbonate to dolomite. Reaction 2 requires an open system in which a one-for-one replacement of calcium by magnesium occurs. Reaction 3 does not require calcium to be removed. The dolomitizing fluid supplies both carbonate and magnesium.

Insights from Alleged Primary Dolomite Sites

Scientists have studied a number of modern sites in which dolomite forms. The dolomite at most of these sites is penecontemporaneous; that is, initially formed calcium carbonate is almost immediately converted into dolomite. Such localities provide clues about the conditions that favor dolomitization. The dolomite produced at these modern sites occurs in local patches; texture is dominantly micritic. Most ancient dolostones consist of sparry dolomite and dolomitized deposits that are hundreds of meters thick and regionally extensive.

Figure 11.11 summarizes several modern dolomitization sites. The hydrologic constraints at each site are crucial. Each has a physical-chemical environment that produces dense brines with an Mg-Ca ratio higher than 8, well above the 5.4 ratio of normal seawater. The brines are produced by evaporation until Ca-rich, Mg-poor minerals such as gypsum and anhydrite form. It is this preferential removal of calcium that changes the Mg-Ca ratio.

Figure 11.11A shows one of the best-known modern dolomite sites, the Coorong coastal plain on the southeastern coast of Australia. Primary dolomitic micrite forms within seepage-fed alkaline lakes and lagoons developed in the interdune flats that separate Quaternary beach barrier sands. During the summer when the region is arid, the pH in the lakes ranges from 8 to 10, and the Mg-Ca ratio in lake wa-

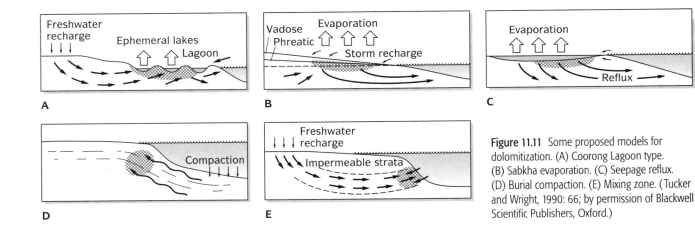

Figure 11.11 Some proposed models for dolomitization. (A) Coorong Lagoon type. (B) Sabkha evaporation. (C) Seepage reflux. (D) Burial compaction. (E) Mixing zone. (Tucker and Wright, 1990: 66; by permission of Blackwell Scientific Publishers, Oxford.)

ter varies from 1 to 20. Precipitation of gypsum and anhydrite preferentially removes Ca^{2+} and SO_4^{2-}. The fluid that remains is the dolomitizing agent, although high alkalinity and the presence of nearby Mg-rich groundwater (generated during the leaching of subjacent volcanic rocks) may help. Although dolomite forms, there is a primary precipitate (the mineral is actually protodolomite) that has the approximate chemical composition of dolomite but lacks the distinctive ordered crystal structure. Therefore, it is unreasonable to apply this model widely. Local patches of ancient laminated micritic dolostone located within outflow zones of extensive carbonate aquifers (for example, within the Eocene Green River Formation, Utah) are probably the best equivalents of Coorong Lakes–style dolostone.

Dolomite also forms as thin (less than 2 m thick), surficial layered patches developed on **sabkhas** (from the Arabic word *sebkha,* for salt flat), the hypersaline tidal and supratidal belts that border the Persian Gulf (Hardie, 1987; Warren, 1991). Figure 11.11B shows an example of this setting. The complex plumbing system is driven by evaporation. Storms and abnormally high tides flood supratidal flats with seawater of normal salinity during the winter and spring. During the long, hot, arid summer, net evaporation leads to the crystallization of Ca-rich evaporite minerals. Upward evaporative pumping causes these dolomitizing fluids to react with aragonite mud. Much of the dolomite formed is a pore precipitate, rather than a direct replacement of pre-existing aragonite mud. Also, most of the dolomite formed is first disordered calcium–magnesium carbonate, rather than ordered dolomite. Over time, it becomes more ordered.

A somewhat similar process occurs locally on the Bahama Banks, where a crust of desiccated, polygonally cracked crust of dolomite mantles the upper surface of supratidal carbonate mud. The dolomitic crust is apparently a residue left by the evaporation of a film of seawater that was left perched above the level of high tide following an occasional storm flood. Again, the precipitation of gypsum is invoked to raise the Mg–Ca ratio, with elevated salinity raising fluid density and promoting its downward movement through readily altered underlying carbonate sediment.

Unfortunately, Coorong Lagoon–type settings and sabkhas generate only local patches of dolostone, and they are typically micritic. How can the broader expanses of predominantly sparry dolostone be explained? Several models have been developed or enhanced since 1990. Each suggests that widespread platform dolostones can be produced from laterally derived—albeit somewhat compositionally modified—normal seawater or from connate pore fluids.

Many of the widespread areas of sparry dolostone are temporally and spatially related to adjacent platform and basinward evaporites. Figure 11.11C illustrates the formation of brine reflux dolomite (for which there are few, if any, modern counterparts). A broad hypersaline basin is shown subjacent to the edge of a carbonate platform. Hypersaline brines generated by evaporation from lagoonal water in that basin become heavier, and (as at Coorong and the Persian Gulf) Mg-rich, thank to the selective removal of calcium to form aragonite, calcite, and gypsum. Such brines eventually become dense enough to displace connate waters, resulting in **reflux;** that is, they slowly seep downward and are flushed through the permeable carbonate sediment, converting it to dolostone. Hydrological modeling suggests that the process can occur across a belt a few tens of kilometers wide in a few million years, with larger platforms requiring either more time or a gradual lateral migration of the evaporite recharge area. Variations in the porosity and permeability of the limestone through which the brines seep will obviously control the texture and extent of secondary dolomitization. An interesting variation on this mechanism (not shown in the figure) proposes that widespread platformal secondary dolostone can also be produced by the brine reflux that should occur beneath broad, deep-water, isolated evaporite basins—the so-called saline giants such as the Messinnian Mediterranean (for which no analogs presently exist).

Figure 11.11D illustrates burial dolomitization, a model that explain the many dolostones that are not intimately associated with evaporites. Burial dolomitization occurs when buried Mg-rich mudrock is compacted and dewatered. Expelled fluids rich in magnesium then invade and dolomitize the adjacent limestone. There are some problems with this model. Although mudrock commonly contains substantial magnesium that could be released during clay mineral diagenesis and burial metamorphism, expelled liquids tend to migrate vertically rather than laterally. Furthermore, mass balance equations suggest that compacting even tremendous volumes of mudrock would generate dolostone equal to as little as 1% of the mudrock volume.

The mixing-zone dolomitization model (Folk and Land, 1975) is a model developed to explain ancient dolostones that show no apparent relationship with hypersaline brine and evaporite (Fig. 11.11E). In this model, seawater of normal salinity and an Mg-to-Ca

ratio of 5.2 is mixed with fresh groundwater, producing brackish groundwater near the coastline. The seawater ostensibly provides enough additional magnesium to supersaturate the brackish groundwater with dolomite. Groundwater circulation is thought to flush the brackish dolomitizing groundwater through subjacent limestone. The validity of this model has been questioned on geochemical grounds. There are no data indicating that significant volumes of dolostone form by mixing seawater with fresh groundwater. Typically, the salinity of fresh groundwater is low; its Mg-to-Ca ratio approaches 1.

The Origin of Dolostone: A Continuing Dilemma

Geochemists continue to develop models for dolomitization. Dense, Mg-rich fluids capable of dolomitization are easy to produce. The problem has always been how and when such fluids can be brought into contact with existing limestone to create regionally extensive belts of dolostone. Depositional settings that produce fluids capable of dolomitization are volumetrically minuscule. Many dolostones show none of the relict textural components (spar, micrite, allochems) associated with the sabkha, alkaline lake, and coastal tidal lagoon settings that most typically form dolomitizing fluids.

The temporal distribution of limestone and dolostone adds a curious twist. Although about one-fifth of all carbonate rocks are dolostone, the proportion of carbonate rock that is dolostone increases progressively with age (Daly, 1907; Ronov, 1972). In Cretaceous and younger carbonate rocks, the limestone-to-dolostone ratio is 80:1; in Paleozoic carbonate rocks, it is 3:1; and in Precambrian carbonate rocks, it is 1:3. These data can be used to support either of two arguments: (1) Depositional settings conducive to the formation of dolostone were more common early in Earth's history; or (2) dolostone is secondary, and older limestones have had more time to alter to dolostone.

The chemistry of seawater has changed over time, especially in the relative abundance of Ca^{2+} and Mg^{2+}. Mid-ocean ridge systems pump calcium into seawater (removing it from newly formed ocean crust and overlying sediment). Magnesium is simultaneously removed from seawater and transferred to rocks. Consequently, during episodes of rapid seafloor spreading (which coincide with a high global sea level), low Mg-Ca ratios produce "calcite seas" in which ooids, marine cements, and reef organisms are dominantly calcite. Conversely, intervals of time (such as the present) characterized by relatively slow rates of seafloor spreading (and lower sea level) generate higher Mg-Ca ratios and "aragonite seas." Aragonite and high-Mg calcite are the dominant minerals in ooids, cements, and reef-making shelled organisms. See Chapter 12 and Stanley and Hardie (1998) for further discussion.

Not surprisingly, there is also a correlation between dolomite abundance and periods of high sea level (Givens and Wilkinson, 1987). Most dolostones occur in rocks formed during greenhouse conditions (warm global temperature, high sea level), but little dolostone is found in rocks deposited during icehouse conditions (cold global temperature, low sea level). Dolomite abundances in deep-sea sediments show the same general pattern (Lumsden, 1988). The exact significance of this crude correlation is not known. Are there unusual characteristics of the planet during greenhouse conditions that promote dolomitization? Or is this pattern a result of the greater abundance of limestone during high sea levels and greenhouse conditions and the scarcity of limestones during low sea levels and icehouse conditions?

Understanding dolomites, especially their origin, is an area of active research. Many unsolved problems continue to challenge us.

CONCLUSIONS

The number of sedimentary geologists who have devoted their careers to the examination, description, classification, and interpretation of limestone and dolostone closely matches the number of petrologists who have specialized in the study of siliciclastic sedimentary rocks. On the basis of the economic importance of carbonate rocks, this attention is clearly warranted. But even more important is the role that these rocks play in deciphering sedimentary environments. The texture and composition of limestone and dolostone are unmatched in their ability to provide reliable insights, both into the physical and chemical conditions that prevailed during deposition and into the geochemical settings that developed long after sedimentation.

FOR FURTHER READING

Bathurst, R. G. C. 1975. *Carbonate Sediments and Their Diagenesis*, 2d ed. Amsterdam: Elsevier.

Bathurst, R. G. C., and L. S. Land. 1986. Carbonate depositional environments. Part 5: Diagenesis. *Colorado School of Mines Quarterly* 81:1–41.

Choquette, P. W., and N. P. James. 1987. Diagenesis 12. Limestones: The deep burial environment. *Geoscience Canada* 14:3–35.

Given, R. K., and B. H. Wilkinson. 1987. Dolomite abundance and stratigraphic age: Constraints on rates and mechanisms of Phanerozoic dolostone formation. *Journal of Sedimentary Petrology* 57:1068–1078.

Hardie, L. A. 1987. Dolomitization: A critical view of some current views. *Journal of Sedimentary Petrology* 57:166–183.

James, N. P., and P. W. Choquette. 1983. Diagenesis 6. Limestones: The sea floor diagenetic environment. *Geoscience Canada* 10:162–179.

James, N. P., and P. W. Choquette. 1984. Diagenesis 9. Limestones: The meteoric diagenetic environment. *Geoscience Canada* 11:161–194.

Land, L. S. 1985. The origin of massive dolomite. *Journal of Geological Education* 33:112–125.

Milliman, J. D. 1974. *Marine Carbonates*. Berlin: Springer-Verlag.

Moore, C. H. 1989. *Carbonate Diagenesis and Porosity*. New York: Elsevier.

Morse, J. W., and F. W. Mackenzie. 1990. *Geochemistry of Sedimentary Carbonates*. New York: Elsevier.

Pray, L. C., and R. C. Murray. 1965. *Dolomitization and Limestone Diagenesis*. SEPM Special Publication 13.

Schneidermann, N., and P. M. Harris. 1985. *Carbonate Cements*. SEPM Special Publication 36.

Scholle, P. A. 1978. *A Color Illustrated Guide to Carbonate Rock Constituents*. American Association of Petroleum Geologists Memoir 27.

Scoffin, T. P. 1987. *An Introduction to Carbonate Sediments and Rocks*. New York: Chapman and Hall.

Stanley, S. M., and L. A. Hardie. 1998. Secular oscillations in the carbonate mineralogy of reef-building and sediment-producing organisms driven by tectonically forced shifts in seawater chemistry. *Palaeogeography, Palaeoclimatology, Palaeoecology.* 144:3–19.

Tucker, M. E., and V. P. Wright. 1990. *Carbonate Sedimentology*. Oxford: Blackwell Scientific Publications.

Warren, J. 1991. Sulfate-dominated sea-marginal and platform evaporative settings. In *Evaporites, Petroleum and Mineral Resources*, ed. J. L. Melvin. Oxford: Blackwell Scientific Publications.

Warren, J. 2000. Dolomite: Occurrence, evolution and economically important associations. *Earth Science Reviews* 52:1–81.

Zenger, D. H., J. B. Dunham, and R. L. Ethington, eds. 1980. *Concepts and Models of Dolomitization*. SEPM Special Publication 28.

12 Carbonate Environments

View of reef corals forming in shallow tropical waters just below the surface. (Fred Braverman/ Minden Pictures.)

ALTHOUGH LIMESTONES CAN FORM IN BOTH NONMARINE lacustrine and pelagic environments, most are produced in shallow marine environments. Limestones make up a major portion of the stratigraphic record on all continents, particularly in ancient epicontinental seaways and shallow shelves.

Carbonate sediments are fundamentally different from clastic sediments in many ways (Table 12.1). As James (1984, p. 209) put it, "Carbonate sediments are born, not made." Whereas terrigenous clastic sediments have been eroded from bedrock and *transported* to the basin of deposition **(allochthonous),** most carbonate sediments form chemically or biochemically *within* the basin of deposition **(autochthonous).** Many carbonate particles undergo little or no hydraulic transport, so the physical sedimentary processes that

TABLE 12.1 Differences Between Siliciclastic and Carbonate Sediments

Carbonate Sediments	Siliciclastic Sediments
Most sediments occur in shallow, tropical environments.	Climate is no constraint, so sediments occur worldwide and at all depths.
Most sediments are marine.	Sediments are both terrestrial and marine.
The grain size of sediments generally reflects the size of organism skeletons and calcified hard parts.	The grain size of sediments reflects the hydraulic energy in the environment.
The presence of lime mud often indicates the prolific growth of organisms whose calcified portions are mud-sized crystallites.	The presence of mud indicates settling out from suspension.
Shallow-water lime sand bodies result primarily from localized physicochemical or biological fixation of carbonate.	Shallow-water sand bodies result from the interaction of currents and waves.
Localized buildups of sediments without accompanying change in hydraulic regimen alter the character of surrounding sedimentary environments.	Changes in the sedimentary environments are generally brought about by widespread changes in the hydraulic regimen.
Sediments are often cemented on the seafloor.	Sediments remain unconsolidated in the environment of deposition and on the seafloor.
Periodic exposure of sediments during deposition results in intensive diagenesis, especially cementation and recrystallization.	Periodic exposure of sediments during deposition leaves deposits relatively unaffected.
The signature of various sedimentary facies is obliterated during low-grade metamorphism	The signature of sedimentary facies survives low-grade metamorphism.

Source: James, 1984.

make cobbles, pebbles, sand, and mud are less important than the composition of the carbonate particle itself, whether precipitated inorganically or formed biogenically. Grain size is chiefly a function of the type of skeletal particle, not of hydraulics.

Another striking difference between carbonate and clastic sediments is the extremely limited conditions under which carbonates constitute the predominant sediment. Clastic sediments are found in virtually every sedimentary environment, both marine and nonmarine, whereas carbonates are common only in clear, warm, shallow, tropical to subtropical seas—the so-called carbonate factory (Fig. 12.1). Most carbonates that do not remain on the shallow shelf are transported only short distances down the slope (basinward) or up the slope (shoreward) into peritidal regions. Freshwater limestones and pelagic oozes are almost the only exceptions to this rule, and they are much less abundant than shelf carbonates.

The environment where carbonates form must meet a number of restrictive conditions that are found only on shallow shelves between 30°N and 30°S lati-

tudes, because carbonate is produced by environmentally sensitive organisms. The shelf waters must be warm—these organisms are tolerant of only a narrow range of temperatures—and shallow—many of these organisms (especially reef-forming corals and lime mud-secreting algae) need light for photosynthesis. Most carbonate production takes place in the upper 10 m of seawater and drops off rapidly below this; below 80 to 100 m, the light is so dim that only a few red algae, ahermatypic corals, mollusks, and echinoderms can live. The water chemistry must be normal marine

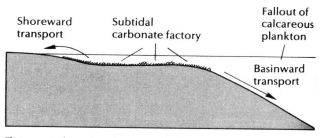

Figure 12.1 The main regions of accumulation of sediments on a carbonate shelf. (After James, 1984: 210.)

because these organisms are tolerant of a very narrow range of salinity. The water must also be clear, free of terrigenous clastics, and only slightly turbulent, because terrigenous quartz sand can abrade the softer calcite, and terrigenous mud can foul gills or digestive tracts, thus hampering the biological activity of most carbonate-secreting organisms. If the water is too turbulent, lime mud is stirred up, which cuts down the light and chokes many marine invertebrates.

With such restrictions, it is not surprising that there are so few active carbonate environments today. Why, then, were they so abundant in the geologic past? In this case, it appears that the present is an imperfect key to the past. Throughout most of geologic history, epeiric seas were widespread and provided enormous areas of the shallow, tropical marine conditions that are ideal for carbonate production. The modern analogs that have been studied the most are the Persian Gulf and the Bahama Platform, but these areas are not nearly so large as the ancient epeiric carbonate seas. One reason is that Pleistocene and Holocene sea levels have been much lower than was typical for most of Earth's history (particularly the Paleozoic and Mesozoic). Another is that Earth's temperatures during the present interglacial period are much lower than was typical of most of the geologic past. Our understanding of ancient carbonate environments is hampered by the scarcity of good modern analogs.

Peritidal Environments

Because carbonates are so depth-restricted, the position of sea level is a natural boundary between carbonate environments. Very different processes occur in peritidal environments and shelf environments; shelf environments are seldom subaerially exposed. The occasional drying that occurs in the intertidal zone means that most organisms cannot survive there. Salinity and water temperature also undergo extreme fluctuations, which further limits the types of organisms that can live there. These chemical conditions have proved to be important for economic reasons as well. Peritidal carbonate flats form laterally persistent, evenly bedded limestones and dolostones, which are host rocks for lead and zinc ores and for petroleum. For this reason, modern carbonate tidal flats have been studied intensively, beginning with Ginsburg's work in the Bahamas in the 1950s (summarized in Ginsburg, 1975) and continuing with studies of tidal flats in Bermuda, Bonaire, Florida, the Persian Gulf, and Shark Bay, Australia.

The shallow intertidal environment is close to optimum for the growth of certain carbonate-forming organisms (especially algal mats), so tidal flats are capable of building up at rates much faster than the normal rates of sea-level rise or shelf subsidence. The classic tidal flat model is a *shallowing-upward* sequence from surf zone to subtidal to intertidal to supratidal deposits to eolian dunes (Fig. 12.2). This sequence can be repeated many times in the rocks because minor fluctuations in relative sea level make big differences in the position of the intertidal zone. Ideally, such repetitive shallowing-upward sequences form on a transgressive carbonate bank, where repeated changes in relative sea level cause the repetition of the sequences. In the past, many parts of the continents were covered with extensive epeiric carbonate seas having hundreds of square kilometers of peritidal zone, so thick tidal flat shallowing-upward sequences are very common in the cratonic record.

Carbonate tidal flats occur on shores that are shielded from daily wave action but are immersed by daily tidal fluctuations and by occasional storms. As on clastic coastlines, the rare storm can have an enormous effect on intertidal environments, producing considerable local erosion. Many tidal flat deposits are formed above mean high tide by storm action. All of the subenvironments in the vertical succession of tidal flats form simultaneously in adjacent belts, so they are another classic example of units that build laterally and are time-transgressive.

The two modern examples of carbonate-forming areas—the arid, evaporitic tidal flats of the Persian Gulf (Fig. 12.3) and the humid, normal marine tidal flats of the Bahamas (see Fig. 12.2)—encompass most of the features seen in ancient carbonate rocks. If the tidal flat is on an extensive shallow shelf, the waves may be slowed at a distant shelf break. In many cases, however, the outermost environment is the surf zone, which absorbs the wave energy. On the *lower foreshore*, the deposits are below the wave swash, so they are poorly sorted, coarse-grained shell deposits with a micritic matrix. Longshore currents produce large-scale festoon cross-bedding. Because the *upper foreshore* is within the swash and backwash zone of the waves, laminated, thick-bedded, well-sorted lime sands and gravels, with planar cross-beds dipping about 15° seaward, are produced here. This environment is subject to frequent subaerial exposure and rapid changes in water chemistry. As a result, the calcite often dissolves slightly and then reprecipitates to cement the pore spaces, forming beach rock.

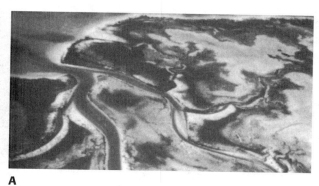

Figure 12.2 (A) The Three Creeks area along the west coast of Andros Island, Bahamas. Tidal channels cut an intertidal marsh composed largely of algae, with some ponds in the intertidal areas. (Stanley, 1989: 1226.) (B) Major features of the peritidal environment. (After Stanley, 1989: 1226.)

A

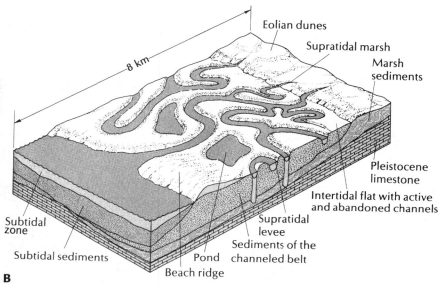

B

Inside the protective barrier is an extensive area of marine subtidal *lagoon* that is never subaerially exposed. Naturally, it is very sheltered, so the main deposits are extensive lime sand and mud banks, with abundant pellets and skeletal fragments. These lagoonal sands and muds are areas of great biotic productivity, so they are often homogenized and bio- turbated by burrowing organisms. Subtidal lagoon lithology is so similar to that of normal marine carbonate shelves that they are difficult to distinguish except for the close association of the lagoon with distinctive intertidal deposits.

The most diagnostic environment is the *intertidal zone* itself (see Figs. 12.2 and 12.3). Here the primary

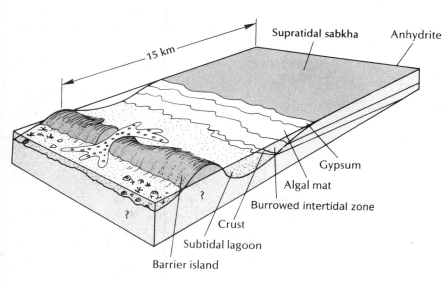

Figure 12.3 The major facies of the regressive tidal flat on the Persian Gulf Trucial Coast. Supratidal sabkha is composed of algal mats with a gypsum crust, which have grown over burrowed subtidal lagoonal sediments. Tidal deltas, composed mainly of ooids, form around inlets cut into small barrier islands composed of mollusk shells, ooids, and coral fragments. Coral reefs can grow seaward of the island, away from the tidal inlets. (Scholle , Bebout, and Moore, 1983: 480; by permission of the American Association of Petroleum Geologists, Tulsa, Okla.)

A

B

Figure 12.4 (A) Hamelin Pool, western Australia, showing rippled intertidal carbonate sands between algal mats. (B) Modern subtidal to intertidal columnar stromatolites at Hamelin Pool. (Courtesy of R. N. Ginsburg.)

factor is the daily alternation of tidal immersion and subaerial exposure. Most of the zone is covered by rippled fine carbonate mudflats, which are often blanketed with algal mats and stromatolites (Fig. 12.4) and by mudcracks (Fig. 12.5) from subaerial exposure. The flats are cut by channels for the outgoing tidewaters. These channels are eventually abandoned and filled

Figure 12.5 Large mudcracks in thick algal mats on the edge of a brine-filled pond, Inagua Island, Bahamas. Gypsum precipitates in the cracks and on the upturned edges. (Courtesy of E. A. Shinn.)

by linear or sinuous stringers of carbonate sand. Sedimentary structures, such as tidal bedding and reactivation surfaces, clearly indicate the reversing currents in the channels. Another characteristic feature of the tidal channels is breccia made of shell debris and rip-up clasts, particularly fragments of laminated mudcracks (Fig. 12.6; see also Fig. 5.6B).

Because the intertidal environment is protected from normal waves, the rare storms tend to have a disproportionately great effect. Storms flood the tidal flat with water carrying much suspended sediment and leave a storm lag of debris and mud, which is eroded and reworked during normal conditions. Shinn, Lloyd, and Ginsburg (1969) have suggested that the tidal flat is like a river delta turned inside out; that is, the sea is the "river" supplying sediment *onshore* to the channeled flats, which are the "delta."

The intertidal environment is distinctive and can be recognized by a number of features. Algal mats and dome- and pillar-shaped stromatolites (see Fig. 12.4) are characteristic, and they produce a rock with even lamination that shows **fenestral,** or sheet-shaped, **porosity** due to gas-filled voids and the shrinking of the desiccated algal mat. Other desiccation features—such as mudcracks, mudchips, channel breccias filled with rip-up clasts and shells, and evaporite minerals—are also characteristic.

The *supratidal zone* is above the high-tide level, so it is immersed only during unusually high spring tides or during storm surges. As a result, desiccation features, including thick layers of mudcracks and algal lamination, are predominant. The algal stromatolites are much more sheetlike or domelike than pillarlike; they form laterally linked hemispheroids. In arid climates (such as the tidal salt flats, or *sabkhas,* of the Persian Gulf; see Fig. 12.3), these intertidal zones are areas where evaporites are deposited, often forming laminae of gypsum or anhydrite and halite. In many cases, the mudcracks fill with evaporites. Sometimes, the evaporite laminae are distorted by soft-sediment deformation or form diapir structures. A particularly characteristic texture, called **birdseye,** results when small ellipsoidal pores form in the algal mat by shrinkage, gas bubbles, or the escape of air during flooding (Fig. 12.7). Above the supratidal salt flat there may be eolian dunes, similar to those found in clastic barrier island complexes except that they are made of carbonate sand.

One of the most striking features of shallowing-upward carbonate sequences is that they often build up into thick sequences of repeated shoaling cycles

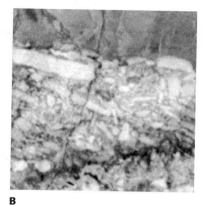

A

B

Figure 12.6 (A) Flat lime mudchip pebbles accumulating on the beach ridge shown in Fig. 12.2B. (B) Ancient flat-pebble limestone conglomerate from the Paleozoic. (D. R. Prothero.)

that can be hundreds of meters thick (Fig. 12.8). Because the rate of carbonate deposition is faster than the rate of platform subsidence or sea-level rise, carbonate sediment accumulates rapidly until it reaches sea level. This explains a single shallowing-upward sequence, but it does not explain the repeated shoaling cycles of carbonate sequences spreading out over vast platforms.

The mechanism for this cyclicity has been widely discussed, and two end-member models have been proposed (Wilkinson, 1982; James, 1984; Hardie, 1986). The **eustatic model** suggests that rapid fluctuations in sea level are responsible for the rapid change in depth of the facies. During periods of stability or slowly rising sea level, the sequence progrades outward in laterally accreted tidal flat sequences (Fig. 12.9, steps 1 and 2). If there is a rapid rise in sea level, the platform floods, which temporarily shuts off tidal flat deposition (step 3). Eventually, new tidal flat sequences pro-

Figure 12.8 Bedded Cambrian carbonates from Fortress Lake, British Columbia, showing dozens of repetitive small-scale shallowing-upward sequences between subtidal-intertidal limestones (dark) and supratidal dolomites (light.) (Courtesy of J. D. Aitken, Geological Survey of Canada.)

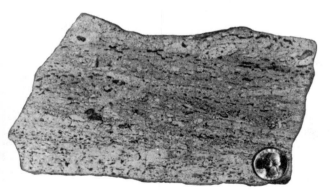

Figure 12.7 Birdseye limestones form in intertidal and supratidal areas where algal layers produce flat mats. Gas bubbles and worm burrows have disrupted the bedding and produced holes that give a bird's-eye appearance to the rock. (Stanley, 1989: 97.)

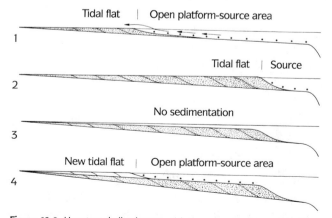

Figure 12.9 How two shallowing-upward sequences can be produced by progradation of a tidal flat wedge. These general conditions apply to both eustatic and autocyclic models. The asterisks show areas of carbonate production. (After James, 1984: 224.)

grade out from the shallow end, building another cycle on top of the last drowned sequence (step 4).

The **autocyclic model** proposed by R. N. Ginsburg (1971; Hardie, 1986) suggests that the cycles result not from sea-level fluctuation but from the rate of carbonate sedimentation as controlled by source area. As in the eustatic model, peritidal sequences prograde outward during periods of stable or slowly rising sea level (steps 1 and 2). But the source area for the carbonate sediments is the subtidal area, which is reduced by the progradation and eventually becomes too small to furnish carbonate for further growth of the prograding wedge. At this point, the lateral growth of the shallowing-upward sequences stops (step 3). Once sea level has risen far enough, however, a new area of subtidal shelf is provided for renewed carbonate production, and the prograding peritidal sequences can build out again across the previous cycle (step 4).

Whichever model prevails for the small-scale cycles, there seems to be no doubt that the larger-scale cycles are controlled by sea-level changes and possibly also by climatic changes (Fischer, 1964; Aitken, 1978; Hardie, 1986). These cycles may also be caused by tectonic changes in the platform. For example, Fischer (1964) was the first to propose that the various frequencies of cyclic carbonate deposition in the Triassic Lofer cyclothems of the Alps might have been caused by climatic changes due to variations in Earth's orbit. New information about the frequency of sea-level changes in geologic history (discussed later) indicates that large-scale eustatic changes could have significant effects.

The main problem with these models is dating. To prove a correlation between facies and orbital variations or particular parts of the sea-level curve, precise dating is needed. So far, this has not been possible in many critical cases because peritidal limestones contain few fossils that are also found in the open ocean and can be correlated worldwide. Nevertheless, this is an area of active interest and research (see Hardie, 1986).

Diagnostic Features of Peritidal Environments

Tectonic Setting Like other carbonate shelves, peritidal environments are restricted to clear, shallow, subtropical to tropical marine shelves, which occur in a stable tectonic setting with little relief. These conditions occur primarily in subsiding passive margins or in epeiric seaways.

Geometry Peritidal environments typically form thin (a few meters thick) but laterally continuous beds representing the various shoreline facies belts, which may build thousands of square kilometers of shelf and form extensive tabular bodies.

Typical Sequence A shallowing-upward sequence from subtidal lagoon muds or beach sands to intertidal rippled, burrowed, and mudcracked flats to supratidal algal flats (with or without evaporites) and eolian dunes is typical.

Sedimentology Features of the intertidal and supratidal environment are very distinctive: algal lamination and stromatolites; mudcracks; tidal channel breccias with shells and mudchip rip-up clasts; birdseye texture; dolomitization; and evaporite minerals (especially gypsum, anhydrite, and halite), which may rise diapirically.

Fossils Stromatolites made by cyanobacteria (blue-green algae) and other resistant types are most common, because of the desiccation and the extreme ranges of temperature and salinity. Some mollusks and burrowing worms can survive on the tidal flat, but most shell debris is washed in by storms.

Subtidal Shelf Carbonates

Shallow carbonate banks are rare today, but judging from the great abundance of shallow marine limestones, they were numerous in ancient epeiric seas. Modern shallow carbonate banks occur on continental shelves with little or no clastic input from rivers (for example, Florida, the Persian Gulf, the coast of Yucatán) or on isolated oceanic platforms (for example, Bermuda, the Bahamas). During major epeiric transgressions, however, the continents must have been covered with a few meters of shallow, warm, clear water, which produced enormous areas of carbonate banks, reefs, and shoals. Box 12.1 describes an ancient example of a huge shallow carbonate bank. As long as subsidence is relatively slow, the carbonate-secreting community of organisms can keep up with it by building up substrate fast enough to remain within the photic zone. If the carbonate bank progrades across an epeiric basin, it can accumulate limestone thicknesses approaching 2000 to 3000 m (for instance, the Cretaceous deposits of central Mexico or the Triassic Dolomites of the Italian Alps).

BOX 12.1 DEVONIAN SHALLOW MARINE CARBONATES OF THE HELDERBERG GROUP, NEW YORK

Shallowing-upward carbonate sequences are common in the stratigraphic record, as reviewed by Hardie (1986). One of the best exposed and most completely described sequences occurs in the Devonian deposits of the Mohawk and Hudson valleys of New York. The Catskill Mountains are rimmed by a resistant limestone cliff extending for over 400 km, known as the Helderberg Escarpment. The Lower Devonian Helderberg Group (Fig. 12.1.1) is made up of a number of carbonate units (described by Rickard, 1962) that represent various shallow subtidal carbonate environments. Some of the most detailed study of the Helderberg Group has concentrated on the basal Manlius Formation, which Laporte (1967, 1969, 1975) has interpreted as a peritidal deposit.

Detailed examination of the Manlius outcrop (Fig. 12.1.2) shows that it is composed of a series of shallowing-upward cycles, averaging 1 to 2 m in thickness (Goodwin and Anderson, 1980). Within these cycles, three facies—supratidal, intertidal, and subtidal (Fig. 12.1.3A)—can usually be recognized, each characterized by a distinctive set of sedimentary structures. Not every cycle includes all three facies, but two out of three are usually present. All three facies occurred at the same time in belts parallel to the ancient shoreline but have been superimposed by the gradual shallowing during each cycle. In the seaward direction, the peritidal deposits of the Manlius intertongue with, and pass into, the shallow marine carbonate bank, represented by the Coeymans Formation, a brachiopod-rich crinoidal calcarenite.

The *supratidal facies* is represented by unfossiliferous, laminated, mudcracked, dolomitic, pelletal carbonate mudstone. The most diagnostic features are the well-preserved desiccation cracks (Fig. 12.1.3B).

Internally, desiccation resulted in irregular spar-filled voids in the algal mat, producing birdseye texture (Fig. 12.1.3C). In detailed structure, the internal lamination is typically wavy and curled upward, due to crinkling and cracking of the algal mat. Between algal laminae are carbonaceous films from the algae and abundant pellets. These so-called bituminous limestones once led geologists to think that the Manlius had been formed in deep, stagnant water.

The supratidal limestones are extensively dolomitized. Dolomite rhombs have replaced the original limestone in many places, and dolomite has also filled the cracks. Laporte (1967) interpreted this dolomite in light of studies of modern supratidal dolomitization (see Fig. 11.11B). Evaporation at the algal mat brought magnesium-rich seawater upward by capillary action. Surface evaporation then precipitated dolomite in a fine-grained layer at the mat surface and spurred replacement of the limestone by dolomite throughout the mat.

The supratidal model is supported by the almost total lack of marine fossils, except for a few skeletal fragments washed in by high tides or storms. Only a few disarticulated ostracod valves, some burrow mottles, and the persistent algal lamination remain as evidence of life on what must have been a hot, hypersaline, often desiccated environment that was hostile to all but certain cyanobacteria.

The *intertidal facies* is represented by sparsely fossiliferous, pelletal carbonate mudstone interbedded with skeletal calcarenite (Fig. 12.1.3D). These 2- to 6-cm-thick beds were once described as ribbon limestones. They contain abundant mudcracks and limestone mudchip conglomerates, indicating episodes of desiccation and erosion of the mudcracked surface.

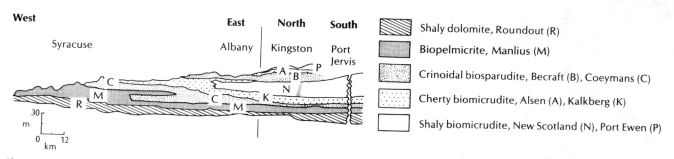

West

Syracuse

East   North   South

Albany   Kingston   Port Jervis

	Shaly dolomite, Roundout (R)
	Biopelmicrite, Manlius (M)
	Crinoidal biosparudite, Becraft (B), Coeymans (C)
	Cherty biomicrudite, Alsen (A), Kalkberg (K)
	Shaly biomicrudite, New Scotland (N), Port Ewen (P)

Figure 12.1.1 Restored section of the Helderberg Group in New York, showing lithology of individual formations. Note the interfingering of lower and middle Helderberg units and the time equivalence of parts of the Manlius, Coeymans, Kalkberg, and New Scotland formations. The Helderberg grades downward into the Upper Silurian–Lower Devonian Roundout Formation. (After Rickard, 1962: 15.)

(continued)

(*Box 12.1 continued*)

Figure 12.1.2 Shoaling-upward sequence within the Manlius Limestone, Helderberg Group, Perryville Quarry, New York. The cycle begins with an erosive base (lower arrow) overlain by reworked stromatoporoid heads. This grades up into a fossiliferous lime wackestone and faintly laminated lime mudstone. Brown dolomitic algal laminites form the top of the sequence just beneath the upper arrow. Scale is about 2 m long. (Courtesy of P. W. Goodwin and E. A. Anderson.)

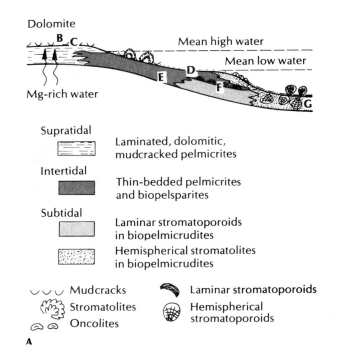

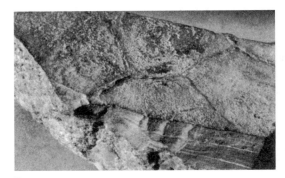

B

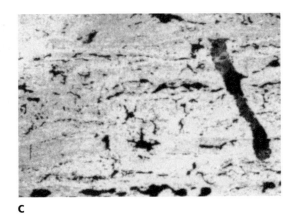

C

Figure 12.1.3 (A) Generalized facies model for the Manlius Formation showing representative rocks of each facies. (B) Supratidal mudcracks, Elmwood Member. (C) Negative print of thin section showing birdseye texture caused by desiccation; larger vertical vug may be a gas trackway. Supratidal facies, middle Thacher Member, Catskill, New York. (D) Negative print of thin section showing skeletal calcarenite lying unconformably on calcareous mud; sand grades up into mud that is truncated above by a second skeletal calcarenite and mudstone. Fossils are abundant and include conical tentaculitids, bryozoans, and ostracods. Intertidal facies, lower Thacher Member. (E) Weathered bedding surfaces strewn with tentaculitids. Intertidal facies, lower Thacher Member, Wiltse, New York. (F) Weathered outcrop showing lateral interfingering of encrusting stromatoporoid masses with skeletal carbonate mudstones. Truncated stratification of nonstromatoporoid beds suggests lateral erosion of sediments followed by accumulation of stromatoporoid masses. Subtidal facies, upper Thacher Member, Clarksville, New York. (G) Outcrop of stratified biopelmicrites (Elmwood and Clark Reservation members) overlain by massive bed of tightly bound stromatoporoids (Jamesville Member.) Subtidal facies. (B–E and G courtesy of L. F. Laporte; F courtesy of L. V. Rickard and the New York State Geological Survey.)

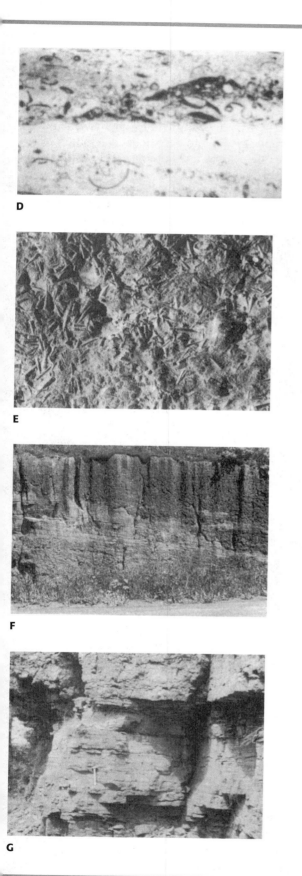

D

E

F

G

There are also abundant algal stromatolites and onco-lites, which are known to prefer intertidal environ-ments. The pelletal carbonate mudstone—with its mudcracks, mudchip conglomerates, stromatolites, and pellets—was often scoured and then covered by a skeletal calcarenite representing submergence of the intertidal flat. The fossil types are few but occur in large numbers. Such a low-diversity, high-abundance pattern is common in highly stressed environments, where only a few tolerant species can survive but their numbers are virtually unchecked by competitors or predators. The fossils include ostracods, tentac-ulitids (Fig. 12.1.3E), small spiriferid brachiopods, tre-postome bryozoans, and spirorbid worms. The more elongate fossils show a strong alignment, indicat-ing that many fossils were reworked by strong tidal currents. The rapid alternation between the mud-cracked, stromatolitic mudstones (indicating relative emergence) and the skeletal calcarenite (indicating submergence) is particularly diagnostic of an environ-ment with rapid alternation of tidal cycles of desicca-tion and submergence.

The *subtidal facies* of the Manlius Formation is characterized by fossiliferous, pelletal carbonate mud-stones interbedded with small biostromes of stromatoporoids. The carbonate mudstones are mas-sive- to medium-bedded and contain a diverse and abundant fauna of rugose corals, brachiopods, ostra-cods, snails, and codiacean algae, which could survive only under continual marine submergence. Biotic ac-tivity is also shown by extensive burrow mottling, which disrupts lamination and gives the mudstones their massive appearance. These mudstones interfin-ger (Fig. 12.1.3F, G) with tabular masses and globular individual heads of stromatoporoids, which appar-ently grew in tidal creeks and channels in the subtidal lagoon. Although the stromatoporoids did not grow on a large enough scale to qualify as a reef, they prob-ably did form wave-resistant structures that sheltered the subtidal lagoon.

When all three of the facies are associated, they clearly indicate peritidal conditions. The repetition of these facies in numerous thick cycles indicates that the Manlius is typical of the shallowing-upward limestone sequences found in many parts of the geologic record.

Overlying and laterally interfingering with the Manlius is the next unit in the Helderberg Group, the Coeymans Formation (see Fig. 12.1.1). It consists of coarse-grained, gray or blue limestone with thin, ir-regular bedding. There is very little clay material. This unit has a normal Devonian marine fauna of brachiopods, high-spired gastropods, straight-shelled

(continued)

(Box 12.1 continued)

Figure 12.1.4 Cross-bedded crinoidal layer of Coeymans-type lithology and stratified beds of Manlius type in the upper Olney Member, near Paris, New York. Subtidal carbonate shelf facies. (Courtesy of L. V. Rickard and the New York State Geological Survey.)

cephalopods, bivalves, crinoids, trilobites, and tabulate corals. Fifty to 80 species are represented, compared with only 5 to 30 species in the more extreme intertidal and shallow subtidal environments represented by the Manlius. The most striking feature is the abundant cross-beds (Fig. 12.1.4), which indicate strong current activity. All these features suggest that there is considerable wave or current activity in a normal marine shelf environment, well below the tide level and just below or at wave base. In some places there are coral bioherms composed of tabulate and rugose corals, crinoid columnals, bryozoans, and stromatoporoids.

Finally, the overlying Kalkberg and New Scotland formations (see Fig. 12.1.1) are composed of fine-grained, irregularly bedded silty limestones, which contain an even more diverse fauna (over 300 species)

of brachiopods, trilobites, scattered tabulate and rugose corals, and especially bryozoans (Fig. 12.1.5). The dominance of fine silt and the irregularity of the bedding suggest very little winnowing by current activity. The abundance of delicate bryozoans is another indicator of relatively quiet conditions. These features, taken together, suggest a shallow marine shelf environment, well below normal wave base.

Thus, the Helderberg facies belts, taken in sequence, represent tidal flat and lagoonal deposits (Manlius), well-agitated subtidal waters near wave base (Coeymans), and open shelf marine waters below wave base (Kalkberg and New Scotland). The overall trend (see Fig. 12.1.1) suggests a sequence that was deepening through continued transgression to the west. These patterns are typical of many carbonate peritidal and shelf sequences throughout the geologic record.

Figure 12.1.5 Print made from acetate peel showing subtidal, burrowed New Scotland biomicrudite with encrusting bryozoans, trilobites, and brachiopods, Bronck's Lake, New York. (Courtesy of L. F. Laporte.)

Shelf deposits are similar in many ways to peritidal and reef carbonates, but there are some major differences. The water is always normal marine, so the organisms found on the shelf tolerate only a narrow range of salinities. The normal marine fauna is much more diverse than the peritidal fauna, however, and a diverse fossil fauna and extensive bioturbation are typical. Because most of the organisms depend directly or indirectly on light, the depths are shallow, ranging from low-tide level down to 100 or 200 m. Like other shallow carbonate environments,

carbonate banks form in a narrow range of temperatures (10° to 30°C), so they are found in tropical or subtropical latitudes. Unlike peritidal lagoons, however, the water is always well-oxygenated. Finally, the carbonate bank is below normal wave base but above storm wave base. Most carbonate shelf sediments are very muddy, but in some areas there is evidence of strong currents or storm activity.

Because the rate of sediment production depends on the growth of organisms and not on hydraulic processes, bedding in carbonates is typically irregular

and discontinuous, with variable thicknesses over a few meters. A local growth spurt or concentration of carbonate particles can build a bed that thickens rapidly over an area of a few meters. Often, there are small hiatuses in the growth pattern, which may result in shaly parting between limestone beds. Some geologists have suggested that these represent episodes of terrigenous sedimentation that temporarily hamper the growth of organisms. Wavy bedding, nodular bedding, flaser bedding, and ball-and-pillow structures are also common but are produced by compaction and not by the hydraulic processes found in terrigenous systems. Stromatolites may also be responsible for wavy bedding. Cross-bedding and ripple marks occur only in areas of moderate current flow, such as exposed banks covered by skeletal sand or ooids. Of course, there are examples of evenly laminated, laterally extensive limestones, but these appear to be the exception. Most lamination is destroyed by the active bioturbation of this environment. Where it survives, even rhythmic lamination is attributable to some unusual fluctuation in climate and water chemistry produced by local conditions.

The sheltered central and inner part of most carbonate banks is an area of low-energy currents and waves; it may be protected by reefs or shoals or lie in coastal lagoons. The bottom is covered with intensely bioturbated peloidal muds. The lime mud is made of aragonite needles precipitated as the skeletons of certain types of calcareous algae. There are sand-sized particles of benthic invertebrates that live and die in the mud. In modern shelves, these are chiefly echinoderms, benthic foraminifers, and mollusks; in ancient seas, brachiopods, bryozoans, and trilobites were also important.

Surrounding the low-energy muds along the outer edge of the platform is a higher-energy environment that is affected by waves and storms. The current action results in coarser carbonate particles, particularly rippled skeletal sands and ooids. Ooids are a particularly good indicator of current energy. They are formed by the rolling and winnowing of small carbonate particles, which build up concentric rings of aragonitic needles as they roll back and forth through the lime mud. The faunas of the exposed banks are less diverse and abundant than those of the quiet lagoons, although an algal mat and a related fauna usually form between storms. Along the edges of the banks are lower-energy environments where seagrasses and burrowing organisms predominate. Skeletal sand and ooids often spill over the banks during storms and become cemented into lumps and grapestone. Wherever plants have formed stabilized mounds locally, the environment is relatively sheltered and a much greater faunal diversity is possible.

One of the best modern analogs of a shallow carbonate bank is the Great Bahama Bank, which covers about 100,000 km² with an average water depth of 5 m or less (Fig. 12.10A). It is surrounded by deep channels and straits 650 m or more deep and is dish-shaped in profile, with a low rim (Fig. 12.10B). Tidal ranges are 0.5 m or less, so wind is the main influence

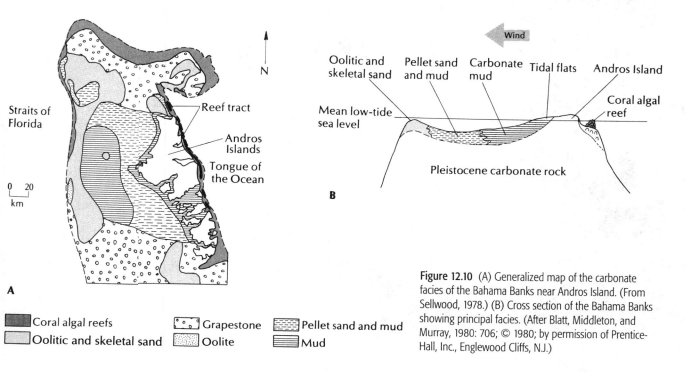

Figure 12.10 (A) Generalized map of the carbonate facies of the Bahama Banks near Andros Island. (From Sellwood, 1978.) (B) Cross section of the Bahama Banks showing principal facies. (After Blatt, Middleton, and Murray, 1980: 706; © 1980; by permission of Prentice-Hall, Inc., Englewood Cliffs, N.J.)

on current flow. The central basin, which has very low energy currents, is covered with pelletal muds. Because the storms and winds come from the north and east, coralgal reefs and skeletal sands rim the higher-energy northern and eastern margins of the bank; behind them are broad banks of grapestone. The quieter northwest margin is rimmed by oolitic shoals, formed in response to the relatively high tidal currents in this area (Fig. 12.11). The edges of the shoals grow very rapidly, despite the higher wave energy, because nutrient-rich waters rise from the deeper marginal channels, making possible the abun-

dant life from which skeletal sands eventually form after they are reworked by waves. The channels themselves (some of which are as deep as 4 km) are floored by pelagic ooze, with some skeletal sand that has spilled over from the shelf.

The Persian Gulf (Fig. 12.12; see also Fig. 12.3) is probably the largest modern example of an inland carbonate sea (as opposed to the offshore bank exemplified by the Bahamas). An area of 350,000 km² is covered with water 20 to 80 m deep. It is almost totally landlocked in a region of desert climate, so evaporitic tidal flats are very common. There is relatively little

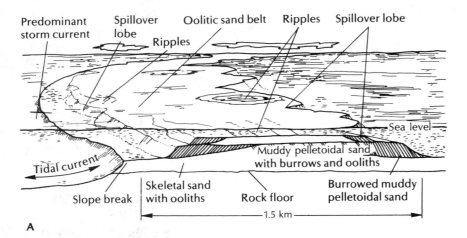

A

B

Figure 12.11 (A) Diagram showing aerial and cross-sectional view of oolitic shoals with tidally generated, bidirectional cross-stratification. (After Davis, 1983: 466.) (B) Aerial view of ooid shoals beneath the clear waters of the Great Bahama Bank near Cat Cay. The strong currents that rolled and produced the ooids have also produced the channels and ripples. (Courtesy of Shell Development Company.)

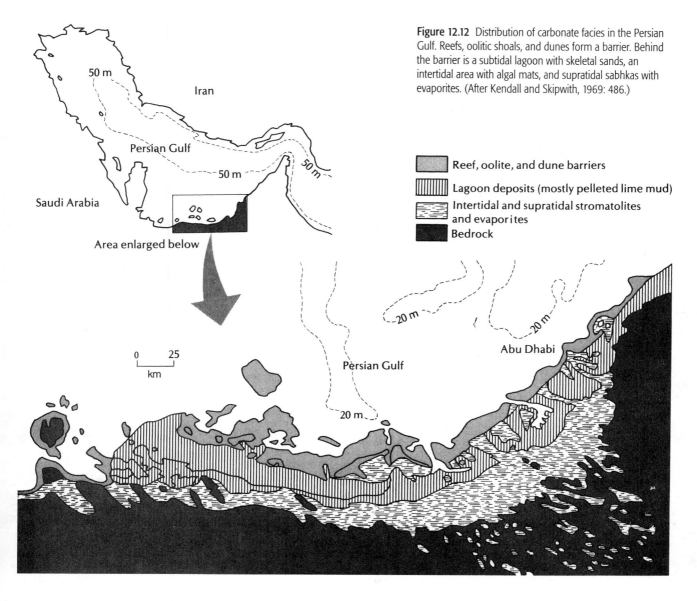

Figure 12.12 Distribution of carbonate facies in the Persian Gulf. Reefs, oolitic shoals, and dunes form a barrier. Behind the barrier is a subtidal lagoon with skeletal sands, an intertidal area with algal mats, and supratidal sabhkas with evaporites. (After Kendall and Skipwith, 1969: 486.)

storm or wave action, and circulation is very restricted, so most of the current activity is caused by tides, which have a range of several meters. Because the current and wave activity is much weaker than at the Great Bahama Bank, skeletal sands are less important than lime mud. The shallow margins of the Persian Gulf are rimmed by muds rich in mollusks. In the deeper waters below wave base in the central gulf, the dominant sediment is dark, argillaceous lime muds. Only the shoals along the southwestern shore are agitated by the winds coming out of the northeast; these winds produce a belt of skeletal sand. Reefs grow just inland of the skeletal sand belt on the southern shore.

There is tremendous lateral variation of facies in carbonate shelves and thus no standard vertical sequence. Under normal conditions, the dominant lithology is pelletal mudstones with abundant fossils. These conditions produce a thick, widespread sheet of limestone with normal marine fossils and remarkable consistency over a wide area. If the shelf is becoming shallower, a shoaling-upward sequence of normal marine and restricted intertidal deposits, and possibly evaporites, is produced (Fig. 12.13). In some shelves, regular influxes of terrigenous mud have produced shaly interbeds within the limestones. Shallow intracratonic basins typically fill with homogeneous limestones that thicken toward the center of the basin. If there is rapid fluctuation of sea level in a shallow basin, a rhythmic alternation of limestones and shales may result.

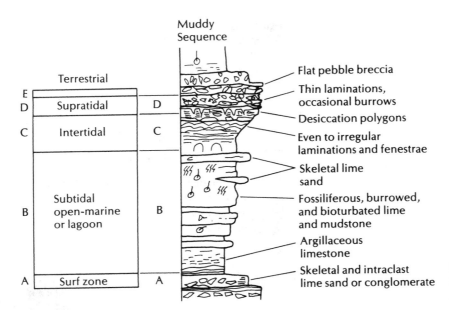

Figure 12.13 Hypothetical shallowing-upward sequence on a low-energy carbonate shelf. (After James, 1984: 218.)

Diagnostic Features of Subtidal Shelf Carbonates

Tectonic Setting All shelf carbonates require warm, clear, shallow, well-oxygenated, normal marine conditions, which typically exist only on continental shelves or epeiric seas in low latitudes with no significant clastic input.

Geometry Subtidal shelf carbonates form homogeneous sheets that can cover thousands of square kilometers and reach hundreds of meters in thickness.

Typical Sequence Under normal conditions, a uniform skeletal pelletal mudstone with remarkable vertical homogeneity is produced. If the sequence shoals upward, the subtidal limestones are capped by intertidal sequences and possibly evaporites.

Sedimentology Although the mineralogy is almost exclusively aragonite, calcite, and dolomite (with rare shales and evaporites), textures are highly variable. Pelletal muds rich in biogenic debris are common, as are ooids, skeletal sands, and bioturbated muds. Bedding of variable thickness, with wedge- and lens-shaped units, is particularly characteristic, as are nodular bedding and flaser bedding caused by compaction.

Fossils The most diagnostic feature is the abundance of fossils of the normal marine fauna, which are tolerant of only a limited range of salinities, light conditions, turbulence, and oxygen content. The shallow carbonate shelf supports by far the greatest diversity of marine faunas of any environment on Earth.

Reefs and Buildups

Unlike any of the systems discussed previously, carbonate reefs and buildups are completely self-generated. They are sediment systems built entirely by the organisms growing in them. The term *reef* has been used so loosely over the years (reviewed by Dunham, 1970, and Heckel, 1974) that the term **buildup** is preferably applied to any body of carbonate rock that has built up topographic relief above the surrounding environment. A reef is a buildup that has grown in the wave zone and has a wave-resistant framework. The term **bioherm** or **biostrome** is generally applied to any in situ accumulation of benthic organisms, whether or not it is topographically high or wave-resistant.

Reefs and buildups are interesting for many reasons. Paleontologists consider them the best examples of a paleoecological community, preserved intact in its own detritus. The presence of a reef implies an adequate supply of nutrients, as well as the usual restrictive conditions of carbonate sedimentation, and so provides a valuable indicator of paleoceanographic conditions. Reefs also give clues to understanding the paleoenvironment at the times in the past when building up was especially common. Buildups also serve as important stratigraphic traps for oil.

A reef community is very complex. The most essential part is the framework builder, and a number of colonial organisms have been responsible for this process throughout geologic history (Fig. 12.14). Today's reef builders may be hermatypic scleractinian corals, coralline algae, bryozoans, or sponges, but in

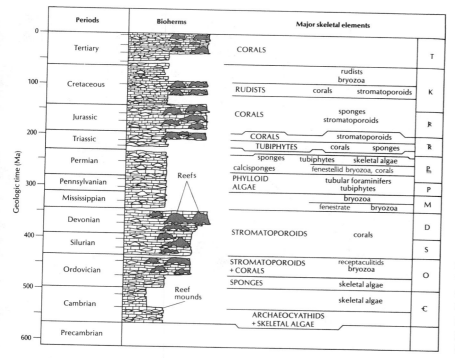

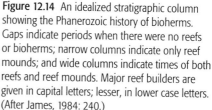

Figure 12.14 An idealized stratigraphic column showing the Phanerozoic history of bioherms. Gaps indicate periods when there were no reefs or bioherms; narrow columns indicate only reef mounds; and wide columns indicate times of both reefs and reef mounds. Major reef builders are given in capital letters; lesser, in lower case letters. (After James, 1984: 240.)

the geologic past they included cyanobacterial stromatolites, archaeocyathids, crinoids and blastoids, stromatoporoids, receptaculitids, tabulate and rugosid corals, rudistid bivalves, and even richthofenid brachiopods. The framework builders are essential to the structure of the reef, making it wave-resistant and forming baffles against strong currents and sediment. As these organisms die, new framework builders grow on top of them, which enables a reef to build upward in response to sea level. The framework organisms are kept in check by a community of bioeroders, including boring algae, worms, sponges, mollusks, and coral-eating fish and echinoids. These organisms continually weaken the framework until storms topple it, thus allowing new growth when sea level is not rising. The framework builders typically make up only 10% of the total volume of the reef. In the many cavities, crevices, and protected interreef patches, lime mud that contains broken skeletal debris and reef fragments accumulates. As this interstitial material fills cavities, the reef becomes even more massive and resistant. Eventually, cementation locks both the framework and interstitial material into a solid, massive limestone unit.

The classification terms for reef limestones reflect the various relationships between framework and interstitial material (Fig. 12.15; see also Fig. 11.5). Embry and Klovan (1971) divided reef carbonates into allochthonous interstitial mud and autochthonous reef rock that grows in situ. If more than 10% of the al-

lochthonous particles are larger than 2 mm in diameter and are matrix-supported, the rock is a **floatstone.** If they are clast-supported, it is a **mudstone. Framestones** are made of autochthonous massive framework builders that grew in situ, whereas **bindstones** are composed of tabular or lamellar fossils that were bound together during deposition. A **bafflestone** is made of in situ stalked organisms that trapped sediment by baffling currents.

The reef organisms themselves give a number of clues about the environment. First, their diversity is sensitive to environmental conditions. A high diversity of both growth forms and taxa indicates nearly optimum conditions, such as a plentiful supply of nutrients and a minimum of physical stresses. During stable, optimum conditions, reef organisms can diversify and specialize, subdividing the niches so that many more organisms can live on the reef. A low diversity, on the other hand, indicates that the environment is unpredictable, stressful, or under initial colonization. Rapid and extreme fluctuations in temperature and salinity, low light levels, or intense wave activity all hamper the growth of reef organisms, so only the hardiest can continue to grow.

Second, the growth form of the framework organisms indicates the wave energy and sedimentation rate (Table 12.2). For example, delicate branching corals can survive only in low-energy wave environments, yet they grow rapidly because they are being buried by sediment due to the high sedimentation

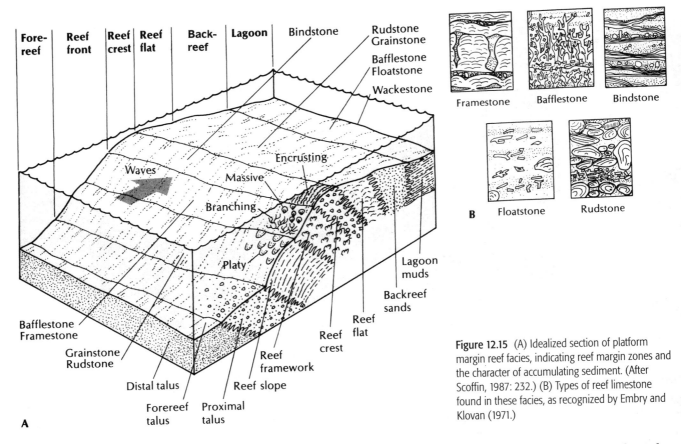

Figure 12.15 (A) Idealized section of platform margin reef facies, indicating reef margin zones and the character of accumulating sediment. (After Scoffin, 1987: 232.) (B) Types of reef limestone found in these facies, as recognized by Embry and Klovan (1971.)

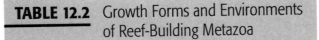

TABLE 12.2 Growth Forms and Environments of Reef-Building Metazoa

Growth Form	Environment	
	Wave Energy	Sedimentation
Delicate, branching	Low	High
Thin, delicate, platelike	Low	Low
Globular, bulbous, columnar	Moderate	High
Robust, dendroid, branching	Moderate to high	Moderate
Hemispherical, domal, irregular, massive	Moderate to high	Low
Encrusting[a]	Intense	Low
Tabular[a]	Moderate	Low

Source: James, 1984.
[a]Encrusting and tabular forms are difficult if not impossible to differentiate in the rock record, yet they indicate very different reef environments.

rate. Encrusting bryozoans or corals, on the other hand, can resist intense wave activity.

Third, the reef is divided into various regions (sometimes called facies) by environmental conditions. For example, a modern reef system shows a spectrum of hydrodynamic conditions, from calm water with periodic subaerial exposure in the backreef, to weakly moving water in the core of the reef, to intense wave and current activity along the wave front. The reef organisms and their growth forms are appropriately zoned to reflect this difference in current activity and strength.

In general, most reefs can be divided into three distinct facies: the reef core, the reef flank, and the inter-reef (Fig. 12.16). The *reef core* is massive, unbedded, and composed of the framework builders filled in by an interstitial matrix of lime mud and skeletal sand. The crest of the reef core is subject to the greatest wave energy, so the growth forms are low and encrusting. If the crest is relatively protected, hemispherical, massive forms with a few interstitial branching forms can occur. The upper part of the reef front is also made of encrusting and massive framework organisms (Box 12.2). Below 30 m or so, the reef flank is below wave base, so the framework builders become much more branching and platelike in response to the quiet water

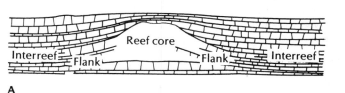

A

B

Figure 12.16 (A) The three major reef facies in cross section. (After Walker, 1984: 229.) (B) Reef and reef-flank deposits over 100 m thick, from the Peechee Formation, Upper Devonian, Flathead Range, Rocky Mountains, Alberta. R denotes reef core. (Courtesy of B. Pratt.)

conditions and the reduced light levels. Hermatypic corals depend on light for their symbiotic algae, as do coralline algae. Modern reefs reach depths of no more than 70 m; below this point, the light level is too low.

Behind the reef crest is the *reef flat,* which is subjected to considerable current and wave activity but is usually sheltered from the worst pounding of the waves. In this region, pavements of skeletal debris and coral rubble accumulate, along with shoals of well-washed lime sand. Because the water is shallow, the framework builders grow rapidly and form irregular clumps and patches separated by current-worked channels of sand and rubble. The most sheltered area of all is the *backreef,* which experiences waves only during storms. It is covered with carbonate sand derived from the biological breakup of carbonate organisms and is moved around by storm waves. In the lagoons, the lime mud settles out of suspension, forming a predominantly muddy matrix. This sheltered environment provides optimum living conditions, limited only by rapid burial in the mud. Among the framework builders, stubby, dendroid forms and large globular forms are common. A great diversity of interstitial organisms live in the mud, especially crinoids, delicate bryozoans, brachiopods, mollusks, ostracods, and calcareous green algae. The diversity of organisms alone is often sufficient to allow one to recognize the sheltered backreef.

Below the growing part of the reef core is the *reef flank.* Debris from the reef washes or slumps down the reef front and into the flank beds, which dip gently away from the core. As a result, the flank beds are usually composed of bedded limestone conglomerates made of reef debris and lime sands that have washed in during storms. Reef-flank beds are another example of deposits that accumulate at an original depositional angle and thus are exceptions to Steno's law of original horizontality. These beds grade into the interreef facies as they get farther from the reef and their dips decrease.

The *interreef facies* is very similar to other shallow-water subtidal carbonate bank deposits. These deposits are usually composed of thin-bedded, pelletal lime muds that are rich in skeletal sands. Often, the only indication of an interreef location is the rare presence of debris from the reef itself, which has been brought there by extreme storm waves. In basins with restricted circulation, evaporite minerals are common in the interreef facies.

One of the most fascinating features of reefs is *reef succession.* Reef communities do not appear on the seafloor fully developed but evolve and change as they colonize and stabilize the substrate (Table 12.3). The *pioneer (stabilization) community* is composed of organisms that encrust or send down roots and stabilize the shoal sands. Calcareous green algae, seagrasses, and crinoids are the most common pioneers, but only a low diversity of them can be maintained in such a turbulent, unstable environment. Once the sediment is stabilized, however, other algae, corals, and sponges take root and grow. On this stabilized framework, the *colonizing community* of the main framework builders takes hold and soon dominates. Typically, a few species of branching or lamellar corals grow very rapidly, outstripping their burial by sediment. After they are well established, the *diversification stage* takes place. At this stage, the bulk of the reef mass forms as the reef grows up to sea level. There is a great diversity of reef-building taxa and growth forms, as well as of interstitial organisms, in response to the diversity of habitats. As the reef community reaches sea level and more turbulent water, the *domination stage* occurs, in which laminated, encrusting forms cover a reef core that is largely filled in.

BOX 12.2 DEVONIAN REEFS OF THE CANNING BASIN, AUSTRALIA

One of the world's best exposed and most spectacular ancient barrier reef complexes is exhumed along the northern margin of the Canning Basin in northwestern Australia. The Canning Basin is the largest sedimentary basin in western Australia, covering an area of 530,000 km² and containing 13,000 m of Ordovician through Cretaceous sediments. The northern margin consists of a belt of Middle and Late Devonian reef complexes about 350 km long and 50 km wide. These reefs grew on the edge of a block-faulted basin that resulted in a relief of several hundred meters above

the basin to the south during the Late Devonian. The barrier reef belt once may have continued another 1000 km to the west and north to connect with similar reefs in the Bonaparte Gulf Basin. Several river gorges expose spectacular sections through these reefs, most notably at Windjana Gorge (Fig. 12.2.1).

As described by Playford (1980, 1981, 1984; Playford and Lowry, 1966), the reef complexes can be divided into three distinct facies and several subfacies (Fig. 12.2.2A): the platform, the marginal slope, and the basin facies. The *reef subfacies* occurs as a narrow rim

Figure 12.2.1 (A) Panoramic photograph of the southeast wall of Windjana Gorge. (Courtesy of P. E. Playford.) (B) Equivalent sketch showing facies relationships in the Frasnian reef platform margin. (After Playford, 1980: 818; by permission of the American Association of Petroleum Geologists, Tulsa, Okla.)

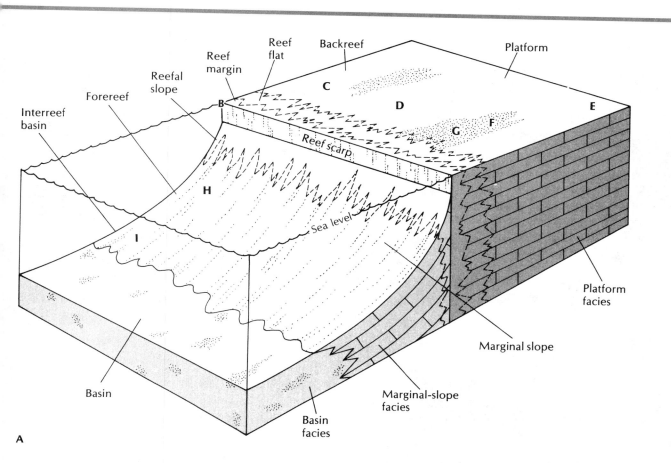

Figure 12.2.2 (A) Location of representative facies in Devonian reefs of the Canning Basin, Australia. (After Playford, 1980: 819; by permission of the American Association of Petroleum Geologists, Tulsa, Oklahoma.) (B) A polished slab from the reef crest showing laminar stromatoporoids and the ?alga *Renalcis* (speckled) surrounded by cavities filled with fibrous calcite. (C) Well-bedded limestone and dolomite in the Pillara Limestone, from the reef flat subfacies. On page 256: (D) Cut and polished slab of a toppled stromatoporoid colony encrusted with the ?alga *Renalcis*. The cavity system has been filled with laminated pelletal mud of the backreef lagoon. (E) Mudcracks in birdseye limestone from the supratidal backreef area. (F) Algal oncolites in fine-grained oolitic pelletoid birdseye calcarenite, from the intertidal-supratidal backreef subfacies. (G) Cross sections of the bivalve *Megalodon* in life position from the backreef immediately behind the reef front. (H) Rolled blocks of reef limestone in the forereef facies. The forereef silty limestones and calcareous siltstones have been distorted by the rolled blocks of limestone from the edge of the reef. (I) Contact between blocks of crinoidal forereef calcarenite in the forereef megabreccia. (B–I courtesy of P. E. Playford.)

(*continued*)

(Box 12.2 continued)

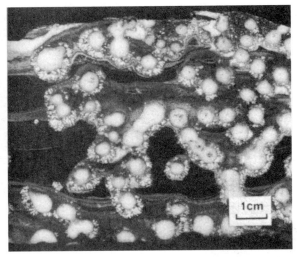

D

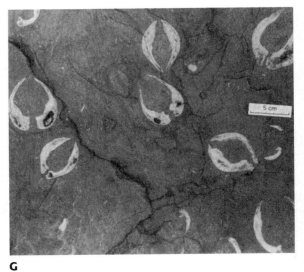

G

E

H

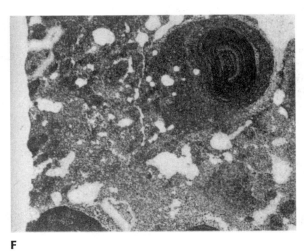

F

I

around the platform and is composed of massive limestone and dolomite built up by colonial cyanobacteria and stromatoporoids. These formed a resistant framework that sheltered the backreef lagoons. The reef framework is typically an interlocking mass of stromatoporoids and the ?alga *Renalcis* (Fig. 12.2.2B), which encrusted all available surfaces. The voids between the framework builders were later filled with fibrous and sparry calcite. Other reef-building organisms included corals, brachiopods, and sponges.

The *backreef subfacies* (Pillara and Nullara limestones) is composed of well-bedded limestone and dolomite that was deposited in the shelf lagoon behind the reef rim (Fig. 12.2.2C) and extends horizontally for tens of kilometers. In some places, these rocks are interbedded with terrigenous sediments. The most abundant rock type of the backreef is dominated by stromatoporoids, which built small

biostromes behind the main reef front. The cavities between the stromatoporoids (Fig. 12.2.2D) are encrusted with the ?alga *Renalcis* and filled with laminated pelletal lime mud. Mudcracked limestones with a birdseye texture are also common, indicating that some parts of the lagoon were emergent at times (Fig. 12.2.2E). In other places, corals such as *Hexagonaria*, rather than stromatoporoids, form small biostromes. Large areas are covered with oolitic limestones, indicating an environment of strong wave and current action. Few organisms were able to live on such a mobile substrate, except for algal oncolites (Fig. 12.2.2F) and the bivalve *Megalodon* (Fig. 12.2.2G). In section, these oolitic limestones exhibit not only oncolites and ooids but also pellets and a pronounced birdseye texture (Fig. 12.2.2F).

The *forereef subfacies* is composed primarily of talus deposits that rolled down the 30° to 35° reef-front

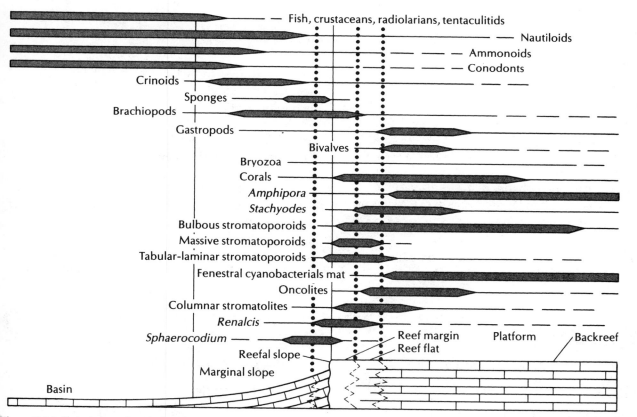

Figure 12.2.3 Ecological zonation of Frasnian (Late Devonian) organisms in the Canning Basin reef complex. The basinal fauna is predominantly free-swimming or floating organisms, such as fish, cephalopods, conodonts, crustaceans, radiolarians, and tentaculitids. The marginal slope is dominated by animals with the ability to anchor to the substrate, such as crinoids, sponges, and brachiopods. The reef crest is made of wave-resistant colonial organisms that build the reef framework, such as corals, massive and laminar stromatoporoids, and algae. The backreef contains organisms that prefer the sheltered conditions and can tolerate elevated temperatures and salinities and occasional desiccation, such as gastropods, bivalves, corals, bulbous stromatoporoids, and stromatolites. (After Playford, 1980: 827; by permission of the American Association of Petroleum Geologists, Tulsa, Okla.)

(continued)

(Box 12.2 continued)

slope. These deposits consist mainly of skeletal sands, conglomerates, and megabreccias. In some places, blocks of reef limestone several meters across have rolled into the forereef deposits and distorted them (Fig. 12.2.2H). Some of these blocks were later colonized on their upper surfaces and eventually buried (Fig. 12.2.2I). Between the blocks of megabreccia, the reef front was colonized by crinoids and brachiopods that lived in the sheltered part of the forereef. Playford (1984) has also described submarine debris flows made of reef talus that cascaded down channels cut into the reef front, producing inversely graded breccia beds similar to those found in clastic alluvial fans (see Chapter 8).

The *basin facies,* found in the basins between reef fronts, is composed of horizontal terrigenous shales, sandstones, and conglomerates interbedded with thin beds of limestone. There is little or no material derived from the limestone platforms; this material apparently was trapped in the forereef slope.

The reef complex shows pronounced ecological zonation (Fig. 12.2.3). The diversity of organisms in the backreef is considerably greater than in the stromatoporoid-dominated community of the reef itself. In addition to stromatoporoids, there are gastropods, bivalves, corals, oncolites, stromatolites, and a number of other algal structures in the backreef. The basin and marginal slopes, on the other hand, were favored by open-ocean organisms such as fish, crustaceans, tentaculitids, nautiloids, ammonoids, and conodonts. The reef slope was colonized by crinoids, sponges, and brachiopods.

In summary, the Devonian reefs of the Canning Basin demonstrate most of the classic features seen in modern reefs: framework-building reef organisms, a sheltered backreef lagoon, a forereef talus, and the open-sea deposits of the basin. Each of these environments had its own range of environmental conditions, resulting in a pronounced ecological zonation of organisms in and around the reef.

Diagnostic Features of Reefs and Buildups

Tectonic Setting Reefs and buildups form at the edges of carbonate banks where upwelling from deeper waters brings up nutrients. They are extremely limited by depth, temperature, salinity, and nutrients. Like other carbonate bank deposits, buildups are found in shallow, low-latitude, passive margins or epiric seas free of clastic input.

Geometry Small, local moundlike or banklike accumulations occur that show rapid lateral changes in facies and thickness form. Patch reefs may be only a few meters high and wide, but some large reef complexes are hundreds of meters thick and kilometers wide.

Typical Sequence Other than reef succession, there is no typical stratigraphic sequence of reefs and buildups. The lateral relationships of the various reef facies are more important than the vertical pattern, although both factors may work in concert to form a complex of massive reefs and bedded interreef limestones.

Sedimentology Framework builders are dominant, so an entire deposit grows and is bound together

TABLE 12.3	Stages of Reef Growth		
Stage	**Type of Limestone**	**Species Diversity**	**Shape of Reef Builder**
Domination	Bindstone to framestone	Low to moderate	Laminated, encrusting
Diversification	Framestone (bindstone), mudstone to wackestone matrix	High	Domal, massive, lamellar branching, encrusting
Colonization	Bafflestone to floatstone (bindstone) with a mudstone to wackestone matrix	Low	Branching, lamellar
Stabilization	Grainstone to rudstone (packstone to wackestone)	Low	Skeletal debris

Source: James, 1984.

in situ. Interstitial lime mud, skeletal fragments, and reef-rock breccias are formed in crevices between the framework organisms. Reefs are formed exclusively of calcite or aragonite, although the reef core is highly susceptible to dolomitization. The back-reef flats may become evaporitic in the appropriate environment.

Fossils   Reefs are formed almost entirely of characteristic fossils whose ecology determines the growth and shape of the reef. Reef organisms are extremely sensitive to temperature, salinity, light, and terrigenous mud, so they are excellent indicators of environmental conditions.

Secular Variation in Carbonates

When we look at modern aragonitic carbonate particles such as ooids, the needles secreted by algae to make lime mud, or cements precipitated in the early stages of diagenesis, our natural assumption is that these sediments have always existed. After all, uniformitarianism suggests that the present is the key to the past, and we need modern analogs and present-day principles of chemistry and mineralogy to understand past conditions. We've already seen in previous chapters, however, that the present is a very imperfect model for the past. For example, the limited shallow marine deposits of the continental shelves are a poor analog for the gigantic epeiric seas that once drowned entire continents (see Chapter 10). In this chapter, we have seen how modern shallow marine carbonate environments (such as the Bahamas or the Persian Gulf) are frustratingly small and limited analogs for the huge limy seas that covered the continents during warm, high-sea-level, greenhouse conditions.

Similarly, we must be cautious about forcing our understanding of ancient carbonate sediments and environments into modern pigeonholes. For example, when we look at ancient ooids made of calcite, we assume that if they do not show evidence of having started out as aragonite needles, they must have been so heavily altered that their original aragonitic texture has been destroyed. Many ancient calcite ooids however, show no sign of diagenetic alteration. Sandberg (1975) first proposed that this absence of diagenetic alteration was not an artifact of poor preservation. He argued instead that in certain times in the geologic past, the chemistry of the seas would have favored precipitation of calcite rather than aragonite. By 1983, Sandberg had developed a model that suggested the

chemistry of the world's oceans had alternated between "aragonite seas," which had a high Mg-Ca ratio in the seawater and precipitated mostly high-Mg calcite and aragonite, and "calcite seas," which had a lower Mg-Ca ratio and precipitated low-Mg calcite (Fig. 12.17). Aragonite seas are known mostly from the icehouse conditions of the Neogene and modern world (since the Antarctic ice caps arose in the Oligocene), from the late Paleozoic, and from the late Proterozoic (which are called Aragonite III, Aragonite II, and Aragonite I, respectively). Between these conditions were the greenhouse worlds of the early and middle Paleozoic (Calcite I) and the Jurassic through Eocene (Calcite II), when the Mg-Ca ratio was lower and the Ca concentration was much higher, so that ooids, lime mud, and early cements were all precipitated directly as low-Mg calcite, not as aragonite.

Stanley and Hardie (1998, 1999) amplified this observation by pointing out a number of other related phenomena. For example, Hardie (1996) noticed that during periods of high Mg-Ca ratios (aragonite seas), there was also abundant precipitation of magnesium sulfate ($MgSO_4$) evaporites. During times of low Mg-Ca ratios, on the other hand, $MgSO_4$ evaporites disappeared and were replaced by sylvite (KCl) and similar evaporites.

The trend is even more obvious when one looks at the kinds of organisms that flourished during calcite seas and aragonite seas (see Fig. 12.17). In aragonite seas, we find that most of the calcareous sediment (especially the micrite) is produced by aragonite-secreting algae, such as *Halimeda* or *Penicillus* today (Aragonite III) or dasycladacean algae during the late Paleozoic (Aragonite II). During calcite seas, however, carbonate sediment is produced by calcite-secreting green algae, such as the receptaculitids of the early Paleozoic (Calcite I) or the huge bloom of nannoplankton (coccolithophorids) that built the thick chalk deposits of the Cretaceous (Calcite II).

Likewise, the dominant groups of reef builders shift with the Mg-Ca ratio. Today, massive reefs are built by hermatypic scleractinian corals made entirely of aragonite. These reef builders first arose in the Triassic, during the end of the last episode of aragonite seas (Aragonite II). A number of other aragonite-secreting reef builders flourished in the early Mesozoic, including a variety of sponges (inozoans, sphinctozoans, and sclerosponges), as did high-Mg-calcite-secreting organisms such as coralline algae (which also flourish today in the Aragonite III world), solenopores, and *Tubiphytes*. By contrast, the reef builders during the early Paleozoic (Calcite I) were all made of calcite.

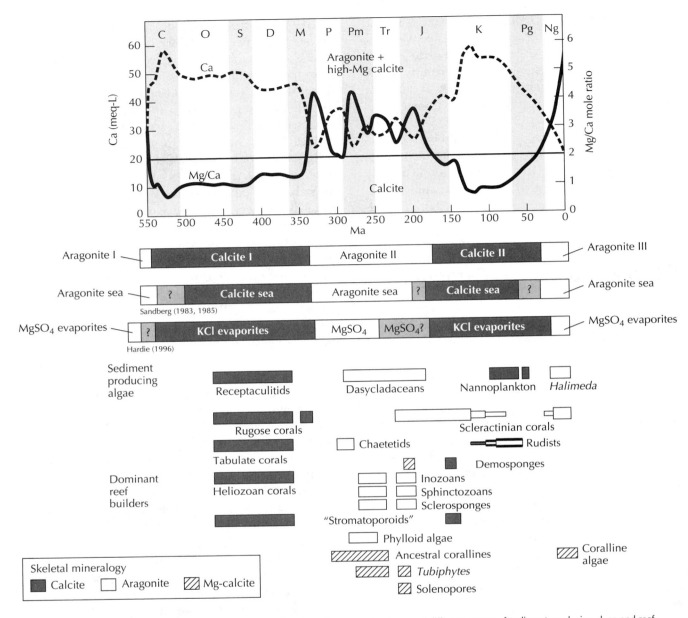

Figure 12.17 Oscillation of the Mg-Ca ratio and Ca concentration through the Phanerozoic, and its correlation with calcite seas and aragonite seas and with other phenomena, such as evaporite precipitation and dominance of different groups of sediment-producing algae and reef-building organisms. (After Stanley and Hardie, 1999.)

These included rugose, heliozoan, and tabulate corals, plus stromatoporoid sponges. The late Mesozoic (Calcite II) saw the return of stromatoporoids and calcite-secreting demosponges. This period also witnessed the replacement of aragonitic coral reefs by huge reefs consisting of the cone-shaped oysters known as rudistids, which are largely made of massive calcite.

Notice that Stanley and Hardie (1998, 1999) are not saying that *every* carbonate-secreting organism was affected or that *everything* switched from low-Mg calcite to aragonite and high-Mg calcite. Most mollusks (which use both calcite and aragonite), brachiopods (which use low-Mg calcite), and echinoderms (which use mostly high-Mg calcite) lived throughout the Phanerozoic without noticeable effects of the Mg-Ca ratio in seawater chemistry. The only organisms that responded were very sensitive to slight changes in seawater chemistry or had poor mechanisms for regulating their carbonate production. These included the sediment-producing algae and the organisms that

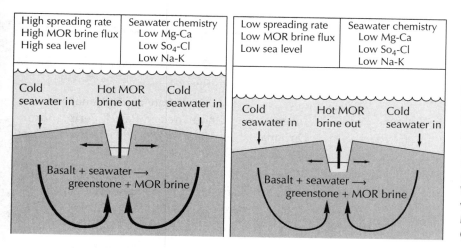

High spreading rate High MOR brine flux High sea level	Seawater chemistry Low Mg-Ca Low So$_4$-Cl Low Na-K

Cold seawater in   Hot MOR brine out   Cold seawater in

Basalt + seawater ⟶ greenstone + MOR brine

Low spreading rate Low MOR brine flux Low sea level	Seawater chemistry Low Mg-Ca Low So$_4$-Cl Low Na-K

Cold seawater in   Hot MOR brine out   Cold seawater in

Basalt + seawater ⟶ greenstone + MOR brine

Figure 12.18 Effect of changing rates of seafloor spreading on the Mg-Ca ratio of seawater. During high rates of spreading (*left*), the high ridge volume raises sea level, and the high volume of MOR (mid-ocean ridge) brine produces rapid conversion of basalt to greenstone, absorbing Mg in the greenschist minerals and lowering the Mg-Ca ratio. When spreading rates are lower (*right*), the ridge volume is less, sea level is lower, and there is more Mg available to change the Mg-Ca ratio in seawater. (After Stanley and Hardie, 1999.)

build huge carbonate skeletons (in a process Stanley and Hardie call hypercalcification), such as the great reef-building organisms. For example, ahermatypic corals, which can live in deep water and do not require symbiotic algae, were unaffected. However, these organisms do not build massive reefs. The hermatypic reef-building corals (such as scleractinians today, and probably tabulates and rugosids in the past) use symbiotic algae in their tissues to precipitate large volumes of carbonate. Therefore, these organisms are highly responsive to the Mg-Ca ratio of seawater.

What drives this oscillation between calcite seas and aragonite seas? Sandberg (1983) first suggested that the oscillation might be a result of the carbon dioxide content of the atmosphere. Greenhouse climates, with their high carbon dioxide levels, might drive down the level of carbonate saturation in the oceans, so that it was below the threshold for aragonite but still capable of producing calcite. And indeed, it appears that the calcite seas are correlated with known greenhouse states and the aragonite seas with icehouse conditions (see Chapter 19). However, Stanley and Hardie (1998, 1999) demonstrated that this does not work geochemically. They also showed that mechanisms that change the Mg-Ca ratio are much more effective than those that change the carbon dioxide–carbonate levels in the ocean. Instead,

they argue that the oscillation may ultimately be controlled by rates of seafloor spreading (Fig. 12.18). During times with high rates of spreading (and also with high sea levels, such as the greenhouse worlds of the early Paleozoic and the Cretaceous), huge volumes of heated brine are released from the volcanic vents along the mid-ocean ridges. This brine is depleted in magnesium, which has been consumed by oceanic basalts as they are converted to greenstone and amphibolites. The magnesium depletion, in turn, leads to higher concentrations of calcium (calcite seas). By contrast, when spreading rates are low (as when the supercontinent of Pangaea formed in the late Paleozoic, or when plate motions slowed down after the collision of India and Africa to close the Tethys Sea in the late Cenozoic), there is reduced consumption of magnesium in the mid-ocean ridges, and therefore the Mg-Ca ratio rises (aragonite seas).

Clearly, we must temper our understanding of ancient carbonate environments with a bit of caution. Not every carbonate facies, or every carbonate-secreting organism, works the same in the modern world as it did in the past. Uniformitarianism is a useful paradigm, but there are conditions from the past that have no modern analogs, reminding us that we must be cautious when examining these environments (see Chapter 19).

CONCLUSIONS

Carbonate facies form a major portion of the stratigraphic record, yet they represent highly restricted environmental conditions. They are very useful for reconstructing ancient environments and ancient latitudes of continents. They are also important for many economic materials, especially the oil and cement on which our society depends.

FOR FURTHER READING

From the references in Chapter 8, the books by Davis, Reading, and Walker contain excellent chapters on carbonate environments. In addition to these, the reader may wish to consult the following sources.

Bathurst, R. G. C. 1975. *Carbonate Sediments and Their Diagenesis*. New York: Elsevier:

Friedman, G. C., ed. 1969. *Depositional Environments in Carbonate Rocks*. SEPM Special Publication 14.

Frost, S. H., M. P. Weiss, and J. B. Saunders, eds. 1977. *Reefs and Related Carbonates: Ecology and Sedimentology*. American Association of Petroleum Geologists Studies in Geology 4.

James, N. P., and I. G. MacIntyre. 1985. Carbonate depositional environments, modern and ancient. Part 1, Reefs. *Quarterly Journal of the Colorado School of Mines* 80(1): 1–70.

Hardie, L. A., and E. A. Shinn. 1986. Carbonate depositional environments, modern and ancient. Part 3, Tidal flats. *Quarterly Journal of the Colorado School of Mines* 81(1): 1–74.

Harris, P. M., C. H. Moore, and J. L. Wilson. 1985. Carbonate depositional environments, modern and ancient. Part 2, Carbonate platforms. *Quarterly Journal of the Colorado School of Mines* 80(4): 1–60.

Laporte, L. F., ed. 1974. *Reefs in Time and Space*. SEPM Special Publication 18.

Mullins, H. T. 1986. Carbonate depositional environments, modern and ancient. Part 4, Periplatform carbonates. *Quarterly Journal of the Colorado School of Mines* 81(2): 1–63.

Sandberg, P. A. 1975. New interpretations of Great Salt Lake ooids and of ancient nonskeletal carbonate minerology. *Sedimentology* 22: 497–537.

Sandberg, P. A. 1983. An oscillating trend in Phanerozoic nonskeletal carbonate mineralogy. *Nature* 305: 19–22.

Scholle, P. A., D. G. Bebout, and C. H. Moore, eds. 1983. *Carbonate Depositional Environments*. American Association of Petroleum Geologists Memoir 33.

Stanley, S. M., and L. A. Hardie. 1999. Hypercalcification: Paleontology links plate tectonics and geochemistry to sedimentology. *GSA Today* 9(2): 1–7.

Wilson, J. L. 1975. *Carbonate Facies in Geologic History*. New York: Springer-Verlag.

13 Other Biogenic Sedimentary Rocks

Signal Hill oil field near Long Beach, California, about 1935. (Courtesy of Chevron, Inc.)

IN CHAPTER 11, WE EXAMINED THE TWO PRINCIPAL, largely organic, carbonate rock types: limestone, which forms as the result of biologic activity (either directly or indirectly), and dolostone, most of which is produced diagenetically from limestone. This chapter covers the other biogenic deposits: (1) siliceous sediment and chert; (2) phosphate sediment and phosphatic rock; and (3) organic-rich materials such as coal and oil, which are concentrated in sediments and sedimentary rocks.

Chert and Siliceous Sediment

Chert is a fine-grained, hard sedimentary rock composed of cryptocrystalline fibrous chalcedony, lesser amounts of microcrystalline and cryptocrystalline quartz, and amorphous silica. It is the product of organic or inorganic precipitation.

Two major types of cherts are found in the geologic record: **bedded (primary) cherts** and **nodular (replacement) cherts.** Each exhibits a distinctive suite of physical characteristics, has its own mode of origin, and occurs within a distinctive lithological and tectonic setting.

Bedded Cherts

Most bedded chert is produced when silica-rich organic oozes deposited on the deep seafloor are recrystallized. Bedded cherts occur as individual bands,

263

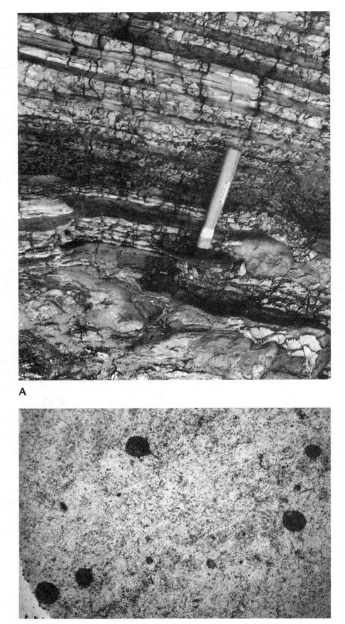

A

B

Figure 13.1 Bedded cherts typically occur within deepwater shale sequences. (A) Finely bedded shales and cherts ("porcellanites") from the middle Miocene Monterey Formation, Mussel Rock, California. (Courtesy of J. Barron.) (B) Most bedded chert is produced by siliceous microfossil oozes accumulating on a deepwater muddy bottom. This thin section of the Ordovician Normanskill chert includes many spiny spherical radiolarian skeletons (about 0.1 to 0.3 mm in diameter.) (D. R. Prothero.)

layers, or laminae that range in thickness from a few millimeters up to several meters (Fig. 13.1). The thicknesses of individual chert layers are often laterally and vertically uniform (ribbon cherts), but they can vary. Internal sedimentary structures are rare. Many bedded cherts occur as part of assemblages of oceanic

crustal rocks and associated deep-sea sediments (ophiolite sequences).

Two categories of bedded chert are recognized based on the presence or absence of fossils.

Bedded fossiliferous cherts contain the more or less obvious remains of such siliceous organisms as diatoms, radiolarians, and sponge spicules. They are obviously biogenic and form when siliceous oozes recrystallize. Classic examples occur in the Miocene Monterey Formation and Jurassic-Cretaceous Franciscan Formation of the California Coast Ranges and in the Ordovician Normanskill Shale of the Hudson Valley in New York.

Nonfossiliferous bedded cherts contain no visible skeletal remains. Most (like the Arkansas Novaculite of the Ouachita Mountains) are probably siliceous oozes so extensively altered that all fossils have been destroyed. When these so-called nonfossiliferous bedded cherts are etched from hydrofluoric acid, however, fossils are invariably seen, so few if any bedded cherts are truly inorganic in origin.

Silica Geochemistry

How is silica dissolved by weathering, transported, and precipitated? Figure 7.4 shows how the solubility of quartz and amorphous silica varies with pH. Most of the silica dissolved in seawater is derived from decomposing rock, but very little of it is produced by the solution of quartz because quartz is almost insoluble (6 to 10 ppm) in natural waters with pH values of less than 11. Most of the dissolved silica is **amorphous silica** that has not yet crystallized into quartz or any other mineral lattice. This chemical phase is quite soluble (100 to 200 ppm) in natural waters with pH values of less than 11. Amorphous silica is not produced by the dissolution of quartz but is the product of chemical weathering, especially hydrolysis, that occurs as feldspar is converted to clay minerals. Although the chemical decomposition of feldspar probably provides more than enough dissolved silica to precipitate the total mass of bedded cherts found in the geologic record, there are additional sources of dissolved silica. Submarine weathering and the diagenesis of basalt and siliciclastic sediment mantling the deep seafloor produce abundant dissolved silica. Submarine geysers pump additional silica-rich fluid directly into seawater at hot spots and along midocean ridges.

Seawater is very undersaturated with silica (normally 1 ppm). Dissolved silica arriving in river water (13 to 14 ppm) is immediately removed by organisms when it reaches the sea. Diatoms, silicoflagellates, and

radiolarians use up nearly all this silica as soon as it is available (see Chapter 10). Striking variations in the amount of dissolved silica in seawater clearly reflect this organic activity. Very low values (<0.01 ppm) in near-surface waters coincide with the photic zone; these values increase progressively to a maximum of 11 ppm where water depth exceeds 2 km.

Microplankton extract dissolved silica to construct minuscule endoskeletons of opaline silica (see Fig. 10.18A). They evolved long enough ago to explain all fossiliferous Phanerozoic bedded cherts. Radiolarians originated in earliest Paleozoic time, diatoms in the middle Mesozoic, and silicoflagellates in the Cretaceous. Cherts with few fossils and the arguably few bedded cherts without fossils probably formed in the same way but diagenesis has probably destroyed all trace of fossils.

What destroys the skeletons? The skeletons of living silica-secreting organisms are covered with a carapace of organic matter that keeps them from dissolving. However, solution begins immediately after death, as the skeletal remains fall slowly to the abyssal ocean floor (see Fig. 10.21) and continues after burial. Most of the silica that dissolves is quickly reprecipitated, leaving no trace of fossils. The precipitation of additional chert from silica dissolved in the seawater trapped in sediment (some of which might be nonbiogenic in origin) further obliterates the skeletal remains.

Precambrian bedded cherts, including many of the ribbonlike bands of chert that occur as components of Precambrian banded iron formation (see Chapter 14), are probably recrystallized from organically produced siliceous oozes. (Others may be produced by replacement.) Ultrahigh magnification with a scanning electron microscope reveals that some of these cherts contain small (micron-sized) spherical and subspherical ornamented structures. At least some of these structures appear to be the remains of photosynthesizing cyanobacteria (algae); others may be fossils of other Precambrian silica-secreting organisms. If not, some other mode of origin for these older deposits must be postulated. No evidence supports the once-popular idea that these deposits result from **syneresis**, the process by which gelatinous masses or flocculated colloidal suspensions of hydrothermally produced silica throw off liquid and dry out to crystallize on the seafloor.

Nodular Cherts

Nodular (also called secondary, or replacement) cherts occur as fist-shaped, spherical, subspherical,

and ovoidal masses of opal, chalcedony, and quartz (Fig. 13.2A) disseminated mainly in shallow-water limestone and dolostone. Nodules vary in size from a few millimeters (pea-sized) to a few centimeters. Individual nodules are often linked together, forming roughly planar bands that create anastomosing networks and lenses of chert. Variations in color from black to white to dull gray probably reflect carbon and water content. Bleached surface rims of distinctly different color with radiating desiccation cracks are common and produce a pattern of internal concentric lamination and color zoning. The sedimentary structures developed in surrounding carbonate rocks often continue right through the nodules. Fossil remains of such silica-secreting organisms as sponges occur along with the silicified remains of organisms that secrete calcareous shells (Fig. 13.2B). In other cases, carbonate allochems such as ooids (see Fig. 11.8) or peloids (Fig. 13.2C) are replaced by silica.

Nodular cherts are clearly of diagenetic origin. They form when silica originally deposited in one place dissolves, migrates, and reprecipitates elsewhere, replacing older material. The dissolved silica is derived from a variety of sources: detrital quartz grains that the wind transported onto carbonate banks, sponge spicules, and microplankton skeletons.

What are the factors that control this process? Silica dissolves in migrating groundwater only if it is very alkaline and the pH is high (see Fig. 7.4), but it is not known exactly why transport of dissolved silica ceases and precipitation of nodular chert begins. Bacterial activity may be crucial to the process (Maliva and Siever, 1989). But in some cases, chert forms inorganically. For example, distinctive modern chert nodules and lenses have been found in cores obtained from the Deep Sea Drilling Project. Dissolved silica disseminated in host mud and chalk simply nucleates at points where magnesium compounds are abundant (Maliva and Siever, 1988).

Phosphorites

Phosphate Geochemistry

Almost all sedimentary rocks contain minor amounts of phosphate. For example, mudrock and limestone typically contain a fraction of a percent of P_2O_5. Sedimentary phosphate deposits (**phosphorites**), on the other hand, are very rare. They are characterized by an abnormally high concentration of P_2O_5 (20% or more). This translates into a phosphate mineral content of roughly 50%.

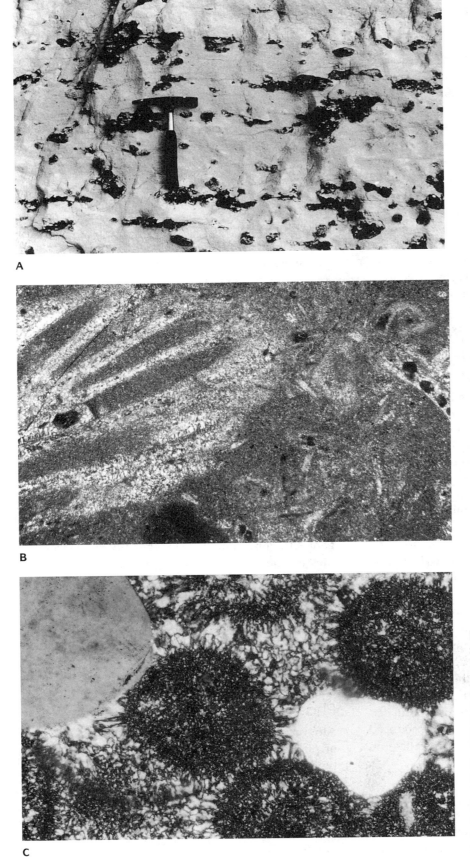

A

B

C

Figure 13.2 Nodular cherts are precipitated by silica-rich groundwaters percolating through limestones. (A) Black chert (flint) nodules from the Upper Cretaceous chalks at Tilleul Plage, Normandy, France. In this case, the boxwork of nodules cuts across the bedding. In other instances, the nodules are concentrated along the burrows of *Thalassinoides.* (Bromley and Ekdale, 1986: 73; courtesy of A. A. Ekdale.) (B) Thin section of Devonian Helderberg chert showing silicified coral and brachiopod fragments. (C) Thin section of Fort Ann chert showing silicified peloids about 0.3 mm in diameter mixed with rounded quartz grains; the voids were filled by drusy and blocky cements, also replaced by silica. See Fig. 11.8. (B and C by D. R. Prothero.)

Phosphate in sedimentary rocks occurs as a variety of minerals, but fluorapatite, $Ca_5(PO_4)_3F$, is the principal species. Hydroxyapatite, $Ca_5(PO_4)_3OH$, is the primary mineral in vertebrate bones and teeth. (The fluoride in drinking water and toothpaste helps convert the hydroxyapatite in our teeth to the more decay-resistant fluorapatite.) Sedimentary apatites of uncertain composition are called **collophane.** In most sedimentary rocks, the minor phosphate typically occurs as detrital clasts of the mineral apatite, as organic fecal matter (coprolites), or as transported bone fragments, all disseminated within limestone, sandstone, or mudrock. Exceptional circumstances are required to form the few phosphorites known. What is the source of the phosphate found in phosphorites? How is it concentrated to generate these unusual deposits?

Phosphate is derived either directly from hydrothermal veins or by chemical decomposition of such phosphate minerals as fluorapatite in igneous and metamorphic rocks. It occurs as ions and as particulate matter adsorbed on organic detritus (Froelich et al., 1982). Phosphate is essential for organisms because it is an integral component of RNA and DNA, the compounds that enable organisms to replicate genetically. This critical nutrient regulates organic productivity.

There are three principal types of phosphorite deposits:

1. Concentrations of nodular phosphorite lie scattered on the floor of some modern outer continental shelves. Well-known ancient analogs of these deposits occur in Morocco and in the Permian Phosphoria Formation of the North American Cordillera.

2. Placer concentrations of transported and reworked organic clasts (bone beds) occur rarely. These include such famous fossil deposits as the Eocene-Oligocene Phosphorites du Quercy of France and the Miocene Bone Valley Formation of Florida.

3. A third type of phosphorite, also extremely rare, is produced by a diagenetic process known as phosphatization. Phosphate-rich fluids leached from **guano** (fecal matter of birds or bats) are concentrated and reprecipitated in limestone.

Origin of Modern and Ancient Phosphorites

Modern *phosphorite nodules* are now concentrated in a few shallow shoals along the outer fringe of continental shelves. Water depth ranges from 40 to 300 m,

and the overall rate of sedimentation is low. Phosphorite minerals occur in gelatinous to solid irregular masses, in spherical clumps, and as slab-shaped chunks. They range in size from bodies a few centimeters in diameter to meter-sized blocks. Coprolites and fish bones are intimately associated with these masses; many contain clusters of phosphate-enriched pellets or ooidlike particles. Relict organic remains suggest that replacement is common.

These curious deposits are apparently generated where cold, nutrient-rich ocean water wells up onto warmer shallow water in shelf areas. The abrupt influx of nutrients leads to extremely high organic productivity at the base of the food chain. Initial bursts of phytoplanktonic activity propagated up the food chain produce large volumes of organic material (waste and remains). Whenever organic productivity exceeds the rate of decomposition, oxygen deficiency results and further decay of phosphate-rich remains ceases.

Ocean water upwelling, heightened organic productivity, and the existence of oxygen-deficient zones (with mass fish kills because of poisoning by phytoplanktonic blooming) are common across the shallow, midlatitude (within 40° of the equator), outer continental shelf areas off western Africa and western South America. Phosphorite is not precipitated directly from seawater but forms in a thin surficial zone of phosphatization developed on the surface of the seafloor. Bacterial decay attacks previously precipitated pellets, bones, and coprolites, releasing free phosphate that replaces pre-existing sediment. Bottom-hugging currents further concentrate phosphate-rich constituents and winnow away non-phosphatic materials.

The best-known ancient analog of modern outer shelf deposits is probably the Permian Phosphoria Formation of Idaho, Montana, and Wyoming (McKelvey et al., 1959). Phosphatic mudrock interbedded with oolitic, pisolitic, and nodular phosphorite occurs in a belt that straddles what was a narrow zone of transition between the shallow shelf and the upper segment of the continental slope (Fig. 13.3). These sediments were deposited on a normal shallow marine shelf that was slowly subsiding. Most of the richest phosphate deposits are thought to have accumulated offshore on the outer shelf and slope, where oceanic upwelling apparently brought up nutrients (especially phosphate) in large quantities. Late Mesozoic to early Cenozoic phosphate deposits of North Africa define a similar band of outer shelf deposition on the southern margin of the Tethys Sea.

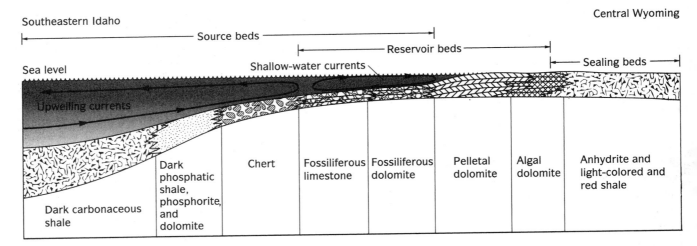

Southeastern Idaho   Central Wyoming

A

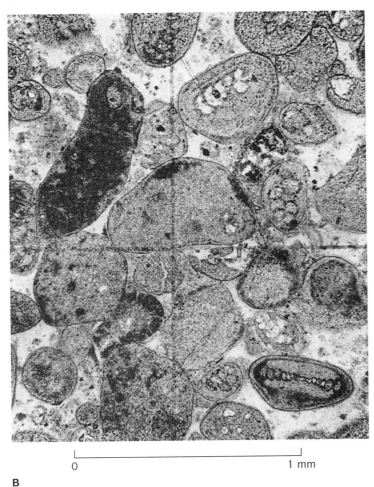

0   1 mm

B

Figure 13.3 The Permian Phosphoria Formation of the northern Rocky Mountains is one of the world's most famous phosphorites. (A) Depositional model for the Phosphoria Formation. Phosphatic beds are thickest in the deep waters near the shelf-slope break, where upwelling currents brought phosphate up from deeper waters and caused a tremendous concentration of phosphate along the submarine slope. (After R. P. Sheldon, 1963: 159; by permission of the U.S. Geological Survey.) (B) Thin section of cherty phosphorite from the Phosphoria Formation containing many foraminiferal tests that became nuclei of ooids. Clear chert fills the interstices between grains, as well as the cracks in grains and the chambers of fossils. (E. R. Cressman, courtesy of U.S. Geological Survey.)

Bone beds are placer concentrations of organic skeletons from which other sediment has been removed. Sediments containing bones, coprolites, and other organic materials are reworked, and the organic remains are concentrated as current lag deposits. The selective dissolution of some phosphate and its recrystallization as cement (a process favored by high bacterial activity and a neutral to slightly acid setting) produce solid phosphorite that replaces mudrock and limestone.

Under exceptional circumstances (geographic isolation is most important), flourishing communities of birds and/or bats confined to islands and peninsulas

or caves generate meters-thick piles of layered guano. Leaching of this excrement concentrates calcium phosphate as an almost insoluble residue. Groundwater circulating downward through this material replaces underlying mudrock or carbonate with phosphate.

Organic-Rich Sediments

Coal, petroleum (oil and natural gas), oil shale, and various types of solid hydrocarbons (such as natural asphalts in tar sands) are accumulations of undecayed carbon-rich organic matter concentrated in sedimentary rocks, not sedimentary rocks per se. Organic matter makes up the bulk of the material in coal, but oil shale, oil, natural gas, and solid hydrocarbons are disseminated in the pore spaces of mudrock, sandstone, and carbonate rocks. These are the **fossil fuels;** they consist of undecayed organic tissue and are burned to produce thermal energy.

Organic-rich deposits are unevenly distributed in time and space because the conditions that favor their formation and preservation occur infrequently and are restricted areally. More than half the world's petroleum reserves occur in the Middle East; North America contains roughly half the world's coal deposits. Few coals are older than Devonian. Coal formation requires the existence of large masses of undecomposed land-based plants. An extensive land plant population evolved only 400 million years ago and did not flourish until Carboniferous time. Oil, oil shale, and natural gas are also almost exclusively of Phanerozoic age; most are restricted to rocks of Mesozoic or Cenozoic age. Precambrian and Paleozoic organisms evidently did not produce sufficient organic material of the kind that can be altered to petroleum. And few settings favor the preservation of the raw material that ultimately is converted to petroleum. Most very old organic material is destroyed either by erosion or by metamorphism and igneous activity.

Carbonaceous deposits in sedimentary rocks are derived from the undecayed remains of plants and minerals. Decay is slow natural combustion, simple oxidation in which organic hydrogen and carbon combine with free oxygen to produce carbon dioxide and water. To be converted into petroleum, coal, or oil shale, organic tissues must not decay. Decay is inhibited only under conditions in which no free oxygen is available, variously defined as anoxic, anaerobic, oxygen-deficient, or reducing. Reducing conditions are promoted by several factors and are especially common on the bottom of deep stagnant bodies of water.

Organic matter preserved in sediment is subdivided into humus, peat, and sapropel. **Humus** is plant matter that accumulates in soil. Most organic tissue decays, producing the organic (humic) acids that play a subordinate role in weathering. Most humus is eventually destroyed completely by decay and decomposition, so it is not an important constituent in organic deposits. The term **peat** is used for the wide variety of incompletely decomposed plant remains that accumulate in fresh and brackish water bogs, marshes, and swamps lacking free oxygen. Peat is also humic organic matter and can be moss-based or leaf-based. **Sapropel** is an inclusive term used for any fine-grained organic material that accumulates subaqueously in anaerobic bodies of water, whether lakes, lagoons, or ocean basins. Most sapropel is composed of the soft organic tissues of marine phytoplankton and zooplankton, but it can include bits and pieces of higher-order plants.

Most coals consist chiefly of humic organic matter (mainly peat). The organic material in shales is believed to be sapropel, but it is sufficiently altered that it can seldom be definitively identified as such. A more general term, **kerogen,** is used for extensively decomposed and altered insoluble organic material of uncertain parentage. Oil and natural gas are believed to be derived from kerogen.

Coal

Coal consists of solid bits and pieces of undecayed plant material, mainly humic organic material but some sapropel. Coal may be widely dispersed but is more typically concentrated as individual layers (coal seams) interbedded with other sedimentary rocks. Coals are classified on the basis of many properties, but only two schemes are widely used: *coal rank* and *coal lithotype.* Each system employs complex nomenclature, but description and classification are straightforward.

Coal Rank   To classify coal by rank, the overall organic carbon content is determined. Coal rank is a measure of the degree of coalification or carbonification; that is, the extent to which impurities, moisture, and gas have been removed from the organic remains, largely by compaction. Peat is not yet coal, even though it consists largely of unconsolidated, undecayed plant remains. It contains 60% carbon along with abundant volatiles and moisture. **Lignite** (brown coal), with roughly 70% carbon and a considerable moisture and volatile content, is the lowest rank of coal. **Bituminous coal** (soft coal) contains

80% to 90% carbon; much of the moisture and many of the volatiles have been removed from the parent organic material by compaction. Finally, **anthracite** (hard coal) contains little moisture or volatiles and is 90% to 100% carbon. Carbon content and combustibility increase with rank. Higher-ranked coals are more valuable fuels because they generate more heat per mass consumed. The economic value of coal is also affected by the content of potential pollutants. Unfortunately, many high-carbon coals of eastern North America are also rich in sulfur (resulting in acid rain) and infusible ash.

The distribution of coals by rank in the geologic record is somewhat systematic. Lignites are young, typically of Mesozoic and Cenozoic age. Bituminous and anthracite coals are somewhat older; those in the northern hemisphere tend to be of late Paleozoic age (mostly Late Carboniferous), and those of the southern hemisphere are Permian and Triassic. Because anthracite coals require extremely intense pressures and elevated temperatures to drive off most of the moisture and volatiles, they develop only where coal-bearing sedimentary rocks have been metamorphosed.

Coal Lithotype   To classify coal by lithotype, the proportion of individual varieties of **macerals** (the microscopic fragments of plant matter) must be examined. These can be studied using polished slab surfaces and a reflected-light microscope equipped with oil-immersion lenses for contrast. The principal maceral groups include *vitrinite,* derived largely from plant cell-wall material and the organic contents of cell cavities; *inertinite,* derived from wood tissue in which the cell structure is preserved; and *liptinite,* macerals derived from megaspores, microspores, and seeds.

The four principal coal lithotypes are known as vitrain, fusain, durain, and clarain. Each consists of differing combinations of the major maceral groups in bands 50 µm or thicker. Coal lithotypes are usually recognized in hand specimen, using color, hardness, luster, and fracture. **Vitrain** (mainly vitrinite) coal layers are shiny black and brittle and show a conchoidal fracture. **Fusain** (composed largely of inertinite) occurs as soft, charcoal-like layers. **Durain** coals (mainly liptinite and inertinite) show a dull luster and irregular fracture. **Clarain** coals (mainly vitrinite and liptinite) exhibit a silky luster, smooth fracture, and internal lamination.

Coal-Forming Conditions and Cyclothems   Three conditions must exist for coal to form. (1) Vegetative growth must be lush. Plant populations flourish best in a warm, moist (tropical) climate. (2) The physical setting must promote the rapid production of large masses of dead vegetation that are quickly sealed off from oxygen so that decay (oxidation) either occurs slowly or not at all. (3) Both of these conditions must occur together and during intervals of time when other sediments are accumulating at a slow rate, so that sediment masking is minimized.

Most coal occurs as regionally extensive blanketlike seams that range in thickness from millimeters to meters or even hundreds of meters. Many seams are components of **cyclothems,** which are rhythmically repeating, vertically ordered sedimentary rock sequences. Other coals are randomly interbedded with siliciclastic sandstone, mudrock, or limestone. Sedimentary rocks associated with coals are deposited in nearshore, transitional depositional settings such as swamps, marshes, river deltas, and coastal plains.

Even though not all cyclothems are coal-bearing, the study of cyclothems has led to a better understanding of coal formation. Figure 13.4A shows the idealized 10-unit coal-bearing cyclothem that serves as a standard for comparison. It begins with three nonmarine sedimentary rocks. Member 1, a basal sandstone unit, rests unconformably on the beds of an underlying cyclothem. This sandstone is overlain by sandy mudrock (member 2) and freshwater limestone or marl (member 3). Member 4 is the underclay, a gray claystone with root casts always found beneath member 5, the coal seam. A series of five marine units overlie the coal. Silty shale (member 6) is overlain by marine limestone (member 7), shale (member 8), still another marine limestone (member 9), and finally silty shale (member 10). This unit is truncated by an erosion surface (an unconformity) that forms the base of the cyclothem above.

A depositional model has been developed to explain the vertical sequence of sedimentary rocks in a cyclothem. The model interprets the three nonmarine members 1, 2, and 3 as fluvial-deltaic sands and muds and interbedded freshwater lacustrine limey marls. Member 4, the underclay, is the soil in which thick vegetative cover flourished, a nearshore forest and coastal plain swamp complex. Member 5, the coal seam itself, formed when marine transgression flooded the vegetative mass, killing and rapidly burying it. Bacterial action, additional burial, and compaction of this undecayed peaty material generates brown coal, bituminous coal, and (with metamorphism) anthracite coal. The destruction of the coastal forest-swamp complex is directly caused by an abrupt

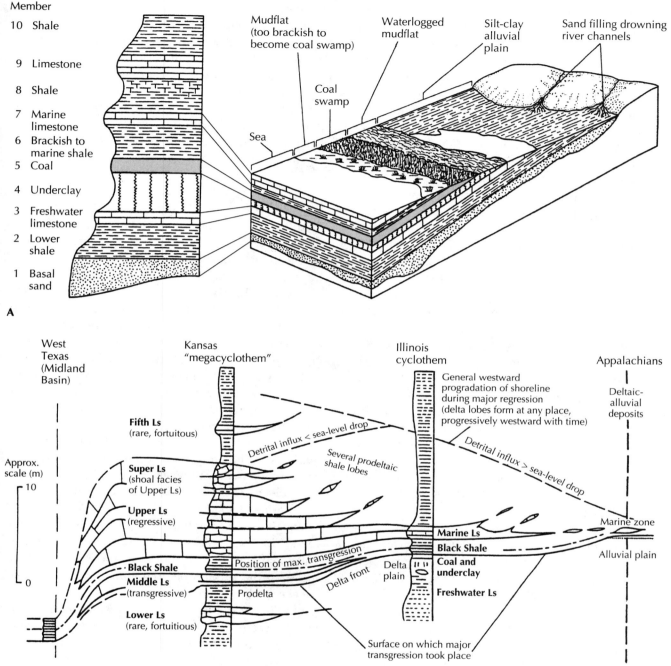

Figure 13.4 (A) The traditional model for the ideal Pennsylvanian coal-bearing cyclothem of the Illinois Basin. (Compare this model with the more recent interpretation of the cyclothem sequence in Fig. 15.5). (Shaw, 1964.) (B) Generalized restored cross section of typical Upper Pennsylvanian eustatic cyclothem along the axis of the midcontinent seaway, showing relationships between members and facies of the Kansas and Illinois cyclothems and with marine deposits in the Appalachians and West Texas. The correlation clearly shows that transgression is relatively abrupt and that most of the sequence is regressive and progradational after the apparently rapid transgression (see Fig. 15.4.) (After Heckel, 1977, 1980.)

marine transgression. The upper, marine portion of a cyclothem, members 6 through 10, records the transgression. The erosion surface marking the top of the cyclothem records a later fall in sea level that led to reemergence of the coastal region and the reestablish-ment of coastal forests and swamps. Natural cyclothems that differ from the idealized 10-member standard presumably are produced when sea-level changes occur more rapidly or when something interrupts the normal sequence of events.

Most coal-bearing cyclothems extend across broad regions of continental interiors. They seldom occur in isolation; most are repetitiously stacked upon one another 40 or 50 at a time, with as many as 100 in some areas. The coal-bearing Carboniferous cyclothems of North America extend over thousands of square kilometers and are remarkably uniform in thickness and internal organization (Fig. 13.4B). During Carboniferous time, much of the interior of North America must have been an extensive, almost featureless, low-lying coastal plain located almost exactly at sea level, much like the present coast of the Gulf of Mexico. The stacking of uniform cyclothems upon one another seen in these Carboniferous deposits means that sea level must have risen and fallen frequently and repeatedly.

Several mechanisms can produce fluctuations in sea level. Tectonism, especially episodic regional upwarping and the subsidence of continental blocks, can cause local marine transgressions and regressions along low-lying coastal plains. But few geologists believe that epeirogenic upwarping and subsidence occur often enough to generate cyclothems on a global scale. Bizarre pulsatory episodes in which continental blocks quickly bobbed up and down seem unlikely and cannot explain the detailed similarity of Permo-Carboniferous cyclothems found on different continents.

Rapid and repeated advances and retreats of the sea are better explained by global changes in sea level (eustacy). Many coal-bearing cyclothems may have been produced by sea-level changes caused by alternating episodes of global cooling and warming. During cold periods, large volumes of seawater are locked up as glacial ice. During warmer interglacial periods, melting ice releases water and causes a rise in sea level. Analyses of Pleistocene ice-water budgets suggest that the net global sea level could fluctuate by roughly 150 m. A sea-level change of this magnitude could easily explain the transgressions and regressions of the Carboniferous cyclothems. Permo-Carboniferous glaciations of this magnitude are well documented on the southern Gondwana continents, and they must have been the controlling factor in late Paleozoic sea-level changes.

Furthermore, the periodicity of global warming and cooling episodes appears to coincide with cyclothem periodicity. Changes in global temperature are caused largely by systematic variations in Earth's orbital trajectory (specifically, the Earth's distance from and orientation to the Sun) that create variations in the amount of solar heating (see Chapter 15).

Random effects of sediment supply can further modify the delicate position of the shoreline along low-lying coasts where lush coal-forming swamps are found. Lateral migration of river channels and deltas results in large variations in the rate of sediment supply and creates additional episodes of marine transgression and regression. Subsiding areas that receive little sediment undergo transgression; those supplied with large volumes of fluvial material emerge.

Petroleum

Petroleum is carbonaceous organic matter that is disseminated in a liquid (oil) and/or gaseous (natural gas) state in the pore spaces of such sedimentary rocks as sandstones and limestone. Although the overall chemistry of petroleum is relatively simple (hydrogen, 85%; carbon, 13%; with subordinate amounts of sulfur, nitrogen, and oxygen), hydrogen and carbon molecules are exceedingly complex. Several hundred organic polymers have been described. Natural gas consists mainly (85%) of methane, CH_4, with subordinate amounts of ethane, propane, and butane. Organic chemists and petroleum engineers are developing a better understanding of how petroleum evolves from organic material buried in sediment, as well as improved techniques for extracting and refining hydrocarbons.

Petroleum **reservoir rocks** are mainly porous shallow marine sandstone, limestone, and dolomite. These lithologies are not ordinarily organic-rich. Petroleum found in these rocks forms elsewhere in **source rocks** and later moves from source rocks to reservoir rocks by the process of **migration.** Most specialists believe that petroleum source rocks are predominantly fine-grained, relatively deepwater marine sediments (Brooks and Fleet, 1987). A recent controversial suggestion that petroleum is inorganic and is derived from the upper mantle is not discussed here (Gold, 1999).

The conversion of undecayed organic matter to petroleum is called **maturation.** Most geologists believe that maturation begins with the diagenetic alteration of sapropelic organic matter to kerogen. Subsequent alteration of kerogen to petroleum (**catagenesis**) takes place only at the elevated temperatures and pressures that accompany burial. Thermal energy converts kerogen to hydrocarbon polymers through thermal degradation and cracking.

The geochemistry of petroleum reflects the primary organic material, the timing of catagenesis, and

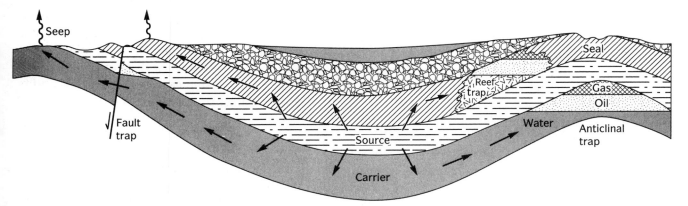

Figure 13.5 The origin and entrapment of oil in sedimentary basins. Organic-rich source sediment is buried and heated sufficiently to form petroleum from the kerogens in the source rock. This oil then migrates upward through permeable strata until it encounters a trap — an anticline, a reef, or a fault — that prevents seepage and allows it to pool. (From Dott and Prothero, 1994: 317; by permission of McGraw-Hill, Inc. New York.)

the specific path and degree of maturation. Telltale organic compounds known as **biomarkers** occur in some crude oils. These compounds resist alteration during maturation and help link a petroleum deposit to its source rock.

Hydrostatic pressure expels mature petroleum compounds from source rocks (**primary migration**). Petroleum expelled by primary migration is propelled by its own buoyancy and eventually rises (**secondary migration**) through permeable carrier rocks, finally coming to rest in reservoir rocks. A structural or stratigraphic trap is necessary to provide an impervious seal (Fig. 13.5). The trap is typically a nonporous, or at least impermeable, sedimentary rock that caps the reservoir and prevents further migration. Many traps are structural. For example, anticlinal folds can be cored with porous reservoir rocks but enveloped by impermeable shales. Normal faults cutting interbedded sandstone and shale can slide shale units in hanging walls directly onto underlying sandstones. Stratigraphic traps also occur. Sandstone reservoir rocks may pinch out landward or seaward beneath overlying mudrock, or marine transgression across the dipping surface of an unconformity may seal off dipping reservoir rocks beneath the unconformity.

Most major oil fields in the Persian Gulf region occur in anticlinal traps developed in folded Mesozoic carbonates. The organic matter accumulated in deepwater sediments deposited on the floor of the Tethys Sea, the Mesozoic ocean that formerly separated Africa and Eurasia. These source rocks are interbedded with and overlain by shallow-water platformal carbonates that are in turn covered by impermeable evaporites. The bedded evaporites were produced by the arid conditions that occurred when the Tethys Sea

was isolated from global seas. Africa and Eurasia then converged, closing the Tethys after maturation by burial metmorphism had converted kerogen in the source rocks to petroleum. Compression folded the sedimentary sequences and produced anticlinical structures sealed by evaporite caps.

Oil Shales

Oil shales are not all shales, and they contain little free oil. As discussed in Chapter 6 and Box 8.4, oil shale is any fine-grained sedimentary rock that yields substantial quantities of oil when heated. Oil shales are exceptionally rich in organic material; their organic content ranges from 5% to 50% by weight and averages roughly 25%. An intensive effort is currently under way to understand better how oil shales originate, because more oil may reside in the world's known oil shale deposits than exists in conventional petroleum reserves.

Like petroleum, oil shales are found almost exclusively in Phanerozoic rocks, mainly in Europe, the former Soviet Union, and western North America. The largest and best known oil shale in North America is the Eocene Green River Formation of southwestern Wyoming, northwestern Colorado, and southeastern Utah (see Box 8.4). The Green River Formation was deposited in a shallow lake. Beds show a delicate rhythmic lamination (see Fig. 8.4.2D). Dark-colored, organic-rich laminae alternate with lighter bands of silt and clay that are traceable over hundreds of kilometers. The rhythmically alternating laminae have been interpreted as varves, or seasonally controlled cyclical sedimentary layers (see Fig. 4.1). Warm summer temperatures produce algal and/or planktonic

blooms that generate abundant organic matter. Seasonal cooling then kills off most of the organisms. Burial of these organic remains by the rapid accumulation of siliciclastic detritus would produce local anoxic conditions, block further decay, and generate the organic-rich laminae. Some large oil shale deposits appear to have been deposited on the floors of lakes. Still others seem to have been deposited on the floors of shallow marine embayments, within lake-swamp-bog complexes, or in isolated basins developed on continental shelves and platforms. These depositional settings must all have had restricted water circulation that created widespread anaerobic bottom waters.

Roughly 75% of the organic matter in oil shales is kerogen; the remainder consists of solid bitumen (organic material of uncertain origin that is soluble in organic solvents). In the Green River Formation, these components appear to originate by chemical degradation and biochemical alteration of undecayed algal remains. Nonorganic material admixed in the millimeter- to centimeter-thick laminae is two to three times more abundant than the carbonaceous component. The nonorganic material is one reason that oil shales are not economically viable at present. To remove it, the oil shale must be heated using a process that requires substantial amounts of water. After the oil has been removed, the volume of processed material exceeds that of the original raw material. These factors raise substantial obstacles to exploiting the Green River Formation, whose basin straddles federally owned land in an area noted for both its aridity and an environmentally sensitive population.

Solid Hydrocarbons

Several types of highly viscous solid hydrocarbons occur in sedimentary rock. Asphalt is semisolid; pyrobitumens and mineral waxes are solid. All are derived from conventional petroleum and occur as surface pools, solid organic dikes and sills, and the fill of porous sandstones. Slow subsurface migration of petroleum accompanied by the gradual loss of gaseous and liquid components and some compositional modifications explain the characteristics and occurrence of these deposits. Examples include the Rancho La Brea tar seep in Los Angeles (well known for its rich late Pleistocene mammalian fossil content) and the Early Cretaceous Athabasca Tar Sands of Alberta in western Canada, a thick sequence of uncemented fluvial and deltaic sandstones containing large volumes of semisolid hydrocarbons (bitumens) with enormous economic potential (Smith, 1987).

CONCLUSIONS

Organisms play a direct or indirect role in forming the sedimentary rocks discussed in this chapter. Bedded cherts form largely by recrystallizing the siliceous shells of microplankton on the deep seafloor, but nodular cherts originate where percolating groundwater remobilizes silica, transports it in solution, and eventually precipitates it within shallow marine carbonate rocks. Sedimentary phosphorites are very scarce in the sedimentary record. Most of these phosphorites probably formed on the outer marine shelf areas where oceanic upwelling promoted high organic productivity and oxygen-deficient conditions.

Of all the rocks described in this book, organic sediments have been the most intensively studied because of their great economic importance. The various fossil fuel deposits, such as coal and petroleum, are unevenly distributed in space and time, because unique geologic conditions are necessary to accumulate organic matter and prevent its subsequent decay. The uneven distribution of these deposits and the unusual conditions required for their formation have important political, economic, and geological consequences.

FOR FURTHER READING

Bentor, Y. K., ed. 1980. *Marine Phosphorites.* SEPM Special Publication 29.

Brooks, J., ed. 1990. *Classic Petroleum Provinces.* Geological Society of London Special Publication 50.

Chilingarian, G. V., and T. F. Yen. 1978. *Bitumens, Asphalts, and Tar Sands.* New York: Elsevier.

Cook, P. J. 1976. Sedimentary phosphate deposits. In *Handbook of Stratabound and Stratiform Ore Deposits,* ed. K. H. Wolf. New York: Elsevier.

Gold, T. 1999. *The Deep Hot Biosphere.* New York: Springer-Verlag.

Hunt, J. M. 1996. *Petroleum Geochemistry and Geology.* New York: W. H. Freeman and Company.

Hyne, N. J. 2001. Nontechnical Guide to Petroleum Geology, Exploration Drilling, and Production, 2d ed. New York: Pennwell.

McBride, E. F., ed. 1979. *Silica in Sediments: Nodular and Bedded Chert.* SEPM Reprint Series 8.

North, F. K. 1990. *Petroleum Geology.* London: Unwin-Hyman.

Nriagu, J. O., and P. B. Moore, eds. 1984. *Phosphate Minerals.* New York: Springer-Verlag.

Rahman, R .A., and R. M. Flores, eds. 1985. *Sedimentology of Coal and Coal-Bearing Sequences.* International Association of Sedimentologists Special Publication 7.

Scott, A. C., ed. 1987. *Coal and Coal-Bearing Strata: Recent Advances.* Geological Society of London Special Publication 32.

Selley, R. C. 1997. Elements of Petroleum Geology, 2d ed. London: Academic.

Sieveking, G., and M. B. Hart, eds. 1986. *The Scientific Study of Flint and Chert.* Cambridge: Cambridge University Press.

Thomas, L. 2002. *Coal Geology.* New York: Wiley.

Tissot, B. P., and D. H. Welte. 1984. *Petroleum Formation and Occurrence,* 2d ed. New York: Springer-Verlag.

Yen, T. F., and G. V. Chilingarian, eds. 1976. *Oil Shale.* New York: Elsevier.

14 Chemical and Nonepiclastic Sedimentary Rocks

Polygonal mudcracks with large salt crystals growing between them in the Devil's Golf Course, Death Valley, California. (© Tom Bean.)

THIS CHAPTER DISCUSSES TWO UNRELATED GROUPS OF rocks: (1) chemical sedimentary rocks that, for the most part, are precipitated from solution inorganically; and (2) a mixed group of miscellaneous sedimentary rocks generated, at least in part, by processes other than the weathering of pre-existing rocks.

The first group, chemical sedimentary rocks, includes (1) iron-rich sedimentary rocks such as banded iron formations and ironstones, and (2) evaporites.

The dissolved components that precipitate to form these rocks are decomposed from pre-existing rocks and minerals. Iron-rich rocks and evaporites make up less than 2% of the total sedimentary rocks on Earth, but they are major sources of iron, many salts, and other chemicals.

The second group includes rocks that are nonepiclastic in origin. Unlike the epiclastic (mainly siliciclastic) sedimentary rocks discussed so far (all of which

are produced by the mechanical disintegration of pre-existing materials), nonepiclastic rocks are generated by processes other than mechanical weathering. For example, although some *volcaniclastic* (literally, rich in fragments of volcanic rock and glass) sedimentary rocks are simply weathered and reworked from volcanic rocks, many are generated by processes other than weathering, such as explosive volcanism (which produces *pyroclastic* material). The term *volcanogenic* is used for all sedimentary deposits generated by volcanic processes. The impact of extraterrestrial objects (bolides) with the surface of the Earth produces **meteoritic** clastic sediment; folding and faulting generate *cataclastic* deposits; and the dissolution and collapse of cavern ceilings and walls produce *solution breccias.*

Solution Geochemistry

To understand chemical sedimentary rocks, we must begin with a general review of chemical solutions. Most chemical and biogenic sedimentary rocks form from seawater. The dissolved constituents in seawater are weathered from the continental blocks and carried to the oceans as runoff. Table 14.1 compares the chemistry of mean river water with that of seawater. **Salinity** is the total amount of dissolved constituents in water. It is expressed either as parts per million (ppm) or as weight percent. The salinity of mean river water is 121 ppm, or 0.012%. The salinity of seawater is much

higher—35,000 ppm, or 3.5%. The major ions dissolved in river water and seawater are also quite different. Modern rivers contain mainly dissolved HCO_3 and CO_3, with minor amounts of Ca, H_4SiO_4, SO_4, Cl, Na, Mg, and K. Seawater contains principally SO_4, Cl, Na, and K. These differences in overall salinity and in the specific proportion of components dissolved in river water and seawater result from modern processes of chemical and biochemical sedimentation. They reflect the degree to which components are extracted from seawater (with or without the assistance of organisms) and incorporated into sediment and sedimentary rock.

The high overall salinity of seawater is due primarily to the buildup of two major constituents over time, sodium and chlorine. These occur in dilute solution in river water and are predominant in seawater because they are removed from solution and incorporated into sedimentary rocks very slowly. In fact, evaporite precipitation is the only way sodium and chlorine leave seawater, and they are the most important constituents of evaporite rocks. Calcium, silica, carbonate, and bicarbonate are abundant in river water but scarce in seawater. They are removed from seawater more rapidly than sodium and chlorine and are incorporated into siliceous and calcareous sediments—chert and limestone (see Chapters 11 and 13).

The tendency of a component to remain in solution or to be selectively removed is expressed by the

---

TABLE 14.1 Relative Abundance of Dissolved Ions in Mean River Water and Seawater

	Mean River Water[a] Total Dissolved Solids (%) (Salinity = 121 ppm or 0.012%)	Seawater[b] Total Dissolved Solids (%) (Salinity = 35,000 ppm or 3.5%)
Bicarbonate $(HCO_3)^-$ and carbonate $(CO_3)^{2-}$	48.6	0.4
Calcium (Ca^{2+})	12.4	1.2
Amorphous silica (H_4SiO_4)	10.8	<0.01
Sulfate $(SO_4)^{2-}$	9.3	7.7
Chlorine (Cl^-)	6.5	55.0
Sodium (Na^+)	5.2	30.6
Magnesium (Mg^{2+})	3.4	3.7
Potassium (K^+)	1.9	1.1
Iron $(Fe^{2+}$ and $Fe^{3+})$	0.6	<0.01
Aluminum $Al(OH)_4^-$	0.2	<0.01
Nitrate $(NO_3)^-$	0.8	<0.01
Total	99.7	99.7

[a]Livingston, 1963.
[b]Mason, 1966.

TABLE 14.2   Selected Ions Dissolved in Seawater: Their Residence Time and Ultimate Fate

Ion	Residence Time (years)[a]	Principal Sediment and Sedimentary Rock Sites
Chlorine	Infinity	Bedded evaporite
Sodium	260,000,000	Bedded evaporite
Magnesium	12,000,000	Dolomite, bedded evaporite
Potassium	11,000,000	Clay mineral diagenesis, bedded evaporite
Sulfate	11,000,000	Bedded evaporite
Calcium	1,000,000	Calcareous ooze, carbonate
Carbonate, bicarbonate	110,000	Calcareous ooze, carbonate
Silica	8000	Siliceous oozes, chert
Manganese	7000	Manganese nodules
Iron	140	Iron-rich sediment
Aluminum	100	Clay minerals, mudrocks

[a]Based on Ross, 1982: Stowe, 1979.

residence time. Residence time is measured in years and can be calculated easily by dividing the total mass of an ion in seawater by its annual mean flux (the amount of ion that enters and leaves the sea yearly). Table 14.2 lists residence times for major ions dissolved in seawater and the sedimentary rock types in which those components are eventually concentrated.

Ions with extremely long residence times are not metabolized by organisms. The principal conditions that favor the inorganic precipitation of these ions require the isolation of seawater in settings where the rate of evaporation exceeds the rate of precipitation and runoff.

Ions such as calcium, carbonate, bicarbonate, silica, and iron have extremely brief residence times. The short residence time reflects the ease with which these ions are either metabolized by organisms and incorporated into shells (silica, bicarbonate, carbonate, and calcium) or extracted inorganically (iron and aluminum). Ions with intermediate residence times, such as potassium and magnesium, must be extracted by slower, more complex processes.

What are the sources of the various constituents? How are they removed from solution and incorporated into sediment? Most of the calcium and some bicarbonate (HCO_3^-), and carbonate (CO_3^{2-}) dissolved in river water are derived from the dissolution of limestone and dolostone. Additional bicarbonate and carbonate come from the dissociation of carbonic acid (H_2CO_3). Carbonic acid is produced as falling rain absorbs atmospheric carbon dioxide. Almost all this dissolved bicarbonate, carbonate, and calcium is rapidly removed from seawater by organisms, forming limestone. Shallow marine organisms such as corals form reef complexes; algae produce lime mud and promote the accretion of ooids. Calcareous oozes are produced when the shells of floating, carbonate-secreting organisms such as foraminiferan, coccoliths, and pteropods sink to the seafloor and, with burial and compaction, recrystallize as pelagic limestone.

Most of the silica dissolved in river water as H_4SiO_4 is a by-product of the weathering of feldspars to clay minerals. Like bicarbonate, carbonate, and calcium, the silica is rapidly removed from seawater by organisms. Pelagic radiolarians and diatoms form their shells (tests) of silica. These shells sink to the ocean floor and accumulate as siliceous oozes. Burial and compaction of these oozes cause their recrystallization as bedded chert.

Other important ions dissolved in river water do not show a dramatic dropoff in abundance in seawater and have residence times of intermediate length. For example, the relative proportion of magnesium dissolved in river water is almost identical to the proportion dissolved in seawater. Dissolved magnesium is derived largely from the weathering of such Mg-bearing minerals as dolomite, pyroxene, and amphibole. The relative proportion of potassium in seawater is about half that dissolved in river water. Most of this potassium is released during the hydrolysis of feldspar and muscovite to clay minerals. The longer residence times for both magnesium and potassium reflect the processes that systematically withdraw them from seawater. Dolomitization eventually con-

sumes magnesium; some potassium is removed through absorption by clay minerals. Absorption converts gibbsite and kaolinite to illite.

The relative abundances of dissolved sulfate (SO_4^{2-}) in river water and seawater are similar. Sulfate has a long residence time because no sedimentation process in the open ocean removes large amounts of it rapidly. Sulfate is eventually extracted from seawater and incorporated into such evaporite minerals as gypsum and anhydrite. Much dissolved sulfate is produced by volcanism and the weathering of sulfide minerals. Some is the undesirable by-product of the burning of sulfur-bearing coal. Burning coal injects sulfur gases into the atmosphere as sulfur dioxide, SO_2 (which oxidizes to sulfur trioxide, SO_3), where it is absorbed by rain and eventually ends up in seawater.

Tiny amounts of iron are dissolved in river water; only a trace is found in seawater. Iron is typically found as a fine colloid rather than in true solution because ferric iron (Fe^{3+}) is extremely insoluble. Colloidal iron is usually flocculated along the coast as rivers enter the sea and is disseminated as a minor constituent in conventional siliciclastic sedimentary rocks. During certain times in the geologic past, however, conditions permitted the formation of unusual iron-rich sedimentary rocks.

Iron-Rich Sedimentary Rocks

Most iron dissolved in river water is derived from the chemical decomposition of iron-bearing minerals such as biotite and pyroxene. Ionic iron occurs in two forms. Ferrous iron (Fe^{2+}) is relatively soluble; ferric iron (Fe^{3+}) is essentially insoluble. In the presence of oxygen, ferrous iron almost immediately oxidizes (rusts) to ferric iron. Because oxygen is abundant in the modern terrestrial atmosphere, the predominant form of iron in most settings is the extremely insoluble ferric iron. As Table 14.1 shows, the already low abundance of iron in river water drops to nearly zero in seawater. The residence time of iron is very brief because it is removed rapidly and concentrated in modern sediment by inorganic flocculation (see Table 14.2).

Flocculation of iron has been important throughout the Phanerozoic. It occurs as soon as dissolved ferric iron (Fe^{3+}) enters the sea. The increased salinity of seawater and the presence of organic matter cause small particles of iron-rich material to crystallize. These particles adhere to one another, accrete, and settle to the seafloor. Ordinarily, only a small amount of iron is generated by flocculation. Flocculated iron particles are incorporated into ordinary sedimentary rocks,

mainly as components of silicate minerals. Almost all sedimentary rocks contain some iron. A typical sandstone contains 2% to 4% iron (weight percent ferrous oxide, FeO, plus ferric oxide, Fe_2O_3), mudrock contains 5% to 6%, and limestone contains less than 1%.

Iron-rich sedimentary rocks are a special category defined by a total iron content that exceeds 15%. The two principal iron-rich sedimentary rock types, **Precambrian banded iron formations** and **Phanerozoic ironstones,** make up less than 1% of the sedimentary rocks in the geologic record but are of enormous importance. Virtually all the world's iron and steel comes from these rocks, particularly the Precambrian banded iron formations. The presence of abundant iron in the "iron ranges" of northern Minnesota and the Upper Peninsula of Michigan was crucial to the industrial growth of the United States in the late 1800s. This iron was shipped in great iron boats through the Great Lakes to steel plants along the waterways of Pennsylvania, Ohio, and Michigan. The other essential ingredient, coal, came from the anthracite mines of the Appalachian belt in Pennsylvania and West Virginia. This is why the steel industry in the United States has long been concentrated in cities such as Pittsburgh and many towns in eastern Ohio and why the auto industry grew up in Detroit.

Precambrian banded iron formations (**BIF** for short) consist of centimeter-thick interlayered alternating bands of chert and iron-rich minerals (Fig. 14.1). These

Figure 14.1 Contorted banded iron formation near Jasper Nob, Ishpeming, Upper Peninsula of Michigan. Note the characteristic alternating bands of chert (light) and iron (dark). (Courtesy of R. H. Dott, Jr.)

formations are direct chemical precipitates, but organisms may have played an important role in their genesis. Their restriction to the earlier Precambrian indicates that iron behaved differently in the first 1800 million years of Earth's history, probably because of lower levels of free oxygen in the early terrestrial atmosphere. In the presence of the modern aerobic (containing abundant free oxygen) atmosphere, soluble ferrous (Fe^{2+}) iron is rapidly oxidized to extremely insoluble ferric iron (Fe^{3+}). As a result, iron can only be transported and deposited as clastic sediment. The texture and mineralogy of banded iron formations clearly indicate that iron was transported in solution and concentrated by precipitation. This means that iron survived over a long interval of time in its soluble ferrous state. Such a condition would occur only in an anaerobic (or at least locally oxygen-poor) Precambrian atmosphere.

The other major iron-rich sedimentary rock, Phanerozoic ironstone, is actually a conventional siliciclastic sedimentary rock (mainly mudrock) in which a high percentage of iron is concentrated. The origin of ironstones is better understood than that of banded iron formations, but their unusual temporal and spatial distribution warrants special explanation.

Banded Iron Formations

No model to explain the origin of banded iron formations has won unanimous acceptance (Trendall and Morris, 1983). The banded appearance is caused by the intimate interbedding of millimeter- to centimeter-thick beds of gray to black chert and red to maroon iron-rich mudrock. Most banded iron formations are early Proterozoic in age (between 1800 and 2400 million years old). Some are Archean, and there are deposits as young as Cambrian. Deposits in the Canadian Shield have been studied the most extensively. Two major categories are recognized, Algoma type and Superior type.

Algoma-Type Banded Iron Formations This variety is almost entirely of Archean age. Individual units are interbedded with deep-water graywackes, mudrocks, and submarine volcanic rocks in greenstone belts. These linear synformal belts of metasedimentary and metavolcanic rocks "float" in the "sea" of granite and high-grade metamorphic rocks that make up most Archean shields. BIF sequences range from a few millimeters to a few meters thick. They are lenticular in shape and of limited areal extent. Belts are a few kilometers wide and long. The principal iron minerals

in these bands of delicately laminated ferruginous mudrock are the oxides hematite and magnetite. Small proportions of iron sulfide, iron carbonate, and iron silicate minerals also occur. Algoma-type deposits are interbedded with deep abyssal plain floor deposits. Their concentration along narrow zones that might coincide with ancient ocean crustal fractures has led to the suggestion that iron-bearing hydrothermal fluids served as the source of iron.

Superior-Type Banded Iron Formations These lower Proterozoic rocks are striking in outcrop because of their spectacular banding (see Fig.14.1). Millimeter- to centimeter-thick layers of reddish iron-rich material alternate with gray to black layers of bedded chert. Sequences of this type are thicker and more extensively developed than Algoma-type deposits. Individual units range from fifty to several hundred meters thick and occupy broad belts a hundred or more kilometers wide and several hundred to more than a thousand kilometers long. BIFs of the Superior type are intimately associated with stable shallow-water shelf sediments such as quartz arenite, limestone, dolostone, and mudrock. Channels, desiccation cracks, crossbedding, and ripple marks are common. Oolites and intraclasts of iron-rich minerals are abundant.

Origin There is considerable difference of opinion about how these deposits form, but most speculation centers on the properties of the early terrestrial atmosphere. Some geochemists argue that initially low levels of atmospheric oxygen (which imply higher partial pressure of atmospheric carbon dioxide) led to more acidic rainfall and lower pH levels in Archean and Proterozoic ocean waters (Garrels, Perry, and Mackenzie, 1973; Clemmey and Badham, 1982). These factors would increase the amount of iron leached during weathering. Lower levels of atmospheric oxygen would also slow the oxidation of soluble ferrous iron to insoluble ferric iron, permitting large volumes of leached iron to be transported to the oceans in solution.

One of the most popular models (Holland, 1973; Klein and Beukes, 1990) suggests that Proterozoic oceans were stratified into (1) deeper masses of anoxic bottom waters containing large amounts of iron, and (2) less dense, more oxic surface waters (Fig. 14.2). This model postulates that periodic overturning of these ocean waters and upwelling onto subjacent shelf areas generated alternating episodes of chert precipitation and deposition of iron-rich mudrock. Although some of this iron may have been weathered from the

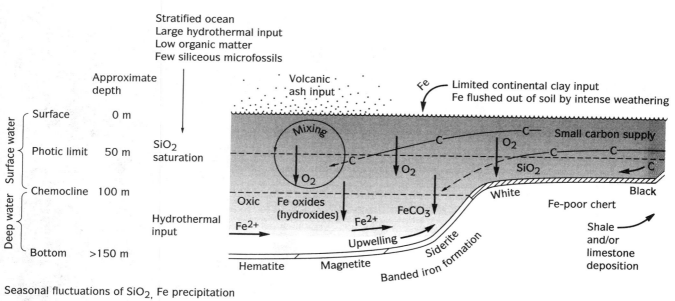

Figure 14.2 Depositional model for banded iron formations. (After Beukes and Klein, 1992: 150.)

continental blocks, the lack of clays in the cherty BIFs suggests that most of the iron was released from submarine vents. Studies of modern hydrothermal vents in mid-ocean ridges show that they contain 10^5 times more iron than typical seawater.

Primitive plants (plus cyanobacteria) and animals probably played an important role in forming both the iron bands and the interbedded chert, so perhaps these are not purely chemical sedimentary rocks after all (Cloud, 1973; Nealson and Meyers, 1990; Lunine, 1999). Garrels (1987) hypothesized that seasonal changes in water temperature and geochemistry controlled organic activity.

Whatever the origin(s) of the BIFs, the conditions required for their formation were vanishing by about 1800 million years ago. The oldest oxidized hematite-rich soil horizons first appear at the same time, and detrital pyrite and uraninite (unstable in an oxidized atmosphere) disappear. These coincidences further suggest that Earth's low atmospheric oxygen content was a major factor in controlling the deposition of BIFs; as oxygen levels began to rise, BIFs rapidly became rare.

Phanerozoic Ironstones (Redbeds)

There are fewer Phanerozoic ironstones or redbeds than banded iron formations, but their origin is better understood. They have a strikingly uneven distribution over time with a few peaks of development. Well-known units that mark these peaks include the

Ordovician Wabana Formation of Newfoundland; the Silurian Clinton Formation in the Appalachians; and various Jurassic ironstones of western Europe, such as the Minette deposits of Luxembourg and Alsace-Lorraine.

The iron in most ironstone deposits is concentrated in ooids (Fig. 14.3). Detrital nuclei are grains of quartz, feldspar, mica, and carbonate. The iron occurs as crystalline iron oxide (especially hematite and goethite) and iron silicate (chamosite) in the mineral layers coating the nuclei. Shells of shallow marine littoral and neritic fossils found within these deposits are partly or completely replaced by iron minerals.

Ironstone beds are a few meters to a few tens of meters thick. Body fossils, cross-bedding, ripple marks, bioturbation, and scour-and-fill structures are common, and ironstones are associated with shallow marine shelf deposits such as limestone, dolomite, mudrock, and sandstone. These features indicate that ironstones accumulated on nearshore shelves. Flocculation of iron-rich clays occurs when detritus comes into contact with seawater, promoting the snowball-like growth of ferruginous oolites. Van Houten (1982) was able to define a series of facies belts using iron minerals. Nearshore shallow-water deposits rich in hematitic oolites grade into an offshore belt of chamosite oolites and eventually into inner and outer shelf deposits characterized by glauconite, an iron-rich clay.

Although the uneven distribution of ironstones is puzzling, their origin has been linked to the erosion

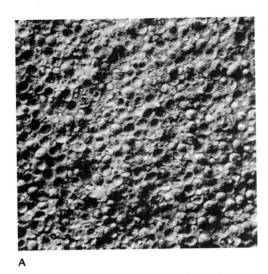

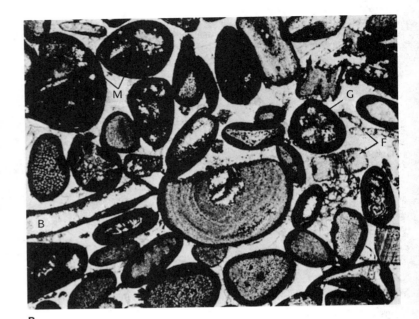

A

B

Figure 14.3 Phanerozoic ironstones and banded iron formations have very different textures. (A) Fractured surface of fresh oolitic ironstone from the Cretaceous of Nigeria. Ooids and matrix consist of siderite, magnetite, and chlorite, with iron hydroxides and clay minerals. Width of photo is 30 mm. (Jones, 1965: 840; by permission of the SEPM.) (B) Oolitic hematite, Silurian Keefer Sandstone, Pennsylvania. Most ooid nuclei are crinoid fragments, but others have nucleated on fragments of brachiopods (B), gastropods (G), and intraclasts made of fossil fragments in a hematite matrix (M), presumably reworked from another part of the ironstone. Hematite permeates the fossil nuclei and their pores, but some of the fossils (F) are only slightly replaced and lack oolitic coatings. The rock has been cemented with clear, nonferruginous calcite. The crinoid in the center is 1 mm in diameter. (Courtesy of R. E. Hunter.)

and redeposition of lateritic soils. Lateritic soils are typically produced when low-lying, iron-rich source rocks such as granite and other crystalline rocks weather in a warm, humid climate (see Chapter 6). This weathering promotes extensive leaching of iron and the buildup of a thick, near-surface soil layer of ferruginous laterite or ferricrete. Well-developed laterites containing more than 50% ferric oxide and hydroxide occur today in the tropical Amazon Basin of South America, where Precambrian shield rocks rapidly decompose to well-leached, ferruginous clayey soils.

The erosion of lateritic soils does not typically generate a coastal belt of ironstone sediment, however. Other special circumstances must be required. When conventional siliciclastic detritus is deposited at a rate equal to or exceeding the rate at which transported lateritic soil debris accumulates, ironstone is prevented from forming by sedimentary masking. The normal background sedimentation overshadows any accumulating lateritic material. Ironstone forms when the depositional site receives only iron-rich muddy detritus; that is, only when there is no supply of normal terrigenous debris.

The restricted distribution of ironstones in time and space confirms that conditions necessary for their formation seldom occurred during the Phanerozoic. Ironstone sedimentation requires the erosion of broad, low-lying source regions mantled with lateritic soil. The eroded material must be transported onto coastal plains that receive little additional siliciclastic detritus. Coincidentally, most of the Phanerozoic ironstones formed during the Ordovician through Devonian and during the Jurassic and Cretaceous. These were periods of greenhouse conditions, when the global climate was warm, ice caps vanished, and sea levels were unusually high (Van Houten and Arthur, 1989). These were also times of black shale deposition, as discussed in Chapter 6. Clearly, warm climates during greenhouse conditions would stimulate the type of deep weathering needed to release so much iron to the sea. Kimberley (1979) points out that minor regressive episodes within the globally high sea level are also needed, so that continental iron-rich muddy sediments can spread over the former seafloor. In this way, iron oxide–rich solutions can percolate down through the marine limestones and replace calcareous ooids.

Other Iron-Rich Sedimentary Rocks

Carbonaceous black shale and limestone containing 10- to 40 cm-thick layers in which crystals of pyrite (iron sulfide) are distributed are another type of iron-rich sedimentary rock. Pyrite forms only in locally reducing (anoxic) environments. The lack of oxygen results from several factors. In the modern Black Sea, for example, the absence of convective overturn prevents the replenishment of oxygen in organic-rich muds accumulating on the bottom. As a result, organic material ceases to decay and any iron present is reduced and recrystallized as pyrite. Ancient pyrite-bearing black shales may have formed in a similar setting.

Modern bog iron deposits represent another iron-rich sedimentary rock type. They consist mainly of the iron minerals goethite and siderite and various manganese minerals such as psilomelane and pyrolusite. They are well developed only in a few high-latitude swamps and lakes in the northern hemisphere. The slow influx of iron- and manganese-bearing acidic groundwater into oxygenated bodies of water seems to promote their formation. Dissolved ferrous iron crystallizes when pH rises.

Finally, irregularly shaped nodules of manganese and iron are scattered across lake bottoms and some portions of the modern deep seafloor. These nodules are rarely larger than a few centimeters in any dimension and are especially prominent in abyssal plain areas where little other sediment accumulates, especially on the flanks of mid-ocean ridges. Iron and manganese nodules result from exceedingly slow concentric accretionary growth. The ultimate source of the manganese and iron (submarine hydrothermal activity?) and the mechanism by which precipitation proceeds, whether organic (bacterial action?) or inorganic, are uncertain (Glasby, 1977).

Evaporites

Evaporites are bedded sedimentary rocks that crystallize from hypersaline solutions known as brines. Brines develop where the amount of water lost through the evaporation of seawater or terrestrial water exceeds freshwater influx from precipitation, surface flow, and groundwater inflow. High temperatures and the lack of precipitation found in arid climates promote brine formation.

More than 100 different minerals occur in bedded evaporite deposits, but few are abundant (Table 14.3). Most consist of highly soluble ions such as sodium and chlorine that have lengthy residence times. This fact underscores the scarcity of brines that are sufficiently saturated to permit minerals composed of these ions to precipitate. The most common evaporite minerals are carbonates (calcite, aragonite, magnesite, and dolomite); sulfates (gypsum and anhydrite); halides (halite, sylvite, and carnallite); and a few borates, silicates, nitrates, and sulfocarbonates.

The few regions of the world that are current sites of evaporite deposition are characterized by very high mean temperatures and an annual rainfall that totals only a few centimeters. Such localities include the semiarid playa lakes scattered across the Great Basin in California and Nevada and the various hypersaline supratidal flats (sabkhas), salt pans, estuaries, and lagoons of the Persian Gulf region in the Middle East. These settings serve as natural laboratories for observing the processes by which evaporites form. But they may not be good analogs of the settings in which many ancient evaporites were deposited.

The few ancient evaporite deposits thought to be nonmarine consist of areally restricted, thin crusts of evaporite minerals. Most ancient sequences extend

TABLE 14.3 The Common Marine and Nonmarine Evaporite Minerals

Common Marine Evaporite Minerals		Common Nonmarine Evaporite Minerals	
Halite	NaCl	Halite, gypsum, anhydrite	
Sylvite	KCl	Epsomite	$MgSO_4 \cdot 7H_2O$
Carnallite	$KMgCl_3 \cdot 6H_2O$	Trona	$Na_2CO_3 \cdot NaHCO_3 \cdot 2H_2O$
Kainite	$KMgClSO_4 \cdot 3H_2O$	Mirabilite	$Na_2SO_4 \cdot 10H_2O$
Anhydrite	$CaSO_4$	Thenardite	$NaSO_4$
Gypsum	$CaSO_4 \cdot 2H_2O$	Bloedite	$Na_2SO_4 \cdot MgSO_4 \cdot 4H_2O$
Polyhalite	$K_2MgCa_2(SO_4)_4 \cdot 2H_2O$	Gaylussite	$Na_2CO_3 \cdot CaCO_3 \cdot 5H_2$
Kieserite	$MgSO_4 \cdot H_2O$	Glauberite	$CaSO_4 \cdot Na_2SO_4$

across broad regions and are hundreds or even thousands of meters thick. They must have been deposited on the floor of broad basins thousands of square kilometers in area.

In 1849, the Italian chemist M. J. Usiglio first evaporated a bucket of seawater and analyzed its contents. Since then, geochemists have used laboratory experiments to trace the steps by which seawater of normal salinity evolves into the hypersaline brines from which the evaporite minerals grow. In some circumstances, evaporitic calcite or aragonite can precipitate when the water volume has been reduced by 50%. Crystallization of evaporite minerals begins only after seawater of normal salinity (35,000 ppm) has been reduced to about 20% of its original volume. The least soluble constituents crystallize first. Calcium and sulfate ions form gypsum and anhydrite. When the volume of seawater is reduced to 10% of the original volume, more soluble components such as potassium, sodium, and chlorine solidify as the minerals halite

and sylvite. Finally, the remaining seawater evaporates almost entirely. This allows the most soluble components to solidify from the last remaining droplets of briny fluid. These last minerals are bitter borates and nitrates.

This orderly process of crystallization as a function of solubility is not always apparent when naturally occurring evaporite mineral sequences are studied. Only some ancient evaporites show a systematic vertical cyclicity in mineralogy that mimics that seen in the laboratory, a pattern that is compatible with progressive crystallization of an increasingly saline brine. Most of these deposits show repetition. Vertical sequences of evaporite minerals are stacked successively on top of one another, suggesting alternating episodes in which brines were repeatedly evaporated and then replenished. In many of these ancient sequences, mineral species appear at random. No apparent relationship exists between a mineral's physical position in the sequence and the solubility of the ions it contains.

A

Figure 14.4 Death Valley, California, is a classic example of a lake basin with concentrically zoned evaporites. (A) Overview of Death Valley from Dante's View (on the southeast rim looking northwest.) The center of the dry lake is halite from the chloride zone, but the arcuate white area in the bottom center of the picture is composed of gypsum. Carbonates and borates are found on the periphery of the gypsum zone at the northern end of the dry lake. (C. B. Hunt, courtesy of U.S. Geological Survey.) (B) Map of Death Valley evaporite zones (*on the facing page*) showing the concentric arrangement of chlorides in the center of the bull's-eye, surrounded by a ring of gypsum and, on the outer fringe, by carbonates or borates. (After Hunt and Mabey, 1966: Plate 1; courtesy of U.S. Geological Survey.)

The natural world of evaporites is far more complex than models based on laboratory experiments.

Evaporite deposits range in age from the Precambrian to the present, but very few are older than Cambrian. Their distribution is uneven. Peaks of development occur during the Cambrian, Permian, and Triassic periods. Presumably, regionally extensive evaporites occur infrequently because the conditions necessary to form them are rare. Widespread aridity must be coupled with a tectonic setting that produces broad, shallow basins or marine estuaries only intermittently connected with the open sea. For this reason, evaporite deposits are typically confined to the high-pressure belts between 10° and 30° latitude, where most of the world's deserts occur. Ancient evaporites

also seem to be concentrated in this paleolatitudinal belt, so evaporite deposits have long been used to reconstruct ancient continental positions and test other hypotheses of climate and paleogeography.

Other types of sedimentation must be minimal so that evaporite deposition is not masked. Because evaporites form in such restricted conditions and tightly constrain ancient paleogeography and climate, it is appropriate to describe the characteristics of the three principal ancient types: nonmarine, shallow marine, and deep marine evaporites.

Nonmarine Evaporites

Nonmarine evaporites accumulate in closed lakes with interior drainage and no external outlets that develop in arid and semiarid regions (Hardie, Smoot, and Eugster, 1978). When rainfall is high, large amounts of dissolved substances enter the lakes. With no outlets and with a hot and dry climate, evaporation reduces the water volume and increases salinity to the point where evaporite minerals begin to crystallize. Continued evaporation shrinks the volume and area of closed lakes. They eventually disappear, leaving a residue of nonmarine evaporites.

Ephemeral dry lakes (playas) occur today in many desert areas. Examples include the Great Salt Lake of Utah, Mono Lake in the Owens Valley of California, and the Dead Sea. Ancient analogs can be identified in a number of ways. Bands of lacustrine evaporites define a bull's-eye pattern (Fig. 14.4). The most soluble minerals, such as halite, are concentrated near the lake center and occur at the top of the evaporite sequence. The least soluble minerals, such as carbonates and sulfates, are concentrated around the rim of the ancient lake and are found at the bottom of the evaporite sequence. A bull's-eye pattern defined on a local scale, together with the association of such evaporites with other demonstrably nonmarine sedimentary rocks, allows this type of evaporite to be identified. Specific mineralogy also helps. Only nonmarine evaporites contain large quantities of borax, epsomite, trona, gaylussite, and glauberite.

Ancient nonmarine evaporites are not particularly common in the stratigraphic record, but some are truly spectacular. One of the most famous is in the Eocene Green River Formation in Wyoming (see Box 8.4). In ancient Lake Goshiute (the Wyoming portion of the Green River system), the Wilkins Peak Member contains many beds of trona (hydrous sodium bicarbonate) and halite totaling more than 50 m in thickness and covering several thousand square kilometers. In

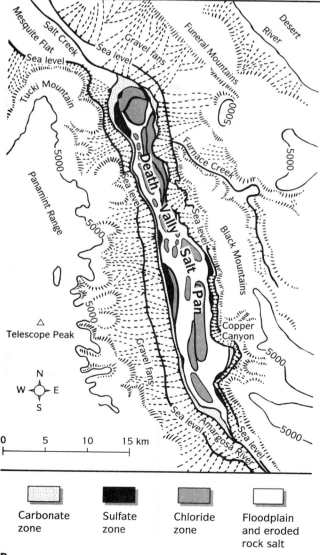

| Carbonate zone | Sulfate zone | Chloride zone | Floodplain and eroded rock salt |

B

some places, the large trona crystals interlock and penetrate upward (see Fig. 8.4.2E). These trona and halite deposits also show the classic bull's-eye pattern, with the most concentrated deposits near the original center of the lake bed in southwesternmost Wyoming.

Shallow Marine Evaporites

Shallow marine evaporites include (1) supratidal and intertidal (coastal sabkha) deposits like those forming today around the perimeter of the Persian Gulf, and (2) truly shallow marine ancient evaporites that formed in what were marine shelf areas with water depths seldom in excess of 5 m. Few if any modern analogs of these latter deposits are known.

Sabkha Deposits   As discussed in Chapter 12, sabkhas are broad coastal supratidal and intertidal flats developed along the margins of arid landmasses. Sediments that accumulate on sabkhas include (1) siliciclastic detritus eroded from adjacent land washed onto the sabkha; (2) offshore deposits of sand and mud that periodic storms wash up and onto the sabkha; and (3) the indigenous sediments of the sabkha itself. Much of the evaporite sediment produced in sabkhas precipitates as saline groundwater seeps into, through, and up and out of the sabkha (Fig. 14.5). Much of this groundwater is seawater that is continually replenished (recharged) beneath the sabkha, but groundwater from the adjacent landmass can

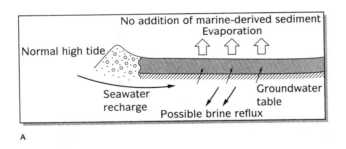

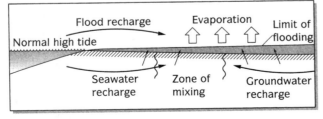

Figure 14.5 Sabkhas receive water from a variety of sources. (A) Sabkha with seawater recharged through the subsurface and with relatively little groundwater influx. (B) Sabkha groundwaters are recharged by a mixture of seawater and groundwater, plus seawater flooding from major storms. (After Walker, 1984: 27; courtesy of Geoscience Canada.)

also feed the system. Groundwater circulation is driven by capillary action and evaporative pumping. Intermittent flooding by the sea also occurs, and beach ridges can trap a reservoir of additional seawater.

Typical sabkha evaporite minerals are anhydrite, gypsum, and dolomite. Much of the anhydrite occurs as irregularly shaped lumps or nodules. These nodules replace altered gypsum crystals originally formed within layers of interbedded carbonate mud or shale. The term **chickenwire structure** is used to refer to this mixture of elongate, irregular clumps of anhydrite separating thin stringers of carbonate and/or siliciclastic mud (Fig. 14.6A, B). This structure is particularly common in sabkha evaporites but is not confined to them.

Cyclicity is common in sabkha evaporite sequences. As deposition proceeds, sabkha deposits naturally prograde oceanward and eventually lie upon intertidal sediments (stromatolites, gypsum, fenestrated birdseye pelleted carbonate mud; see Chapter 12). These in turn rest on oolitic and bioclastic subtidal carbonate rocks of the subtidal zone (Fig. 14.6C). Ancient sabkha evaporites of Permian age have been recognized in the Delaware Basin of Texas and New Mexico (Warren and Kendall, 1985).

Shallow Marine Shelf and Basin Deposits   The classic ancient evaporite deposits of the world appear to be of shallow marine origin. Examples include the Permian Zechstein deposits of northern Europe, much of the Permian Delaware Basin deposits of West Texas (Fig. 14.7), Silurian-Devonian deposits of the Elk Point Basin in western Canada and the Salina Basin of Michigan and upstate New York, and Devonian-Jurassic evaporites of the Williston Basin of Montana. The great volume of these deposits is particularly striking and puzzling. Each consists of hundreds of meters of bedded evaporite minerals, especially halite and gypsum, that extend across thousands of square kilometers. This is surprising because evaporation of a single column of seawater 1 km thick yields only a 15-m sequence of evaporite. Of this 15 m, 3 m would be gypsum and another 1.5 m would be halite.

How can we explain the classic ancient sequences consisting of meters and even several tens of meters of gypsum and anhydrite and overlying beds of halite several meters thick? Does their existence imply evaporation of an enormous volume of seawater? Precipitation of thick sequences consisting only of gypsum dictates a level of salinity produced by evaporating at least 70% but less than 90% of the original volume of

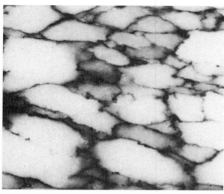

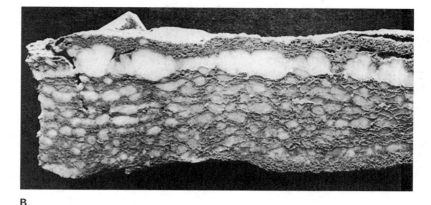

A

B

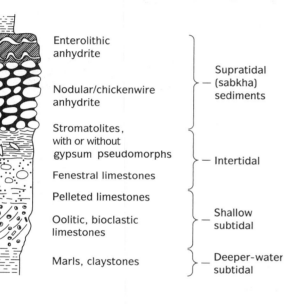

Enterolithic anhydrite

Nodular/chickenwire anhydrite

— Supratidal (sabkha) sediments

Stromatolites, with or without gypsum pseudomorphs

Fenestral limestones

— Intertidal

Pelleted limestones

Oolitic, bioclastic limestones

— Shallow subtidal

Marls, claystones

— Deeper-water subtidal

C

Figure 14.6 Sabkhas produce a number of distinctive structures. (A) Mosaic anhydrite (chickenwire structure) commonly formed when anhydrite nodules coalesce, shown at actual size. (Courtesy of R. P. Glaister.) (B) Nodular gypsum is common just below the surface of the sabkha. (Courtesy of L. A. Hardie.) (C) Typical vertical cycle of sabkha sediments. Such cycles range from several meters to several tens of meters in thickness. (After Tucker, 1991: 179; by permission of Blackwell Scientific Publications, Oxford.)

seawater, or else halite and other salts would also be found. Such brines must have been replenished continually with additional seawater so that salinity stayed in an equilibrium between these narrow limits for long periods of time. The repeated vertical cycles (see Fig. 14.7) of evaporite minerals (for example, calcite and anhydrite) must also be explained.

Obviously, these sequences do not form in a single episode during which tremendous volumes of seawater are evaporated. To produce a thick, regionally extensive evaporite sequence, a basin must somehow be periodically isolated from the open ocean (Clark and Tallbacka, 1980; Hardie and Eugster, 1970). A physical barrier, perhaps a structural ridge or growing reef strategically positioned at the basin mouth, can function as a screen or filter, making such a **barred basin** an evaporating "pan." At high tides or during storm surges, seawater of normal salinity enters the basin and replenishes its stock of normal seawater. During low tides, especially those that coincide with daily temperature highs, rapid evaporation of seawater can

raise salinity to the saturation point. Crystallization thus proceeds in a quasi-orderly cyclical fashion (Fig. 14.8). Barred basins permit partial or incomplete sequences of evaporite minerals to develop. For example, layers of halite and potash salts that would normally overlie basal carbonate and gypsum-anhydrite beds would not form if denser, late-developing hypersaline brines escaped by reflux. And the timing of high tides, low tides, and peaks of evaporation might be out of sync.

This is a delicate system, requiring both the barrier and the basin to persist over a long time. Most North American shallow marine evaporites probably accumulated in this kind of broad, barred cratonic or intracratonic basin, with the evaporite minerals forming in several environments: on the floor of shallow brine pools, on salt flats, and in estuaries and lagoons. Depths must have been very shallow (rarely more than a few meters, routinely perhaps only a few centimeters), so that high salinities in seawater could be reached repeatedly within hours. Curiously, delicate

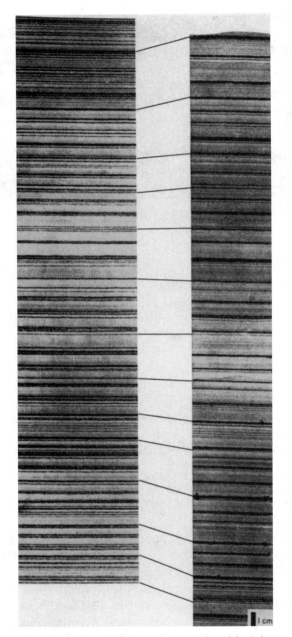

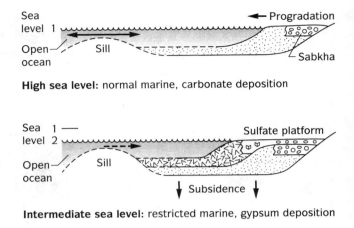

High sea level: normal marine, carbonate deposition

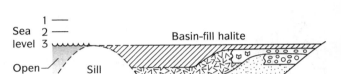

Intermediate sea level: restricted marine, gypsum deposition

Low sea level: salt lake – saline pan, halite precipitation

Figure 14.8 Model depicting the filling of a barred evaporite basin. Repeated refilling of the basin with normal seawater, followed by repeated evaporation, generates enormously thick evaporite sequences. (After Walker, 1984: 271; courtesy of Geoscience Canada.)

Figure 14.7 The Permian Castile Formation evaporites of the Delaware Basin of West Texas show spectacular fine-scale lamination, presumably from fluctuating climatic cycles. These subsurface cores from areas 95 km apart show the possible correlation between laminae of calcite (dark) and anhydrite (light.) (Dean and Anderson, 1982: 340.)

internal laminations extending for tens of kilometers in these sequences imply that this extremely shallow sediment surface remained remarkably undisturbed.

Deep Marine Evaporites

Ancient evaporite deposits that lack the delicate, regionally extensive laminations are interpreted as deep-water deposits. They contain transported size-

graded clasts of gypsum and anhydrite (slope-driven turbidites?) and broken and contorted masses of slumped evaporite (Schmaltz, 1969). The exact depth at which they formed is uncertain. Estimates based on slope requirements and paleotopography range from as little as 40 m (Davies and Ludlum, 1973) in the Devonian Elk Point Basin of Montana up to 600 m for the Pleistocene Dead Sea (Katz, Kolodny, and Nissenbaum, 1977).

Figure 14.9 illustrates general basin models for deep-water evaporite deposits as well as models that combine shallow- and deep-water origins. The major problem with deep-water, deep-basin models is that it's hard to *keep* the water deep if evaporation is occurring fast enough to precipitate minerals. To circumvent this problem, it is suggested that most evaporites originally form along the shallow margins of the basin but then are transported by gravity sliding to the center of the basin.

Most authors suggest that deep basins with evaporite deposits were filled with shallow water. Under these circumstances, a large sill would allow repeated inflow of seawater down into the basin where it would then evaporate. The famous late Miocene (Messinian Stage) evaporites of the Mediterranean

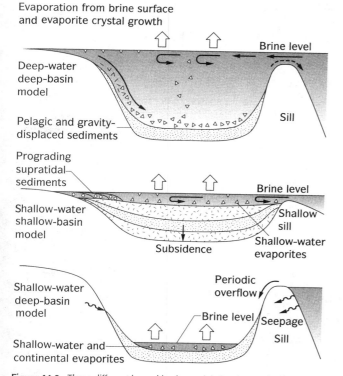

Figure 14.9 Three different barred basin models for the production of evaporite sequences in deep waters and deep basins, shallow waters and shallow basins, and shallow waters and deep basins. (After Walker, 1984: 290; Courtesy of Geoscience Canada.)

now the eastern end of the Mediterranean. Only the western connection through the Straits of Gibraltar remained open, as it does today. Because of high mean temperatures, the Mediterranean today (and in the past) evaporates at a very fast rate, and the few rivers that drain into it are insufficient to maintain it. Thus, the huge one-way flow of water from the Atlantic through the Gibraltar Straits is essential to maintaining the Mediterranean at normal marine salinity.

At the end of the Miocene, however, Africa also began to collide with Spain, forming the Atlas Mountains and restricting the Gibraltar Straits. A pulse of Antarctic glaciation in the latest Miocene triggered a global drop in sea level, and suddenly the Atlantic no longer flowed into the Mediterranean. Within a few thousand years, 2.5 million km^3 of water slowly evaporated, forming a gigantic version of Death Valley or the Dead Sea, complete with a bull's-eye pattern of evaporites in several of the subbasins of the Mediterranean. Unlike Death Valley or the Dead Sea, however, this basin was 4000 m (13,000 feet) deep!

This staggering discovery was made in 1970, when the Deep Sea Drilling Project drilled core after core in the Mediterranean and found thick deposits of upper Miocene evaporites in the basin center and thick alluvial fan conglomerates and sands around the margins. The entire basin had evidently become one huge, deep salt pan with 3000 m of relief on its flanks. Seismic imaging of the Nile Delta region showed that the Nile is underlain by an ancient canyon almost 2500 m deep, now filled by river sediments. This could occur only if the Mediterranean had dried up completely, incising the river valleys as much as 3000 m lower than

are an example of this process (Fig. 14.10). Throughout the Cenozoic, the great Tethys seaway that once ran from the Mediterranean to Australia became fragmented and destroyed. India collided with Asia in the Eocene, and Africa collided with the Arabian Peninsula during the middle Miocene, shutting off what is

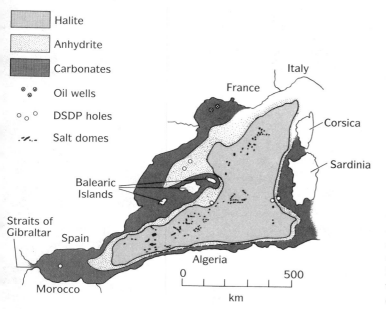

Figure 14.10 Concentrically zoned marine evaporites occur in the Balearic Basin of the western Mediterranean Sea, immediately east of the Straits of Gibraltar. Apparently, this basin was completely dried up and 4000 m below sea level during the peak of the Messinian event. Carbonates and sulfates (anhydrite) form concentric rings around the coastal region. The center of the basin is occupied by halite; and potash salts, such as sylvite, occur in the very center of the bull's-eye. (After Hsü, 1972: 390; by permission of Elsevier Science Publishers.)

their present level. Because the best-known outcrops of these late Miocene evaporites occur on the island of Sicily near the Straits of Messina, this great evaporation event is known as the **Messinian event,** after the name for the final stage of the Miocene.

Multiple cycles of evaporites in these Messinian deposits imply that the Mediterranean refilled and dried up many times. Some halite and gypsum deposits reach a staggering 2000 m in thickness. Since a single episode of complete evaporation would produce only 100 m of halite and gypsum, at least 40 separate drying events in less than a million years would be required to produce all these evaporites.

Nonepiclastic Sedimentary Rocks

Four processes other than the weathering of preexisting rocks produce sediment and sedimentary rocks. Table 14.4 shows how nonepiclastic sedimentary rocks are classified. The most important of these classes are the **volcanogenic sedimentary rocks.** These rocks are formed directly by volcanic processes and are closely linked temporally and spatially with active volcanism. **Pyroclastic sedimentary rocks** are produced by explosive activity; other kinds of

TABLE 14.4   Classification of Nonepiclastic Sedimentary Rocks

Type and Mechanism of Origin	Subcategories
I. Volcanogenic Directly produced by processes associated with the extrusion and cooling of igneous magmas	A. Pyroclastic: Composed of tephra—fragments of rock, mineral, and glass generated by volcanic explosions. Pyroclastic sedimentary rocks are subdivided on the basis of tephra size into *agglomerate* and *breccia* (>64 mm), *lapilli stone* (2–64 mm), and coarse ($\frac{1}{16}$–2 mm) and fine ($<\frac{1}{16}$ mm) *tuff*. 　Pyroclastic deposits are subdivided on the basis of origin into 1. Airfall deposits 2. Volcaniclastic flow deposits 　a. Ignimbrites 　b. Pyroclastic surge deposits 　c. Lahars (mudflows) B. Autoclastic: Produced by the uneven cooling and brecciation of hot lavas in contact with the atmosphere C. Hyaloclastite: Produced by the uneven cooling and brecciation of hot lavas in contact with water
II. Cataclastic Directly related to the crushing, grinding, and brecciation associated with intense folding and faulting	A. Fault breccia and mylonite B. Fold breccia and mylonite C. Crush conglomerate
III. Collapse or founder breccia Produced by the collapse or foundering of rocks into open spaces generated when underlying material is removed	A. Collapse breccia: The mechanism responsible for removal of underlying material is indeterminate B. Solution breccia: Selective dissolution of carbonate or evaporite causes collapse
IV. Meteoritic or impact or fallback breccia Produced by the impact of high-velocity extraterrestrial bodies (bolides) of indeterminate origin (from the asteroid belt or the Oort cometary cloud?)	

volcanogenic rocks are produced by less spectacular processes related to the cooling of lava. Any sedimentary rock containing a large number of volcanic fragments, without regard to its origin or environment, is called **volcaniclastic**. Thus, sedimentary rocks weathered from volcanic source rocks are volcaniclastic, rather than volcanogenic.

Three other kinds of nonepiclastic, nonvolcanogenic sedimentary rocks are recognized. They are generally coarse-grained and much less common than volcanogenic rocks. **Cataclastic rocks** form as the result of tectonic activity, mainly fragmentation and brecciation associated with folding and faulting. **Collapse** or **solution breccias** develop where solution or stress cause overlying rock units to collapse into open spaces below. **Meteoritic clastic sediments** (also called impact or fallback breccias) are produced by the impact of extraterrestrial bodies with Earth's surface.

Volcanogenic Sedimentary Rocks

Most sedimentary geology courses and textbooks focus on sandstones and carbonates and rarely discuss volcanogenic sedimentary rocks in any detail. Yet in some parts of the world, sedimentary rocks of volcanic origin are far more abundant than epiclastic sandstones or other types of sedimentary rock. In the geologic past, volcanogenic and volcaniclastic sedimentary rocks have made up a significant percentage of the total rock record (typically about 25% of the sedimentary record). In some areas, volcanogenic and volcaniclastic sediments are at least twice as abundant as epiclastic sandstones (which rarely make up more than 10% of the sedimentary record, according to Ronov, 1968) or carbonates (which make up less than 15%, on the average).

Volcanogenics are especially important for understanding the stratigraphy of active mountain belts. They were common in the Archean, when most sediments were composed of fresh or reworked volcanic material, ranging from lithic graywackes to pillow lava breccias. Volcanogenics also have economic importance; many of them are associated with hydrothermal mineralization, which makes them major sources of gold, uranium, and sulfide minerals. Finally, volcanogenics have a disproportionate effect on human affairs, as demonstrated by the fatal eruptions at Mount Vesuvius in Italy, Krakatoa and Tambora in Indonesia, Mount St. Helens in Washington State, Nevado del Ruiz in Colombia, Mount Unzen in Japan, and Mount Pinatubo in the Philippines.

Given the great geologic and economic importance of volcanogenic sediments and the fact that volcanoes have always inspired awe, fear, and fascination, it is surprising that these sediments have been so neglected. Although sedimentary petrologists concede that volcanogenic sedimentary rocks are voluminous, few have studied them. There are many reasons for this neglect. Distinguishing the various volcanogenic components in hand specimens or thin sections is difficult. Volcanic constituents are quite susceptible to chemical decomposition and textural diagenesis. For example, shards of volcanic glass older than Tertiary are almost unknown because they devitrify into clay minerals and zeolites.

Much of the neglect of volcanogenic rocks by sedimentary petrologists stems from deliberate bias, however. Traditionally, volcanogenics are the domain of igneous petrologists, and sedimentary geologists are reluctant to trespass in unfamiliar territory. Because volcanogenic rocks are hybrids of igneous and sedimentary processes, sedimentary geologists have long avoided them.

By historical accident, stratigraphy was first studied and thus much better understood in the stable continental interior (the domain of quartz arenites and carbonates) than in mountain belts, with their structural complexity and abundant volcanogenic sediments. Consequently, sedimentary petrologists have hesitated to work in geologic terrains where large volumes of volcanogenic sedimentary rocks occur, preferring instead to study sedimentary rock sequences that lack volcanogenic components. Plate tectonics and terrane tectonics now demonstrate that ancient mobile belts incorporate volcanic arcs and subjacent forearc and back-arc basins (see Chapter 19). Because immense volumes of volcanogenic rocks occur in these domains, the renewed interest of sedimentary petrologists in these areas is greatly advancing our understanding and interpretation of these fascinating deposits. The short section that follows cannot begin to do justice to the subject. We strongly recommend the excellent texts by Fisher and Schmincke (1984) and Cas and Wright (1987), or the chapter by Lajoie and Stix (1992), for a detailed discussion of the subject.

Pyroclastic Rocks Pyroclastic rocks are produced when tephra lithifies. **Tephra** is material of any size or composition ejected by volcanic explosions. Several size classes of tephra similar to the major size classes of siliciclastic sedimentary rocks are used to categorize pyroclastic sedimentary rock types without regard to composition (Table 14.5). Coarse tuff consists predominantly of sand-sized tephra (lithified coarse ash); fine tuff is silt-sized tephra (lithified fine ash) (Fig. 14.11A).

TABLE 14.5 Size Classification of Tephra and Pyroclastic Sedimentary Rocks

Size Class Range	Siliclastic Sediment Size Class	Tephra Name	Pyroclastic Sedimentary Rock Type
>64 mm	Cobbles and boulders	Blocks[a] (angular)	Volcanic breccia[b]
		Bombs[a] (rounded)	Agglomerate[b]
2–64 mm	Granules and pebbles	Lapilli	Lapilli stone
$\frac{1}{16}$–1 mm	Sand	Coarse ash	Coarse tuff
<$\frac{1}{16}$ mm	Silt and clay	Fine ash	Fine tuff

[a]Blocks are angular to subangular clasts of pre-existing congealed lava; bombs are rounded blobs of lava that congealed in flight.
[b]If very coarse clasts that are bombs dominate, the pyroclastic rock is agglomerate. If very coarse clasts that are blocks dominate, the pyroclastic rock is volcanic breccia.

A

C

B

D

Figure 14.11 Pyroclastic tephra deposits are classified by size and shape. (A) Thick sequence of sand- and dust-sized volcanic ash deposits. (D. R. Prothero.) (B) Popcorn-sized lapilli made of pumice, from the Jemez volcanic field, New Mexico. (D. R. Prothero.) (C) Agglomerate is composed of volcanic bombs. This example is from Karymsky volcano, Kamchatka Peninsula, Russia. (Courtesy of R. Hazlett.) (D) Volcanic breccia, Augustine volcano, Alaska. (Courtesy of H. U. Schmincke.)

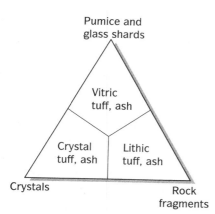

Figure 14.12 Triangular diagram for classifying tephra of various sizes using three end members: ash (<2 mm), lapilli (2 to 64 mm), and blocks and bombs (>64 mm.) (After Friedman, Sanders, and Kopaska-Merkel, 1992: 41; by permission of Merrill, New York.)

Lapilli (2–64 mm) stone contains mainly pebble-sized tephra (Fig. 14.11B). Very coarse-grained pyroclastic deposits are either volcanic **agglomerates** (mainly bombs) (Fig. 14.11C) or **volcanic breccias** (mainly blocks) (Fig. 14.11D). **Blocks** and **bombs** are very coarse (>64 mm grain diameter) tephra. Blocks are large, angular, sharp-edged fragments of lava that congealed prior to explosive volcanism. Bombs are the same size as blocks but are rounded, bloblike masses of lava that congealed during flight after being erupted. Different size classes of tephra are typically intermixed.

Some classification schemes use differences in the mineralogy and petrology of tephra to categorize pyroclastic sedimentary rocks. (In fact, agglomerates and volcanic breccias are separated using compositional distinctions that are related to the contrasting origin of blocks and bombs.) Figure 14.12 illustrates a classification scheme for subdividing tuff based on the proportion of three principal components: (1) pumice and shards of glass, (2) crystals, and (3) rock fragments.

The physical characteristics used to distinguish these three components are shown in Figure 14.13. Tephra in tuff derived directly from magma is composed of volcanic glass. Volcanic glass forms mainly

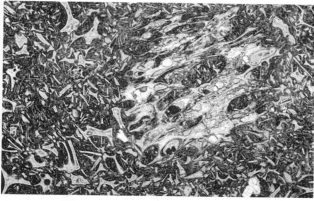

A

B

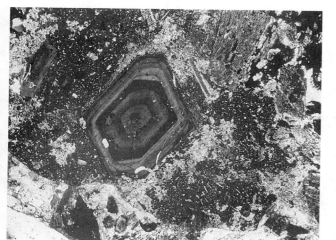

C

Figure 14.13 In thin section, volcanogenics exhibit a variety of textures. (A) A vitric tuff is composed largely of altered volcanic glass (obsidian) that shows the characteristic triangular and concave shape from having formed around the walls and junctures between bubbles of gas. In the center is a large pumice fragment, showing the distinctive vesicular "bubble" texture. Field of view is 6 mm across. From the lower Miocene Lospe Formation, Santa Maria Basin, California. (Courtesy of R. Cole.) (B) A crystal tuff is composed of euhedral phenocrysts. Here, feldspars and hornblende phenocrysts are suspended in an altered glassy groundmass with a distinct flow texture (note the horizontal layering). This example is known as "peperino," from Naples, Italy. (D. R. Prothero.) (C) A lithic tuff is made of volcanic rock fragments (dark clasts with light plagioclase phenocrysts.) This example also has a beautifully zoned plagioclase. (Courtesy of R. H. Dott, Jr.)

from the disaggregation of highly vesicular pumice or scoria. It appears as sickle-shaped or lunate glass shards that are pale white to yellow and isotropic in thin section (Fig. 14.13A). If crystallization occurred before the eruption, most tephra is composed of mineral crystals (Fig. 14.13B). Components identified as crystals are fragments of quartz and feldspar or, less commonly, pyroxene and amphibole. Because these crystals grow in a fluid medium before they are ejected, they tend to be euhedral. Zoning is common. Tephra can also consist of any rocks present in the immediate vicinity of the volcanic eruption (Fig. 14.13C). Country rocks quite often include large volumes of pre-existing congealed lava, but no igneous, metamorphic, or sedimentary rock type is excluded.

The name assigned to a pyroclastic deposit incorporates both tephra size and the relative percentages of volcanic glass, mineral crystals, and rock fragments. These are measured in the field with pebble counts (if overall grain size is sufficiently coarse) or by point-counting thin sections. The following categories are distinguished: (1) glass-rich vitric tuff or lapilli vitric tuff, (2) crystal-rich crystal tuff or lapilli crystal tuff, and (3) lithic tuff or lapilli lithic tuff. An alternative scheme designates the rock as either tuff or lapilli stone. A prefix added to this term lists end-member components in order of increasing abundance. A rock fragment, crystal lapilli stone would be a pyroclastic sedimentary rock in which granule- and pebble-sized tephra are composed mainly of mineral crystals with some rock fragments.

Pyroclastic sedimentary rocks originate in several ways. Most are pyroclastic airfall deposits or various kinds of volcaniclastic flow deposits.

Pyroclastic airfall deposits form when ejecta thrown into the air by a volcanic eruption fall back to the surface, mainly in the immediate vicinity of the volcanic vent or fissure. Not surprisingly, these deposits rapidly coarsen and thicken toward the source, with large, poorly sorted, angular blocks and bombs immediately circumscribing the explosive center. A widespread thin blanket of better-sorted, better-rounded ash can be spread downwind (see Fig. 15.26). The extent of a pyroclastic air-fall deposit largely reflects the volume of the eruption, the strength and frequency of prevailing winds, and the volume and texture of the tephra produced. Individual falls may exhibit primitive size grading. Clasts can be welded together, depending on the temperature of the material as it falls and the rate at which the falling material accumulates. Individual falls accumulate as uniformly thick blankets of tephra that mantle topographic highs and lows (Fig. 14.14). Air-fall deposits that land in bodies of water are more complex.

Volcaniclastic flow deposits form when tephra is remobilized and moves downslope as a pyroclastic flow (producing ignimbrites), a pyroclastic surge, or a lahar.

Ignimbrites are produced by *nuée ardente* (literally "glowing cloud") eruptions (Fig. 14.15A). These are gravity-propelled clouds of ground-hugging hot tephra and gas that move downslope at velocities of up to 200 km/hr (125 mph). They are basically hot, subaerial density currents. Ignimbrite deposits are not well organized internally, but coarser clasts may show upward size grading (Fig. 14.15B). In this respect, they are similar to the inverse grading seen in debris flows (see Fig. 3.10). Like debris flows, pyroclastic flows are nonturbulent fluidized sediment gravity flows that

Figure 14.14 Ashfalls often cover pre-existing topography and can accumulate on relatively steep slopes. This example of mantle bedding from O-shima Island, Japan, shows multiple ashfall layers that have covered an irregular topographic surface like a blanket. (Courtesy of S. Aramaki.)

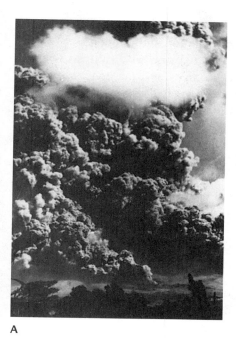

A

B

C

Figure 14.15 (A) Nuées ardentes, or glowing clouds of hot volcanic ash, dust, and gas, plummeted down the slopes of Mayon in the Philippines at about 160 km/hr during the eruption of 1968. (Courtesy of W. Melson, Smithsonian Institution.) (B) Inversely graded pyroclastic flow deposit, Tenerife, Azores. (Courtesy of R. V. Fisher.) (C) The hottest parts of the pyroclastic flow become welded tuffs, or ignimbrites. This specimen shows dark, lens-shaped, collapsed pumice fragments enclosed in a light matrix of welded glass shards. (R. G. Schmidt and R. L. Smith, courtesy of U.S. Geological Survey.)

carry a great range of particle sizes. Particles are suspended by upward-escaping gases and by grain-to-grain contacts. The concentration of particles is high and there is little pore space, so dilation must occur for the particles to move past one another and flow. Because the shear effects are greatest at the base of the flow, the smaller ash and lapilli are found there; larger lapilli, blocks, and bombs are driven toward the top of the flow, where there is less shear (as in aqueous debris flows).

The interior portions of ignimbrite deposits are insulated because they lie between cooler basal and surface layers. Consequently, they retain heat and often alter to welded tuff (Fig. 14.15C). Ignimbrites have a distribution quite unlike that of air-fall deposits. They follow and fill pre-existing drainage corridors (Fig. 14.15B) but do not mantle intervening topographic divides.

Nuée ardentes and their ignimbrites can be truly catastrophic. The most famous and deadly of these was the May 8, 1902, eruption of Mount Pelée on the Caribbean island of Martinique. Although the volcano had been rumbling and producing ash clouds for weeks, the island's inhabitants did not evacuate. Scientists and photographers recorded the growth of a huge spine of rock plugging the vent as it gradually rose out of the summit. Suddenly, the mountain exploded and blew off its volcanic plug, and a nuée ardente rushed down the flanks at speeds up to 160 km/hr (100 mph). In a few minutes and with no sound, the searing emulsion of gas, ash, and dust enveloped the city of St. Pierre and killed 29,000 people

almost instantly. The only survivor of the eruption was a prisoner in a deep dungeon, who was protected from the blast by his cell. He later became a sideshow celebrity in the Barnum & Bailey Circus.

Pyroclastic surge deposits are produced by rapid, episodic, or discontinuous downslope movements of pyroclastic material; gas; and, in some cases, water. Individual deposits are thinner and finer-grained than ignimbrite units and are often richer in crystals and lithic fragments than pyroclastic flows. Surge deposits differ from ignimbrites because they have well-defined internal organization expressed by planar lamination, trough cross-bedding, and planar cross-bedding (Fig. 14.16). Like ignimbrites, pyroclastic surge deposits mantle both topographic highs and lows, but these deposits systematically thicken in valleys and thin in drapelike fashion over topographic divides.

Pyroclastic surges are low-concentration density currents with the particles supported mainly by turbulence (rather than by gas fluidization, as in ignimbrites). The surge is driven by the momentum of the expanding gas, by the momentum of the particles, and by gravity (depending on the slope). Particle concentration occurs at the base of the surge, so sedimentation takes place in the density-stratified bed-load region. As in aqueous flows, many of the particles are moving by saltation and traction.

Lahars are mudflows formed of saturated volcanic material. They can occur when large amounts of precipitation fall onto slopes that are mantled with unconsolidated ash or when a volcanic eruption mixes with water from a river or melting snowcap (Fig. 14.17A). Except for the abundance of volcanic materials, lahars resemble conventional mudflows. They are very poorly sorted with a matrix-supported fabric and little or no internal bedding. Like conventional mudflows and debris flows, they often exhibit reverse graded bedding (Fig. 14.17B) or reverse grading at the base of the flow changing to normal grading at the top (called reverse-to-normal grading).

Although lahars superficially resemble reverse-graded pyroclastic surges, lahars can be distinguished by evidence (such as abundant wood fragments) that they were water-saturated, not composed of superheated gases. Indeed, lahars can be full of tree trunks and branches (Fig. 14.17A), which are oriented subparallel to the bedding surface and may point downstream.

Although wet, muddy lahars may not seem as terrifying as hot nuées ardentes roaring down the sides of a volcano, they can be just as destructive. The

Figure 14.16 Pyroclastic surge deposits show well-developed internal bedding features, unlike most pyroclastic flows. In this example from the Bandelier Tuff, Jemez Mountains, New Mexico, a pumice lapilli tuff (see Fig. 14.11B) exhibits faint cross-bedding and scouring into the underlying columnar-jointed ignimbrite flow. (D. R. Prothero.)

A

B

Figure 14.17 Lahars are debris flows composed of volcanic material. They usually form when a volcano melts its snowcap or heavy rains mobilize loose volcanic ash and debris. (A) The valleys emanating from Mount St. Helens were filled with thick lahar deposits, characteristically covered by trees and boulders that were carried along. (Courtesy of R. Hazlett.) (B) Outcrop of inversely graded lahar deposits, from the 1929 eruption of Mount Pelée, Martinique. (Courtesy of R. V. Fisher.)

eruption of Mount St. Helens on May 18, 1980, produced many lahars formed when the snowcap melted during the eruption. They swept down the flanks of the volcano and flooded the river valleys, causing more damage and loss of life than the actual blast or ash cloud. Lahars can be very voluminous and extensive, and they make up more total material than all other products of a volcano combined. One formation in the Sierra Nevada of California contains 8000 km^3 of lahar material, enough to cover Delaware with a deposit more than 1 km thick.

The most famous of recent lahars struck the towns below Nevado del Ruiz in Colombia after its November 5, 1985, eruption. Formed from the melted snowcap and from rivers swollen by recent rains, lahars rushed down 11 flank valleys of the volcano at speeds of 35 km/hr (20 mph). A number of smaller villages were inundated, and thousands of people were killed. The greatest loss of life occurred when a wall of mud 5 m high swept through the city of Armero at 11 P.M., while most people were asleep. Of the 25,000 residents of the city, 20,000 died in minutes. Survivors reported that the initial wall of mud was cold but became increasingly hot, eventually "smoking" or "scalding." Most of the victims were buried under 5 m of hot mud and could not be located unless a pool of blood seeped to the surface. The majority were never found.

Base surges are sediment gravity flows that form when steam-saturated eruption columns collapse and travel outward across the ground surface. They were originally described from a surge developed at the base of the mushroom cloud over the underwater nuclear explosion at Bikini Atoll in 1947, and they have since been seen emanating from explosive volcanic plumes. Base surges are composed of a turbulent mixture of water vapor or condensed droplets and solid particles. The surge may start out dry if the water is vaporized, but if water condenses during transport, the wet mixture will behave as a fluid. The surface tension of the water on the particles binds them together and causes the mass to exhibit plastic flow.

Base surges are moderately to poorly sorted, with a rapid decrease in grain size and thickness away from the source. They commonly exhibit sedimentary structures, such as cross-bedding and fine lamination, that resemble nonvolcanic flows (Fig. 14.18).

Other Volcanogenic Sedimentary Rocks Other volcanogenic sedimentary rock types owe their origin to igneous activity other than volcanic eruption. For example, lava flows typically cool in an orderly fashion. The basal portion of a flow is chilled against underlying bedrock. Simultaneously, the upper portion cools as thermal energy is radiated into the atmosphere. A delicate and brittle crust is generated by cooling of the upper lava surface. Because it rests on underlying material that is moving, this crust easily breaks into angular, pancakelike blocks. The rock that results is **autoclastic breccia**. Continued flow of lava further fragments these brecciated slabs; the slabs slide off the nose of the flow and become incorporated into its base. As a result, upper and lower

Figure 14.18 Base surge deposits are much better sorted than pyroclastic flows and may exhibit sedimentary structures. This example from the Tule Lake area, northern California, shows both well-developed cross-bedding and parallel lamination. Note hammer at lower right for scale. (Courtesy of R. V. Fisher.)

Figure 14.19 Hyaloclastites are produced when submarine pillow lavas break up. This pillow breccia from the Oregon coast shows a few remnant pillows but is made mostly of loose angular fragments. (Courtesy of R. V. Fisher.)

zones of autoclastic breccia enclose a massive or flow-banded central portion.

Hot lava flows also come into direct contact with water, causing rapid and dramatic cooling that produces thin, brittle, easily brecciated chilled rinds enveloping still-moving masses of magma. These breccias are known as **hyaloclastites.** Many pillow lavas show these well-developed rinds of breccia (Figure 14.19).

Nonepiclastic, Nonvolcanogenic Sedimentary Rocks

Cataclastic Rocks   These rocks are produced by the mechanical fragmentation of pre-existing rocks due to intense folding and faulting. The movement of one crustal block relative to another grinds and pulverizes narrow zones of the rocks into fine-grained, banded, powdery **cataclastic breccia** or **microbreccia (mylonite)** that is extremely granulated and sheared. There are three principal varieties. **Fault breccias** consist of blocks of country rocks floating in finer-grained powdery fault gouge. When mapped, brec-

ciated zones are largely confined to the belt of faulting. **Fold breccias** form where layers of brittle sedimentary rock such as sandstone lie within less competent, plastic lithologies such as mudrock. As the sequence bends, brittle rock layers are brecciated and jostled together within the surrounding carapace of plastic material, especially where folding creates sharp anticlinal crests and synclinal troughs. **Crush breccias** develop where brittle rocks are cut by a series of close joints. The intersection of joints produces cubic blocks of material that, with further deformation, rotate against one another and generate breccia

Collapse (Founder) and Solution Breccias   These rocks form when material underlying a rock unit is selectively removed. The unsupported rock ceiling collapses, and large angular blocks tumble down or free-fall into the underlying space. Solution breccias are a specific type of collapse breccia. They form when the underlying material is excavated by solution. Solution breccias are associated with readily soluble subterranean bedded evaporite and limestone units.

Meteoritic, Impact, or Fallback Breccias   These breccias are produced when bodies of extraterrestrial origin (**bolides**) impact Earth's surface at high velocity. The tremendous kinetic energy produced by the impact fragments the solid material at the target. Material excavated from craters is injected into the atmosphere and stratosphere. The bulk of the material produced falls back to Earth near the point of impact.

Few impact or fallback breccias are well preserved. The material produced is notoriously susceptible to the ravages of weathering and erosion. Also, breccias produced by impact events are largely confined to the local region surrounding the target point. Finally, Earth rarely collides with bolides massive enough to generate sizable amounts of fallback material. These bolides are of uncertain origin, but they are probably derived from the asteroid belt or from clouds of cometary material far out in space. Chapter 5 briefly discusses well-known breccias generated at the Cretaceous-Tertiary boundary and in mid-Tertiary time. These breccias are produced not only as the direct result of impacts but also by giant water waves (supertsunamis) generated by impact.

Identifying a deposit as an impact breccia requires careful examination of a number of features. Defining characteristics include close association of the deposit with **tektites** ("splash droplets"; see Fig. 15.27), the presence within the deposit of such high-pressure polymorphs of SiO_2 minerals as coesite and stishovite, and the presence of a widespread contemporary layer of fine-grained ejecta that is enriched in platinum group metals such as iridium.

Bolide impacts may be an important process that has initiated significant global change over the course of geologic time. Some geologists and astronomers believe that impacts have caused mass extinctions periodically throughout Earth's history.

CONCLUSIONS

Iron-rich sediments are our primary sources of iron and steel and yield important clues about the Precambrian atmosphere and oceans. Evaporites are literally the "salt of the earth," essential in all cultures not only for diet but for a wide range of industrial uses. Evaporites also reveal fascinating aspects of the geologic past, such as the great salt basins associated with the Silurian-Devonian reefs, the amazing evaporite

episodes in the great Permian basins, and the staggering implications of the Mediterranean as a desert. Volcanogenic deposits are becoming increasingly important as stratigraphers begin to decipher mountain belts. Impact deposits capture the imagination, and some geologists think they hold the key to understanding great mass extinctions, such as the one that wiped out the dinosaurs.

FOR FURTHER READING

Cas, R. A. F., and J. V. Wright. 1987. *Volcanic Successions: Modern and Ancient*. London: George Allen and Unwin.

Dean, W. E., and B. C. Schreiber, eds. 1978. *Notes on a Short Course on Marine Evaporites*. SEPM Short Course No. 4.

Fisher, R. V., and H. U. Schmincke. 1984. *Pyroclastic Rocks*. Berlin: Springer-Verlag.

Fisher, R. V., and H. U. Schmincke. 1994. Volcaniclastic sediment transport and deposition. In *Sediment Transport and Depositional Processes*, ed. K. Pye. Boston: Blackwell Scientific Publications.

Hardie, L. A. 1984. Evaporites: Marine or non-marine? *American Journal of Science* 284:193–240.

Hardie, L. A. 1991. On the significance of evaporites. *Annual Reviews of Earth and Planetary Sciences* 19:131–168.

Hardie, L. A., and H. P. Eugster. 1971. The depositional environment of marine evaporites: A case for shallow, clastic accumulation. *Sedimentology* 16:187–220.

Hsü, K. J. 1983. *The Mediterranean Was a Desert*. Princeton, N.J.: Princeton University Press.

Kendall, A. C. 1984. Evaporites. In *Facies Models*, 2d ed., ed. R. G. Walker. Toronto: Geological Association of Canada.

Lajoie, J., and J. Stix. 1992. Volcaniclastic rocks. In *Facies Models Response to Sea Level Change*, ed. R. G. Walker and N. P. James. Toronto: Geological Association of Canada.

Lunine, J. I. 1999. *Earth: Evolution of a Habitable World*. New York: Cambridge University Press.

Park, R. G. 1988. *Foundations of Structural Geology*. Glasgow: Blackies.

Raup, D. M. 1985. *The Nemesis Affair: A Story of the Death of Dinosaurs and the Ways of Science.* New York: W. W. Norton.

Schreiber, B. C., ed. 1988. *Evaporites and Hydrocarbons.* New York: Columbia University Press.

Sharpton, V. L., and P. D. Ward, eds. 1990. *Global Catastrophes in Earth History: An Interdisciplinary Conference on Impacts, Volcanism, and Mass Mortalities.* Geological Society of America Special Paper 247.

Silver, L. T., and P. H. Schultz, eds. 1982. *Geological Implications of Impacts of Large Asteroids and Comets on the Earth.* Geological Society of America Special Paper 190.

Sonnenfeld, P. 1984. *Brines and Evaporites.* New York: Academic Press.

Trendall, A. F., and R. C. Morris, eds. 1983. *Iron Formations, Facts and Problems.* Amsterdam: Elsevier.

Warren, J. K. 1989. *Evaporite Sedimentology.* Englewood Cliffs, N.J.: Prentice-Hall.

Young, T. P., and W. E. G. Taylor, eds. 1989. *Phanerozoic Ironstones.* Geological Society of London Special Publication 46.

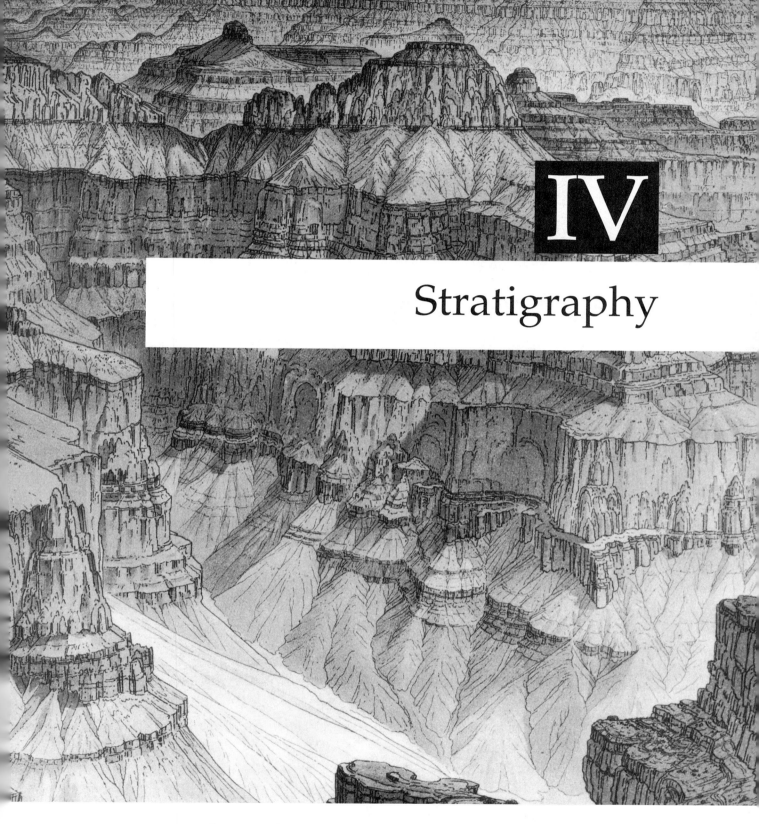

IV

Stratigraphy

The panorama from Point Sublime in the Grand Canyon, looking east. Drawn by W. H. Holmes. (From Dutton, 1882; courtesy of U.S. Geological Survey.)

15 Lithostratigraphy

The Pulpit, composed of *Protoceras* sandstone channels, Poleside Member of the Brule Formation, Badlands National Park, South Dakota. (Photo taken by N. H. Darton in 1898; courtesy of U.S. Geological Survey.)

SO FAR, WE HAVE EXAMINED HOW SEDIMENTARY ROCKS form, and we have attempted to infer ancient depositional environments from local outcrops. Now we will integrate information and inferences from outcrop evidence into a larger context.

As we begin to examine the rock record, we discover that it is preserved and exposed only in certain places,

and it is full of gaps. This fact leads to such questions as: What conditions are ideal for the formation and preservation of stratigraphic sequences? How completely do these sequences record geologic history? The limited exposure of outcrops on the modern surface of the Earth means that many deposits of the same age are no longer connected. To reconstruct the

geologic history that these outcrops represent, we must ask certain questions: How are these outcrops and environments related in space and time? How can they be correlated, and what is the time significance of these correlations? In addition to correlating outcrops, we can learn much by mapping and plotting the geometry of the rock units under study.

Matching the sedimentary properties of rock units is only one way to correlate them. Another important method is biostratigraphy, or correlation by the fossil content of the rocks. Biostratigraphy requires another level of understanding, because the biological properties of the fossils are as important as the physical properties of the rocks that contain them. In recent years, the geophysical properties of strata have become very important in correlation, particularly in the subsurface. Well-logging uses physical properties of the rock that can be detected by devices pulled through drill holes and is the primary tool of modern subsurface correlation. Seismic stratigraphy provides information about subsurface units that could not otherwise be detected. Changes in Earth's magnetic field over geologic time have made it possible to correlate rocks by the way they record these changes, giving us the discipline of magnetostratigraphy.

The geochemical properties of strata are equally important. The regular pattern of change of certain stable isotopes over geologic time has allowed correlation by this geochemical pattern. However, it is the decay of unstable (radioactive) isotopes that gives us numerical age estimates, rather than mere relative correlation. Integrating these numerical ages with the relative sequence of events recorded in strata is often the most difficult and challenging part of stratigraphy. The geologist not only must develop a correlation framework and put dates into it but also must be aware of the relative strengths and pitfalls of each method. Integrated chronostratigraphy requires the skillful and well-informed use of many disciplines, a training that all geologists should have.

Finally, building on the depositional framework of local basins, geologists must consider the larger framework of plate tectonics. What factors ultimately control the crustal movements that form basins in the first place? The stratigraphy of basin formation and tectonics, has been completely revolutionized since the 1960s. For the first time, we now have a global model that explains and predicts *why, when,* and *where* sedimentary basins form. The enormous ramifications of this understanding are just beginning to change the field of stratigraphy. The stratigraphy of the twenty-first century is likely to be very different from stratig-

raphy as it was understood through most of the twentieth century. Indeed, it is a fortunate time to be a stratigrapher interpreting the rock record!

Facies

"Layer Cake" Geology and Facies Change

In the late 1700s and early 1800s, catastrophist geologists thought that the rock record had been laid down in uniform sheets over the whole world during Noah's Flood. If this had been the case, the same rock layers would be evenly distributed over the Earth everywhere, with little difference in lithology or thickness. This concept of the rock record is often called "layer cake" geology, and it is surprising how many modern geologists are still influenced by this kind of thinking. The impressive sequences of rock layers in such places as the Grand Canyon do tempt us to imagine that such sequences might extend indefinitely without change in thickness or lithology, each layer representing the same period of time everywhere.

One of the first applications of uniformitarianism to geologic thinking was a comparison of modern depositional conditions with their ancient rock equivalents. Any examination of modern depositional environments shows that most do not extend for great distances laterally but eventually change into other depositional environments. Many sedimentary rock types are being deposited simultaneously in different areas, and no single rock type is deposited over a very wide area at one time. Rock types are diagnostic of local environments, not of ages. These ideas first appeared in the geologic literature in the late 1700s. For example, the famous French chemist Antoine Lavoisier (1743–1794) produced a diagram (Fig. 15.1) in 1789 showing that gravels should occur near the shore but fine clays could be expected far from shore. Other geologists had noticed that the lithology of rock formations changes over distance. In 1839, Sedgwick and Murchison named the Devonian Period after the rocks of Devonshire, because although they were the same age as the Old Red Sandstone, their lithology differed.

In 1669, Nicholas Steno coined the term **facies** (from the Latin word meaning "aspect" or "appearance") for the entire aspect of a part of Earth's surface during a certain interval of geologic time (Teichert, 1958). The Swiss geologist Amanz Gressly (1814–1865) first expounded the modern concept of facies in 1838. In his study of the Mesozoic sequences of the Jura Mountains of Switzerland (the type area of the Jurassic), Gressly went beyond previous stratigraphers who

Figure 15.1 Antoine Lavoisier's diagram of the relationships of coarse littoral *(Bancs Littoraux)* and finer pelagic *(Bancs Pelagiens)* sediments to the northern French coastline. Lavoisier recognized that gravel can be moved only by waves near the shore whereas fine sediments can be carried into deeper water. He also saw that distinctive organisms inhabited each environment. If sea level rose and flooded the land *(la Mer montante),* both littoral and pelagic sediments would migrate landward. If sea level fell *(la Mer descendante),* they would shift seaward. (Lavoisier, 1789.)

named rock units based on local isolated sections. He traced rock units laterally and found that they changed in appearance. Gressly's use of the term *facies* designated lateral changes in the appearance of a rock unit and emphasized the fact that rock units are not uniform in lithology but change as their depositional environment changes. Continental geologists subsequently standardized this meaning in their literature. According to Haug (1907), "A facies is the sum of lithologic and paleontologic characteristics of a deposit in a given place," and this definition prevails among many stratigraphers today. It implies that facies change over distance, as Gressly originally noted.

The word *facies* was so handy, and its use became so widespread, that its meaning broadened beyond Gressly's original intention. Outside Europe, it came to denote major stratigraphic sequences, whereas Gressly's concept of facies was restricted to lateral changes within a single rock unit. Some geologists used the term for just about any similar association of rocks occurring in certain geographic, oceanographic, or inferred tectonic settings. It has even been used by metamorphic petrologists to denote metamorphic rocks of similar assemblages of minerals (Eskola, 1915). At one time, there was much discussion about whether this broad usage of *facies* had made the term so vague as to be useless (Moore, 1949; Teichert, 1958; Weller, 1958; Markevich, 1960; Krumbein and Sloss, 1963).

In recent years, however, a broad definition along the lines of "a body of rock with specified character-

istics" (Reading, 1986, p. 4) or "associations of sedimentary rock that share some aspect of appearance" (Blatt, Middleton, and Murray, 1980, p. 618) has prevailed and seems to be understood by most geologists. Specifically, a **lithofacies** denotes a consistent lithologic character within a formation (for example, *shale lithofacies* or *evaporite lithofacies*) and excludes its fossil character (which was included in Gressly's concept). Similarly, the term **biofacies** applies to the consistent biological character of a formation (for example, *oyster bank facies*). In this sense, a biofacies is also an ecological association.

Because of the emphasis during the past two decades on depositional environments, the word *facies* has become closely associated with genetic interpretations. Currently, it is most often used in the environmental sense, as the lithologic product of a depositional system such as *fluvial facies* or *shallow marine facies* (see, for example, Reading, 1986; Walker, 1984). This is the way we have used the word in the first three parts of this book. Most modern geologists prefer to retain this long-established and useful word and to indicate its meaning by context, rather than coin a new terminology to separate Gressly's meaning from broadly descriptive facies terms (for example, *sandstone facies*, strictly speaking a lithofacies) and from more interpretive terms (for example, various environmental or tectonic facies). *Sandstone facies* is clearly a descriptive term, but *fluvial facies* is just as clearly an interpretive term indicating the product of a depositional environment.

Sedimentary environments that produce changes of facies occur in natural associations. The facies that are formed next to one another in a vertical section of rock will be the same as those found next to one another in a modern depositional system. This fact was described by Johannes Walther (1894), who stated, "Only those facies and facies-areas can be superimposed, primarily, which can be observed beside each other at the present time." The principle that *facies that occur in conformable vertical successions of strata also occur in laterally adjacent environments* is known as **Walther's law of correlation of facies.**

Walther's law is a useful tool for stratigraphers. In many instances, the interpretation of the environments that deposited a few units will be clear, but a conformable unit between two of them, or one that lies above or below them, cannot be interpreted so easily. Walther's law can be used to limit the possible interpretations. For example, in the traditional model of the coal cyclothem sequence (see Fig. 13.4A), there are shales and limestones both above and below the coal. Environmental associations and Walther's law suggest that the lower shale and limestone formed in freshwater lakes or fluvial settings, but the upper shales and limestones were marine. Marine and nonmarine fossils in the limestones confirm this interpretation, but there are fewer fossils in the shale. Here Walther's law helps to rule out unlikely interpretations.

Nevertheless, Walther's law must be used with caution. Environmental changes can be extremely rapid and leave no record, so there might be a gap in the sequence of facies. A classic example of such a gap is found in the Upper Devonian–Lower Mississippian Chattanooga and New Albany shales found widely in the Appalachians and the Midwest (Lineback, 1970). Because these shales overlie shallow marine carbonates, they had long been interpreted as shallow lagoonal deposits, despite many indicators of deep-water deposition. Closer examination showed that a subtle unconformity exists between the shales and the carbonates (Cluff, Reinbold, and Lineback, 1981; Conkin, Conkin, and Lipschutz, 1980), which is indicated only by pyrite and phosphate mineralization. The Chattanooga and New Albany shales were in fact deep-water, basinal black shales (described in Chapter 6) deposited unconformably on shallow marine limestones. The basin dropped before the shales were deposited and left a record of this rapid deepening only a few centimeters thick. The apparent "sequence" of these depositional environments is not a true conformable sequence after all, because it is interrupted by a major unconformity. Walther's law does not apply across unconformities!

Transgression and Regression

Certain facies associations are common in the rock record. For example, most clastic shorelines show a series of depositional environments that are progressively finer-grained in the offshore direction (Fig. 15.2). Thus, there are facies belts of coarse sands and gravels (mostly fluvial in origin), finer sands (beach and shallow offshore environments), and silts and clays (shallow shelf muds, prodeltaic clays). If the relative sea level changes (either by eustatic sea-level change or by local uplift or subsidence), deposits of these facies belts accumulate. Three facies patterns are possible.

Facies belts could pile up vertically (Fig. 15.2A) if the relative rate of sea-level rise is exactly balanced

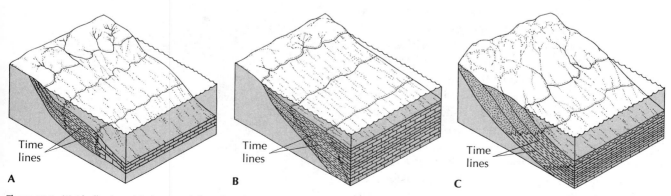

Figure 15.2 (A) Idealized model of accumulation of sandy, muddy, and carbonate facies belts, assuming no relative change of sea level. (B) The same facies belts during transgression. The time lines cut across the rock units, and the sequence fines upward. (C) Regressive or progradational facies pattern caused by a relative retreat of sea level. Once again, the time lines cut across the rock units, but in this case the local stratigraphic sequence coarsens upward.

by the sedimentary output of the land, which would cause the shoreline to remain in approximately the same place. This condition is so unusual that it is almost never encountered in the stratigraphic record.

Facies show a **transgressive** pattern (Fig. 15.2B) when the sediment supply is overwhelmed by a relative rise in sea level, or when the land subsides tectonically. Both cause the shoreline to move landward.

Facies show a **regressive** pattern (Fig. 15.2C) when the shoreline moves seaward due to an excess sediment supply from the land (**progradation**), when the land is tectonically uplifted and the sea retreats, or when there is a relative lowering of sea level. Because it is difficult to tell whether sediment supply, tectonic uplift, or sea-level change is primarily responsible for this pattern, many geologists use the term *regression* to mean any movement of the shoreline away from the land.

Transgressive-regressive facies patterns form most of the marine stratigraphic record, so it is important to recognize and understand them. Each process leaves a distinctive pattern of facies, both vertically and laterally. Because transgression brings deeper-water deposits—with their progressively finer grain sizes—landward, the stratigraphic column becomes *finer upward* at every point. Conversely, during regression or progradation, coarser shallow-water deposits can build out (prograde) over the finer deposits of the former marine shelf; regressive sequences thus become *coarser upward*. There are many exceptions to this generalization, of course. We have seen that offshore sand bars do occur on the continental shelf (see Chapter 10) amid fine-grained shales. In many cases, however, transgression and regression can be inferred from a single vertical section by this simple rule of thumb.

Let us look at one more aspect of transgressive-regressive sedimentary packages. An ancient depositional surface of a given age existed in several different environments simultaneously. Thus, in the stratigraphic record, the nearshore sandstones are equivalent in age to some of the offshore shales and limestones. As the various transgressions and regressions moved in and out, they deposited a continuous spectrum of facies across these depositional surfaces. Thus, *the time lines* (which represent depositional surfaces of equal age) *cut across the facies boundaries*. The rock units or facies, such as the shale or sandstone, are of *different ages in different places*; that is, they are **time-transgressive** (see Fig. 15.2).

Other lateral relationships between rock bodies are more descriptive. A rock body can become thinner laterally to a point where it vanishes; it is said to **pinch out**. Pinch-outs are important in oil geology because they are often natural *stratigraphic traps* (see Fig. 13.5). If a porous sandstone reservoir rock pinches out between two impermeable shales, oil may be trapped at the thinnest part of the pinch-out. Typically, the boundary between two laterally adjacent rock bodies is very jagged due to rapid changes in the shoreline; the result is a series of pinch-outs that form an **intertonguing** pattern (Fig. 15.3). Rock bodies that do not have distinct boundaries between them are said to **grade** into one another, or intergrade. Lateral grada-

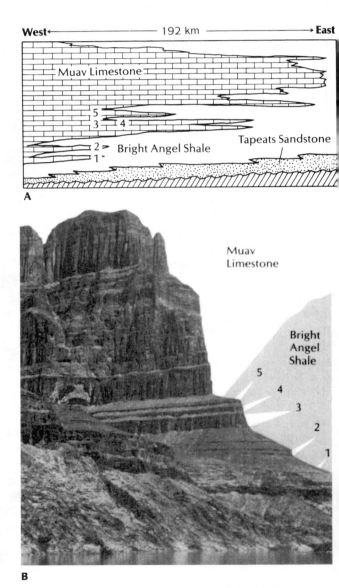

Figure 15.3 (A) Summary of the intertonguing relationships between Cambrian Bright Angel Shale and Muav Limestone in the western Grand Canyon. (After McKee and Resser, 1945.) (B) Outcrop showing intertonguing; the shale members of the Bright Angel Shale are shaded. (Courtesy of P. W. Huntoon.)

tion from porous sandstone to impermeable shale, if overlain by another impermeable layer, can also create a stratigraphic trap. Oil may accumulate where the zone of "shale-out" is relatively narrow.

Asymmetry of Transgressive and Regressive Cycles

The symmetrical facies record of marine transgressions and regressions (Fig. 15.4A) long taught in introductory historical geology classes is now being reconsidered. For years, stratigraphers have noticed that the classic fining-upward sequence of transgressive facies is extremely rare and that coarsening-

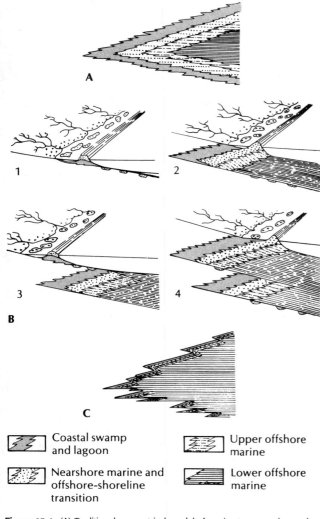

Coastal swamp and lagoon

Nearshore marine and offshore-shoreline transition

Upper offshore marine

Lower offshore marine

Figure 15.4 (A) Traditional symmetrical model of marine transgression and regression. (B) Stages of deposition of shoreline deposits showing reworking and apparent rapid transgression followed by thick progradational sequences. (C) The pattern that results from this process. Even the basal transgressive sequence is composed of numerous small regressive wedges interrupted by thin, apparently rapid, transgressions. (Ryer, 1977: 175, 185, 186; by permission of the Geological Society of America.)

upward patterns are very common, even when sea level is rising. Fischer (1961) and Swift (1968) pointed out that during transgression, the reworking effect of the rising sea level tends to obliterate facies that had been deposited previously. The preservation of a transgressing shallow marine or coastal sequence is much less likely than was once thought, because these deposits are reworked by shoreface and shallow marine processes before they are buried. As a result, unless subsidence is very rapid, transgression leaves a very thin sequence, making the sea level *appear* to have risen very rapidly.

Most of the deposits formed during a sea-level rise are *regressive* sequences that have built out and coarsened upward during the progradation of nearshore facies (Fig. 15.4B). These regressive sequences are less likely to be reworked before being buried by later depositional processes and so are preserved more often. Therefore, frequent changes in relative sea level have an asymmetrical effect. Minor episodes of regression during a large-scale transgression produce almost all the stratigraphic record. The result is an asymmetrical cycle of progradational wedges during the transgressive phase and a similar sawtooth pattern during the regressive phase (Fig. 15.4C).

This revised concept of how transgression and regression are recorded in the rock record has far-reaching implications. For example, the classic model of the fining-upward transgressive coal cyclothem of Illinois (see Fig. 13.4) may be obsolete. This model placed the base of each cycle at the base of the channel sandstone, which represents a fluvial channel sand (Fig. 15.5, right-hand side); transgressive deposits covered the fluvial sands and muds with progressively deeper-water facies. If the revised concepts are correct, the natural break in the sequence is not the channel sandstone but the lowest marine shale or limestone above the coal (Fig. 15.5, left-hand side). This stratum represents the rapid transgression of the marine facies across the basin. The entire sequence of marine limestones and shales, deltaic distributary channel sands, and interdistributary subaerial shales and coals overlies this surface. This is exactly what one would expect of a prograding delta sequence. Contrary to what was previously thought, the cyclothem sequence is not a series of transgressive episodes interrupted by rapid sea-level falls but a series of frequent thin, rapid transgressions followed by prograding deltaic basin fills (see Fig. 13.4B).

Although this asymmetrical model has gained widespread acceptance, it must be tested further

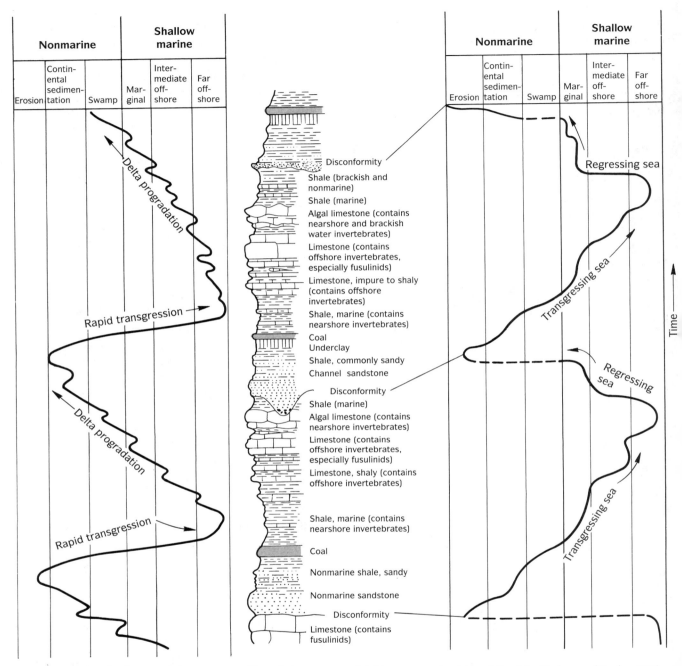

Figure 15.5 Classic Carboniferous cyclothem sequence with two facies interpretations that depend on whether one adopts the symmetrical or the asymmetrical transgression-regression model. At right is the traditional model (after Crowell, 1978), which treats the disconformity beneath the channel sandstone as a rapid regression followed by slow transgression. At left is a more modern interpretation (after Friedman and Sanders, 1978), which places a rapid transgression at the base of the first marine limestone and considers the rest of the sequence to be a regressive, progradational delta front. Correlation of these coal-bearing cyclothems with the more open marine cyclothems supports the second model (see Fig. 13.4).

because it does not explain all examples of apparently symmetrical transgressive-regressive cycles. Bourgeois (1980) has pointed out that transgressive shelf sequences are well documented on tectonically active margins, where rapid subsidence of the basin produced transgression despite high rates of sedimentation. In essence, these basins subside so rapidly that the transgressive sediments sink out of harm's way before they can be reworked. Most of the examples of thin transgressive sequences just discussed are from passive margins or cratonic seaways, which have less subsidence and sediment supply.

Asymmetrical cycles also seem to have been produced in areas of relatively high sediment supply.

For example, Brett and Baird (1986) found that the Devonian Hamilton Group of upstate New York shows both symmetrical and asymmetrical patterns in the same cycles. The onshore deposits are asymmetrical and shallow upward because sediment aggraded upward to the limit imposed by wave energy, in a region of abundant sediment supply. When sea level rose higher, the initial sediment starvation was followed by inundation and deep-water sedimentation. The offshore shales, on the other hand, show symmetrical cycles and thinner deposits because their sediment supply was always relatively low. The geologist must keep all these possibilities in mind when examining the local outcrops for evidence of transgression or regression.

A Framework for Accumulation

We have seen that sediments can accumulate due to relative changes in sediment supply or sea level, but what conditions are necessary for their accumulation and preservation in the rock record? How complete is the record they preserve? Standing on the rim of the Grand Canyon, you see below you a sequence of sediments almost a mile thick. A casual tourist might get the impression that all of geologic time is represented by some sort of sediment in the walls of the canyon. It seems natural to assume that such sediments are accumulating continuously in the permanent rock record, because sedimentation goes on all around us.

Yet a more critical look at the canyon reveals that the vast sedimentary sequence is visible only because the Colorado River has carved the canyon and carried away all the sediments that once filled it. During a raft trip down the Grand Canyon, you pass rockfalls and landslides, shoot rapids over and around boulders that have fallen into the river, and gulp mouthfuls of Colorado River water, muddy and gritty with sediments that are being actively transported to Lake Mead and Hoover Dam. Indeed, the entire Colorado Plateau is a region of rapid uplift and erosion, which is the reason for the depth of the gorge in the first place. Very little permanent sedimentary record is accumulating there to represent the Holocene.

Most of the terrestrial environments discussed in Chapters 8 and 9 will meet a similar fate. Any feature above sea level is doomed to be eroded eventually, although this fate may be forestalled temporarily if the feature is uplifted rapidly enough. Any

sediments that accumulate above sea level will be reworked and redeposited somewhere downhill or downstream unless the proper circumstances entomb them. The term **base level of erosion** has long been used (Powell, 1875; Gilbert, 1914) to describe the *level below which erosion* (chiefly stream erosion) *cannot occur.* In most cases, base level is sea level, although mountain lakes and other ephemeral features can establish temporary local base levels that are above sea level. Joseph Barrell (1917) applied this concept more broadly than his predecessors. To him, base level was the *level above which sediments cannot accumulate permanently.* Dunbar and Rodgers (1957) called this the **base level of aggradation.** It is an imaginary surface of equilibrium between the forces of erosion and deposition. Barrell's definition emphasizes the fact that most terrestrial sedimentary environments are above base level and will ultimately be reworked or eroded. The concept of base level was developed to explain why terrestrial deposits add relatively little to the sedimentary record and have low preservation potential. The main exception to this generalization occurs in nonmarine basins that drop down so far that they accumulate a thick pile of sediment and establish their own local base level (Fig. 15.6A).

Because sedimentary environments below base level are much more likely to be preserved, we would expect marine sedimentary deposits to dominate the stratigraphic record, and they do (Fig. 15.6B). Still, the shallow marine environment is subject to fluctuations in sea level and to strong coastal currents, so erosion also can prevail there. The lowest point below base level is, of course, the deep marine environment. Here the sedimentary record is the most continuous and has the best chance of preservation. Yet the deep-sea record has limitations, too. Rates of sedimentation tend to be extremely slow, so the strata are very thin; deep-sea currents can locally scour or dissolve the seafloor sediment; and most of the deep-sea record has not been uplifted onto the land where it can be studied by geologists. The growth of marine geology over the past three decades, however, has enabled us to study this superlative sedimentary sequence for the first time.

In summary, *preservation of the sedimentary record is the rare exception rather than the rule.* Indeed, special circumstances are needed for sediments to become part of the permanent rock record. Every stratigraphic sequence we examine is composed of small packages of sediment that have been preserved only by luck; the other sediments formed at the same

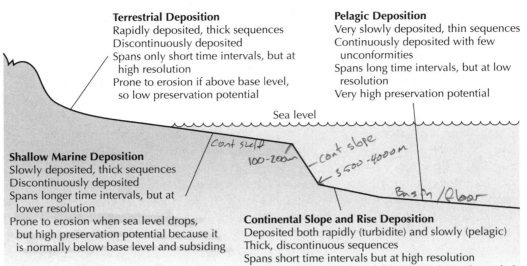

A

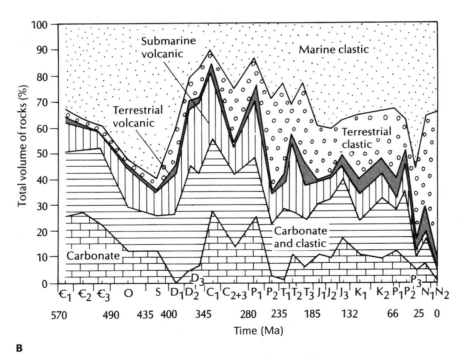

B

Figure 15.6 (A) The patterns of sedimentation in different environments. (B) Relative percentages of various lithofacies through the Phanerozoic. Notice that marine clastic and carbonate rocks typically make up 60% to 80% of the rock record throughout geologic history. (After Ronov et al., 1980: 319; by permission of Elsevier Science Publishers.)

time were destroyed. A rise in sea level helps to preserve many packages of sediment, but ultimately, preservation depends on one thing: *net subsidence.* The rate of basin subsidence must exceed both the rate of sea-level drop and the rate of erosion, so that the subsiding basin remains below base level until— again by luck—it is exposed for geologists to examine and interpret.

Many of the best examples of nonmarine depositional sequences discussed in Chapter 8 were trapped in rapidly downdropped basins, especially fault grabens. Rising sea level can build up a series of deltaic, barrier island, and shallow shelf sediments, which have a good chance of preservation if they sink into some sort of basin before the next sea-level drop can expose them to erosion.

Gaps in the Record

Diastems and the Completeness of the Rock Record

In the early 1900s, the discovery of radiometric dating forced stratigraphers to examine the discrepancy between the age of the Earth determined by radiometric methods and the age estimates based on the thickness of the sedimentary record. If the radiometric dates were correct, then the average rate of accumulation of sediments was on the order of meters per thousands of years. Yet studies of modern rates of sedimentation indicated that typical sedimentation rates are on the order of meters per year to meters per hundreds of years, implying that the stratigraphic record is very incomplete. To use Ager's phrase (1981), it is "more gaps than record." There must be far more unrecorded time in the stratigraphic record than is inferred from the major unconformities.

In 1917, Joseph Barrell first articulated the idea that every bedding plane could account for some of this unrepresented time. He concluded that there are many obscure small-scale unconformities in the stratigraphic record, which he called **diastems.** Barrell used the concept of base level to explain the intermittent nature of sedimentation. As sediments are washed toward the sea, they move steadily into deeper water until they reach depths (according to grain size) at which the water energy is no longer sufficient to transport them. Below base level, sediment will accumulate (and be preserved) whenever it is available; above this level it cannot accumulate permanently and eventually will be transported farther until it reaches a site below base level. Base level is not always fixed, however, but can fluctuate due to grain size, scouring bottom currents, changes in wave base, and marine slumping.

Base level also fluctuates with changes in the environment and the basin. At first glance, the history of an aggrading sedimentary basin might appear to be a continuous rise in base level as the basin is filled (Fig. 15.7). Superimposed on this upward trend, however, are oscillations of global sea level as well as even smaller-scale local trends that temporarily lower base level and cause erosion. The net result is that sediment can accumulate and be preserved during only part of the net aggradational phase of the major oscillations. At all other times, there is either erosion of the recently accumulated record or nondeposition. The effect of this oscillation of base level is that geologic time is represented by sediment only sporadically. Out of the total span of time, only short intervals had a chance to

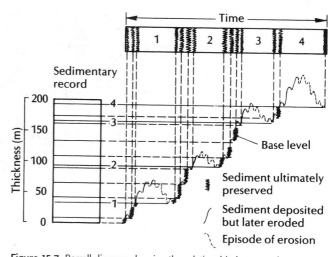

Figure 15.7 Barrell diagram showing the relationship between the sedimentary record and the actual time it represents. Because base level fluctuates up and down, most geologic time is not represented by the sedimentary record. Some sediments are never deposited because the basins have a low base level (dashed lines), and others are destroyed by erosion after being deposited (solid lines.) It is only during rises in base level that are not immediately followed by erosion that the sedimentary record accumulates (wiggly lines), forming a temporal record (*top*) made up of only these relatively short intervals. Barrell concluded that there must be many cryptic unconformities between bedding planes, which he called *diastems.* Major unconformities are numbered 1–4. (After Barrell, 1917: 796; by permission of the Geological Society of America.)

be preserved. Thus, the sedimentary record is not a continuous record of Earth's history but a series of thick packages representing very small portions of the total time elapsed.

The concept of shifting base level has also been applied to nonmarine environments. For example, sediments can be trapped in a terrestrial basin and remain there until the basin is filled and buried. In this case, the basin floor is a local base level. Although such basins are above the ultimate base level near sea level, a local disequilibrium can be maintained over the short term. These terrestrial deposits show the same episodic accumulation as those formed in the marine environment. Loope (1985) has drawn a Barrell diagram (Fig. 15.8) for the Permian dune deposits of the Cedar Mesa Sandstone in the Canyonlands of Utah. What appears to be a continuous record of eolian deposition is actually the record of a few episodes of dune migration that were triggered by short regressive events in the adjacent marine environment. The bulk of geologic time is represented by the bedding planes between dune sequences, when episodes of deflation and soil formation eroded and stabilized the dune sequence.

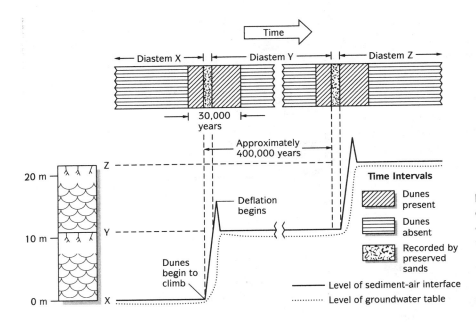

Figure 15.8 Modern application of the Barrell diagram to the Cedar Mesa Sandstone of Utah, an eolian deposit. Preservation of sediments occurs only while the sand dunes climb; after the dunes stabilize with a high groundwater table, no further accumulation occurs, and deflation begins. (After Loope, 1985: 75; by permission of the Geological Society of America.)

Unconformities

Gaps in the rock record come in many different sizes, from barely detectable bedding-plane diastems to huge unconformities. Dunbar and Rodgers (1957) defined an **unconformity** as "a temporal break in a stratigraphic sequence resulting from a change in regimen that caused deposition to cease for a considerable span of time. It normally implies uplift and erosion with the loss of some of the previously formed record." There are four types of unconformities (Fig. 15.9):

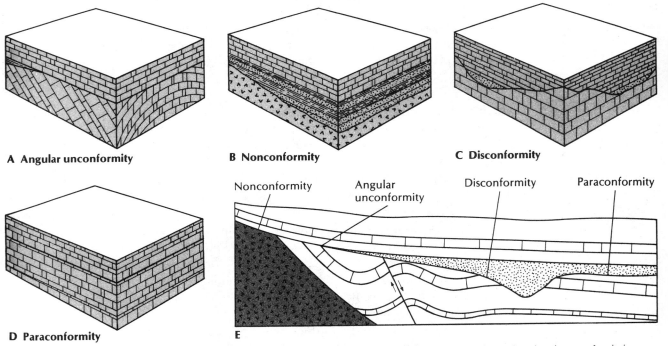

Figure 15.9 Four major types of unconformity. (A) Angular unconformity between tilted and undeformed sediments. (B) Nonconformity between crystalline (igneous or metamorphic) and sedimentary rocks. (C) Disconformity between two parallel bodies of sediment, with some evidence of an erosional gap between them, such as a channel or other erosional surface. (D) Paraconformity, in which the sediments are parallel but there is no direct physical evidence of erosion; the unconformity is detected by determining the ages of the existing units (usually by biostratigraphy.) (E) Changing character of unconformities over distance if the younger unit laps across different types of bedrock. (After Dunbar and Rodgers, 1957.)

Angular unconformity—This familiar type of unconformity occurs when the sequence beneath the unconformity is tilted at some angle to the strata above it.

Nonconformity—The erosional surface truncates massive plutonic igneous or metamorphic rocks and is covered by sediment.

Disconformity—Sedimentary sequences above and below the unconformity are parallel, but there is an erosional surface between them (for example, a river channel base, an ancient karst surface, or a soil horizon).

Paraconformity (obscure unconformity)—The sequences above and below the unconformity are parallel, and there is a normal bedding-plane contact with no obvious erosional surface between them. In this case, the unconformity is recognized by other evidence, such as fossils, indicating that the beds are of significantly different ages. Some authors doubt the need for the paraconformity category, arguing that most alleged paraconformities, if traced laterally, eventually exhibit some physical evidence of erosion and thus prove actually to be disconformities (Davis, 1983).

The Grand Canyon section illustrates these categories (Fig. 15.10). For example, all of the upper Proterozoic and Paleozoic sedimentary rocks lie nonconformably on Precambrian schists and granites (Fig. 15.10E). The upper Proterozoic Unkar and Chuar groups are tilted and overlain by the Cambrian sequence in an angular unconformity (Fig. 15.10D). The thick Cambrian succession of Tapeats Sandstone, Bright Angel Shale, and Muav Limestone is overlain by a disconformity that spans the entire Ordovician and Silurian and most of the Devonian (Fig. 15.10A). Channeled into the Muav Limestone are thin sequences of the Devonian Temple Butte Limestone (Fig. 15.10B), which in turn were eroded and filled in by the Mississippian Redwall Limestone. The recently named Pennsylvanian Surprise Canyon Formation (Fig. 15.10C) is disconformably eroded into the Redwall Limestone. There are many other gaps in the remaining Pennsylvanian and Permian parts of the sequence (Fig. 15.10A, B), some of which have been considered to be paraconformities. Most of these gaps, however, show some evidence of erosional breaks and are disconformities. This apparently complete record, as impressive as it seems, is indeed "more gaps than record." As the geologic time column (Figs. 15.10A, B) shows, only small parts of the Late Cambrian, Late Devonian, mid-Mississippian, and Permo-Pennsylvanian are represented. Most of the Proterozoic, all of the Ordovician and Silurian,

and all of the Mesozoic and Cenozoic have been eroded away or were never deposited. In numerical terms, less than 40% of the Paleozoic is actually represented by rock in this classically "complete" sequence; the duration of the Ordovician-Silurian-Devonian unconformity alone is longer than the total time represented by the rocks in the canyon.

The most important criteria for recognizing unconformities in the field are usually sedimentary. These include a basal conglomerate above the unconformity; a deeply weathered erosional surface, sometimes with a clear soil horizon; truncation of bedding planes or of clasts in the lower sequence; surfaces that have been postdepositionally altered by burrowing, boring, or cementation into hardgrounds; and surfaces that show relief due to an erosional episode.

Unconformities can also be recognized from paleontological clues. For example, if one or more faunal zones is missing, an unconformity is present. Many authors have suggested that gaps in the evolution of a lineage are evidence of an unconformity. Recently, however, paleontologists (Eldredge and Gould, 1972) have argued that the gaps may be a result of the rapid evolution of lineages, not missing sections. If several lineages show an abrupt change at the same level, however, it is likely that the gap represents an unconformity rather than simultaneous rapid evolution.

Structural criteria can also help to identify an unconformity. Strata are not the only geologic phenomena that are truncated by an erosional surface; occasionally, dikes or faults may be truncated. If the dips above and below a surface are greatly discordant, the surface is an angular unconformity.

All these criteria must be used with caution. Large-scale dune cross-beds, for example, may give the appearance of an angular unconformity in a local outcrop. If the dipping flank beds of a reef or the dipping foreset beds of a delta are truncated by a horizontal erosional surface, the disconformity can appear to be an angular unconformity. Compaction of saturated sediments can cause soft-sediment deformation between layers, which gives the appearance of structural deformation. In a limited outcrop, the discordance between the deformed and undeformed layers might be erroneously interpreted as an unconformity.

Obviously, then, regional studies are necessary to decide whether an apparent erosional surface is an unconformity and, if so, how significant it is. In some cases, regional studies reveal that a disconformity changes into a nonconformity or angular unconformity over distance (see Fig. 15.9E). In the Grand

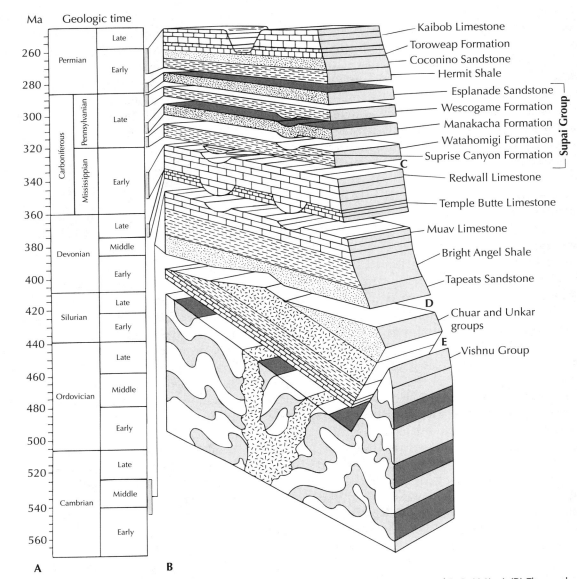

Figure 15.10 Although the Grand Canyon section may appear to the casual viewer to be a complete record of geologic time, it is remarkably incomplete. *Above:* (A, B) The actual temporal span of the Paleozoic units of the Grand Canyon that are shown in B. Less than 10% of Paleozoic time is actually represented, including parts of the Middle Cambrian, a bit of Late Devonian, some Early Mississippian, and three short slices of the Pennsylvanian—but no Ordovician or Silurian. Most of the Early Permian, but no Late Permian, is represented. *Facing page:* (C) The disconformity between the Redwall Limestone and the overlying Pennsylvanian rocks shows a clear paleovalley fill. (Courtesy of E. D. McKee.) (D) The angular unconformity between the tilted Precambrian sediments and the overlying Cambrian deposits is shown. Some of the Precambrian units formed resistant paleotopography in the Cambrian, so that the Cambrian deposits lap over them irregularly. (N. W. Carkhuff; courtesy of U. S. Geological Survey.) (E) Elsewhere in the Grand Canyon, the Cambrian sediments lap nonconformably over Precambrian schists and granites (rocks with vertical foliation.) (D. R. Prothero.)

Canyon, the Cambrian Tapeats Sandstone laps nonconformably over the basement schists (see Fig. 15.10E) and forms an angular unconformity with the tilted Proterozoic metasediments (see Fig. 15.10D). In other cases, a conformable sequence changes into an unconformity over distance, especially as it thins out toward the margin of a basin (Fig. 15.11A). Although relatively continuous deposition takes place in se-

quences A and B in the center of the basin through times T_1 to T_{10}, the margin of the basin is subject to rapid fluctuations in base level (Fig. 15.11B). At about time T_5, base level begins to move downward and seaward (that is, to the right), so that deposition of sequence A ceases on the basin margin. When base level reaches point P at time T_7, it begins to move upward and landward again, preventing strata B_1

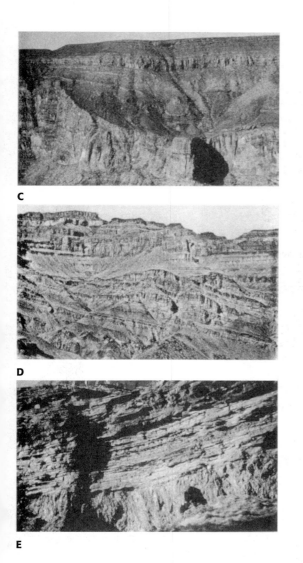

and B_2 from being deposited toward the margin of the basin. Also at time T_7, there is slight uplift of the landward part of sequence A, so that strata E_6 through E_9 are progressively removed by erosion. Eventually, depositional surface NP comes to lie directly on erosional surface LP. Between them is a large time gap in the rock record (a **lacuna**) composed of time intervals when strata were never deposited (a **hiatus** in the sense of Wheeler, 1964) and when strata were removed after deposition (the **degradational vacuity** of Wheeler, 1964). Figure 15.11C shows an example of a lacuna composed of both a hiatus in the Niobrara Chalk and a degradational vacuity in the underlying Carlile Shale.

Catastrophic Uniformitarianism

In his provocative book *The Nature of the Stratigraphical Record* (1981), Derek Ager debunks many working assumptions about the rock record that are

common among geologists, especially the "myth of continuous sedimentation." Areas that have apparently continuous records (see Fig. 15.6A) also have very low rates of accumulation (millimeters per year). At these rates, it would take 200 years to bury a typical sea urchin found in the Cretaceous chalks. The continuity in such stratigraphic sections is clearly an illusion because sections that span the same time interval in other basins are often orders of magnitude thicker. One centimeter in a so-called continuous section can be represented by tens of meters of section in some other basin. Each interval of time is represented by a very thin section or, more likely, by no rock at all. Ager shows that even "classic" sections have obvious and obscure unconformities, and he presents an unconventional picture of what even the "best" sections represent (Fig. 15.12). His metaphor for the accumulation of the sedimentary record is not the conventional "gentle, continuous rain from heaven" but is more like the child's definition of a net: "A lot of holes tied together with string. The stratigraphical record is a lot of holes tied together with sediment" (Ager, 1981, p. 35). It is like the life of a soldier, "long periods of boredom and brief periods of terror" (Ager, 1981, pp. 106–107).

Part of the reason for the discontinuities in the record is the nature of sedimentation itself. The extreme gradualist bias that geologists inherited from Lyell caused them to overemphasize the gradual accumulation of sediment from daily processes and to ignore the large-scale catastrophic events. Perhaps the only place where sedimentation bears any resemblance to the conventional concept of a "gentle, continuous rain from heaven" is the deep seafloor, where planktonic microfossils and clay particles rain down continuously from the surface waters. Even here, though, scouring by bottom currents and dissolution caused by oceanographic changes are taking place (see Chapter 10). In terrestrial and shallow marine realms, permanent preservation occurs only when a local basin drops below the base level (see Fig. 15.6A). Most sedimentary environments do not show permanent net accumulation of sediment; exceptions are environments that have a high chance of preservation, such as deltas, downdropping basins and grabens, and growing reef complexes. The importance of gradual accumulation is probably overemphasized because rates of accumulation are so low. Ager argues that gradualistic biases have led geologists to neglect the importance of rare catastrophic events. For example, during normal conditions, the Gulf of Mexico coastal region accumulates sediment

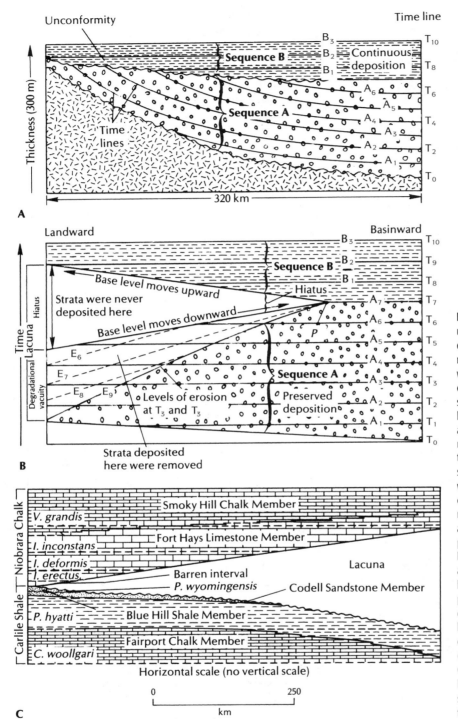

Figure 15.11 The origin and history of an unconformity are revealed in cross sections at the edge of a sedimentary basin. (A) Physical relationships between hypothetical rock units, in which sequence A rests nonconformably on eroded igneous rocks and is in turn overlain by sequence B. There is continuous deposition between A and B in the basin center (*right*), but an unconformity develops toward the basin margin. (B) The thickness axis is replaced by a time axis, so the curved time lines in A become straight. As deposition of sequence A continues without interruption from times T_0 to T_7 in the center of the basin, uplift on the basin margin moves the base level downward and erodes strata along erosional surfaces E_6 through E_9. This produces a degradational vacuity. At point P, the base level reverses direction and starts to rise again. However, little of sequence B is preserved because its base level starts out so low. This produces a hiatus. The combination of a degradational vacuity and a hiatus produces a lacuna. (After Wheeler, 1964: 606; by permission of the Geological Society of America.) (C) Example of a stratigraphic lacuna from the Cretaceous of Kansas. Time planes are provided by biostratigraphic zone boundaries. The stratigraphic gap becomes progressively larger toward the right. (After Hattin, 1975.)

at an average rate of 10 cm per 1000 years. Yet any given segment of coast has a 95% probability of being hit every 3000 years by a major hurricane, which scours much deeper than the 30 cm accumulated since the last hurricane. Much of the stratigraphic record in this region may be accumulated storm deposits. The daily, uniformitarian processes of the marine environment serve only to rework the top of the last storm deposit, and the effects of these processes are usually wiped out by the next storm. Ager cites many other examples of catastrophic events that have disproportionate effects on the stratigraphic record. Among the best-known examples are turbidites and olisthostromes (discussed in Chapter 10), which slump or flow down the continental slope in a matter of minutes or hours yet make up most of the strati-

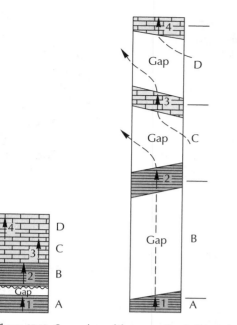

Figure 15.12 Comparison of the conventional picture of a part of the stratigraphic record (*left*) with the probable true picture (*right*). The gaps also conceal the immigration and emigration of fossil taxa (1, 2, 3, 4), giving the impression that they range continuously through time (left). (After Ager, 1981.)

ural catastrophism. Ager (1981) emphasized the importance of rare, high-energy events, but Dott pointed out that even normal fair-weather processes are inherently discontinuous and episodic. For example, a prograding shoreline (Fig. 15.13A) deposits a relatively continuous sequence offshore; but as the shelf, foreshore, beach, and dunes make their way across a given area, they are subject to more energetic and variable conditions. Hence, their record is more episodic. Similarly, the distal parts of the submarine fan (Fig. 15.13B) are influenced by the almost steady rain of pelagic sedimentation. Because channels of the inner fan are products of episodic turbidity current activity, however, there are significant gaps between them. The geologist must be alert to possible gaps wherever there is a small break in the bedding of an outcrop. Continuity may well be an illusion.

Correlation

There is no place on Earth's surface where the entire geologic record is represented by a single section. Even such classic exposures as the Grand Canyon lack all but small parts of the Precambrian and Paleozoic (see Fig. 15.10). The classic stratigraphic column of England and Wales is slightly more complete, but many important intervals are missing. In most areas, the stratigraphic record must be patched together from many short, local sections. The process of demonstrating the equivalence or correspondence of geographically separated parts of a geologic unit is called **correlation.**

The word *correlation*, like the word *facies*, has had a number of contradictory and confusing meanings. Before the 1960s, geologists often used the term

graphic record of those marine environments. As Ager (1981, p. 50) points out, "The periodic catastrophic event may have more effect than vast periods of gradual evolution." He calls this the Phenomenon of Quantum Sedimentation, or "catastrophic uniformitarianism." It is uniformitarian in that natural laws and processes are invoked (rather than the supernatural processes of biblical catastrophists), but it denies Lyell's gradualism of rates.

Dott (1983) suggested that the term **episodic sedimentation** be used to avoid confusion with supernat-

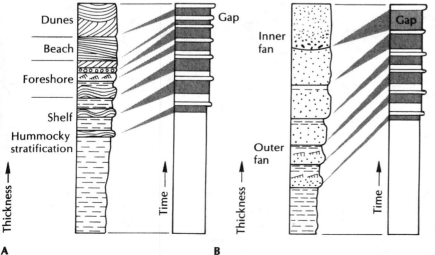

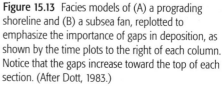

Figure 15.13 Facies models of (A) a prograding shoreline and (B) a subsea fan, replotted to emphasize the importance of gaps in deposition, as shown by the time plots to the right of each column. Notice that the gaps increase toward the top of each section. (After Dott, 1983.)

correlation to mean *time equivalence*. As we will see, geologists today think that very few rock units are time-equivalent over significant distances. To avoid the problems of proving time equivalence, most modern geologists define correlation as the equivalence of lithologic units, without time implication. Usually, it is clear from the context whether lithologic correlation or time correlation is meant. In this section, we concentrate on lithologic correlation.

Establishing lithostratigraphic correlation can be simple, but it is usually very complex. The simplest method is to establish physical continuity between exposures, either by "walking it out" in the field or by tracing it on maps or cross sections. Outcrops are rarely extensive enough to allow us to trace beds continuously, however, so other procedures are necessary. Some rock units have distinctive diagnostic features that make them easy to recognize in different outcrops. If there is a strong lithologic similarity between two discontinuous exposures that lie in the same position in the sequence, they can be tentatively correlated.

Sometimes the sequence itself is distinctive enough that clear correlations can be made, even though some units may no longer show lithologic similarity because of facies change. In this case, their position in the sequence is adequate for correlation (Fig. 15.14).

Other criteria have been used in the absence of better means of correlation. Unconformities can be obvious markers that span long distances, and there is evidence (discussed later) that many regional unconformities are worldwide and synchronous. Some unconformities are excellent tie points between sections, but unfortunately, the rocks above and below an unconformity are probably not correlative. As we saw earlier, the lacunae above and below an unconformity can increase in size over distance (see Fig. 15.11B). If the erosional surface cuts through a sequence of units, then the units below the unconformity are of different ages in different places. In some regions, structural deformation and metamorphism occurred at restricted times in geologic history. In such cases, the degree of structural develop-

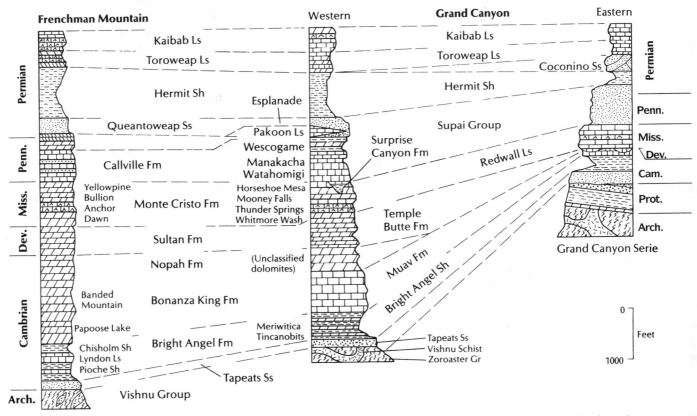

Figure 15.14 Correlation between the eastern and western Grand Canyon Paleozoic sequences (*right*) and the comparable sequence at Frenchman Mountain, just east of Las Vegas, Nevada. Note that most units can be correlated by similarity in lithology, even if they change names and thicknesses over distance. Other units can be correlated by similarity in position between units that do not change from one exposure to the other. For example, the Pennsylvanian rocks are limestones in the west and sandy shales of the Supai Group in the Grand Canyon, but they are overlain by the Hermit Shale and underlain by the Mississippian limestones in both places, which establishes their approximate correlation. (From Bachhuber, Rowland, and Huntoon, 1987.)

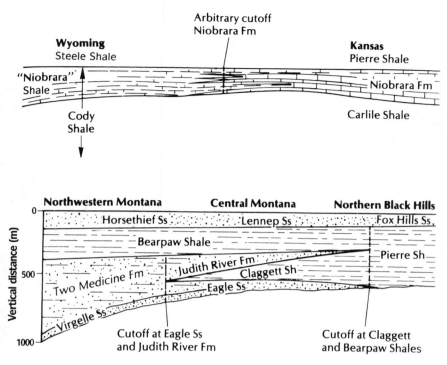

Figure 15.15 Facies change and stratigraphic cutoff. In the top diagram, the chalk and limestone of the Niobrara Formation of Kansas are shown to pass, through intertonguing and gradation, into a shale in Wyoming. Because the two lithologies are not given the same name, there must be an arbitrary cutoff along the line at which the unit loses its predominantly carbonate lithology. In the lower diagram, the intertonguing interval between the Pierre and Bearpaw shales is subdivided into distinct formations. This requires an arbitrary cutoff between the Claggett, Bearpaw, and Pierre shales, and between the Judith River, Eagle, and Two Medicine sandstones. (After Krumbein and Sloss, 1963.)

ment or metamorphism can be used for tentative correlation in the absence of better indicators.

Correlation is seldom simple or straightforward. Some of the problems are purely artificial. For example, when geologists have different concepts of what constitute good marker beds on which to base their boundaries between units, the boundaries can change arbitrarily. This can lead to absurdities. Suppose that two state geological surveys define the boundary of a rock unit differently. The boundaries are mapped consistently within both states, but at the state line, the boundary is "offset" by the difference in definition. Regional maps that attempt to compile information from several states often show these "state-line faults." An unwary map user might be misled into thinking that real faulting had occurred precisely on the state line.

Other problems arise from real changes in the rocks. A facies change can cause one lithology to change gradually into another, and the boundary between them, the **stratigraphic cutoff,** must be arbitrary. Sometimes facies changes and intertonguing cause groups of units to merge or split; then the scheme of stratigraphic cutoffs must be even more arbitrary and complex (Fig. 15.15).

Time Correlation

Catastrophist geologists saw each rock layer as a phase of the receding Noachian floodwaters and therefore considered a layer to be isochronous, or time-equivalent, wherever it was recognized. The concept of facies change showed that rock types are not good time markers, but layer-cake notions still persisted. When facies analysis and biostratigraphy were applied to some classic sections, surprising conclusions often emerged. For example, the famous Catskill sequence of upstate New York (Fig. 15.16) was once divided into units based on gross lithology. The "Catskill Group" was coarse-grained sands, gravels, and redbeds; the "Chemung Group" was all the sandy shales; and so on. These rock units were placed in vertical succession, and each was thought to represent a different time period. Detailed tracing of the rock units and their fossils, however, later demonstrated that the time lines cut across the lithologic units. Each of the old "groups" was markedly time-transgressive, spanning much of the Devonian.

In North America, layer-cake thought persisted well into the 1950s, primarily because of the influence of E. O. Ulrich (1857–1944). As described by Dunbar and Rodgers (1957, pp. 284–288), Ulrich considered the European ideas of facies changes to be relatively unimportant in his work on the Paleozoic of North America. He viewed each rock unit and its distinctive fauna as the product of one isolated advance of the shallow seas over the continental interior. Thus, he interpreted the units in the Devonian Catskill sequence as the products of several separate advances rather than as one continuous facies

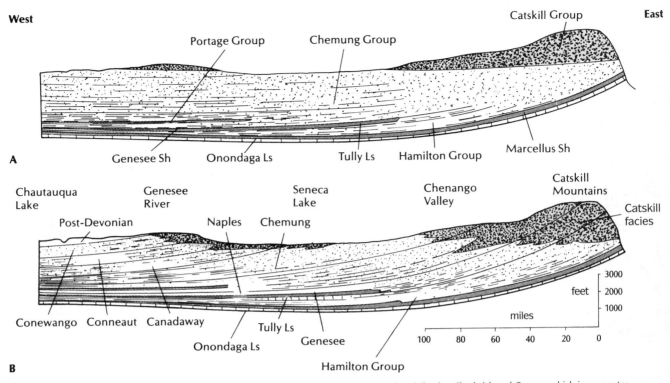

Figure 15.16 East-west cross sections across the Devonian Catskill sequences of southern New York. (A) Traditional interpretation (pre-1930), which treated the units as a layer-cake sequence of Hamilton and Portage shales, Chemung sandstones, and Catskill redbeds. (B) Present interpretation, following Chadwick and Cooper, which incorporates modern concepts of facies change. Time planes are shown by the curved lines; each unit consists of Catskill redbeds in the east, sandy facies in the center, and shales in the west. (After Dunbar and Rodgers, 1957.)

sequence. For him, any difference in fauna or lithology, no matter how trivial, was proof of difference in age and evidence of an unconformity.

Naturally, this reasoning produced correlations and paleogeographic reconstructions that seem bizarre today. In every example in Ulrich's 1916 paper, for instance, he was exactly wrong. It took decades for North American geologists to undo the damage his erroneous correlations had done. His influence was so great that his layer-cake ideas still lurk behind the work of North American geologists, more than 50 years after his work was discredited.

Contrary to Ulrich's ideas, rock units can be noticeably diachronous, even over short distances. Gazing into the Grand Canyon, one sees cliffs of Tapeats Sandstone, slopes of Bright Angel Shale, and more cliffs of Muav Limestone above the Precambrian basement (see Fig. 15.10). From the canyon rim, these units look continuous and uniform in thickness, so it seems reasonable to think in terms of "the time when the Tapeats Sandstone was deposited." The stratigraphic distribution of trilobites, however, shows that even over a distance of only 180 km, all these formations are markedly time-transgressive (Fig. 15.17). The

Muav Limestone in the west is equivalent in age to the Tapeats Sandstone in the east.

In his influential book, *Time in Stratigraphy* (1964), Alan Shaw attacked the problem of the time significance of rock units. He developed a model for the sedimentation in the broad, epeiric seas that produced most of the stratigraphic record of continental cratons. These epeiric seas must have been very broad and shallow, extending for hundreds of kilometers with slopes of only centimeters per kilometer. Restricted water circulation due to limited wave, current, and tide action would have produced a regular series of facies belts (Fig. 15.18). During transgression or regression, these facies belts would shift back and forth relative to the shoreline, each belt producing a characteristic rock unit that is time-transgressive throughout. From this, Shaw (1964, p. 71) concluded that "all laterally traceable nonvolcanic epeiric marine sedimentary rock units must be presumed to be diachronous."

Ager (1981, p. 62) presents a modern analog of the formation of diachronous deposits. On the coast of the Persian Gulf, the facies belts consist of subtidal lagoonal muds, intertidal algal mats, and supratidal

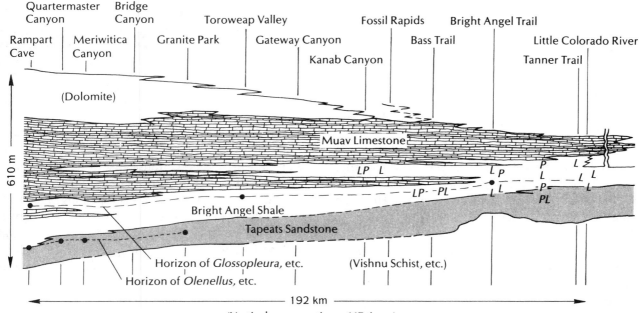

West

East

Quartermaster Canyon

Bridge Canyon

Rampart Cave

Meriwitica Canyon

Granite Park

Toroweap Valley

Gateway Canyon

Kanab Canyon

Fossil Rapids

Bass Trail

Bright Angel Trail

Little Colorado River

Tanner Trail

(Dolomite)

610 m

Muav Limestone

Bright Angel Shale

Tapeats Sandstone

Horizon of *Glossopleura*, etc.

Horizon of *Olenellus*, etc.

(Vishnu Schist, etc.)

192 km

(Vertical exaggeration = 117 times)

Figure 15.17 Time relationships of transgressing Cambrian marine sediments in the Grand Canyon. The dotted lines indicate biostratigraphic time planes showing that the Tapeats Sandstone in the east is equivalent in age to the Muav Limestone in the west; L and P denote *Lingulella* and *Paterina* fossils. (After McKee and Resser, 1945.)

sands and evaporites of the sabkha, or supratidal salt flats (Fig. 15.19). A pit dug through the sabkha about 6.5 km from the shore revealed an algal mat similar to the one forming today at the tide line. Radiocarbon dating of this mat gives ages of 4000 years, or approximately 460 years of diachronism per kilometer (a little more than 2 m/yr). Ager emphasizes that the layer-cake metaphor of sedimentation as a "continuous rain from heaven" must be replaced with the concept of **lateral sedimentation,** or "the moving finger writes." In Ager's words (1981, p. 61), "If it looks the same it must be diachronous."

Another study of the lateral migration of coastal environments (sand spits, bay bottoms, and tidal marshes) showed different results (LeFournier and Friedman, 1974). Based on the changing positions of dikes in France, rates of lateral migration of 10 m/yr were estimated. This is five times faster than the rates

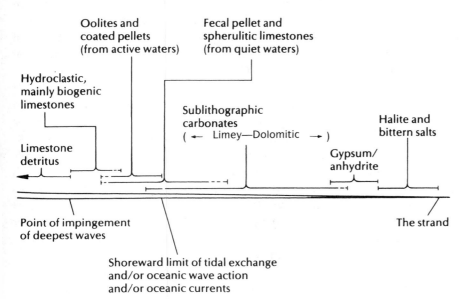

Oolites and coated pellets (from active waters)

Fecal pellet and spherulitic limestones (from quiet waters)

Hydroclastic, mainly biogenic limestones

Sublithographic carbonates (← Limey—Dolomitic →)

Halite and bittern salts

Limestone detritus

Gypsum/ anhydrite

Point of impingement of deepest waves

The strand

Shoreward limit of tidal exchange and/or oceanic wave action and/or oceanic currents

Figure 15.18 Shaw's model of the standard facies belts in a limey epeiric sea, with vertical scale greatly exaggerated. Because the lithologic types are contemporaneous and laterally adjacent, each could be expected to form time-transgressive rock units as the seas transgress and regress. (After Shaw, 1964.)

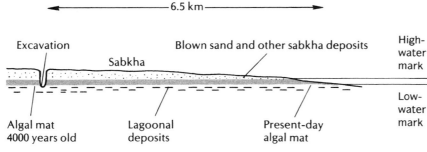

Figure 15.19 Ager used the example of an algal mat in the Persian Gulf to demonstrate diachronism. The mat is forming on the surface today, but its lateral equivalent — buried beneath the sabkha — can be radiocarbon dated at 4000 years old. The algal mat transgressed only 6.5 km in 4000 years. (After Ager, 1981.)

calculated by Ager (1981). Rates calculated from the migration of sandy spits in the barrier island sequence on the Atlantic shores of Long Island and New Jersey are even faster (Kumar, 1973; Sanders, 1970). The Long Island spit complex migrated at 65 m/yr (see Fig. 9.20), and the point of Sandy Hook in New Jersey has moved about 12 m/yr. The Mississippi Delta migrates at a rate of 75 m/yr (Gould, 1970). The rates are several orders of magnitude faster than the rates calculated by Ager and six to seven times faster than the migration of French coastal facies.

On the scale of historic time, a lateral migration of millimeters to meters per year seems very slow, and sedimentary units are significantly diachronous. But on the scale of geologic time, these rates are so fast as to be nearly instantaneous. As we saw in the previous section, events separated by a few thousand years to hundreds of thousands of years can seldom be resolved in the geologic record. The geologist must keep two nearly contradictory concepts in mind when thinking about the time significance of a lithologic unit. On the scale of a local depositional environment and a short time span, rock units are markedly time-transgressive. Point-bar sequences, delta fronts, prograding barrier islands, shifting tidal channels, and spits are but a few of the many examples of depositional systems that migrate laterally and produce bodies of uniform character that transgress time. On a scale of millions of years, however, this amount of change cannot be resolved. With such poor time resolution, most rock units seem roughly isochronous.

Most of the ancient examples of diachronism that have been studied, however, show that over long distances rock units transgress large amounts of time, usually millions of years. This apparent discrepancy with modern rates of migration can be explained by the zigzag pattern of transgression. Episodes of transgression can indeed be as rapid as modern rates suggest. As shown by Barrell's diagram (see Fig. 15.7), however, transgressions need not be continuous. Instead, a short episode of transgression can be followed by a regression that reduces the net lateral shift

of the shoreline. As this zigzag "dance" of transgression and regression continues to move landward, it produces the typical intertonguing found at the base of many rock units (see, for example, Fig. 15.17). The basal deposits of a single transgressive episode over a very short distance (less than a few kilometers) might indeed be synchronous to the extent that the dating can resolve the age of these deposits. Over any significant distance, however, the base of the unit is significantly younger and therefore time-transgressive.

When approaching a contact between formations in an outcrop for the first time, a geologist cannot be sure whether the contact is roughly synchronous or significantly transgresses time. Ideally, the geologist should use whatever means are available to obtain an estimate of local rates of lateral migration for use on the problem at hand. The beds may be significantly time-transgressive, or the scale of resolution may be so coarse and the distances so short that their ages are indeed "close enough that the difference doesn't matter" (Shaw, 1964).

The Nature of the Control

Transgressive-regressive packages bounded by unconformities make up the sedimentary record of most basins on cratons and continental margins. Some of these unconformities are small; others span great distances and long periods of time. Major unconformities divide the stratigraphic record of every continent into discrete packages, which Sloss, Krumbein, and Dapples (1949) and Sloss (1963) called **sequences** (Fig. 15.20). These major unconformity-bounded packages of sediment are thought to represent Earth's large-scale tectonic or eustatic events, each of which persisted for tens of millions of years. The sequences unite natural packages of sedimentary history that may span system boundaries. For example, the Lower Ordovician in most of North America is the culmination of the long Cambrian transgressive episode, which Sloss called the "Sauk Sequence." The rest of the Ordovician record is part of another transgressive

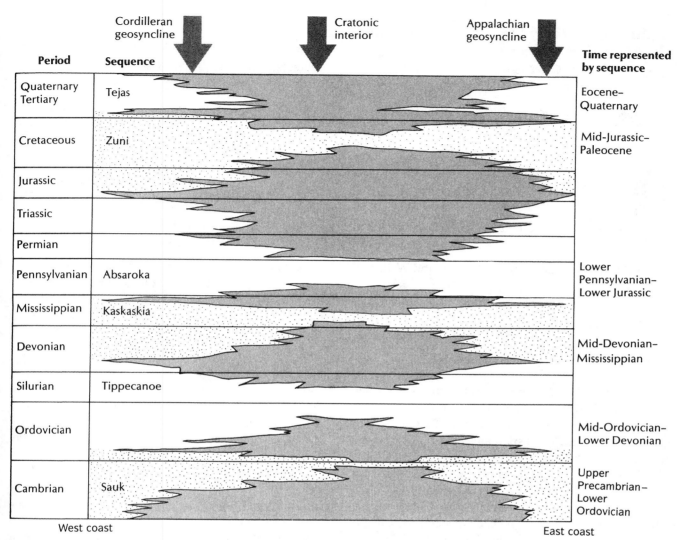

Figure 15.20 Sloss's diagram of the time-stratigraphic relationships of unconformity-bounded sequences in North America. Dark areas represent large gaps in the stratigraphic record, which become smaller toward the continental margins. Light areas represent strata. These "Sloss sequences" have also been recognized in Russia and South America. (After Sloss, 1963: 110; by permission of the Geological Society of America.)

episode that continues into the Silurian, called the "Tippecanoe Sequence." The Sauk-Tippecanoe "boundary" represents a major event in North American history, but the Cambrian-Ordovician and Ordovician-Silurian system boundaries are not marked by any major tectonic or eustatic events in North America. Therefore, the sequences are more useful for describing the geologic history of North America than are European terms such as Cambrian, Ordovician, and Silurian.

Further analysis of these sequences reveals that they are not restricted to North America. A similar pattern of unconformities is found on the Russian Platform (Sloss, 1972, 1978) and in South America (Soares, Landim, and Fulfair, 1978). There is also a spectrum of small-, medium-, and large-scale unconformities preserved in the record of each continent. In some cases

(Busch and Rollins, 1984; Ross and Ross, 1985), these unconformities can be matched up between continents. According to these authors, the Carboniferous and Permian records of most of the major continents show a similar pattern of cyclic transgression and regression that can be matched, cycle for cycle, around the globe. These global similarities in transgressive-regressive cycles are unlikely to have been controlled by local tectonics but must have a global cause. The only reasonable controlling agent for this kind of change is global sea level.

Similarly, the seismic evidence from the passive margins of the world (discussed in Chapter 17) shows a similar pattern of small-, medium-, and large-scale sea-level cycles of onlap and offlap. As defined by Peter Vail and others (Vail et al., 1977a, b; Haq, Hardenbol, and Vail, 1987, 1988), there are four types of

TABLE 15.1 Stratigraphic Cycles and Their Probable Causes

Type[a]	Other Terms	Duration (million years)	Probable Cause
First-order	Supercycle (Fischer, 1981)	200–400	Major eustatic cycles caused by formation and breakup of super-continents
Second-order	Sequence (Sloss, 1963) Synthem (Ramsbottom, 1979)	10–100	Eustatic cycles induced by volume changes in global mid-ocean ridge system
Third-order	Mesothem (Ramsbottom, 1979)	1–10	?Crustal flexures or changes in the geoid
Fourth-order	Cyclothem (Wanless and Weller, 1932)	0.2–0.5	Rapid eustatic fluctuations induced by the growth and decay of continental ice sheets, growth and abandonment of deltas

[a]Vail, et al., 1977a, b.

cycles (Table 15.1). **First-order cycles** (the **supercycles** of Fischer, 1981) span hundreds of millions of years. **Second-order cycles** (the **synthems** of Chang, 1975, and Ramsbottom, 1979) span tens of millions of years and produce the major unconformities shown in Fig. 15.20 and Sloss's sequences. **Third-order cycles** last only a few million years. **Fourth-order cycles,** last hundreds of thousands of years and correspond to the classic cyclothems of the late Paleozoic. Busch and Rollins (1984) and Busch and West (1987) recognize even smaller-scale patterns. In their scheme, fourth-order cycles—the **mesothems** of Chang (1975) and Ramsbottom (1979)—have durations of 600,000 years to 3 million years; **fifth-order cycles** (traditional cyclothems) span 300,000 to 500,000 years; and **sixth-order cycles** (punctuated aggradational cycles of Goodwin and Anderson, 1985) range from 50,000 to 130,000 years long.

First-Order Cycles

It seems clear that first-order cycles are related to major plate movements (Worsley, Nance, and Moody, 1984). For example, the suturing of all continents into the Pangaea supercontinent in the Permian coincides with a long period of continental emergence, and the rapid breakup and separation of continents in the Cretaceous corresponds to a long period of continental inundation. Fischer (1981, 1982, 1984) suggested that Earth has gone through two complete 300-million-year supercycles between what he calls ice-

house and greenhouse states (Fig. 15.21). During icehouse states (late Precambrian, late Paleozoic, and late Cenozoic), there were extensive polar ice caps, lower sea levels, and steep temperature gradients between the poles and the equator. Although there was a thin layer of warm water at the surface, the bulk of the ocean was composed of cold water masses produced at the poles, and the mean temperature of the ocean was about 3°C.

Greenhouse states, on the other hand, had no polar ice caps, and so sea level was higher. This implies that the temperature gradient between poles and equator was much less steep and that the ocean was also less stratified, with an average temperature of about 15°C (almost 10°C warmer than today). Fischer suggested that the greenhouse phase is triggered by increased carbon dioxide in the atmosphere (the greenhouse effect), which traps solar radiation and warms the Earth (Barron and Washington, 1982, 1985). He pointed out that peaks of volcanism are correlated with the greenhouse phase of the cycle (see Fig. 15.21), which suggests that increased mantle activity during these periods releases large amounts of greenhouse gases from the mantle into the atmosphere by means of increased volcanism. When this mantle "overturn" slows, weathering on the land withdraws carbon dioxide from the atmosphere until its rate of supply by volcanism is balanced by the rate of withdrawal by the lithosphere. This low-level balance corresponds to the icehouse phase of the supercycle. The growth of a

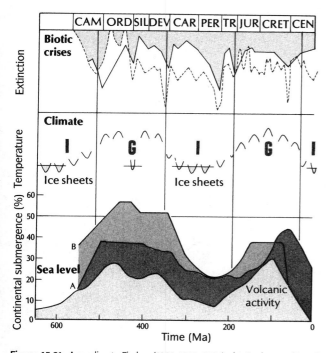

Figure 15.21 According to Fischer (1981, 1982, 1984), the Earth goes through 300-million-year supercycles between icehouse (I) and greenhouse (G) states. These first-order cycles are shown by the alternating phases of climate. The greenhouse phases correlate with peaks in volcanism and global sea-level highs, shown in two versions in curves A and B. Two versions of the curve of extinctions of marine organisms are also shown (solid and dashed lines). Many mass extinction events ("biotic crises" along the top) also appear to be associated with the transitions between phases of the supercycles. (After Fischer, 1982b: 132; © 1984 by Princeton University Press.)

glacial armor over the land (preventing further withdrawal of carbon dioxide by lithospheric weathering) and possibly increased mantle convection have apparently prevented the Earth from becoming a ball of ice like some other planets.

Fischer's models have many appealing features, but they are based on only the last 700 million years of Earth's history. With only two complete icehouse-greenhouse cycles, it is difficult to prove that a cyclic causation mechanism really exists. Two greenhouse phases alternating with three periods of glaciation could easily result from coincidental causes that have no underlying cyclic mechanism. Crowell and Frakes (1970) have shown that most of the icehouse phases of Earth's history coincide with times when the continents were near the poles. In addition, there are exceptions that are difficult to explain by these cycles, such as the Late Ordovician glaciation, which occurred near the peak of a supposed greenhouse phase. Recent studies of the Ordovician glaciation, however, have shown that it was very short and was restricted to the North African part of Gondwanaland, so it may not

invalidate the large-scale greenhouse concept (Crowley and Baum, 1991; Brenchley et al., 1994).

Second-Order Cycles

Many of the second-order cycles are thought to be related to changes in mid-ocean ridge volume and seafloor spreading rates. During periods of rapid seafloor spreading, the greater volume of recently erupted and intruded oceanic crust that has not yet cooled and contracted increases the ridge volume (Fig. 15.22). During periods of slow seafloor spreading, the crust has more time to cool and occupies less volume. There seems to be good correspondence between spreading rates and the frequency and magnitude of sea-level changes during the Cretaceous (Hallam, 1963; Hays and Pitman, 1973; Pitman, 1978). However, ridge volume changes are too slow to account for cycles of much less than 10 million years.

Sheridan (1987a, 1987b) suggested that the cycles of fast and slow seafloor spreading may be controlled by cyclic changes of mantle convection, operating on a scale of 30 to 60 million years (shorter than the cycles postulated by Fischer). He pointed out that the periods of fast spreading and high sea level are correlated with periods of high magnetic field activity, which is controlled by the Earth's core and mantle (Fig. 15.23A). Most of the periods of high sea level since the Cambrian are correlated with periods of stable magnetic field polarity (Fig. 15.23B), with a phase lag of about 10 million years. Vogt (1975) and Sheridan (1983, 1987a, 1987b) proposed a model (Fig. 15.23C) in which Earth alternates between two phases. When its core is relatively cool and quiet, there is a weak magnetic field and few reversals; to compensate for this decreased core energy and to maintain the energy balance of the Earth, the mantle undergoes rapid, smooth convection in a series of well-developed convection cells and mantle plumes (Fig. 15.23C, time = T_1). As the core heats up and becomes more turbulent, it produces many magnetic field reversals and a strong magnetic field; the mantle, on the other hand, becomes relatively cooler, and the convection and plate motion slow down, with few discrete convection cells (Fig. 15.23C, time = T_3).

Mound and Mitrovica (1998) suggested another, possibly related, mechanism for the second-order cycles. They proposed that true polar wander (long-term wander of Earth's rotational pole) occurs over very long time spans (tens of millions of years). When Earth's rotational pole changes, the crust must respond by slow viscoelastic motions upward and downward, which would cause eustatic changes. In

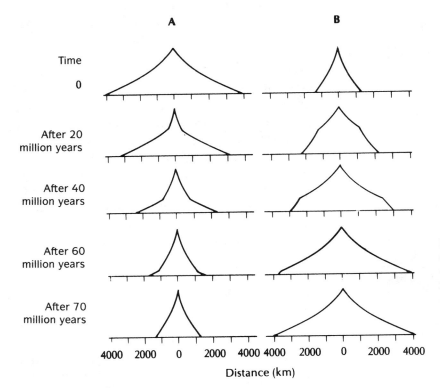

Figure 15.22 Profiles of fast- and slow-spreading mid-ocean ridges through 70 million years of spreading. (A) Ridge that had been spreading at 6 cm/yr slows to 2 cm/yr. After 70 million years, it has one-third its original volume, and epeiric seas return to the ocean basins. (B) Ridge that had been spreading at 2 cm/yr changes to 6 cm/yr, greatly increasing its volume, displacing water, and causing a sea-level rise. (After Pitman, 1978: 1391; by permission of the Geological Society of America.)

many ways, this hypothesis is comparable to the postulated "inertial true polar wander" event of the Early Cambrian (Kirschvink et al., 1997), which suggests that there was a mass imbalance of continents on the Cambrian Earth, resulting in extraordinarily fast plate motions as the masses redistributed to more stable equatorial positions. However, the scale of the late Mesozoic-Cenozoic polar wander postulated by Mound and Mitrovica (1998) is not so extreme as the Cambrian event.

Third-Order Cycles

Cycles with durations of 1 to 10 million years (but typically less than 3 million years) are known as third-order cycles. Although third-order cycles are ubiquitous in the Phanerozoic, it is not yet established that they are globally synchronous, since they are usually shorter than the resolution of biostratigraphy. If they are not synchronous, they might be caused by local tectonic processes; in this case, a mechanism that causes global sea-level change might not be required to explain these cycles.

Haq, Hardenbol, and Vail (1988) and Vail et al. (1977a, b) attributed third-order cycles to the waxing and waning of glaciers, but most glacial changes occur faster than this and are probably responsible for fourth-order cycles (discussed shortly). Kauffman (1984) noticed that third-order marine transgressions in the Cretaceous strata of the Western Interior corre-

sponded to tectonism and volcanism, whereas regressions were marked by tectonic and volcanic quiescence. Since then, a number of authors have suggested that short-term changes in the Earth's surface might be responsible for cycles on a scale of a few million years. For example, Cloetingh (1988) argued that episodic changes in the Earth's crustal stress field resulting from the jostling of tectonic plates could produce uplifts and subsidence of tens of meters in a few million years. Mörner (1981, 1987a, 1987b) pointed to satellite evidence that Earth's shape, or geoid, has "sags" and "bulges" with an amplitude of up to 180 m. The migration of these sags and bulges could result in global sea-level changes. As Devoy (1987) and Christie-Blick, Mountain, and Miller (1990) point out, however, the geoidal sags and bulges move too slowly for third-order cycles and would lift both the continents and ocean basins, producing no net change in sea level.

Clearly, the third-order cycles have not yet been fully explained, although many interesting ideas have been proposed. If these cycles are not truly global, then local tectonic and crustal changes may be all that is necessary to explain them.

Fourth-Order Cycles

Global sea-level changes on the scale of a few hundred thousand years (fourth-order cycles) are thought to be caused by changes in global ice volume. During the maximum Pleistocene ice advance, so much seawater

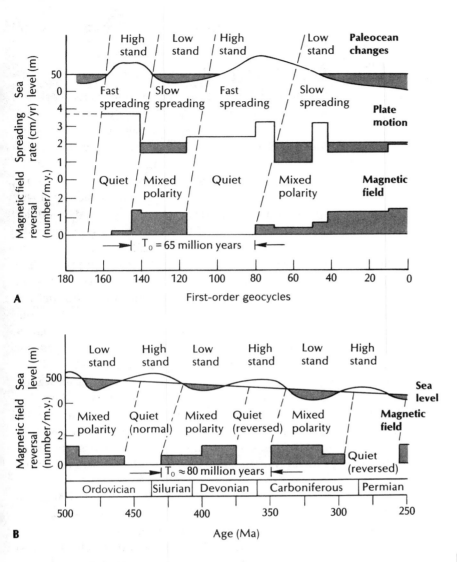

A First-order geocycles

B Age (Ma)

C Fundamental period: $T_0 = 65$ million years
Phase lag: $\Delta T = T_1 - T_3 = 10$ million years

Figure 15.23 Sheridan (1987a, 1987b) has shown a correlation between the activity in Earth's core (as inferred from magnetic field behavior) and mantle and the changes in spreading rate and sea level. (A) The correlations among sea level, spreading rate, and magnetic field behavior over the last 180 million years. (B) Although there is no way to calculate spreading rate prior to the Jurassic, the correlation between field behavior and sea level can be traced back to the Ordovician. In both A and B, the cycles take about 65 to 80 million years and have a phase lag of about 10 million years between the magnetic field behavior change and the corresponding sea-level change. (C) Sheridan's model for these correlations. At time T_1, the core is cooler, with a weak magnetic field and few reversals. To compensate, the mantle becomes hotter and develops strong convection and faster plate motion. By time T_3, the core heats up and develops strong convection, resulting in many field reversals and a stronger magnetic field. The mantle, by contrast, convects less and plate motion is reduced. (After Sheridan, 1987a: 64–67; © American Geophysical Union.)

was trapped in the ice caps that sea level was 100 m lower than it is now. If the present ice caps melted, sea level would rise by 40 to 50 m (Donovan and Jones, 1979). That adds up to about 150 m of net sea-level change at rates as rapid as a few centimeters per year. It seems clear that the fourth-order cycles are glacially controlled. Indeed, the evidence for glacial control of late Paleozoic global sea level is strong (Crowell, 1978), and the glacial control of late Cenozoic sea-level changes is equally well documented (Kennett, 1982). Nevertheless, global changes in ice volume do not explain the well-known sea-level cycles of the Jurassic and Cretaceous, when there were no ice sheets, so far as we know. Several authors (Barron, Arthur, and Kauffman, 1985; Fischer, Herbert, and Premoli-Silva, 1985) have proposed models that explain these cycles in terms of alternating wet (possibly cooler) and dry (possibly hotter) climates.

But what controls ice volume and glacial-interglacial cycles? Since 1976, we have known that most of the Pleistocene temperature variations were controlled by changes in the amount of sunlight Earth receives. These changes are caused by variations in Earth's orbital geometry that affect its distance from and orientation to the Sun (Hays, Imbrie, and Shackleton, 1976; Imbrie and Imbrie, 1979; Denton and Hughes, 1981; Fischer, 1986). The effects of these **Milankovitch cycles** were once known exclusively from the Pleistocene, but analysis of the classic Carboniferous coal-bearing cyclothems has shown a periodicity that may be explainable by the Milankovitch cycles (Heckel, 1986). Milankovitch cycles have also been invoked to explain the periodicities found in Triassic lake sediments (Olsen, 1984) and in Cretaceous marine sediments (Fischer, Herbert, and Premoli-Silva, 1985; Fischer, 1986). Goodwin and Anderson (1985) attributed the periodicity

of cycles in the Devonian of New York to the Milankovitch cycles.

Event Stratigraphy

Whether or not Milankovitch mechanisms work for most other short-term cycles, one thing is becoming clear: *cycles controlled by eustatic changes have time significance.* If transgressions and regressions are induced by simultaneous worldwide changes in sea level, then the peaks (maxima) of transgressions and regressions are more or less synchronous worldwide. Israelsky (1949) first introduced the idea that the correlation of lithologies by their position in a cycle can have time significance, and Ager (1981, p. 70) called such correlation **event stratigraphy.** During a typical transgressive-regressive cycle (Fig. 15.24), lithostratigraphic units are clearly diachronous. But different lithologies that are formed at the same point in a cycle (such as the deposits during peak transgression) are approximately isochronous. The correlation of *different* lithologies that represent the *same* point in the cycle (the event) has a greater chance than any other technique of indicating time significance. (The fossils in this instance are either scarce or highly facies-controlled.) Event stratigraphy offers a method of worldwide correlation of eustatic cycles, with some assurance that they are isochronous. However, correlations of beds other than those at the peaks of cycles are much less certain.

If the correlation of the peak of a single transgression or regression with other such peaks is uncertain, the larger-scale pattern may suggest a correlation. Busch and Rollins (1984) and Busch and West (1987) argue that the pattern of cycles nested within (or superimposed upon) larger-scale cycles can be correlated over large areas. For example, the Carboniferous of the North American midcontinent (particularly Kansas) and that of the northern Ap-

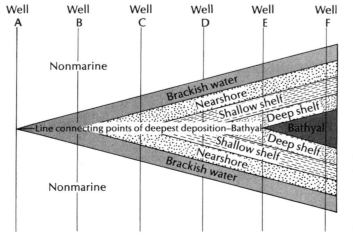

Figure 15.24 Israelsky's (1949) diagram of transgressive-regressive facies cycles, showing that the point of maximum transgression in each facies is essentially time-equivalent. Each lithostratigraphic unit, however, is markedly diachronous. At peak transgression, bathyal deposits shown at right are synchronous with nonmarine deposits shown at left. Correlation by events such as transgressive or regressive maxima and minima has come to be known as *event stratigraphy* (Ager, 1973, 1981).

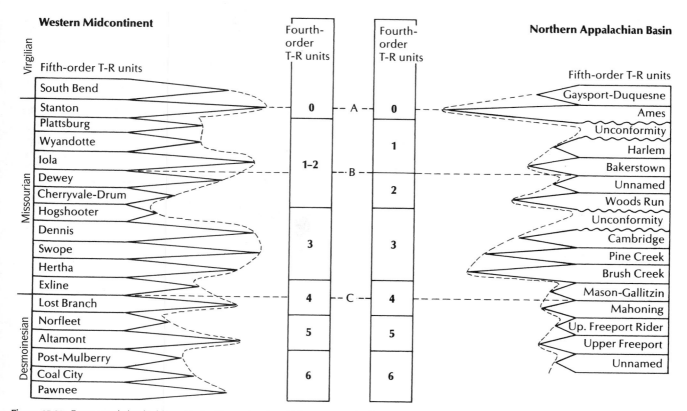

Figure 15.25 Event correlation by hierarchical patterns of cycles within cycles. Both the western midcontinent and the northern Appalachians show a series of fifth-order transgressive-regressive cycles (T-R) clustered into larger fourth-order cycles (numbered). These cycles can be checked by biostratigraphic markers (levels A, B, and C), which show that the cycles are isochronous over 1200 km. (After Busch and West, 1987: 160; © American Geophysical Union.)

palachians (mostly Pennsylvania, Ohio, and West Virginia) show a similar pattern of fifth-order transgressive-regressive cyclothems clustered into larger fourth-order cycles (Fig. 15.25). The peaks of unusually large transgressive-regressive cycles (such as the Pine Creek–Brush Creek of the Appalachians, the Hertha-Swope-Dennis of the midcontinent, or the Ames and the Stanton) serve as the major tie points for lining up the pattern of cycles. These peaks have been checked by biostratigraphic markers (see Fig. 15.25) and have been shown to be isochronous.

In summary, we have seen how early stratigraphers tended to think of rock units as having time significance, an inheritance of the old layer-cake ideas that dated back to flood geology. In the 1930s and 1940s, the pendulum began to swing the other way, and the importance of lateral facies changes began to be emphasized. Shaw (1964) and Ager (1973, 1981) represented some of the more extreme proponents of lateral sedimentation: "if it looks the *same*, it must be *different* in age." Because most sedimentation is controlled by local facies patterns, these ideas about lateral sedimentation are still valid. We cannot consider

the lithostratigraphic boundaries between rock units to have time significance, especially over long distances. With the recent emphasis on eustatically controlled cycles and their effects on sedimentation, one can find many places in the world where lithostratigraphic units do appear to have time significance. It must be emphasized, however, that these correlations are rare exceptions. The assumption of synchrony can be justified only by independent evidence that these cycles are indeed global.

Geologically Instantaneous Events

How do we establish time equivalence in the stratigraphic record? In most instances, radiometric dates are rare or unavailable and have estimated measurement errors that are too large for this purpose. The ideal time marker should be *widespread*, *distinctive*, and *geologically instantaneous*. Generally, events that were separated by years to tens of years (or less) cannot be resolved in the stratigraphic record and are therefore considered instantaneous with respect to geologic time scales.

The primary tool for inferring time relationships in the geologic record is biostratigraphy (discussed in detail in Chapter 16). In the absence of fossil evidence, a number of rock types might be considered to be formed instantaneously (in the geologic sense). The most common of these rock types are volcanic deposits of a single eruption, such as ash layers and their diagenetically altered equivalents, bentonites (see Chapter 6 and Fig. 6.10). Ash deposits fall or flow in a matter of hours or days and cover wide areas during that time. If ash deposits can be correlated precisely and possibly even dated, they provide unique time planes. The use of ash layers to mark geologic time is called **tephrostratigraphy** or **tephrochronology.**

In areas of frequent volcanic activity, tephrostratigraphy can be a powerful tool. Determining which ash beds match up requires detailed petrographic analysis of the ash and its geochemical "fingerprinting" by comparing the amounts of stable, heavy trace elements that are present. Other methods, such as biostratigraphy or magnetic stratigraphy, can be used to check the correlations. Once the ash layers have been correlated, this method offers advantages over most other methods of correlating sections. For one thing, ash layers are geologically instantaneous and so provide a network of "time planes" among the sections. Also, many ash layers can be dated radiometrically, so it is possible to obtain a numerical age in a sedimentary sequence.

Many explosive volcanic events spread huge clouds of ash over enormous areas, making it possible to correlate sequences across great distances and a variety of sedimentary environments. For example, the Long Valley Caldera near Mammoth Mountain in California erupted about 740,000 years ago (Gilbert, 1938; Dalrymple, Cox, and Doell, 1965; Izett et al., 1970; Izett and Naeser, 1976; Izett, 1981) and spread an ash cloud all the way to eastern Nebraska, forming the Bishop Tuff (Fig. 15.26). The eruption was so massive that 150 km³ of ash were displaced up to 2000 km, with pyroclastic deposits up to 125 m thick near the source and several centimeters thick 1500 km away from the source. This ash layer, along with many others, has been used to correlate Pleistocene sequences all over the western United States (Izett, 1981). By combining the data from this ash with data from other ashes that happened to be blown westward (Sarna-Wojcicki et al., 1987), one can construct a network of ash datum levels that correlate terrestrial and marine sections from the North Pacific to Nebraska. Terrestrial and marine sections have also been correlated in this way on the Pacific coast of Central America (Bowles,

Figure 15.26 (A) Outcrop of the Bishop Tuff, an immense ash-flow sheet that erupted from Long Valley Caldera, California, about 740,000 years ago. The finely laminated basal pumice layer is shown here; note that the size of pumice fragments increases up the section. (P. C. Bateman, courtesy of U.S. Geological Survey.) (B) Extent of the Bishop ash cloud, which traveled more than 1000 km to eastern Nebraska. (After Izett and Naeser, 1976: 587; by permission of the Geological Society of America.)

Jack, and Carmichael, 1973) and in the Mediterranean (Keller, Ryan, and Ninkovitch, 1978). Such precise correlations between the terrestrial and marine environments are very seldom achieved because the lithofacies and the fossils of the two environments are completely different. Magnetic stratigraphy is the only other method that can achieve this kind of correlation across facies.

Other rock types that are less common and widespread than volcanic ash but are formed rapidly are also useful as time markers. Any deposit formed by a single catastrophic slide or flow—such as a turbidite, olistostrome, or debris flow—can be considered geologically instantaneous. Some lake sediments are deposited by a uniform rain of silt and clay from suspension and therefore form along time planes. This is

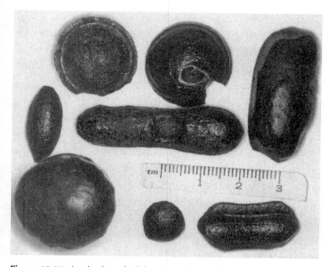

Figure 15.27 A selection of tektites, aerodynamically shaped because they cooled while flying through the atmosphere. (Courtesy of B. Mason.)

particularly true of climatically controlled laminations and varve deposits because each layer reflects a climatic change and is therefore isochronous (see Fig. 4.1). The varve deposits of lakes that show migrating facies belts, however, cannot be considered good time markers (Buchheim and Biaggi, 1988).

Certain features within sedimentary beds can also be used as time markers. In Chapter 17, we will discuss correlation by magnetic reversals and isotopic shifts, which appear to be synchronous. Other geochemical signatures of unique events, such as the iridium anomaly that is believed to represent the impact of an extraterrestrial body (Alvarez et al., 1980), can also be used as time markers. **Tektites** (Fig. 15.27) are glassy spherules that are thought to have been scattered around the Earth by the impacts of meteorites. If a series of tektite layers can be clearly identified as derived from the same event, the layers are useful time markers. Tektites must be used with caution, however, because there are many pitfalls in interpretation.

Clearly, geologists must use any and all available correlation tools and know the strengths of each. When possible, one method of correlation should be checked against another. In Chapter 18, we will look at several examples of cross-checks. The investigator should keep several pitfalls in mind, particularly the easy mental habit of layer-cake thinking. *Rock units are probably diachronous unless proven otherwise.*

Time, Time-Rock, and Rock Units

Throughout the early history of geology, there was continual controversy about the relationship be-

tween rock and time units and about the time significance of rock bodies. Many early geologists thought that every rock unit was deposited during a discrete period of time, such as before, during, or after Noah's Flood. Werner felt that the formation of a specific rock type, such as a schist or limestone, was unique to one of his four great divisions of Earth's history. In his scheme, one could recognize a particular geologic time by a specific lithology. As the principle of faunal succession came to be applied widely, geologists realized that rocks of similar lithology could differ in age. The classic stratigraphic column of England is full of shales, limestones, and sandstones of various ages. These rocks can be distinguished only by their fossil content, because their lithologies are so similar. In addition, it was finally recognized that rock units could be of similar age and yet have very different lithologies, as Sedgwick and Murchison showed when they named the Devonian. The Old Red Sandstone is very different in lithology from the graywackes of Devonshire, yet the fossils showed that both units belong between the Silurian and Carboniferous in the rock column.

The battle between Sedgwick and Murchison over the Cambrian and Silurian is a classic demonstration of this conceptual conflict. Sedgwick proceeded in the old-fashioned way, describing local lithologies and attaching names to them. Murchison used faunal succession to define the Silurian so that it could be recognized internationally. Murchison's fossils enabled him to trace his "Silurian" into rocks that Sedgwick had defined as Cambrian on the basis of lithology.

Faunal succession was developed by men who thought that the Earth is very young. They believed that fossils had been laid down in successive flood deposits spanning very little time. The publication of Darwin's *On the Origin of Species* was still 50 years in the future, so faunal succession had no evolutionary connotation then. Yet the sequence of faunas and their utility in distinguishing strata of similar lithology suggested that fossils were an independent, more reliable marker of time than rocks. Geologic time and observable rock sequences had become separate but related concepts.

The difference between time and rock units still confuses people today. Because observable rock units are not unique to a time and can often transgress time, formally defined *rock units* (**lithostratigraphic units**) have no explicit time connotation. Like time on the human scale, geologic time is an abstract concept that can be marked by many different methods:

changes in fossil faunas, radioactive decay, changes in oxygen isotopes, reversals of Earth's magnetic field, or whatever is convenient. Therefore, *geologic time units* (**geochronologic units**) are also purely abstract concepts. Time and rock units are not completely separate, however. Many geologic events have left evidence in the rock record, and the rock record is our only concrete representation of geologic time. For this reason, geologists have developed a hybrid unit, the *time-rock* (or time-stratigraphic, or **chronostratigraphic**) unit. A time-rock unit is the sum total of the rocks formed worldwide during a specified interval of geologic time (as recognized by fossils, radiometric dates, or other means). A time-rock unit has physical reality even though it cannot be seen in its entirety in any one section and its boundaries are based on the abstract concept of time. Those boundaries are established by biostratigraphy (see Chapter 16).

Schenck and Muller (1941) formalized the subdivisions of the three categories of time, time-rock, and rock units and recognized the hierarchy of units that had developed over the years. Their scheme is shown in Fig. 15.28. Schenck and Muller emphasized that rock units have no explicit time connotation by showing them at right angles to the time and time-rock units. The equivalence of time units and their corresponding time-rock units is shown by the dotted lines. Thus, the Jurassic System consists of all the rocks deposited during the Jurassic Period.

The important distinction between time and rock units is often confused, even by geologists who should know better. It is easy to fall into the layer-cake thinking of our catastrophist forebears and refer to "Tapeats Sandstone time" or "the thickness of the Late Jurassic" or events that occurred "during the Upper Cretaceous." As Owen (1987) points out, geologists who

never refer to "upper Tuesday" or "late peninsula of Michigan" will easily confuse time and place words in geology. Such confusion occurs on an even more subtle level. For example, the phrase "pre-Dakota unconformity" implies that the Dakota Sandstone is a geochronologic unit; it should be "sub-Dakota unconformity." This may seem like nitpicking, but as Owen (1987, p. 367) puts it, "authors who are careless in usage of time and place words run the risk of implying that their carelessness may extend to data collection, analysis, and conclusions as well."

Each unit in the sequence is composed of one or more units of the level beneath it in the hierarchy. For example, the Phanerozoic Eon contains three eras (Paleozoic, Mesozoic, and Cenozoic). The Cenozoic Era contains two periods (Tertiary and Quaternary). The Tertiary Period contains five epochs (Paleocene, Eocene, Oligocene, Miocene and Pliocene). The Oligocene Epoch contains two internationally recognized ages (Rupelian and Chattian), and the corresponding stages are based on biostratigraphic zones. For example, in the deep sea, the Chattian Stage is presently based on two successive planktonic foraminiferal biostratigraphic zones: the *Paragloborotalia opima opima* Zone and the *Globigerina ciperoensis* Zone. Other zonations can be used, but the foraminiferal zones have achieved international consensus in this instance.

Rock units have a similar hierarchy. For example, the Oligocene White River Group is composed of two formations in South Dakota: the Chadron Formation and Brule Formation. The Brule Formation, in turn, is made up of the Scenic Member and the Poleslide Member. Within the Scenic Member are several local units, such as the lower *Oreodon* beds and the *Metamynodon* sandstones.

Instead of building up from the bottom (as in the case of zones), the fundamental unit in lithostratigraphy is the **formation.** A formation is characterized by two important properties: It must have *identifiable and distinctive lithic characteristics,* and it must be *mappable on the Earth's surface or traceable in the subsurface.*

Formations can be lumped into groups and supergroups or subdivided into members and beds. This classification is not obligatory, however. Many formations have no members and are not part of any larger group. Whether a rock unit merits the rank of member, formation, or group is up to the discretion of the geologist who describes it. For example, in the Paleozoic sequence in the Grand Canyon (see Fig. 15.10), the top four units are formations (Kaibab Limestone, Toroweap Limestone, Coconino Sandstone, and

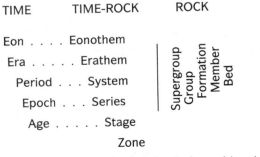

TIME	TIME-ROCK	ROCK
Eon	Eonothem	
Era	Erathem	Supergroup Group Formation Member Bed
Period . . .	System	
Epoch . . .	Series	
Age	Stage	
	Zone	

Figure 15.28 Schenck and Muller's (1941) scheme of time, time-rock, and rock units. The hierarchies of time and time-rock units are parallel; that is, a given time unit has an equivalent time-rock unit. However, the hierarchy of rock units is placed at right angles to the time and time-rock hierarchy, because rock units have no time connotation.

Hermit Shale). Below it are four formations that are hard to distinguish and so are usually united into the Supai Group (McKee, 1982). Beneath the Supai is the Redwall Limestone (McKee and Gutschick, 1969), a formation with four *members* (from bottom to top: Whitmore Wash, Thunder Springs, Mooney Falls, and Horseshoe Mesa), which are comparable in thickness to the *formations* above them. The Redwall Limestone (a *formation*) is approximately equal in thickness to the Supai *Group.* The decision about whether units deserve member, formation, or group rank is based not on thickness but on other criteria such as their distinctiveness, lateral continuity, and ease of recognition.

The criterion of mappability or traceability is one of convenience. In some areas that have thousands of meters of section (see Box 15.1), formations can be hundreds of meters thick and can be recognized on large-scale maps, aerial photographs, and even satellite photos. In other areas, formations are very thin and can be mapped only on very small scales. Clearly, if a formation is mappable, there must be some lithic criterion by which it can be distinguished and mapped. But lithic distinctiveness does not necessarily imply lithic uniformity. Formations can include many different rock types and can change lithology over distance. All that is required is some sort of distinctiveness on which the boundaries, or **contacts,** can be based and mapped. Because formations are units of convenience, one geologist's formation might be only a member to a stratigrapher working in an area with different geology.

The Stratigraphic Code

The inconsistencies and conflicts of terminology and concepts were major sources of confusion to early geologists. Eventually, it became necessary to standardize usage and erect a set of guidelines for stratigraphers to follow. The first American Commission on Stratigraphic Nomenclature met in the early 1930s (Ashley et al., 1933). The American code has been revised many times since then, and its most recent revision (1983) appears at the end of the book as Appendix A. International usage is not as unanimous as North American nomenclature, but the essential features have been codified by the International Subcommission on Stratigraphic Classification in the International Stratigraphic Guide (Hedberg, 1976; Salvador, 1994).

The North American Stratigraphic Code states the criteria that are to be used to distinguish a *formal* rock unit from an *informal* one. The name of a formal geo-

logic unit is capitalized (for example, Morrison Formation) and must be created under strict rules (articles 3 through 21). The creation of a formal unit must be published in a recognized scientific medium, and the original description must list a number of important defining criteria. For example, a formation must be named after a local geographic feature, not a person or some combination of Greek and Latin roots. There are rules about priority of names and compendia to determine if a name has already been used. In the United States, this duty is carried out by the frequent publications of the Geologic Names Committee of the U.S. Geological Survey. The boundaries, dimensions, and age of the formation must be given clearly in the original description. Most important, a type section, or **stratotype,** must be designated so that later geologists can determine what the name originally referred to and what the author's conception of the name was. The type section should be the best reference section available. Ideally, it should be well exposed, and both the top and bottom contacts of the formation should be visible. It should be as complete and continuous as possible, without faulting or long covered intervals. Additional **reference sections** are also helpful in establishing the variability of the formation or giving a backup section in case the type is covered, built over, bulldozed, or otherwise lost (or poorly chosen in the first place). In many cases, the formation can be studied in the subsurface as well. It is best to designate the type section on a surface outcrop and to designate subsurface sections as reference sections, because the surface data are accessible to more geologists. Some units, however, are known only from boreholes.

When a new or revised description of a unit is published, the information should include the following elements:

1. Name and rank of unit

2. Locations of type and reference sections, including a map or air photo

3. Detailed description of the unit in the stratotype, including the character and height of the section or the well depth of contacts

4. Comments on the local or regional extent of the unit and its variability

5. Graphic log of the unit (including geophysical logs of subsurface units)

6. Location of curated reference material

7. Discussion of the unit's relationship to other contemporaneous stratigraphic units in adjacent areas

BOX 15.1 MEASURING AND DESCRIBING STRATIGRAPHIC SECTIONS

The fundamental data for nearly all stratigraphic studies are measured stratigraphic sections. Accurate measurement, precise description, and careful collection of rock samples and fossils are essential before any correlation or interpretation can be made. A stratigraphic column provides a one-dimensional sequence of units; two columns, placed side-by-side and correlated, produce a two-dimensional cross section; three or more columns, placed on a map and correlated in three dimensions, result in a spatial representation of all the rock units.

Each section presents its own peculiarities. In reconnaissance mapping, it is often necessary to measure a number of sections in minimal detail. For a careful bed-by-bed analysis, it may take days to measure one section. The section(s) to be measured must be chosen carefully. In some regions, only partial exposures are available, and every decent outcrop must be measured. Arid regions with high relief, on the other hand, usually have excellent outcrops everywhere. In this case, geologists prefer the sections that are the thickest, the most uncomplicated structurally, somewhat distant from other sections, and the most accessible and climbable. Most geologists learn by experience to recognize outcrops that will enable them to concentrate on measuring and describing rather than on keeping their footing.

Section measurement begins with recording all the pertinent information, such as date and precise topographic location, in the field notebook (Fig. 15.1.1). Photographs of the outcrop can supplement field sketches. The geologist must also determine if the beds have any dip and which measurement technique (discussed later) is appropriate to the thickness, exposure, and attitude of the outcrop. Sections are always measured from the bottom of the exposure because, by superposition, the lowest beds accumulated first and are the oldest. The first step is to describe and subdivide the strata in terms of homogeneous, informal units that are natural for the exposures in the outcrop. These may or may not correspond to formally named formations or members. In many cases, the geologist does not know in advance where the formal boundaries lie but must determine this later. Each informal unit is numbered, and a complete sedimentary rock description is recorded in the field notebook. The dominant lithology, texture, color (on both weathered and fresh surfaces), macroscopic mineralogy, induration, and outcrop exposure and resistance to weathering are the most important properties to be noted in the field. The contact relationships (sharp or gradational) between units are also noted, as are any visible structural features (folding, brecciation, or fracturing). Sedimentary structures are informative, and if they give any indication of the locations of tops of deposition beds or of current directions, these should be measured and recorded.

All of these features can be summarized by symbols to produce what is known as a **graphic log** (Fig. 15.1.2). An example was shown in Figure 9.2.3. The graphic log has the great advantage of reducing long verbal descriptions to a visual plot that becomes easy to read with practice. Grain sizes can be shown by the relative width of the stratigraphic profile.

Unit#	Section 8 — 7-17-60, E.D. McKee Kaibab Trail (North) Section 11-12, T32N R3E, Coconino Co., AZ Slope and cliff units of Esplanade Ss. measured on N. side of Roaring Springs Cyn. Lwr. part of Supai Group measured on S. side.	Thickness (ft.)
	Redwall Ls - Horseshoe Mesa Mbr. Ls., olive gray, aphanitic; contains much red-brown mudst. Filling sol'n. cavities; forms top of main cliff. Unconformity inferred Surface even and flat where observed. Supai Group - Watahomigi Fm.	(Not meas.)
1	Lower slope conglom., mod. → dark reddish-brown, locally disturbed, as pockets in basal mudst.	2.5
2	Siltst., grayish-red, muddy, thin-bedded (0.5 - 2')	5.0
3	Concealed slope.	22.0
4	Ls., grayish-red-purple, aphanitic; massive, thick	3.0
	-cross-bedded (planar) in upper part; mudst. partings.	227.5
23	Mudst., light-brown, calcareous, shaly	11.0
24	Ss. pale reddish-brown, v.f. grained; flat-bedded, thin → med. bedded (6-30"), forms receding ledges.	19.0
25	Mudst. grayish-red brown, calcareous; structureless conchoidal fracture, forms recess	5.5
26	Ss., grayish-orange v.f. grained; hard, massive, thick-bedded (4 - 8') weathers to rough gray surface. Top Supai Group Unconformity Erosion Channels shallow, filled with thin-bedded shaly mudst. Hermit Sh.	21.0
	Concealed slope.	13.0

Figure 15.1.1 Example of field notes taken in measuring section. (Data from McKee, 1982.)

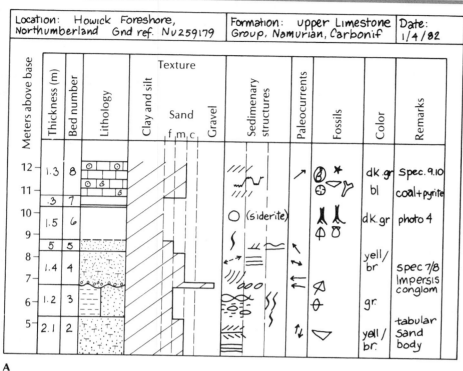

Figure 15.1.2 (A) Example of field notes taken while measuring a section and constructing a graphic log. (B) Summary draft of graphic log, based on data in A.

Lithology is indicated by a series of standardized patterns. Sedimentary structures and fossils can be shown symbolically either directly on the column or in a series of parallel columns. Paleocurrent directions can be plotted and placed in the column where they were measured. Some features still require verbal description, so a separate column for notes is appropriate.

Any fossils that occur should be noted and identified as well as field conditions permit. If the specimens are important or should be studied further, they should be carefully collected and their *exact* position on the column recorded. Fossils without precise stratigraphic data are almost always useless for stratigraphy. Only amateurs and rockhounds pick up fossils in the field without carefully recording

(continued)

(Box 15.1 continued)

C

Figure 15.1.2 *(continued)* (C) Standard symbols for lithology, sedimentary structures, and fossils used in measuring sections and constructing graphic logs. (After Tucker, 1982.)

where they are found. Trained geologists have no excuse! In most studies, samples of the rock units are needed for laboratory analysis or detailed study. These samples need not be large (a few rocks can make a backpack heavy), but it's better to be safe than sorry. Some problem with the description or correlation may arise later in the laboratory, and it might be impossible to return to the outcrop.

The method of measuring stratigraphic thickness varies with the attitude, exposure, and total thickness of the beds to be measured. The simplest measurements are of horizontal and well-exposed beds, typical of many sequences in the U.S. midcontinent. In such cases, the fastest and easiest method is to use a sighting hand level and measure in increments of one's own eye height (Fig. 15.1.3A). Starting from the base of the outcrop, the geologist sights to a point that is

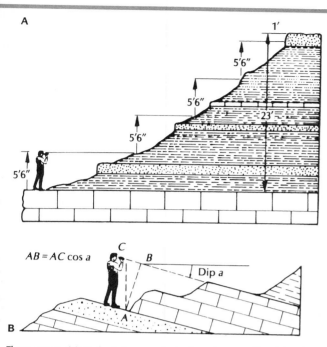

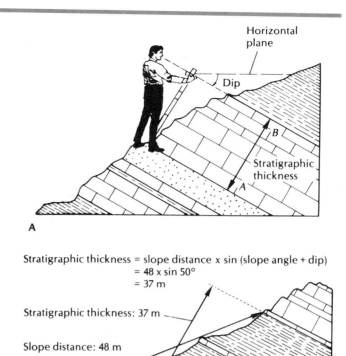

Figure 15.1.3 (A) Method of measuring horizontal strata with a sighting hand level, using eye height as the standard dimension. (After Krumbein and Sloss, 1963.) (B) Hewett method of measuring the thickness of dipping strata, using a clinometer set at the angle of dip. In this method, stratigraphic thickness is equal to eye height times the cosine of the dip angle. (After Kottlowski, 1965.)

Figure 15.1.4 (A) Measuring dipping strata with a Jacob's staff. (After Kottlowski, 1965.) (B) Measuring dipping strata using a measuring tape across a sloping surface. In this method, both the dip and slope angle must be measured, in addition to determining the slope distance by measuring tape. (After Krumbein and Sloss, 1963.)

level with eye height and mentally notes its location. Then the geologist walks to that point and repeats the process, describing the beds between sightings along the way. If the level has a clinometer (such as is found in a Brunton transit or an Abney hand level), it is possible to measure dipping beds using the **Hewett method**. The clinometer is set at the angle of dip, and the geologist sights for the next point to stand on, moving perpendicular to the strike (Fig. 15.1.3B). In this case, the eye height must be multiplied by the cosine of the dip angle to get the true stratigraphic thickness. If the dip is consistent, this correction can be made once at the beginning, and then each increment on the section in the field notebook is the corrected eye height. Most standard field notebooks contain trigonometric tables for making such corrections.

The best way to measure dipping beds, especially when the exposures are steep or discontinuous, is to use a simple device called a **Jacob's staff** (Fig. 15.1.4A). A Jacob's staff is a lightweight pole of some convenient standard length (such as 1.5 m) with smaller increments marked on the side. A horizontal sighting bar is usually attached to the top, and a good Jacob's staff also has a clinometer. Expensive Jacob's staffs hold a Brunton transit in a clamp on the top and even

telescope shut when not in use. A simple Jacob's staff can be made from a pole with a rotating plate on the top (attached by a wing nut) onto which is glued a protractor and an inexpensive line level. It is not necessary to spend a lot of money to get an accurately measured section.

The Jacob's staff can be used to measure horizontal beds in the same manner, except that the increment of measurement is the length of the Jacob's staff (a convenient standard unit) rather than eye height. Few geologists have a convenient eye height; thus, measuring a section with the Jacob's staff is sometimes faster than hand leveling. The Jacob's staff is by far the easiest and most accurate method for measuring inclined bedding. The staff is held perpendicular to the bedding surface, and the geologist sights along the top, perpendicular to the strike, to find the next point of measurement (Fig. 15.1.4A). The clinometer at the top is set at the dip angle to ensure that the staff is exactly perpendicular to bedding. Because good outcrops are

(continued)

(*Box 15.1 continued*)

seldom exposed in a single straight transect, the geologist must often move laterally along a bedding plane and continue to measure the section where the exposures are better. Wherever the transect begins, it must be kept perpendicular to the strike of the beds, or else a trigonometric correction is necessary. For this reason, the geologist should continually keep track of the strike and dip of the beds, especially if the attitude is very steep or changes rapidly. A single geologist can measure a section quite rapidly with a Jacob's staff and then go back and fill in the details and make collections. Two geologists, however, can do the job in less than half the time because one can measure while the other writes descriptions in the field notebook and collects samples.

Two geologists are necessary for methods that require measuring tape or a similar device. If the total section is not very thick, or if the individual beds are very thin and each requires detailed study, a measuring tape is preferable. One can use a measuring tape and simple trigonomentry to measure inclined beds along a slope (Fig. 15.1.4B). Many other methods of measuring dipping beds in unusual ways, or with unusual tools, are possible. The excellent little book by Kottlowski (1965) is highly recommended for the details of section measuring techniques.

Informal units are also common, and the code recognizes their role in "works in progress" and in innovative approaches to stratigraphy. For example, Sloss's "sequences" (discussed above) are informal units. Many of the working units in economic geology—such as coal beds, oil sands, aquifers, and quarry beds—are informal. Informal units serve a valuable purpose in enabling geologists to articulate new concepts outside the restrictions of the code. As informal units become more widely accepted, they may eventually become formal. Many of the new units and categories in the code were once informal units. Informal units do not have the status of the formal units of the code, however, and are not capitalized. As more detailed studies are done, many more units are recognized than require formal definition. Most of these units will remain informal because establishing all of them would cause an overwhelming proliferation of names. As the authors of the code put it, "No geologic unit should be established and defined, whether formally or informally, unless its recognition serves a clear purpose."

One of the most common errors made by geologists is to confuse informal with formal units in print or in slides and illustrations. All names in a formally defined unit are capitalized, but only the geographic term is capitalized in informal names. For example, the Tapeats Sandstone of the Grand Canyon is a formally defined unit, but the *Metamynodon* sandstones of the Big Badlands of South Dakota are informal. This distinction becomes tricky when referring to time and time-rock terms. For example, the subdivisions (periods/systems) of the Paleozoic and Mesozoic eras have been formally defined, but those for the Cenozoic have not. Thus, it is correct to refer to the Late Devon-

ian Epoch or the Lower Jurassic Series, but the upper or late Eocene must always remain uncapitalized. There are more subtle pitfalls, as well. For example, the Cretaceous has only two formal subdivisions, so it is correct to refer to Early/Lower and Late/Upper Cretaceous, but not to "Middle" Cretaceous (except as an informal term). Owen (1987) clarifies many of these confusing points, and the geologic time scale (Appendix B) is a useful guide to determining which subdivisions have attained formal recognition.

Many striking features of the new North American Stratigraphic Code do not appear in any of its predecessors. The widening domain of stratigraphy is reflected in classifications for rock bodies that are not layered. The new units are divided into the following two classes and are defined in Appendix A (new categories are in italics):

Material categories based on content or physical limits: Lithostratigraphic, *lithodemic, magnetopolarity,* biostratigraphic, pedostratigraphic, *allostratigraphic*

Categories expressing or related to geologic age: Chronostratigraphic, *polarity-chronostratigraphic,* geochronologic, *polarity-chronologic, diachronic, geochronometric*

The most remarkable of the new categories are the **lithodemic units,** which apply to unstratified igneous and metamorphic rocks. With formal recognition of this category, the rules of stratigraphy now apply to all rocks on the Earth's surface. Because lithodemic units are unstratified, their boundaries are based on the positions of lithologic changes. Most such boundaries are intrusive contacts, but there can be gradational contacts. Metamorphic rocks often show gradational changes in which the degree of deformation, or the presence of indicator minerals, changes gradually

away from the source of temperature or pressure. The basic unit, a **lithodeme,** is comparable in concept to a sedimentary formation. Ideally, a lithodeme has some sort of lithologic consistency and is mappable. The rock name in a lithodeme is capitalized, as in San Marcos Gabbro or Pelona Schist. A group of lithodemic units can be united into a formal unit called a **suite,** as long as all the included lithodemes are either igneous or metamorphic. If two or more classes of rock (that is, igneous, metamorphic, or sedimentary) are lumped together into a formal unit, it is called a **complex.** The Franciscan Complex, for example, is composed mostly of metamorphic rocks but also contains undeformed sediments and igneous intrusions.

Units that are defined and identified on the basis of the discontinuities that bound them are called **allostratigraphic units.** The lithologies of allostratigraphic units are often very similar, but the thin discontinuities between them prevent them from being considered a single formation. The code cites terrace gravel deposits on opposite sides of valley walls as an example of allostratigraphic units. Such deposits may be either physically overlapping or separated, but they are usually very different in age and are bounded by a distinct unconformity. In the case of terrace deposits, each may have been deposited by a different glacial episode. The unconformity-bounded seismic sequences of Vail et al. (1977a, b) and the sequences of Sloss (1963) are also examples of allostratigraphic units, although they are not yet formally defined. Allostratigraphic units have the same criteria for recognition and formal definition as other formal units.

Previously, only time (geochronologic) and time-rock (chronostratigraphic) units existed as units that depend on geologic age. The new code recognizes **diachronic units,** which are spans of time represented by rock bodies that are known to be diachronous. This is a remarkable reversal from the old concept that rock bodies were usually time-equivalent. Now there is a formal method by which geologists can refer to the time represented by rock bodies that are explicitly *not* time-equivalent.

The new code also distinguishes between geochronologic units, which are ultimately based on chronostratigraphic units, and **geochronometric units,** which are expressed in years and have no material referent. This unit has become necessary, especially in Precambrian geology, in which events are referred to by their numerical, not relative, age.

The code is clearly written, so it is unnecessary to paraphrase it further here. Most geologists have agreed to follow the code, and most geologic publications insist on adherence to it. If one wants to be understood, and to understand what other people think and write, knowing the code is a necessity.

CONCLUSIONS

Stratigraphy has come a long way from the layer-cake thinking of the late nineteenth and early twentieth centuries. For a modern geologist, it is no longer sufficient to merely describe the rocks. The geologist must also keep the larger interpretive framework in mind. How do these rocks correlate with others? How can their age be established? Do they fit into the framework of transgressions and regressions? How completely do they represent the geologic record? These and many other similar questions show that there is much more to interpreting a stratigraphic sequence than most nongeologists realize.

FOR FURTHER READING

Ager, D. V. 1981. *The Nature of the Stratigraphical Record,* 2d ed. New York: John Wiley.

Dott, R. H., Jr. 1983. Episodic sedimentation: How normal is average? How rare is rare? Does it matter? *Journal of Sedimentary Petrology* 53:5–23.

Einsele, G., W. Ricken, and A. Seilacher, eds. 1991. *Cycles and Events in Stratigraphy.* Berlin: Springer-Verlag.

Hallam, A. 1992. *Phanerozoic Sea-Level Changes.* New York: Columbia University Press.

Kottlowski, F. E. 1965. *Measuring Stratigraphic Sections*. New York: Holt, Rinehart, and Winston.

Miall, A. D. 1999. *Principles of Sedimentary Basin Analysis*, 3d ed. New York: Springer-Verlag.

Owen, D. E. 1987. Usage of stratigraphic terminology in papers, illustrations, and talks. *Journal of Sedimentary Petrology* 57:363–372.

Salvador, A. 1994. *International Stratigraphic Guide*, 2d ed. Boulder, Colo.: Geological Society of America.

Schoch, R. M. 1989. *Stratigraphy: Principles and Methods*. New York: Van Nostrand Reinhold.

Shaw, A. B. 1964. *Time in Stratigraphy*. New York: McGraw-Hill.

Wilgus, C. K., et al., eds. 1988. *Sea Level Changes: An Integrated Approach*. SEPM Special Publication 42.

16 Biostratigraphy

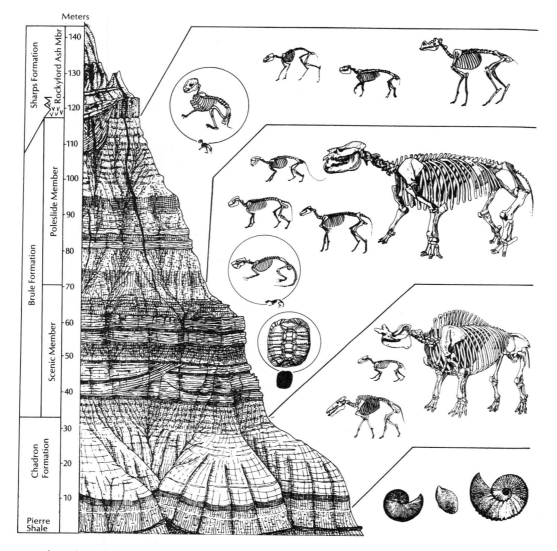

Biostratigraphic succession of fossil mammals and turtles from the Big Badlands, South Dakota. (Courtesy of G. J. Retallack.)

AFTER THE THE PRINCIPLE OF FAUNAL SUCCESSION WAS discovered, geologists slowly came to appreciate the significance of fossils in correlating and dating rocks. Indeed, in the last two centuries, fossils have proved to be the only widely applicable tool of time correlation in sedimentary rocks, although they must be used with caution. Their use for stratigraphic correlation is called **biostratigraphy.**

Once the principle of faunal succession was widely accepted, further refinements of its use soon followed. The French geologist Alcide d'Orbigny, in his comprehensive study of the Jurassic (*Terrains Jurassiques,* 1842), showed that fossil **assemblages** are the keys to correlation. No matter what the lithology of a rock unit, or where it is found, it can be recognized by its characteristic assemblage of fossils. D'Orbigny called

341

all of the strata defined by one fossil assemblage a **stage,** a term that has now been included in the formal time-stratigraphic hierarchy. Of course, d'Orbigny's stages were catastrophist in meaning; that is, each stage represented a separate creation and flood. As d'Orbigny's 10 stages were studied and subdivided by later stratigraphers, the number of stages increased far beyond what any catastrophist geologist could accept. For other reasons, this multiplicity of stages was also unacceptable to the growing number of uniformitarian geologists. "The naming of new stages became more a sport than a scientific endeavor" (Berry, 1987, p. 125).

In Germany, Friedrich Quenstedt found that d'Orbigny's stages did not work well outside France. Quenstedt had studied the Jurassic succession in minute detail, measuring sections and recording the position of each fossil. He found that d'Orbigny's stages were too broadly and vaguely defined for his use and began to divide his sections according to the stratigraphic ranges of individual fossil taxa. His methods were amplified and popularized by one of his students, Albert Oppel (1831–1865). Oppel traveled widely, applying Quenstedt's methods to the Jurassic rocks of France, Switzerland, and England. After plotting hundreds of vertical ranges, he realized that the patterns were repeated on a fine scale all over Europe. These patterns could be broken into distinct aggregates, bounded at the bottom by the appearance of certain taxa and at the top by the appearance of other taxa. Between 1856 and 1858, Oppel described the diagnostic aggregates, or "congregations," of fossils in his scheme of **overlapping range zones.** His scheme was the first to allow the clear-cut, nonarbitrary recognition and comparison of time units. Oppel also grouped his zones into "stages," but these differed from d'Orbigny's stages by being based on smaller units with distinct boundaries. Eventually, most of d'Orbigny's stages were redefined in this manner.

Controlling Factors: Evolution and Paleoecology

Oppel's zonation scheme, based on Jurassic ammonites, did not turn out to be applicable worldwide, partly because of changes in taxonomy. Modern ammonite workers tend to combine (or "lump") species that Oppel considered distinct. The major limitation on his scheme, though, was that the ammonites he studied were not found worldwide. In Europe, for example, ammonites occur in two biogeographic provinces, one in northwestern Europe (Oppel's area) and one near the Mediterranean. Efforts to use Oppel's scheme in the Mediterranean province failed because few of the ammonites he recognized were found there. It soon became clear that zonation schemes must first be worked out locally and then compared between regions to see how they correlate.

The failure of Oppel's scheme shows how evolution and paleoecology control biostratigraphic distributions. The *evolution* of organisms is the enabling factor, providing the progressive changes in species through time that make biostratigraphy possible. Unlike any other correlation method, biostratigraphy is based on the unique, sequential, nonrepeating appearance of fossils through time. The presence of a single fossil can often be used to determine the age of a rock very accurately. This is not true of the lithology of the rock, its magnetic polarity, its seismic velocity, or its isotopic composition; none of these properties are unique and none can be used alone. If the rock can be radiometrically dated, then another criterion of age can be used; except for volcanic ashes and lava flows, however, few stratified rocks can be radiometrically dated.

Paleoecology is the limiting factor on biostratigraphic distributions of organisms because no organism has ever inhabited every environment on Earth. The distribution of organisms is also restricted on a local scale. **Facies-controlled organisms** are restricted to particular sedimentary environments (Fig. 16.1). Such organisms manage to migrate as depositional environments change and thus to find their old environment in a

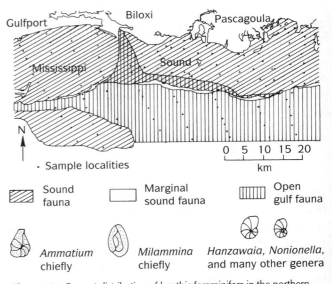

Figure 16.1 Present distribution of benthic foraminifers in the northern Gulf of Mexico showing their strong ecological separation. This type of facies-controlled distribution is likely to create problems for the biostratigrapher. (After Eicher, 1976: 9.)

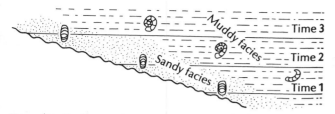

Figure 16.2 Significance of contrasting rates of evolution and shifting environments. The brachiopod *Lingula* lives in the sandy nearshore facies and evolves extremely slowly. It is a poor biostratigraphic indicator because it tracks its environment and thus changes little. Ammonoids that swam in the open ocean, however, are excellent for biostratigraphy. They evolved rapidly and were not tied to any particular facies because they are free-swimming. (After Dott and Prothero, 1994: 86.)

new place; they have evolved very slowly, if at all. For example, there is a distinct ecological community dominated by *Lingula* (a genus of inarticulate brachiopod) that has persisted unchanged in Europe since the Cambrian, while other invertebrate communities changed rapidly. This *Lingula* community appears to follow its accustomed environment, avoiding evolutionary change, and thus has no geological time significance (Fig. 16.2). An unwary biostratigrapher might use the first appearance of a slowly changing, facies-controlled fossil as a time indicator, when it is really an indicator of time-transgressive facies change.

In the biostratigraphy of the Gulf Coast, there are assemblages of robust, heavy-walled agglutinated foraminifers that do not occur in the Gulf of Mexico today but are known elsewhere at bathyal and abyssal water depths (Poag, 1977). These foraminifers were

used widely in zonation of the Oligocene and Miocene sediments of the Gulf and Caribbean. When offshore drilling began to recover younger deposits, however, these assemblages were found to extend to the top of the Pleistocene. It turns out that this agglutinated foraminiferal assemblage is more useful for paleoecology than for biostratigraphy. The only way to determine whether facies control is a factor in the distribution of fossils is to compare the biostratigraphy of the suspect group with the biostratigraphy of other groups in the area or to check it with physical stratigraphic methods. Although fossils are better time indicators than any other stratigraphic tool, they too must be used with caution.

Another example of facies control was documented by McDougall (1980). For years, benthic foraminifers were used to construct a biostratigraphic zonation of the Eocene marine sequence along the North American Pacific coast. The zonation in the state of Washington (Rau, 1958, 1966) was worked out independently of the zonation in California (Schenck and Kleinpell, 1936; Mallory, 1959; Donnelly, 1976). When McDougall examined the foraminifers that were found in both environments and analyzed their ecological associations, she found that foraminifers that marked one "time" interval in Washington occurred at a completely different "time" period in California. For example, the *Sigmomorphina schencki* zone, which was used to mark the early Refugian in Washington (Fig. 16.3A), was based on deep-water forms that occurred throughout

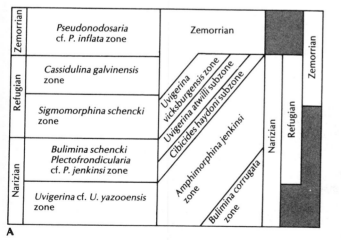
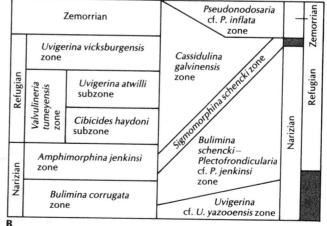

Figure 16.3 Correlation of the late Eocene benthic foraminiferal zonations between California and Washington, showing the problems caused by the strong facies control on these organisms. (A) Correlation scheme assuming that the Washington zonation is valid. Under this scheme, the California zonation is time-transgressive, and the stages that are based on California forms overlap. (B) Correlation assuming that the California zonation is valid. In this case, the Washington zonation is time-transgressive, and the

stages based on the Washington foraminifers overlap. The discrepancy between these zonations occurred because most of these taxa were strongly facies-controlled; they disappeared from one area simply because they had migrated north or south with the facies changes and varying water depth. For example, the *Sigmomorphina schencki* zone appears in the late Narizian of California but migrates north to Washington by the early Refugian. (After McDougall, 1980: 15.)

344

the Narizian and Refugian in California (Fig. 16.3B). Similarly, the *Cibicides haydoni* subzone, used to mark the early Refugian in California, was based on shallow-water forms that spanned the late Narizian and entire Refugian in Washington. It turned out that the Narizian, Refugian, and Zemorrian Pacific coast benthic foraminiferal stages were not sequential chronostratigraphic units, as had been thought. Instead, they overlapped or duplicated one another in various ways, depending on whether the stages were defined based on the foraminifers found in California or those found in Washington. Such problems are enough to give a biostratigrapher gray hairs! The latest recalibration of these stages has been summarized by Prothero (2001).

The distribution of fossils in the rock record is controlled by two primary factors: evolution and paleoecology. Evolution is the key to telling time, and paleoecology is often useful in determining sedimentary environments. Untangling the two factors is not always simple, but we are fortunate to have both sources of data.

Biostratigraphic Zonation

As Quenstedt and Oppel showed, the key to using fossils for telling time is careful plotting of the vertical stratigraphic distribution, or **range,** of every fossil in a local section. Fossils whose original stratigraphic positions were not recorded at the time of collection (usu-

ally fossils collected by amateurs or geologists who are not stratigraphers) are useless for this purpose. The first step, then, is to obtain a reliable database: a large suite of various fossils from one or more sections of the local area, with *accurate stratigraphic data that were gathered at the time of collection of each specimen.*

After all the fossils have been cleaned up, they must be identified. This is not a trivial task. Sometimes distinctions between one fossil species and another are so subtle that only a specialist can tell them apart. In other cases, a group may not have been studied in years, so there is no clear understanding of which species are valid or when they lived. Identification can also be hampered by poor preservation. An incorrect identification can lead to serious problems if the stratigrapher follows it to incorrect correlations.

The next step is to compile the ranges into a **zonation.** Shaw (1964, p. 102) pointed out that every extinct organism divides geologic time into three parts: *the time before it appeared, the time during which it existed, and the time since its disappearance.* All the strata that actually contain a given fossil species are said to be in its **range zone.** In a local section, the observed range of a fossil is its *partial range zone,* or **teilzone.** Teilzones are the empirical database from which all biostratigraphy is derived, but no teilzone represents the total range in time and space of a species. The various classes of zones are called **biozones.**

The absence of a fossil in an area can result from a number of factors (Fig. 16.4). A species first appears on

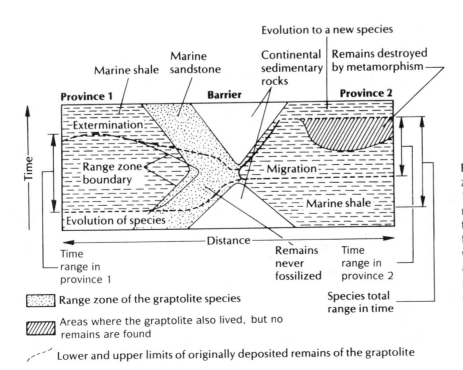

Figure 16.4 Time-space relationships of the range zone of a fossil. The species evolves first in province 1, but the barrier to province 2 is present, so its migrational first appearance in province 2 is later than its true first appearance. Likewise, it dies off first in province 1 but lasts longer in province 2, where it eventually disappears by evolving into another species. In addition to these factors, which prevent synchronous first and last appearances in different basins, the range zone can be shortened by nonpreservation or by erosion or metamorphism. The total species range in time is shorter than its local range in any one region. (After Eicher, 1976: 10.)

Earth when it evolves from an ancestral form (evolutionary first occurrence), but it first appears in regions away from its origin due to immigration. Various factors may have limited its immigration so that in some areas a taxon appears in the record very much later than when it first evolved in its home range. Likewise, the last appearance of a species on Earth is its final extinction event, but it may have been exterminated in many local regions before its final extinction. It may also have emigrated out of a region long before it became extinct worldwide. The first appearance of a fossil, whether by evolution or immigration, is called a **first appearance datum (FAD)**, or first occurrence datum (FOD). Similarly, the last appearance, whether by extinction or emigration, is called a **last appearance datum (LAD)**, or last occurrence datum (LOD).

If the fossil record were perfect everywhere, biostratigraphic ranges would be simple functions of these four types of events: evolution, extinction, immigration, and emigration. Other factors are involved, however. Fossils of a species will seldom be found in rocks deposited in environments that were unsuitable for that species (see Fig. 16.4). In many environments where the species did occur, the rocks may not be suitable for preserving fossils or may have been eroded away along with their fossils. Also, the rocks may have been heated and deformed by a metamorphic event, so that few or no recognizable fossils remain. The rocks can also be inaccessible for sampling, or a particular fossil might accidentally be missed in the sampling. Thus, the final, observed distribution of a fossil group is affected by many factors that have nothing to do with its biology. Much of the original "signal" has been destroyed by the "noise" of sampling, nonpreservation, and destruction. Biostratigraphic techniques are designed to reconstruct as much of the signal as possible.

The nomenclature of zonation has been confused since its beginning, but the most recent stratigraphic codes (Hedberg, 1976; Appendix A) have attempted to standardize it. **Interval zones** form the first class of zones. They are bounded by two specific first or last occurrences of taxa. The most important type of interval zone is the **concurrent range zone** (Fig. 16.5A), defined by the overlap of two or more taxa. A **taxon range zone** (Fig. 16.5B) is based on the first and last occurrence of a single taxon. A **lineage zone** (Fig. 16.5C) is based on successive evolutionary first or last occurrences within a single lineage. An **interval zone** (Fig. 16.5D) (not to be confused with interval zones as a class of zones) is defined by two successive first or last occurrences of unrelated taxa.

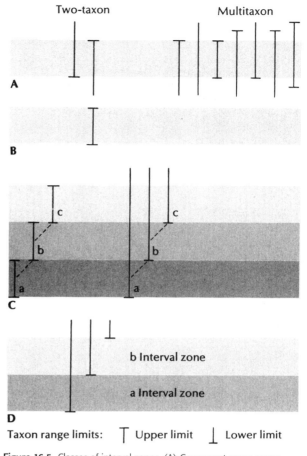

Figure 16.5 Classes of interval zones. (A) Concurrent range zones, defined by the first and last appearance of two or more taxa with overlapping ranges. (B) Taxon range zone, defined by the first and last appearance of a single taxon. (C) Lineage zones, or phylozones, defined by the evolutionary first appearance of successive taxa in a lineage. (D) Interval zones, defined by two successive first or last occurrences of partially overlapping ranges. (After Hedberg, 1976: 51.)

Assemblage zones, which are based on the association of three or more taxa, form the second class of zones. Normal assemblage zones are characterized by numerous taxa (Fig. 16.6A); thus, their boundaries are slightly vague and are recognizable only when a sufficient number of their characterizing taxa are present. This can be an advantage, however. Zones based strictly on one or two taxa are unrecognizable when those defining taxa are absent, whereas if any of the characterizing taxa of an assemblage zone are present, the zone can be recognized. An assemblage zone can be given more definite boundaries by using two or more first or last occurrences to define it (it can still be recognized by its characterizing taxa, even if the defining taxa are absent). The North American Stratigraphic Code and the International Stratigraphic Guide call this type of zone an **Oppel zone** (Fig. 16.6B) because it seems to be most similar to Oppel's original

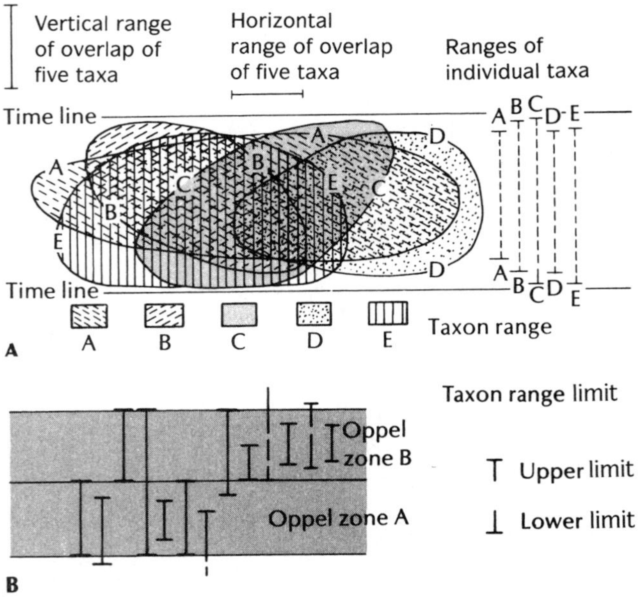

Figure 16.6 Classes of assemblage zones. (A) Typical assemblage zones are defined by a suite of taxa, so the upper and lower boundaries are vague, depending on how many taxa are used to define the zone. The actual zone of overlap of all five taxa in this example is considerably less than the total assemblage zone. (B) Oppel zones, defined by the overlap of several taxa. (After Hedberg, 1976: 57.)

taxa. Because certain fossils can overwhelm a fauna or flora at times, it is tempting to correlate peaks in the abundance of these taxa. Such an abundance change is usually the result of local ecological factors, however, and probably has little time significance. When peaks in abundance are the result of time-significant changes, such as worldwide climatic change, abundance zones can serve as time markers. For example, an opportunistic coccolith called *Braarudosphaera* bloomed in the deep sea in tremendous numbers at times of climatic stress and formed distinctive *Braarudosphaera* chalks. Because these chalks were controlled by climate, they have been used successfully to correlate rock units.

Other changes in fossil lineages can have biostratigraphic utility. For example, the coiling direction in the planktonic foraminifer *Globorotalia truncatulinoides* shifts from right-coiling in warm waters to left-coiling in cool waters (Fig. 16.7). The rapid temperature shifts in the world ocean during the Pleistocene show up in the changing coiling ratios of this microfossil. Because the temperature shifts are controlled by global climatic change, the coiling ratios are good time markers.

There are many practical problems with applying the theory of biostratigraphy. All formal biostratigraphic zonations should conform to the stratigraphic procedures outlined by the code. A zone must be formally named, usually by the species names of the two taxa that define its upper and lower limits or by the

usage (although he actually used several different types of zones; see Hancock, 1977, and Berry, 1987).

Abundance zones, also known as "peak" or "acme" zones, form the third class of zones. Abundance zones are characterized by the relative abundance of certain

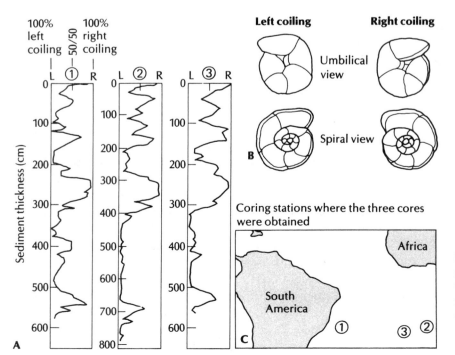

Figure 16.7 Correlation based on climatically controlled changes in coiling direction in the foraminifer *Globotruncana truncatulinoides* in the South Atlantic. This animal is predominantly right-coiling during interglacial periods but switches to left-coiling during glacial periods. (After Eicher, 1976: 115.)

name of one particularly abundant species that first appears in it. There must also be a description of the taxonomic content of the zone (including both the defining and the characterizing taxa). A type section must be designated, and the known geographic extent of the zonation must be given.

As stratigraphers have become more careful in documenting ranges and describing type sections, inevitable disputes have arisen over boundaries. The type sections or areas of two successive biostratigraphic units are usually in two separate areas, and there may be no overlap. Often, neither section preserves the boundary between the two units, so biostratigraphers must search for a third area where the transition is recorded. Ideally, this section should be as continuous and fossiliferous as possible, with several taxonomic groups to compare.

After the detailed local biostratigraphy of the available groups has been worked out, the stratigrapher must decide which biostratigraphic event(s) should serve as boundaries between units. At this point, philosophical disputes become acute. Some workers argue that boundaries should be drawn at mass extinctions and other faunal breaks because the high turnover of taxa would make the boundary easy to recognize. Others argue that boundaries should never be based on faunal breaks because the abrupt truncation of many ranges of taxa along a narrow zone probably indicates a major hiatus or unconformity.

Drawing the boundary between two time-stratigraphic units along a major gap is indeed risky. If that gap should be filled later by some other section located elsewhere, the boundary may become blurred. Some stratigraphers advocate drawing the boundary at the position of the evolutionary first occurrence of a single, distinctive taxon. If the evolutionary sequence of that taxon is relatively complete, it is less likely that a section is missing. Ager (1964, 1973, 1981) suggests that once such a boundary is agreed upon, some physical marker (metaphorically called "the Golden Spike") should be driven into the outcrop at the precise level of the boundary. Although this procedure is more arbitrary than using natural breaks, there should be no more argument once it is done. To date, a number of major boundaries have been defined internationally in this manner (Bassett, 1985; Cowie, Ziegler, and Remane, 1989): the Silurian-Devonian boundary (see Box 16.1); the base of the Cambrian (base of the *Phycodes pedum* trace fossil zone near the town of Fortune, southeastern Newfoundland, Canada); the Ordovician-Silurian boundary (base of the *Parakidograptus acuminatus* graptolite

zone at Dob's Linn, Scotland); the Devonian-Carboniferous boundary (base of the *Siphonodella sulcata* conodont zone in a trench at La Serre, southern France); the Cretaceous-Tertiary boundary (the iridium-bearing boundary clay at El Kef, Tunisia); and the Eocene-Oligocene boundary (last appearance of hantkeninid planktonic foraminifera at a quarry in Massignano, near Ancona, Italy).

As of this writing, over 40 boundaries have been formally defined by biostratigraphic criteria. Each time a committee meets to establish a boundary, it follows the procedures described in Box 16.1. New boundaries are established almost every month, and the latest status of the international committees and their decisions are listed in the quarterly journal *Episodes* (http://www.iugs.org/iugs/pubs/pubs. htm). A quick glance at the time scale published by the International Union of the Geological Sciences (IUGS) shows the current status of the zonation of the Phanerozoic (current status listed on the Micropaleontology Press website at http://micropress.org/stratigraphy/gssp.htm, and available as an Adobe Acrobat file at http://micropress.org/stratigraphy/chus.pdf. Once the boundary has been decided by international agreement, it is designated as the **GSSP** (Global Standard Stratotype Section and Point). Further details are given at the site for the International Stratigraphic Chart (http://micropress.org/stratigraphy/over.htm).

The choice of which boundary types to recognize can also vary with the biostratigraphic method used. In outcrops, many stratigraphers mark zones from the bottom (first occurrence) upward because they measure a section from the base. These "topless" zones have the advantage that if a boundary in the stratotype happens to be drawn on a local hiatus, and geologists find a new section that spans this hiatus elsewhere, the new section can be assigned automatically to the zone below (Fig. 16.8). Geologists working with subsurface data tend to use the tops of range zones.

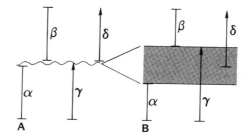

Figure 16.8 (A) Zones defined by both bottoms and tops of ranges (α and β) and "topless" zones (γ and δ). (B) Restoration of the section removed by the unconformity in A results in a gap between α and β, but no gap results from topless zones.

BOX 16.1   THE "GOLDEN SPIKE" AT THE SILURIAN-DEVONIAN BOUNDARY

The establishment of the Silurian-Devonian boundary is a good example of how such boundaries are established internationally. The boundary has been defined as the first occurrence of the graptolite *Monograptus uniformis uniformis* in bed 20 (Fig. 16.1.1) of the section at Klonk, the Czech Republic (Chlupác, Jaeger, and Zikmundova, 1972; McLaren, 1973, 1977). The name *Klonk* is a fortunate coincidence, because the sound of it suggests the driving of the "Golden Spike." But the area was not chosen because of its name. A number of areas around the world were considered by a group of stratigraphers who were familiar with faunas that spanned the Silurian-Devonian boundary. As reviewed by McLaren (1973, 1977), the Silurian-

Devonian Boundary Committee of the Commission of Stratigraphy, IUGS, was organized in Copenhagen in 1960 and gave its final report and recommendations in 1972.

The committee's first problem was that the type areas of both systems were inadequate. Murchison based the Silurian on marine rocks in the Welsh Borderland found under the nonmarine Devonian Old Red Sandstone. When Murchison and Sedgwick found marine Devonian rocks in southwestern England, they eventually recognized that this sequence was equivalent to the Old Red Sandstone and was younger than the Silurian. Unfortunately, the base of the marine type Devonian is not exposed in Britain, and the top of the marine type Silurian is capped by nonmarine rocks. Because neither Wales nor Devon nor any other area in Britain contained a good sequence spanning the Silurian-Devonian boundary, the old type area had to be set aside and other areas considered.

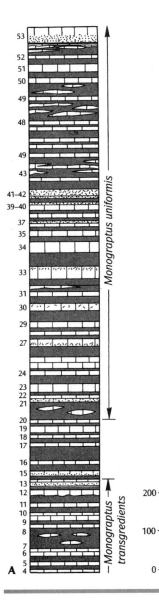

B

Figure 16.1.1 (A) Type section of the Silurian-Devonian boundary, near Klonk, Czech Republic. The boundary is defined as the first appearance of the graptolite *Monograptus uniformis,* which occurs in bed 20 in this section. (After Chlupác, 1972.) (B) Photograph of the Klonk section. Bed 20 is visible just above the point of the hammer. (Courtesy of I. Chlupác.)

Another important consideration came from studies in Bohemia (now the Czech Republic) of sequences first described by the pioneering geologist Joachim Barrande in 1846. These sequences had been important in Murchison and Sedgwick's disputes over the Silurian because they were much thicker, more complete, and more fossiliferous than the type areas in Britain (Secord, 1986). By 1960, it was apparent to the committee that the "Barrandian" sections included a latest Silurian stage, the Lochkovian, that was missing from the British sequence. The Barrandian was thus one of the prime candidates for a type section.

Nevertheless, the committee looked into several other candidate areas and considered many different fossil assemblages that could be used to define the boundary. After discussing graptolites, conodonts, corals, brachiopods, trilobites, ostracods, crinoids, vertebrates, and plants, they settled on graptolites. Because graptolites were pelagic, their fossils occur worldwide and make the best zonal indicators of all the candidates. By 1968, the committee had chosen the base of the *Monograptus uniformis* zone as the base of the Devonian.

Now the task was to determine the place where this zone is best exposed. They sought a well-studied rock section with good exposures. The fossils of this section should be abundant at every level, thoroughly studied, and well documented. Ideally, the best section would be the one that is most complete, with no evidence of missing or truncated zones. If it should contain a gap, and a section were found later that spanned the gap, then the sharp boundary could become blurred. Finally, the type section should be easy to reach and should remain accessible to geologists of any nationality who wanted to study the section and collect fossils, then and in the future.

In addition to the Barrandian in the Czech Republic, sections were studied in three places in the former Soviet Union; the Carnic Alps of Austria and Italy; the Holy Cross Mountains of Poland; Aragon in Spain; the Roberts Mountains in Nevada; the Gaspé in Quebec, and additional areas in Algiers, Morocco, Australia, Thailand, the Yukon, and the Canadian Arctic. By 1971, the choice was narrowed to four candidates: Podolia in the former Soviet Union, Morocco, Nevada, and the Barrandian in the Czech Republic.

In August 1972, the committee met during the International Geological Congress in Montreal. They weighed the pros and cons of each section and finally chose the Barrandian by a vote of 31 to 1. The Barrandian was selected because it was thoroughly described and had a rich and uninterrupted succession of well-studied graptolites, conodonts, and other fossils. There were enough facies changes within the Barrandian sections to make positive correlations with other benthic fossil assemblages. Two Barrandian sections were voted on, and the one at Klonk was chosen. Another vote was taken on whether to draw the boundary at bed 20, the base of the *M. uniformis* zone (base of the Devonian), or at bed 13, the top of the *M. transgrediens* zone (top of the Silurian). There was a 170-cm gap between these beds, with no fossils. After discussion, the base was drawn at the bottom of bed 20 (see Fig. 16.1.1).

McLaren (1973, 1977) reviewed some of the lessons learned from this experience. Although the procedure sounds very legalistic, such a process is necessary if stratigraphers are to communicate effectively. As McLaren (1973, p. 23) put it, "Definition is not science, but a necessary prerequisite to scientific discussion." If scientists cannot agree on what a name means or how to recognize it, then they end up talking about different things. This can hinder their ability to describe what they are thinking about and to communicate ideas.

The committee's work also encouraged international cooperation and discussion and cleared up many misconceptions and miscorrelations that would have taken years to discover. One of the early misconceptions was the idea that rock units could be used as time markers. Some geologists wanted the boundary to be drawn at the Ludlow Bone Bed in the Welsh Borderland, the traditional top of the type Silurian. As we saw in Chapter 15, rock units rarely are worldwide and synchronous enough to make good chronostratigraphic boundaries.

McLaren (1973) commented on the remarkable collegiality of the delegates to the committee, who did not form into warring factions representing particular nations or schools of thought (as they easily could have). The democratic procedure used to debate and decide the matter made the results accessible and acceptable to the entire geological community.

Finally, we have seen how the priority of the name or type area, although important, may not always be the deciding factor. The type areas of the Silurian and Devonian in England were both inadequate for the purpose of defining their mutual boundary. Similarly, the type area for the base of the Cambrian is now in Canada, not in Sedgwick's sections in Wales. The type section for the base of the Ordovician is not in Murchison's sections in Wales but is in Scotland because Lapworth (1879) had found better fossils there and had defined the Ordovician in 1879 with primary reference to Scotland.

350

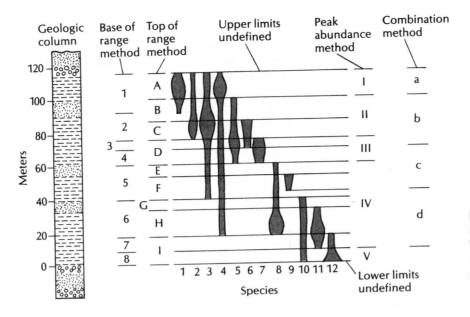

Figure 16.9 Four methods of recognizing concurrent range zones in a single section. Widths of lines indicate relative abundances of specimens. Note that each method gives a slightly different zonation. (After Eicher, 1976: 109.)

Not only are they accustomed to drilling downward, but the drilling mud may continue to bring up fossils from the borehole wall long after the base of a particular biostratigraphic zone has been passed. Other geologists have tried to use one of these methods in combination with abundance data, which are particularly striking in some sections (Fig. 16.9).

The decision about whether to use the tops or bottoms of ranges depends upon a number of factors. The first appearance datum in a local section or core rarely coincides precisely with its time of evolution. The actual first occurrence is likely to be somewhat later because it takes time for the organism to disperse from its site of evolution to the sampling site. Its first occurrence may be lower in the section than it really should be, however, because of misidentification, downhole caving, or reworking. Its first occurrence may be much later than its actual evolution because of incomplete sampling or because the local environment was unsuitable for preservation. Similar arguments apply to the probability of identifying a fossil's last occurrence. The actual last occurrence might be earlier than it appears to be because of misidentification, reworking, or uphole contamination. The actual last appearance might be later than it appears as a result of incomplete sampling or local extinction due to facies changes. The theoretical probability curves are likely to be modified by local circumstances. For example, the characteristics of an organism (its ability to disperse or its facies sensitivity) can change the shape of the curve; so can the type of sample (core, outcrops, or cuttings). The

practicing biostratigrapher soon gains a sense of which problems are most serious in a given situation as he or she learns the idiosyncrasies of each particular geologic settings.

The Time Significance of Biostratigraphic Events

What types of biostratigraphic events make the best boundaries? Many authors (for example, Murphy, 1977) consider the evolutionary first occurrence (lineage zone) to be the best marker because the evolution of a taxon is an unrepeatable event that happens at a unique time and place. There are drawbacks to this method, however. Evolutionary first occurrences are rare, and they may not appear in the sections of interest. If Eldredge and Gould (1972) are right in their theory of punctuated equilibrium, there are few examples of gradualistic change in the fossil record. Gould and Eldredge (1977) claimed that most "gradualistic" evolutionary changes in taxa do not hold up under close scrutiny. Many cases of apparent gradualism may be a result of repeated immigration events rather than of the evolutionary change of a lineage in situ. If gradual change actually does occur and is used to determine boundaries, then problems of definition arise. Does one define the evolutionary first occurrence of a species as the first appearance of the character that defines the species, or as the point when 50% of the population has developed this char-

acter? In this case, there is a conflict between species defined for biostratigraphic purposes and the biological meaning of species. The definition of species by the 50% criterion, for example, would be considered unnatural by a biologist.

Some geologists (for example, Repenning, 1967) have argued that immigrational or emigrational events should be preferred for boundaries. Immigrational first occurrences, in particular, can be abrupt and unambiguous. The dispersal of most marine organisms, particularly those with planktonic larval stages, is very rapid by geologic standards (Scheltema, 1977). This is also true of land vertebrates (Lazarus and Prothero, 1984; Flynn, MacFadden, and Mckenna, 1984). The main problem with this marker is that organisms immigrate to different places at different times. The notorious "*Hipparion* datum," based on the first appearance of the three-toed horse *Hipparion* in the Old World, was once used as the base of the Pliocene (Colbert, 1935). Recent work (Woodburne, 1989) has shown that the *Hipparion* immigration "event" was at least two different immigrations of different hipparionine horses at different times during the Miocene.

Extinctions are probably less reliable boundaries because it is well known that taxa can linger in a refuge long after they have disappeared elsewhere. Even so, extinctions can be attractive boundaries because they often cluster at horizons that represent mass extinctions, which are major episodes in the history of life.

Stratigraphers have long assumed that biostratigraphic events are reasonably synchronous within the resolving power of the stratigraphic record. This assumption was criticized by Thomas Henry Huxley (1862), who pointed out that biostratigraphers have demonstrated only *similiarity in order of occurrence* (**homotaxis**), not synchrony. In recent years, other, independent techniques have been employed to test the time significance of biostratigraphic events. Hays and Shackleton (1976) used oxygen-isotope stratigraphy to show that the extinction of the radiolarian *Stylatractus universus* was globally synchronous. Haq and others (1980) conducted a similar test using carbon isotopes. Berggren and Van Couvering (1978) stated that asynchrony, or **heterochrony,** of biostratigraphic datum planes is rare in the marine realm. Yet Srinivasan and Kennett (1981) showed that the last appearances of some marine microfossils can be heterochronous. Some marine microfossils, such as the benthic foraminifers, are notorious for being facies-controlled and heterochronic, as we saw in the example of the North American Pacific coast Eocene benthic foraminifers (McDougall, 1980).

In land sections, Prothero (1982) and Flynn, MacFadden, and McKenna (1984) tested mammalian biostratigraphy against magnetostratigraphy and found that mammal *assemblages* do not show significant heterochrony as a rule, although the stratigraphic occurrence of *individual taxa* can be significantly time-transgressive. These studies demonstrated that the basic assumptions of biostratigraphy must be constantly reexamined when new tools of correlation are developed. Most of the methods and assumptions of biostratigraphy have proved to be quite robust. This is hardly surprising, because biostratigraphy has been used successfully since the days of William Smith. For these reasons, biostratigraphy will probably remain the primary tool for correlation and for determining relative age.

Index Fossils

Some fossils are so abundant and characteristic of key formations that they are known as **index fossils,** and many formations in North America can be recognized by such fossils. Information about index fossils was cataloged in Shimer and Shrock's *Index Fossils of North America* (1944). Most fossil groups (including index fossils) that make good biostratigraphic indicators are distinctive, widespread, abundant, independent of facies, rapidly changing, and short-ranging. From these properties, one can see that the best biostratigraphic indicators are pelagic organisms that evolve rapidly. Pelagic organisms live in the surface waters of the oceans, so they can disperse quickly around the world. They are unaffected by the various bottom facies that control benthic organisms. Practical experience has shown that ammonoids, graptolites, conodonts, foraminifers, and other planktonic microfossils give the best results, although some benthic groups are also useful.

There are also some problems with using index fossils for correlation. Too often, the index fossil is equated with the formation, and no attempt is made to document the *actual range* of the fossil *within* the formation. This results in a loss of resolution of the data. The stratigraphic range of the fossil is often reported to be the same as the total thickness of the formation, which may extend the range artificially. In addition, the use of index fossils can cause confusion of rock units with time-rock units and imply that rock units are time-equivalent. Too much reliance on one index

fossil can seriously limit a biostratigrapher who works in areas where the index fossil does not occur. The absence of a key index fossil does not necessarily mean that the rocks are not of the age of that index fossil. Finally, Shaw (1964, p. 99) pointed out that a reliance on index fossils can lead to other abuses. For example, the rote memorization of index fossils and their formations ("*Spirifer grimesi* is the index fossil of the Burlington Limestone") can lead a geologist to recognize formations by their fossils rather than by their lithologies. The tedium of having to memorize lists of index fossils also tends to drive many good young minds away from paleontology.

North American Land Mammal "Ages" and Biochronology

Land mammal fossils have long been used as stratigraphic markers in Cenozoic terrestrial deposits. Most fossil mammals occur in isolated quarries and pockets, however, and are seldom distributed evenly through thick stratigraphic sections, as are marine invertebrates. As a result, mammalian biostratigraphers have not always followed classic Oppelian biostratigraphic procedures. A few thick sections that show superposition of faunas are known, but as a whole, the sequence of mammalian faunas must be worked out indirectly. The result has been called **biochronology**. Williams (1901) defined a **biochron** as a unit of time during which an association of taxa is interpreted to have lived (Woodburne, 1977). A biochron is equivalent to the biozone of an assemblage of taxa, except that it is not directly tied to any actual stratigraphic sections.

Tedford (1970) showed that classic biostratigraphic procedures were used by early North American mammalian paleontologists (for example, the "life zones" of Osborn and Matthew, 1909). These procedures fell into disuse, and a biochronological sequence was built up by piecing together a sequence of faunas based on their stages of evolution. In 1941, the sequence of mammalian fauna was formally codified by a committee chaired by Horace E. Wood II (Wood et al., 1941). The Wood Committee set up a series of provincial land mammal "ages" that were based on the classic sequence of mammal fauna in North America. These came to be known as North American Land Mammal "Ages." However, they are not true ages in the formal time-stratigraphic sense because they are not built from biostratigraphic zones based on actual rock sections.

The Wood Committee attempted to define each unit unambiguously by listing index fossils, first and last occurrences, characteristic fossils, and typical and correlative areas. These multiple criteria have since led to conflicts that would not have occurred if classic biostratigraphic methods had been followed originally. For example, the late Eocene Chadronian Land Mammal "Age" is defined by two criteria: (1) the co-occurrence of the horse *Mesohippus* and titanotheres and (2) the limits of the Chadron Formation. At the time of the Wood Committee's work, the last occurrence of titanotheres was thought to coincide exactly with the top of the Chadron Formation, so there was no conflict. Since then, titanotheres have been found in rocks above the Chadron Formation (Prothero, 1982; Prothero and Whittlesey, 1998). Now we must choose between conflicting criteria. Is the end of the Chadronian marked by the last appearance of titanotheres or by the top of the Chadron Formation?

In recent years, mammalian paleontologists have made efforts to return to classic biostratigraphic principles (reviewed by Woodburne, 1977, 1987). Much of the mammalian time scale is being restructured in terms of biostratigraphic range zones tied to local sections (Woodburne, 2003). There are parts of the Tertiary, however, that may never be zoned, so biochronology is still widely used. Its robustness was demonstrated when radiometric dates first became available for the land mammal record and showed that the biochronological sequence was in the correct order (Evernden et al., 1964). European paleomammalogists, who have fewer good stratigraphic sections to work from, use an explicitly biochronological approach. Each fauna is placed in order according to its stage of evolution. *Niveaux repères*, or reference levels, of classic faunas are used in place of type sections (Jaeger and Hartenberger, 1975; Thaler, 1972).

Quantitative Biostratigraphy

A biostratigraphic database can become so large that it is impossible for a biostratigrapher to find the pattern among so many stratigraphic sections with so many taxonomic ranges. In recent years, techniques have been introduced to quantify and handle large databases, mostly by computer.

The most popular method, **graphic correlation** or "Shaw plots," was introduced by Alan Shaw in his book *Time in Stratigraphy* (1964). Much of the work can be done on graph paper with the aid of a pocket calcu-

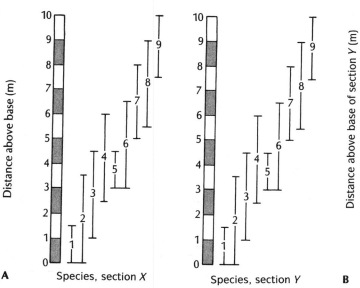

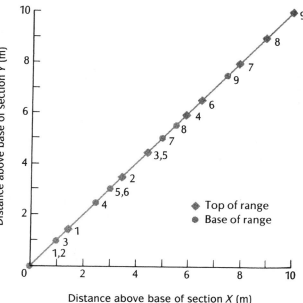

Figure 16.10 Shaw's graphic correlation method. (A) Two sections with the stratigraphic distribution of nine taxa. (B) Correlation between sections *X* and *Y*. Because the ranges in both sections are identical, the line of correlation between the two sections is straight and has a 45° slope. (After Eicher, 1976: 112, 113.)

lator, although the entire procedure is usually done on a computer. In essence, it uses the statistician's meaning of the word *correlation*. When two variables are correlated, they form a cluster of points on a bivariate data plot. The statistician then attempts to fit a line of correlation to the cluster. First and last occurrences of fossil taxa become data points along an axis that represents the stratigraphic section. If two stratigraphic sections are placed along the two bivariate axes (abscissa and ordinate), the data points in both sections can be plotted in bivariate space (Fig. 16.10). The line connecting these points is the line of correlation.

The line of correlation is the most powerful aspect of Shaw's method. If all the points fit exactly on the line, there is perfect correlation; outlying points immediately become apparent and can be examined to see if they represent artificial or real range extensions or truncations. Real data sets seldom show perfect correlation, and the scatter of points near the line is a measure of the scatter of the data.

If the two sections are plotted to the same scale, the slope of the line is particularly informative. A line with a perfect 45° slope (Fig. 16.10B) indicates that the two sections have identical rates of rock accumulation and identical distributions of fossils. (Shaw speaks of "rock accumulation" because postdepositional compaction can shorten sections and distort true rates of sedimentation.) Usually, the rates of accumulation

in two sections are not identical, in which case the ranges in one section will be proportionately shorter (Fig. 16.11A). The plot of these two sections will not result in a line with a 45° slope but will be a line inclined toward the axis that represents the section with the higher rate of accumulation (Fig. 16.11B). If the rate of accumulation during deposition changes in one section relative to the other, the slope will change, giving a "dogleg" pattern (Fig. 16.12). If there is a hiatus in one of the sections, the ranges will tend to truncate at that level (Fig. 16.13A); the resulting plot will have an obvious "terrace" representing the missing section (Fig. 16.13B). The graphic correlation method is a powerful tool for spotting bad range data, unconformities, and missing sections and for interpreting rates of rock accumulation.

Shaw's method has applications beyond the correlation of two sections. It can be used as the basis for a large-scale correlation scheme among multiple sections. Shaw recommends that the stratigrapher begin by correlating the two best sections in the study area. Any outliers on the line representing true range extensions are added to the best reference section, producing a **composite standard.** The composite standard can then be correlated with each additional section, one at a time, and any further range extensions can be added to it. The result is a composite standard that is not a real section but a synthesis of all the information

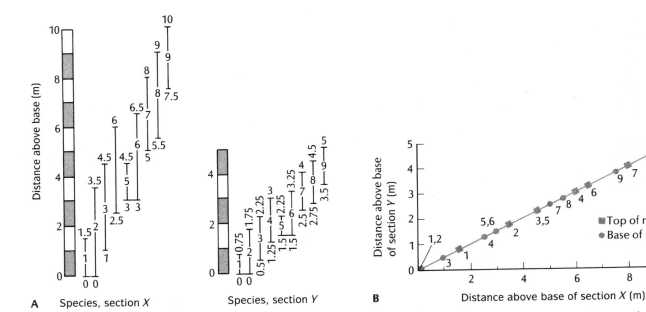

Figure 16.11 (A) Two sections that have the same relative spacing of ranges but different rates of rock accumulation. (B) Graphic correlation of these two sections. The points form a straight line, but the slope is deflected toward the axis of the section with the higher accumulation rate. (After Shaw, 1964.)

from a group of sections. The composite standard ranges are not teilzones but are the maximum ranges of each taxon in the area. If this were done for every section in the world, the total range zone would be approximated. The subdivisions of the composite standard are no longer measured in thicknesses of section but are abstract units that approximate geologic time planes. These can be used to correlate and align sections, producing results that show patterns more clearly than correlations based on lithologic boundaries or on individual biostratigraphic planes.

Because Shaw's methods are easily learned and are applicable to a variety of biostratigraphic problems, they have had widespread use in the oil industry (Miller, 1977). Simple graphic correlation plots can be used routinely when plotting any two sets of stratigraphic data, including magnetostratigraphic data and isotopic data. A number of other methods of quantitative biostratigraphy have been developed, but the graphic correlation method is by far the most commonly used. Further information on other quantitative methods can be found in Prothero (1990), Cubitt and Reyment (1982), and Kauffman and Hazel (1977).

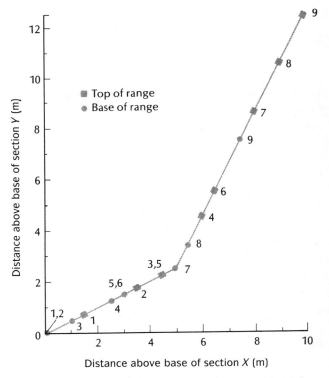

Figure 16.12 Graphic correlation of two sections that show a change in the rate of rock accumulation. This correlation forms the characteristic "dogleg" kink in the slope of the line. (After Eicher, 1976: 114.)

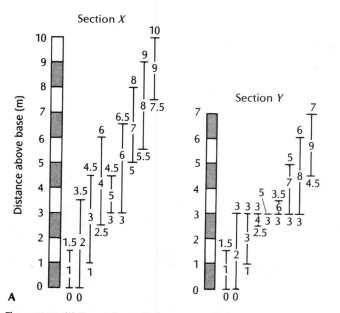

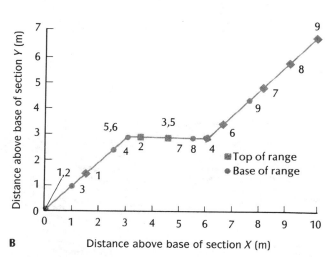

Figure 16.13 (A) Two sections with the same taxa, but one section has an unconformity that truncates the ranges of some taxa. (B) Graphic correlation plot of these two sections. The unconformity causes a "step" or "plateau." The section with the unconformity (*Y*) shows no change, whereas the complete section (*X*) continues to spread out the points. (After Eicher, 1976: 113.)

CONCLUSIONS

Since the early days of geology and the principle of faunal succession, geologists have recognized that fossils are the only practical means of determining the age of most sedimentary rocks. As long as geologists need to know the age of their rocks, there will always be a need for paleontologists. Yet biostratigraphy is not as simple as identifying the fossil and pinning down the age. The distribution of fossils can be affected by many factors that make them difficult to use, or it may give misleading results. Understanding these fundamental problems, assumptions, and limitations of biostratigraphy is essential for any geologist who wants to know the reliability of a biostratigraphic age estimate.

FOR FURTHER READING

Aubry, M.-P., W. A. Berggren, J. A. Van Couvering, and F. Steininger. 1999. Problems in chronostratigraphy: Stages, series, unit and boundary stratotypes, global stratotype section and point, and tarnished golden spikes. *Earth-Science Reviews* 46: 99–148

Cubitt, J. M., and R. A. Reyment, eds. 1982. *Quantitative Stratigraphic Correlation.* New York: John Wiley and Sons.

Eicher, D. L. 1976. *Geologic Time,* 2d ed. Englewood Cliffs, N.J.: Prentice-Hall.

Gradstein, F. M., J. P. Agterberg, J. C. Brower, and W. J. Schwarzacher, eds. 1985. *Quan-titative Stratigraphy.* Dordrecht, Netherlands: D. Reidel.

Griffiths, C., and F. Gradstein. 1997. *Essentials of Quan-titative Stratigraphy.* London: Chapman and Hall.

Kauffman, E. G., and J. E. Hazel. 1977. *Concepts and Methods of Biostratigraphy.* Stroudsburg, Pa.: Dowden, Hutchinson, and Ross.

Mann, K. O., and H. R. Lane, eds. 1995. *Graphic Correlation.* SEPM Special Publication 53.

Prothero, D. R. 2001. Chronostratigraphic calibration of the Pacific Coast Cenozoic: A summary. *Pacific Section SEPM Book* 91: 377–394.

Shaw, A. B. 1964. *Time in Stratigraphy.* New York: McGraw-Hill.

Woodburne, M. O., ed. 2003. *Late Cretaceous and Cenozoic Mammals of North America.* New York: Columbia University Press.

17 Geophysical and Chemostratigraphic Correlation

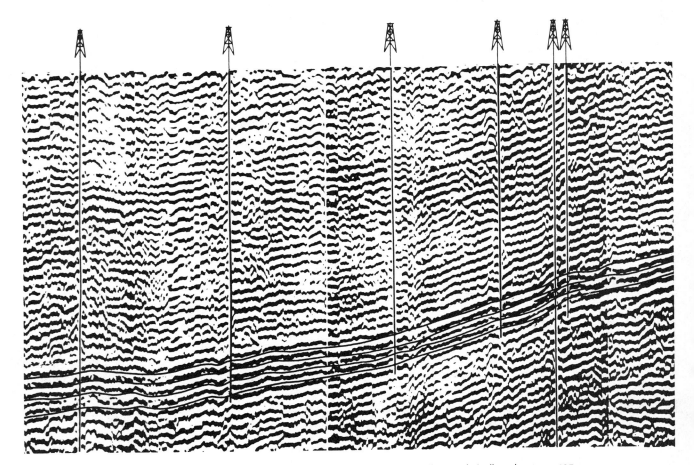

Seismic section from a Tertiary basin in South America, showing well control. (Vail et al., 1977a: 105; courtesy of American Association of Petroleum Geologists, Tulsa, Okla.)

WE HAVE EXAMINED THE CORRELATION OF SECTIONS based on the lithologic features of the rocks (Chapter 15) or on their fossil content (Chapter 16). In this chapter, we discuss methods of correlation that depend on the geophysical or geochemical properties of rocks. Many such correlation techniques have been developed relatively recently and are just proving their stratigraphic potential. Each method has certain strengths and limitations. These methods include examining the geophysical and geochemical properties of rocks in the walls of a drillhole (well logging), bouncing seismic waves off subsurface layers to determine their structure (seismic stratigraphy), measuring the magnetic properties of rocks to correlate their changes in polarity (magnetic stratigraphy), and measuring the changes in ratios of stable isotopes (chemostratigraphy).

Well Logging

One of the oldest and most widely used methods that depends on the geophysical properties of rocks is subsurface **well logging.** Many geologic problems cannot be solved on the basis of surface outcrops. This is particularly true in the areas where outcrops are limited, such as the midcontinent of North America. In some places, tens of meters or more of glacial debris or soil horizons cover nearly all outcrops. Seismic reflection

can be used to determine the subsurface structure of these covered areas, but a direct sample of the formation is needed to be sure of the lithology of the rocks from which the seismic reflections are coming. The most practical way to obtain this subsurface information is to drill a well and record all the useful information possible from a core of the drilled sequence.

Continuous core recovery from most wells is far too expensive (thousands to hundreds of thousands of dollars) for more than occasional use, except when great detail is needed. In modern drilling, a lubricating mud is pumped down through the drill pipe (Fig. 17.1) to cool and lubricate the drilling area. The drilling mud (usually a mixture of bentonitic clay and oil or water, plus barite to regulate the density) is then forced up the well to the surface so that mud constantly flows toward the mud tanks, carrying the cuttings, or chips, away from the drilled formation to prevent the well from clogging. These chips are a primary source of information about the subsurface unit. The site geologist usually keeps a continuous log and samples of the chips as they come up and are screened

out of the drilling mud. The chips are the only direct record of the lithology of the sequence as it is encountered by the drill bit. If the density of one lithology greatly differs from the next, however, the chips may rise to the surface at different rates and give an erroneous impression of the sequence. The driller also keeps a log of the drilling rate, which indicates the degree of induration of the formation.

The second and more important source of stratigraphic information comes after the well has been drilled and the drill pipe removed. A **caliper** lowered down the hole measures the hole diameter as it is pulled upward. These measurements not only indicate the presence of shales, which are prone to caving, but also are important for calculating the response of the logging tools. An exploring device, called a **sonde,** is then lowered down the hole; as it is raised, various electrical and other properties of the hole are measured (Fig. 17.2). A whole suite of measurements can be logged, depending on what information the driller is seeking. Usually, several measurements are made simultaneously, and the results are combined for a better understanding of the logged lithostratigraphic sequence. Grouped by the properties that they measure best, the various types of well logs include the following:

Bulk properties of the rock:	Dipmeter and sonic logs
Compositional aspects:	Gamma ray log
Fluid properties of pores:	Spontaneous potential (SP) log, resistivity (R) log, and compensated neutron log (CNL)
Gases in the well:	Gas detector and gas chromatograph

Spontaneous Potential and Resistivity Logs

Of the methods just listed, the two most commonly used are the spontaneous potential and resistivity logs. These properties are usually measured together, with spontaneous potential conventionally plotted on the left and resistivity on the right (Fig. 17.3).

The **spontaneous potential (SP) log** measures the difference in electrical potential between two widely spaced electrodes, a grounded electrode at the top of the well and an electrode on the sonde. Drilling mud invades the pore spaces of the rock adjacent to the well. When the drilling mud and the natural pore

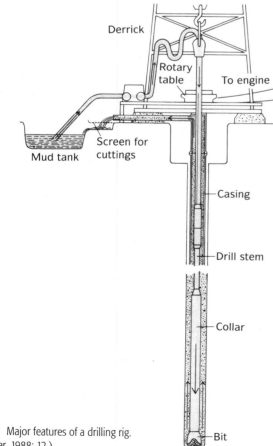

Figure 17.1 Major features of a drilling rig. (After Siever, 1988: 12.)

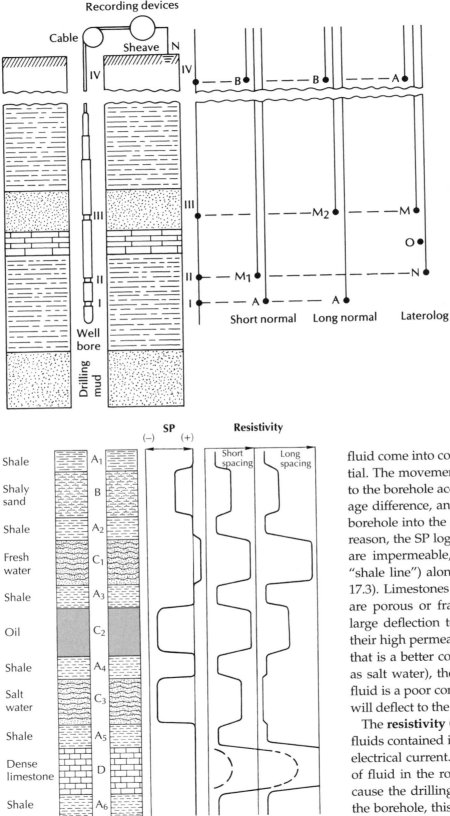

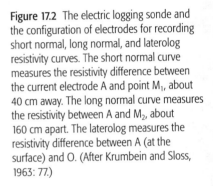

Figure 17.2 The electric logging sonde and the configuration of electrodes for recording short normal, long normal, and laterolog resistivity curves. The short normal curve measures the resistivity difference between the current electrode A and point M_1, about 40 cm away. The long normal curve measures the resistivity between A and M_2, about 160 cm apart. The laterolog measures the resistivity difference between A (at the surface) and O. (After Krumbein and Sloss, 1963: 77.)

Figure 17.3 Idealized spontaneous potential and resistivity curves for various combinations of rock types and contained fluids. (After Krumbein and Sloss, 1963: 78.)

fluid come into contact, they set up an electrical potential. The movement of ions from the drilled formation to the borehole accounts for 85% of the measured voltage difference, and invasion of drilling mud from the borehole into the formation accounts for 15%. For this reason, the SP log is a measure of permeability. Shales are impermeable, so their reading is near zero (the "shale line") along the right edge of the log (see Fig. 17.3). Limestones are low in permeability unless they are porous or fractured. Sandstones usually show a large deflection toward the negative pole because of their high permeability. If the sonde encounters a fluid that is a better conductor than the drilling mud (such as salt water), the curve will deflect to the left; if the fluid is a poor conductor (such as fresh water or oil), it will deflect to the right.

The **resistivity (R) log** measures the resistivity of the fluids contained in the surrounding rock to an applied electrical current. The resistivity indicates the amount of fluid in the rock and therefore the pore space. Because the drilling mud invades the porous rock from the borehole, this log actually measures the difference in resistivity between the invaded zone (mud has a high resistivity) and the uninvaded zone (natural pore fluids have a lower resistivity). Resistivity increases

with decreasing pore space; 10% porosity is about 10 times more resistive than 30% porosity.

Dense rocks with no pore space (such as limestones and quartzites) have very high resistivities, so they usually deflect the record to the right, even off scale. Off-scale peaks are usually truncated, and their tops begin again at the zero line. In addition, the R profile for limestone is typically jagged because of solution cavities or fractures. Sandstones filled with a nonconducting fluid (such as oil or fresh water) also deflect to the right. Shales, on the other hand, have low resistivities and deflect to the left.

The spacing of the electrodes is critical. For example, because drilling muds are usually less saline and more resistive than pore waters, a salt-water-bearing unit infiltrated by drilling mud shows a high resistivity and can be mistaken for an oil sand. To obtain the true resistivity beyond the invaded zone, the spacing between electrodes must be increased. A typical R log has two curves, one for **short normal spacing** (about 40 cm of separation) and one for **long normal spacing** (about 160 to 180 cm of separation). Even longer spacing can be obtained by placing a third electrode (O in Fig. 17.2) between the measuring electrodes and recording the resistivity between O and the surface (A in Fig. 17.2). This extralong spacing (called **laterolog**) measures the resistivity at a long distance from the bore and is least sensitive to mud invasion. The differ-ence between the short and long spacing of R logs is shown in Fig. 17.3. Fresh water gives a large deflection to the right with long spacing, but oil shows almost no change. A permeable salt-water-bearing unit, on the other hand, deflects to the left on the long spacing and to the right on the short spacing.

Other Logs

Other types of logs measure other properties of the borehole. The **dipmeter** has four electrodes spaced 90° apart that measure the resistivity simultaneously in four directions. A rock bed has the same resistivity on all sides of the borehole. Therefore, if the bed is horizontal, the resistivities in the four directions will be the same (Fig. 17.4A). If the bed dips, however, one or more of the electrodes may encounter a different bed with some other resistivity value. The computer selects similar resistivities on the four different curves to calculate a dip. The continuous log of changes in dips is used not only to determine regional structure and unconformities but also to detect changes in the bedding attitudes of sedimentary structures. Certain assemblages of dips indicate a diagnostic suite of sedimentary structures that aid in paleoenvironmental interpretation (Fig. 17.4B).

The **gamma-ray log** measures the natural radioactivity of the strata. Unlike the other methods, it can be used even after the borehole has been cased in

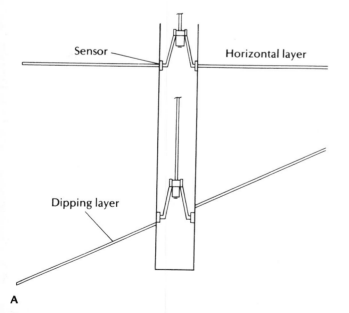

A

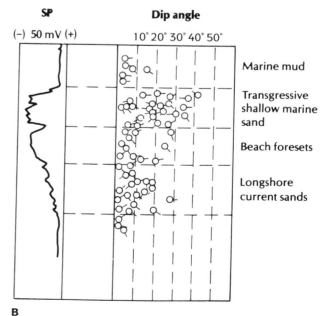

B

Figure 17.4 (A) A dipmeter, which is used to determine the dip of rock layers in a well. If the bed is horizontal, all the electrodes detect the bed simultaneously; if the bed dips, then one electrode detects it before the others. (From Hyne, 1984.)

(B) Environmental interpretation using a high-resolution dipmeter. The consistency or variability of the dips and the steepness of their dip angle can be characteristic of certain facies. (After Gilreath, Healy, and Yelverton, 1969: 108.)

cement to prevent it from collapsing. Because most of the gamma radiation comes from the decay of ^{40}K in the minerals of the surrounding rock, this method is sensitive to rocks that are high in potassium. Potassium feldspars, micas, clays, and organic material have the highest potassium content; quartz and calcite have none. Thus, shales and arkosic, lithic, or muddy sandstones will give high gamma-ray values, and limestones and quartz-rich sandstones will give very low ones.

Sonic, or acoustic, **velocity logs** use the methods of seismology on a very small scale. By measuring the

TABLE 17.1 Wire–Line Well Logs

Name	Measures	Primary Uses	Comments
Electrical log			Older type of log
Spontaneous potential (self-potential), SP	Permeable vs. shale beds	Correlation, location of reservoir rocks	Reservoir rocks kick to left
Resistivity, R	Electrical resistivity		
Short normal	Adjacent to well bore	Tops and bottoms	Oil and gas gives kick to far right
Long normal	Away from well bore	Identification of reservoir fluids	Higher oil saturations kick further to right
Induction log, dual induction log	Measures SP, R, and conductivity	Same as electrical log; shallow induction same as short normal and deep induction same as long normal	Used in well filled with any type of drilling mud or air; No SP measurement in air or oil drilling mud; Modern type of log
Gamma-ray log	Natural radioactivity of rocks	Correlation, tops and bottoms, location of reservoir rocks, shale content	Shale kicks to right; used in cased and uncased wells
Neutron porosity log, neutron log	Hydrogen atom density	Porosity	Reads low on gas reservoirs; used in cased and uncased wells
Formation density log, density log	Density of rock	Porosity	Must know lithology (matrix)
Acoustic velocity log, sonic log, velocity log	Sound velocity through rock layer; measures interval transit time Δt	Porosity and correlation	Must know lithology (matrix)
Caliper log	Size of well bore	Engineering calculations; Calibration of other logs	Thick filter cake (small hole) indicates permeable zone
Dipmeter, dip log	Orientation of subsurface rocks	Interpretation of structure and depositional environment	Uncased well

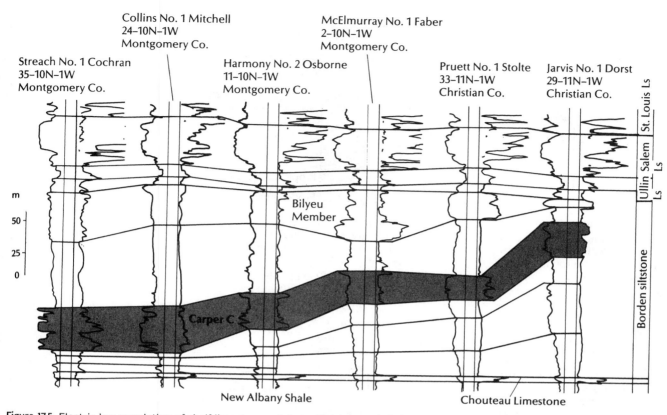

Figure 17.5 Electric log correlation of shelf limestones, deltaic siltstones and shales, and delta-front turbidite sands (Carper C) from the Mississippian of central Illinois. (From Lineback, 1968: 15; by permission of the Illinois Geological Survey.)

velocities of sound waves traveling through rock, geologists can estimate the density and porosity of the rock. The logging tool has a sound transmitter at one end and two receivers spaced along it. Both receivers pick up pulses sent out by the transmitter, with the more distant receiver picking them up slightly later. From the difference in receiving times and the distance between the receivers, the sonic velocity through the rock can be calculated. Shale has the lowest sonic velocity, followed by sandstone, limestone, and dolostone, which has the highest. As porosity increases, sonic velocity decreases because the gas or liquid in the pore spaces has a much lower sonic velocity than the rock. If the lithology is known, the porosity can be calculated.

There are a number of other well-logging methods of lesser importance, and new methods and refinements on old methods are developed continuously. Table 17.1 summarizes some of the more important methods and their relative strengths and weaknesses. After a geologist uses these methods, they become almost second nature. For example, the lithologic interpretation depicted in Figure 17.5 shows how limestone, shale, and sandstone curves can be matched for

a considerable distance. In many cases, the shapes of the curves indicate not only the lithology and pore water but also the geometry of the unit. Many of the classic stratigraphic sequences that are characteristic of particular depositional environments have characteristic shapes on the SP log (Fig. 17.6). Well-log interpretation is not always straightforward, however. There are many subtle variations in rocks and in the mechanical behavior of the logging method that can be misleading. It always pays to have a good understanding of *why* a particular method works in case the rules of thumb break down.

Seismic Stratigraphy

Waves in the Earth

One of the most powerful and rapidly growing fields of stratigraphy and geophysics is **seismic stratigraphy.** Long used to determine the deep structure of the Earth and the shallow surface structure for oil exploration, seismic stratigraphy has reached new levels of sophistication and information content since about 1985. Now it is possible not only to recognize

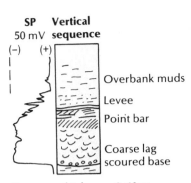

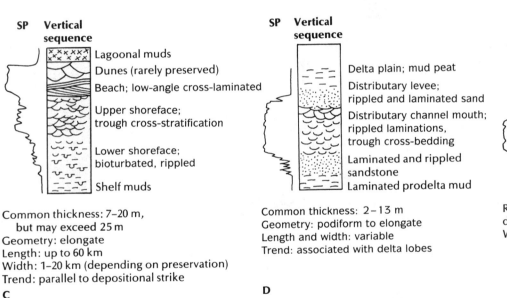

A

Common thickness: 2–10 m
Geometry: multistory sheet or shoestring
Width and length: extremely variable
Trend: roughly perpendicular to
    depositional strike

B

Common thickness: 3–13 m
Geometry: sheet or shoestring
Length: up to tens of kilometers
Width: variable
Trend: parallel to depositional strike

C

Common thickness: 7–20 m,
    but may exceed 25 m
Geometry: elongate
Length: up to 60 km
Width: 1–20 km (depending on preservation)
Trend: parallel to depositional strike

D

Common thickness: 2–13 m
Geometry: podiform to elongate
Length and width: variable
Trend: associated with delta lobes

Real example of preserved
distributary channel mouth:
Wilcox Group, Gulf Coast

Figure 17.6 Idealized examples of stratigraphic motifs and spontaneous-potential (SP) logs for (A) a fluvial meander, (B) a coastal barrier, (C) a barrier island, and (D) a distributary channel mouth. Note that the log shape is an approximation of the grain size and sand content. For example, the fining-upward fluvial sequence also tapers upward on the SP curve; the coarsening-upward barrier island sequence also widens upward. (After Galloway, 1978: 180.)

stratigraphic horizons but also to recognize the shape of stratigraphic sequences and interpret their depositional history, to recognize unconformities and reconstruct the transgressional-regressional history of an area, and even to detect the fluid content of rocks and identify hydrocarbon accumulations. Seismic stratigraphy allows two- and three-dimensional analyses of subsurface geology with resolutions of tens to hundreds of meters. The technique has also become less expensive as it has been refined; as a result, running a seismic line is much cheaper than drilling a dry hole.

In simple terms, reflection seismology is the process of making a loud bang and then listening for echoes. It is similar to a ship's sonar, which listens for the returning echoes of sonar "pings" to estimate the depth of the seafloor. On land, the loud bang is usually produced by an explosion of buried dynamite or by some sort of mechanical pounder or thumper. The most widely used thumper method is called Vibroseis® (developed by Conoco); an array of large trucks produces vibrations in the ground (Fig. 17.7). At sea, a device called an air gun explodes a bubble of gas underwater to create a noise.

Figure 17.7 COCORP seismic reflection trucks in Wyoming. Each truck has a large plate in back that is lowered and pressed to the ground until it supports the truck. Hydraulic motors vibrate the truck on the plate, creating a rhythmic pulse that is ideal for deep seismic reflection. (Courtesy of S. Kaufman.)

The sound waves travel in all directions, but only those that travel almost directly downward can be reflected off structures underneath the explosion (Fig. 17.8). On land, the returning sound wave is detected by a listening device called a **geophone** (or "jug"; the

geophone crew members are called "jug hustlers"). At sea, the reflected sound is recorded by a hydrophone. A geophone is a remarkably simple device. It consists of a casing that encloses a spring-mounted magnetic coil wired to the sound truck. When the ground vibrates, the casing vibrates with it, but the internal magnet, hanging from its spring, remains motionless due to inertia. The relative motion between the magnet and the geophone casing is recorded as electrical current fluctuations, which become the signals transmitted to the receiving equipment.

A single shot recorded by a single geophone is recorded on a strip chart as a long line, with a pulse representing the time when the reflected sound wave returned. If more than one reflecting surface is encountered, there will be more than one pulse on the trace. Therefore, the vertical scale on most seismic lines is **two-way travel time** in seconds. (Knowledge of the average seismic velocity of various rock types enables geologists to calculate the actual depth of the reflecting layer in meters rather than in seconds, but this step is necessary only when they are trying to match seismic records with well logs or surface outcrops.) A single strip chart is as one-dimensional as a borehole record. To record two-dimensional layers and other structures, an **array** of recording points is needed (see Fig. 17.8). Each point records the shot separately and produces a strip chart. When all the charts are lined up, the pulses, or reflections, from the same layer also line up from one chart to the next, and the first approximate measurement of the layer is made. This is called "picking" the marker reflections.

The geophones in an array are laid out at regular intervals in a traverse away from each recording truck, which contains the electronics to record the signals and the computer to process and interpret them (Fig. 17.9). (At sea, a ship tows a string of regularly spaced hydrophones underwater to achieve the same effect.) The pulse of sound is then generated. Every geophone in the linear array picks up the direct reflection at a time that depends on its distance from the

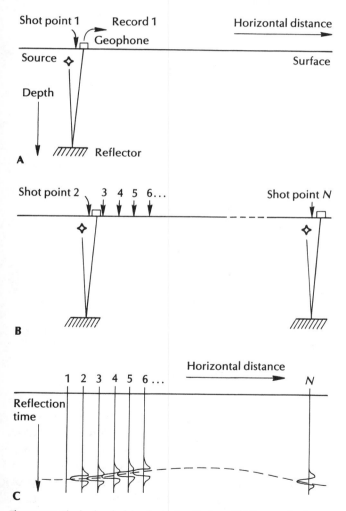

Figure 17.8 The basic principle of seismic reflection. (A) When a seismic wave is generated, the geophone picks up the wave's two-way travel time down to a reflecting layer and back. (B) Moving the shot point and geophone generates a series of reflections of the layer. (C) These reflections show up as a wiggly trace on the seismic record, which can be correlated across the profile. (After Anstey, 1982: 4.)

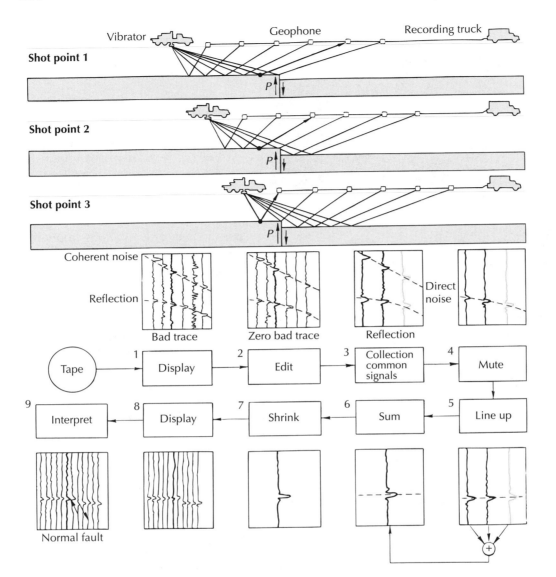

Figure 17.9 The generation and processing of seismic data require many steps. The vibrator trucks (see Fig. 17.7) generate a low-frequency signal that is reflected off an underground layer. A series of geophones picks up the arrivals of waves reflected at various angles from the source. The trucks and geophones are then moved, enabling every geophone to pick up reflections from various angles (note angles of reflection from point P). These reflections are recorded on tape and the recordings (1) displayed. Computer editing (2) removes any noisy traces, and the reflection signals from a common depth point are collected (3). High-amplitude noise is muted (4), and common signals are lined up (5) to compensate for the various angles of the geophones. The pulse is then summed (6), shrunk (7), and displayed (8) along with records from other positions on the seismic line. These can then be interpreted (9) as a geologic feature (in this case, a normal fault.) (After Cook, Brown, and Oliver, 1980: 160.)

acoustic source. To improve the signal-to-noise ratio, the vibrator truck and its array are moved a short distance downline after a shot is made, so that the reflections from the same layer will be picked up by the geophones from slightly different positions. This is called the **common-depth-point method.** By repeating the shots and measuring from slightly different positions along a linear transect, the signals are recorded many times and can be collected and stacked. This procedure amplifies the signals and screens out the noise because the noise has no regular pattern along the line. Finally, all the signals from the single traverse are collected, screened by the computer, and printed out as a **seismic profile.** The dense clusters of vertical traces from the many subsurface reflector horizons form an almost continuous pattern of lines and pulses. If all the complicated signal processing is done correctly, the reflections from distinctive horizons align and look like real cross sections of units.

Seismic Profiles

Seismic profiles are not stratigraphic cross sections in the strict sense. The vertical scale is two-way travel

time, not actual thickness. If the seismic velocity is the same in all the underlying rocks, then the travel time and thickness are linearly related, but variations in lithology and seismic velocity almost always confuse this relationship. Moreover, the reflector horizons are not necessarily lithologic boundaries. A seismic reflection is produced by any abrupt change in seismic velocity, usually the result of a sharp contrast in density, or **acoustic impedance.** Most bedding horizons produce reflections, but if the boundary is subtle and lies between two lithologies of the same density, the signal may be very weak. On the other hand, lithologic features that have little stratigraphic significance can produce strong reflections. Dense chert horizons, for example, produce reflections as strong as or stronger than major unconformities in the continental margins. For a long time, these chert horizons—which do not correspond to any formation boundary—confused marine geologists. Many other lithologic phenomena can also produce spurious reflections, and these are often difficult to screen out. Great care must be taken in interpreting seismic profiles because they look deceptively easy to read, but they are not.

Nevertheless, the structures of many subsurface layers and their approximate depths can be revealed by seismic profiling. If outcrop or well data are available, the prominent reflectors can be matched up with known lithologic horizons. In doing this, however, the relative scales must be kept in mind. A single seismic "wiggle," or deflection, usually represents tens to hundreds of meters of section, which is orders of magnitude coarser than the resolution of a well log (Fig. 17.10). Only the most prominent lithologic breaks appear on the seismic record, and the actual deflection may represent a zone that is meters or

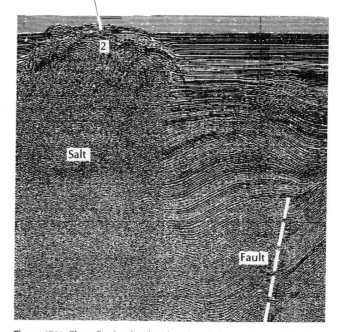

Figure 17.11 The reflection-free interior of a salt dome. (Anstey, 1982: 39; courtesy of U.S. Geological Survey.)

more in thickness (typically, 10 m or more per impedance "wiggle"), where the density contrast is greatest. Layered stratigraphic sequences are the most obvious and easy to interpret, and structural features such as dipping beds, folds, and faults are usually apparent. Unlayered structures do not show any coherent pattern of reflections but instead are characterized by an almost random scatter of reflections. Massive features such as salt domes, reefs, and hardrock basement usually show this random pattern (Fig. 17.11). Even subtle differences in the texture of the reflections can be meaningful. For example, nonmarine beds typically have jagged reflections, with many discontinuous horizons; quiet marine deposition produces reflections that are smooth, continuous, and homogeneous (Fig. 17.12). The presence of gas (which has a very low seismic velocity) inside a reservoir rock changes the seismic reflections considerably, producing "bright spots." This is one of several instances in which hydrocarbon accumulations produce direct seismic traces. In most cases, however, economically valuable hydrocarbon deposits must be found by analyzing the total stratigraphic pattern.

The COCORP Project

In recent years, seismic stratigraphy has made discoveries that have implications far beyond local geology and economic concerns. One of the most important

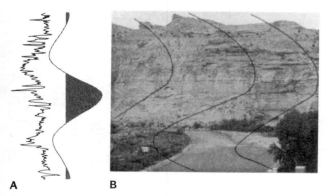

Figure 17.10 (A) The relative dimensions of typical stratification (as seen on an electric log) and the size of a seismic pulse. (After Anstey, 1982: 53.) (B) The scale of the seismic pulse compared with an outcrop about 150 m thick. Obviously, seismic stratigraphy is suitable only for coarse-scale resolution of stratigraphic features. (Photo by D. K. Prothero; figure based on an idea from Miall, 1984.)

A   B

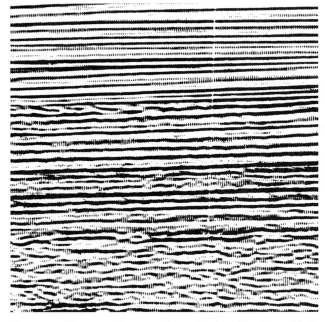

Figure 17.12 The contrast between calm marine deposition (upper third) and nonmarine deposition (lower third.) The middle third shows some open marine and some shallow-water deposition. (Anstey, 1982: 25; courtesy of Merlin Geophysical Company.)

developments has been the Consortium on Continental Reflection Profiling (**COCORP**) project, developed at Cornell University with the aid of a consortium of oil companies. The COCORP project has concentrated on producing detailed, high-resolution profiles with considerable depth penetration (up to 50 km); the ultimate aim is to produce complete, deep-continental profiles across most of North America. The project uses a large array of Vibroseis® trucks and stacks a much larger sequence of seismic arrays than was possible before the advent of high-powered computers. The COCORP project has made several key traverses through mountain ranges and basins, with some startling results. For example, its profile of the Wind River thrust of Wyoming showed it to be much more deeply seated than previously thought. Even more surprising are the profiles across the Appalachians, which have revealed that these mountains are underlain by deep, horizontal thrust faults (Fig. 17.13) that have hundreds of kilometers of offset. Contrary to the old notion of shallow, steep thrusts that die out at depth, it appears that the Appalachians are produced by "thin-skinned"

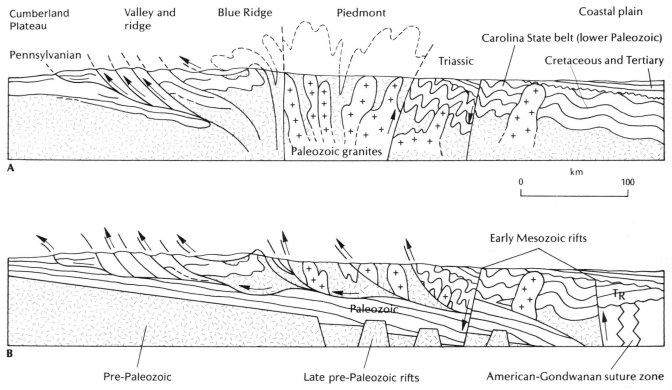

Figure 17.13 (A) The traditional interpretation of the basement structure of the southern Appalachians. The Piedmont and Blue Ridge were thought to have deep crustal roots, with intrusions coming from deep in the crust. (B) The COCORP seismic reflections showed that the southern Appalachians are actually made of thin-skinned thrusts; thus, the crystalline basement is part of a thinthrust slice and has no roots. Instead, it is thrust over Paleozoic sediments. (After Dott and Prothero, 1994: 347.)

thrusting that has transported Precambrian basement hundreds of kilometers over Paleozoic shelf sediments. This radically changes the proposed tectonic models for this area. It is also possible that hydrocarbons are trapped below these thrusts in regions that have never before produced oil.

The Exxon-Vail Curve

Probably the most widely discussed result of seismic stratigraphy has been the interpretation of unconformities. These show up on seismic records as places where beds abruptly truncate or gently pinch out, and they often produce strong reflections. Each unconformity-bounded package on a seismic profile is called a **seismic sequence** (Fig. 17.14A). The sequences are

bounded by curved lines on the seismic profile, but with the biostratigraphy of multiple wells, it is possible to determine the ages of the beds. Assuming that these minor reflectors within sequences are isochronous, one can estimate the ages and magnitudes of the unconformities that bound them. The reflectors then can be placed in a chronologic framework (Fig. 17.14B). From the ages and dimensions of these plots, the ages and the magnitudes of changes in coastal onlap and offlap that formed these seismic packages can be interpreted (Fig. 17.14C).

The seismic profiles of a number of passive continental margins have been studied in this fashion, and each was analyzed for the ages and magnitudes of its major unconformities. In the northwest African

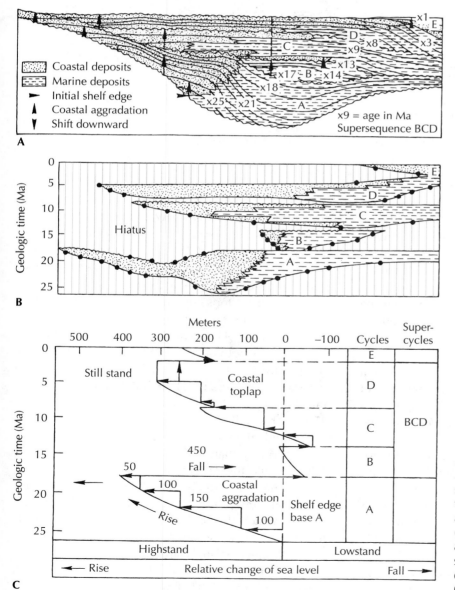

Figure 17.14 (A) Seismic sequences are analyzed by marking the unconformities in a seismic section and determining their degrees of onlap or offlap and their ages. (B) These can then be plotted as a series of unconformity-bounded sequences with a time axis. The lateral extent of the units is shown, and the vertically striped regions are absent units. (C) From this diagram, a curve of relative onlap and offlap can be constructed for this particular section. (After Vail et al., 1977a: 78; by permission of the American Association of Petroleum Geologists, Tulsa, Okla.)

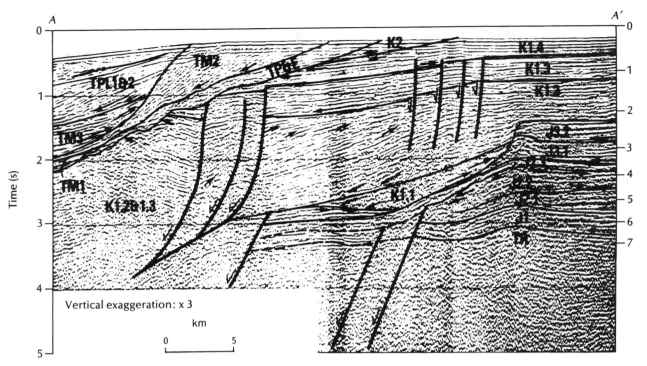

Figure 17.15 Seismic profile across the passive margin of northwestern Africa, showing sequences as defined by seismic reflections. In this view, the fault-bounded Triassic basement (TII) can be clearly seen and is overlain by several small Jurassic packages (J1–J3.2). The bulk of the section is composed of thick Cretaceous shelf sequences (K1.1–K2), which are cut by normal faults. There are thin Paleocene and Eocene deposits (TP&E) and Miocene deposits (TM1–TM3), but the Oligocene is completely absent due to an enormous offlap event (see Figs. 17.16 and 17.17). Major unconformities also separate the Jurassic from the Cretaceous, the K1.1 from the K1.2 sequences, and the Cretaceous from the Paleocene. Based on this and similar sequences from other passive margins, Vail et al. (1977b: 137) constructed their famous "sea-level" curves. (After Mitchum, Vail, and Sangree, 1977; by permission of the American Association of Petroleum Geologists, Tulsa, Okla.)

margin, for example, there are major unconformities between the Jurassic and the Cretaceous (J3.2 and K1.1 in Fig. 17.15), between the Cretaceous and the Tertiary (K2 and TP&E), between the Eocene and the Miocene (TP&E and TM1, eliminating the entire Oligocene), and between the middle and upper Miocene (TM2 and TM3). There are minor unconformities between units within the Jurassic and within the Cretaceous. A significant number of normal faults are also apparent in the seismic section.

After a number of these curves were interpreted, it became clear that most passive margins have major unconformities of approximately the same magnitude and the same age. From this, Vail, Mitchum, and Thompson (1977), who were working for Exxon, suggested that sea level had fluctuated on a worldwide basis, producing unconformities of approximately the same magnitude on all continental margins at the same time. The product of their work was originally called the **Vail sea-level curve** (Fig. 17.16); it purports to show the eustatic changes through the last 200 million years.

Since the initial publication of the Exxon-Vail curve, however, there has been considerable controversy over what controls the shape and apparent magnitude of the cycles in the curve. As we saw in Chapter 15, the large-scale first-order cycles (Fig. 17.17) appear to have been caused by major continental movements. But the jagged, sawtooth shape of the second- and third-order cycles—with their curved bottoms and flat, abrupt tops—led to controversy. This asymmetry could not be explained by the known fluctuations of sea level, and now it appears that the sea-level fall was not that rapid. When sea level falls, it is normal for the shallow marine coastal sequence to prograde out across the shelf, building up a sedimentary package even though sea level is dropping (Fig. 17.18). This "fills in" the peak during the regressive phase and gives the appearance that sea level falls more abruptly than it rises. Because sea level can be falling while the prograding wedge is still building up, it is necessary to distinguish between the descriptive terms **onlap** and **offlap** (which describe the degree to which the package laps onto the continent) and the true rise and fall of global sea level. Vail and his colleagues (Vail, Hardenbol, and Todd, 1984; Haq, Hardenbol, and Vail, 1987, 1988; Posamentier, Jervey, and Vail, 1988) have differentiated between

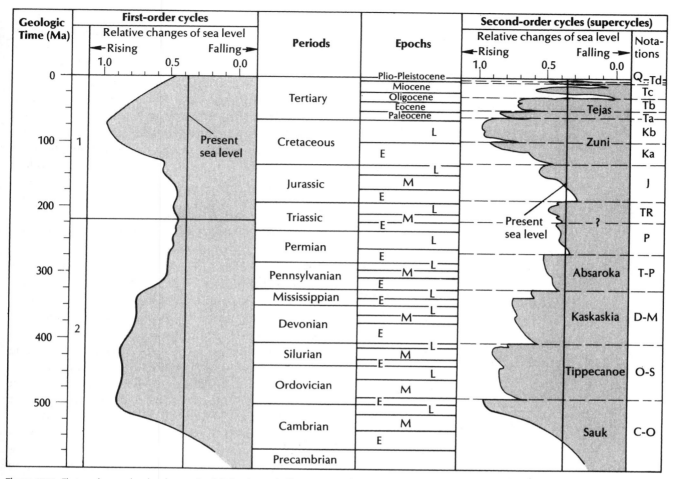

Figure 17.16 First- and second-order changes in global onlap and offlap (called "sea-level changes" in this diagram) through the Phanerozoic. Note that there have been two first-order peaks of major transgression (during the early Paleozoic and the Cretaceous) and peaks of regression during the late Precambrian, Permo-Triassic, and late Cenozoic. These are punctuated by second-order cycles that span tens of millions of years. The Paleozoic second-order cycles correspond to Sloss's (1963) sequences (see Fig. 15.20.) (After Vail et al., 1977a: 84; by permission of the American Association of Petroleum Geologists, Tulsa, Okla.)

their sawtooth curve of coastal onlap and offlap and a true sea-level curve, which has a smoother, more rounded shape (see Fig. 17.17 and Appendix B).

The Exxon-Vail curve has been widely and enthusiastically used by many geologists to correlate their strata to the global time scale. Indeed, the Exxon team asserts that "sequence stratigraphy offers a unifying concept to divide the rock record into chronostratigraphic units, avoids the weaknesses and incorporates the strengths of other methodologies, and provides a global framework for geochemical, geochronological, paleontological, and facies analysis" (Baum and Vail, 1988, p. 322). The latest tests of the Exxon-Vail curve were published in the volume edited by Wilgus et al. (1988). A number of authors (for example, Olsson, 1988; Baum and Vail, 1988) identified many of the key unconformities predicted by the Vail model in sections exposed on land. Poag and Schlee

(1984) did the same with well data from the North American Atlantic margin. But some geologists are concerned that these identifications involve circular reasoning. Rather than test the Exxon-Vail curve to see if it is valid, some seismic stratigraphers have assumed that this "sea-level" curve is superior to other methods of correlation. When studies of continental-margin unconformities agree with the Exxon-Vail curve, they are treated as confirmation; if they do not, they are dismissed as some local tectonic signal or attributed to sloppy dating.

Nevertheless, some sequence stratigraphers have used the Exxon-Vail curve as their primary tool of correlation. For example, Campion et al. (1994) recognized three sequence boundaries in the middle-late Eocene strata of the Transverse Ranges in coastal California. They confidently correlated these sequence boundaries to sea-level changes on the

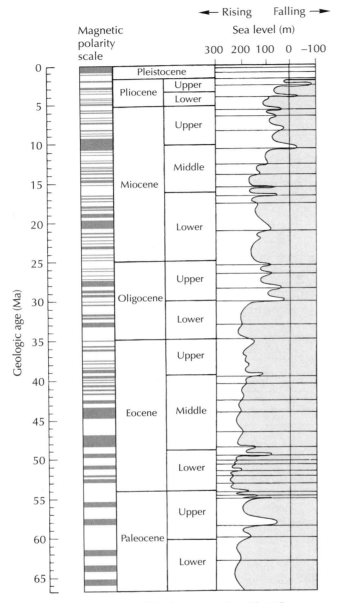

Figure 17.17 Updated version of the Cenozoic portion of the Vail curve generated by Haq, Hardenbol, and Vail (1987, 1988.) This curve differs from the Vail et al. (1977a: 85) "sea-level" curve in showing a smoother, true eustatic curve rather than the sawtooth onlap-offlap curve (see Fig. 17.16 and Appendix B.)

Haq, Hardenbol, and Vail (1987) curve and assigned ages to the strata based on this correlation. However, when a detailed study of the biostratigraphy and magnetostratigraphy was undertaken (Prothero, 2001), it turned out that most of their correlations were off by 3 to 6 million years, and none of their sequence boundaries matched eustatic events on the Haq, Hardenbol, and Vail (1987) curve. Sequence stratigraphic correlations might be sufficient in the absence of any other time indicator, but they are no substitute for good biostratigraphy.

In a particularly scathing review, Miall (1992) pointed out that there are so many cycles and unconformities in the Exxon-Vail curve that *any* record of unconformities in a section could be made to match. He pointed to a classic paper by Zeller (1964) entitled "Cycles and Psychology," which used artificially generated "stratigraphic sequences." Geologists, trained in pattern recognition, were able to "see" correlations where none existed, simply because there were so many events that some matched up by random chance. Miall (1992) argued that the Exxon-Vail curve could be "matched up" with 77% of the events in a completely random, artificial sequence.

Another serious criticism of the Exxon-Vail curve is its apparent chronostratigraphic precision (Miall, 1991, 1993; Hallam, 1992; Dickinson, 1993; Snyder and Spinosa, 1993). Each of the unconformities and cycle boundaries is confidently labeled with a numerical age, usually given to the nearest 100,000 years. Most of the third- and fourth-order cycles have durations on the order of 2 million years or less, yet the dating methods have error estimates of ±5 million years—longer than most of the cycles. Even between different figures in the same paper, there are discrepancies. For example, Haq, Hardenbol, and Vail (1988) give the age of the top of the nannoplankton NP16 zone as 44 Ma in their figure 11 but 41 Ma in their figure 14; similarly, the top of the NP15 zone is quoted as 46 Ma in figure 11 and 43.1 Ma in figure 14 (Miall, 1991). Such large internal errors of as much as 3 million years are longer than most of the individual cycles, and such errors could cause miscorrelation by one whole sequence or more.

Aubry (1991) demonstrated the difficulty of dating these unconformities. She studied deep-sea cores and sections all over Eurasia, North Africa, and North America and found that there are at least two major regional unconformities that might correlate with the lower-middle Eocene boundary, or 49.5 Ma "event" in the Haq, Hardenbol, and Vail (1988) curve. The 49.5 Ma "event" might correspond to one, two, or more local unconformities, with or without regional tectonic influences. The 49.5 Ma "event" appears to be a result of miscorrelation of biostratigraphic data and conflation of several different events. And this is only one out of 121 third-order sequences in the Haq, Hardenbol, and Vail (1988) curve! If this kind of analysis were done for every "event," even more confusion would result. As Miall (1993) put it, "The more detailed the study, the more confused the resulting patterns of correlations, and the less support that remains for the so-called global cycle chart." Miall (1991)

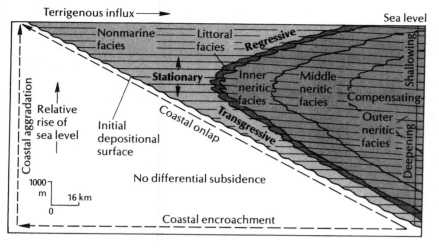

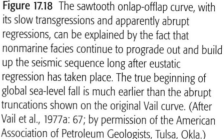

Figure 17.18 The sawtooth onlap-offlap curve, with its slow transgressions and apparently abrupt regressions, can be explained by the fact that nonmarine facies continue to prograde out and build up the seismic sequence long after eustatic regression has taken place. The true beginning of global sea-level fall is much earlier than the abrupt truncations shown on the original Vail curve. (After Vail et al., 1977a: 67; by permission of the American Association of Petroleum Geologists, Tulsa, Okla.)

wrote, "Stratigraphic correlation with the Exxon chart will almost always succeed, but it is questionable what this proves. The occurrence of numerous global third-order cycles remains to be demonstrated. The existing Exxon cycle chart should be abandoned—it is too flawed to fix."

Many geologists (Bally, 1980; Watts, 1982; Thorne and Watts, 1984; Watts and Thorne, 1984; Parkinson and Summerhayes, 1985; Summerhayes, 1986; Miall, 1986, 1991, 1992; Burton, Kendall, and Lerche, 1987; Johnson, 1987; Carter, 1988; Hubbard, 1988; Kendall and Lerche, 1988; Christie-Blick, Mountain, and Miller, 1990; Jordan and Flemings, 1990; Reynolds, Steckle, and Coakley, 1990; Christie-Blick, 1991; Christie-Blick and Driscoll, 1995; Fulthorpe, 1991; Hallam, 1992) have challenged the idea of eustatic control of the second- and third-order cycles. They point out that these cycles have been measured exclusively from passive continental margins. Hubbard (1988) studied three of the major basins that were responsible for much of the sea-level curve proposed by Vail et al. (1977a,b) and found that most of the sequence boundaries were not synchronous but were related to local tectonic activity. Only in basins of the same age with identical subsidence history and sediment input rates did the second-order cycles appear to be synchronous. Hubbard (1988) suggested that the global second- and third-order cycles may be an illusion created by the fact that the Exxon-Vail curve was constructed from basins that were very similar in age and tectonic history. If this is so, then the cycles do not have global chronostratigraphic significance and should not be used in correlation. Likewise, Williams (1988) found that the oxygen isotope record of the oceans agreed with the Exxon-Vail curve for some events but not for others. In short, the Vail second- and third-order cycles

probably have a significant eustatic component, but there is also much local tectonic "noise" that has not yet been filtered out of the eustatic "signal." Indeed, some authors (Burton, Kendall, and Lerche, 1987; Kendall et al., 1992) question whether we can determine eustatic sea level at all. The Exxon-Vail curves should not be used uncritically. Although Fig. 17.17 and Appendix B show the Haq, Hardenbol, and Vail (1987, 1988) "sea-level" curves, not all the fluctuations have been clearly documented as eustatic events.

To their credit, the Exxon authors (Posamentier, Jervey, and Vail, 1988; Posamentier and Vail, 1988) have begun to refine their models to incorporate the effects of local tectonism. These new models attempt to separate the parts of the Exxon-Vail curve that are eustatically controlled from those that are due to local tectonism or progradation. Further research on the sequences themselves should begin to filter out local tectonics and produce a true global sea-level curve.

Sequence Stratigraphy

As Weimer and Posamentier (1993) point out, regardless of one's position on the eustatic versus tectonic origin of the second- and third-order cycles, a much larger, more interesting topic has arisen from the Exxon seismic work: **sequence stratigraphy.** It began in the 1980s (summarized in the papers collected in Wilgus et al., 1988) and has swept through parts of the profession like a hurricane. As of 2003, sequence stratigraphy is extremely influential in the major oil companies, and many of the talks at the meetings of the American Association of Petroleum Geologists and the SEPM are framed in sequence stratigraphic terms. Many books and papers on the subject (for example, Van Wagoner et al., 1990; Loucks and Sarg, 1993; Weimer and Posamentier, 1993; Posamentier et al.,

1993; Emery and Myers, 1996) are available, because it has touched nearly every aspect of stratigraphy. Sloss (1993) wrote that "the emergence of the New Sequence Stratigraphy has been the single most vitalizing force propelling students in post-Pangaea stratigraphy, sedimentation, and sedimentary-tectonic synthesis since the development of the plate-tectonic paradigm." Many other geologists are more skeptical, although they have not been as vocal or prolific as the proponents (see Miall, 1997, for a review).

In Chapter 15, we saw how Sloss (1963) divided cratonic sediments into unconformity-bounded packages he called "sequences" (see Fig. 15.20). In the 1970s, the Exxon-Vail group extended this idea to their seismic profiles of the Atlantic passive margins (see the papers in the volume edited by Payton, 1977). Although sequence stratigraphic concepts were originally developed from seismic reflection profiles, they have since been extended to any stratigraphic succession with a substantial marine component. Mitchum (1977) and Van Wagoner et al. (1988) define a **sequence** as *a relatively conformable succession of genetically related stata bounded by unconformities and their correlative conformities.* The sequence is further subdivided into **systems tracts,** which represent stratal wedges produced during various episodes of high or

low sea level (Fig. 17.19). Systems tracts, in turn, are built up of many small-scale **parasequences,** which are relatively conformable successions of genetically related beds bounded by surfaces that represent abrupt changes in water depth (**marine flooding surfaces**). Most parasequences are small enough in scale (meters to tens of meters thick) that they can be seen in outcrop; they are the smallest units that the long-amplitude seismic waves can resolve.

Three principal types of stacking arrangements are commonly found in sequence stratigraphy (Fig. 17.20): progradational, retrogradational, and aggradational. They are controlled by the balance between the depositional rate and the **accommodation rate,** or the rate of subsidence of the basin that allows the accumulation of sediment. If the accommodation rate is much higher than the sediment supply rate, there is a net transgression, or a **retrogradational** stacking pattern. If the sediment supply outstrips accommodation, the parasequence stacks build outward in a net regressive, or **progradational,** pattern. If the two factors are nearly in equilibrium, parasequences will be stacked vertically in an **aggradational** pattern.

An idealized continental margin profile (see Fig. 17.19, nicknamed the "sea slug" model) comprises numerous systems tracts that are stacked on top of and

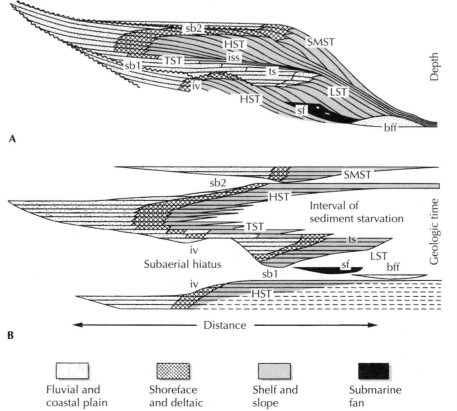

Figure 17.19 Conceptual cross section in relation to depth (A) and geologic time (B) showing stratal geometry, systems tracts, and the distribution of siliciclastic facies within unconformity-bounded sequences deposited in a basin with a shelf break. Systems tracts: SMST, shelf margin; HST, highstand; TST, transgressive; LST, lowstand. Sequence boundaries: sb1, type 1; sb2, type 2. Other abbreviations: iss, interval of sediment starvation; ts, transgressive surface (corresponding to the time of maximum regression); iv, incised valley; sf, slope fan; bff, basin floor fan. (After Vail, 1987: 13; by permission of the American Association of Petroleum Geologists, Tulsa, Okla.)

Fluvial and coastal plain

Shoreface and deltaic

Shelf and slope

Submarine fan

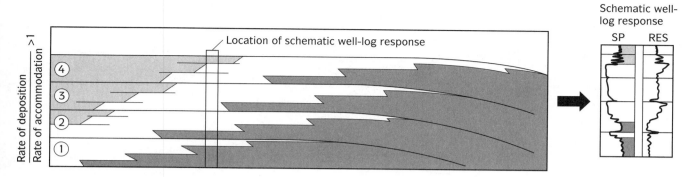

Schematic well-log response

Progradational parasequence set

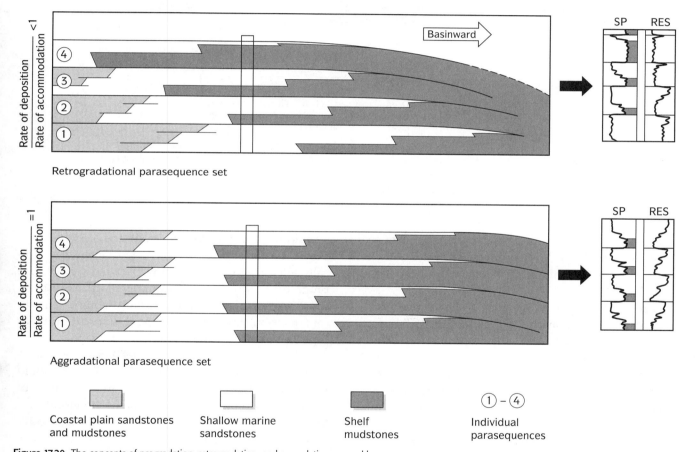

Retrogradational parasequence set

Basinward

Aggradational parasequence set

Coastal plain sandstones and mudstones

Shallow marine sandstones

Shelf mudstones

(1)–(4) Individual parasequences

Figure 17.20 The concepts of progradation, retrogradation, and aggradation as used by sequence stratigraphers. (After Vail et al., 1989: 137.)

adjacent to one another. For example, the late stages of rise in relative sea level might produce a **highstand systems tract** (Fig. 17.21A; HST in Fig. 17.19). This package is produced after maximum flooding, when sets of aggrading parasequences cover an older unconformable surface and are eventually succeeded by prograding parasequences as the sediments build outward. If relative sea level then drops rapidly, the HST will be truncated and incised by an unconformity, and

deposition will shift to the **lowstand systems tract** (Fig. 17.21B; LST in Fig. 17.19). LSTs accumulate as submarine fan and slope deposits near the base of the continental margin, whereas the older HST is actively eroded as it is exposed. When a relatively rapid sea-level fall occurs and erodes deeply into the continental margin, it produces a **type 1 unconformity** or **type 1 sequence boundary** (sb1 in Fig. 17.19). At the end of the sea-level fall, or at the beginning of the subsequent

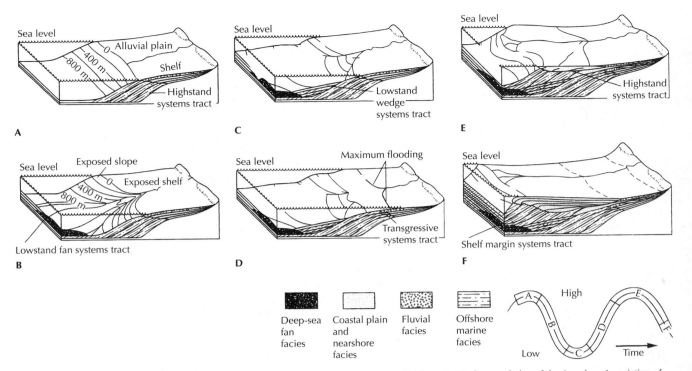

Figure 17.21 Characteristic systems tracts and their relationship to the eustatic curve, according to sequence stratigraphy. These block diagrams are idealized and have a strong vertical exaggeration. (After Posamentier, Jervey, and Vail, 1988: 112; by permission of the American Association of Petroleum Geologists, Tulsa, Okla.)

sea-level rise (that is, at times when the change of sea level is very slow), **lowstand wedge systems tracts** can prograde over the LST or onlap onto the slope (Fig. 17.21C).

A relatively rapid rise in sea level will produce a **transgressive systems tract** (Fig. 17.21D; TST in Fig. 17.19). TSTs produce a rapidly transgressing, or retrogradational, set of parasequences that onlap until they reach a point of maximum flooding. As the rate of relative sea-level rise slows, the parasequences change from retrogradational to aggradational. While sediments are aggrading during maximum flooding on the continent, sediment starvation and **condensed sections** are found in the deeper-water deposits (see Fig. 17.19B). Thus, the TST is represented by an unconformity in the deep water but can be traced to a correlative conformable surface in the aggrading nearshore deposits. Once maximum flooding has been reached, TSTs are succeeded by prograding HSTs, producing another thick continental margin wedge of strata (Fig. 17.21E). If, however, there is a gradual fall of sea level, sediments will shift outward to the edge of the continental shelf and aggrade at the edge, forming a **shelf margin systems tract** (Fig. 17.21F; SMST in Fig. 17.19). Unlike the deep marine turbidites and shales of the LST (which are formed by

a rapid drop in relative sea level), the SMST is composed mostly of shallow-water deposits and is typically adjacent to the HST. However, the shift from a prograding HST to an aggrading SMST is also responsible for a subtle unconformity without significant incision, known as a **type 2 unconformity** or **type 2 sequence boundary** (sb2 in Fig. 17.19).

As Weimer and Posamentier (1993) point out, however, the "sea slug" model (see Fig. 17.19A) should not be taken too literally. It is really an idealization of numerous sedimentary basins around the world, cartooned to show every possible feature. No real stratigraphic package matches it in every detail, but portions of most sedimentary basins show some features of the model. Certain features are common in certain settings and are not found in others. For example, thick TSTs are rare in siliciclastic settings but are fairly common in carbonate margins (Loucks and Sarg, 1993), with their repetitive shallowing-upward cyclicity (see Figs. 12.8 and 12.9). Deep-water submarine fans and shales of the LST, on the other hand, would rarely show up in shallow cratonic basins.

At this point, most novices are staggered by all the new terminology, some of which seems to unnecessarily rename such old concepts as transgression, regression, and progradation. Indeed, the jargon has

alienated many geologists as well, who find new names applied to familiar, well-studied outcrops and ask if any new insight has really been gained. Sequence stratigraphers regard the terminology as an essential part of their way of looking at the world, because many of the concepts they have introduced had not been discussed before. Indeed, this is typical of any radical new movement in science—jargon is essential to differentiate the movement from the old orthodoxy and make its concepts distinct by giving them new names, even if they superficially resemble some of the old ways of thinking.

Although sequence stratigraphy arose from the study of passive continental margins as a way of seeing strata, it has now been enthusiastically applied to any unconformity-bounded stratal package. Many authors (Einsele, 1985; Hallam, 1992) question whether sequence stratigraphic concepts are appropriate to cratonic basins, which have extensive stretches of very shallow seas, no shelf-slope break, and very low sedimentation and accommodation rates. On the other hand, Weimer (1992) made a strong case for the sequence stratigraphic approach in the epeiric seas of the Western Interior Cretaceous (Fig. 17.22; see Box

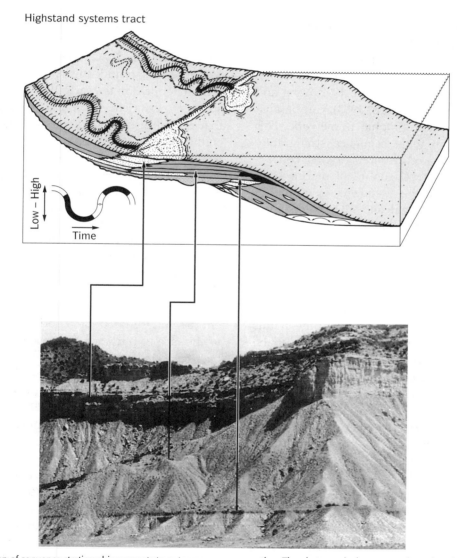

Figure 17.22 Application of sequence stratigraphic concepts to outcrops from the Western Interior Cretaceous in the Book Cliffs, Douglas Creek Arch, Colorado. In this example, the rate of eustatic rise is at a minimum and falls slowly in the late highstand. The rates of deposition are greater than the rate of sea-level rise, so parasequences build basinward in aggradational to progradational parasequence sets of the highstand systems tract. In addition, parasequences downlap onto the condensed section. The photograph shows a condensed section (phosphatic oolites) and a progradational parasequence set, as well as a highstand systems tract, consisting of the Castlegate Sandstone (top of cliffs), and the Buck Tongue and Sego Members of the Price River Formation. (After Van Wagoner et al., 1990: 42; by permission of the American Association of Petroleum Geologists, Tulsa, Okla.)

10.2). Tye et al. (1993) and Hewlett and Jordan (1993) were able to apply some sequence stratigraphic concepts to active-margin basins.

Some unconformity-bounded nonmarine deposits have been analyzed in a sequence stratigraphic framework. This approach may seem plausible in areas where fluvial base level might rise or fall in response to sea-level changes. But as one proceeds upstream, such local factors as climate-induced changes in fluvial discharge and sediment flux begin to take over (Posamentier and James, 1993; Blum, 1993). Indeed, these factors may be much more important than sea-level changes in determining fluvial, deposition; such factors would produce terrestrial deposits that are out of phase with marine deposits (Schumm, 1993; Koss, Ethridge, and Schumm, 1994).

Even regions that are far from either sea-level effects or base level control of deposition, such as eolian sand dunes (Kocurek and Havholm, 1993; Yang and Nio, 1993) or lake deposits (Liro, 1993), can be analyzed in a sequence stratigraphic framework. This raises the question of what sequence stratigraphy *means*, once it is so divorced from its original conceptual foundation of sea-level control of marine sedimentary packages. Are the "sequence-stratigraphy-colored glasses" so powerful and all-encompassing that *any* set of strata can be made to fit this world view? Are we mindlessly matching up unconformities to a preconceived conceptual framework without looking hard at whether the data support this framework? Some geologists (see Miall, 1997) worry that sequence stratigraphers have become near-religious in their zeal to explain every unconformity in this manner without adequate independent tests of the age and origin of the unconformities.

Currently most of the controversy over sequence stratigraphy still centers on the issue of tectonic versus eustatic control of stratal packages and whether the sequence stratigraphic framework is the best way to explain relative onlap and offlap and changes in sedimentary basin fill (Jordan and Flemings, 1990; Reynolds, Steckler, and Coakley, 1990; Christie-Blick, Mountain, and Miller, 1990; Walker, 1990; Christie-Blick, 1991; Christie-Blick and Driscoll, 1995; Hallam, 1992; Miall, 1991; Schumm, 1993; Koss, Ethridge, and Schumm, 1994). For many practicing geologists, however, these debates are irrelevant. All that matters is that sequence stratigraphy *works*. Unconformity-bounded packages and parasequences with sharp tops representing sudden changes in relative sea level are widespread in many marine strata and are easy to recognize in the field and in seismic section (see Fig. 17.22). As Weimer and Posamentier (1993) argue, "From the perspective of the oil exploration industry, the key question regarding the sequence stratigraphic concepts is this: can these concepts be used to decrease risk in exploration, or applied to better develop an oil reservoir?"

Since the furor over sequence stratigraphy in the 1980s and 1990s, the geological community has largely settled into an uneasy truce. The detailed critiques of sequence stratigraphy have been published in many places (summarized by Miall, 1997), but the practitioners of sequence stratigraphy (largely those working for the oil companies, and a few working within academia) are apparently no longer listening. The last decade of critiques has not affected the way "Sequence Stratigraphy: The Next Generation," operates. Instead, sequence stratigraphic theory has largely frozen in the form it adopted in the early 1990s, and it no longer responds to criticism. Miall and Miall (2001) point out that sequence stratigraphy has achieved an almost cultlike status. Its adherents no longer question the tenets of their paradigm and no longer respond to suggestions from the rest of the geological community. Miall and Miall (2001) also note that recent volumes (such as de Graciansky et al., 1998) employ circular reasoning. Such volumes assume eustatic control of their cycles and no longer attempt to examine any other sources of data that might test whether the cycles were in fact eustatic. Such insularity is a worrisome trend, because the more isolated a research community becomes, the more likely it is to waste its efforts on misguided ideas and to avoid the cold slap in the face of outside peer review that could snap it back to reality.

Another sign of change can be seen in professional journals and in the programs of scientific meetings. From their peak in the mid-1990s, sequence stratigraphic talks are becoming much less common in meetings of such groups as the Geological Society of America. They still predominate in the meetings of industry groups, such as the American Association of Petroleum Geologists, but they are definitely receding from the meetings and journals of the SEPM (Society for Sedimentary Geology). Clearly, the explosive growth phase of interest in sequence stratigraphy is over. Sequence stratigraphy is largely entrenched in certain subsets of the profession but is receding from other areas it once dominated. Only time will tell whether the approach will continue to be fruitful or its limitations will eventually outstrip its advantages.

Magnetostratigraphy
Rock Magnetism

Another geophysical correlation technique that has achieved great importance over the past 20 years is magnetic polarity stratigraphy, or **magnetostratigraphy.** Unlike the methods described previously, magnetostratigraphy has been used primarily to correlate surface outcrops, although it also has been used to correlate subsurface cores. Magnetostratigraphy correlates rocks by means of the similarities of their magnetic patterns. It has some unique advantages, including the possibility of recognizing time planes that are worldwide and independent of facies. Its main limitation is that it seldom works independently but must be used with some other geochronologic method.

Most rocks contain minerals that are naturally magnetic, such as the iron oxides magnetite (Fe_3O_4) and hematite (Fe_2O_3). In cooling igneous rock, these minerals do not lock into a permanent magnetic **remanence** until they cool to a temperature called the **Curie point.** For hematite, the Curie point is about 650°C, and for pure magnetite (with no titanium), it is 578°C. There are three types of remanent magnetization: thermal, detrital, and chemical.

Once a magnetic mineral cools below the Curie point, its magnetization is locked into the crystal, aligned in the direction of the Earth's magnetic field at the time. Magnetization formed by cooling below the Curie point is called **thermal remanent magnetization (TRM).** Some igneous rocks, especially basaltic lava flows, contain large amounts of magnetic minerals and are strongly magnetized.

When an igneous rock is weathered and eroded, its magnetic minerals, along with the rest of the rock, become sedimentary particles. The small magnetic grains (only a few micrometers in diameter) behave as tiny bar magnets, aligning themselves with the magnetic field of the Earth when the grains are deposited as sediments. The magnetization of these sediments when they have lithified is called **detrital remanent magnetization (DRM).** DRM is usually two or three orders of magnitude weaker than TRM, depending on how much magnetic mineral the sedimentary rock contains.

After weathering, iron is dissolved from the rock and moves through the groundwater. Eventually, it can precipitate in another place, usually as hematite. As chemically precipitated iron minerals nucleate and grow, they acquire a remanence that is parallel to the magnetic field direction of the Earth at the time they form. This is called **chemical remanent magnetization (CRM).** Because the hematite weathering product could have been acquired at any time, the CRM may have little to do with the magnetic field prevailing when the rock formed. In this case, it is irrelevant to the stratigraphic age of the unit and may form an overprint of noise over the primary signal that would only hinder magnetostratigraphic investigations. There are examples, however, of sedimentary rocks whose chemical remanence can be shown to have formed at the same time as the rock unit itself.

Sampling, Measurement, and Analysis

One or more of these three types of remanence may be present in a rock as it occurs in the field. To measure the magnetization, however, samples must be taken back to the laboratory. In well-indurated rocks, such as lava flows, samples must be collected with a portable coring drill (Fig. 17.23). The drill produces a small core, which must be oriented with respect to true north before it is taken from the outcrop. Samples of softer sedimentary rocks can be collected with simple hand tools; again, the orientation must be carefully recorded. Even soft muds can be sampled if they are enclosed in a container with the field orientation marked on it. Three or more samples are usually collected at each stratigraphic level or site, so that they can be averaged. A larger number of samples also allows the calculation of simple statistics. The most important statistical calculation is the variation around a mean direction. This is a measure of how well the samples cluster and can be used to determine whether the magnetic vectors differ significantly from a random scatter.

A **magnetometer** is a device that measures both the intensity and the direction of the magnetic vector of the sample. Samples that are strongly magnetized can be measured with a spinner magnetometer (Fig. 17.24A). This device rotates the sample rapidly around an axis within a coil and measures the electrical current created in the coil by the spinning sample. By spinning the sample in various orientations, the direction and intensity of the magnetic vector can be determined. Most samples are too weakly magnetized or too poorly cemented together to be measured by a spinner magnetometer.

In the mid-1970s, the cryogenic magnetometer (Fig. 17.24B) was developed to measure such specimens. The cryogenic magnetometer uses liquid helium at 4 K (kelvin) (4 degrees above absolute zero) to create a superconducting region around the sensors. Electrical currents move with almost no resistance at such low

A

B

Figure 17.23 (A) Paleomagnetic sampling of lava flows with a coring drill. (Courtesy of S. Bogue.) (B) An orientation device is placed over the drilled core to transfer the compass azimuth and dip to the core before it is broken free and removed. (D. R. Prothero.)

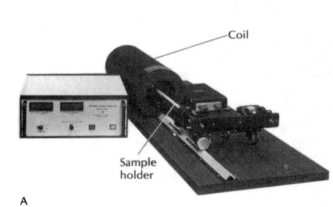

A

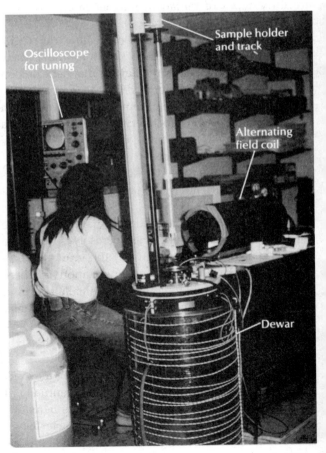

B

Figure 17.24 (A) Spinner magnetometer. The long white rod spins the sample within the sensing coil (black cylinder.) (Courtesy of Schonstedt Instrument Company.) (B) Cryogenic magnetometer. The sample is placed at the end of the long white plastic tube and lowered on the track to the left into the dewar, where it is measured in the superconducting sensing area in the bottom. The black cylinder to the right is the coil of the alternating field demagnetizer. (D. R. Prothero.)

temperatures. When a magnetized specimen is brought into the superconducting sensor area, the specimen's magnetic field sets up a current in the superconducting coil, which then can be measured. The cryogenic magnetometer has many advantages over older methods. It is more sensitive by almost three to four orders of magnitude and thus can measure even the most weakly magnetized lithologies. It responds almost instantly, in contrast to the many minutes required to measure each sample with a spinner magnetometer. In addition, the sample does not have to be spun, so even liquid suspensions and live animals can be measured. Many laboratories now have cryogenic magnetometers that allow the sample to be rotated while it is in the sensing area. All measurements are taken, averaged, and corrected by a microcomputer that interfaces with the magnetometer electronics. The most advanced labs even have an automated sample changer to minimize the tedious labor of moving samples.

When a sample is first measured in the laboratory, all the magnetization it has acquired since it was first formed is still present. This initial magnetization is called the **natural remanent magnetization (NRM).** Often, the primary (or original) magnetization of the sample is overprinted by a younger magnetization direction that was picked up from the Earth's present magnetic field. To get rid of younger overprinting, each sample must be partially demagnetized, or "cleaned," until only the primary component remains. The younger overprinting is usually the easiest to erase because it was acquired more recently and usually is not strong enough to realign the primary magnetic fields of the mineral grains.

Demagnetization can be done in two ways, and sometimes both are used on the same sample. In **alternating field (AF) demagnetization,** the sample is placed in a strong alternating field, and the weaker components of magnetization are eliminated as they oscillate back and forth in response to the rapid change of field direction. With stronger alternating fields, more and more of the magnetization is destroyed until a cleaning field is reached in which only the primary component remains. In **thermal demagnetization,** the sample is heated to higher and higher temperatures in multiple steps, after which it is allowed to cool in an area shielded from the Earth's magnetic field; the magnetization is then measured. The weaker, overprinted components of magnetization disappears at the lower temperatures to which a sample is heated; the primary component is elimi-

nated only when the sample reaches the Curie point of the magnetic mineral.

Demagnetization can be a complicated, tedious procedure, and sometimes good results cannot be obtained. Problems with demagnetization can lead to erroneous results. For example, most early paleomagnetic studies were done with AF demagnetization alone because it was the easiest to use. It was later discovered that many terrestrial mudstones have an overprinting due to chemical remanent magnetization from goethite and other iron hydroxides, which does not respond to AF treatment. Thermal treatment at 200°C, which dehydrated the goethite and eliminated its overprinted component, radically changed the results of several geologists, who had to revise their earlier interpretations.

Field Reversals and the Polarity Time Scale

The result of all this sampling, cleaning, and measurement is a series of rock samples that show the direction and approximate intensity of the Earth's magnetic field at the time they were lithified. As early as 1906, the French physicist Bernard Brunhes noticed that some volcanic rocks were magnetized not in the direction of the Earth's present magnetic field but 180° in the opposite direction, or *reversed* from the present *normal* direction. Our compass needles would have pointed south 780,000 years ago, before the Earth's field last reversed. The observations of Brunhes and others of reversed rock magnetization were not followed up until the early 1960s, when a group of scientists led by Allan Cox, Richard Doell, and G. Brent Dalrymple at the U.S. Geological Survey began to study the magnetization of ancient rocks systematically. By chance, one of the earliest samples studied, the Haruna dacite of Japan, has the virtually unique property of being self-reversing. As a result, many geologists doubted that the new evidence of reversely magnetized rocks meant that the entire magnetic field of the Earth had reversed.

Between 1963 and 1969, Cox, Doell, and Dalrymple continued to provide new data, in friendly competition with the McDougall and Chamalaun group at the Australian National University. To convince skeptical geologists, these paleomagnetists needed to show not only that the magnetization of their samples was reversed but also that all rocks of the same age were of the same polarity worldwide. Only worldwide data could not be explained by local self-

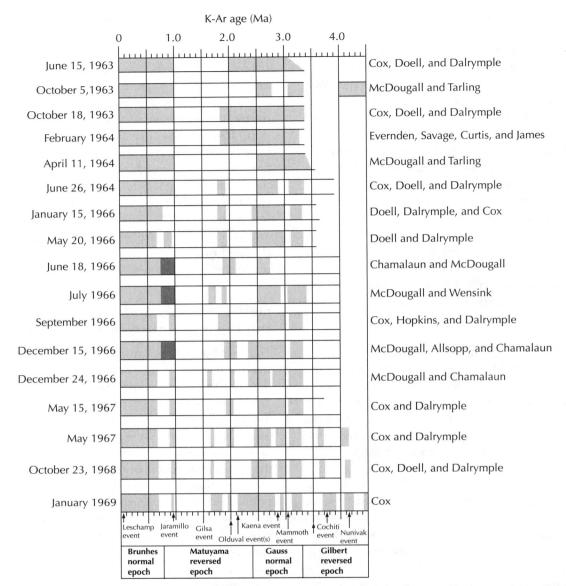

Figure 17.25 Successive versions of the geomagnetic reversal time scale as determined from the potassium-argon dating of volcanic rocks from various continents. Shaded areas represent periods of normal polarity, and unshaded areas are periods of reversed polarity. The original 1963 time scale of Cox, Doell, and Dalrymple had only three polarity episodes in the last 3 million years, but the competition between them and McDougall and Chamalaun in Australia produced so much data that the 1969 version of the time scale contained more than 20 events in the last 5 million years. (After Dalrymple, 1972: 214.)

reversal. They sampled lava flows of various ages from all over the world, not only because such rocks are strongly magnetized but also because they could be dated directly by the newly refined method of potassium-argon dating. As the samples accumulated, the pattern became clear. Regardless of where they came from, rocks of the same age had the same magnetic polarity (Fig. 17.25). Not only had the Earth's magnetic field reversed in the geologic past, but it had done so repeatedly at irregular intervals. The data produced a distinctive pattern of long and short reversed and normal episodes that could be correlated worldwide. We now call this the **magnetic polarity time scale.**

Magnetostratigraphic Correlation

The polarity history of the Earth presented stratigraphers with a powerful tool for interregional correlation. The pattern of magnetic polarity reversals is irregular and nonperiodic, and distinctive long episodes can be recognized. Once a local polarity record has been matched to the global polarity time scale, it is possible to make extremely precise correlations. First, polarity reversals happen *worldwide* and

are *independent of facies or lithology*. No other method of correlation can make this claim. The magnetic record of a rock can be measured whether it is a basalt, a siltstone, a limestone, or whatever. The magnetic pattern allows correlation between the deep sea and the terrestrial record where no such correlation was possible before. Second, polarity reversals are *geologically instantaneous*. Studies of sequences that were deposited at high sedimentation rates have shown that a typical polarity reversal takes only 4000 to 5000 years to occur. For all intents and purposes, polarity zone boundaries are isochronous.

Magnetostratigraphy has its own peculiar limitations as well. Polarity events are not unique. Rocks of normal and reversed polarity occur throughout the geologic record, so a single sample cannot be dated by its polarity. Only a long sequence of rocks with polarity zones of distinctive lengths can be correlated. Even then, if there are significant hiatuses or rapid changes in sedimentation rates, the interpretation can be erroneous. Even if the polarity reversal pattern appears to be distinctive, it cannot be correlated by itself unless the top of the section is recent and the first zone down marks the present episode of normal polarity. In all other sections, some other independent method

(usually biostratigraphy or radiometric dating) is necessary to place the section in its approximate geochronologic position. After this is done, the pattern can be matched to the polarity time scale at that age to see what interpretation gives the best fit. There may be several reasonable interpretations, and in such cases the correlation is ambiguous. Thus, magnetostratigraphy, despite its strengths, has important limitations. It is not independent but requires other dating techniques. It requires long sections of favorable lithology that produce a correlative pattern. If there are unconformities or fluctuations in sedimentation rate, the pattern may be distorted or incomplete.

As the polarity time scale continues to develop, general patterns are emerging (Fig. 17.26). Most of the Cenozoic and Late Cretaceous was a period of rapidly changing, or mixed, polarity, which makes it suitable for magnetostratigraphy. The same is true of much of the Triassic, the Jurassic, and the Early Carboniferous. On the other hand, some parts of geologic history were characterized by long periods of stability. During the Late Cretaceous, the Earth's field was continuously normal for 36 million years. During the Late Carboniferous and Permian, the Earth's field was reversed for 70 million years. Obviously, during long

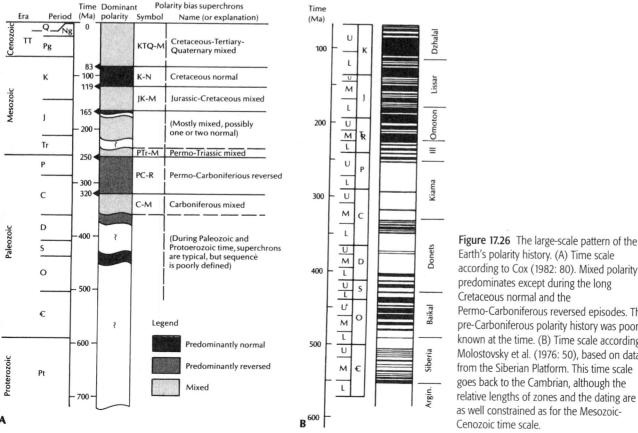

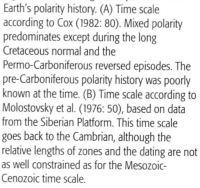

Figure 17.26 The large-scale pattern of the Earth's polarity history. (A) Time scale according to Cox (1982: 80). Mixed polarity predominates except during the long Cretaceous normal and the Permo-Carboniferous reversed episodes. The pre-Carboniferous polarity history was poorly known at the time. (B) Time scale according to Molostovsky et al. (1976: 50), based on data from the Siberian Platform. This time scale goes back to the Cambrian, although the relative lengths of zones and the dating are not as well constrained as for the Mesozoic-Cenozoic time scale.

periods of field stability, there is little potential for magnetostratigraphy. Pre-Carboniferous magnetic stratigraphy is still too poorly known to generalize about the dominant patterns. Work on the Siberian Platform (Molostovsky et al., 1976) has produced a tentative Paleozoic magnetic polarity time scale, but it is not as easy to calibrate as the Mesozoic-Cenozoic seafloor spreading record (Fig. 17.26B).

Despite these limitations, many classic studies have revealed the power of the method. For example, the relative thicknesses of polarity zones may fluctuate somewhat, but in many cases they are remarkably consistent. As another example, the polarity zones in Antarctic deep-sea cores (Fig. 17.27) are all different in total thickness, but each core has zones of the same relative length. Even in the more episodic sedimen-

tary conditions of terrestrial fluvial sequences in Pakistan, there have been good results. For example, the sections in the Potwar Plateau region of Pakistan (Fig. 17.28A) produce polarity zones of consistent relative thicknesses. The mammalian biostratigraphy and dated volcanic tuffs place the sequence in the magnetic polarity time scale. A Shaw graphic correlation plot of one of the sections against the polarity time scale (Fig. 17.28B) shows that the slope is relatively straight, with linear changes. If any of the polarity zones had irregular thicknesses, they would stand out as outlying points far from the line.

The North American Stratigraphic Code (see Appendix A) has standardized the terminology of magnetostratigraphy. In the early days of magnetic stratigraphy (see Fig. 17.25), polarity events were called

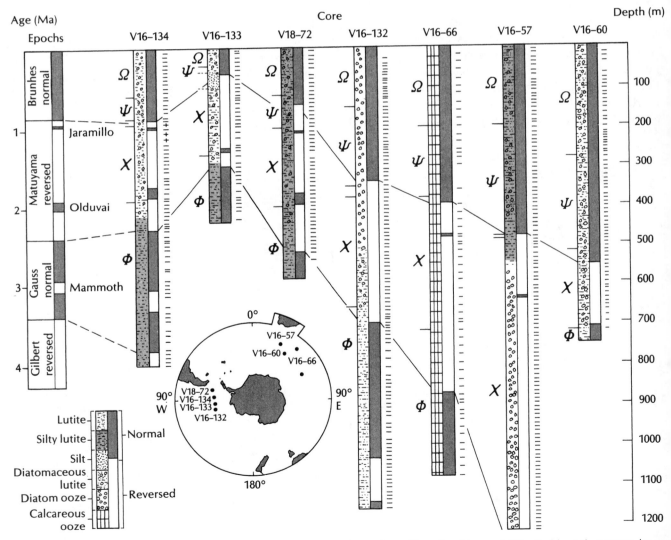

Figure 17.27 One of the earliest examples of the magnetostratigraphic correlation of sediments. The polarity zones found in Antarctic deep-sea cores clearly agree with the biostratigraphy of the radiolarians. (Greek letters denote biostratigraphic zones.) (After Opdyke et al., 1966: 351, by permission of N. Opdyke and *Science;* © 1966 by the AAAS.)

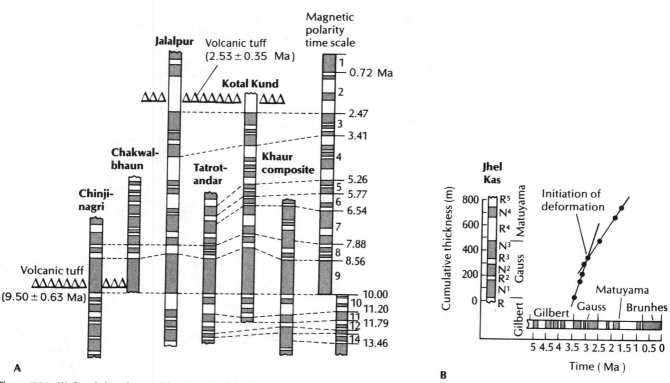

Figure 17.28 (A) Correlation of terrestrial sections from the Miocene Siwalik Group of Pakistan with the magnetic polarity time scale (shown at right.) The long normal interval with the tuff dated at 9.50 ± 0.63 Ma ties the section to Epoch 9 (old terminology; now called Chron C5.) The date of 2.55 ± 0.35 Ma at the base of a reversed interval ties the upper part of the section to the Matuyama reversed chron (Epoch 2 in the old terminology.) From these correlations, the rest of the pattern in the area can be internally correlated and each interval matched to the polarity time scale. (B) The relative spacing of events in individual sections (ordinate) matches the polarity times (abscissa) quite well, as shown by the graphic correlation plot. (After Johnson et al., 1982: 34, 38; by permission of Elsevier Science Publishers.)

"epochs" and were named after great scientists associated with magnetism (Brunhes, Matuyama, Gauss, and Gilbert). Shorter changes in polarity within the epochs were called "events" and were named after the place they were found (for example, Olduvai Gorge in Tanzania, Mammoth Mountain in the Sierra Nevada, and Jaramillo Creek in New Mexico). Paleomagnetists soon ran out of names of great magnetists, so they began numbering instead. Using the first polarity interval prior to the Gilbert as Epoch 5, they numbered both normal and reversed intervals back to Epoch 23 in the early Miocene (Hays and Opdyke, 1967; Opdyke, Burckle, and Todd, 1974). Stratigraphers soon pointed out that these intervals were not the same as epochs of the geochronologic hierarchy (see Chapter 15). In addition, the numbering scheme was somewhat arbitrary, because not all normal and reversed boundaries were "epoch" boundaries.

Eventually, magnetostratigraphers began to work with polarity events before the Miocene and needed more labels. They adopted the numbers of the positive magnetic anomalies assigned by Heirtzler et al. (1968) to the seafloor spreading profiles. Each episode of nor-

mal polarity was numbered the same as its positive anomaly on the seafloor. Events were called magnetic polarity **chrons** to prevent confusion with the epochs of geochronology. According to LaBrecque et al. (1983), a chron begins at the top of the normal interval that corresponds to the numbered positive magnetic anomaly and ends just above the beginning of the next oldest normal interval below it. To prevent confusion with the older numbering scheme, the chrons of LaBrecque et al. (1983) are labeled with the prefix C. The suffixes N and R have also been used to subdivide the normal and reversed part of the same chron, and some chrons have been further subdivided with additional numbers and letters (Cox, 1982). Part of the old magnetic Epoch 7 became Chron C4, Epoch 13 became Chron C5A, and so on (Tauxe et al., 1987). The current terminology is shown in Appendix B.

In the new code, the chrons can be subdivided into subchrons or lumped into superchrons. The traditional terminology for the last 5 million years of polarity history is retained, but now these periods are called the Matuyama Chron, the Olduvai Subchron, and so on. The code makes a distinction between

magnetostratigraphic units (bodies of rock distinguished by any difference in magnetic properties) and magnetopolarity units (bodies of rock distinguished by differences in magnetic polarity). By this definition, magnetostratigraphic units include not only magnetopolarity units but also units distinguished by slight changes in field direction, as happens during reversal transitions or as a result of secular variation. Like other formal stratigraphic units, the magnetic polarity units require a type section and adherence to all the other recommended procedures of formal designation and publication.

Chemostratigraphy

Although isotopes formed by radioactive decay have become important in dating (see Chapter 18), there are many isotopes that are stable and do not decay over time. These isotopes exist in well-defined ratios in the ocean and atmosphere. When oceanographic and climatic changes occur, some isotopes can become more or less abundant with respect to others, due to **fractionation,** or separation of isotopes by their differences in atomic weight. These fluctuations of certain isotope ratios have become a powerful tool for stratigraphy in recent years.

Oxygen Isotopes

Most of the oxygen in the Earth (99.756%) is in the form of ^{16}O, which has eight protons and eight neutrons; a slightly heavier stable isotope, ^{18}O, has two more neutrons and is relatively rare (0.205%). Usually, these two isotopes are present in the ocean in this ratio. In 1947, Harold Urey and Cesare Emiliani found that these two isotopes fractionated with changes in temperature. By examining the oxygen in the calcite of the shells of foraminifers, Emiliani found that the oxygen isotope ratio seemed to fluctuate in response to the changes of temperature caused by the ice ages.

Later work showed that the picture was more complicated. In addition to the temperature effect, there was also an effect due to ice volume. Water that is richer in ^{16}O, which makes it lighter, evaporates more readily than water with more ^{18}O. In the unglaciated world (Fig. 17.29), this water rains into the ocean or onto the land and travels back to the ocean via the rivers, so there is no change in the ratio of the oxygen isotopes. During glaciation, the clouds precipitate their ^{16}O-rich water on the ice caps as snow, where it remains locked up for a long time. As a result, the oceans become relatively depleted in ^{16}O and enriched in ^{18}O during periods of glaciation. It is now apparent

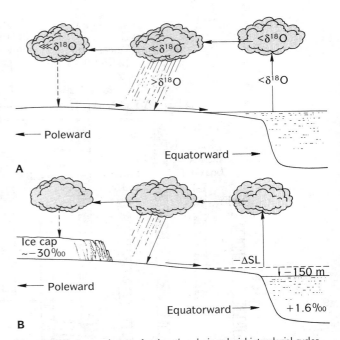

Figure 17.29 Oxygen isotope fractionation during glacial-interglacial cycles. (A) Water carrying the lighter isotope ^{16}O is preferentially evaporated to form clouds. As the clouds move landward and rain out, they become even more ^{18}O-depleted. During interglacial periods, however, this ^{18}O-poor water returns to the sea, and there is no net change. (B) During glacial periods, the ^{18}O-depleted water is trapped in the ice caps, which have $^{18}O/^{16}O$ ratios of -30 parts per thousand (‰). The ocean, as a consequence, is relatively enriched in ^{18}O ($+1.6$‰). (After Matthews, 1984: 89.)

that the change in oxygen isotope ratio during the ice ages is primarily related to changes in ice volume, with a minor effect due to temperature.

Oxygen isotopes are measured with respect to an arbitrary laboratory standard called **PDB,** after the Pee Dee belemnite. Calcite from this abundant cephalopod in the Cretaceous Pee Dee Formation of South Carolina is used to calibrate the mass spectrometer. The ratio is calculated by the following equation:

$$\delta^{18}O = \frac{[(^{18}O/^{16}O)_{sample} - (^{18}O/^{16}O)_{standard}]}{(^{18}O/^{16}O)_{standard}} \times 1000$$

A shell that has a $\delta^{18}O$ value of 3 parts per thousand (parts per mil, or ‰) to PDB means that the CO_2 derived from that shell is 3 parts per mil richer in that isotope than PDB. Positive $\delta^{18}O$ values are enriched in ^{18}O, indicating increased ice volume and cooling; negative values are enriched in ^{16}O, indicating decreased ice volume and warming.

The result of these changes in ice volume and temperature is a distinct pattern of fluctuating oxygen isotopes in the Pliocene and Pleistocene (Fig. 17.30). This oxygen isotope record has been divided into dis-

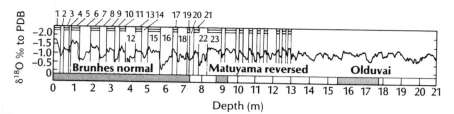

Figure 17.30 Oxygen isotope fluctuations in the carbonate of planktonic foraminifers for the last 2 million years. The more negative values (upward fluctuations) are interglacial periods; the more positive values (downward fluctuations) are glacial periods. The major glacial-interglacial isotopic stages were given numbers by Emiliani (1955, 1966) and Shackleton and Opdyke (1973), and their correspondence to the magnetic polarity time scale is shown. (After Shackleton and Opdyke, 1976: 451; by permission of the Geological Society of America.)

tinct "stages," which are numbered back from the present. Because oxygen isotopes in the ocean respond to climatic changes, these "stages" are worldwide. Ice caps advance and retreat so rapidly that oxygen isotope changes can be considered geologically instantaneous. Like paleomagnetic samples, the ratio of a single sample is not sufficient to determine its age, because each value of $\delta^{18}O$ has occurred many times in geologic history.

A series of samples is needed to match a pattern to the global oxygen isotope record (Fig. 17.31). Biostratigraphic control is necessary to determine which part of the total pattern is being matched. Like magnetic stratigraphy, isotope stratigraphy is not independent but relies on another method of determining time. Unlike magnetic stratigraphy, however, oxygen isotope stratigraphy is dependent on lithology. Oxygen isotopes are measured in the shells of marine organisms, because the isotope record occurs primarily in ocean water. Not all marine organisms give good results, either. Some organisms fractionate oxygen isotopes in their own distinctive ways by physiological means and thus are useless for this kind of analysis. A large number of organisms must be sampled to screen out those that might give spurious results.

Carbon Isotopes

Like oxygen, carbon has more than one stable isotope in the Earth's oceans and atmosphere. Most of the carbon (98.89%) is in the form of ^{12}C, which has six protons and six neutrons. However, 1.11% is the heavier isotope ^{13}C, which has an extra neutron. These two isotopes circulate through the ocean and are incorporated into the calcite of organisms in much the same way as oxygen isotopes. Typically, the two are measured together during an isotopic analysis, and the formula for calculating $\delta^{13}C$ is the same as the formula for $\delta^{18}O$, with the appropriate substitutions.

The carbon system is not controlled by ice volume and temperature but by oceanic circulation. Organic materials tend to be low in ^{13}C, so when they decay, they release not only excess ^{12}C but also less ^{13}C, so the value of $\delta^{13}C$ decreases in the water. Deep ocean waters, in particular, are traps for organic nutrients and CO_2, which are relatively depleted in surface waters as a result of photosynthesis, the sinking of organic debris, and respiration by bottom-dwelling organisms. This means that deep-ocean waters trap carbon that is depleted in ^{13}C and rich in ^{12}C. When major changes in oceanic circulation occur, these bottom waters exchange with the ocean surface waters and release their ^{12}C, which makes the $\delta^{13}C$ value more negative.

In general, the ratios of the carbon isotopes are primarily a reflection of oceanographic and climatic changes, which happen on a global scale in a geologic instant. Thus, carbon isotopes, like oxygen isotopes,

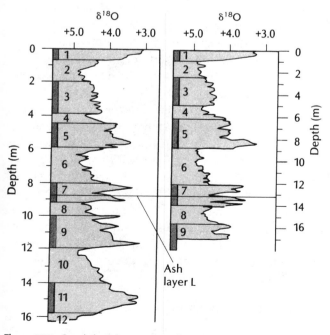

Figure 17.31 Correlation of oxygen isotopic oscillations between two deep-sea cores from the eastern equatorial Pacific. The two cores are scaled by the correlation along ash layer L. Isotopic stages 1 through 12 are indicated. (After Ninkovitch and Shackleton, 1975: 32; by permission of Elsevier Science Publishers.)

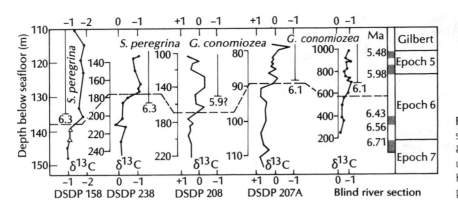

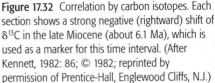

Figure 17.32 Correlation by carbon isotopes. Each section shows a strong negative (rightward) shift of $\delta^{13}C$ in the late Miocene (about 6.1 Ma), which is used as a marker for this time interval. (After Kennett, 1982: 86; © 1982; reprinted by permission of Prentice-Hall, Englewood Cliffs, N.J.)

can be good time markers. Distinctive $\delta^{13}C$ "spikes" (Fig. 17.32) have been used in correlation, although there is no global $\delta^{13}C$ curve like that for oxygen isotopes. Like the use of oxygen isotopes, the use of carbon isotope stratigraphy requires a continuous record with a distinct pattern and an independent method of time correlation (usually biostratigraphy). Carbon isotopes also depend on the organisms that trap them.

Until recently, neither oxygen nor carbon isotopes could be used in fresh water or on land because the systems depend on the organism's being in physiological contact with normal ocean water. Recent studies, however, have shown that terrestrial oxygen isotopes can also be used. For example, Koch, Zachos, and Gingerich (1992) showed that the striking change in carbon isotopes at the Paleocene-Eocene boundary in the deep-sea record could also be detected in the carbon isotopes of terrestrial carbonate in soils, and even in the carbon component of fossil mammal teeth. This not only provided a firm correlation between marine and land records but also demonstrated that the global carbon changes at the Paleocene-Eocene boundary must have been extreme.

An even more striking example of terrestrial isotopes comes from the late Miocene, about 7 Ma. Carbon isotope analyses of soils and mammalian teeth from North America, Pakistan, South America, and East Africa also show the same strong positive enrichment from about −7 per mil to as great as +5 per mil (Quade, Cerling, and Bowman, 1989; Cerling et al., 1997). The prevailing interpretation of this trend has been the increase of plants that use the C4 (Hatch-Slack) photosynthetic pathway, which today is found only in temperate and tropical grasses. Most other plants (including high-latitude and high-altitude grasses) use the more conventional C3 (Calvin) photosynthetic pathway. Such a dramatic

change in global isotopes suggests a huge expansion of temperate and tropical grasslands at 7 Ma, producing large savannas in most parts of the world in the late Miocene.

The fact that grasslands expanded in the Miocene is not surprising. What is surprising is that the teeth of herbivores—such as horses, camels, and antelopes—became high-crowned for eating gritty vegetation (such as grasses) at 15 to 16 Ma, almost 7 to 8 million years before the carbon isotope signal of grassland expansion. Retallack (1997) has argued that there must have been an expansion of C3 grasslands in the middle Miocene, but if so, we have no modern analogs for such a habitat, because all modern C3 grasses occur only in high altitudes or high latitudes. Others have suggested that the high-crowned teeth of horses and other herbivores might have developed in response to grittier vegetation from windblown dust on their plant fodder, rather than from the expansion of a particular kind of grass. Unfortunately, both of these hypotheses are hard to test.

These isotopic data provide another surprise as well. Contrary to expectation, there is no great faunal change at 7 Ma when the C4 grasslands expand—no great increase in grazing mammal diversity, nor any decline in leaf-eating browsers. In fact, there is not much faunal change at all (Prothero, 1999). This result contradicts long-held notions that mammals must respond to every small change in climate or vegetation. Terrestrial isotopic studies of this kind are just beginning, but they are already producing interesting results that contradict many long-held notions!

Strontium Isotopes

Recently, refinements in the measurement of strontium isotopes have made possible a new form of isotope stratigraphy. Strontium, along with calcium, is an alkali earth element on the periodic table and

readily forms divalent cations. Its ionic radius is only slightly larger than that of calcium, so it often fills the calcium sites in trace amounts in many minerals. The common isotope of strontium in the oceans is ^{86}Sr. ^{87}Sr occurs in lesser abundance; the normal ^{87}Sr/^{86}Sr ratio in modern oceanic waters is around 0.7090. As ^{87}Sr is produced by the decay of ^{87}Rb, this ratio fluctuates through geologic time, with a steady, almost linear increase since the Jurassic. Recent work (DePaolo and Ingram, 1985; Elderfield et al., 1982; Palmer and Elderfield, 1985; Elderfield, 1986; Hess, Bencher, and Schilling, 1986) has shown that the rate of change is continuous and linear since the late Eocene (Fig. 17.33). If the ^{87}Sr/^{86}Sr ratio of calcite in a marine organism can be measured precisely enough, the organism can be placed on the curve and dated (McArthur, Howarth, and Bailey, 2001).

This method can date single samples, unlike the other stable isotope methods. Like the other methods, however, it is presently limited to normal marine organisms. In addition, it does not appear that it will work for samples older than Eocene, so another time control is necessary to obtain a preliminary estimate of age. It is a technique with great potential, however, particularly if other dating techniques cannot be used.

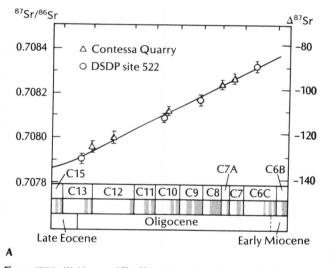

A

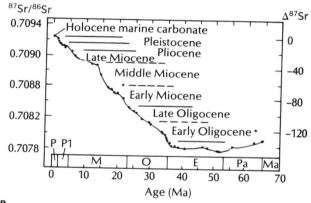

B

Figure 17.33 (A) Measured ^{87}Sr/^{86}Sr values of late Eocene and Oligocene chalks calibrated against the magnetostratigraphy. The vertical error bars are remarkably small, allowing for very tight correlation. (B) The change in strontium isotopic ratios has been almost linear since the early Tertiary, allowing a single unique determination of strontium isotopes to date a sample with fairly high precision. (From DePaolo and Ingram, 1985: 939, by permission of the authors and Science; © 1985 by the AAAS.)

CONCLUSIONS

The geophysical and geochemical methods of correlation outlined in this chapter are among the newest techniques in stratigraphy. In many cases, they offer insights about the stratigraphic record that are not possible by any other method. For example, all we know about subsurface geology comes from well logging and seismic stratigraphy. Magnetic stratigraphy allows highly precise numerical dating and correlation of rocks across facies in a way that was never possible in the past. Oxygen and carbon isotopes have become the principal tool of paleoceanography and paleoclimatology, and strontium isotopes have allowed precise dating of Cenozoic marine rocks. Finally, sequence stratigraphy has been hailed as a scientific revolution, and it has given us many new insights about previously studied stratigraphic sequences.

FOR FURTHER READING

Anstey, N. A. 1982. *Simple Seismics.* Boston: International Human Resources Development Corporation.

Asquith, G. B. 1982. *Basic Well Log Analysis for Geologists.* Tulsa, Okla.: Amer. Assoc. Petrol. Geol. Methods in Exploration Series.

Butler, R. F. 1992. *Paleomagnetism, Magnetic Domains to Geologic Terranes.* Cambridge, Mass.: Blackwell.

Cerling, T. E., J. M. Harris, B. J. MacFadden, M. G. Leakey, J. Quade, V. Eisenmann, and J. R. Ehleringer. 1997. Global vegetation change through the Miocene/Pliocene boundary. *Nature* 389:153–158.

Christie-Blick, N., and N. W. Driscoll. 1995. Sequence stratigraphy. *Annual Reviews of Earth and Planetary Sciences* 23:451–478.

Cox, A. V., ed. 1973. *Plate Tectonics and Geomagnetic Reversals.* San Francisco: W. H. Freeman and Company.

de Graciansky, P.-C., J. Hardenbol, T. Jacquin, and P. R. Vail, eds. 1998. *Mesozoic and Cenozoic Sequence Stratigraphy of European Basins.* SEPM Special Publication 60.

Dobrin, M. B. 1976. *Introduction to Geophysical Prospecting,* 3d ed. New York: McGraw-Hill.

Doveton, J. H. 1994. *Geologic Log Analysis.* SEPM Short Course Notes 29.

Emery, D., and K. Myers, eds. 1996. *Sequence Stratigraphy.* Cambridge, Mass.: Blackwell.

Hallam, A. 1992. *Phanerozoic Sea-Level Changes.* New York: Columbia.

Hyne, N. J. 1984. *Geology for Petroleum Exploration, Drilling and Production.* New York: McGraw-Hill.

Kennett, J. P. 1980. *Magnetic Stratigraphy of Sediments.* Stroudsburg, Pa.: Dowden, Hutchinson and Ross.

Koch, P. L., J. C. Zachos, and P. D. Gingerich. 1992. Correlation between isotopic records in marine and continental carbon reservoirs near the Paleocene/Eocene boundary. *Nature* 358:319–322.

Loucks, R., and J. F. Sarg, eds. 1993. *Carbonate Sequence Stratigraphy: Recent Developments and Applications.* Amer. Assoc. Petrol. Geol. Memoir 57.

McArthur, J. M. 1994. Recent trends in strontium isotope stratigraphy. *Terra Nova* 6:331–358.

McArthur, J. M., R. J. Howarth, and T. R. Bailey. 2001. Strontium isotope stratigraphy: LOWESS version 3: Best fit to the marine Sr-isotope curve for 0–509 Ma and accompanying look-up tables for deriving numerical age. *Journal of Geology* 109:155–170.

McElhinny, M. W. 1973. *Palaeomagnetism and Plate Tectonics.* Cambridge: Cambridge University Press.

Miall, A. D. 1991. Stratigraphic sequences and their chronostratigraphic correlation. *Journal of Sedimentary Petrology* 61:497–505.

Miall, A. D. 1992. Exxon global cycle chart: An event for every occasion? *Geology* 20:787–790.

Miall, A. D. 1997. *The Geology of Stratigraphic Sequences.* New York: Springer Verlag.

Miall, A. D., and C. E. Miall. 2001. Sequence stratigraphy as a scientific enterprise: The evolution and persistence of conflicting paradigms. *Earth-Science Reviews* 54:321–348.

Payton, C. E., ed. 1977. *Seismic Stratigraphy: Applications to Hydrocarbon Exploration.* Amer. Assoc. Petrol. Geol. Memoir 26.

Posamentier, H. W., C. P. Summerhayes, B. U. Haq, and G. P. Allen, eds. 1993. *Sequence Stratigraphy and Facies Associations.* International Association of Sedimentologists Special Publication 18.

Prothero, D. R. 1999. Does climatic change drive mammalian evolution? *GSA Today* 9(9): 1–5.

Prothero, D. R. 2001. Magnetostratigraphic tests of sequence stratigraphic correlations from the southern California Paleogene. *Journal of Sedimentary Research B* 71:525–535.

Quade, J., T. E. Cerling, and J. R. Bowman. 1989. Development of the Asian monsoon revealed by marked ecological shift during the latest Miocene in northern Pakistan. *Nature* 342:163–166.

Retallack, G. J. 1997. Neogene expansion of the North American prairie. *Palaios* 12:380–390.

Sheriff, R. E. 1978. *A First Course in Geophysical Exploration and Interpretation.* Boston: International Human Resources Development Corporation.

Sheriff, R. E. 1980. *Seismic Stratigraphy*. Boston: International Human Resources Development Corporation.

Tarling, D. H. 1983. *Palaeomagnetism*. London: Chapman and Hall.

Van Wagoner, J. C., R. M. Mitchum, K. M. Campion, and V. D. Rahmaninan, 1990. *Siliciclastic Sequence Stratigraphy in Well Logs, Cores, and Outcrops*. Amer. Assoc. Petrol. Geol. Methods in Exploration 7.

Weimer, P., and H. Posamentier, eds. 1993. *Siliciclastic Sequence Stratigraphy: Recent Developments and Applications*. Amer. Assoc. Petrol. Geol. Memoir 58.

Wilgus, C. K., B. S. Hasting, C. G. St. C. Kendall, H. W. Posamentier, C. A. Ross, and J. C. Van Wagoner, eds. 1988. *Sea Level Change: An Integrated Approach*. SEPM Special Publication 42.

Williams, D. F., I. Lerche, and W. E. Full. 1988. *Isotope Chronostratigraphy, Theory and Methods*. San Diego, Calif.: Academic Press.

18 Geochronology and Chronostratigraphy

An example of cross-cutting relationships that establish relative ages. The igneous dike cuts through the shales but is truncated by an erosional surface on which is overlain the massive sandstone. A radiometric date on the dike would give a minimum age for the shales and a maximum age for the sandstone. (D. R. Prothero.)

ALL OF THE DATING METHODS DISCUSSED SO FAR CAN determine the **relative ages** of rocks, fossils, and geologic events. There was no way of knowing the **numerical ages** of events until the discovery of radioactive decay in the early twentieth century. Radioactive decay is one of the few predictable, clocklike processes in nature, and understanding how it works is essential to obtaining and interpreting numerical ages. Note that we say "numerical" rather than "absolute" dating, as you might find in older textbooks. Radiometric dates are merely numerical estimates of the age in years, but they are not truly "absolute," because they are subject to all sorts of errors and problems, as we will discuss shortly. To call them

"absolute" dates gives them a false connotation of superiority or truth. Indeed, the 1983 revision of the North American Stratigraphic Code recommends abandoning the term "absolute dates."

Although numerical dating is virtually always done on igneous or metamorphic rocks, we include it in a textbook on sedimentary rocks because an understanding of geochronology is essential to stratigraphy. The 1983 revision of the North American Stratigraphic Code (Appendix A) now includes categories for igneous and metamorphic rock bodies, so all types of rock that show relative ages are now within the purview of stratigraphy.

Geochronology

Radioactive Decay

During the process of radioactive decay, unstable parent atoms change into more stable daughter atoms by emitting subatomic particles and energy. If a neutron is lost, the atomic mass (given as a superscript before the symbol of the element; for example, ^{238}U) changes, producing a different isotope of the same element (for example, ^{235}U). If an atom loses a proton, both the atomic number (usually given as a subscript before the symbol of the element, for example, $_7N$) and the atomic mass change, and a different element results (for example, $_6C$). Sometimes both protons and neutrons are given off as alpha particles, thereby changing the atomic number by two and the atomic mass by four. An example might be

$$_{92}^{238}U \rightarrow {}_{90}^{234}Th + {}_2^4He \text{ (an alpha particle, } \alpha)$$

In beta decay, the parent nucleus emits an electron, and an uncharged neutron in the nucleus changes into a positively charged proton. The latter event increases the atomic number by one, changing the element but not the mass. For example:

$$_{19}^{40}K \rightarrow {}_{20}^{40}Ca + e^- \text{ (a beta particle, } \beta)$$

In electron capture, a proton in the nucleus picks up an orbital electron and becomes a neutron, giving the atom a lower atomic number, but the mass remains unchanged. For example:

$$_{19}^{40}K + \text{orbital electron} \rightarrow {}_{18}^{40}Ar + \text{gamma particle } (\gamma)$$

Figure 18.1 shows a typical radioactive decay series, including changing atomic numbers and masses and alpha and beta decay products. Because radioactive decay takes place entirely within the nucleus of an atom, it is unaffected by chemical changes (such as oxidation and reduction); such changes affect only the

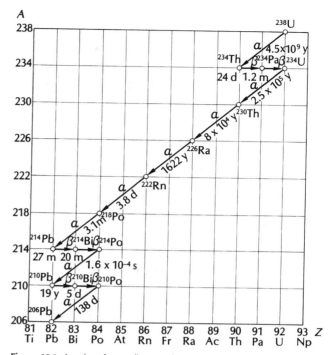

Figure 18.1 A series of naturally occurring unstable nuclides having ^{238}U as parent. Half-lives are given in years (y), days (d), hours (h), minutes (m), or seconds (s); α denotes alpha emission and β denotes beta emission. (After Faure 1986: 285.)

orbital electrons. The ratio of parent to daughter atoms of a radioactive element present in a crystal is determined only by the elapsed time since the radioactive element and its decay products were locked into the crystal (assuming neither have escaped). If we take N to represent the number of radioactive nuclei present in a sample, then

$$-\frac{\delta N}{\delta t} \propto N$$

Rearranging gives $-\delta N / N = \lambda \delta t$, where λ is a previously determined **decay constant.** By integrating both sides, one obtains

$$-\int \frac{\delta N}{N} = \lambda \int \delta t$$

$$-\ln N = \lambda t + C$$

where C is the constant of integration and can be evaluated at time $t = 0$. Therefore, if $N = N_0$, then $C = -\ln N_0$ (since $t = 0$), and

$$-\ln N = \lambda t - \ln N_0$$

$$\ln N - \ln N_0 = -\lambda t$$

$$\ln \frac{N}{N_0} = -\lambda t$$

$$\frac{N}{N_0} = e^{-\lambda t}$$

$$N = N_0 e^{-\lambda t}$$

The preceding equation, $N = N_0e^{-\lambda t}$, gives the radioactive decay curve shown in Fig. 18.2. The decay process is exponential, so the amount of parent material decreases exponentially and the amount of daughter product increases exponentially. Another way of looking at this process is shown in Fig. 18.3. At each equal increment of time (called a **half-life**), half of the parent material decays to daughter atoms. If there are eight parent atoms initially, there are four parents and four daughters after the first half-life, two parents and six daughters after two half-lives, and one parent and seven daughters after three half-lives. The half-life is thus defined as *the increment of time needed for half the parent atoms to decay to daughter products*. The half-life has a constant value for each different set of radioactive parents and daughters. It can be calculated from the preceding formulas. If we represent half-life by $t_{1/2}$, when $t = t_{1/2}$, $N = 1/2\,N_0$; thus

$$\tfrac{1}{2}N_0 = N_0e^{-\lambda t_{1/2}}$$

$$-\ln\left(\tfrac{1}{2} = e^{-\lambda t_{1/2}}\right)$$

$$\ln 2 = \lambda t_{1/2}$$

That is, half-life $= t_{1/2} = \ln 2/\lambda = 0.693/\lambda$.

In other cases, we may want to rearrange the equation so that we can solve for t, the age, in terms of the known half-life, decay constant, and numbers of parent (p) and daughter (d) atoms. Since

$$N_0 = p + d = pe^{\lambda t}$$

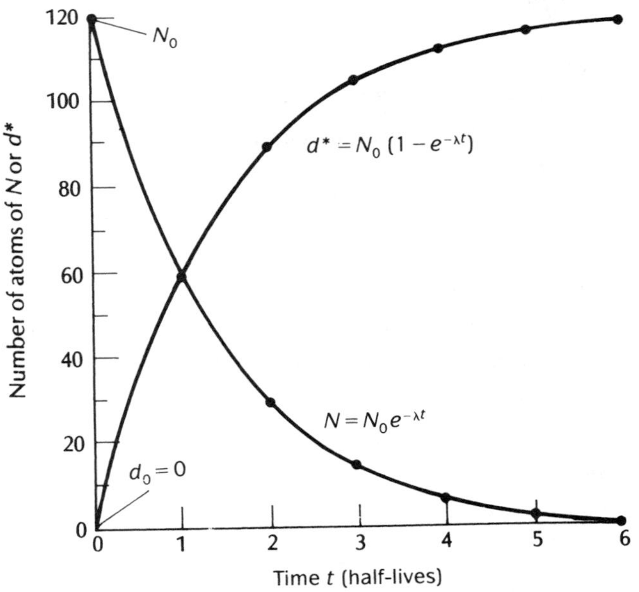

Figure 18.2 Decay curves of a hypothetical radionuclide N to a stable radiogenic daughter d^* as a function of time measured in half-lives. Note that $N \rightarrow 0$ as $t \rightarrow \infty$, whereas $d^* \rightarrow N_0$ as $t \rightarrow \infty$. (After Faure, 1986: 42.)

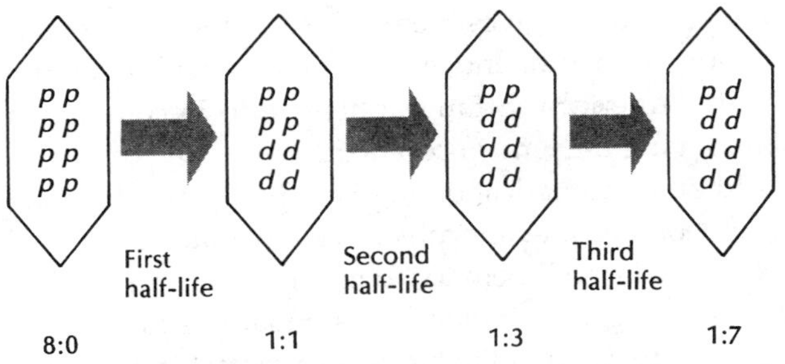

Figure 18.3 Radioactive decay of a hypothetical crystal with eight atoms of an unstable parent isotope p. After one half-life, half the parent material has decayed to the daughter product d, resulting in four parent and four daughter atoms. After the second half-life, half the remaining parent material again decays, leaving two parents and six daughter atoms. After the third half-life, one of the two remaining parent atoms decays, leaving only one parent and seven daughters. The change in the $p{:}d$ ratio from 8:0 to 1:1 to 1:3 to 1:7 is a measure of the age of the crystal in terms of half-life.

and since $N/N_0 = e^{-\lambda t}$, then $N\,e^{\lambda t} = N_0$ and $p = N$. In this case,

$$\frac{p + d}{p} = e^{\lambda t}$$

$$t = \frac{1}{\lambda}\ln\left(\frac{p + d}{p}\right) = \frac{1}{\lambda}\ln\left(\frac{d}{p} + 1\right)$$

This last equation is the simplest equation for calculating the age of most radiogenic systems.

Measurement

The standard tool of geochronology and isotope geochemistry is the **mass spectrometer** (Fig. 18.4), a device that separates and counts atoms of different masses or charges in a sample substance. The mass spectrometer, first developed by Nier in 1940, consists of three essential parts: (1) a source of a positively charged, monoenergetic beam of ions; (2) a magnetic analyzer; and (3) an ion collector. The entire system is evacuated to pressures of 10^{-6} to 10^{-9} mm of mercury. The solid sample is heated in the ion source until it emits ions, which then are accelerated and collimated into a narrow beam. This ion beam passes through the magnetic analyzer (Fig. 18.5), a curved track surrounded by a magnet. The magnetic field forces the ions to move in curved paths through the analyzer. The heavier the mass of an ion, the less it responds to the magnetic field and the less curved its path. By the time the ions have gone through the analyzer, they have been sorted by their masses into separate beams. The lightest ions show the most magnetic deflection, and the heaviest ions show the least. The collector picks up these separated ion

Figure 18.4 The modern Nier-type mass spectrometer, the standard tool for nearly all studies of isotope geochemistry. The ion source (cylindrical device on the top left of the cabinet) bombards the sample with electrons to ionize it. The ions travel the curved path through the magnetic field (*center*), where they are separated according to their mass. The ion collector at the right side of the curved path detects the ions of isotopes of various masses. The results are then analyzed by computer to determine the amounts of each isotope. (Courtesy of VG Isotopes Ltd.)

beams and records the relative strengths of their peaks (Fig. 18.6), which measure the relative abundance of each isotope in the sample.

Even the most sensitive modern mass spectrometers have limitations, the most important of which is **measurement error.** This is usually about $\pm 0.2\%$ to $\pm 2.0\%$, which means as much as ± 2 million years for an age of 100 million years or ± 20 million years for an age of 1 billion years. Thus, under the best conditions (fresh specimens, great analytical care, and material that is old enough to contain a measurable amount of daughter product), a sample said to be 100 million years old is really 98 to 102 million years old. The age cannot be resolved better than that, because of the limits of the instrument. The magnitude of this error limits spectrometry to estimating the age of a sample; the relative size of the error tells how reliable the estimation is. Because of measurement error and bad samples, we are hesitant to refer to a date as "accurate" or "close to the truth." We *can* make some statement about its **precision,** or reproducibility, which is indicated by the size of the error and by subsequent analyses (see Fig. 18.16).

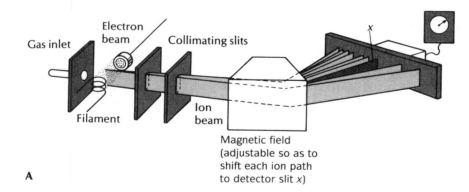

A

B

Figure 18.5 (A) The working parts of the mass spectrometer. (B) Close-up of the curved ion path between the poles of the magnet. (Courtesy of VG Isotopes, Ltd.)

Criteria for Usefulness

The short-lived, unrenewable radioactive isotopes that formed at the beginning of Earth's history 4.5 billion years ago have all decayed. Only isotopes with long half-lives (Table 18.1) are still present on Earth to serve as geological clocks. Many of these have limited geological application and are not discussed here. A handful of isotopes occur in great enough abundance in the right types of rocks to be geologically useful (Table 18.2).

Besides an appropriate half-life, there are many other criteria for whether a system is geologically useful. The radiometric measurement of a sample gives the time of its cooling, when solidification locked the parent and daughter atoms into the mineral. Whether this "date" is also a meaningful age must be determined from geological relationships. For example, a detrital particle records the date of its cooling in its igneous parent but gives no indication of the date at which it became a sedimentary particle or the date at which it was incorporated into sedimentary rock. This can lead to serious problems. A volcanic ash flow may pick up a small amount of detrital feldspar or biotite that is much older than the ash flow. If such particles are inadvertently included in the analysis, an erroneous age estimate can result. Except for volcaniclastic rocks, *very few sedimentary rocks can be dated directly.* They can be dated only by indirect methods, such as their relationship with associated lava flows or with cross-cutting dikes. Finding enough places where these relationships exist has taken decades, but each success has refined the numerical dating of the geologic time scale (see Box 18.1).

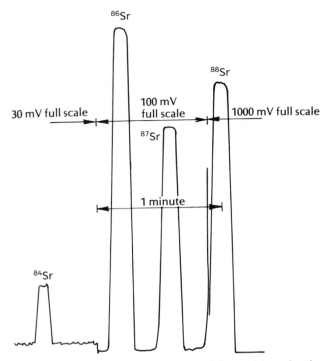

Figure 18.6 Display of a mass spectrometer. Each isotope (of strontium, in this case) produces a peak, and the amounts of each isotope are gauged by the height of the peak. (After Gast, 1962; 932; © 1962, Pergamon Press, Inc.)

In the past few years, enormous improvements have been made in reducing the error estimates on samples. New technology, such as the Super-High-Resolution Ion Microprobe (SHRIMP), allows measurement errors of $\pm 0.1\%$ or less. Thus, even Precambrian dates of 3 billion years may have measurement errors of only ± 1 million to ± 5 million years or less—extraordinary precision for events of that age.

TABLE 18.1 Decay Schemes					
Parent Isotope	Daughter Isotope	Parent's Natural Abundance (%)	Half-Life	Parent's Decay Constant	Practical Dating Range
^{40}K	^{40}Ar	0.01167	1.250×10^9 yr	$\lambda_\beta = 4.692 \times 10^{-10}$/yr	1 to >4500 Ma
	^{40}Ca			$\lambda_\epsilon = 0.581 \times 10^{-10}$/yr	>1500 Ma
^{87}Rb	^{87}Sr	27.835	48.8×10^9 yr	1.42×10^{-11}/yr	10 to >4500 Ma
^{147}Sm	^{143}Nd	14.97	1.060×10^9 yr	6.54×10^{-12}/yr	>200 Ma
^{176}Lu	^{176}Hf	2.59	3.50×10^9 yr	1.94×10^{-11}/yr	>200 Ma
^{232}Th	^{208}Pb	~100	14.010×10^9 yr	4.9475×10^{-11}/yr	10 to >4500 Ma
^{235}U	^{207}Pb	0.72	0.7038×10^9 yr	9.8485×10^{-10}/yr	10 to >4500 Ma
^{238}U	^{206}Pb	99.28	4.468×10^9 yr	1.55125×10^{-10}/yr	10 to >4500 Ma
^{14}C	^{14}N	—	5370 yr	1.29×10^{-4}/yr	<80,000 yr

After Parrish and Roddick (1985).

TABLE 18.2	Materials Used for Geochronology
Dating Method	**Materials That Can Be Dated**
K-Ar	Hornblende, muscovite, biotite-phlogopite, feldspars, glauconite, whole-rock volcanics, some glasses
Rb-Sr	Micas, K-feldspar, cogenetic whole rocks that have a dispersion of Rb/Sr ratios, apatite and sphene for initial $^{87}Sr/^{86}Sr$
Sm-Nd	Pyroxene, plagioclase, garnet, apatite, sphene, other phases; whole rocks with a dispersion of Sm/Nd ratios
Lu-Hf	Much the same as Sm-Nd, zircon for initial Hf isotopic composition
U-Th-Pb	Zircon, monazite, xenotime, baddeleyite (ZrO_2), sphene, apatite, allanite, pyrochlore, U or Th minerals
Pb-Pb	Galena or other Pb minerals, K-feldspar, tellurides, carbonates in carbonatites
^{238}U fission track	Zircon, apatite, sphene, garnet, epidote, volcanic glass

After Parrish and Roddick (1985).

The loss of parent or daughter atoms as a result of heat or weathering creates other problems. For example, during metamorphism, the original mineral can be recrystallized, and some of the parent or daughter atoms may escape, thus completely eliminating any possibility that the original parent rock can be dated. Recrystallization, however, resets the clock so that the decay process begins all over again. The dates then will reflect the time of metamorphism. This process is called **metamorphic resetting.** Weathering can cause a similar problem by allowing some of the parent or daughter atoms to escape; once again, any date derived from the remaining atoms is meaningless. These kinds of problems are not easy to overcome, so they are best avoided. Great care must be taken to get the freshest samples possible and to check every date several times.

Another major limitation of radiometric methods is that there must have been enough of a radioactive element in the rock for there to be measurable amounts of parent and daughter atoms. Most radioactive elements are too rare in terrestrial rocks to be useful, with the exception of potassium, rubidium, and uranium. Also, the half-life of the parent element must be short enough for there to be a measurable amount of daughter product; this places a lower limit of about 100 million years on the usefulness of the Rb-Sr, U-Pb, and Th-Pb systems (see Table 18.1). On the other hand, systems with half-lives that are too short are radioactively "dead." Carbon-14, for example, has such a short half-life that the duration of its usefulness is about 60,000 to 80,000 years. In specimens older than this, there is no parent material left to measure.

Potassium-Argon Dating

The decay series most commonly used for dating is potassium-argon. Potassium, the seventh most abundant element in the Earth's crust, is especially

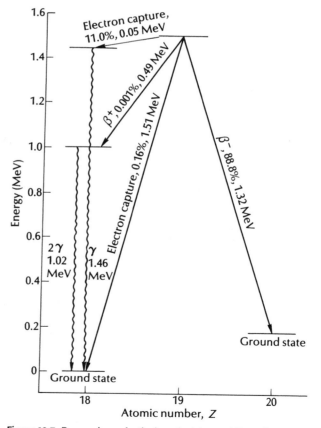

Figure 18.7 Decay scheme for the branched decay of ^{40}K to ^{40}Ar by electron capture and by positron (β^+) emission, and to ^{40}Ca by emission of negative beta particles (β^-). The percentage of the parent material in each decay path and the energy changes are also shown. (After Dalrymple and Lanphere, 1969.)

BOX 18.1 BRACKETING THE AGE OF THE SILURIAN-DEVONIAN BOUNDARY

In Chapter 16 (Box 16.1), we saw how the Silurian-Devonian boundary was established biostratigraphically. Although this gives us a standard for its recognition and relative age, it says nothing about its numerical age. The type section in the Czech Republic has no igneous intrusives or extrusives, so the boundary cannot be directly dated there. Geologists have had to search all over the world for places where rocks of late Silurian or early Devonian age (as established by their fossils) occur in relationship to datable igneous rocks. Very few of the rocks found in such relationships have preserved the actual Silurian-Devonian boundary, and the dates of the igneous rocks that are in relation to boundary rocks may not be close to the boundary. The entire process requires considerable interpolation between dated horizons and those that have the diagnostic fossils.

We are dealing with ages on the order of 400 million years, so the typical error for these samples is on the order of ±5 million to ±15 million years.

Even if we had a lava flow perfectly interbedded between the Silurian and Devonian, its error would

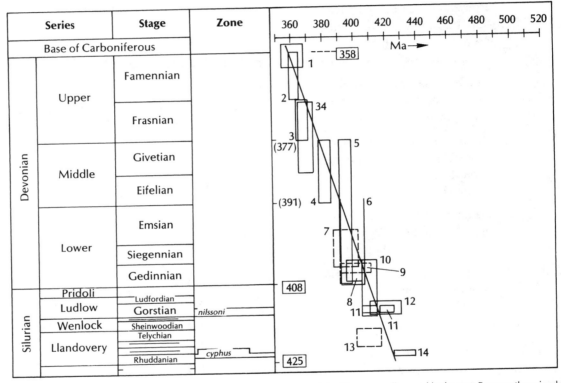

Figure 18.1.1 Radiometric constraints on rocks of Late Silurian and Devonian age. Each date has some error due to the imprecision of radiometrics (shown by the horizontal widths of the boxes, typically ±4 million to ±5 million years) and some error due to difficulties in locating the date in a stratigraphic horizon (vertical height of the box.) Some of the numbered boxes are discussed in the text. Because there is a large margin for error in every date, McKerrow, Lambert, and Cox (1985) chose to fit a line through the centers of most of the boxes to interpolate the ages of the boundaries. (After Snelling, 1985: 86; by permission of Blackwell Scientific Publishers.)

common in potassium feldspars, micas, hornblende, and clays, which are found in many geological situations. The most common isotope of potassium is ^{39}K (92% of total potassium), and the next most common is ^{41}K (7.9%). The rarest isotope is the radioactive form ^{40}K (0.01167%); because it is an unstable atom, most of Earth's original ^{40}K has already decayed, enhancing its rarity.

Potassium-40 has a peculiar branching decay path (Fig. 18.7). Eleven percent of any given amount of ^{40}K decays to ^{40}Ar by electron capture and gamma decay. The rest decays to ^{40}Ca by beta decay. Large amounts of nonradiogenic ^{40}Ca also occur in crustal rocks, so this decay path is not often used for dating. On the other hand, ^{40}Ar is formed only by radioactive decay, and so this is the system that is used. Because there are

(Box 18.1 continued)

allow us to give the date only to the nearest 5 million years or so. Experience has also shown that there may be several conflicting dates, even when the sampling and measurements are done as carefully as possible. Consequently, geochronologists (Odin, 1982; McKerrow, Lambert, and Cox, 1985; Gale, 1985) have used a suite of dates from different areas (Fig. 18.1.1). Instead of treating each date as a single point, they represented each date by a rectangle. The horizontal dimension of a rectangle represents the measurement error (plus or minus) in millions of years. The vertical dimension represents the stratigraphic error, or the degree to which the date can be placed precisely in the fossil sequence. Some dates have huge stratigraphic errors because they occur well above or below the sequence of interest and thus provide only a maximum or minimum age. Dates with no upper or lower stratigraphic limit may be indicated by open rectangles.

Only a few of the dates shown in Fig. 18.1.1 are relevant to the age of the Silurian-Devonian boundary. Date 5 (the Shap Granite in England) intrudes Upper Silurian rocks but may be as young as Late Devonian. Date 6 (the Skiddaw Granite, also in England) is post–Late Silurian, but there is no younger limit on its stratigraphic age. Both dates fix the curve only crudely, and their relationship to the line of correlation through them suggests that they are Early or Middle Devonian. Date 8 (the Lorne Lavas of England, interbedded with lowest Devonian rocks) gives an Rb-Sr age of 399 ± 5 Ma for the earliest Devonian. Date 9 comes from the Hedgehog Volcanics of Maine, which are interbedded with Lower Devonian rocks and produce Rb-Sr and whole-rock ages of 400 ± 10 Ma for the earliest Devonian. Date 10 is from the Gocup Granite of Australia, which could range in age from the Middle Silurian to Early Devonian. It gives a date of 409 ± 5 Ma by K-Ar on muscovite.

Dates 11 and 12 occur in the Late but not latest Silurian (Ludlovian), so they help fix the line but are not close to the Silurian-Devonian boundary. Date 11 is from the Laidlaw Volcanics of Australia, which are interbedded with deposits containing early Ludlovian fossils. The spread on the dates from this area is discouraging: 409 ± 5 Ma by K-Ar on sanidine; 420 ± 5 Ma by K-Ar on biotite; 425 ± 17 Ma by Rb-Sr on whole rock; and 421 ± 5 Ma by Rb-Sr on biotite. Date 12 is a little more useful because it comes from the type Ludlovian in the Welsh Borderland. It gives a date of 419 ± 10 Ma by K-Ar on biotite. Not shown in Fig. 18.1.1 are new dates from the Wormit Bay Lavas of Scotland, which are interbedded with rocks that may be Ludlovian or Pridolian in age. These give a date of 408 ± 5 Ma by Rb-Sr on biotite; this is the first date listed that is truly latest Silurian.

From the preceding numbers, it is apparent how difficult it is to get a decent date with good stratigraphic constraints anywhere near the desired target of the Silurian-Devonian boundary. The only reasonable procedure is to fit a line through the spread of the rectangles, as shown in Fig. 18.1.1. *The age of the boundary is not directly dated anywhere, but it is estimated by interpolation.* Naturally, the method of interpolation can vary from author to author, and each estimate may be modified slightly when new dates are available. Consequently, various authors give different estimates for the desired age. A typical range of estimates from the last few decades for the age of the Silurian-Devonian boundary includes:

411 Ma: McKerrow et al., 1980

401 Ma: Jones, Carr, and Wright, 1981

400 Ma: Odin, 1982, 1985

408 Ma: Harland et al., 1982

412 Ma: McKerrow, Lambert, and Cox, 1985

408 Ma: Gale, 1985

This range illustrates the uncertainty inherent in the whole problem. Whenever you see a numerical time scale or run across a geologist confidently giving an "absolute" age, it pays to keep in mind the real limitations of those "solid" numbers.

two decay paths, there are two decay constants, λ_ϵ for the K-Ar decay path, and λ_β for the K-Ca decay path.

The main problem with the potassium-argon system is that the daughter product, argon, is an inert gas that can leak out of a crystal. If this has happened in a sample, one will find too little daughter product, and the date obtained will be too young. Argon-40, which also occurs in the atmosphere, can adhere to the sample and contaminate it, making the date too old. To check for this, it is necessary to analyze also for ^{36}Ar, which has a known ratio to ^{40}Ar in the atmosphere ($^{40}Ar/^{36}Ar = 296$). The formula for correcting

TABLE 18.3   Critical Table for Conversion of K-Ar Ages from Western Constants to New IUGS Constants

Age	F	Age	F	Age	F	Age	F	Age	F	Age	F
0	1.0268	182	32	385	96	606	1.0160	847	24	1111	88
5	66	193	1.0230	397	94	619	58	861	22	1127	86
15	64	204	28	409	92	632	56	875	1.0120	1142	84
25	62	215	26	421	1.0190	645	54	889	1.0118	1158	82
35	1.0260	226	24	433	88	658	52	903	16	1174	1.0080
45	58	237	22	445	86	671	1.0150	918	14	1190	78
56	56	248	1.0220	457	84	684	48	932	12	1206	75
66	54	259	1.0218	469	82	697	46	947	1.0110	1222	74
76	52	271	16	481	1.0180	710	44	961	08	1238	72
87	1.0250	282	14	493	78	724	42	976	06	1254	1.0070
97	48	293	12	506	76	737	1.0140	990	04	1270	1.0068
108	46	305	1.0210	518	74	750	38	1005	02	1287	66
118	44	316	08	530	72	764	36	1020	1.0100	1303	64
129	42	327	06	543	1.0170	778	34	1035	1.0098	1320	62
139	1.0240	339	04	555	1.0168	791	32	1050	96	1336	1.0060
150	38	350	02	568	66	805	1.0130	1065	94	1353	58
161	36	362	1.0200	580	64	819	28	1081	92	1370	56
172	34	374	1.0198	593	62	833	26	1096	1.0090	1387	54

Note: To convert an age based on old western constants to one based on new IUGS constants, multiply by the indicated correction factor F. Ages are in 10^6 yr. Old western constants: $\lambda_\epsilon + \lambda_\epsilon^1 = 0.585 \times 10^{-10}$ yr^{-1}, $\lambda_\beta = 4.72 \times 10^{-10}$ yr^{-1}, ^{40}K/K$_{total} = 1.19 \times 10^{-4}$ mol/mol; new IUGS constants: $\lambda_\epsilon + \lambda_\epsilon^1 = 0.581 \times 10^{-10}$ yr^{-1}, $\lambda_\beta = 4.962 \times 10^{-10}$ yr^{-1}, ^{40}K/K$_{total} = 1.167 \times 10^{-4}$ mol/mol.

After Dalrymple (1979).

for atmospheric argon is

$$^{40}\text{Ar (measured)} - 296(^{36}\text{Ar}) = {}^{40}\text{Ar (radiogenic)}$$

To obtain useful dates, the crystal lattice of a mineral must be tight enough to retain the argon gas. At depths of 5 km or more or at temperatures above 200°C, most minerals leak argon, so a date from a rock that has been at these or higher pressures or temperatures has probably been metamorphically reset. If the mineral has been altered or weathered, it almost certainly has leaked some argon. As the sample is broken down and prepared in the laboratory, it must also resist dissolution, or more argon will leak out.

Nonetheless, the potassium-argon system is very useful for many geological problems because potassium is abundant in many rocks. There is no upper age limit on this system because its half-life of 1.3 billion years is long enough for use on Precambrian rocks and nearly everything younger. The lower limit depends on atmospheric argon contamination and the sensitivity of the mass spectrometer used. Biotite that is younger than a million years old cannot usually be dated, but sanidines as young as 10,000 years have been dated successfully.

The age equation for potassium-argon dating is more complex than the standard formula because

Age	F	Age	F	Age	F	Age	F	Age	F	Age	F	Age	F
1404	52	1731	16	2099	0.9980	2520	44	3008	08	3585	72	4283	38
1421	1.0050	1750	14	2121	78	2546	42	3038	06	3620	0.9870	4326	36
1439	48	1770	12	2143	76	2571	0.9940	3068	04	3656	0.9868	4370	34
1456	46	1789	1.0010	2165	74	2597	38	3098	02	3692	66	4414	32
1474	44	1809	08	2188	72	2622	36	3128	0.9900	3728	64	4459	0.9830
1491	42	1829	06	2210	0.9970	2648	34	3159	0.9898	3765	62	4505	28
1509	1.0040	1849	04	2233	0.9968	2675	32	3189	96	3802	0.9860	4551	26
1527	38	1869	02	2256	66	2701	0.9930	3221	94	3840	58	4597	24
1545	36	1889	1.0000	2279	64	2728	28	3252	92	3878	56	4645	22
1563	34	1909	0.9998	2302	62	2755	26	3284	0.9890	3916	54	4693	0.9820
1581	32	1930	96	2326	0.9960	2782	24	3316	88	3955	52	4741	0.9818
1599	1.0030	1950	94	2350	58	2809	22	3348	86	3994	0.9850	4790	16
1618	28	1971	92	2373	56	2837	0.9920	3381	84	4034	48	4840	14
1636	26	1992	0.990	2397	54	2865	0.9918	3414	82	4074	46	4891	12
1655	24	2013	88	2422	52	2893	16	3448	0.9880	4115	44	4942	0.9810
1674	22	2035	86	2446	0.9950	2922	14	3482	78	4156	42	4994	08
1693	1.0020	2056	84	2471	48	2950	12	3516	76	4198	0.9840	5049	06
1712	1.0018	2078	82	2495	46	2979	0.9910	3550	74	4240			04

there are two decay constants:

$$t = \frac{1}{\lambda_\epsilon + \lambda_\beta} \ln\left[\frac{(^{40}Ar_{rad})}{(^{40}K)} \left(\frac{\lambda_\epsilon + \lambda_\epsilon}{\lambda_\epsilon} \right) + 1 \right]$$

The decay constants λ_ϵ and λ_β were corrected in 1976 (Steiger and Jäger, 1977), so any date printed before then must be corrected. Table 18.3 provides an easy method to correct any older date for the new constants. For a potassium-argon date to be reliable, a number of criteria must be met:

1. No argon should have escaped or leaked.

2. The crystal must have cooled rapidly so that argon is trapped.

3. No extra argon should have been added. Correction must be made for ^{40}Ar contamination from the atmosphere.

4. The system must have been closed to potassium gain or loss during its history.

5. There should be no unusual isotopic fractionation of potassium.

The suitability of various minerals for potassium-argon dating is highly variable.

Micas contain 7% to 9% potassium and are common in felsic igneous rocks and in metamorphic rocks. Biotite and muscovite are both reasonably reliable but are very susceptible to weathering and reheating. They are common as detrital grains, so one must be wary of detrital contamination.

Hornblende contains 0.1% to 1.5% potassium and is very common in felsic igneous rocks and in certain high-grade metamorphic rocks (especially amphibolite grade). Hornblende holds tenaciously to argon, so it is good for dating metamorphic terranes.

Potassium feldspars contain 7% to 12% potassium and occur in a wide variety of rocks. Plutonic igneous feldspars show large variations in argon loss and are seldom reliable. Sanidine, the high-temperature potassium feldspar common in volcanic rocks, cools rapidly and has a different structural state from low-temperature potassium feldspars. As a result, it traps argon well and produces dates with small error bars. It is especially good for dating young, fresh volcanics, but it alters easily and so seldom survives in its fresh state from rocks as old as the Mesozoic. Contamination from detrital potassium feldspars can also be a big problem.

Plagioclases contain 0.01% to 2% potassium and are abundant in nearly all igneous rocks, but their potassium content is so low that they are rarely usable.

Pyroxenes contain 0% to 1% potassium and are common in basic igneous rocks, but their low percentage of potassium limits their use. At depth, there is apparently significant contamination of pyroxenes by ^{40}Ar. Pyroxenes in shallow eruptive basalts have been successfully used for dating, however.

When a rock is too fine-grained for a geologist to pick out individual crystals in order to measure the age of the minerals, the whole rock can be broken down and dated (called **whole-rock analysis**). If the rock is fresh and has a tight, nonporous groundmass that traps all the argon, it may be suitable for dating. Whole-rock dating is not suitable for plutonic igneous rocks that have been heated above 200°C and so have lost argon. However, it is very good for fresh basalts, especially because the individual minerals in basalts tend to be extremely fine grained.

Glauconites (discussed in Chapter 6) are found in many shallow marine environments, where they grow authigenically at low temperatures in sediments. They have been widely used for K-Ar dating in Mesozoic and Cenozoic sediments, since they contain a few percent potassium. Ideally, they would seem to be the exception to the rule that sediments cannot be dated directly. Experience has shown, however, that they are susceptible to both argon loss and potassium gain, making their dates 10% to 20% too low (Thompson and Hower, 1973; Berggren et al., 1978; Berggren, Kent, and Flynn, 1985; Obradovich, 1988). Some scientists have chosen to rely heavily on glauconites for dating the geologic time scale and have come up with calibrations that differ by several million years from those derived from high-temperature potassium-argon dates. When this conflict arises, the glauconite dates should be viewed with suspicion because they have a poor track record.

In general, then, potassium-argon dating is excellent for most geologic problems, but it must be used with common sense. A single age determination is usually insufficient. Samples should be tested several times to reduce the error estimates. The date is even more reliable if two minerals from the same rock, two specimens from the same unit, or two radiogenic minerals are used. Agreement of dates within 5% is generally considered concordant. Anomalous ages can occur for a number of reasons. Ages that are too young probably result from bad samples, reheating, alteration, weathering, or potassium gain. Ages that are too old can result from atmospheric ^{40}Ar contamination or inherited ^{40}Ar. Random errors can be produced by analytical problems such as those discussed earlier, as well as by faulty sampling, mislabeled samples, or an incorrect preliminary age estimate based on mistakes in determining field relationships.

Argon-Argon Dating

Another technique that is being used increasingly is ^{40}Ar/^{39}Ar dating, in which potassium-rich samples are bombarded with neutrons in a nuclear reactor to convert ^{39}K to ^{39}Ar. Argon-39 is unstable, with a half-life of 269 years, and decays back to potassium-39 by beta decay. The ratio of ^{39}K to ^{40}K is known, so ^{39}Ar serves as a proxy for ^{40}K. Argon-40, of course, is the daughter product we are trying to measure. After irradiation, the argon gas is extracted from the sample as in conventional K-Ar methods and is analyzed in the mass spectrometer. Using stable ^{36}Ar to correct for atmospheric and other interferences, one can calculate an age once the conversion rate of ^{39}K to ^{39}Ar is known. This rate is determined by irradiating a standard of known age along with the sample. This technique has many advantages:

1. It does not require a separate analysis of potassium, so it is unnecessary to split the sample to retrieve the solid potassium; only argon gas need be measured.

2. It is so sensitive that very small samples can be used.

3. Because only one element is measured, the uncertainties of sample weighing and concentration measurement are eliminated.

4. With the proper techniques, atmospheric argon contamination can be removed; in fact, the method of measurement called step heating not only eliminates this problem but also allows correction for partial loss of argon from a sample that has been disturbed since cooling.

Figure 18.8 shows how the argon-argon system works. If the ^{39}Ar was uniformly distributed at potassium sites within the mineral before decay, then it should diffuse out of all parts of the mineral at the same rate. Therefore, measurements of the ^{40}Ar/^{40}K ratio taken after step heating from the center to the edge of the sample should produce constant values and a flat curve (Fig. 18.8A). If the sample has been heated during its history, however, some of the ^{40}Ar will have diffused from the edges, but the ^{40}Ar in the center, which was sealed in, will have kept the original ^{40}Ar/^{40}K ratio. Thus a ^{40}Ar/^{40}K plot of an altered sample (Fig. 18.8B) will be low in ^{40}Ar near the margin relative to the center.

As argon continues to decay, it will be trapped throughout the crystal as long as the crystal remains a closed system. This results in an increased ^{40}Ar/^{40}K ratio in both the center and the edge of the crystal (Fig. 18.8C). The ^{40}Ar/^{40}K ratio in the center of the crystal

reflects its original age because the center was relatively immune to alteration. The age estimate from the edge of the crystal gives the time elapsed since partial outgassing caused by geologic heating. Thus, this method is one of the few that gives dates for both the original cooling and later thermal events.

During analysis, this cooling history is reconstructed by stepwise heating of the crystal. At the lowest temperature, the argon from the edge diffuses out first and is measured. At each higher temperature step, argon from deeper into the crystal is released and measured. If the fraction of ^{39}Ar released stays on a plateau through most of the heating steps, then this plateau represents the age of the argon in the unaltered center of the crystal (Fig. 18.9). If the plateau is relatively smooth all the way across (Fig. 18.10A), then an age can be estimated by averaging all the values for the ^{39}Ar on the plateau to produce an **integrated age.** The more irregular the steps in the plateau (Fig. 18.9B), the greater the loss of argon has been and therefore the greater the measurement error.

Another method of obtaining argon-argon dates is the laser-fusion method. Instead of the crystal being slowly heated and its argon measured to obtain a plateau, the crystal is melted by a laser. As the crystal fuses, it releases its argon, which is measured as was

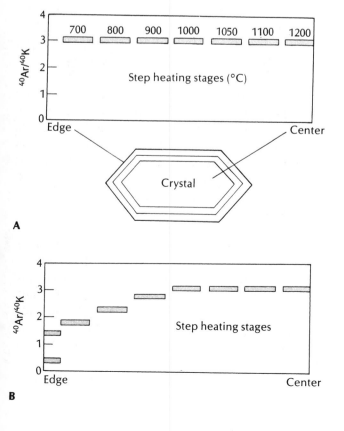

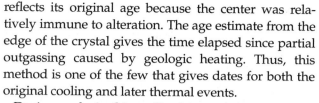

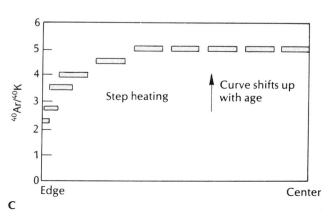

Figure 18.8 The argon-argon dating method. (A) Plot of the ^{40}Ar/^{40}K ratio with step heating of an unaltered crystal. If there has been no leakage, the outer rim of the crystal will release the same amount of material at low temperatures as the center releases at high temperatures. (B) An altered crystal, however, will show lower ratios at the lower temperatures due to leakage of argon from the edge. The plateau at the higher temperatures should give the true age of the unaltered center of the crystal. (C) As time passes and the crystal gets older, the entire curve shifts upward, but the plateau remains as long as no further alteration takes place.

A

B

C

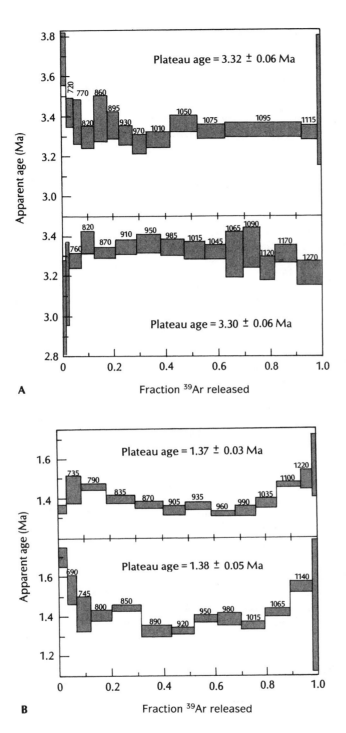

Figure 18.9 Examples of real argon-argon data. (A) Age spectra of feldspars from pumice clasts in the Toroto Tuff, Lake Turkana. Uncertainty of the age for each heating step at the level of one standard deviation is shown by the thickness of the bar. Temperature at which gas was released is shown by numbers over the bar. Both samples have very well defined plateaus, with only minor edge alteration at 760°C and below. (B) Age spectra from feldspars from the Chari Tuff, Lake Turkana. The plateaus are less well defined than in the previous example. (After McDougall, 1985: 165, 171; by permission of the Geological Society of America.)

just described. The advantage of this method is that the laser can be aimed very precisely, so that the argon can come from a single crystal, or even from part of a single crystal. By analyzing a large number of carefully chosen crystals, the investigator can determine which crystals appear altered or contaminated and which ones give good agreement on their ages. Thus, both high reproducibility and high precision are possible. Unlike in the step-heating method, however, one cannot plot a plateau that separates the altered edges from the original date obtained from the plateau. Nevertheless, by fusing enough carefully inspected crystals, the investigator can determine if the fresh crystals cluster around a concordant value and can assess the reliability of the date in terms of its scatter (the usual "plus or minus" value).

The main limitation of the argon-argon dating system is that minerals can absorb atmospheric argon. This contamination is difficult to detect except by the irregularities in the plateau of the curve. The best results have been obtained on hornblende and potassium feldspars; biotites tend to produce very irregular age spectra. The major advantage of this system over the potassium-argon system is its potential for detecting and dating thermal disturbance with a single sample. For this reason, it will probably be used more often in the future. Indeed, most of the good argon-argon dating labs are now so backlogged with research commitments that it can take years to get a date analyzed.

Rubidium-Strontium Dating

Rubidium-87 decays to strontium-87 in a single step by beta decay. Twenty-eight percent of the rubidium in the Earth's crust is ^{87}Rb, but rubidium is not a common element in the crust as a whole, occurring in abundances of about 3 to 140 ppm. Rubidium is an alkali element that forms a univalent cation with an ionic radius slightly larger than that of potassium, so it substitutes readily for potassium in many minerals. Strontium, however, forms a divalent cation, so it prefers calcium sites. As a result, when ^{87}Rb decays to ^{87}Sr, there is a strong tendency for the daughter strontium to migrate out of the host mineral. If whole-rock dating is used, however, the radiogenic strontium can be found in the calcium sites. Rubidium, like potassium, occurs primarily in micas, feldspars, and amphiboles. In most minerals, some inherited ^{87}Sr has been trapped in the crystal. To correct for this, ^{86}Sr in rock samples is also mea-

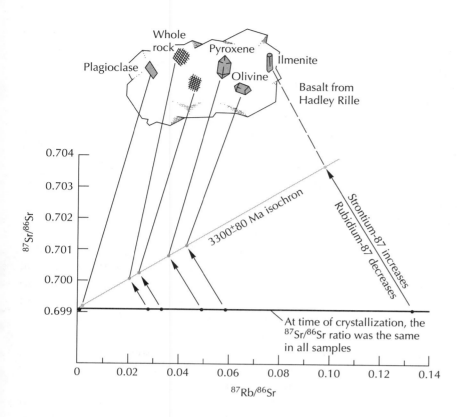

Figure 18.10 Rubidium-strontium isochron dating of a lunar basalt. The isochron method depends on analysis of samples with differing rubidium content. At the time of crystallization, the $^{87}Sr/^{86}Sr$ ratio was the same for all minerals in the sample. In minerals with low rubidium content, such as plagioclase or the whole rock, this ratio changed relatively little; in minerals with high rubidium content, such as ilmenite, it changed much more over the same amount of time. The slope of the isochron increases with time and provides the date. (After Eicher, 1976: 128.)

sured because its ratio to ^{87}Sr is known for most of Earth's history. Therefore, the age formula has the correction factor built in:

$$t = \frac{1}{\lambda} \ln \left\{ 1 + \left(\frac{^{86}Sr}{^{87}Rb} \right) \left[\underbrace{\left(\frac{^{87}Sr}{^{86}Rb_{present}} \right) - \left(\frac{^{87}Sr}{^{86}Rb_{initial}} \right)}_{\text{radiogenic Sr connection}} \right] \right\}$$

Rubidium-strontium dating works best in older samples because rubidium is so rare that a long decay period is necessary to accumulate measurable quantities of strontium. The correction for the $^{87}Sr/^{86}Sr$ ratio can introduce huge errors because ^{86}Sr is so abundant, especially in younger rocks. For example, corrections for $^{87}Sr/^{86}Sr$ are good to one part in 100 million, but in many samples, one part in 100 million is all the ^{87}Sr that 3 million years of decay of ^{87}Rb will produce. A young sample could have 100% error, but Precambrian samples have enough radiogenic strontium that the error is only on the order of 0.3%.

The most reliable technique of rubidium-strontium dating is called the **isochron method.** A single rock contains several minerals with different initial concentrations of rubidium. Yet at the time of crystallization, the $^{87}Sr/^{86}Sr$ ratio throughout the rock was constant (See Fig. 18.10). As the ^{87}Rb in the rock decayed, the

$^{87}Sr/^{86}Sr$ ratios in the minerals that contained more of the parent ^{87}Rb changed more than the ratios in the minerals with less ^{87}Rb. Therefore, measurements of the strontium isotopes present in the various minerals of a sample differ and fall along different paths in the plot shown in the figure. At any given time, the isotope ratios of all the minerals should fall along a common line, called an isochron. The slope of the isochron line is the ratio of radiogenic ^{87}Sr to ^{87}Rb in the sample and thus is a measurement of the age of the sample; older samples have a progressively steeper slope.

The rubidium-strontium method is useful in both igneous and metamorphic rocks that are old enough. It has proved especially useful for lunar basalts, meteorites, and metamorphic rocks, providing both the age of original crystallization and the date of metamorphism. Because the migrating ^{87}Sr is trapped in calcium sites, the whole-rock age represents the original date of crystallization. Individual minerals will have lost strontium during metamorphism; therefore, their rubidium-strontium age is a measure of their metamorphic resetting date. The **blocking temperature** (the temperature at which metamorphic resetting takes place) also varies in the different minerals, giving a rough index of the maximum temperature reached by the rock. The whole rock has a blocking temperature of 700°C, but the

muscovite in it is stable only at 500°C and the biotite only at 280° to 320°C. If resetting has occurred in the biotite but not the muscovite, for example, the maximum temperature can be bracketed.

Uranium-Lead Dating

Uranium-lead dating is often the most reliable dating technique, especially for old igneous rocks, because uranium is much more abundant than rubidium. The system is made even more robust by the fact that three uranium-lead isotopic decay series are available for cross-checking. The Earth's uranium is 99.27% ^{238}U and 0.72% ^{235}U. These uranium isotopes decay to different isotopes of lead, ^{206}Pb and ^{207}Pb, respectively. A third system, ^{232}Th, decays to ^{208}Pb, so all three related systems can be used for internal consistency. Each isotope decays to lead through a series of alpha and beta decays and produces a number of short-lived intermediate products and helium at each alpha decay step (see Fig. 18.1). The $^{207}Pb/^{206}Pb$ ratio in a rock also gives an age estimate, which can be used to check for concordance. Inherited lead can throw measurements off, so the nonradiogenic ^{204}Pb must also be measured. The normal ratio of ^{204}Pb to ^{207}Pb and to ^{206}Pb is assumed to be equivalent to the primordial ratio measured in meteorites, and this amount can be subtracted to correct for inherited lead.

Minerals that are rich in uranium are rare, but many minerals contain traces of it. Less common minerals—such as zircon, sphene, and monazite—are often used because they are impervious to weathering or other alteration and are extremely stable. They are also very easy to isolate by standard separation techniques, followed by immersion in a bath of hydrofluoric acid until the entire rock except the zircons has dissolved. These minerals are also very stable detrital particles, so care must be taken to avoid detrital grains in a sample. Some zircons are known to have been picked up by magmas and crystallized into much younger rock, but these can be recognized by their reworked texture. A reworked zircon seldom shows a euhedral habit.

Because three systems are involved, the best way to determine the age is a **concordia plot** (Fig. 18.11). On a plot with the $^{207}Pb/^{235}U$ ratio as the abscissa and the $^{206}Pb/^{238}U$ ratio as the ordinate, all the sample data points should fall along a smooth curve called the **concordia curve.** This is the theoretical curve that data from the sample should continue to follow as the uranium decays. When the material of the sample cooled and crystallized, its lead-to-uranium ratios would have been at the origin of the plot, and as the sample aged, the ratios would have moved up the curve. The age of the sample can be read directly from the position of the sample on the curve. If the ratios of isotopes in the sample fall off the curve, they are **discordant** (Fig. 18.12). This discordance may be due to metamorphism, variable lead loss, or continuous lead loss by diffusion. Discordant results plot below the concordia curve, usually in straight lines. A regression line fitted to these discordant points intercepts the concordia curve in two places. The upper intercept should give the original age of cooling, and the lower intercept (chronologically younger) gives the time of metamorphism or lead loss. Even discordant data yield valuable information.

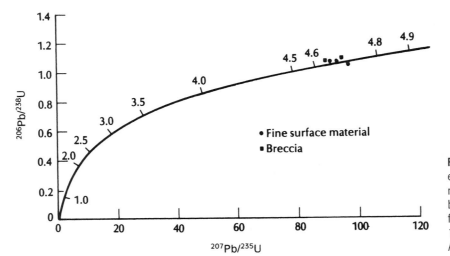

Figure 18.11 Uranium-lead concordia plot showing essentially concordant ages of lunar fine surface material and breccia collected by *Apollo 11.* Dates between 4.6 and 4.7 billion years agree with the data for the age of the solar system. (After Rankin et al., 1969: 743; by permission of the authors and the AAAS; © 1969 by the AAAS.)

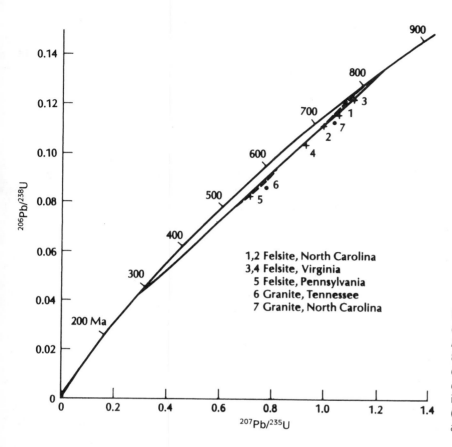

1,2 Felsite, North Carolina
3,4 Felsite, Virginia
5 Felsite, Pennsylvania
6 Granite, Tennessee
7 Granite, North Carolina

Figure 18.12 A concordia plot of a series of discordant uranium-lead ages from the Appalachians indicates a major igneous episode at 820 Ma (upper intersection of straight line with concordia curve) and metamorphic reheating during the Appalachian orogeny at 240 Ma (lower intersection of straight line with concordia curve.) (After Wetherill, 1971: 386; by permission of the authors and the AAAS; © 1971 by the AAAS.)

Fission-Track Dating

Spontaneous radioactive fission of ^{238}U produces not only daughter isotopes but also two high-energy charged particles that shoot out in opposite directions from the decaying nucleus. As they are emitted, these particles may disrupt the crystal lattice of their host mineral and leave tracks that can be seen under a microscope (Fig. 18.13). These **fission tracks,** about 10 to 20 μm long, can be seen in many crystals. Because the number of tracks is a function of the original amount of radioactive parent in the crystal, counting the tracks should give the age of the material. The older a crystal (given equal amounts of parent material), the more fission tracks will be formed. The method works best on substances with hard, resistant crystal lattices that are translucent and can be studied easily. Fresh volcanic glass, zircon, sphene, apatite, muscovite, biotite, and tektites have been successfully dated in this way. The sample is cut to expose a cross section, etched with hydrofluoric acid or other appropriate etchant to make the tracks visible, and then polished. The tracks are visible under a petrographic microscope at 800 to 1800 power and are counted using an eyepiece graticule to get a **spontaneous track density.**

The tracks are products of the fission of ^{238}U. To find out how much parent ^{238}U is present in the crystal, the sample is then placed in a reactor and bombarded with neutrons. This bombardment generates new particle tracks from the thermal-neutron-induced decay

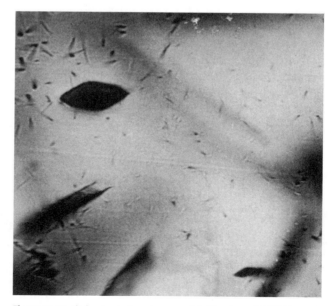

Figure 18.13 Fission tracks in a zircon crystal. (Courtesy of C. W. Naeser.)

of ^{238}U. The more original ^{238}U there is, the more tracks will be produced. These are counted in the same way as the spontaneous tracks, giving an **induced track density**. The age can then be calculated as a ratio of the spontaneous-to-induced track density:

$$A = \ln \left[1 + \left(\frac{\rho_s}{\rho_i} \right) \left(\frac{\lambda_D \Phi \sigma I}{\lambda_F} \right) \right] \frac{1}{\lambda_D}$$

where ρ_s is the spontaneous track density; ρ_i is the induced track density; ϕ is the neutron dose in neutrons/cm^2; and the constants currently in use (Naeser, 1979) include $I = 7.252 \times 10^{-3}$, $\lambda_D = 1.551 \times 10^{-10}$/year, $\lambda_F = 7.03 \times 10^{-17}$/year, and $\sigma = 580 \times 10^{-24}$ cm^2.

Fission-track dating has its stringent requirements. The tracks must be stable and etchable. The uranium must be uniformly distributed through the crystal and must be concentrated enough to give a minimum of 10 tracks/cm^2. The crystal must also be free of lattice defects, distortions, and other flaws. The main problem with the procedure is that heating can anneal fission tracks, eliminating some of them and giving ages that are too young. The tracks of various minerals anneal at different temperatures. In a whole rock, the discordant fission-track ages can be used to estimate the ages of both the primary cooling and any later metamorphic events. For example, apatite has a blocking temperature of only 74° to 120°C, but sphene anneals at 197° to 350°C and zircon at 150° to 210°C. Thus, the apatite might give the age of latest cooling, whereas the sphene is more likely to retain the original age.

As long as the mineral to be dated can be recovered, weathering is not a problem with fission tracks. In general, fission-track ages yield error estimates that are larger than those produced by potassium-argon dating. These larger error estimates are a result of problems with counting the large number of tracks necessary to achieve errors of 1% to 2%. To achieve errors of less than 2%, more than 10,000 tracks would have to be counted for both the spontaneous track density and the induced track density. Fewer than 2000 tracks are usually counted for practical reasons. Fission-track dating is excellent for dating young Tertiary and Quaternary volcanic rocks or for determining the provenance of detrital zircons. In addition, it is less expensive than most other dating techniques.

Carbon-14 Dating

Carbon-14 dating was first developed in the 1940s for use in archeology. It is the only isotopic decay system that occurs naturally in sedimentary rocks and fossils, so it is an exception to the rule that sedimentary rocks cannot be dated directly. The half-life of the system is so short (5730 years) that all the primordial carbon-14 decayed long ago. However, ^{14}C is continually produced in the atmosphere by the bombardment of ^{14}N by cosmogenic neutrons. The ^{14}N loses a proton, thereby becoming ^{14}C. This atmospheric radioactive carbon is assimilated into the carbon cycles of plants and animals while they live. The decay clock starts at the moment the tissue is no longer living and stops exchanging carbon with the environment.

The nuclear reaction is

$$\text{neutron} + {}^{14}_{7}N \rightarrow {}^{14}_{6}C + {}^{1}_{1}H$$

and the corresponding decay reaction is

$${}^{14}_{6}C \rightarrow {}^{14}_{7}N + \beta + \text{energy}$$

The age equation is

$$t = 19.035 \times 10^3 \log \frac{A}{A_0} \text{ in years}$$

where A is the measured activity of the sample in disintegrations per minute per gram of carbon (dpm/g), and A_0 is the initial activity. The currently accepted value for A_0 is 13.56 ± 0.07 dpm/g.

Carbon-14 dating is based on two assumptions: first, that the rate of ^{14}C production in the upper atmosphere is constant; second, that the rate of assimilation of ^{14}C by organisms is rapid relative to its rate of decay. Both of these assumptions appear to be valid.

The procedure for measuring carbon-14 in a sample is quite different from that for any other isotopic system. First, the sample is burned in a vacuum chamber to produce CO_2, part of which is composed of ^{14}C. The carbon dioxide gas is run through a chamber from which most of the other atmospheric gases have been removed. It is then concentrated in a copper counting tube, where a Geiger counter measures the decay activity. During measurement, the sample must be shielded from atmospheric gamma rays and neutron bombardment. In many cases, only a few grams of material are needed, but in others, several kilograms may be required (Table 18.4). Even with sufficient material, it takes 12 to 24 hours of counting per sample to get a measurement. For precision, multiple samples are counted and recounted to screen out background noise and to get consistent measurements with small error estimates.

The main limitation of carbon-14 dating for geology is its short half-life. The practical upper age limit is about 40,000 years, although with extraordinary care

TABLE 18.4 Material Suitable for Dating by the Carbon-14 Method

Material	Amount Required[a]	Comments
Charcoal and wood	25 g	Usually reliable, except for finely divided charcoal, which may adsorb humic acids; removable by treatment with NaOH. Subject to "post-sample-growth error"; that is, the difference in time between growth of a tree and use of the wood by humans.
Grains, seeds, nutshells, grasses, twigs, cloth, paper, hide, burned bones	25 g	Usually reliable. These materials are short-lived and have negligible post-sample-growth errors.
Organic material mixed with soil	50–300 g	Should contain at least 1% organic carbon in the form of visible pieces. Efforts should be made to remove as much soil as possible in the field.
Peat	50–200 g	Often reliable, but intrusive roots of modern plants must be removed. The coincidence of peat formation with the occupation of archeological sites requires careful consideration.
Ivory	50 g	Often well preserved and reliable. Interior of tusks is younger than the exterior. Some ivory tools may have been carved from old rather than contemporary material.
Bones (charred)	300 g	Heavily charred bones are reliable. Lightly charred bones are not because exchange with modern radiocarbon is possible.
Bones (collagen)	1000 g or more	Organic carbon in bones, called collagen, is reliable. However, the organic carbon content is low and decreases with age to less than 2%.
Shells (inorganic carbon)	100 g	The carbon in the calcite or aragonite of shells may exchange with radiocarbon in carbonate-bearing groundwater. Shell isotope carbon may be initially enriched in ^{14}C relative to wood due to fractionation. It may also be depleted in ^{14}C due to incorporation of "dead" carbon derived by weathering of old carbonate rocks. The reliability of shell dates is therefore questionable.
Shell (organic carbon)	Several kg	Organic carbon is present in the form of conchiolin, which makes up 1% – 2% of modern shells. Dates may be subject to systematic errors due to uncertainty of initial ^{14}C activity of this material.
Lake marl and deep-sea or lake sediment	Variable	Such materials are datable on the basis of the radiocarbon content of calcium carbonate. Special care must be taken to evaluate errors due to special local circumstances.
Pottery and iron	2–5 kg	Pottery sherds and metallic iron may contain radiocarbon that was incorporated at the time of manufacture. Reliable dates of such samples have been reported.

[a]The approximate amounts of material required for dating are based on the assumption that 6 g of carbon should be available, which is sufficient to fill an 8-liter counter with CO_2 at a pressure of 1 atm.

After Ralph (1971).

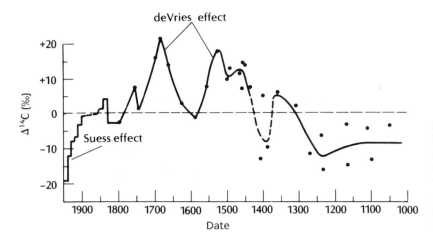

Figure 18.14 Deviation of the initial radiocarbon activity in wood samples of known age. The decline in radiocarbon content starting at about the year 1900 results from the increase in all CO_2 from burning fossil fuels (the Suess effect). The anomalously high radiocarbon activity around 1500 and 1710 is known as the de Vries effect, but its causes are not understood. (After Faure, 1986: 392.)

and extremely sensitive equipment, marginal dates of 60,000 to 80,000 years have been produced. Another problem has been uncovered by analyzing the carbon in dated tree rings. These studies show that the relative amount of atmospheric ^{14}C has fluctuated tremendously in historic times, decreasing significantly in the last 100 years as a result of the burning of trees and fossil fuels (Fig. 18.14). Nuclear explosions have also added slight amounts of atmospheric ^{14}C. A correction must be added to the date to compensate for these effects.

There are other pitfalls as well. Marine shells are notorious for picking up carbon that has been dissolved in seawater from ancient carbonate rocks. This ancient carbon is radioactively "dead" and depresses the ^{14}C count, giving the impression that the shell is much older than it really is. This is why an occasional living shell gives a very ancient date. In general, however, carbon-14 is the primary dating tool for Quaternary geology and archeology. It is possible to date directly any organic material that was once living, such as shells, wood, peat, bones, baskets, cloth, paper, food materials, and even some pottery and iron objects. Carbon-14 dating proved critical in dating the ice ages as well; undoubtedly, new uses will continue to be found for it.

Other Dating Methods

Some relatively minor isotopic decay systems have significant applications. **Thorium-230** has been used to date deep-sea cores that are too old for carbon-14 dating (40,000 years) and too young for the first magnetic reversal boundary (780,000 years) and that have no potassium for K-Ar dating. Thorium-230, part of the ^{238}U decay series (see Fig. 18.1), has a half-life of 75,000 years. When it forms from the decay of ^{238}U, it

is fractionated in seawater, with the uranium remaining in solution and the thorium being incorporated into the bottom sediments. If the sedimentation rates and the ^{230}Th precipitation rates are approximately constant, the thorium can be measured from the top of a long core downward to obtain the age.

The most important new discovery concerns the recalibration of radiocarbon with other systems, such as ^{234}U-^{230}Th (Bard et al., 1990). A core drilled through coral reefs in Barbados produced a continuous record for the last 30,000 years, with both carbon-14 and U-Th dates from the same samples. The two sets of ages deviated in a consistent way, with the radiocarbon dates on the young side. When the more recent data were compared with tree ring dates, it turns out that the U-Th dates were more reliable, and the radiocarbon is too young in any date older than 9000 years. Thus, nearly all published radiocarbon ages need to be recalibrated. This can have significant effects. For example, recalibrating the radiocarbon time scale puts the peak of the last glacial maximum at 21,500 years ago, not 18,000 years ago as had long been thought. This recalibration changes not only the textbooks but also the calculations of the Milankovitch models of orbital forcing of the ice ages. Fortunately, the new dates make the Milankovitch model fit better than previously reported.

Another system used in similar situations is **thorium-230–protactinium-231.** Produced from the decay of ^{235}U, ^{231}Pa is also found in marine sediments. It has a half-life of only 37,000 years. If this isotope is found in combination with ^{230}Th, the two have a combined half-life of 57,000 years. The ratio of these two isotopes thus changes constantly through time, because they decay at different rates. By combining the two isotopes, one can obtain a date that is independent of sedimentation rates.

Therefore, this method is more versatile than using ^{230}Th alone. Because of the short half-life, however, its maximum limit is about 150,000 years.

Another system that has been used in recent years is **samarium-neodymium** (Sm-Nd). Both are rare Earth elements with high atomic numbers on the periodic table. The ^{147}Sm changes to ^{143}Nd by alpha decay with a half-life of 106 billion years. Both elements are found widely in the Earth's crust in trace amounts of less than 10 ppm. Phosphate minerals such as apatite and monazite, and alkalic igneous rocks, tend to have higher concentrations. Mafic and ultramafic igneous rocks, however, have the highest Sm/Nd ratios and are the most easily dated. Whole-rock dating, or the isochron method of dating various minerals in the same rock, is most commonly used. Sm-Nd dates are less susceptible to alteration by metamorphism than are Rb-Sr dates. They are especially useful for dating rocks that were initially low in rubidium, such as mafic and ultramafic rocks.

Chronostratigraphy

The "Web of Correlation"

Because most real-world situations present us with a frustrating and incomplete mix of data from various sources, and because every method of dating and correlating has its strengths and weaknesses, most stratigraphic problems are tackled using a variety of techniques. Usually, this approach requires collaboration among specialists in various branches of geology. Once this information from several techniques is gathered, however, it is possible to integrate the results of the various methods into a combined picture, cross-checking one set of correlations or dates against the others to spot possible problems (Fig. 18.15). The result is a correlation and a chronology of geologic events that are more rigorously tested and have greater precision and resolution than any single method could produce. This "web of correlation" (to paraphrase Berry, 1987) is often complex and difficult to construct, but the value of the system is correspondingly greater. *Establishing the time relationships among geologic units* by means of integrated methods is called **chronostratigraphy.**

Sometimes, two data sources (for instance, a biostratigraphic identification and a radiometric date) may disagree. The specialists who produced each result must reexamine their data to see if the conflict can be resolved. This process is healthy because the flaws in the data often become apparent, and the result is greater refinement. Ideally, every episode of conflict and resolution produces a smaller and smaller range of disagreement, and the corresponding chronology of events needs fewer and fewer changes. The resulting chronology has greater **accuracy;** it is closer to that abstract, unattainable ideal we call "truth" (Fig. 18.16).

As we check the results of one method against another, we must avoid circular reasoning. Each method must retain its own integrity, or else it tells

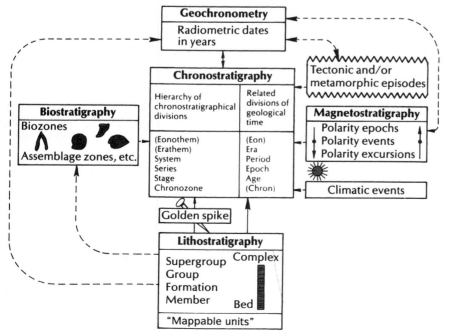

Figure 18.15 Flowchart of the procedures of stratigraphy and their relationships. The chronostratigraphical record is the central repository for the data derived from the procedures and phenomena indicated around them. (After Dineley, 1984: 13.)

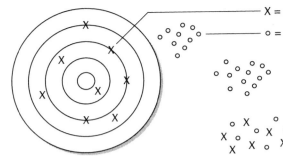
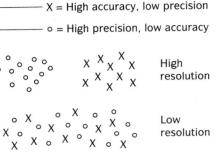

Figure 18.16 The relationships among precision, accuracy, and resolution. In this analogy of three marksmen shooting at a target, one cluster of shots has high precision (high reproducibility) but is off the mark (not accurate), whereas another is more accurate (closer to the bull's-eye) but not precise (poorly clustered.) If the two different clusters are easily distinguished, this permits higher resolution than if the two clusters overlap. (Based on an idea from Griffiths, 1967.)

us only what we already know (see Box 18.2). A classic example concerns a certain biostratigrapher who implicitly incorporated the stratigraphic ranges of his fossils into his species definitions. In other words, he defined species not only by their anatomical features but also by the formations in which they occurred. Each formation had a different species, even if the fossils had not actually changed in any recognizable way. The result was that the formations were correlated by their fossils, and the fossils were recognized by the formation they came from. Another paleontologist called this man's bluff by handing him a loose specimen of one of the fossils. Without knowing the formation from which it had come, he could not identify its species.

The methods of dating and correlation that we have examined fall into two categories that correspond broadly to the concepts that underlie Gould's 1987 book, *Time's Arrow, Time's Cycle*. Some methods of dating are based on irreversible processes that operate continuously in one direction ("time's arrow"). These include biostratigraphy, which is based on the unidirectional evolution of organisms through time, and geochronology, which is based on the unidirectional process of radioactive decay. Both of these methods can be used independently of any other method, although they seldom provide maximal information alone.

Other methods are based on processes that have several possible states and fluctuate or cycle among them ("time's cycle"). Correlation is based on the overall *pattern* of fluctuations of these states and depends on biostratigraphy or geochronology to place this pattern within the matrix of geologic history. Most other correlation methods fall into this second category. For example, most lithologic correlation is based on a pattern of fluctuations among certain rock types: sandstone, limestone, shale, and so on. Recognition of

the pattern comes from the *sequence* of rock types and the *relative thickness* of each type. A particular unit (or sample) cannot be dated by itself but depends on the sequence and ultimately on biostratigraphy and/or geochronology. By contrast, a single key fossil or a single sample with decaying radioisotopes in the minerals can be dated in isolation, after the necessary background work is done (for example, the biostratigraphic sequence of an area or the decay constants of an isotopic system).

Other stratigraphic methods also depend on pattern recognition. Magnetostratigraphy, for example, is based on recognition of a pattern of fluctuation between two stable states: normal and reversed polarity. A sample can be measured by itself, but it tells us only that the rock cooled or was deposited during a time when the Earth's polarity was in one of the two possible states. A reversely magnetized rock could have been formed during the most recent episode of reversed polarity, the Matuyama, which ended about 780,000 years ago, or in any one of the hundreds of reversed episodes that preceded it. A magnetostratigraphic sequence with an identifiable pattern of polarities is needed to correlate the rocks to the polarity time scale. Usually, such a sequence also needs some independent means of dating (typically biostratigraphy or geochronology) to give us some idea of how old the pattern is. Then it can be matched with the appropriate part of the magnetic polarity time scale to see if there are characteristic polarity intervals that match in both duration and sequence.

A similar approach is required for most other means of correlation. Seismic correlations are based on changes in density of the material through which the seismic wave has passed and make sense only in the context of the entire pattern. Stable isotope correlations are based on the fluctuation of carbon or oxygen isotopes and make sense only if a complete

BOX 18.2   THE KBS TUFF AND THE PITFALLS OF "ABSOLUTE" DATING

The continental deposits of the East African Rift have become world famous as sites of important discoveries of human fossils. These include Olduvai Gorge in northern Tanzania, the areas around Lake Turkana (formerly Lake Rudolf) in northern Kenya, the Omo River drainage in southern Ethiopia, and the Hadar area in the Afar Triangle of northern Ethiopia. This last area is the home of one of the earliest hominid relatives, *Australopithecus afarensis*. The most famous specimen of this species is a nearly complete skeleton of a female nicknamed "Lucy."

Because of the interest in these hominid fossils and their ages, there have been many efforts to determine the stratigraphy as precisely as possible. These rift grabens provided suitable homes for our early ancestors and fossilization sites for their bones; fortunately, they also provided the depositional environment needed for exceptional stratigraphic detail. Rift valleys tend to subside continuously for long periods of time, producing unusually thick terrestrial sequences. Their frequent faulting and tilting mean that many of these sequences are well exposed today, although detailed mapping is required to work out the faulting. The rift valley was and is a site of active volcanism, so there are also many layers of airfall and ash-flow tuffs, which are suitable for both radiometric dating and tephro-stratigraphic correlation. Where the sequences are thick and fine-grained enough, they are also suitable for magnetic stratigraphy. Finally, these deposits entombed not only the hominids but also the entire evolving East African Miocene-to-Recent mammal fauna, some of which are excellent biostratigraphic indicators. All these techniques have been applied to the East African sequence, and the cross-checking, correction of bad correlations and dates, and ultimate refinement of the system are exceptionally thorough.

The discovery of Olduvai Gorge in 1911 began the study of the geology of the region, and the fossil faunas were examined by a number of people for the next 50 years (reviewed by Leakey, 1978). The area received worldwide attention in 1959 with the discovery of the first good hominid specimens and again in 1961 with the publication of a potassium-argon date of 1.8 million years on Olduvai Bed I (Leakey, Evernden, and Curtis, 1961; Evernden and Curtis, 1965). This was one of the first potassium-argon dates to be analyzed, and it radically altered the prevailing view of the age of human origins by proving them to be much older than had been thought. In subsequent years, the stratigraphy of the Olduvai section was worked out in great

detail, using dates from other ashes and flows (Hay, 1975). Work in the Omo Valley produced a longer and more complete sequence, with many radiometric dates going back to more than 3 million years ago.

Spectacular hominid fossils also came from the Koobi Fora area on the east shore of Lake Turkana. Of these, Richard Leakey's best *Homo habilis* skull, known by its catalogue number, 1470, was the most significant. Not only was it much more advanced than the australopithecines that had been found previously, but it was associated with stone tools. The specimen and the tools were found below the KBS Tuff, which was dated by the $^{40}Ar/^{39}Ar$ method (Fitch and Miller, 1970) at 2.61 ± 0.26 million years.

As described in Don Johanson and Maitland Edey's 1981 book *Lucy* and in Roger Lewin's 1987 *Bones of Contention*, this date caused severe problems for the paleoanthropologists. The 1470 skull was not only very advanced but almost a million years older than would have been predicted from other areas. The chief evidence that the date was anomalously old came from the mammalian biostratigraphy. As worked out by a number of scientists, the fauna found with 1470 were much more like those known from about 2 million years ago at Omo and Olduvai. Of the mammals, the elephants and especially the pigs (Maglio and Cooke, 1972; Cooke, 1976) proved to be the most useful. Maglio's (1972) "Faunal zone a," which included the pig *Mesochoerus limnetes* and the elephant *Elephas recki*, occurred in association with dates of around 2 million years at Omo but below the KBS Tuff at Koobi Fora. At a 1975 conference in London, the dispute between Leakey and the adherents to the old KBS Tuff date, and those who doubted it, reached a climax. As described by Johanson and Edey (1981, p. 239), one of Leakey's associates arrived wearing a hat he called a "pig-proof helmet" to protect against the "pig men." Basil Cooke, the chief "pig man," wore a tie with "MCP" woven into it; most of the audience assumed that this abbreviation meant "male chauvinist pig." After Cooke made a strong case for the pig biostratigraphy and against the old KBS Tuff date, he said, "You may think you know what MCP stands for, but you don't. It really stands for *Mesochoerus* correlates properly."

The mammalian biostratigraphy was double-checked by Tim White and John Harris (1977), who also came to the conclusion that the date on the KBS Tuff was too old. Their detailed studies of the fauna even showed that there was a significant gap in the

(continued)

(Box 18.2 continued)

Koobi Fora sequence that was not apparent from the lithostratigraphy or troublesome K-Ar dates. Several more attempts to date the KBS Tuff produced new dates on *two different* KBS tuffs of 1.6 and 1.8 Ma by K-Ar (Curtis et al., 1975), and then a better date of 1.8 ± 0.1 Ma by K-Ar (Drake et al., 1980). The $^{40}Ar/^{39}Ar$ dates on the KBS Tuff that first produced the notorious date of 2.61 Ma (Fitch and Miller, 1970) later gave a date of 2.42 Ma (Fitch, Hooker, and Miller, 1976). When reanalyzed by others (McDougall et al., 1980; McDougall, 1985), the KBS Tuff produced a $^{40}Ar/^{39}Ar$ date of 1.88 ± 0.02 Ma. The early fission-track dates on the KBS Tuff first gave dates of 2.42 Ma (Hurford, Gleadow, and Naeser, 1976) but eventually produced ages of 1.87 ± 0.04 Ma (Gleadow, 1980). Ironically, because of the initial bad date on the KBS Tuff and its tremendous scientific importance, the KBS Tuff is now one of the best- dated units in the world.

In concert with the dates on this most controversial unit, many of the other tuffs at Omo, Koobi Fora, and other areas produced dates that provided a total framework for the area. Another approach to these volcaniclastic units was tephrostratigraphy by chemical fingerprinting (Cerling and Brown, 1982). Eight of the tuffs at Koobi Fora could be matched with those at Omo based on their trace and minor elements. Chemical fingerprinting showed that the Tulu Bor Tuff in the Koobi Fora Formation had been misidentified. This method made it possible to use the tuffs as synchronous surfaces to separate the Koobi Fora Formation into a set of sequential, isochronous units (Findlater, 1978) and to match these with the units at Omo.

Magnetostratigraphy of the Koobi Fora Formation was first attempted by Brock and Isaac (1974), but many of the specimens apparently had a normal overprint that was not removed during demagnetization. This, compounded with the fact that Brock and Isaac were trying to calibrate their magnetostratigraphy with the erroneous dates of the KBS Tuff, meant that the magnetostratigraphy had to be reinterpreted. Originally, Brock and Isaac (1974) correlated the Koobi Fora Formation with the polarity time scale from just above the Olduvai event to the top of the Gilbert reversed interval. Because of the anomalously old date on the KBS Tuff, they placed a disconformity above it to account for the "missing" lower Matuyama and upper Gauss intervals. As revised by Hillhouse et al. (1977), the sequence spanned from above the Olduvai down to the middle of the Gauss, eliminating the disconformity that was demanded by the bad date on the KBS Tuff. The most recent interpretation added much additional sequence to the section and showed that the sequence spans part of the Matuyama, with a big disconformity below the KBS Tuff and above another sequence that runs from the top of the Gauss to the middle of the Gilbert. Clearly, the paleomagnetic interpretation is susceptible to the revisions in radiometric dating and does not reveal the disconformities by itself. On the other hand, the paleomagnetic reversal boundaries provide additional time planes for even

sequence has been analyzed to determine the characteristic peaks and valleys in the pattern. The method of strontium isotopes, however, is an exception: it appears to be directional rather than cyclic (see Chapter 17). Since the Eocene, the unidirectional change in the $^{87}Sr/^{86}Sr$ ratio from about 0.707 to 0.709 means that any given sample can be dated within the level of precision of measurement.

Resolution and Precision

Other factors that become important when combining stratigraphic methods are their relative **precision**—the inherent error of a system, which indicates how tightly clustered or repeatable the measurements are (see Fig. 18.16) and **resolution**—the ability of a system to discriminate between two closely spaced events in geologic time. Every stratigraphic method makes compromises between the two qualities. The precision of radiometric dating is stated in terms of the percent analytical error (given by the "plus or minus" after a date). This precision remains fairly constant in terms of the error percentages, but the resolution decreases with increasing sample age (Fig. 18.17). For example, a date of 10 million years with an analytical error of $\pm 1\%$ means that radiometry can resolve events that are separated by about 100,000 years. The same 1% error on dates of around 100 million years means that the method can only resolve events that are 1 million years or more apart. A more typical 5% error on a date of 100 million years means that the method can only resolve events that are 5 million or more years apart.

Precision in biostratigraphy is limited by a number of factors (discussed in Chapter 16), such as upward or downward mixing or transport of fossils, improper identification or poor preservation of fossils, and unfossiliferous intervals. In the best cases, however, the

better resolution of events than was possible with the ashes or biostratigraphy alone.

The example of the Koobi Fora and related formations of East Africa points out a number of problems typical of this kind of study. The radiometric dates provide the numerical ages, but they are subject to many types of error and so had to be redone many times in several laboratories by three methods over 15 years before all the results were consistent and undisputed. Because the KBS Tuff date supported Leakey's contention that the genus *Homo* is very old, there was much political and sociological maneuvering before the problem was corrected. Leakey and his associates placed too much importance on the 2.6 Ma dates, which caused much bitterness in paleoanthropology for almost a decade (Lewin, 1987).

As discussed by Lewin (1987), the mystery of why the original KBS feldspar crystals gave the anomalously old date has never been solved. Every time they were reanalyzed by Fitch and Miller, their ages came out at around 2.4 to 2.6 Ma. Lewin (1987, p. 252) found some of the original crystals and had them reanalyzed by the K-Ar method; their dates came out 1.87 ± 0.04 Ma. Even methods that are supposedly objective and "absolute," such as radiometric dating, can be biased by the expectations of the experimenter. For example, Lewin (1987, p. 246) quotes Hurford on running the fission-track method:

You can bias your results ten percent either way, easily. You go crystal by crystal, and you begin to see where the rolling average is going. If you need the count to be higher with the crystal you're working on, so that it will fit in, you might include something that is a double track. If you want the count to be lower, you don't include it. That was poor practice.

Biostratigraphy proved to be the most reliable method, least subject to errors. It also revealed gaps in the sequence that were not made apparent by other methods. It is only locally applicable to that part of East Africa, however, and does not correlate with other continents or the deep sea. Tephrochronology also improved correlation within the Turkana Basin but, like the evidence of the mammals, it was not useful outside the areas of the volcanic ashfall.

Finally, magnetostratigraphy provided the finest resolution of time and the best correlation with other parts of the world, but it was subject to problems with the magnetic signal and had to be interpreted in the framework of the available radiometrics and biostratigraphy. As each of these sources of data was reinterpreted, the magnetic story changed too. As of 1995 however, most of the controversies had apparently subsided, and now the hominid-bearing sequences of East Africa are among the best-dated terrestrial sequences in the world.

precision is extremely good and is limited only by the rate of sediment accumulation. For example, some deep-sea cores have nearly continuous records of microfossils, with no large gaps, so the first appearance of a key microfossil can be very precisely located if the preservation and identification of the fossils are good. If the core also happens to represent a time of rapid sedimentation, a precision of thousands to tens of thousands of years is possible. If there were annual climatic varves or some similar phenomenon, a precision of hundreds of years to less than a year is possible.

Resolution in biostratigraphy, on the other hand, is limited primarily by the rate of evolution of the species that make up the fossil assemblages or individual lineages. If they evolve slowly, then every zone is long, and resolution is low. If they evolve rapidly, the zones can be short, and excellent resolution is possible. Resolution varies from group to group among the biostratigraphic indicator fossils (Fig. 18.18).

For example, 117 ammonite zones and subzones are currently recognized in the 70-million-year span of the Jurassic (House, 1985), so each zone or subzone averages only 600,000 years in length. Ramsbottom (1979) reported a zonation of the Namurian (Early Carboniferous) down to 25,000-year increments using goniatitic ammonoids. As is apparent from Fig. 18.18, not all fossil groups have resolution this fine, but many have zones that are shorter than 3 million years in average duration. Because radiometric precision decreases with increasing age, the relative resolution of the biostratigraphy compared to the numerical age gives a rough measure of the equivalence of biostratigraphic resolution. In Fig. 18.18, for example, Cambrian trilobites have zonal lengths of about 3 million years, but the radiometric precision is typically ± 5 million to ± 10 million years. Cretaceous zonal lengths are only a million years or less, but dates typically have errors in the range of ± 2 million to ± 4

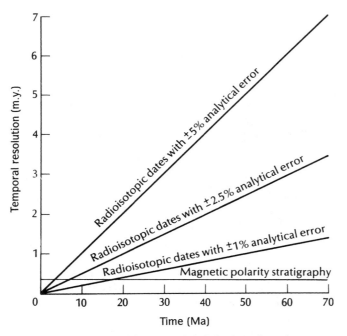

Figure 18.17 Comparison of the temporal resolution for radiometric dating and magnetic polarity chronology for the Cenozoic. Analytical errors of 1%, 2.5%, and 5% are shown for radiometric dating. The maximum temporal resolution is twice the value of the standard error. The average frequency of reversals for the Cenozoic is 3 per million years. Notice that the resolution of magnetic stratigraphy is constant through time whereas the resolution of radiometrics decreases with increasing age. (After Flynn, MacFadden, and McKenna, 1984: 689; © 1984; by permission of the University of Chicago Press.)

million years. Scaling the zonal lengths to the error in numerical age puts Cretaceous ammonite subzones on approximately the same 1:200 line on which Cambrian trilobites fall. From Fig. 18.18, it appears that the finest resolution available is in the Devonian and Carboniferous, in which the zones are extremely short despite the great age of the system. For most of these examples, good biostratigraphy has much better resolution than radiometric dates.

Let us consider one more system. The precision of magnetic stratigraphy is limited by the quality of the magnetic signal, the density of sampling, the completeness of the stratigraphic record, and the time it takes to make a transition from one polarity to the other. In exceptional cases, however, a detailed record of the transition interval itself can provide finer precision than the 4000 to 5000 years needed for a polarity transition. The resolution, on the other hand, is limited by the length of the polarity interval. For the Cenozoic, the frequency of polarity changes ranges from 3.84 per million years in the Neogene to 1.85 per million years in the Paleogene (Flynn, MacFadden, and Mckenna, 1984). The average length of a polarity zone is about 3

million years or less. *This does not change with increasing age* (see Fig. 18.17), so the resolution relative to the numerical age is much better than even the best radiometric dates for anything older than the Quaternary. During the mixed polarity of the Jurassic and early Cretaceous, the paleomagnetic resolution is again on the order of less than a million years, whereas the resolution of radiometric ages is typically 3 to 8 million years, or worse.

Magnetostratigraphy, then, offers resolution on the same order of magnitude as biostratigraphy (or better in some cases), but it is applicable to just about any rock type, terrestrial or marine. High-resolution biostratigraphy, on the other hand, is possible only in environments in which rapidly evolving organisms are preserved. In the Jurassic, for example, the marine zonation based on ammonites has very high resolution, better than the paleomagnetic record. There is no comparable terrestrial biostratigraphy, however, so magnetic stratigraphy offers the potential of greater resolution in that environment.

The resolving power of geochronologic systems in the Cenozoic is shown in Fig. 18.19. On the vertical axis, the age of the event increases logarithmically upward. On the horizontal axis, the time separation of

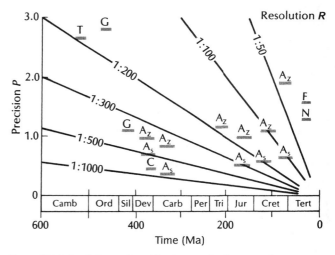

Figure 18.18 Resolution and precision in some fossil groups of biostratigraphic importance. Abbreviations: A, Ammonoidea; C, conodonts; G, graptolites; N, nannoplankton; T, trilobites. Subscripts: z, zones; s, subzones. Precision P is given on the vertical axis and refers to discrimination in years of average zonal or subzonal units. If t is the span of years by numerical dating, and z is the number of zones or subzones over that interval, then $P = t/z$. Resolution R represents the resolving power for a given value of P for the numerical time span of the zone considered. If T is the mean numerical age for the time span t of z zones, then $R = T/P$. Radiating lines of equal resolution are shown in terms of $1/R$. (After House, 1985: 275; by permission of Blackwell Scientific Publishers.)

events increases logarithmically to the right. Naturally, events of very short duration, from 1 hour (tsunamis and gravity flows) to about 1 year, cannot be resolved except in the most unusual circumstances. They fall in the simultaneous-instantaneous zone, the zone in which events can be considered geologically instantaneous. The range of accuracy of ^{14}C dating is from tens of years to about 80,000 years, whereas the accuracy of K-Ar dating begins in the range of 500,000 years and older. The zone of resolution for these methods is much more limited. Potentially, ^{14}C dating can resolve events spaced just years apart, although it is much more practical for events spaced ten to tens of thousands of years apart. Because the outside limit is less than 100,000 years, however, ^{14}C dating cannot be used for events further apart than 10,000 years.

Potassium-argon dating, on the other hand, cannot discriminate events separated by tens of thousands of years except in the Pleistocene. Resolution decreases with increasing age, so during the Tertiary, events separated by a million years are at the lower limit of resolution (as we have already seen). Notice that at this point the zone of biochronological discrimination begins to overlap that for potassium-argon (see Fig. 18.19). Although the resolution of potassium-argon continues to get worse in older systems (hence the slope), the resolution of biochronological discrimination remains on the order of 1 million to 10 million years (the vertical boundary). This reinforces the point that biostratigraphic resolution is dependent on the rate of turnover of the fossil species in question and not on their numerical ages.

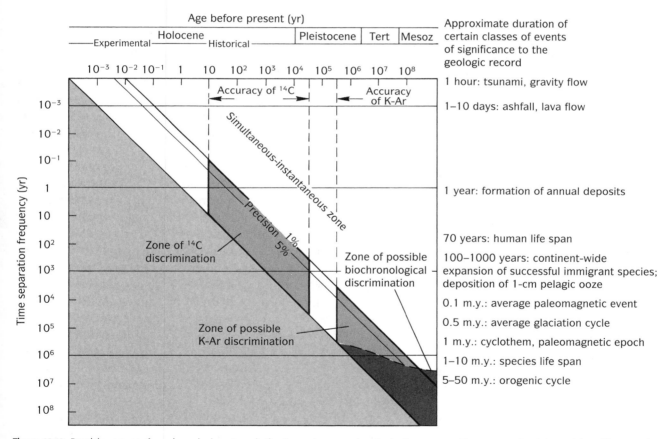

Figure 18.19 Resolving power of geochronologic systems in the Cenozoic. The horizontal axis shows age increasing exponentially to the right, from hours to hundreds of millions of years. The vertical axis shows the time separation of events, from hours to hundreds of millions of years. The range is from events of short duration (tsunamis, turbidity currents, ashfalls, lava flows) to events of long duration (species life spans, orogenic cycles). Carbon-14 dating is useful only in the area represented by the parallelogram, ranging from tens of years to less than 100,000 years. In the most recent materials, ^{14}C dating can resolve events that are years to tens of years apart. At its older limit, it can only resolve events that are hundreds of years apart. Events more closely spaced than this cannot be resolved and fall in the simultaneous-instantaneous zone. Similarly, K-Ar dating can date materials less than a million years old, but its resolution is at best only tens of thousands of years. The light gray area indicates the zone in which Cenozoic biochronology is effective. Note that although biostratigraphy can seldom resolve events spaced less than a million years apart, this resolution does not decrease with increasing age. (After Berggren and Van Couvering, 1978: 43; by permission of the American Association of Petroleum Geologists, Tulsa, Okla.)

Constructing the Geologic Time Scale: An Example from the Eocene-Oligocene

So far, we have looked at local or regional units that have been correlated to a standard geologic time scale. But how is that time scale constructed? How can we say, for example, that the Eocene/Oligocene boundary is 38 or 34 or 32 million years old? How do we recognize it in our local outcrops?

The global geologic time scale is constructed by integrating stratigraphies and dates from all over the world. Combining data of unequal quality from so many sources makes worldwide correlation much trickier than local or regional correlation because there are many more possible sources of error. In some cases, the data have yielded relatively uncontroversial results; in other cases, the data are so poor that there is little confidence in the numbers and little grounds for strong argument. We have seen such a case with the dating of the Silurian-Devonian boundary (see Box 18.1). In a few other cases, however, there are multiple sources of data, but their interpretation is subject to alternative points of view and strong controversy. These last examples are the most revealing because they show us most clearly the steps by which most of the time scale has been constructed. Let us look at one part of the time scale that has been particularly controversial: the Eocene/Oligocene boundary.

The names *Eocene* and *Miocene* were coined originally by Charles Lyell (1833) to divide the Tertiary according to faunal percentages. Unlike most stratigraphic names, they were based on an abstract concept, the "ticking" turnover of the molluscan "clock," rather than on a typical area or type section. Clearly, Lyell conceived of time units as independent of local stratigraphy, but most of Lyell's contemporaries and successors were more interested in designating stages based on local lithostratigraphic sequences, which may or may not represent all of Tertiary time. Despite his interest in abstract time terms, Lyell designated the Paris and London basins as the typical areas for his Eocene. Lyell (1833, pp. 57–58) explicitly recognized that there might be further subdivisions of his units but did not try to designate them himself.

Naturally, geologists soon began to find areas with faunas that were intermediate between Lyell's Eocene and Miocene. The upper part of the Eocene is sparsely fossiliferous in the Paris Basin, but in northern Germany and Belgium an extensive marine transgression produced many fossils that are younger than those in the Eocene of the Paris Basin. These northern rocks were called "lower Miocene" by some and "upper Eocene" by others, until von Beyrich (1854) coined the term *Oligocene* for them. In 1856, von Beyrich more precisely defined the units included in the Oligocene, and its upper and lower boundaries.

Long before this, however, geologists had named parts of their local successions of fossiliferous strata. As reviewed in detail by Berggren (1971), these names and their type areas proliferated and soon became a confusing mess. In many cases, two successive stages were not in the same basin or otherwise did not lie in superposition to one another, so it was impossible to determine whether they were really successive or overlapped because of lateral facies changes. In other cases, it became clear that a stage, as originally defined, left a gap between it and the next stage. Then arguments ensued as to whether to draw the boundary at the base of the upper stage or at the top of the lower stage or to create another stage in between.

The primary cause of the whole controversy was that these Tertiary stages were created for local sequences within various European basins, which had undergone complicated local histories of transgression and regression throughout the Tertiary (Fig. 18.20). The type area of each stage is thus part of a time-transgressive sedimentary cycle that is separated from adjacent cycles by unconformities above and below it. When a geologist tried to trace any of these stages laterally from its type area, problems were encountered. These problems were compounded by the fact that most of the faunas were endemic, shallow marine mollusks. These were often difficult to match from one European basin to another, let alone with other parts of the world. The result was almost a century of argument, much of it trivial or fruitless. When someone wanted to call some rocks in North America "Eocene" or "Oligocene," it was almost impossible for anyone else to substantiate or refute the claim. Not only were there few fossils in North America that could be correlated to the type areas in Europe, but there was considerable confusion as to what was Eocene or Oligocene even within the type areas in Europe.

By the 1960s and 1970s, a solution began to appear. What was needed were the most continuous possible sequences, with few gaps, containing an uninterrupted record of fossils that could be correlated around the world. The shallow marine basins of Europe failed to provide this, but the deep-sea record, with its globally distributed, rapidly evolving planktonic microfossils, did. Deep-sea sequences were studied first from uplifted exposures on land in New

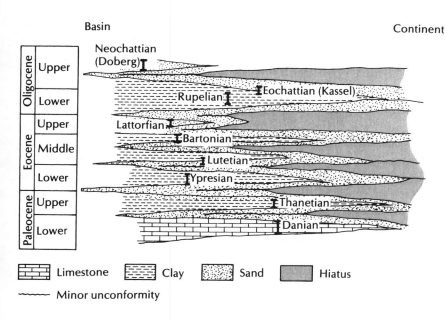

Figure 18.20 Depositional history of the type areas of the Paleogene stages and ages in northwestern Europe. Notice that the section is incomplete and is composed of irregular transgressive and regressive sequences with large hiatuses. Each of the stratotype sections spans only a small part of the total age or epoch. This incompleteness is responsible for much of the confusion and dispute about stratigraphic terminology in the Paleogene over the last century. (After Berggren and Van Couvering, 1978: 216; by permission of the American Association of Petroleum Geologists, Tulsa, Okla.)

Zealand and Italy and then from oceanic cores recovered by the Deep Sea Drilling Project. In particular, the changes in planktonic foraminifers provided clear, nearly global datum levels that could be correlated from core to core with great confidence. Because these changes were generally controlled by climatic changes in global water masses, they proved to be synchronous within the resolution of the system in nearly all examples. The planktonic foraminiferal and calcareous nannofossil zones became the worldwide standard of correlation for the Tertiary for lower and middle latitudes. In areas of siliceous sedimentation, diatom and radiolarian zones were used. A few key cores had both calcareous and siliceous sediments and provided the necessary connection between the two stratigraphies.

The micropaleontologist, therefore, was in the key position to correlate strata in the Tertiary. Berggren (1971) and many other scientists examined samples from the type areas in Europe to find the microfossils needed to tie the Tertiary epochs and stages there to the deep-sea record. This was not an easy task, because the shallow-water European sequences contained few open-ocean planktonic microfossils, and often these were poorly preserved. This research showed that some long-accepted beliefs and standards were wrong and that some entrenched names were useless or overlapped other names. Naturally, this led to more controversy, but most of it has died down since the studies of the early 1970s.

Yet the deep sea, which provided the continuous, fossiliferous record needed to correlate the European type areas on a worldwide basis, could not provide numerical ages for these epochs and stages. Radiometric dates were needed, but unfortunately, they were not ideal, either. Virtually no ashfalls occurred in the deep marine sequences, and few occurred in the shallow marine sequences. Most numerical dates were obtained from glauconites, which grew diagenetically in the shallow marine environment. Dates from glauconites are subject to problems that do not affect dates from high-temperature minerals deposited by volcanic events. Glauconites are notorious for leaking argon, and most of the original glauconite dates have been rejected for this reason. Other glauconite dates were rejected because of equipment insensitivity to atmospheric argon. There are problems with even the freshest glauconites, however. Because they form diagenetically, they can crystallize long after they are buried and thus can lock in a much younger age. They occur as rounded pellets on the seafloor, so they are very easily transported and reworked into younger deposits, giving spuriously old ages. Other problems have been documented as well. In short, many isotope geologists regard glauconite dates with suspicion, especially when they clash with high-temperature dates that are not subject to these problems. Nevertheless, glauconites are abundant and have been used by default for dating the Tertiary (Hardenbol and Berggren, 1978, Fig. 6; Odin, 1978, 1982).

Besides the steady rain of microfossils onto the deep seafloor, there is another steady deep-sea process that records events. This process is seafloor spreading. As the ridges spread apart, they are intruded by magmas from the mantle that record the prevailing direction of the magnetic field as they cool. This cooled mid-ocean ridge crust is then split apart by further spreading, and the magnetic record is transported away from the ridge in both directions. As the Earth's magnetic field

flips back and forth between normal and reversed polarity, the continuous process of spreading and cooling of magma records the polarity history away from the ridge. Thus, the spreading seafloor acts like a tape recorder of the Earth's magnetic field.

How do we calibrate this tape recording? The first test was Leg 3 of the Deep Sea Drilling Project. The researchers drilled holes on both sides of the Mid-Atlantic Ridge in the South Atlantic and found that the cores got progressively older with distance from the ridge. The basalts of the deep seafloor were so altered that they could not be radiometrically dated, but the sediments overlying the basalts had microfossils that yielded a biostratigraphic age. To obtain reliable numerical estimates, the youngest part of the paleomagnetic time scale was constructed by dating many fresh lava flows all over the world and then measuring their polarities. As described in Chapter 17, these measurements eventually produced a magnetic polarity time scale for the last 5 million years. Basalts in Iceland can be dated as far back as 13 Ma (Harrison, McDougall, and Watkins, 1979), providing both polarity and direct basalt dates, but for the older dates, the radiometric error has already exceeded the precision achievable by other methods. The time scale earlier than 13 Ma must be constructed by extrapolation of the seafloor spreading record itself.

The first historic attempt at this extrapolation was made by Heirtzler et al. (1968), who compared the spreading profiles from the South Atlantic, South Pacific, North Pacific, and Indian oceans and found that the South Atlantic had the least variation in spreading rate and that its spreading history extended back to about 80 Ma. Then they did something daring. Using a spreading rate calculated from the dated time scale from 0 to 4 Ma, they extrapolated back to 80 Ma and produced the *entire* time scale. We now know that their extrapolated time scale is within 10% of the currently accepted version, a remarkable achievement considering the lack of control points. The assumption of constancy of the rate of seafloor spreading must have been a good one for them to come so close on the first attempt.

The process has been refined many times since the Heirtzler et al. (1968) time scale. Alvarez et al. (1977) analyzed a sequence of Cretaceous pelagic limestones near Gubbio, Italy, that produced a continuous fossil and paleomagnetic record for the Late Cretaceous and Paleocene. Using this tie point and a date for the Late Cretaceous, LaBrecque, Kent, and Cande (1977) produced a new version of the time scale that was tied at the Late Cretaceous and at the Quaternary. All of the

time scale between these tie points had to be interpolated. Ness, Levi, and Couch. (1980) revised the time scale yet again because new K-Ar decay constants (Steiger and Jäger, 1977) had just been published. Unfortunately, Ness, Levi, and Couch (1980) simply revised the ages of the interpolated boundaries rather than going back to the original dates to recalculate them to see how they changed the time scale.

Lowrie and Alvarez (1981) once again modified the time scale according to the Gubbio sequence (Lowrie et al., 1982), which now had almost continuous magnetic stratigraphy and marine microfossils for the Late Cretaceous through the Miocene. They used the paleomagnetic signature of eight stratigraphic boundaries that were based on microfossil evidence to add more calibration points between the original two points of LaBrecque, Kent, and Cande (1977). Unfortunately, they assumed that the ages of these biostratigraphic boundaries were well constrained, and so they fit the spreading profile between them. This forced fit caused the spreading profile to change by rapid jerks between each pair of data points, which goes against everything that geophysicists know about seafloor spreading. Although the spreading rate varies from ocean to ocean and changes slowly over time, there is no geophysical mechanism that could change it by rapid spurts and stops every few million years. These rapid fluctuations were an artifact of the uncertainties of the numerical ages of the eight paleontological datum levels, which were less reliably known than the rate of seafloor spreading.

The Gubbio section of Lowrie et al. (1982) and Leg 73 of the Deep Sea Drilling Project (Poore et al., 1982) both provided continuous pelagic records with good microfossils and magnetic data. It was clear from these records that the microfossils used to recognize the Eocene/Oligocene boundary occurred between normal Chrons C13 and C15 of the polarity time scale, in the upper third of the reversal between them, known as C13R. (Early in the study of marine magnetics, anomaly 14 was dropped from the seafloor spreading profiles as a mistake.) At that time, however, neither of these sections could be radiometrically dated. What was needed were sections with good high-temperature dates and magnetic stratigraphy that could be tied directly to the polarity time scale. These data were first provided by Prothero, Denham, and Farmer (1982, 1983; Prothero, 1985a) in a section at Flagstaff Rim, near Casper in central Wyoming (Fig. 18.21). This section spanned more than 200 m of volcaniclastic sediments that were full of Chadronian (then thought to be early Oligocene) mammals. More important, there

A

Figure 18.21 (A) The classic section of the White River Formation at Flagstaff Rim, near Alcova, Wyoming. The section contains mammal fossils that span most of the Chadronian land mammal "age" (Emry, 1973.) It also contains ash layers (labeled B, F, G, and J) that were K-Ar dated by Evernden et al. (1964) and recently redated by Ar-Ar methods. (Courtesy of R. J. Emry.) (B) Old and new interpretations of the time scale, based on the change in dating methods. On the left is the time scale of Berggren, Kent, and Flynn (1985), which was partially based on the old Flagstaff Rim K-Ar dates. In the center is the new adjustment to the time scale suggested by the new Flagstaff Rim Ar-Ar dates, plus Ar-Ar dates from previously undated ashes at Dilts Ranch and elsewhere in the Brule Formation (Swisher and Prothero, 1990; Prothero and Swisher, 1992.) On the right are the revisions in the North American Land Mammal "Ages" (NALMA) suggested by the new dates.

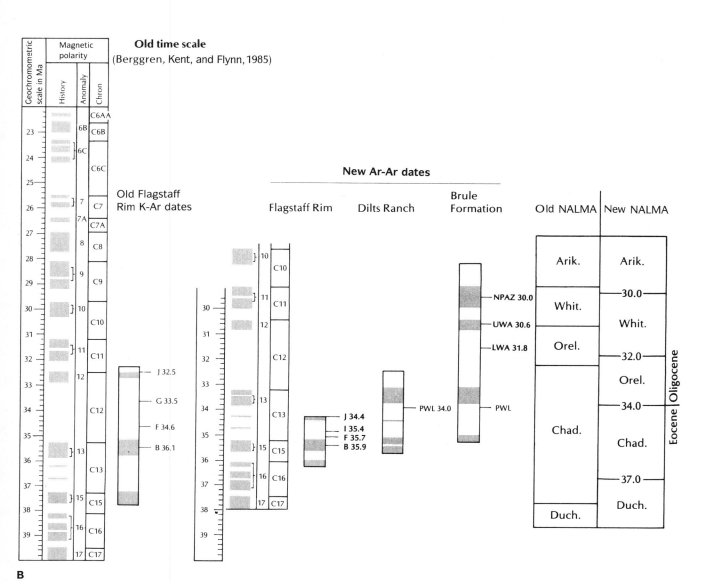

B

were already K-Ar dates in four places in the section. An unusually long episode of reversed polarity was found between dates of 32.5 Ma (Ash J) and 36.1 Ma (Ash B). The only reversal in the Oligocene part of the polarity time scale that approaches this almost 3-million-year duration was the reversal between Chrons C12 and C13, which is more than 2 million years long (see Fig. 18.21). Other dates (Ashes F and G) established intermediate points in this correlation. The mammal record showed no evidence of a major unconformity that might reflect missing sections and therefore missing polarity intervals. On this basis, direct radiometric calibration points (Prothero, Denham, and Farmer, 1982, 1983; Prothero, 1985a) were provided between the calibration points from the Neogene and the Late Cretaceous. Other calibration points (Prothero 1985a; Prothero and Armentrout, 1985; Flynn, 1986) came from terrestrial and shallow marine sequences in North America with good magnetic stratigraphy and direct high-temperature radiometric dates.

With these and other new data, another version of the time scale (Berggren, Kent, and Flynn, 1985) was established using these direct calibrations of the magnetic polarity record. Avoiding the sudden, unwarranted jerks in spreading rate postulated by Lowrie and Alvarez (1981), Berggren, Kent, and Flynn (1985) fit three line segments through the calibration points that were based on high-temperature (not glauconite) dates found in sequences with clearly identifiable polarity records. The two inflection points of these three line segments correspond to periods of well-established changes in spreading rates, based on comparison with other spreading profiles. The Berggren, Kent, and Flynn (1985) time scale does not assume constant spreading for the whole time interval, but only between times where it has been independently established that there were changes in spreading rates. This is preferable to the Lowrie and Alvarez (1981) approach, which assumed that the biostratigraphic levels were well dated (the biostratigraphers knew they were not) and then produced geophysically impossible fluctuations in spreading rate. Berggren, Kent, and Flynn, on the other hand, assumed that spreading rate is relatively constant (except where demonstrated otherwise) compared to the large known errors of biostratigraphic age assignments and radiometric dates. Based on these methods, Berggren, Kent, and Flynn estimated the age of the Eocene/Oligocene boundary at 36.5 Ma.

Other scientists, however, constructed their own time scales, which give radically different versions of the Eocene/Oligocene boundary and other important

events. The most persistent of these was Odin (1978, 1982), who used glauconites from the European marine sequences. Odin produced a time scale that consistently gave dates much younger that those from high-temperature methods. Curry and Odin (1982) and Odin and Curry (1985) placed the Eocene/Oligocene boundary at 32 Ma or younger, and their Paleocene/Eocene boundary differs from the estimate of Berggren, Kent, and Flynn (1985) by almost 4 million years! Such a huge difference cannot be attributed to the normal error in the radiometric dates, which is typically less than a million years for the early Tertiary.

For anyone trying to use a time scale and make calculations based on these numbers, this discrepancy is so big that it cannot be ignored. Each reader must look over the extensive documentation of each time scale and judge the assumptions that were made and the quality of data that were used to construct it. Aubry et al. (1988) did so and found some severe problems with Odin's glauconite dates. Many of Odin and Curry's (1985) dates were mislocated stratigraphically (Aubry et al., 1988, Fig. 4). When they are replotted, they are more consistent. In other cases, there was a wide range of glauconite dates from the same bed, and Odin and Curry (1985) arbitrarily picked certain ones and ignored others that tended to give older ages. In other cases, they selectively ignored dates that did not agree with their preconceptions, even though these came from areas with well-established stratigraphy. Obradovich (1988) showed that although some glauconites gave ages that appear to be reliable, others did not, and there was no way of knowing in advance which were more reliable.

A more fundamental problem is the reliability of the glauconite dates themselves. As discussed earlier, there are many reasons that their reliability is generally inferior to that of high-temperature dates. Aubry et al. (1988, Fig. 6) replotted the dates from Odin and Curry (1985) and found that although all the high-temperature volcanic dates gave a consistent line of correlation, some of the glauconite dates agreed with it but most were too young. There seems to be systematic bias toward ages from glauconites that are too young, so that when glauconite dates are used in the time scale, the time scale is also too young. As Obradovich (1988) pointed out, what is being measured is actually the date when the glauconite crystals are "closed" and no longer exchange materials with the surrounding sediment. Even Odin (1982, p. 728) admits that this closure may occur millions of years after the glauconite grain was deposited, so that most glauconite ages are minimum estimates.

Nevertheless, a number of divergent estimates have appeared for the age of the Eocene/Oligocene boundary. Harris and Zullo (1980) and Harris, Fullagar, and Winters (1984) dated the Eocene/Oligocene boundary at 34 Ma, based on an Rb-Sr date from glauconites in the Castle Hayne Limestone in North Carolina. Berggren and Aubry (1983) and Berggren, Kent, and Flynn (1985), however, have shown that this date was misplaced stratigraphically; it was not even in the right microfossil zone, but significantly older. Harris and Fullagar (1989) have dated a bentonite, rather than unreliable glauconites, and their dates (both K-Ar and Rb-Sr) suggest that the Eocene/Oligocene boundary is younger than 36 Ma.

Glass and Crosbie (1982) estimated the age of the Eocene/Oligocene boundary at 32.3 Ma based on K-Ar dates and fission-track dates of microtektite layers from the Caribbean. But Berggren, Kent, and Flynn (1985) have shown that there are problems with the biostratigraphic age assignment of this date. More recently, Glass, Hall, and York (1986) dated microtektites with laser-fusion argon-argon dating and obtained an age for the late Eocene of 35.4 ± 0.6 Ma. By extrapolation, they placed the Eocene/Oligocene boundary at 34.4±0.6 Ma. However, it is not certain how early or late in the Priabonian (late Eocene) their microtektites are located, so this extrapolation does not place very precise limits on the Eocene/Oligocene boundary. It could be only slightly younger, or much younger, than their 34.4 Ma date.

Next, the Italian Apennine sections originally analyzed by Lowrie et al. (1982) provided high-temperature volcanic dates that are closely tied to marine biostratigraphy and magnetostratigraphy. In several places in the classic Gubbio sections, there are ash partings that were dated by both K-Ar and Rb-Sr methods by Montanari et al. (1985). Their first publications gave an estimate of 35.7±0.4 Ma for the Eocene/Oligocene boundary. Although this is a bit younger than the 36.5 Ma estimate of Berggren, Kent, and Flynn (1985), most of the dates of Montanari et al. (1985) fit quite well on the Berggren time scale line. However, additional dates from Gubbio and another section near Massignano have caused Montanari et al. (1988) to reject one contaminated date of 35.4±0.4 Ma for the base of Chron C12R (early Oligocene), and accept dates of 36.0±0.4 Ma and 36.4±0.3 Ma for the late Eocene. By refitting their line to these dates, they estimated the age of the Eocene/Oligocene boundary at 33.7±0.5 Ma.

By 1989, it was becoming clear that something was wrong with the Berggren, Kent, and Flynn (1985) esti-mate of 36.5 Ma for the Eocene/Oligocene boundary. Then a breakthrough occurred. The volcanic ashes from Flagstaff Rim, Wyoming (see Fig. 18.21), which had been originally dated by K-Ar, were redated by laser-fusion Ar-Ar methods. Carl Swisher of the Berkeley Geochronology Center found that many of the old K-Ar numbers were unreliable, probably due to contamination of the samples when they were originally analyzed in 1963. By isolating single, unweathered crystals of biotite and sanidine and analyzing dozens of them with laser-fusion methods, Swisher produced dates that differed radically from the old estimates (see Fig. 18.21). For example, the old Flagstaff Rim dates ranging from 32.5 Ma (Ash J) to 36.1 Ma (Ash B) has been replaced by dates of 34.4 Ma (Ash J) to 35.8 Ma (Ash B). By shortening the span of the Flagstaff Rim section from more than 4 million years to less than 2 million years, it is clear that the reversed interval between Ashes F and J cannot be the long Chron C12R reversed interval, as Prothero, Denham, and Farmer (1982, 1983) and Prothero (1985a) thought. Instead, it appears that Flagstaff Rim spans Chron C13R and C15N, and possibly also C16N (see Fig. 18.21).

Where did the long C12R reversal go? Fortunately, the new Ar-Ar technique has made it possible to date many other ashes that were previously undatable by K-Ar methods. The Dilts Ranch section, near Douglas, Wyoming (see Fig. 18.21), produced a number of Chadronian dates that agreed with the new Flagstaff Rim dates. A date of 34.0 Ma on the Persistent White Layer (PWL) occurs just below the Chadronian-Orellan boundary. In addition, new dates on the Lower and Upper Whitney Ashes produced the first reliable dates from the Orellan or Whitneyan. From these new dates, it is apparent that Chron C12R must be the long reversal that spans the early Orellan to the middle Whitneyan. It is bracketed by dates of 34.0 Ma (the PWL) and 30.8 Ma (the Upper Whitney Ash), giving a duration of about 2.5 to 3.0 million years, in good agreement with the known duration of Chron C12R.

The implications of these new dates for the time scale are staggering. The old Berggren, Kent, and Flynn (1985) time scale is shown at the left in Fig. 18.21, and the new adjustments are shown in the middle. The new Ar-Ar dates suggest a shift of almost 2 million years for Chrons C12 through C16! More important, this places the Eocene-Oligocene boundary (upper third of Chron C13R) around 34.0 to 33.5 Ma, in good agreement with many of the new dates discussed above. These new dates were fitted to a new time scale by Berggren et al. (1992) and Berggren, Kent, and Hardenbol (1995). Any Paleogene time scale (including

the Haq, Hardenbol, and Vail 1987, 1988 Exxon cycle chart) published before 1990 is now grossly out of date.

The dispute over the age of the Eocene-Oligocene boundary now appears to be over, with a general agreement on dates of around 33.5 to 34.0 Ma. Yet this is no mere tempest in a teapot, interesting only to specialists. The Eocene-Oligocene transition was one of the most crucial climatic and faunal extinction crises in the entire Tertiary (Cavelier et al., 1981; Prothero, 1985b, 1994a, 1994b), so dating it is very important. For example, Prothero (1985b) labeled the Chadronian/ Orellan extinctions the "mid-Oligocene event" and

thought that the Terminal Eocene Event occurred at the beginning of the Chadronian. The new calibrations (see Fig. 18.21) make it clear that the Chadronian-Orellan transition is near the Terminal Eocene Event and the mid-Oligocene event is much later.

Clearly, a time scale is not an easy thing to construct, and there are a whole host of assumptions and decisions made in producing the clean, simple results that most people use. A good geologist or stratigrapher, like an intelligent shopper, must do the homework. The old motto, *Caveat emptor* ("Let the buyer beware"), applies to all scientific work, including time scales.

CONCLUSIONS

Radiometric dating provides the only practical method of assigning numerical dates to events in the geologic past. A number of different radioisotopic systems can be used, but each is appropriate only under certain circumstances. A good geologist knows which isotopic system can be used and what its strengths and limitations are. Even more fundamental is an understanding of the geological context of radiometric dates and their relationships to other dating methods (such as biostratigraphy, isotope stratigraphy, and magnetic stratigraphy). Establishing a chronostratigraphic framework for a rock sequence is not simply a matter of grabbing a sample and getting an answer out of a black box in a lab. It requires a careful judgment of the relative strengths

and weaknesses of each dating method, as well as an assessment of which methods are giving more reliable answers and which are not. Many scientists have been fooled by a naive reliance on radiometric dates (such as the KBS Tuff example in Box 18.2) without understanding their potential weaknesses or geological context. In other cases, whole time scales have been changed by erroneous dates (such as the glauconite dates in the Eocene/Oligocene boundary example) or by updated methods (such as when argon-argon dating radically revised the old K-Ar dates). All of these pitfalls should make it clear that geochronology and chronostratigraphy are not simple exercises but require extensive geological training and common sense.

FOR FURTHER READING

Berggren, W. A., D. V. Kent, and J. Hardenbol, eds. 1995. *Geochronology, Time Scales, and Stratigraphic Correlation.* SEPM Special Publication 54.

Dalrymple, G. B. 1991. *The Age of the Earth.* Palo Alto, Calif.: Stanford University Press.

Dalrymple, G. B., and M. A. Lanphere. 1969. *Potassium-Argon Dating.* San Francisco: W. H. Freeman and Company.

Eicher, D. L. 1976. *Geologic Time,* 2d ed. Englewood Cliffs, N.J.: Prentice-Hall.

Faure, G. 1986. *Principles of Isotope Geology,* 2d ed. New York: John Wiley.

Harland, W. B., A. V. Cox, P. G. Llewellyn, C. A. G. Pickton, A. G. Smith, and R. Walters. 1989. *A Geologic Time Scale.* Cambridge: Cambridge University Press.

Johanson, D., and M. Edey. 1981. *Lucy: The Beginnings of Humankind.* New York: Simon and Schuster.

Lewin, R. 1987. *Bones of Contention: Controversies in the Search for Human Origins.* New York: Simon and Schuster.

McDougall, I., and T. M. Harrison. 1988. *Geochronology and Thermochronology by the $^{40}Ar/^{39}Ar$ Method.* New York: Oxford University Press.

Parrish, R., and J. C. Roddick. 1985. *Geochronology and Isotope Geology for the Geologist and Explorationist.* Geol. Assoc. Canada, Cordilleran Section, Short Course 4.

Prothero, D.R. 1994. *The Eocene-Oligocene Transition: Paradise Lost.* New York: Columbia University Press.

York, D., and R. M. Farquhar. 1972. The *Earth's Age and Geochronology.* Oxford: Pergamon Press.

19 Sedimentary Rocks in Space and Time

Rifting between Africa and the Arabian Peninsula, forming the Red Sea (*foreground*), the Gulf of Aqaba (*center right*), and the Sinai Peninsula. (Stanley, 1989: 164.)

THIS BOOK BEGAN WITH A DISCUSSION OF THE DETAILED aspects of sedimentary rocks: their texture, composition, classification, nomenclature, structures, diagenesis, depositional settings, and overall genesis. Subsequent chapters considered the principal aspects of stratigraphy: facies, unconformities, biostratigraphy, and the means by which stratified rocks can be correlated with one another and tied to a numerically calibrated relative standard geologic time scale.

Now we turn to the bigger picture and address the larger questions about sedimentary rocks. What factors control the overall distribution of sedimentary rocks in time and space? How are sedimentary basins initially produced? How do such basins develop through time? What roles do weathering, erosion, and sedimentation play in the growth of terrestrial continental crust and the origin and evolution of individual orogenic belts? What important secular changes occur

within the sedimentary rock record? What insights do such variations through time imply about short-term and long-term global evolution?

Basin Analysis

Several tools enable geologists to analyze the three-dimensional geometry and depositional history of an ancient sedimentary basin. Many types of evidence can be developed that indicate the source of sediments and their direction of transport; the architecture and thickness of rock units and their relative ages; and, ultimately, the tectonic factors that controlled the geologic history of the basin. This kind of research is called **basin analysis.** Most of the world's nonrenewable fuel resources are found in sedimentary basins, and predicting the occurrence, abundance, and cost of recovery of oil, gas, and coal requires the most complete understanding of these basins possible.

Many of the fundamental methods of basin analysis are discussed throughout this book. The determination of the source and direction of transport of the

basinal sediments, however, is particularly common in basin analysis. One finds this information by examining the basin sediments to determine their provenance, or source area. In field studies, the geologist must be alert for any rock type or clast that is megascopically distinctive and should also study the area surrounding the basin for possible sources of the sediment.

If the rocks contain conglomerates with megascopic pebbles and cobbles, it is possible to determine provenance right at the outcrop. More often than not, however, provenance studies require detailed petrographic analysis. The most useful diagnostic grains are often visible only in thin section. If a sandstone has few lithic fragments or other polycrystalline grains that are diagnostic by themselves, then the percentage of certain key mineral grains may be important. Sometimes the assemblages of quartz, feldspars, and lithic fragments can be very distinctive and can allow quite elaborate reconstructions to be made of ancient current directions and drainage basins, as well as of the uplifted source areas (Fig. 19.1).

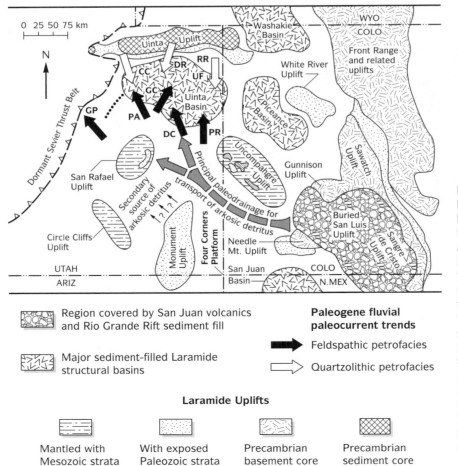

Figure 19.1 Map showing major Laramide basins of Colorado and Utah and the uplifted basement rock during the Cretaceous through Eocene. Using distinctive petrographic assemblages in the sandstone, Dickinson, Lawton, and Inman (1986) recognized a "quartzolithic petrofacies" in the Uinta Basin in northeastern Utah derived from the quartzites of the Uinta Mountains to the north, and a "feldspathic lithofacies" eroded from uplifts to the south. Because the uplifts in that region are mantled with Mesozoic strata and have few feldspar-rich Precambrian basement source rocks, however, Dickinson and others reconstructed a paleodrainage across western Colorado that supplies the arkosic debris to the "feldspathic lithofacies." The Monument Uplift in southeastern Utah is also a possible source of the arkose. (After Dickinson, Lawton, and Inman, 1986: 290; by permission of the SEPM.)

Region covered by San Juan volcanics and Rio Grande Rift sediment fill

Major sediment-filled Laramide structural basins

Paleogene fluvial paleocurrent trends

Feldspathic petrofacies

Quartzolithic petrofacies

Laramide Uplifts

Mantled with Mesozoic strata

With exposed Paleozoic strata

Precambrian basement core

Precambrian sediment core

Most diagnostic are the heavy minerals (with densities greater than the ubiquitous quartz and feldspars), which can be separated from others by settling crushed samples in a dense liquid such as bromoform. Assemblages of dense minerals can often be traced to distinctive source terranes. For example, some minerals—such as garnet, epidote, staurolite, and kyanite-sillimanite-andalusite—are characteristic of metamorphic terranes. Others may indicate the presence of certain igneous rocks. Tourmaline, beryl, topaz, and monazite, for example, are usually derived from pegmatites. The most stable heavy minerals— such as zircon, tourmaline, and rutile—can maintain their integrity through many sedimentary cycles. Thus, their abundance increases as sediments mature. The relative abundance of these three minerals is often called the **ZTR index** and is used as an indicator of sediment maturity.

One also can map the textural trends of the particles in a basin. For example, the roundness of many sedimentary grains usually increases with distance from the source, so a map of grain roundness (Fig. 19.2A) can indicate source and direction of transport. Similarly, grain size usually decreases with distance from the source, so a map of maximum clast diameter (Fig. 19.2B) can reveal much information about the source area.

The most direct method of determining direction of transport, however, is to measure sedimentary structures that show clear directional trends. Cross-beds, ripple marks, and the various directional bottom marks discussed in Chapter 4 all give clues to the flow direction of a depositional event. These structures can be measured in the field and used to plot paleocurrents (see Box 4.1).

Paleocurrent analysis can be a powerful tool, but a reliable study may require hundreds of measurements. Many of the sedimentary environments discussed earlier have highly variable current directions. For example, the currents in a meandering

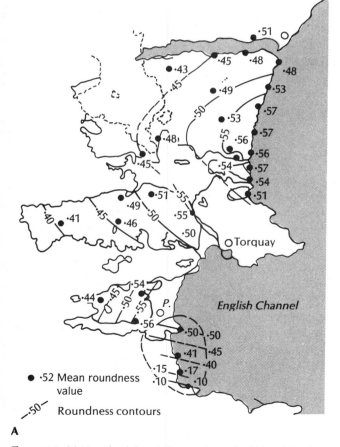

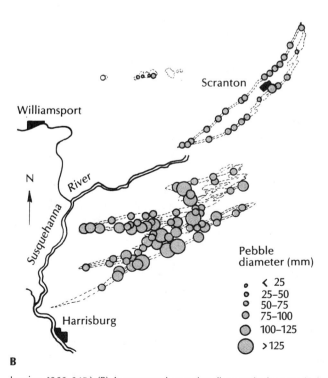

Figure 19.2 (A) Map of variations in the roundness of pebbles of the Triassic Lower New Red Sandstone of England. The highest roundness values are found north of Torquay, indicating a souce that is farther inland. (After Laming, 1966: 945.) (B) Average maximum clast diameter in the Pottsville Formation of Pennsylvania. Note that the coarsest material is in the southeast, indicating a source area in this direction. (After Meckel, 1967: 225.)

river can point in all directions of the compass, depending on the part of the meander loop from which they come. Measurements from a restricted outcrop might be from a part of the meander loop that happened to be flowing opposite to the overall flow direction of the river. Reliance on such measurements could result in a very erroneous paleogeographic reconstruction of the river system. In beach environments, there can be multiple paleocurrent directions. The foreshore strata could be seaward-dipping, the backshore strata could be landward-dipping, and there could also be shore-parallel structures from longshore currents. All these strata could be preserved in the same limited outcrop. A field geologist who lacked an understanding of the paleoenvironment would be baffled.

In any paleocurrent study, the geologist must take as many measurements as is practical and make field observations to determine from which part of the sedimentary environment these currents might have come. In addition, paleocurrent measurements from dipping beds must be corrected to the horizontal. This is easy to do on a stereonet, although many calculators and small computers can be programmed to perform such corrections. To display a large sample of directional data and get a sense of the overall trend, the geologist can plot paleocurrent data on a rose diagram, as discussed in Box 4.1 (see also Box 19.1 and Fig. 19.1.4).

Stratigraphic Diagrams and Maps

Provenance studies and paleocurrent analyses each give a good sense of the two-dimensional areal distribution of features in and around a sedimentary basin. For a fuller understanding, however, a three-dimensional view of the basin is essential. Such a view can be conveyed in a number of different types of cross sections and maps, some of which are unique to stratigraphy.

Stratigraphic Cross Sections

Once a series of stratigraphic sections have been measured in an area, they can be drafted side by side and correlated by the procedures described in Chapter 15. The result is a two-dimensional representation of the Earth known as a **stratigraphic cross section.** It differs from a normal geologic cross section in that no topography is shown, and structural deformation is either corrected for or shown diagrammatically.

Stratigraphic cross sections are usually aligned along a shared stratigraphic datum or by the topographic elevation of their beds if they are horizontal. After lines of correlation are drawn, patterns of pinch-out and other differences in thickness, intertonguing, facies changes, unconformities, and biostratigraphy can be shown and their relationships delineated. A clear stratigraphic cross section often carries more information about the geologic history of a region than any other type of diagram.

But a stratigraphic cross section is only two-dimensional. One can obtain an even greater understanding of the geometry of lithologic units by drawing a series of cross sections on an isometric map base to form a type of three-dimensional cross section known as a **fence diagram** (Fig. 19.3). Fence diagrams can make some patterns and relationships very clear, but they must be drafted carefully or they become confusing. Reducing cross sections to fences of the appropriate scale for the isometric diagram also loses much of the detail. If the geology is too complicated, it is impossible to represent it on a few fences; but the more fences that are drawn, the more difficult the diagram is to read. Computer programs are now available that eliminate some of the mechanical problems of drawing cross sections and fence diagrams (particularly in isometric drawing). Correlation is not an automatic procedure, however, so geologists must use their own judgment in deciding what is geologically reasonable.

Stratigraphic Maps

A map is a two-dimensional representation of points in a three-dimensional space. The purpose of many maps, such as roadmaps, is to show horizontal distances and directional relationships, but maps can also reduce three-dimensional objects to two dimensions by shadowing and shading (which requires considerable artistic talent) or by contouring (which is more mechanical). Topographic maps, the most familiar example of contouring, reduce the three-dimensional surface of the Earth to two dimensions. Contouring can also be used to represent three-dimensional surfaces of an underground rock bed (structure contour maps) or three-dimensional thicknesses of rock (isopach maps). A stratigraphic map shows the areal distribution, configuration, or aspect of a stratigraphic unit or surface. The simplest stratigraphic maps, such as isopach and structure contour maps, show the external geometry of a rock body. More complex maps also depict internal variations in the composition of a

BOX 19.1 BASIN ANALYSIS OF THE RIDGE BASIN, CALIFORNIA

A particularly well exposed and thoroughly studied basin is the Ridge Basin in the Transverse Ranges north of Los Angeles, California (summarized in Crowell and Link, 1982). Small in area, it is now up-lifted and deeply dissected, so that most of the features can be studied in outcrop (Fig. 19.1.1). The Ridge Basin formed between two major fault systems, the San Gabriel and San Andreas (Fig. 19.1.2A). These systems produced a very narrow (10 to 15 km wide) but extremely deep basin that accumulated a thickness of 13,500 m (more than 8 miles) of sediments over about 8 million years of the late Miocene and early Pliocene. Because the basin is so narrow and deep, facies changes across it are very rapid (Fig. 19.1.2B).

The fault-bounded basin margins are characterized by coarse fanglomerates and breccias, which pass laterally into the sandstones and shales of the basin center. The Ridge Basin began accumulating sediment in the late Miocene when a marine incursion deposited the Castaic Formation, up to 2800 m of marine mudstones (Figs. 19.1.2B, 19.1.3). Although the stratigraphic thickness of the Ridge

Basin itself is thousands of meters, the water depth in the Castaic Formation was only in the range of 45 to 90 m (Stanton, 1960, 1966). Most of the thickness of the stratigraphic units in the Ridge Basin resulted from continual subsidence along the faults and not from catastrophic deepening. After the Castaic marine incursion, the Ridge Basin continued to fill with lacustrine muds of the Peace Valley Formation. These mudstones interfinger with fluvial and deltaic sandstones and conglomerates of the Ridge Route Formation on the northeastern margin and the coarse Violin Breccia on the southwestern margin (see Figs. 19.1.2B and 19.1.3). The Peace Valley–Ridge Route sequence is capped by the upper Miocene–lower Pliocene Hungry Valley Formation, 1200 m of fluvial and alluvial sandstones, conglomerates, and mudstones.

Basin analysis of the Ridge Basin included detailed sedimentological studies of each of the stratigraphic units shown in Fig. 19.1.2B and the study of their depositional environments. In addition, fossils (trace, plant, invertebrate, and mammal) were used to determine the age and depositional environments

Figure 19.1.1 Aerial photograph of the northern part of the Ridge Basin, showing the thick sequence of lacustrine shales and deltaic and fluvial sandstones dipping away from the viewer. Frenchman Flat and Highway 99 at lower left, Hungry Valley and Frazier Mountain in the background. (Courtesy of J. C. Crowell.)

(continued)

(Box 19.1 continued)

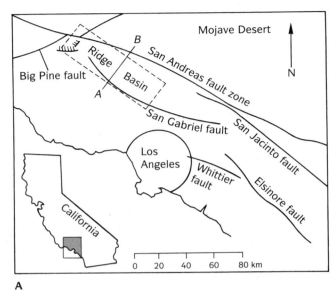

A

Figure 19.1.2 (A) Location of the Ridge Basin, southern California. Note location of cross section *A–B*. (B) Diagrammatic cross section of the Ridge Basin showing the major stratigraphic and structural relationships. Section located along line *A–B*. (After Crowell and Link, 1982: 1, 2.)

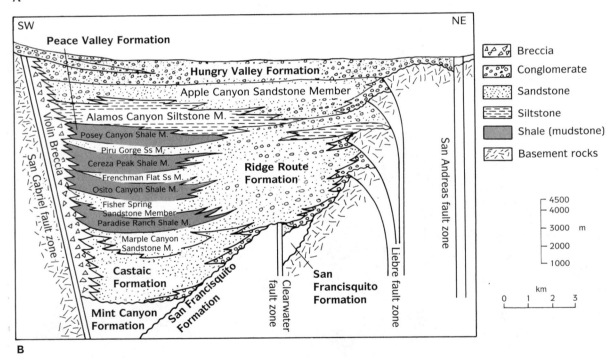

B

of each unit. Parts of the sequence have also been dated with magnetic stratigraphy (Ensley and Verosub, 1982). Some of the units have also been analyzed geochemically.

Because the basin is so narrow and well exposed, with such obvious source areas, it is ideal for studies of source and direction of transport. The large number of paleocurrent measurements is summarized in

Fig. 19.1.4. Most of the bidirectional paleocurrent marks (such as ripple marks, cross-beds, and groove casts) show strong east-west orientation, roughly perpendicular to the axis of the basin. These marks correspond to currents that transported sediments from the source areas in the east and west. Some of the paleocurrents, particularly those in the center of the basin, show a northwest-southeast orientation,

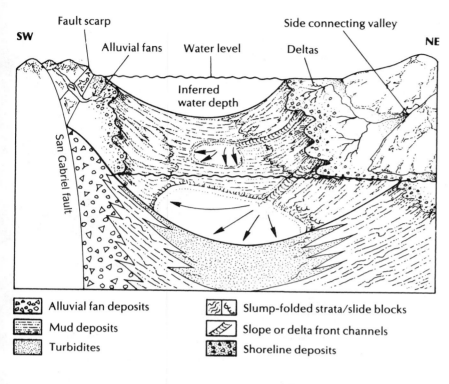

Fault scarp   Side connecting valley

SW   Alluvial fans   Water level   Deltas   NE

Inferred water depth

San Gabriel fault

▱ Alluvial fan deposits
☰ Mud deposits
⣿ Turbidites

▨ Slump-folded strata/slide blocks
▧ Slope or delta front channels
▦ Shoreline deposits

Figure 19.1.3 Depositional interpretation of Ridge Basin showing the breccias and alluvial fans produced along the fault scarps. These grade rapidly into fluvial-deltaic deposits, then drop off sharply into deep-water shales and turbidites. (After Crowell and Link, 1982: 87.)

indicating that the transport in the center of the basin was southeasterly. Apparently, the basin drained to the southeast, so the paleoslope for the deeper basin sediments was in that direction.

Provenance studies of the Ridge Basin are complicated by the variety of source areas available on each side of the basin. Pebble counts in the Hungry Valley Formation (Fig. 19.1.5) show that the most common rock types are granitics and gneisses, which are exposed widely to the west and north. Volcanic rocks are locally abundant near the Big Pine fault in the northwest and in the southeast margin of the basin. Certain distinct rock types—such as pink granites, augen gneisses, and fault-zone mylonite—are derived only from the Alamo Mountain–Frazier Mountain area to the west and do not occur to the north or east. Marbles, on the other hand, are exposed only north of the San Andreas fault and appear to be from a northeasterly source area. These studies show that the bulk of the Ridge Basin sedimentary fill came from the north and east (Fig. 19.1.6), even though the basement southwest of the San Gabriel fault scarp was a closer source for sediments on the southwestern side of the basin.

Provenance analysis of the cobbles in the Violin Breccia reveals an unusual mode of deposition. The Violin Breccia lithologies are restricted to gneisses, diorites, and granitic rocks, which have a very limited source area on the southwest side of the fault. No sedimentary clasts are found from prospective source areas just a few kilometers to the west or from more distant areas. All these lines of evidence suggest that the Violin Breccia source area was not very extensive and was drained by small streams with limited drainage basins. Yet the Violin Breccia has a total stratigraphic thickness of 13,400 m. How could such an enormous thickness of conglomerate have formed from such a small, poorly drained source area in only 7 million years?

Crowell (1982) pointed out that the characteristics of a basin trapped between two strike-slip faults provide a mechanism for sediment accumulation. The southwestern source area moved northwestward relative to the basin as the San Gabriel fault continued to move in the late Miocene. In effect, the alluvial fans accumulating in the Ridge Basin were transported southeastward, away from their source, like cars on a southeastbound coal train (Fig. 19.1.7A). As each alluvial fan

(continued)

(*Box 19.1 continued*)

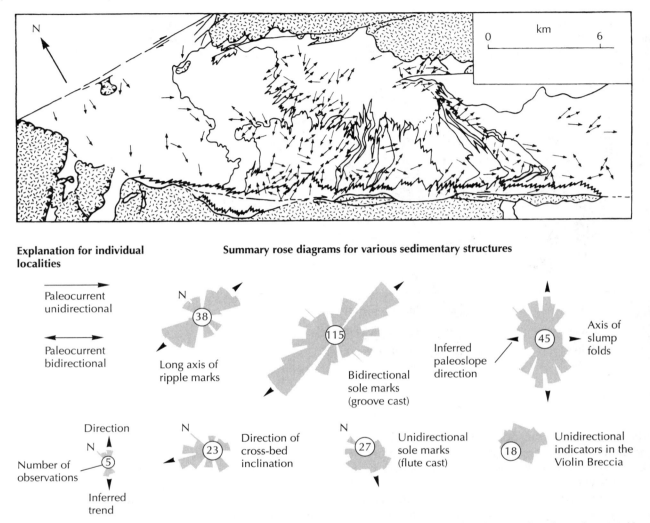

Figure 19.1.4 Paleocurrent rose diagrams for a variety of directional sedimentary current indicators in the Ridge Basin. Notice that the bulk of the unidirectional indicators are out of the north-northeast, even on the southwestern edge of the basin where the souces from the southwest would have been much closer. The basin ultimately drained toward the southeast, as indicated by the paleocurrents. (After Crowell and Link, 1982: 270.)

deposit moved south from the source area, a new "coal car" (alluvial fan deposit) was formed behind it to the north. Consequently, the thick Violin Breccia is actually made of a sequence of shingled fan deposits, with the oldest lying farthest south (Fig. 19.1.7B) and each progressively younger fan overlapping the one before it on its northern end. Thus, the vertical stratigraphic thickness in any one place is less than a third of the total stratigraphic thickness, and the formation is time-transgressive over about 7 million years.

The peculiarities of the Ridge Basin are a result of its unusual geologic setting. Unlike most basins in the craton or mountain belts, the Ridge Basin formed

as a chasm or gap opened between two curved segments of a strike-slip fault (see Fig. 19.1.6). This accounts for the extremely narrow yet deep basin, with an enormous thickness of sediment accumulated over 8 million years. These strike-slip basins are not common in the geologic record, but they produce a very distinctive result: an extraordinarily thick but narrow sedimentary package with coarse conglomerates on the margins and very rapid facies changes to lacustrine and marine shales and sandstones in the middle. As the Ridge Basin example becomes better known, more such examples are being found in ancient sediments.

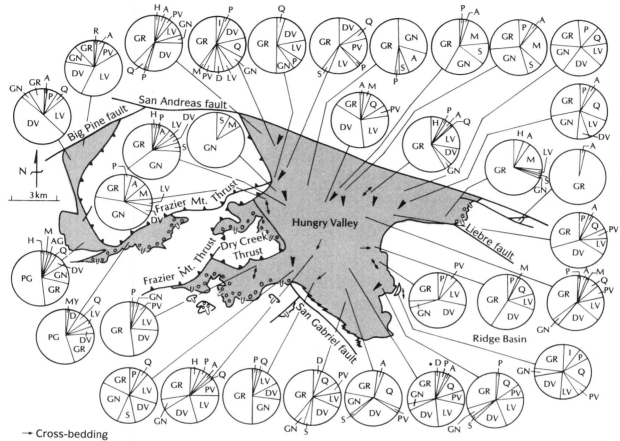

→ Cross-bedding
↔ Channel orientation
► Imbrication

Figure 19.1.5 Provenance of pebble types from the Hungry Valley Formation, as analyzed by R. G. Bohannon. Relative percentages of types are indicated by pie diagrams. Abbreviations: A, amphibolite; AG, augen gneiss; D, diorite; DV, dark volcanics; GR, granitics; GN, gneiss; H, hornfels; I, intraclasts, such as rip-up clasts of underlying strata of shale and sandstone; LV, light-colored volcanics; M, marble; MY, mylonite; P, pegmatite; PG, pink granite; PV, purple volcanics; Q, quartz; S, schist. (After Crowell and Link, 1982: 147.)

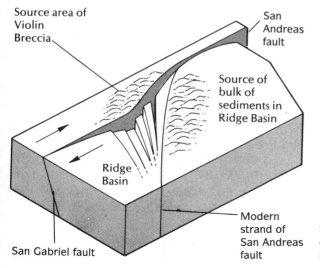

Figure 19.1.6 Block diagram showing the origin of the Ridge Basin as a sigmoidal bend in the San Andreas fault. This peculiar origin explains the extremely steep yet narrow geometry of the basin. (After Crowell and Link, 1982: 150.)

(*continued*)

(Box 19.1 continued)

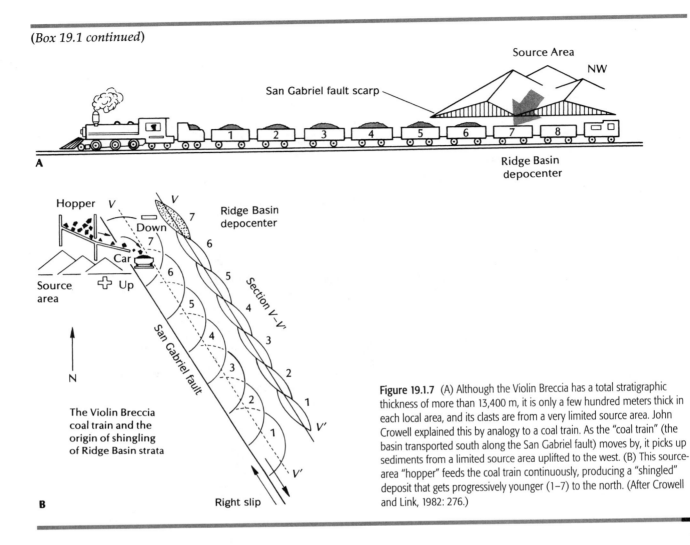

Figure 19.1.7 (A) Although the Violin Breccia has a total stratigraphic thickness of more than 13,400 m, it is only a few hundred meters thick in each local area, and its clasts are from a very limited source area. John Crowell explained this by analogy to a coal train. As the "coal train" (the basin transported south along the San Gabriel fault) moves by, it picks up sediments from a limited source area uplifted to the west. (B) This source-area "hopper" feeds the coal train continuously, producing a "shingled" deposit that gets progressively younger (1–7) to the north. (After Crowell and Link, 1982: 276.)

rock body, its vertical variability, and its internal geometry. Such maps are usually classed as facies maps. Finally, the whole geologic picture can be integrated into interpretive maps such as paleotectonic maps and paleogeographic maps.

The **structure contour map** is similar to a topographic contour map except that the surface contoured is not topography but the surface of an underground lithologic unit (Fig. 19.4). Like other contour maps, it is made by placing as many data points as possible on the map base and contouring between those that are different. This procedure requires an extensive array of subsurface points, as well as all the outcrop elevations available. If few data points are available, the contour lines are best drawn by hand using a bit of interpretation to generate as reasonable a picture as possible. When the density of data points is great enough, the contours can be drawn mechani-

cally. Computer programs are now available that can accomplish this at minimal cost. Still, every geologist should have some practice with the art of manual contouring. There is no better way to understand contour lines and to learn their limitations.

Structure contour maps are not difficult to interpret. Structural highs such as domes and anticlines show up clearly, as do low spots such as basins and synclines. Structure contours that are tightly bunched along a straight line usually indicate faulting, and the contours themselves can show the offset. The structure contour map is a primary tool for finding structural traps for hydrocarbons, such as anticlines and faults.

Isopachs are points of equal thickness of a rock unit, and an isopach map contours these points. This procedure requires data points along the top and the bottom of the bed. If there is any dip, the apparent thickness of a bed may be greater than its true thickness.

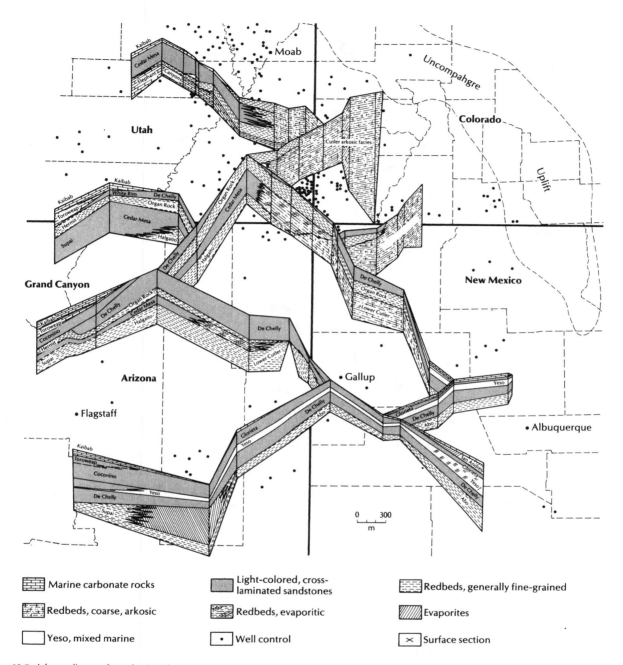

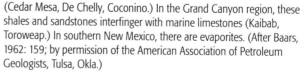

Figure 19.3 A fence diagram from the Permian rocks of the Four Corners region. By means of this three-dimensional projection of a series of cross sections, complex facies relationships can be seen clearly. The arkoses from the Uncompahgre Uplift in Colorado (Cutler Formation) interfinger with nonmarine shales (Organ Rock, Halgaito) and eolian sandstones (Cedar Mesa, De Chelly, Coconino.) In the Grand Canyon region, these shales and sandstones interfinger with marine limestones (Kaibab, Toroweap.) In southern New Mexico, there are evaporites. (After Baars, 1962: 159; by permission of the American Association of Petroleum Geologists, Tulsa, Okla.)

The apparent thickness of a unit is shown by an **isochore map.** Many so-called isopach maps are actually isochore maps because they are not corrected for structure. Isopach data usually come from wells scattered around the area of a map base, but they can also be drawn from two structure contour maps by subtracting the elevations of the lower surface from those of the upper surface. The database is contoured by the usual methods. The result, however, is not a surface but must be visualized as an irregular lens, or pillow, or bowl, or shoestring, or whatever three-dimensional shape seems appropriate. The contours are not on the top surface of the structure (which is considered horizontal by convention), but along the bottom surface.

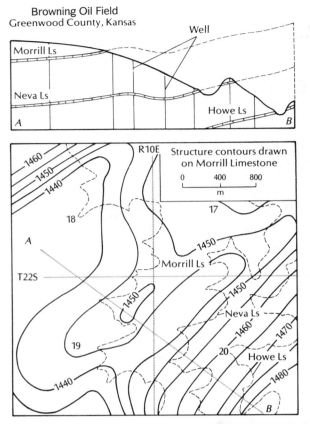

Figure 19.4 *Below:* Structure contour map showing the relationship of structure contours to outcrop patterns. *Above:* The projection of the contours on the Morrill Limestone beyond its outcrop edge through the use of lower marker beds and known intervals between them. (After Krumbein and Sloss, 1963: 438.)

They indicate where the body bulges up or down, pinches out, or is truncated (Fig. 19.5).

Isopach and isochore maps help locate features that are found in no other type of map. A concentric bull's-eye pattern with increasing thicknesses toward the center clearly shows a basin, with the deepest part having the thickest isopachs. Abrupt changes in isopach spacing can indicate the boundary between thin shelf sequences and thick basin deposits. The boundary between the shelf and the basin is called the **tectonic hinge.** Unusual flexures or breaks in the isopach pattern usually indicate buried folded or faulted uplifts. Where isopachs thin to zero, stratigraphic pinch-outs or abrupt fault boundaries are indicated by tight contour spacing. Zero isopachs may indicate erosion rather than depositional thinning, however. In such cases, the unit has been carved out from above, and the zero isopachs represent "holes" in its upper surface. Further evidence is then needed to determine the true significance of the peculiar patterns on the isopach map.

A chronologic series of isopach maps indicates the time when basins were subsiding and accumulating sediment and when they were shallow or being eroded. A trend toward zero isopach, which represents a pinch-out, may imply an ancient shoreline. The thickness and dimensions of a basin are often critical in determining whether it has hydrocarbon reservoir potential or where stratigraphic and structural traps lie. For these reasons, isopach maps have long been used in petroleum geology.

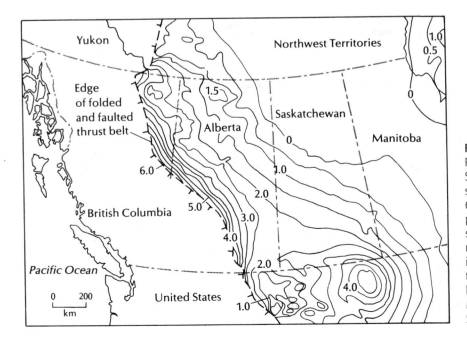

Figure 19.5 Isopach contours of sediments in the Rocky Mountain foreland basin of Alberta, Saskatchewan, and Montana. Thickness increases southwestward, with a maximum thickness of 6 km right against the fold-and-thrust belt, which produced the downwarping and supplied the sedimentary fill. The bull's-eye pattern in the lower right indicates a bowl-shaped closed basin. The irregular contours with the 1-km thickness just to the left of the basin indicate thin spots across basement uplifts. (After Porter, Price, and McCrossan, 1982: 165; by permission of the Royal Society.)

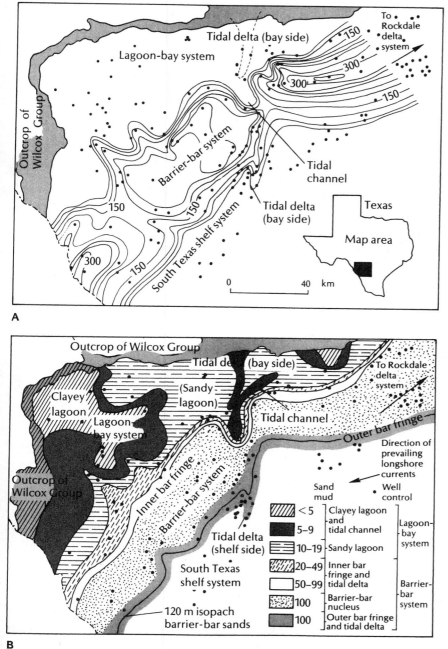

A

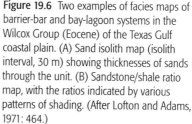

Figure 19.6 Two examples of facies maps of barrier-bar and bay-lagoon systems in the Wilcox Group (Eocene) of the Texas Gulf coastal plain. (A) Sand isolith map (isolith interval, 30 m) showing thicknesses of sands through the unit. (B) Sandstone/shale ratio map, with the ratios indicated by various patterns of shading. (After Lofton and Adams, 1971: 464.)

A **facies map** shows the areal variation of some aspect of a stratigraphic unit. Variations in lithologic aspects and attributes are shown by **lithofacies maps**; variations in faunal aspects are shown by **biofacies maps**. Facies maps may show the thicknesses of a single component of a unit, such as the sandstone thickness in a formation. An isopach map of a single rock component in a unit is called an **isolith map** (Fig. 19.6A). The actual thickness of a component can be converted to its percentage of the thickness of the whole stratigraphic unit, producing a percentage map. More than one component can be shown in a lithofacies map; for example, the ratios of two components (such as sand to shale) can be calculated and contoured. Such a **ratio map** (Fig. 19.6B) gives a good idea of where a unit is changing in grain size or where various environments are found. If three components are represented, a triangular diagram for the ratios of the three components must be used as a key (Fig. 19.7). The resulting **triangle facies map** is a bit more difficult to read but often gives a good sense of the distributions of sedimentary environments and facies changes. Further, the degree of mixing of the three components can be calculated mathematically,

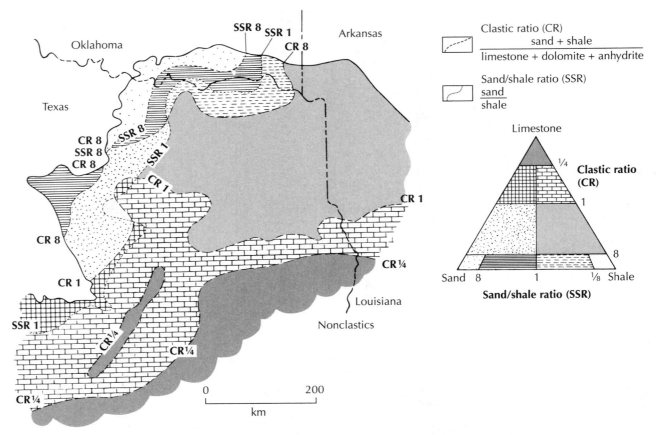

Figure 19.7 Triangle facies map of the Cretaceous Trinity Group. Relative percentages of the three components (sand, shale, and nonclastics) are shown by the different patterns, as keyed in the three-component triangle. (After Krumbein and Sloss, 1963.)

and these values can be contoured to form an **entropy map.** This kind of map is a good indicator of how mixed or homogeneous a rock unit is. Many other types of lithofacies maps have been created for specific purposes, but they are less commonly used.

Most facies maps are based on two-dimensional representations of three-dimensional features. Computer technology has made isometric projections of isopach, isolith, and other facies data much easier to plot. Figure 19.8 shows a perspective diagram of the isopachs and the sandstone, siltstone, and shale isoliths for Upper Cretaceous formations in Alberta. The computer perspective diagram is much easier to visualize than a two-dimensional map and conveys information very clearly. A high density of data points is necessary for the computer to plot this kind of diagram accurately, however.

An ordinary geologic map shows the distribution of rock outcrops on the Earth's present topographic surface. One can use subsurface information to strip away younger units and to reconstruct the pattern of outcrops before they were buried. This kind of map is

called a *paleogeologic map,* or a **subcrop** (that is, *sub*surface out*crop*) **map.** If one could use a time machine to visit the world as it was before the younger units were deposited, the paleogeologic map could be used as an ordinary geologic map. Conversely, if one mapped the subcrop pattern of the *base* of a sequence that covers an ancient surface, one would have a geologic map of units as seen from below, a so-called **worm's-eye view map,** or **supercrop map.**

Paleogeologic maps are particularly useful in areas such as the craton, where broad regional unconformities occur and younger transgressive sequences lap over older beds that have been eroded to expose many other units. The paleogeologic map may show structural features that are not visible because they are covered by younger features that were deposited after structural deformation ceased. A series of paleogeologic maps is often one of the best tools for determining the timing and extent of episodes of structural deformation. In addition, paleogeologic maps give a good sense of the relative magnitudes of unconformities, transgressions, and regressions. They can also

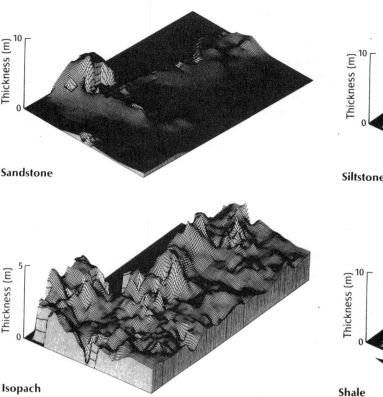

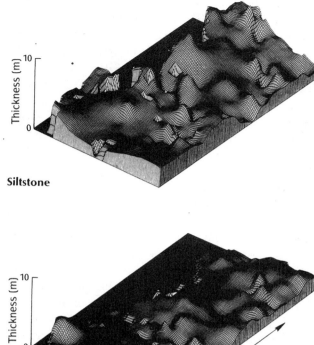

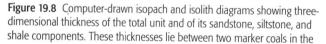

Figure 19.8 Computer-drawn isopach and isolith diagrams showing three-dimensional thickness of the total unit and of its sandstone, siltstone, and shale components. These thicknesses lie between two marker coals in the Cretaceous of central Alberta. (After Hughes, 1984: 70; by permission of the Minister of Supplies and Services, Canada.)

have economic applications. If, for example, a region is covered by an impermeable caprock, the paleogeologic map will show where the permeable reservoir rocks occur unconformably beneath it—a classic stratigraphic trap for hydrocarbons.

The stratigraphic maps described so far are based on such objective attributes of lithologic units as elevation, thickness, geologic age, percentage of components, and homogeneity. Sometimes it is useful to display subjective interpretations of these lithologies on a geologic map. From inferences based on lithofacies distribution and paleoenvironmental evidence, the geologist can construct a **paleogeographic map,** which shows the distribution of ancient environments at a given time in geologic history (see, for example, Figs. 8.5.1 or 9.1.1). Paleogeographic maps show the shorelines, the areas of marine deposition, the eroding uplands, the rivers, and sometimes the climatic and vegetational belts at some instant in geologic history. Historical reconstructions of this kind are often the ultimate goal of historical geology. Naturally, such interpretation can be highly speculative, especially if there are missing areas as a result of unconformities. Such

information can be economically useful, however, in predicting the distribution of shorelines, deltas, or other stratigraphic traps, or in getting a good perspective of the total paleoenvironmental picture.

Tectonics and Sedimentation

For almost two centuries, we have known that particular sedimentary rocks and basin geometries are associated with particular types of tectonic situations. But not until the development of plate tectonics in the 1960s was there an adequate explanation of *why* these sedimentary rocks and sequences formed during certain times and places. In the 1970s, enormous scientific breakthroughs related sedimentary sequences to specific motions of Earth's crustal plates. Now it is possible to explain sedimentary sequences with plate-tectonic models and even to infer the motions of ancient plates from the sedimentary record they leave behind.

Cratonic Sedimentation

As early as 1859, the pioneering American geologist James Hall recognized that there is a fundamental

difference between sedimentation in the center of a continent and that on its margins. The center of a continent appears to be relatively stable, with thin sedimentary sequences punctuated by unconformities. As one approaches the continental margins, however, one finds an increasing association between deep sedimentary troughs that have accumulated thousands of meters of sediment and the mountain belts that exist on or near many continental margins. The stable center of the continent was named the **craton** by Stille in 1936. *Kraton* is the Greek word for "shield," and the cores of the continents are indeed broadly convex upward like the shields of ancient warriors. The marginal troughs that parallel the mountain belts were called **geosynclines** by James Dwight Dana in 1873. We will discuss geosynclines in the next section.

The craton is characterized by thin sedimentary strata that unconformably overlie Precambrian basement. Most continents have a Precambrian core or shield area to which mountain belts have been added by later accretion along the margins. Because cratons are usually positive features (they are often above base level), much of their post-Precambrian history is represented by erosion surfaces and unconformities. Limited vertical movement in the craton has formed shallow **basins** and **arches,** which have a few tens of

meters of relief and which have sunk or risen very slowly and steadily over long spans of time (Fig. 19.9). The cratonic sedimentary cover on most continents is primarily Paleozoic and Mesozoic; averages about a kilometer in thickness; and is composed almost entirely of shallow marine sandstones, limestones, and shales, as well as fluvial and deltaic sequences. The sandstones are usually clean and very mature, indicating that the grains have gone through multiple cycles of reworking before final deposition.

Cratonic basins are typically shallow and bowl-shaped, with the units thickening gradually toward the center (Fig. 19.10). The isopachs of many basins show a bull's-eye pattern, indicating a very regular thickening toward the center of subsidence. Judging from the sedimentary records of such basins, they remained structural and topographic lows for long periods of time. Yet their fill is not continuous. During some periods, these basins underwent gentle downwarping; during others, they were relatively stable. As a consequence, their sedimentary record consists of unconformity-bounded packages that represent intervals of major transgression across the entire craton. These packages were called sequences by Sloss, as discussed in Chapters 15 and 17. Even during the peak of a transgression, however, the cratonic basin fill was

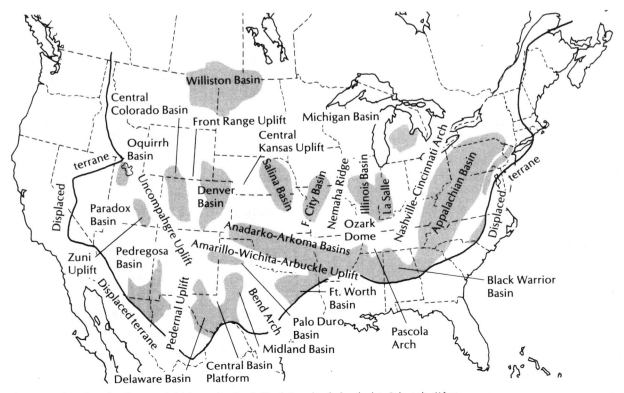

Figure 19.9 Location of major cratonic basins and arches in North America during the late Paleozoic. (After Sloss, 1982: 36; by permission of the Geological Society of America.)

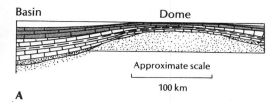

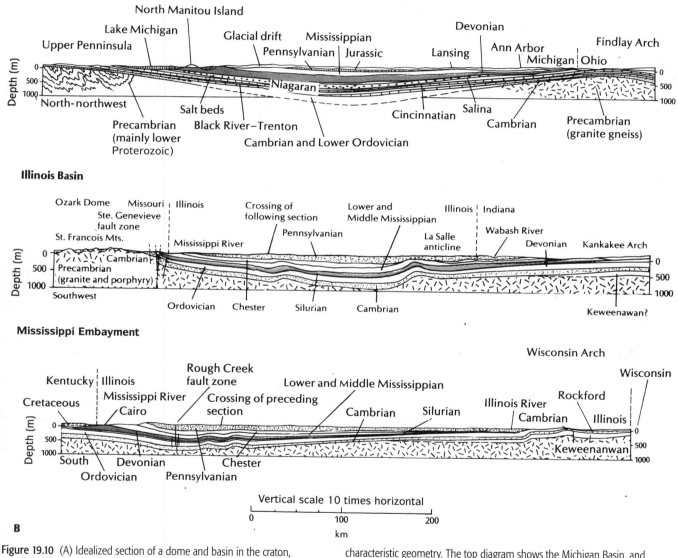

Figure 19.10 (A) Idealized section of a dome and basin in the craton, showing thinning of all units away from the basin and toward the dome. This results partly from deposition of a greater thickness of each unit in the basin and partly from truncation and overlap along unconformities. (B) Cross sections of two representative cratonic basins showing the characteristic geometry. The top diagram shows the Michigan Basin, and the lower two diagrams show east-west (*middle*) and north-south (*bottom*) cross sections of the Illinois Basin. (After King, 1977: 28, 35; © 1977 Princeton University Press; by permission of Princeton University Press.)

thin and discontinuous compared with the sediment found along continental margins.

The cause of this gentle cratonic basinal downwarping has been argued for a long time. Lately, it has become apparent that these basins are not in fact very different in origin from those on continental margins.

Many cratonic basins apparently resulted from deep intracontinental rifts that were interrupted and became deeply buried. Even though these ancient rift valleys did not split the craton, they remain zones of crustal thinning and weakness and so are subject to further subsidence. An example is the bull's-eye pat-

tern seen in the Michigan Basin, which is centered in southern Michigan. Maps of this basin show smoothly concentric isopachs, indicating an almost perfectly bowl-shaped depression. Recent gravity data, however, show a pronounced linear trend in the depression, suggesting a deeply buried northwest-southeast trending graben underneath the Paleozoic cover (Byers, 1982). It appears that the Michigan Basin is not a simple bowl but is underlain by a narrow graben that is part of an ancient Proterozoic rift system found beneath Lake Superior, Minnesota, and much of the midcontinent. Discoveries such as this promise to radically change many of our traditional notions about cratonic sedimentation and tectonics.

Sedimentation in Orogenic Belts: The Classic Geosynclinal Interpretation

Orogenic systems such as the Appalachians, Alps, Himalaya, and Cordillera expose inordinately thick sequences of sedimentary rock that have been folded, faulted, metamorphosed, and ultimately thrust upward as topographically spectacular mountain belts. Throughout the nineteenth century and well into the twentieth, few geologists accepted the notion of continental drift. Ocean basins were thought to be relatively ancient features; continents were believed to be younger and of variable age. Most important, the positions of continental blocks and ocean basins were regarded as fixed.

The permanence of continental blocks and ocean basins made it difficult to explain the origin and evolution of mobile belts. Hall (1859) recognized a genetic connection between thick sedimentary sequences and elongate mountain systems, and he argued that the subsidence of elongate segments of crust adjacent to continental blocks was caused by the weight of the sediments that accumulated there. Dana (1873) coined the term *geosyncline* for such elongate sediment-filled troughs and suggested that their subsequent deformation was due to lateral compression, not sediment loading.

The American model for geosynclines was naturally the orogenic belts that border the eastern and western margins of the North American continent: the Appalachian and Cordilleran orogenic systems (Fig. 19.11A). By 1900, European geologists had applied Dana's concept of geosynclines to the thick, compressed sedimentary sequences that form the Alps. Based on the Alpine example, Europeans regarded

geosynclines not as asymmetrical, subsiding troughs on the continental margins but as symmetrical features that formed between continents (Europe and Africa) and straddled oceanic crust, with thick, deep marine sediment fill. European and American geologists argued back and forth throughout the first half of the twentieth century, debating geosynclinal terminology; the nature and origin of sediment sources; and the precise mechanisms by which geosynclinal troughs first formed, subsequently filled, and were finally compressed and uplifted.

In 1936, the German geologist Hans Stille published a watershed paper on geosynclines. In the present context of plate tectonics, his paper was prophetic. He subdivided geosynclinal belts into two components, the **miogeosyncline** and the **eugeosyncline.** The miogeosyncline consisted of a sequence of shallow marine sandstone and limestone that thinned toward the craton, thickened toward the bounding ocean basin, and generally mimicked cratonic sequences in lithological character. Oceanward, the miogeosynclinal ("near geosyncline") belt or trough ended, passing abruptly into a eugeosynclinal ("true syncline") belt in which sediment fill was thicker and of deeper marine origin: shale, sandstone, volcanic rock, and chert. The eugeosynclinal belt contained abundant submarine volcanics and volcaniclastic debris. Much of the sediment seemed to be derived either from now-missing source areas within the ocean basin (so-called volcanic islands and tectonic lands) or from distant, now-missing borderlands. (Robert S. Dietz, 1963, later provided a modified model for these geosynclinal components, shown in Fig. 19.11C, in which the miogeosyncline is replaced with a modern passive continental shelf and the eugeosyncline with a present-day passive continental slope–rise–abyssal plain complex.)

As the eugeosyncline and miogeosyncline were subjected to compression (of uncertain origin), the more intensely deformed and metamorphosed eugeosyncline was commonly thrust continentward up and over the miogeosyncline. Inherent in geosynclinal theory was the related concept of tectonic cycles. Sediment filling a geosynclinal belt varied as a result of mobility or tectonism. Relatively clean, well-sorted quartzose sands and interbedded carbonates typified an initial stage in which a continental margin (or intracontinental basin) slowly subsided (the *preorogenic stage*). As the pace of subsidence accelerated, a thick sequence of deep marine shale, turbidite wacke sandstone, and chert (referred to as **flysch** by the Swiss) was deposited (the *synorogenic stage*). Finally, as mountains were deformed and uplifted, a thick sequence of

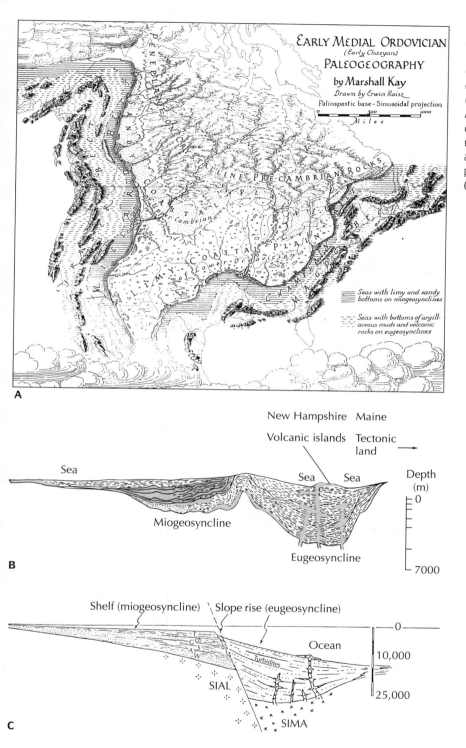

A

B

C

Figure 19.11 (A) Paleogeographic reconstruction of Early Ordovician North America showing the bordering Cordilleran and Appalachian Geosynclines with paired miogeosynclinal and eugeosynclinal belts (From Kay, 1951: Plate1.) (B) Classic profile of a geosyncline based on the Appalachians. (After Kay, 1951: 26; by permission of the Geological Society of America) (C) Dietz's reconstruction across the Appalachians showing analogy to present passive Atlantic margin with a paired continental shelf and slope-rise area. (After Dietz, 1963: 320.)

mixed marine and nonmarine—but largely terrestrial (fluvial and lacustrine)—conglomerate, sandstone (often redbeds), and mud was eroded and deposited along their flanks (**molasse** of the *postorogenic stage*). Marshall Kay's classic monograph *North American Geosynclines* (1951) summarized the nomenclature applied to geosynclines and proposed a uniform scheme for their classification.

A number of questions about geosynclines remained unresolved, however. What and where were modern analogs to ancient geosynclines? Why did geosynclinal subsidence begin? Why were some geosynclinal belts (such as the Appalachians) confined to continental margins, whereas others (the Alps, Urals, and Himalaya) lay between continental crustal blocks? What caused the initial tensional

stress regime necessary to initiate subsidence to develop gradually into laterally applied compressional stress? By the mid-1950s, it became obvious that myriad schemes for classifying and naming geosynclines and fitting their sediment fill into an unrealistic, idealized sedimentary tectonic framework was no substitute for true understanding. Explanation had to be sought elsewhere.

Plate Tectonics and Sedimentary Basins

In the late 1950s and 1960s, geology underwent a scientific revolution as profound as those of Copernican astronomy, Einsteinian physics, or Darwinian biology. The discovery of seafloor spreading showed that the Earth's crust is not the static foundation we had assumed it to be. Instead, it is composed of a group of lithospheric plates that move relative to one another over the surface of the globe. Plate-tectonic theory developed rapidly after it was widely accepted in the late 1960s and early 1970s, but stratigraphers were relatively slow to recognize its implications for the formation of sedimentary basins. A landmark paper by John Dewey and John Bird in 1970 first related the formation of mountain belts and geosynclinal basins to plate-tectonic models, and by the late 1970s most mountain belts and sedimentary basins were being explained by plate tectonics.

Plate-tectonic theory contained many revolutionary implications. Geosyncline theory became obsolete, and its terms survive today with only part of their original meanings. Even more startling are the implications of plate tectonics for the tempo and scale of geologic events. All of the modern seafloor has been created since the Jurassic, and all older seafloor has been subducted and lost beneath the trenches. Ocean basins up to 500 km wide can form and disappear in 50 to 100 million years. This cycle of opening and closing of an ocean basin is called a **Wilson cycle,** and it appears that there have been many Wilson cycles in the history of each continent. This means that no sedimentary basin with a long history of deposition is likely to have remained static or in its original position, except on cratons.

On continental margins that are associated with subduction zones, it appears that much of the geology consists of terranes from elsewhere (**exotic,** or **allochthonous, terranes**), that have been accreted by the collisions of plates. Seventy percent of the Western Cordillera of North America, for example, is estimated to be made of blocks that were not part of North America in the Precambrian but came from other parts of the Pacific Basin during the Phanerozoic. Most of California, Oregon, Washington, and some of Idaho and Nevada are essentially foreign—a fact that has profound implications for geologic studies in this region. Every time you cross a major geologic boundary, such as a fault zone or volcanic arc, you may be stepping onto rocks that have no genetic relationship to the terrane you just left. This sense of lateral discontinuity radically changes the way stratigraphers view these regions (Byers, 1982).

Plate tectonics has given us a fairly coherent picture of both *how* and *why* the major crustal plates move. Most of the major plates have a core of *continental crust,* which is sialic in composition, much less dense than the mantle, and about 30 to 50 km thick. This core of continental crust is surrounded by a thin (5 km) layer of *basaltic oceanic crust,* which is much denser than continental crust. Oceanic crust arises as magma along the mid-ocean ridges as the seafloor spreads when continents drift apart. Then, because ocean crust is thin and dense, it slides under the edge of the buoyant continental crust and sinks into the Earth's mantle. When continents collide, however, neither can sink under the other, so they produce an uplifted mountain belt such as the modern Himalaya.

Three major types of margins between lithospheric plates are known:

1. If two plates are separating, they form a **divergent** (passive) **margin.** Divergent margins are characterized by extensional features, especially seafloor spreading, extensional grabens, and normal faulting.

2. If two plates are moving toward each other, they form a **convergent** (active) **margin.** Such margins are characterized by compressional tectonics, especially folding and thrusting. Except when the converging plates are continental crustal plates, there is subduction of one plate (usually oceanic) beneath the other. Thus, a convergent margin includes a subduction zone and all its associated features.

3. If two plates are sliding past each other, neither separating nor converging, they form a **transform margin,** which is characterized by horizontal shear and strike-slip faulting.

Many plates are bordered by combinations of all three types of margins, or individual margins change from one type to another during their geologic history, so these categories are necessarily oversimplified. Nonetheless, they are useful models for

understanding the processes of mountain building and sedimentary-basin development.

Divergent Margins

The evolution of divergent margins might be subtitled "how to build an ocean." In the first stage (Fig. 19.12A), continental crust begins to rift apart, and hot mantle plumes upwell underneath the rift soon afterward. Upwarping produces domed uplifts that shed coarse, immature alluvial and fluvial deposits onto their flanks. Mantle upwelling can also result in volcanic eruptions, which contribute alkalic volcaniclas-

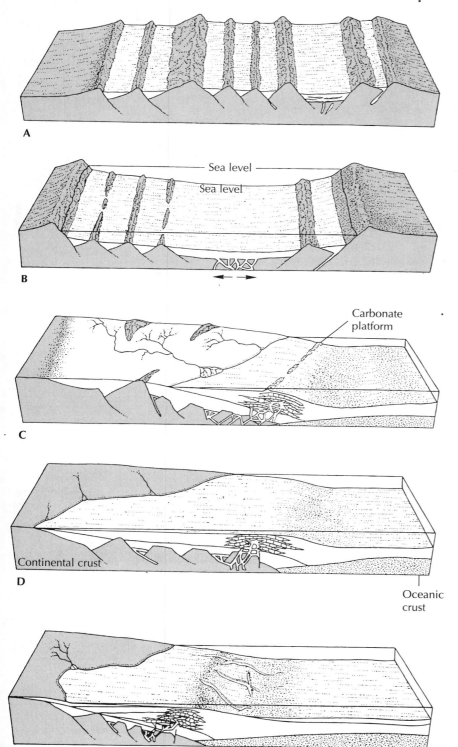

Figure 19.12 Cross sections of a passive margin showing four phases of development. (A) Rift valley phase. (B) Protooceanic gulf phase. (C) Early postrift phase. (D) Late postrift phase, when clastic sediments overwhelm the carbonate platform and build up the shelf-slope-rise complex. (E) Late postrift phase, when low sea level incises submarine canyons into the shelf edge. (After Schlee and Jansa, 1981: 78.)

Figure 19.13 Aerial photograph of the East African Rift Valley in the Afar Triangle, Ethiopia. Note the massive, tilted, normal fault blocks. The depressions are coated with white evaporites from dry lakes. (Tazieff, 1970: 48.)

tic sediments to the sedimentary piles on either side of the rift.

In the second stage, uplift and extension breaks apart the crust, and a crustal block drops down to form a fault graben called a **rift valley.** This stage is occurring today in the East African Rift Valley (Fig. 19.13). Continued spreading results in the development of a progressively wider rift graben, with numerous downdropped normal fault blocks along both flanks of the rift (see Fig. 19.12A). These fault grabens form basins for the sediment that erodes from the upthrown areas. Most of this sediment is coarse, immature alluvial debris and lesser fluvial deposits. In the center of the basin, small lakes may form lacustrine shales or limestones and even evaporites. Frequently, the faults become vents for alkalic volcanics and produce ashfalls and ash flows, as well as abundant volcaniclastic sediment. The famous hominid-bearing beds in East Africa (discussed in Box 18.2) are a good example of this stage of rift development. A classic ancient example of a rift graben sequence is the Triassic-Jurassic Newark Supergroup of the eastern margin of North America (Fig. 19.14). Fault grabens were formed by the Triassic opening of the Atlantic, and these basins were filled by a thick sequence of arkoses, lithic arenites, lacustrine shales, and volcanics such as the Palisades Sill and the Watchung lava flows.

Not all rift valleys become ocean basins. Hot spots arise at isolated points and produce a triradiate set of rift valleys, which is one class of triple junction (Fig. 19.15A). Two of the three arms of the triple junction link with rifts of nearby triple junctions to form the

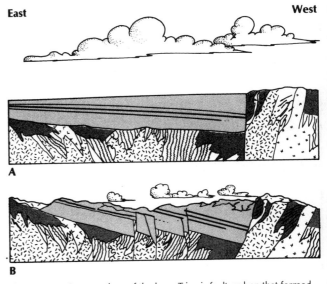

Figure 19.14 Cross sections of the large Triassic fault graben that formed during the early opening of the Atlantic. It is filled with the redbeds and lacustrine shales of the Newark Supergroup. (A) Early stage of filling of the basin. As the basin subsided, lavas welled up to form dikes and sills, and alluvial fans spread from the uplands to the east. (B) Eventual destruction of the basin by further normal faulting. (After Stanley, 1989: 459.)

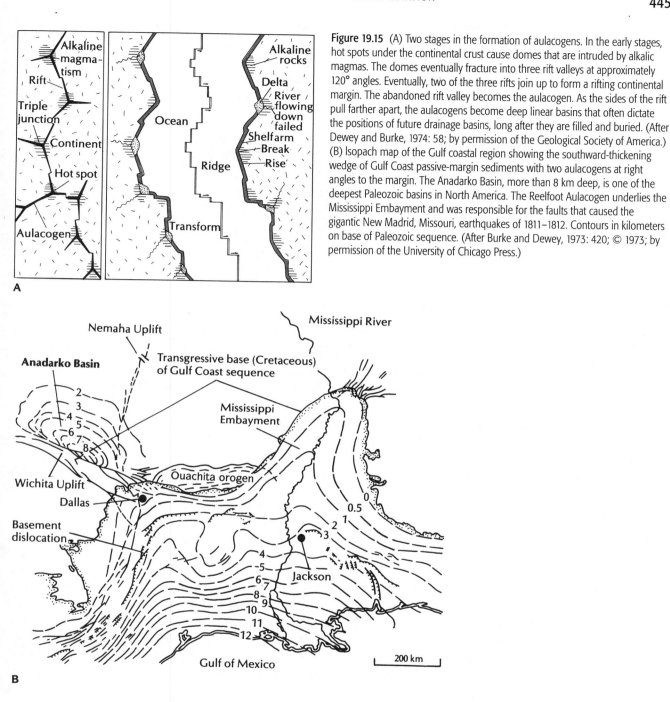

Figure 19.15 (A) Two stages in the formation of aulacogens. In the early stages, hot spots under the continental crust cause domes that are intruded by alkalic magmas. The domes eventually fracture into three rift valleys at approximately 120° angles. Eventually, two of the three rifts join up to form a rifting continental margin. The abandoned rift valley becomes the aulacogen. As the sides of the rift pull farther apart, the aulacogens become deep linear basins that often dictate the positions of future drainage basins, long after they are filled and buried. (After Dewey and Burke, 1974: 58; by permission of the Geological Society of America.) (B) Isopach map of the Gulf coastal region showing the southward-thickening wedge of Gulf Coast passive-margin sediments with two aulacogens at right angles to the margin. The Anadarko Basin, more than 8 km deep, is one of the deepest Paleozoic basins in North America. The Reelfoot Aulacogen underlies the Mississippi Embayment and was responsible for the faults that caused the gigantic New Madrid, Missouri, earthquakes of 1811–1812. Contours in kilometers on base of Paleozoic sequence. (After Burke and Dewey, 1973: 420; © 1973; by permission of the University of Chicago Press.)

eventual continental margin. The third arm opens for a while but stops spreading when the other two arms link. This failed third arm of a triple junction, called an **aulacogen,** eventually fills with sediments. Aulacogens form at high angles to the eventual continental margin and are zones of subsidence. River systems become entrenched in the abandoned trough and eventually fill it with fluvial and deltaic deposits. Nearly all of the larger rivers that drain into the modern Atlantic (including the Mississippi, the Amazon, the Congo, and the Niger), and several lesser rivers, have developed in aulacogens formed during Triassic and Jurassic rifting. After deltaic and coastal-plain sequences bury the aulacogens, a deep linear basin remains that can be detected in the subsurface. For example, the Anadarko Basin and the Mississippi Embayment are believed to be ancient aulacogens now buried under the coastal plain (Fig. 19.15B). Many other deep, narrow continental rifts, such as the Precambrian Keweenawan basins beneath Lake Superior and Wisconsin and Minnesota, are also believed to be aulacogens.

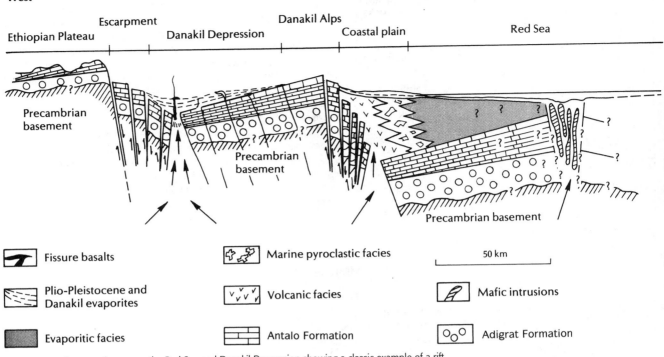

Figure 19.16 Cross section across the Red Sea and Danakil Depression showing a classic example of a rift valley filling with a proto-oceanic gulf sequence. (After Hutchinson and Engels, 1970: 328; by permission of the Royal Society.)

The third stage might be called the **proto-oceanic gulf stage** (see Fig. 19.12B). The two successful arms of the triple junction continue to spread and eventually are invaded by marine waters. They have not yet spread far enough, however, for oceanic crust to appear at the center. The fault grabens of the rift valley margin are buried under a sequence of shallow marine shales, limestones, and pelagic oozes. Many of these basins happen to be located in tropical areas and so produce sabkha dolostones and evaporites. Because the rift is still narrow at this stage, the marine waters are probably restricted in their circulation, which may result in thick sequences of marine evaporites. The best modern analog for this stage is the Red Sea (Fig. 19.16), which is 2000 km long, 200 to 350 km wide, and only 1 to 2 km deep. Here, the graben sequences are capped by alkalic volcanics, shallow marine shales and limestones, supratidal dolostones and sabkha deposits, and thick evaporite sequences. In the center of the Red Sea, marine oozes accumulate. The rifting of the North Atlantic reached this stage in the Jurassic, and the Gulf of Mexico became a restricted evaporitic basin that produced a thick unit known as the Louann Salt, which is responsible for most of the salt domes and much of the oil wealth of the Gulf Coast.

The fourth stage of development is **seafloor spreading** (see Fig. 19.12C–E). Continental separation reaches the point where oceanic crust forms at the spreading ridge (see Fig. 19.15A). As the spreading continues, the continental margins separate and sink down the ridge flanks. Pelagic oozes and shales accumulate over most of the seafloor, and the turbidite prism of the continental slope and rise forms near the continental margin. Above the shelf-slope break, the proto-oceanic gulf sediments become the basement for a thick sequence of shelf sediments and deltaic deposits, building up the continental terrace and eventually the continental embankment (see Fig. 19.12C–E). This sequence of clean marine sandstones and limestones progrades outward from the continent and is sometimes bounded by a reef at the shelf-slope break (Fig. 19.17A). The modern passive margin sequence of the North Atlantic shows many variations on this geometry (see Fig. 19.24A, B), but the basic pattern is the same. The total sequence is almost 10 km thick, with a basement of Triassic-Jurassic grabens, proto-oceanic Jurassic evaporites, and Lower Cretaceous shallow marine sediments. The rest of the shelf built outward during the Cretaceous and Tertiary, providing the bulk of the sequence that now underlies our present At-

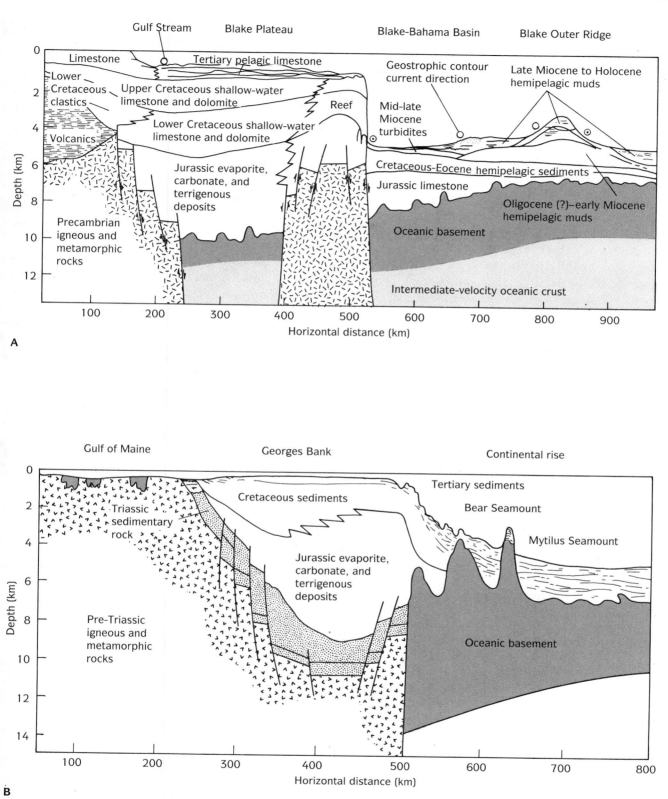

Figure 19.17 (A) Cross section of the Florida Platform, Blake Plateau, and Blake-Bahama Basin, showing a classic example of a thick passive-margin continental shelf-slope-rise sequence. Notice the buried rift grabens, the Jurassic proto-oceanic gulf evaporites, and the Cretaceous shelf-edge reefs and shallow shelf limestones. (B) Cross section of the Atlantic margin through the Gulf of Maine. Similar Triassic grabens, Jurassic evaporites, and Cretaceous shelf-slope-rise sediments can be seen. (After Sheridan, 1974: 400.)

lantic continental shelf and coastal plain. This is the climax of the evolution of passive margins, and this geometry continues to develop unless the plate motions change so that the passive margin becomes a convergent margin or becomes involved in a continental collision.

Convergent Margins

When two plates converge, several types of mountain belts can be produced. If both plates contain continental crust, neither can be completely subducted, so they become uplifted and deformed and eventually suture along the line of collision, like the modern Himalaya. The suture belt itself is an area of erosion; it sheds coarse clastic debris and fluvial deposits off its flanks into adjoining plains. The best known example of a suture belt is the molasse sequences of the Siwalik Hills of Pakistan and India, which are a product of Himalayan uplift since the Miocene.

If two oceanic plates converge, either can be subducted beneath the other, forming the classic **island arc complex** (Fig. 19.18) on the overriding plate. These intraoceanic arcs (Fig. 19.18A) are well known along the Pacific Rim today and include many familiar modern analogs, such as Japan, Indonesia, and the Philippines. Because of the predominance of oceanic crust in these islands, the volcanics are predominantly basalts. When the downgoing slab is melted in the subduction zone, it forms a small island arc by volcanic eruption on the overriding plate. The products of submarine volcanism (pillow lavas) and submarine sedimentation (turbidites, graywackes, shales, pelagic oozes) are predominant because this arc setting is mostly under water. Some intraoceanic arcs exhibit **backarc spreading,** producing an interarc basin behind the active arc (Fig. 19.18A). As backarc spreading continues, a **remnant arc** bordered by both active and inactive interarc basins may be left behind.

When oceanic crust subducts beneath continental crust, a continental-margin arc forms (Fig. 19.18B). The best modern example of a continental-margin arc is the Andes, so the arc complex is sometimes called an Andean-type arc (in contrast to the intraoceanic, or Japan-type, arc).

Because the rising magma in a continental-margin arc complex must penetrate a thick layer of continental crust, it becomes differentiated in the classic calc-alkaline trend of andesite-dacite-rhyolite. The

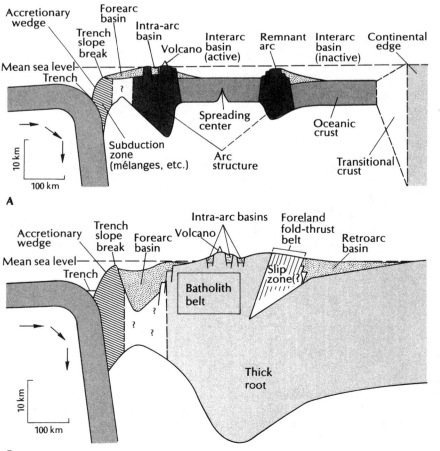

Figure 19.18 Schematic diagrams of sedimentary basins associated with arcs. (A) Intraoceanic (Japan-type) arc with backarc spreading. (B) Marginal (Andean-type) arc. (After Dickinson and Yarborough, 1976: 31; by permission of the American Association of Petroleum Geologists, Tulsa, Okla.)

A

batholith underneath the volcanic arc is also felsic and follows the calc-alkaline trend, with quartz monzonites and granites predominant. Most of the volcanics are silicic, so they tend to erupt explosively and to form ignimbrites and ash flows. Because the region is typically epicontinental and subaerial, the sediments filling the basins are mostly immature alluvial and fluvial sandstones and shales. The continental-margin arc may develop a foreland fold-thrust belt and retroarc basin, which are discussed shortly.

The two types of arcs have certain features in common. Where two plates meet, a **trench** is formed. Some of the material lying on the subducted plate goes down the "hole" without leaving a trace. Most of the pelagic sediments are scraped off the subducted plate onto the overriding plate, and these sediments accumulate along the arc to form an **accretionary wedge** just below the trench-slope break (see Fig. 19.18). The accretionary wedge is built up by continual underplating of material scraped off the downgoing slab, so new material is added to the *bottom* of the

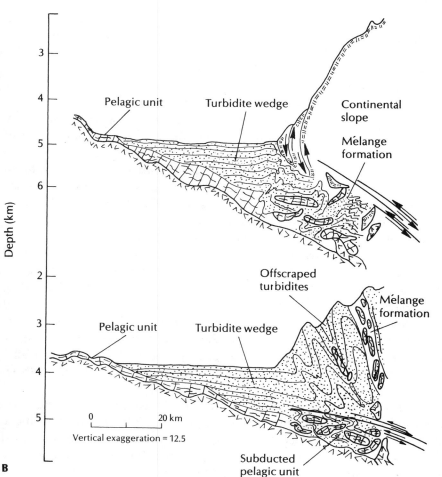

Figure 19.19 (A) Partially rounded exotic blocks in Franciscan mélange, near San Simeon, Coast Ranges of California. Largest blocks are 1 m across. (D. R. Prothero.) (B) Mode of formation of mélange as scrapings from a subducting plate (terrigenous and pelagic sedimentary blocks and oceanic ophiolites) accumulate and are intensely folded, sheared, and metamorphosed. (Scholl and Marlow, 1974: 200.)

pile constantly. The accreted slices are mostly oceanic sediment and pieces of oceanic crust (pillow lavas, sheeted dikes, and layered gabbros) known as **ophiolites,** which were formed by eruption at the mid-ocean ridge.

The most characteristic rock type of the accretionary wedge is **mélange** (French, "mixture"), a mass of chaotically mixed, brecciated blocks in a highly sheared matrix (Fig. 19.19A). This deformation and pervasive shearing and brecciation are due to the tremendous compressional and shear forces generated by the downgoing slab (Fig. 19.19B). Mélange is so mixed up that it shows no stratigraphic continuity or sequence, and blocks and boulders from many sources are mixed together. Some are exotic blocks from terranes no longer present in the vicinity. Mélanges can look something like submarine landslide deposits (or olisthostromes, discussed in Chapter 10). Unlike olisthostromes, however, mélanges are not associated with undeformed deep marine turbidites and shales but with tectonic deformation. A mélange terrane is regional in scale, with various degrees of deformation in various parts of the wedge. Mélanges are usually much more sheared and fractured than olisthostromes, and the included blocks are themselves fractured. In some cases, the sheared matrix of a mélange has actually undergone ductile flow. A summary of the major differences between mélanges and olisthostromes is shown in Table 19.1. The distinction between them is important because their presence implies completely different tectonic and sedimentary settings.

The metamorphism of the accretionary prism is also unique. Because the downgoing slab is cold relative to the mantle, metamorphism takes place at high pressure but at relatively low temperature. This is called blueschist metamorphism, because the predominant mineral is the blue amphibole glaucophane. The presence of glaucophane blueschist is diagnostic of a subduction zone.

On top of the slabs of ophiolite, mélange, blueschists, and deformed oceanic shales and cherts of the accretionary wedge is a thin mantle of pelagic sedimentary cover. In some places, there are even basins on top of the accreted material (Fig. 19.20). The sediment trapped in these basins has a good chance of becoming entrained during the deformation process and eventually becoming deformed and metamorphosed.

Above the trench-slope break is a basin in the arc-trench gap, the **forearc basin** (see Fig. 19.18). The forearc basin can be shallow marine or subaerial; most of its sediments are volcanic debris shed from the arc or from the accretionary wedge. Many forearc basins are

TABLE 19.1 The Differences Between Mélange and Olistostrome

Criterion	Mélange	Olistostrome
Clast character	Angular, fractured, sheared; may be deformed into boudins or smeared phacoids	Angular to rounded, may be fractured but not sheared
Clast source	Overlying and/or underlying unit	Overlying unit only
Matrix	Sheared, plastic intrusion into sheared	Not necessarily sheared
Contacts with other units	Sheared	May have sedimentary contacts with "normal" slope or trench sediments

From Hsü (1974).

filled with coarse shallow marine, deltaic, and fluvial deposits. In other instances, the forearc basin is very deep and accumulates turbidites. The classic ancient example is the Great Valley Group, which underlies the Central Valley of California. It is a basin almost 70 km wide and 12 km deep, filled with volcaniclastic submarine fan turbidites. On the margin, shallower marine deposits occur.

The **arc** itself is generally made of stratified volcanics and volcaniclastic debris. Many arcs have small graben basins at their crests, forming **intra-arc basins** (see Fig. 19.18). The best known of these basins are found in the Altiplano of the Andes of Peru and Bolivia. An example of one such basin is that filled by Lake Titicaca, the world's highest lake. Intra-arc basins are filled primarily with coarse, immature alluvial volcaniclastic detritus and occasionally with lacustrine clays.

The two backarc settings (see Fig. 19.18) produce different sedimentary products. **Interarc basins** are made of oceanic or thinned, "oceanized" continental crust and are completely marine. Their main sediments are volcaniclastic turbidites from the arc, montmorillonitic

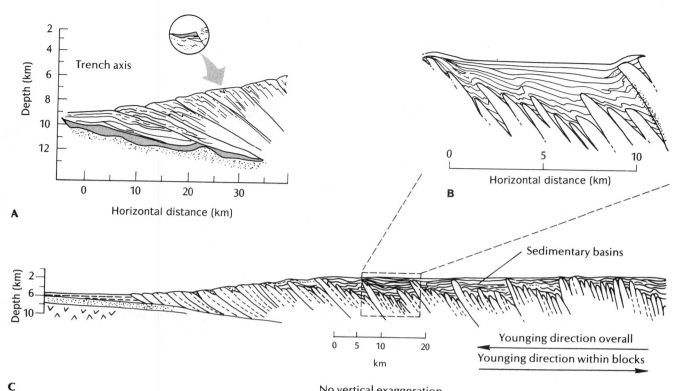

Figure 19.20 (A) Geometry of an accretionary wedge showing the thick slabs of crustal material that are sheared off and stacked up from the bottom. (B) Enlargement of the circled inset in A showing the shingling of the accreted slabs, which get younger toward the bottom of the pile.

(C) Enlargement of the accretionary wedge basins showing the fill of pelagic sediment when they become tectonically deformed with further accretion. (After Karig and Sharman, 1975; Moore and Karig, 1976.)

clays from weathered volcanics, biogenic ooze, and small amounts of eolian dust from the continent. The continent sheds debris into the proximal part of the basin. **Retroarc basins** are created by the compression of the intracontinental arc, which forms the **foreland fold-thrust belt** (see Fig. 19.18B). This thrusting and basement uplift typically exposes metamorphic infrastructure, which consists of high-grade metamorphic terranes from deep in the continental basement. The sediments that fill the retroarc basin are shed not only from the arc and foreland fold-thrust belt but also from epicontinental seaways. There can be a complete mixture of shallow marine, deltaic, and fluvial sandstones and shales, which are relatively mature except near a faulted upland source.

Transform Margins

In transform margins, the dominant motion is neither compressional nor extensional but strike-slip. Theoretically, there should be simple translation of crust with no opportunity for basins to form. In the real world, however, transform margins are seldom straight; instead they curve and are composed of nu-

merous parallel shear zones. This is well demonstrated by the San Andreas fault zone in California. Because of irregularities in the sliding plates, there is local compression and extension, forming deep, narrow, fault-bounded troughs. If they are caused by an extensional gap, they are called **transtensional,** or "pull-apart," **basins.** A transtensional basin has steep, fault-bounded walls and can drop precipitously to great depths. A well-known modern example is the Salton Trough along the southern San Andreas fault. If the transform basins are caused by a compressional downwarping, they are known as **transpressional basins;** these can also be very deep and narrow and accumulate thick piles of sediment. For example, the Mio-Pliocene Ridge Basin of California (discussed in Box 19.1) accumulated more than 13,500 m of breccias, alluvial fan deposits, lacustrine silts, marine turbidites, and evaporites. Other such basins are starved of clastics and accumulate deep pelagic silts and oozes, such as the Miocene Monterey Formation, a cherty shale in California. Such a thick, narrow, and immature sedimentary accumulation can occur only in basins on transform margins.

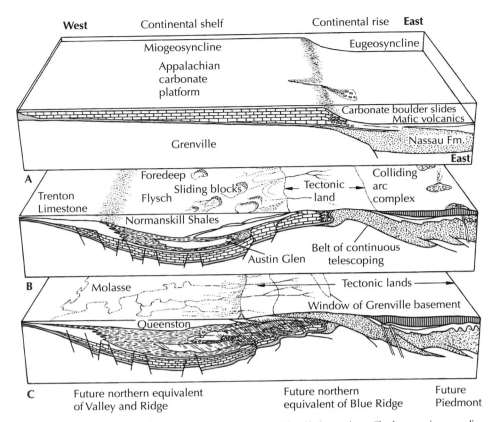

Figure 19.21 The Cambro-Ordovician "geosyncline" in western New England. (A) During the Cambrian, passive margin conditions persisted, with a thick shelf-platform sequence (miogeosyncline) and slope-rise package (eugeosyncline.) (B) In the Middle Ordovician, the passive margin changed to an active margin when an island arc complex from the proto-Atlantic began to collide with the continent. The former miogeosyncline was downbuckled into a eugeosynclinal trough, trapping flysch and huge submarine slides shed from the upwarped "tectonic land" to the west. (C) In the Late Ordovician, the collision stopped and the basins filled with nonmarine molasse from the remnant "tectonic land." (After Dewey and Bird, 1970: 1043.)

Geosynclines and Plate Tectonics

The elements of the classic geosyncline can now be reinterpreted in the light of plate tectonics. The classic eugeosyncline-miogeosyncline association is the normal product of passive-margin sinking. The miogeosynclinal sequence is produced by continental shelf sedimentation, and the eugeosyncline accumulates on the continental slope and rise (Fig. 19.21A; also see Fig. 9.12).

To complicate matters, some elements of the eugeosyncline are a result of active-margin tectonics. The abundant volcanics result from the approach of an island arc during the closing of an ocean (Fig. 19.21B, C). The debris shed from an upland on the seaward side of the trough was once thought to be derived from a mysterious "tectonic land," but this upland was the arc or continent that finally collided during the last phases of closing. For example, much of the debris that filled the Appalachian eugeosyncline during the Ordovician Taconic orogeny and the Devonian Acadian orogeny was shed from Europe or Africa,

which are now no longer in the vicinity. The characteristic compressive deformation and overthrusting of the eugeosyncline onto the miogeosyncline are also caused by the continental collision in the last phases of ocean closure. Sometimes this collisional suture zone includes elements of the accretionary prism, such as ophiolites and mélanges.

Thus, the complex association of geologic features that puzzled the geosyncline theorists for so long is not attributable to a single cause but to a combination of features formed by active-margin tectonics superposed on rocks formed by passive-margin sedimentation.

Tectonics and Sandstone Petrology

We have seen how plate-tectonic models can be used to predict what types of sediments will be found in various types of basins. The final step is to reverse the process and infer ancient plate configurations from the types of sediments found. The overall geometry and lithology of a stratigraphic sequence

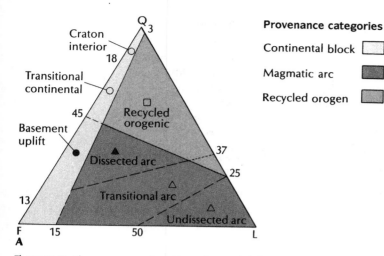

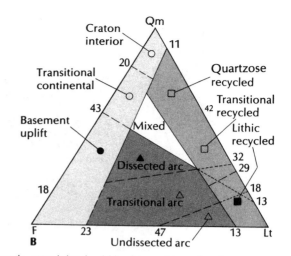

Figure 19.22 The provenance of sandstones is strongly influenced by the type of tectonic sources that were present. On (A) a QFL (quartz, feldspar, lithics) plot and (B) a QmFLt (monocrystalline quartz, feldspar, lithics including polycrystalline quartz) plot, pure quartz sandstones typically come from the cratonic interior, feldspathic sandstones from basement uplifts, and lithic sandstones from eroded arcs. Finer subdivisions are also shown. (After Dickinson et al., 1983: 223; by permission of the Geological Society of America.)

are the most diagnostic features. For example, a complex of coarse fanglomerates and turbidites, thousands of meters thick, could form only in a strike-slip pull-apart basin. Rock associations such as mélanges and ophiolites are also diagnostic of their tectonic settings. In a few instances, a single mineral, such as glaucophane, is diagnostic of a particular tectonic regime—namely, blueschist metamorphism and a subduction zone. For the vast majority of sedimentary rocks, however, the clues are more subtle.

In a series of classic papers, Dickinson and colleagues (Dickinson and Suczek, 1979; Dickinson and Valloni, 1980; Dickinson, 1982; Dickinson et al., 1983) used triangle diagrams of quartz, feldspar, and lithic fragments (**QFL plots**) to interpret the tectonic provenance of sandstones (Fig. 19.22). A slightly different form of the classic QFL plot is the **QmFLt plot,** which has poles at monocrystalline quartz (Qm), feldspar, and all polycrystalline lithic fragments (Lt), including polycrystalline quartz. In these plots, samples from known tectonic regimes were plotted with respect to the three end-member components. A clear pattern of fields representing distinct tectonic terranes shows up on the plots (Fig. 19.23). Cratonic-interior sandstones are usually very mature, pure monocrystalline quartz, so they plot very close to the Q or Qm pole. Basement uplifts typically produce considerable feldspar from unroofed plutons, resulting in arkosic sandstones that cluster along the Q–F join. Undissected-arc terranes generate much lithic debris, so they cluster near the L or Lt pole. As arc terranes are progressively dissected, they move toward the center of the plot. Orogenic debris that has been recycled plots along the Q–L join.

Although tectonic sources influence sandstone composition, they are not the only influence. Potter (1978) did a statistical study of the sediments issuing from major rivers around the world. Most sands plotted correctly according to the schemes just outlined, but the Ganges, Indus, and Brahmaputra sands (which are derived from the continent-continent collision that formed the Himalaya) were classified as a "trailing coast" rather than a "collisional coast." Suttner, Basu, and Mack (1981) have shown that climate can also influence sandstone composition (Fig. 19.24). For example, metamorphic source terranes in humid climates produce a weathered quartz-rich sandstone, whereas plutonic igneous source terranes in humid climates are more feldspar-rich. Arid climates preserve the lithic fragments produced from metamorphic terranes and both the feldspar and the lithics produced from plutonic terranes. Thus, caution must be used when interpreting sandstones for tectonic influences. Ideally, the petrologic data should be complemented by data from other rock types and total field relations, so the effects of tectonic source and climate can be separated.

Another complicating factor is diagenesis. Some sandstones have undergone such extensive diagenetic alteration that the proportions of the components in the triangle diagrams have been completely changed; this situation leads to an erroneous tectonic interpretation. For example, Milliken, McBride, and Land (1989) studied the Oligocene Frio Formation from the Texas Gulf coastal plain and found that it

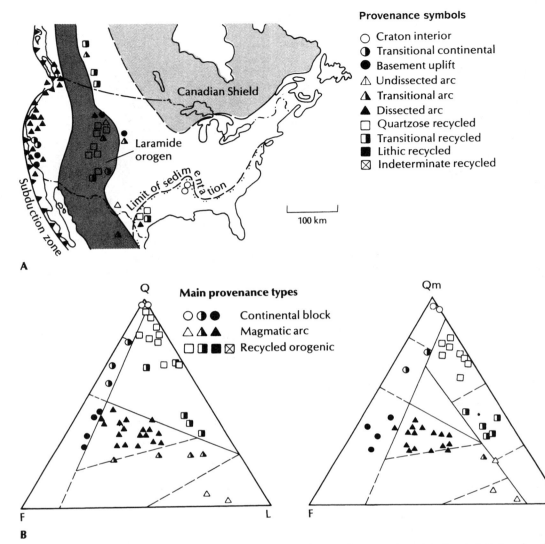

Figure 19.23 (A) Paleotectonic map showing the tectonic source of selected Cretaceous-Paleogene sandstones (indicated by circles, triangles, and squares.) (B) QFL and QmFLt plots for these sandstones showing how the tectonic sources cluster into discrete fields based on their composition. (After Dickinson et al., 1983: 232; by permission of the Geological Society of America.)

had undergone extensive diagenetic alteration and loss of feldspars. The altered remnants of the feldspar grains could be seen, however, so the original feldspar content could be determined. By this method, Milliken, McBride, and Land (1989) estimated that 50% to 80% of the feldspars had been lost. Such a drastic change in composition moves the Frio Formation from the "magmatic arc" field of the triangle to the "recycled orogen" portion. Actually, the Frio Formation is found on a passive margin, so it fits neither of these tectonic settings. It is unusual among passive-margin sandstones in having a high content of volcaniclastic debris, an artifact of the extraordinary Oligocene volcanic eruptions that occurred from Nevada to Colorado to New Mexico to Trans-Pecos Texas.

Secular Changes in the Sedimentary Record

Uniformitarianism has long been the most fundamental concept of geology. For many geologists, uniformitarianism makes two basic assumptions. First, natural laws (and consequently geologic processes) have operated in a *consistent manner* throughout time (this uniformity of processes is known as **actualism**). Second, the geologic processes have *always* operated with *essentially the same intensity* (this uniformity of rates is known as **gradualism**). The second assumption, so strongly entrenched, is simply not true. We believe actualism, not gradualism, to be a more accurate characterization of the Earth because, as we have

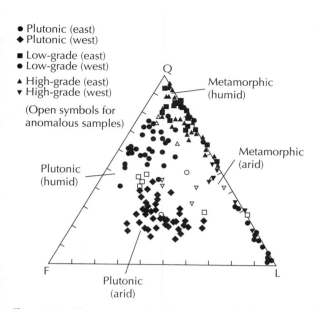

- Plutonic (east)
- Plutonic (west)
- Low-grade (east)
- Low-grade (west)
- High-grade (east)
- High-grade (west)

(Open symbols for anomalous samples)

Figure 19.24 Climate can strongly influence the composition of sandstones. Here first-cycle sand from a modern river with known sources is plotted on a QFL diagram. Metamorphic and plutonic sources clustered very differently on the plot, depending on whether they came from a humid region with active chemical weathering or an arid region with limited chemical weathering and breakdown of feldspars and lithic fragments. (After Suttner, Basu, and Mack, 1981: 1240.)

emphasized throughout this text, from the perspective of the terrestrial sedimentary rock record, *the present is often a very imperfect key to the past, and vice versa.*

Actualism accepts the uniformitarian premise that natural laws and the manner in which geologic processes operate remain constant, but it realistically acknowledges that Earth has evolved continuously. Not only has the intensity of geologic processes varied greatly over time, but even the very nature of those processes has changed as our world has evolved. For example, the history of the Moon clearly implies that a process no longer important on Earth (the impacting of extraterrestrial bodies) must have been a recurrent and common phenomenon during the first 500 million years of Earth's history.

Actualism has important implications for sedimentary geology. Applying it reinforces our conviction that many modern sedimentary environments may not be very good analogs for earlier stages of geologic history. The Earth is now in the middle of a brief interglacial age just 20,000 years after the peak of the last glaciation. Under today's somewhat unusual conditions, modern continental shelves are probably poor analogs for the great epicontinental seas that prevailed during much of the Phanerozoic. Sediment textures and primary depositional structures found on the surface of existing continental shelves are largely inher-

ited from lower sea-level conditions (which exposed the continental shelf to subaerial processes) that prevailed during the preceding glacial age. Consequently, although the modern Bahama Banks and Persian Gulf duplicate many conditions in the epeiric seas (water depth and temperature, salinity, current frequency and velocity), the scale is not comparable. We can probably begin to understand limestone depositional environments from these modern analogs, but we should continually remind ourselves that they do not begin to approach the scale of the huge epicontinental carbonate banks of the Paleozoic.

The use of modern analogs is even more problematic when other aspects of sedimentary rocks are taken into account. Consider the following likely possibilities.

- Weathering, erosion, and sediment transport would operate quite differently in a world without land plants.

- The present icehouse state produces atmospheric and ocean water temperatures and chemistry, as well as sea levels, markedly different from conditions in a greenhouse state (see Chapter 11).

- Banded iron formations (as well as pyrite- and uraninite-bearing fluvial conglomerates) suggest an early terrestrial reducing atmosphere that contained little or no free oxygen; this is quite a different situation from the present Earth.

- An early terrrestrial atmosphere with less oxygen and more carbon dioxide probably produced early seas with a higher pH and fewer marine organisms capable of depositing limestone.

- Plate-tectonic mechanisms perhaps did not operate or may have operated differently, particulary if the main driving force behind seafloor spreading and continental drift is thermal energy released from the terrestrial interior. If most of Earth's internal heat is generated by radioactive decay, half-life decay would have generated much higher rates of heat flow early in terrestrial history. Some geologists argue that this means a thinner lithosphere and shallower, more rapidly moving convection cells in the Archean asthenosphere. This might have led to smaller, less stable protocontinental blocks that were less likely to be episodically covered with shallow seas.

- If the Moon was closer to the Earth during the Precambrian and even into the Paleozoic (as coral growth rings suggest), there would have been much larger tides than now occur.

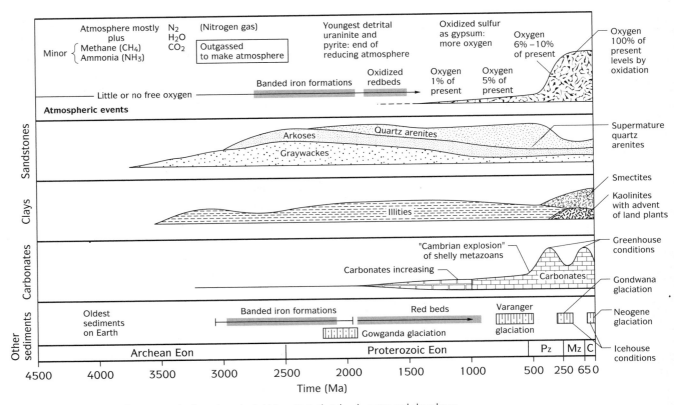

Figure 19.25 Trends in sedimentary rocks through geologic history. Note that the character and abundance of sandstones, limestones, and shales have changed greatly over the last 4.5 billion years, so the present is a very imperfect "key to the past."

In short, the past serves as a rather unreliable guide to the present, *and vice versa*. It is anything but a perfect key. Historical geologists must learn to concede that the Earth has evolved continuously. The secular changes in Earth's sedimentary rock record discussed in the next sections provide some of the clearest insights into the nature of these terrestrial changes (Fig. 19.25).

Sandstones Through Time

Secular changes in the kinds of sandstone we find clearly indicate an evolving Earth. Most Archean feldspathic arenite (arkose) deposits contain more plagioclase feldspar than potassium feldspar, the reverse of Phanerozoic sandstones. And a majority of lower and middle Archean sandstones are mud-rich lithic graywackes. They occur as thick turbidites that are invariably immature compositionally and texturally. Clasts of such ultramafic rocks as dunite and peridotite are much more common than granitic rock fragments. Detrital grains of olivine, pyroxene, and plagioclase feldspar are abundant; clasts of quartz and potassium feldspar

are scarce. All these clasts are mainly angular and poorly sorted and show little evidence of abrasion during reworking by strong shallow-water transport. These characteristics are inconsistent with the existence of broad granitic continental crustal blocks episodically covered with shallow seas. Although some terrigenous material is clearly derived from granite and high-pressure, high-temperature granulite, slabs of sialic continental crust were evidently thinner and restricted areally. Many of the eroding emergent areas were probably ultramafic and mafic magmatic arcs; some surface exposures of mantle, or at least very deep crust, were eroded. Most sandstones were rapidly transported short distances directly into deep-water ocean trenches and forearc and backarc basins (which may survive as greenstone belts). There was little reworking because no broad, shallow shelves existed; sedimentary recycling was uncommon. The absence of land plants combined with a corrosive, reducing atmosphere intensified weathering on the land and accelerated denudation.

Arkoses composed of potassium, rather than plagioclase feldspar, and quartz-rich siliciclastics increase to-

ward the end of the Archean Eon. This trend accelerates during the Proterozoic Eon. The sheer volume of lower Proterozoic pure quartz arenite (now quartzite) is puzzling. It implies extensive weathering and erosion of very large volumes of granite and rhyolite as well as wacke and mudrock. The cumulative volume of quartz in just a few of the classic lower Proterozoic quartzites of the Canadian Shield requires the erosion of at least 10,000 km^3 of granitic rock. This value must be multiplied many hundredfold if the total quartz of all early Proterozoic quartzites is to be explained.

During the Proterozoic, continental blocks apparently became more granitic and stable enough that more mature quartz sandstones begin to appear (perhaps recycled and cannibalized from older sandstones). Tectonism became modern in most aspects. Proterozoic mobile belts exhibit an internal structural and stratigraphic framework much like that of Phanerozoic orogens. This is consistent with a world in which seafloor spreading, continental drift, subduction, collision, and suturing occurred in the same fashion as they do now. This tectonism permitted shallow epicontinental seas to cover large, relatively stable continental blocks from time to time. The pattern of sandstone deposition better fits such modern tectonic settings as rift systems, continental shelves, and continental rises.

Lower Paleozoic sandstones, like lower Proterozoic quartzites, are largely supermature quartz arenite. They typically consist of 99% quartz grains that are invariably well-rounded and well-sorted, convincing evidence that weathering was intense and that transport occurred by agents that are efficient abraders (such as wind, surf, and strong longshore currents). Sand grains must have been reweathered and eroded repeatedly. Chemical decomposition under tropical conditions can, in a few cases, produce pure quartz arenite in a single cycle, but these deposits are angular and poorly sorted. To achieve the textural and compositional supermaturity of such deposits as the Cambrian Tapeats Sandstone in the Grand Canyon and the Ordovician St. Peter Sandstone of the upper Midwest, multiple recycling and eolian reworking are probably necessary (Dott and Prothero, 1994). The lack of land plant cover until the Silurian or Devonian also must have helped to produce these supermature quartz arenites, which virtually disappear after the early Paleozoic.

A final, surprising Phanerozoic secular trend in sandstones is obviously the result of diagenesis. Modern turbidite sands, if lithified, would be arenite, rather than graywacke. Yet almost all of the Phanerozoic (and Precambrian) graded, sole-marked turbidite sands are composed of graywacke. Metamorphism, disaggregation of rock fragments, and degradation of feldspar all helped to produce this change.

Carbonates Through Time

The present is also not a very reliable key to the past from the perspective of carbonates (Tucker, 1992). The mineralogy of limestone clearly suggests subtle changes in the chemistry of seawater during the Phanerozoic and more radical changes prior to that. Many of these changes are probably related to alternating icehouse and greenhouse states, which control ocean water level, temperature, and chemistry (Fig. 19.26). During most of the Phanerozoic, when the Earth experienced warm, carbon-dioxide-rich greenhouse conditions, sea level was high and the continents were flooded with shallow seas in which limestone accumulated easily. During the icehouse

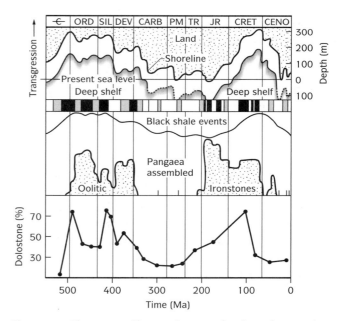

Figure 19.26 The two great Phanerozoic supercycles of greenhouse and icehouse conditions seem to explain many secular changes in sedimentary rocks (compare with Fig. 15.21). During greenhouse conditions, when the planet was warmer and atmospheric carbon dioxide and sea levels were higher (drowning the continents), certain sedimentary rocks types are typical. Black shales and oolitic ironstones are restricted to the early Paleozoic and Jurassic-Eocene greenhouse conditions. Limestone and dolostone abundances and the predominance of calcitic, rather than aragonitic, ooids are also characteristic. During icehouse conditions (Carboniferous-Triassic, Oligocene to present), sea levels were lower and conditions were colder, producing glacial deposits on the poles and evidence of climatic extremes (abundant evaporites, dune sands, oxidized terrestrial redbeds instead of epicontinental marine limestones). (After Van Houten and Arthur, 1989: 48.)

conditions (late Paleozoic–Triassic and Oligocene–present), limestone deposition is confined largely to the open ocean basins where it accumulates as calcareous planktonic ooze. (There are significant exceptions, such as the Bahama Banks.) During the Phanerozoic, at least, the rate at which limestone forms has remained relatively constant, but the site has shifted back and forth between shallow and deep marine settings.

Precambrian limestones are much less abundant. The absence of broad, stable continental blocks covered with shallow seas partly explains their comparative scarcity. However, much of the lack of Precambrian carbonates has to do with organic evolution. Stromatolites indicate that carbonate-secreting cyanobacteria evolved as early as 3.5 billion years ago, but they did not flourish until Proterozoic time, held back perhaps by seawater chemistry and/or the absence of shallow marine platforms. Or perhaps Precambrian organisms just weren't very efficient carbonate producers. Thick and abundant limestone production appears to be most closely tied to the important Cambrian "explosion," when hard-shelled animals suddenly developed the ability to secrete calcium carbonate efficiently. (An important side effect of organic evolution is the role plants and animals play in regulating carbon dioxide in the atmosphere and in ocean water.)

Atmospheric composition might also control dolomite formation. Low carbon dioxide content promotes calcite precipitation; higher levels of carbon dioxide favor precipitation of both calcium and magnesium, which boosts dolomite production. Dolostones do seem to coincide with greenhouse conditions (see Fig. 19.26). The generally uneven distribution of dolostones through time probably reflects a variety of factors: geochemistry, tectonism, and the age of the carbonate deposit. The older a limestone, the more likely it is that it will be dolomitized sooner or later.

As we discussed in Chapter 12, the nature of carbonate deposition has also changed through time. During the icehouse worlds of the late Cenozoic and late Paleozoic, ocean chemistry favored precipitation of lime mud and ooids as aragonite. During the greenhouse worlds of the early to mid-Paleozoic and later Mesozoic, however, calcite was secreted as lime mud and made up most carbonate particles, such as ooids. This alternation between "calcite seas" and "aragonite seas" seems to be controlled largely by the Mg-Ca ratio in the world's oceans, which in turn is largely a result of changes in spreading rate and production of

Mg in the world's spreading ridges (see Chapter 12 for further details).

The kinds of allochems found in carbonates vary with age. The predominant skeletal fragments change because of organic evolution. For example, crinoidal limestones are common in Paleozoic (especially Mississippian) rocks but are rare after crinoids were decimated by the Permo-Triassic extinctions. Likewise, limestones made of brachiopods, trilobites, or bryozoans are also typical of the Paleozoic but not of younger rocks, since those groups were also decimated at the end of the Permian. Ammonitic limestones are found only in the late Paleozoic and Mesozoic. Chalks made of calcareous nannoplankton and foraminifers do not become abundant until the Cretaceous, when both groups emerged. Modern limestones are made mostly of the fragments of bivalves and gastropods, calcareous algae, and occasional echinoids, because these have been the predominant groups of carbonate-secreting benthic organisms in the Cenozoic.

But even nonskeletal grains such as ooids and peloids have an uneven distribution. Episodes of rising and falling sea level appear to promote ooid formation, for unknown reasons. The mineralogy of ooids has also changed. Ooids precipitated during high-carbon-dioxide (greenhouse) conditions are calcitic; aragonite is found in ooids forming presently under existing icehouse conditions.

Mudrocks Through Time

The abundance of shales has been fairly constant, which probably reflects the constancy of hydrolysis in converting feldspars to clays. Yet the character of mudrocks evolves. Archean and lower Proterozoic mudrocks contain mostly reduced iron and other minerals; oxidized minerals occur only in rocks from the upper Proterozoic. Organic carbon and carbonate become more abundant in upper Proterozoic and Phanerozoic mudrocks. This is not surprising, because life became more diverse and carbon was more likely to be trapped in sediment. Efforts to identify more detailed secular variations in both major and trace elements are beginning to show even more significant and promising results.

Most Precambrian and Paleozoic shales are composed largely of illitic clays, but as we saw in Chapters 6 and 7, this is a result of diagenesis. Illite is the most stable clay during burial and diagenetic alteration. Expandable smectitic clays such as montmorillonite are rare in rocks older than late Cenozoic, be-

cause diagenesis almost always drives out water, changing the structure and chemistry of the clay. Kaolinite is also scarce in rocks older than the mid-Paleozoic, probably because no land plants existed before that time to produce the intense leaching necessary for their formation. Finally, black shales seem peculiar to greenhouse conditions and are particularly common at the beginning of major transgressions in areas that are sediment-starved, producing deep, mostly anoxic conditions (see Fig. 19.26).

Iron-Rich Rocks and Evaporites Through Time

Iron-rich sedimentary rocks provide the best means of tracing atmospheric evolution. Debate over the details continues (Holland, 1984; Gregor et al., 1988), but most scientists agree that the abundance of banded iron formations before about 1.8 billion years ago must mean relatively low atmospheric oxygen levels. After banded iron formations disappear forever, iron in sediment accumulates as oxidized redbeds and episodic, oolitic ironstones. Ironstones apparently coincide with tropical lateritic weathering and low rates of supply of other sediment, and they are restricted to greenhouse conditions (see Fig. 19.26). Clearly, the level of atmospheric oxygen rose, but argument continues over just how rapidly.

The first abundant evaporites (especially gypsum) appear shortly after 1.2 billion years ago. Of course, evaporite formation could not occur until the advent of broad, stable landmasses that could be episodically covered by thin films of evaporating seawater. But the composition of the atmosphere probably also controlled evaporite formation. Geochemical calculations show that sulfur under reducing conditions precipitates as pyrite. Sulfate minerals such as gypsum cannot form until the atmosphere is fairly oxidizing. Gypsum, like iron-rich sedimentary rocks, is quite sensitive to an evolving atmosphere.

CONCLUSIONS

The distribution of sedimentary rocks in space and time is not random. One of the long-standing puzzles of geology is the problem of why sediments accumulate in basins in some places and times but not in others. Through plate tectonics, we now understand why many sedimentary basins form. Similarly, the characteristics of the sedimentary record have changed through time. These secular changes in the composition and abundance of sandstones, limestones, mudrocks, iron-rich rocks, and evaporites tell us much about how Earth's atmosphere and oceans have evolved.

FOR FURTHER READING

Allen, P. A., and J. R. Allen. 1990. *Basin Analysis: Principles and Applications*. Oxford: Blackwell Scientific Publications.

Brenner, R. L., and J. A. McHargue. 1988. *Integrative Stratigraphy*. Englewood Cliffs, N.J.: Prentice-Hall.

Busby, C. J., and R. V. Ingersoll, eds. 1995. *Tectonics of Sedimentary Basins*. Cambridge, Mass.: Blackwell Science.

Condie, K. C. 1989. *Plate Tectonics and Crustal Evolution*. Oxford: Pergamon Press.

Conybeare, C. E. B. 1979. *Lithostratigraphic Analysis of Sedimentary Basins*. New York: Academic Press.

Crowell, J. C., and M. H. Link, eds. 1982. *Geologic History of the Ridge Basin, Southern California*. SEPM Pacific Section, Field Trip Guidebook 22.

Dickinson, W. R., ed. 1974. *Tectonics and Sedimentation*. SEPM Special Publication 22.

Dott, R. H., Jr., and R. H. Shaver, eds. 1974. *Modern and Ancient Geosynclinal Sedimentation*. SEPM Special Publication 19.

Gregor, C. B., R. M. Garrels, F. T. Mackenzie, and J. B. Maynard, eds. 1988. *Chemical Cycles in the Evolution of the Earth*. New York: John Wiley.

Howell, D. G. 1990. *Tectonics of Suspect Terrane*. New York; Chapman and Hall

Ingersoll, R. V. 1988. Tectonics of sedimentary basins. *Geological Society of America Bulletin.* 100:1704–1719.

Kleinspehn, K. L., and C. Paola, eds. 1988. *New Perspectives in Basin Analysis.* New York: Springer-Verlag.

Kottlowski, F. E. 1965. *Measuring Stratigraphic Sections.* New York: Holt, Rinehart, and Winston.

Krumbein, W. C., and L. L. Sloss. 1963. *Stratigraphy and Sedimentation.* San Francisco: W. H. Freeman, and Company.

Langstaff, C. S., and D. Morrill. 1981. *Geologic Cross Sections.* Boston: International Human Resources Development Corporation.

LeRoy, L. W., and J. W. Low. 1954. *Graphic Problems in Petroleum Geology.* New York: Harper and Brothers.

Levorsen, A. I. 1967. *Geology of Petroleum.* San Francisco: W. H. Freeman and Company.

Lowe, D. R. 1980. Archean sedimentation. *Annual Review of Earth and Planetary Sciences* 8:145–167.

Miall, A. D. 1990. *Principles of Sedimentary Basin Analysis,* 2d ed. New York: Springer-Verlag.

North, F. K. 1985. *Petroleum Geology.* Boston: Allen and Unwin.

Pettijohn, F. J., P. E. Potter, and R. Siever. 1987. *Sand and Sandstone,* 2d ed. New York: Springer-Verlag.

Potter, P. E., and F. J. Pettijohn. 1977. *Paleocurrents and Basin Analysis,* 2d ed. New York: Springer-Verlag.

Prothero, D. R., and R. M. Dott, Jr. 2003. *Evolution of the Earth,* 7th ed. New York: McGraw-Hill.

Ronov, A. B. 1964. Common tendencies in the chemical evolution of the Earth's crust, ocean, and atmosphere. *Geochemistry International* 1:713–737.

Taylor, S. R., and S. M. McClennan. 1985. *The Continental Crust: Its Composition and Evolution.* Oxford: Blackwell Scientific Publications.

APPENDIXES

Appendix A

North American Stratigraphic Code[1]
North American Commission on
Stratigraphic Nomenclature

FOREWORD

This code of recommended procedures for classifying and naming stratigraphic and related units has been prepared during a four-year period, by and for North American earth scientists, under the auspices of the North American Commission on Stratigraphic Nomenclature. It represents the thought and work of scores of persons, and thousands of hours of writing and editing. Opportunities to participate in and review the work have been provided throughout its development, as cited in the Preamble, to a degree unprecedented during preparation of earlier codes.

Publication of the International Stratigraphic Guide in 1976 made evident some insufficiencies of the American Stratigraphic Codes of 1961 and 1970. The Commission considered whether to discard our codes, patch them over, or rewrite them fully, and chose the last. We believe it desirable to sponsor a code of stratigraphic practice for use in North America, for we can adapt to new methods and points of view more rapidly than a worldwide body. A timely example was the recognized need to develop modes of establishing formal nonstratiform (igneous and highgrade metamorphic) rock units, an objective which is met in this Code, but not yet in the Guide.

The ways in which this Code differs from earlier American codes are evident from the Contents. Some categories have disappeared and others are new, but this Code has evolved from earlier codes and from the International Stratigraphic Guide. Some new units have not yet stood the test of long practice, and conceivably may not, but they are introduced toward meeting recognized and defined needs of the profession. Take this Code, use it, but do not condemn it because it contains something new or not of direct interest to you. Innovations that prove unacceptable to the profession will expire without damage to other concepts and procedures, just as did the geologic-climate units of the 1961 Code.

This Code is necessarily somewhat innovative because of: (1) the decision to write a new code, rather than to revise the old; (2) the open invitation to members of the geologic profession to offer suggestions and ideas, both in writing and orally; and (3) the progress in the earth sciences since completion of previous codes. This report strives to incorporate the strength and acceptance of established practice, with suggestions for meeting future needs perceived by our colleagues; its authors have attempted to bring together the good from the past, the lessons of the Guide, and carefully reasoned provisions for the immediate future.

Participants in preparation of this Code are listed in Appendix I, but many others helped with their suggestions and comments. Major contributions were made by the members, and especially the chairmen, of the named subcommittees and advisory groups under the guidance of the Code Committee, chaired by Steven S. Oriel, who also served as principal, but not sole, editor. Amidst the noteworthy contributions by many, those of James D. Aitken have been outstanding. The work was performed for and supported by the Commission, chaired by Malcolm P. Weiss from 1978 to 1982.

This Code is the product of a truly North American effort. Many former and current commissioners representing not only the ten organizational members of the North American Commission on Stratigraphic Nomenclature (Appendix II), but other institutions as well, generated the product. Endorsement by constituent organizations is anticipated, and scientific communication will be fostered if Canadian, United States, and Mexican scientists, editors, and administrators consult Code recommendations for guidance in scientific reports. The Commission will appreciate reports of formal adoption or endorsement of the Code, and asks that they be transmitted to the Chairman of the Commission (c/o American Association of Petroleum Geologists, Box 979, Tulsa, Oklahoma 74101, U.S.A.).

Any code necessarily represents but a stage in the evolution of scientific communication. Suggestions for future changes of, or additions to, the North American Stratigraphic Code are welcome. Suggested and adopted modifications will be announced to the profession, as in the past, by serial Notes and Reports published in the *Bulletin* of the American Association of Petroleum Geologists. Suggestions may be made to representatives of your association or agency who are current commissioners, or directly to the Commission itself. The Commission meets annually, during the national meetings of the Geological Society of America.

1982 NORTH AMERICAN COMMISSION
ON STRATIGRAPHIC NOMENCLATURE

[1]Reprinted by permission from American Association of Petroleum Geologists Bulletin, v. 67, no. 5 (May, 1983), pp. 841–875.
Copies are available at $1.00 per copy prepaid. Order from American Association of Petroleum Geologists. Box 979, Tulsa, Oklahoma 74101.

CONTENTS

PART I. PREAMBLE

BACKGROUND

PERSPECTIVE

Codes of Stratigraphic Nomenclature prepared by the American Commission on Stratigraphic Nomenclature (ACSN, 1961) and its predecessor (Committee on Stratigraphic Nomenclature, 1933) have been used widely as a basis for stratigraphic terminology. Their formulation was a response to needs recognized during the past century by government surveys (both national and local) and by editors of scientific journals for uniform standards and common procedures in defining and classifying formal rock bodies, their fossils, and the time spans represented by them. The most recent Code (ACSN, 1970) is a slightly revised version of that published in 1961, incorporating some minor amendments adopted by the Commission between 1962 and 1969. The Codes have served the profession admirably and have been drawn upon heavily for codes and guides prepared in other parts of the world (ISSC, 1976, p. 104–106). The principles embodied by any code, however, reflect the state of knowledge at the time of its preparation, and even the most recent code is now in need of revision.

New concepts and techniques developed during the past two decades have revolutionized the earth sciences. Moreover, increasingly evident have been the limitations of previous codes in meeting some needs of Precambrian and Quaternary geology and in classification of plutonic, high-grade metamorphic, volcanic, and intensely deformed rock assemblages. In addition, the important contributions of numerous international stratigraphic organizations associated with both the International Union of Geological Sciences (IUGS) and UNESCO, including working groups of the International Geological Correlation Program (IGCP), merit recognition and incorporation into a North American code.

For these and other reasons, revision of the American Code has been undertaken by committees appointed by the North American Commission on Stratigraphic Nomenclature (NACSN). The Commission, founded as the American Commission on Stratigraphic Nomenclature in 1946 (ACSN, 1947), was renamed the NACSN in 1978 (Weiss, 1979b) to emphasize that delegates from ten organizations in Canada, the United States, and Mexico represent the geological profession throughout North America (Appendix II).

Although many past and current members of the Commission helped prepare this revision of the Code, the participation of all interested geologists has been sought (for example, Weiss, 1979a). Open forums were held at the national meetings of both the Geological Society of America at San Diego in November, 1979, and the American Association of Petroleum Geologists at Denver in June, 1980, at which comments and suggestions were offered by more than 150 geologists. The resulting draft of this report was printed, through the courtesy of the Canadian Society of Petroleum Geologists, on October 1, 1981, and additional comments were invited from the profession for a period of one year before submittal of this report to the Commission for adoption. More than 50 responses were received with sufficient suggestions for improvement to prompt moderate revision of the printed draft (NACSN, 1981). We are particularly indebted to Hollis D. Hedberg and Amos Salvador for their exhaustive and perceptive reviews of early drafts of this Code, as well as to those who responded to the request for comments. Participants in the preparation and revisions of this report, and conferees, are listed in Appendix I.

Some of the expenses incurred in the course of this work were defrayed by National Science Foundation Grant EAR 7919845, for which we express appreciation. Institutions represented by the participants have been especially generous in their support.

SCOPE

The North American Stratigraphic Code seeks to describe explicit practices for classifying and naming all formally defined geologic units. *Stratigraphic procedures* and principles, although developed initially to bring order to strata and the events recorded therein, are applicable to all earth materials, not solely to strata. They promote systematic and rigorous study of the composition, geometry, sequence, history, and genesis of rocks and unconsolidated materials. They provide the framework within which time and space relations among rock bodies that constitute the Earth are ordered systematically. Stratigraphic procedures are used not only to reconstruct the history of the Earth and of extra-terrestrial bodies, but also to define the distribution and geometry of some commodities needed by society. *Stratigraphic classification* systematically arranges and partitions bodies of rock or unconsolidated materials of the Earth's crust into units based on their inherent properties or attributes.

A *stratigraphic code* or guide is a formulation of current views on stratigraphic principles and procedures designed to promote standardized classification and formal nomenclature of rock materials. It provides the basis for formalization of the language used to denote rock units and their spatial and temporal relations. To be effective, a code must be widely accepted and used; geologic organizations and journals may adopt its recommendations for nomenclatural procedure. Because any code embodies only current concepts and principles, it should have the flexibility to provide for both changes and additions to improve its relevance to new scientific problems.

Any system of nomenclature must be sufficiently explicit to enable users to distinguish objects that are embraced in a class from those that are not. This stratigraphic code makes no attempt to systematize structural, petrographic, paleontologic, or physiographic terms. Terms from these other fields that are used as part of formal stratigraphic names should be sufficiently general as to be unaffected by revisions of precise petrographic or other classifications.

The objective of a system of classification is to promote unambiguous communication in a manner not so restrictive as to inhibit scientific progress. To minimize ambiguity, a code must promote recognition of the distinction between observable features (reproducible data) and inferences or interpretations. Moreover, it should be sufficiently adaptable and flexible to promote the further development of science.

Stratigraphic classification promotes understanding of the *geometry* and *sequence* of rock bodies. The development of stratigraphy as a science required formulation of the Law of Superposition to explain sequential stratal relations. Although superposition is not applicable to many igneous, metamorphic, and tectonic rock assemblages, other criteria (such as cross-cutting relations and isotopic dating) can be used to determine sequential arrangements among rock bodies.

The term *stratigraphic unit* may be defined in several ways. Etymological emphasis requires that it be a stratum or assemblage of adjacent strata distinguished by any or several of the many properties that rocks may possess (ISSC, 1976, p. 13). The scope of stratigraphic classification and procedures, however, suggests a broader definition: a naturally occurring body of rock or rock material distinguished from adjoining rock on the basis of some stated property or properties. Commonly used properties include composition, texture, included fossils, magnetic signature, radioactivity, seismic velocity, and age. Sufficient care is required in defining the boundaries of a unit to enable others to distinguish the material body from those adjoining it. Units based on one property commonly do not coincide with those based on another and, therefore,

distinctive terms are needed to identify the property used in defining each unit.

The adjective *stratigraphic* is used in two ways in the remainder of this report. In discussions of lithic (used here as synonymous with "lithologic") units, a conscious attempt is made to restrict the term to lithostratigraphic or layered rocks and sequences that obey the Law of Superposition. For non-stratiform rocks (of plutonic or tectonic origin, for example), the term *lithodemic* (see Article 27) is used. The adjective *stratigraphic* is also used in a broader sense to refer to those procedures derived from stratigraphy which are now applied to all classes of earth materials.

An assumption made in the material that follows is that the reader has some degree of familiarity with basic principles of stratigraphy as outlined, for example, by Dunbar and Rodgers (1957), Weller (1960), Shaw (1964), Matthews (1974), or the International Stratigraphic Guide (ISSC, 1976).

RELATION OF CODES TO INTERNATIONAL GUIDE

Publication of the International Stratigraphic Guide by the International Subcommission on Stratigraphic Classification (ISSC, 1976), which is being endorsed and adopted throughout the world, played a part in prompting examination of the American Stratigraphic Code and the decision to revise it.

The International Guide embodies principles and procedures that had been adopted by several national and regional stratigraphic committees and commissions. More than two decades of effort by H. D. Hedberg and other members of the Subcommission (ISSC, 1976, p. VI, 1, 3) developed the consensus required for preparation of the Guide. Although the Guide attempts to cover all kinds of rocks and the diverse ways of investigating them, it is necessarily incomplete. Mechanisms are needed to stimulate individual innovations toward promulgating new concepts, principles, and practices which subsequently may be found worthy of inclusion in later editions of the Guide. The flexibility of national and regional committees or commissions enables them to perform this function more readily than an international subcommission, even while they adopt the Guide as the international standard of stratigraphic classification.

A guiding principle in preparing this Code has been to make it as consistent as possible with the International Guide, which was endorsed by the ACSN in 1976, and at the same time to foster further innovations to meet the expanding and changing needs of earth scientists on the North American continent.

OVERVIEW

CATEGORIES RECOGNIZED

An attempt is made in this Code to strike a balance between serving the needs of those in evolving specialties and resisting the proliferation of categories of units. Consequently, more formal categories are recognized here than in previous codes or in the International Guide (ISSC, 1976). On the other hand, no special provision is made for formalizing certain kinds of units (deep oceanic, for example) which may be accommodated by available categories.

Four principal categories of units have previously been used widely in traditional stratigraphic work; these have been termed lithostratigraphic, biostratigraphic, chronostratigraphic, and geochronologic and are distinguished as follows:

1. A *lithostratigraphic unit* is a stratum or body of strata, generally but not invariably layered, generally but not invariably tabular, which conforms to the Law of Superposition and is distinguished and delimited on the basis of lithic characteristics and stratigraphic position. Example: Navajo Sandstone.

2. A *biostratigraphic unit* is a body of rock defined and characterized by its fossil content. Example: *Discoaster multiradiatus* Interval Zone.

3. A *chronostratigraphic unit* is a body of rock established to serve as the material reference for all rocks formed during the same span of time. Example: Devonian System. Each boundary of a chronostratigraphic unit is synchronous. Chronostratigraphy provides a means of organizing strata into units based on their age relations. A chronostratigraphic body also serves as the basis for defining the specific interval of geologic time, or geochronologic unit, represented by the referent.

4. A *geochronologic unit* is a division of time distinguished or the basis of the rock record preserved in a chronostratigraphic unit. Example: Devonian Period.

The first two categories are comparable in that they consist of material units defined on the basis of content. The third category differs from the first two in that it serves primarily as the standard for recognizing and isolating materials of a specific age. The fourth, in contrast, is not a material, but rather a conceptual, unit; it is a division of time. Although a geochronologic unit is not a stratigraphic body, it is so intimately tied to chronostratigraphy that the two are discussed properly together.

Properties and procedures that may be used in distinguishing geologic units are both diverse and numerous (ISSC, 1976, p. 1, 96; Harland, 1977, p. 230), but all may be assigned to the following principal classes of categories used in stratigraphic classification (Table 1), which are discussed below:

I. Material categories based on content, inherent attributes, or physical limits,

II. Categories distinguished by geologic age:

    A. Material categories used to define temporal spans, and

    B. Temporal categories.

Table 1. Categories of Units Defined*

MATERIAL CATEGORIES BASED ON CONTENT OR PHYSICAL LIMITS

Lithostratigraphic (22)
Lithodemic (31)**
Magnetopolarity (44)
Biostratigraphic (48)
Pedostratigraphic (55)
Allostratigraphic (58)

CATEGORIES EXPRESSING OR RELATED TO GEOLOGIC AGE

Material Categories Used to Define Temporal Spans
  Chronostratigraphic (66)
  Polarity-Chronostratigraphic (83)
Temporal (Non-Material) Categories
  Geochronologic (80)
  Polarity-Chronologic (88)
  Diachronic (91)
  Geochronometric (96)

*Numbers in parentheses are the numbers of the Articles where units are defined.
**Italicized categories are those introduced or developed since publication of the previous code (ACSN, 1970).

Material Categories Based on Content or Physical Limits

The basic building blocks for most geologic work are rock bodies defined on the basis of composition and related lithic characteristics, or on their physical, chemical, or biologic

content or properties. Emphasis is placed on the relative objectivity and reproducibility of data used in defining units within each category.

Foremost properties of rocks are composition, texture, fabric, structure, and color, which together are designated *lithic characteristics*. These serve as the basis for distinguishing and defining the most fundamental of all formal units. Such units based primarily on composition are divided into two categories (Henderson and others, 1980): lithostratigraphic (Article 22) and lithodemic (defined here in Article 31). A lithostratigraphic unit obeys the Law of Superposition, whereas a lithodemic unit does not. A *lithodemic unit* is a defined body of predominantly intrusive, highly metamorphosed, or intensely deformed rock that, because it is intrusive or has lost primary structure through metamorphism or tectonism, generally does not conform to the Law of Superposition.

Recognition during the past several decades that remanent magnetism in rocks records the Earth's past magnetic characteristics (Cox, Doell, and Dalrymple, 1963) provides a powerful new tool encompassed by magnetostratigraphy (McDougall, 1977; McElhinny, 1978). *Magnetostratigraphy* (Article 43) is the study of remanent magnetism in rocks; it is the record of the Earth's magnetic polarity (or field reversals), dipole-field-pole position (including apparent polar wander), the non-dipole component (secular variation), and field intensity. Polarity is of particular utility and is used to define a *magnetopolarity unit* (Article 44) as a body of rock identifed by its remanent magnetic polarity (ACSN, 1976; ISSC, 1979). Empirical demonstration of uniform polarity does not necessarily have direct temporal connotations because the remanent magnetism need not be related to rock deposition or crystallization. Nevertheless, polarity is a physical attribute that may characterize a body of rock.

Biologic remains contained in, or forming, strata are uniquely important in stratigraphic practice. First, they provide the means of defining and recognizing material units based on fossil content (biostratigraphic units). Second, the irreversibility of organic evolution makes it possible to partition enclosing strata temporally. Third, biologic remains provide important data for the reconstruction of ancient environments of deposition.

Composition also is important in distinguishing pedostratigraphic units. A *pedostratigraphic unit* is a body of rock that consists of one or more pedologic horizons developed in one or more lithic units now buried by a formally defined lithostratigraphic or allostratigraphic unit or units. A pedostratigraphic unit is the part of a buried soil characterized by one or more clearly defined soil horizons containing pedogenically formed minerals and organic compounds. Pedostratigraphic terminology is discussed below and in Article 55.

Many upper Cenozoic, especially Quaternary, deposits are distinguished and delineated on the basis of content, for which lithostratigraphic classification is appropriate. However, others are delineated on the basis of criteria other than content. To facilitate the reconstruction of geologic history, some compositionally similar deposits in vertical sequence merit distinction as separate stratigraphic units because they are the products of different processes; others merit distinction because they are of demonstrably different ages. Lithostratigraphic classification of these units is impractical and a new approach, allostratigraphic classification, is introduced here and may prove applicable to older deposits as well. An *allostratigraphic unit* is a mappable stratiform body of sedimentary rock defined and identified on the basis of bounding discontinuities (Article 58 and related Remarks).

Geologic-Climate units, defined in the previous Code (ACSN, 1970, p. 31), are abandoned here because they proved to be of dubious utility. Inferences regarding climate are subjective and too tenuous a basis for the definition of

formal geologic units. Such inferences commonly are based on deposits assigned more appropriately to lithostratigraphic or allostratigraphic units and may be expressed in terms of diachronic units (defined below).

Categories Expressing or Related to Geologic Age

Time is a single, irreversible continuum. Nevertheless, various categories of units are used to define intervals of geologic time, just as terms having different bases, such as Paleolithic, Renaissance, and Elizabethan, are used to designate specific periods of human history. Different temporal categories are established to express intervals of time distinguished in different ways.

Major objectives of stratigraphic classification are to provide a basis for systematic ordering of the time and space relations of rock bodies and to establish a time framework for the discussion of geologic history. For such purposes, units of geologic time traditionally have been named to represent the span of time during which a well-described sequence of rock, or a chronostratigraphic unit, was deposited ("time units based on material referents," Fig. 1). This procedure continues, to the exclusion of other possible approaches, to be standard practice in studies of Phanerozoic rocks. Despite admonitions in previous American codes and the International Stratigraphic Guide (ISSC, 1976, p. 81) that similar procedures should be applied to the Precambrian, no comparable chronostratigraphic units, or geochronologic units derived therefrom, proposed for the Precambrian have yet been accepted worldwide. Instead, the IUGS Subcommission on Precambrian Stratigraphy (Sims, 1979) and its Working Groups (Harrison and Peterman, 1980) recommend division of Precambrian time into *geochronometric units* having no material referents.

A distinction is made throughout this report between *isochronous* and *synchronous*, as urged by Cumming, Fuller, and Porter (1959, p. 730), although the terms have been used synonymously by many. *Isochronous* means of equal duration; *synchronous* means simultaneous, or occurring at the same time. Although two rock bodies of very different ages may be formed during equal durations of time, the term *isochronous* is not applied to them in the earth sciences. Rather, isochronous bodies are those bounded by synchronous surfaces and formed during the same span of time. *Isochron*, in contrast, is used for a line connecting points of equal age on a graph representing physical or chemical phenomena; the line represents the same or equal time. The adjective *diachronous* is applied either to a rock unit with one or two bounding surfaces which are not synchronous, or to a boundary which is not synchronous (which "transgresses time").

Two classes of time units based on material referents, or stratotypes, are recognized (Fig. 1). The first is that of the traditional and conceptually isochronous units, and includes *geochronologic units*, which are based on *chronostratigraphic units*, and *polarity-geochronologic units*. These isochronous units have worldwide applicability and may be used even in areas lacking a material record of the named span of time. The second class of time units, newly defined in this Code, consists of *diachronic units* (Article 91), which are based on rock bodies known to be diachronous. In contrast to isochronous units, a diachronic term is used only where a material referent is present; a diachronic unit is coextensive with the material body or bodies on which it is based.

A *chronostratigraphic unit*, as defined above and in Article 66, is a body of rock established to serve as the material reference for all rocks formed during the same span of time; its boundaries are synchronous. It is the referent for a *geochronologic unit*, as defined above and in Article 80. Internationally accepted and traditional chronostratigraphic units were based initially on the time spans of lithostratigraphic

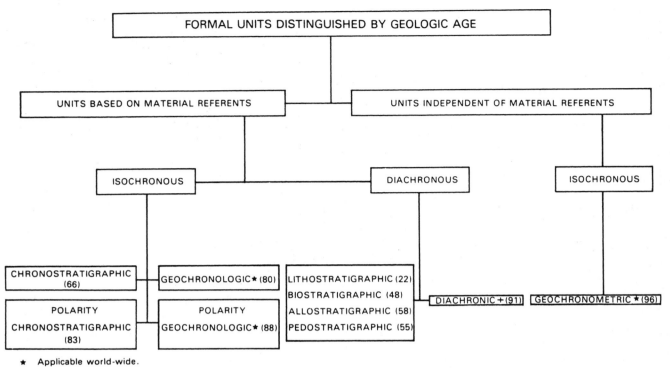

★   Applicable world-wide.
+   Applicable only where material referents are present.
( )   Number of article in which defined.

FIG. 1.—Relation of geologic time units to the kinds of rock-unit referents on which most are based.

units, biostratigraphic units, or other features of the rock record that have specific durations. In sum, they form the Standard Global Chronostratigraphic Scale (ISSC, 1976, p. 76–81; Harland, 1978), consisting of established systems and series.

A *polarity-chronostratigraphic unit* is a body of rock that contains a primary magnetopolarity record imposed when the rock was deposited or crystallized (Article 83). It serves as a material standard or referent for a part of geologic time during which the Earth's magnetic field had a characteristic polarity or sequence of polarities; that is, for a *polarity-chronologic unit* (Article 88).

A *diachronic unit* comprises the unequal spans of time represented by one or more specific diachronous rock bodies (Article 91). Such bodies may be lithostratigraphic, biostratigraphic, pedostratigraphic, allostratigraphic, or an assemblage of such units. A diachronic unit is applicable only where its material referent is present.

A *geochronometric* (or chronometric) *unit* is an isochronous direct division of geologic time expressed in years (Article 96). It has no material referent.

Pedostratigraphic Terms

The definition and nomenclature for pedostratigraphic[2] units in this Code differ from those for soil-stratigraphic units in the previous Code (ACSN, 1970, Article 18), by being more specific with regard to content, boundaries, and the basis for determining stratigraphic position.

The term "soil" has different meanings to the geologist, the soil scientist, the engineer, and the layman, and com-

monly has no stratigraphic significance. The term *paleosol* is currently used in North America for any soil that formed on a landscape of the past; it may be a buried soil, a relict soil, or an exhumed soil (Ruhe, 1965; Valentine and Dalrymple, 1976).

A *pedologic soil* is composed of one or more soil horizons.[3] A *soil horizon* is a layer within a pedologic soil that (1) is approximately parallel to the soil surface, (2) has distinctive physical, chemical, biological, and morphological properties that differ from those of adjacent, genetically related, soil horizons, and (3) is distinguished from other soil horizons by objective compositional properties that can be observed or measured in the field. The physical boundaries of buried pedologic horizons are objective traceable boundaries with stratigraphic significance. A buried pedologic soil provides the material basis for definition of a stratigraphic unit in pedostratigraphic classification (Article 55), but a buried pedologic soil may be somewhat more inclusive than a pedostratigraphic unit. A pedologic soil may contain both an 0-horizon and the entire C-horizon (Fig. 6), whereas the former is excluded and the latter need not be included in a pedostratigraphic unit.

The definition and nomenclature for pedostratigraphic units in this Code differ from those of soil stratigraphic units proposed by the International Union for Quaternary Research and International Society of Soil Science (Parsons, 1981). The pedostratigraphic unit, geosol, also differs from the proposed INQUA-ISSS soil-stratigraphic unit, pedoderm, in several ways, the most important of which are: (1) a geosol may be in any part of the geologic column, whereas a pedoderm is a surficial soil; (2) a geosol is a buried soil,

---

[2]From Greek, *pedon*, ground or soil.

[3]As used in a geological sense, a *horizon* is a surface or line. In pedology, however, it is a body of material, and such usage is continued here.

whereas a pedoderm may be a buried, relict, or exhumed soil; (3) the boundaries and stratigraphic position of a geosol are defined and delineated by criteria that differ from those for a pedoderm; and (4) a geosol may be either all or only a part of a buried soil, whereas a pedoderm is the entire soil.

The term *geosol*, as defined by Morrison (1967, p. 3), is a laterally traceable, mappable, geologic weathering profile that has a consistent stratigraphic position. The term is adopted and redefined here as the fundamental and only unit in formal pedostratigraphic classification (Article 56).

FORMAL AND INFORMAL UNITS

Although the emphasis in this Code is necessarily on formal categories of geologic units, informal nomenclature is highly useful in stratigraphic work.

Formally named units are those that are named in accordance with an established scheme of classification; the fact of formality is conveyed by capitalization of the initial letter of the *rank* or *unit* term (for example, Morrison Formation). Informal units, whose unit terms are ordinary nouns, are not protected by the stability provided by proper formalization and recommended classification procedures. Informal terms are devised for both economic and scientific reasons. Formalization is appropriate for those units requiring stability of nomenclature, particularly those likely to be extended far beyond the locality in which they were first recognized. Informal terms are appropriate for casually mentioned, innovative, and most economic units, those defined by unconventional criteria, and those that may be too thin to map at usual scales.

Casually mentioned geologic units not defined in accordance with this Code are informal. For many of these, there may be insufficient need or information, or perhaps an inappropriate basis, for formal designations. Informal designations as beds or lithozones (the pebbly beds, the shaly zone, third coal) are appropriate for many such units.

Most economic units, such as aquifers, oil sands, coal beds, quarry layers, and ore-bearing "reefs," are informal, even though they may be named. Some such units, however, are so significant scientifically and economically that they merit formal recognition as beds, members, or formations.

Innovative approaches in regional stratigraphic studies have resulted in the recognition and definition of units best left as informal, at least for the time being. Units bounded by major regional unconformities on the North American craton were designated "sequences" (example: Sauk sequence) by Sloss (1963). Major unconformity-bounded units also were designated "synthems" by Chang (1975), who recommended that they be treated formally. Marker-defined units that are continuous from one lithofacies to another were designated "formats" by Forgotson (1957). The term "chronosome" was proposed by Schultz (1982) for rocks of diverse facies corresponding to geographic variations in sedimentation during an interval of deposition identified on the basis of bounding stratigraphic markers. Successions of faunal zones containing evolutionarily related forms, but bounded by non-evolutionary biotic discontinuities, were termed "biomeres" (Palmer, 1965). The foregoing are only a few selected examples to demonstrate how informality provides a continuing avenue for innovation.

The terms *magnafacies* and *parvafacies*, coined by Caster (1934) to emphasize the distinction between lithostratigraphic and chronostratigraphic units in sequences displaying marked facies variation, have remained informal despite their impact on clarifying the concepts involved.

Tephrochronologic studies provide examples of informal units too thin to map at conventional scales but yet invaluable for dating important geologic events. Although some such units are named for physiographic features and places where first recognized (e.g., Guaje pumice bed, where it is not mapped as the Guaje Member of the Bandelier Tuff), others bear the same name as the volcanic vent (e.g., Huckleberry Ridge ash bed of Izett and Wilcox, 1981).

Informal geologic units are designated by ordinary nouns, adjectives or geographic terms and lithic or unit-terms that are not capitalized (chalky formation or beds, St. Francis coal).

No geologic unit should be established and defined, whether formally or informally, unless its recognition serves a clear purpose.

CORRELATION

Correlation is a procedure for demonstrating correspondence between geographically separated parts of a geologic unit. The term is a general one having diverse meanings in different disciplines. Demonstration of temporal correspondence is one of the most important objectives of stratigraphy. The term "correlation" frequently is misused to express the idea that a unit has been identified or recognized.

Correlation is used in this Code as the demonstration of correspondence between two geologic units in both some defined property and relative stratigraphic position. Because correspondence may be based on various properties, three kinds of correlation are best distinguished by more specific terms. *Lithocorrelation* links units of similar lithology and stratigraphic position (or sequential or geometric relation, for lithodemic units). *Biocorrelation* expresses similarity of fossil content and biostratigraphic position. *Chronocorrelation* expresses correspondence in age and in chronostratigraphic position.

Other terms that have been used for the similarity of content and stratal succession are homotaxy and chronotaxy. *Homotaxy* is the similarity in separate regions of the serial arrangement or succession of strata of comparable compositions or of included fossils. The term is derived from *homotaxis*, proposed by Huxley (1862, p. xlvi) to emphasize that similarity in succession does not prove age equivalence of comparable units. The term *chronotaxy* has been applied to similar stratigraphic sequences composed of units which are of equivalent age (Henbest, 1952, p. 310).

Criteria used for ascertaining temporal and other types of correspondence are diverse (ISSC, 1976, p. 86–93) and new criteria will emerge in the future. Evolving statistical tests, as well as isotopic and paleomagnetic techniques, complement the traditional paleontologic and lithologic procedures. Boundaries defined by one set of criteria need not correspond to those defined by others.

PART II. ARTICLES

INTRODUCTION

Article 1.—**Purpose.** This Code describes explicit stratigraphic procedures for classifying and naming geologic units accorded formal status. Such procedures, if widely adopted, assure consistent and uniform usage in classification and terminology and therefore promote unambiguous communication.

Article 2.—**Categories.** Categories of formal stratigraphic units, though diverse, are of three classes (Table 1). The first class is of rock-material categories based on inherent attributes or content and stratigraphic position, and includes lithostratigraphic, lithodemic, magnetopolarity, biostratigraphic, pedostratigraphic, and allostratigraphic units. The second class is of material categories used as standards for defining spans of geologic time, and includes chronostratigraphic

and polarity-chronostratigraphic units. The third class is of non-material temporal categories, and includes geochronologic, polarity-chronologic, geochronometric, and diachronic units.

GENERAL PROCEDURES

DEFINITION OF FORMAL UNITS

Article 3.—**Requirements for Formally Named Geologic Units.** Naming, establishing, revising, redefining, and abandoning formal geologic units require publication in a recognized scientific medium of a comprehensive statement which includes: (i) intent to designate or modify a formal unit; (ii) designation of category and rank of unit; (iii) selection and derivation of name; (iv) specification of stratotype (where applicable); (v) description of unit; (vi) definition of boundaries; (vii) historical background; (viii) dimensions, shape, and other regional aspects; (ix) geologic age; (x) correlations; and possibly (xi) genesis (where applicable). These requirements apply to subsurface and offshore, as well as exposed, units.

Article 4.—**Publication.**[4] "Publication in a recognized scientific medium" in conformance with this Code means that a work, when first issued, must (1) be reproduced in ink on paper or by some method that assures numerous identical copies and wide distribution; (2) be issued for the purpose of scientific, public, permanent record; and (3) be readily obtainable by purchase or free distribution.

Remarks. (a) **Inadequate publication.**—The following do not constitute publication within the meaning of the Code: (1) distribution of microfilms, microcards, or matter reproduced by similar methods; (2) distribution to colleagues or students of a note, even if printed, in explanation of an accompanying illustration; (3) distribution of proof sheets; (4) open-file release; (5) theses, dissertations, and dissertation abstracts; (6) mention at a scientific or other meeting; (7) mention in an abstract, map explanation, or figure caption; (8) labeling of a rock specimen in a collection; (9) mere deposit of a document in a library; (10) anonymous publication; or (11) mention in the popular press or in a legal document.

(b). **Guidebooks.**—A guidebook with distribution limited to participants of a field excursion does not meet the test of availability. Some organizations publish and distribute widely large editions of serial guidebooks that include refereed regional papers; although these do meet the tests of scientific purpose and availability, and therefore constitute valid publication, other media are preferable.

Article 5.—**Intent and Utility.** To be valid, a new unit must serve a clear purpose and be duly proposed and duly described, and the intent to establish it must be specified. Casual mention of a unit, such as "the granite exposed near the Middleville schoolhouse," does not establish a new formal unit, nor does mere use in a table, columnar section, or map.

Remark. (a) **Demonstration of purpose served.**—The initial definition or revision of a named geologic unit constitutes, in essence, a proposal. As such, it lacks status until use by others demonstrates that a clear purpose has been served. A unit becomes established through repeated demonstration of its utility. The decision not to use a newly proposed or a newly revised term requires a full discussion of its unsuitability.

Article 6.—**Category and Rank.** The category and rank of a new or revised unit must be specified.

---

[4]This article is modified slightly from a statement by the International Commission of Zoological Nomenclature (1964, p. 7–9).

Remark. (a) **Need for specification.**—Many stratigraphic controversies have arisen from confusion or misinterpretation of the category of a unit (for example, lithostratigraphic vs. chronostratigraphic). Specification and unambiguous description of the category is of paramount importance. Selection and designation of an appropriate rank from the distinctive terminology developed for each category help serve this function (Table 2).

Article 7.—**Name.** The name of a formal geologic unit is compound. For most categories, the name of a unit should consist of a geographic name combined with an appropriate rank (Wasatch Formation) or descriptive term (Viola Limestone). Biostratigraphic units are designated by appropriate biologic forms (*Exus albus* Assemblage Biozone). Worldwide chronostratigraphic units bear long established and generally accepted names of diverse origins (Triassic System). The first letters of all words used in the names of formal geologic units are capitalized (except for the trivial species and subspecies terms in the name of a biostratigraphic unit).

Remarks. (a) **Appropriate geographic terms.**—Geographic names derived from permanent natural or artificial features at or near which the unit is present are preferable to those derived from impermanent features such as farms, schools, stores, churches, crossroads, and small communities. Appropriate names may be selected from those shown on topographic, state, provincial, county, forest service, hydrographic, or comparable maps, particularly those showing names approved by a national board for geographic names. The generic part of a geographic name, e.g., river, lake, village, should be omitted from new terms, unless required to distinguish between two otherwise identical names (e.g., Redstone Formation and Redstone River Formation). Two names should not be derived from the same geographic feature. A unit should not be named for the source of its components; for example, a deposit inferred to have been derived from the Keewatin glaciation center should not be designated the "Keewatin Till."

(b) **Duplication of names.**—Responsibility for avoiding duplication, either in use of the same name for different units (homonymy) or in use of different names for the same unit (synonymy), rests with the proposer. Although the same geographic term has been applied to different categories of units (example: the lithostratigraphic Word Formation and the chronostratigraphic Wordian Stage) now entrenched in the literature, the practice is undesirable. The extensive geologic nomenclature of North America, including not only names but also nomenclatural history of formal units, is recorded in compendia maintained by the Committee on Stratigraphic Nomenclature of the Geological Survey of Canada, Ottawa, Ontario; by the Geologic Names Committee of the United States Geological Survey, Reston, Virginia; by the Instituto de Geologia, Ciudad Universitaria, México, D.F.; and by many state and provincial geological surveys. These organizations respond to inquiries regarding the availability of names, and some are prepared to reserve names for units likely to be defined in the next year or two.

(c) **Priority and preservation of established names.**—Stability of nomenclature is maintained by use of the rule of priority and by preservation of well-established names. Names should not be modified without explaining the need. Priority in publication is to be respected, but priority alone does not justify displacing a well-established name by one neither well-known nor commonly used; nor should an inadequately established name be preserved merely on the basis of priority. Redefinitions in precise terms are preferable to abandonment of the names of well-established units which may have been defined imprecisely but nonetheless in conformance with older and less stringent standards.

(d) **Differences of spelling and changes in name.**—The geographic component of a well-established stratigraphic name is not changed due to differences in spelling or changes in the name of a geographic feature. The name Bennett Shale, for example, used for more than half a century, need not be altered because the town is named Bennet. Nor should the Mauch Chunk Formation be changed because the town has been renamed Jim Thorpe. Disappearance of an impermanent geographic feature, such as a town, does not affect the name of an established geologic unit.

Table 2. Categories and Ranks of Units Defined in This Code*

A. Material Units

LITHOSTRATIGRAPHIC	LITHODEMIC	MAGNETOPOLARITY	BIOSTRATIGRAPHIC	PEDOSTRATIGRAPHIC	ALLOSTRATIGRAPHIC
Supergroup	Supersuite				
Group	Suite	Polarity Superzone			Allogroup
Formation	*Lithodeme*	*Polarity zone*	*Biozone* (Interval, Assemblage or Abundance)	*Geosol*	*Alloformation*
Member (or Lens, or Tongue)		Polarity Subzone	Subbiozone		Allomember
Bed(s) or Flow(s)					

(LITHODEMIC column: Supersuite / Suite / Lithodeme bracketed as "Complex")

B. Temporal and Related Chronostratigraphic Units

CHRONO-STRATIGRAPHIC	GEOCHRONOLOGIC GEOCHRONOMETRIC	POLARITY CHRONO-STRATIGRAPHIC	POLARITY CHRONOLOGIC	DIACHRONIC
Eonothem	Eon	Polarity Superchronozone	Polarity Superchron	
Erathem (Supersystem)	Era (Superperiod)			
System (Subsystem)	*Period* (Subperiod)	*Polarity Chronozone*	*Polarity Chron*	*Episode*
Series	Epoch			Phase
Stage (Substage)	Age (Subage)	Polarity Subchronozone	Polarity Subchron	Span
Chronozone	Chron			Cline

(DIACHRONIC column units bracketed as "Diachron")

*Fundamental Units are italicized.

(e) **Names in different countries and different languages.** — For geologic units that cross local and international boundaries, a single name for each is preferable to several. Spelling of a geographic name commonly conforms to the usage of the country and linguistic group involved. Although geographic names are not translated (Cuchillo is not translated to Knife), lithologic or rank terms are (Edwards Limestone, Caliza Edwards; Formación La Casita, La Casita Formation).

Article 8. — **Stratotypes.** The designation of a unit or boundary stratotype (type section or type locality) is essential in the definition of most formal geologic units. Many kinds of units are best defined by reference to an accessible and specific sequence of rock that may be examined and studied by others. A stratotype is the standard (original or subsequently designated) for a named geologic unit or boundary and constitutes the basis for definition or recognition of that unit or boundary; therefore, it must be illustrative and representative of the concept of the unit or boundary being defined.

Remarks. (a) **Unit stratotypes.** — A unit stratotype is the type section for a stratiform deposit or the type area for a nonstratiform body that serves as the standard for definition and recognition of a geologic unit. The upper and lower limits of a unit stratotype are designated points in a specific sequence or locality and serve as the standards for definition and recognition of a stratigraphic unit's boundaries.

(b) **Boundary stratotype.** — A boundary stratotype is the type locality for the boundary reference point for a stratigraphic unit. Both boundary stratotypes for any unit need not be in the same section or region. Each boundary stratotype serves as the standard for definition and recognition of the base of a stratigraphic unit. The top of a unit may be defined by the boundary stratotype of the next higher stratigraphic unit.

(c) **Type locality.** — A type locality is the specified geographic locality where the stratotype of a formal unit or unit boundary was originally defined and named. A type area is the geographic territory encompassing the type locality. Before the concept of a stratotype was developed, only type localities and areas were designated for many geologic units which are now long- and well-established. Stratotypes, though now mandatory in defining most stratiform units, are impractical in definitions of many large nonstratiform rock bodies whose diverse major components may be best displayed at several reference localities.

(d) **Composite-stratotype.** — A composite-stratotype consists of several reference sections (which may include a type section) required to demonstrate the range or totality of a stratigraphic unit.

(e) **Reference sections.** — Reference sections may serve as invaluable standards in definitions or revisions of formal geologic units. For those well-established stratigraphic units for which a type section never was specified, a principal reference section (lectostratotype of ISSC, 1976, p. 26) may be designated. A principal reference section (neostratotype of ISSC, 1976, p. 26) also may be designated for those units or boundaries whose stratotypes have been destroyed, covered, or otherwise made inaccessible. Supplementary reference sections often are designated to illustrate the diversity or heterogeneity of a defined unit or some critical feature not evident or exposed in the stratotype. Once a unit or boundary stratotype section is designated, it is never abandoned or changed; however, if a stratotype proves inadequate, it may be supplemented by a principal reference section or by several reference sections that may constitute a composite-stratotype.

(f) **Stratotype descriptions.** — Stratotypes should be described both geographically and geologically. Sufficient geographic detail must be included to enable others to find the stratotype in the field, and may consist of maps and/or aerial photographs showing location and access, as well as appropriate coordinates or bearings. Geologic

information should include thickness, descriptive criteria appropriate to the recognition of the unit and its boundaries, and discussion of the relation of the unit to other geologic units of the area. A carefully measured and described section provides the best foundation for definition of stratiform units. Graphic profiles, columnar sections, structure-sections, and photographs are useful supplements to a description; a geologic map of the area including the type locality is essential.

Article 9.—**Unit Description.** A unit proposed for formal status should be described and defined so clearly that any subsequent investigator can recognize that unit unequivocally. Distinguishing features that characterize a unit may include any or several of the following: composition, texture, primary structures, structural attitudes, biologic remains, readily apparent mineral composition (e.g., calcite vs. dolomite), geochemistry, geophysical properties (including magnetic signatures), geomorphic expression, unconformable or cross-cutting relations, and age. Although all distinguishing features pertinent to the unit category should be described sufficiently to characterize the unit, those not pertinent to the category (such as age and inferred genesis for lithostratigraphic units, or lithology for biostratigraphic units) should not be made part of the definition.

Article 10.—**Boundaries.** The criteria specified for the recognition of boundaries between adjoining geologic units are of paramount importance because they provide the basis for scientific reproducibility of results. Care is required in describing the criteria, which must be appropriate to the category of unit involved.

Remarks. (a) **Boundaries between intergradational units.**—Contacts between rocks of markedly contrasting composition are appropriate boundaries of lithic units, but some rocks grade into, or intertongue with, others of different lithology. Consequently, some boundaries are necessarily arbitrary as, for example, the top of the uppermost limestone in a sequence of interbedded limestone and shale. Such arbitrary boundaries commonly are diachronous.

(b) **Overlaps and gaps.**—The problem of overlaps and gaps between long-established adjacent chronostratigraphic units is being addressed by international IUGS and IGCP working groups appointed to deal with various parts of the geologic column. The procedure recommended by the Geological Society of London (George and others, 1969; Holland and others, 1978), of defining only the basal boundaries of chronostratigraphic units, has been widely adopted (e.g., McLaren, 1977) to resolve the problem. Such boundaries are defined by a carefully selected and agreed-upon boundary-stratotype (marker-point type section or "golden spike") which becomes the standard for the base of a chronostratigraphic unit. The concept of the mutual-boundary stratotype (ISSC, 1976, p. 84–86), based on the assumption of continuous deposition in selected sequences, also has been used to define chronostratigraphic units.

Although international chronostratigraphic units of series and higher rank are being redefined by IUGS and IGCP working groups, there may be a continuing need for some provincial series. Adoption of the basal boundary-stratotype concept is urged.

Article 11.—**Historical Background.** A proposal for a new name must include a nomenclatorial history of rocks assigned to the proposed unit, describing how they were treated previously and by whom (references), as well as such matters as priorities, possible synonymy, and other pertinent considerations. Consideration of the historical background of an older unit commonly provides the basis for justifying definition of a new unit.

Article 12.—**Dimensions and Regional Relations.** A perspective on the magnitude of a unit should be provided by such information as may be available on the geographic extent of a unit; observed ranges in thickness, composition, and geomorphic expression; relations to other kinds and ranks of

stratigraphic units; correlations with other nearby sequences; and the bases for recognizing and extending the unit beyond the type locality. If the unit is not known anywhere but in an area of limited extent, informal designation is recommended.

Article 13.—**Age.** For most formal material geologic units, other than chronostratigraphic and polaritychronostratigraphic, inferences regarding geologic age play no proper role in their definition. Nevertheless, the age, as well as the basis for its assignment, are important features of the unit and should be stated. For many lithodemic units, the age of the protolith should be distinguished from that of the metamorphism or deformation. If the basis for assigning an age is tenuous, a doubt should be expressed.

Remarks. (a) **Dating.**—The geochronologic ordering of the rock record, whether in terms of radioactive-decay rates or other processes, is generally called "dating." However, the use of the noun "date" to mean "isotopic age" is not recommended. Similarly, the term "absolute age" should be suppressed in favor of "isotopic age" for an age determined on the basis of isotopic ratios. The more inclusive term "numerical age" is recommended for all ages determined from isotopic ratios, fission tracks, and other quantifiable age-related phenomena.

(b) **Calibration.**—The dating of chronostratigraphic boundaries in terms of numerical ages is a special form of dating for which the word "calibration" should be used. The geochronologic time-scale now in use has been developed mainly through such calibration of chronostratigraphic sequences.

(c) **Convention and abbreviations.**—The age of a stratigraphic unit or the time of a geologic event, as commonly determined by numerical dating or by reference to a calibrated time-scale. may be expressed in years before the present. The unit of time is the modern year as presently recognized worldwide. Recommended (but not mandatory) abbreviations for such ages are SI (International System of Units) multipliers coupled with "a" for annum: ka, Ma, and Ga[5] for kilo-annum (10^3 years), Mega-annum (10^6 years), and Giga-annum (10^9 years), respectively. Use of these terms after the age value follows the convention established in the field of C-14 dating. The "present" refers to 1950 AD, and such qualifiers as "ago" or "before the present" are omitted after the value because measurement of the duration from the present to the past is implicit in the designation. In contrast, the duration of a remote interval of geologic time, as a number of years, should not be expressed by the same symbols. Abbreviations for numbers of years, without reference to the present, are informal (e.g., y or yr for years; my, m.y., or m.yr. for millions of years; and so forth, as preference dictates). For example, boundaries of the Late Cretaceous Epoch currently are calibrated at 63 Ma and 96 Ma, but the interval of time represented by this epoch is 33 m.y.

(d) **Expression of "age" of lithodemic units.**—The adjectives "early," "middle," and "late" should be used with the appropriate geochronologic term to designate the age of lithodemic units. For example, a granite dated isotopically at 510 Ma should be referred to using the geochronologic term "Late Cambrian granite" rather than either the chronostratigraphic term "Upper Cambrian granite" or the more cumbersome designation "granite of Late Cambrian age."

Article 14.—**Correlation.** Information regarding spatial and temporal counterparts of a newly defined unit beyond the type area provides readers with an enlarged perspective. Discussions of criteria used in correlating a unit with those in other areas should make clear the distinction between data and inferences.

Article 15.—**Genesis.** Objective data are used to define and classify geologic units and to express their spatial and temporal relations. Although many of the categories defined in this Code (e.g., lithostratigraphic group, plutonic suite)

[5]Note that the initial letters of Mega- and Giga- are capitalized, but that of kilo- is not, by SI convention.

have genetic connotations, inferences regarding geologic history or specific environments of formation may play no proper role in the definition of a unit. However, observations, as well as inferences, that bear on genesis are of great interest to readers and should be discussed.

Article 16.—**Subsurface and Subsea Units.** The foregoing procedures for establishing formal geologic units apply also to subsurface and offshore or subsea units. Complete lithologic and paleontologic descriptions or logs of the samples or cores are required in written or graphic form, or both. Boundaries and divisions, if any, of the unit should be indicated clearly with their depths from an established datum.

Remarks. (a) **Naming subsurface units.**—A subsurface unit may be named for the borehole (Eagle Mills Formation), oil field (Smackover Limestone), or mine which is intended to serve as the stratotype, or for a nearby geographic feature. The hole or mine should be located precisely, both with map and exact geographic coordinates, and identified fully (operator or company, farm or lease block, dates drilled or mined, surface elevation and total depth, etc).

(b) **Additional recommendations.**—Inclusion of appropriate borehole geophysical logs is urged. Moreover, rock and fossil samples and cores and all pertinent accompanying materials should be stored, and available for examination, at appropriate federal, state, provincial, university, or museum depositories. For offshore or subsea units (Clipperton Formation of Tracey and others, 1971, p.22; Argo Salt of McIver, 1972, p. 57), the names of the project and vessel, depth of sea floor, and pertinent regional sampling and geophysical data should be added.

(c) **Seismostratigraphic units.**—High-resolution seismic methods now can delineate stratal geometry and continuity at a level of confidence not previously attainable. Accordingly, seismic surveys have come to be the principal adjunct of the drill in subsurface exploration. On the other hand, the method identifies rock types only broadly and by inference. Thus, formalization of units known only from seismic profiles is inappropriate. Once the stratigraphy is calibrated by drilling, the seismic method may provide objective well-to-well correlations.

REVISION AND ABANDONMENT OF FORMAL UNITS

Article 17.—**Requirements for Major Changes.** Formally defined and named geologic units may be redefined, revised, or abandoned, but revision and abandonment require as much justification as establishment of a new unit.

Remark. (a) **Distinction between redefinition and revision.**—Redefinition of a unit involves changing the view or emphasis on the content of the unit without changing the boundaries or rank, and differs only slightly from redescription. Neither redefinition nor redescription is considered revision. A redescription corrects an inadequate or inaccurate description, whereas a redefinition may change a descriptive (for example, lithologic) designation. Revision involves either minor changes in the definition of one or both boundaries or in the rank of a unit (normally, elevation to a higher rank). Correction of a misidentification of a unit outside its type area is neither redefinition nor revision.

Article 18.—**Redefinition.** A correction or change in the descriptive term applied to a stratigraphic or lithodemic unit is a redefinition which does not require a new geographic term.

Remarks. (a) **Change in lithic designation.**—Priority should not prevent more exact lithic designation if the original designation is not everywhere applicable; for example, the Niobrara Chalk changes gradually westward to a unit in which shale is prominent, for which the designation "Niobrara Shale" or "Formation" is appropriate. Many carbonate formations originally designated "limestone" or "dolomite" are found to be geographically inconsis-

tent as to prevailing rock type. The appropriate lithic term or "formation" is again preferable for such units.

(b) **Original lithic designation inappropriate.**—Restudy of some long-established lithostratigraphic units has shown that the original lithic designation was incorrect according to modern criteria; for example, some "shales" have the chemical and mineralogical composition of limestone, and some rocks described as felsic lavas now are understood to be welded tuffs. Such new knowledge is recognized by changing the lithic designation of the unit, while retaining the original geographic term. Similarly, changes in the classification of igneous rocks have resulted in recognition that rocks originally described as quartz monzonite now are more appropriately termed granite. Such lithic designations may be modernized when the new classification is widely adopted. If heterogeneous bodies of plutonic rock have been misleadingly identified with a single compositional term, such as "gabbro," the adoption of a neutral term, such as "intrusion" or "pluton," may be advisable.

Article 19.—**Revision.** Revision involves either minor changes in the definition of one or both boundaries of a unit, or in the unit's rank.

Remarks. (a) **Boundary change.**—Revision is justifiable if a minor change in boundary or content will make a unit more natural and useful. If revision modifies only a minor part of the content of a previously established unit, the original name may be retained.

(b) **Change in rank.**—Change in rank of a stratigraphic or temporal unit requires neither redefinition of its boundaries nor alteration of the geographic part of its name. A member may become a formation or vice versa, a formation may become a group or vice versa, and a lithodeme may become a suite or vice versa.

(c) **Examples of changes from area to area.**—The Conasauga Shale is recognized as a formation in Georgia and as a group in eastern Tennessee; the Osgood Formation, Laurel Limestone, and Waldron Shale in Indiana are classed as members of the Wayne Formation in a part of Tennessee; the Virgelle Sandstone is a formation in western Montana and a member of the Eagle Sandstone in central Montana; the Skull Creek Shale and the Newcastle Sandstone in North Dakota are members of the Ashville Formation in Manitoba.

(d) **Example of change in single area.**—The rank of a unit may be changed without changing its content. For example, the Madison Limestone of early work in Montana later became the Madison Group, containing several formations.

(e) **Retention of type section.**—When the rank of a geologic unit is changed, the original type section or type locality is retained for the newly ranked unit (see Article 22c).

(f) **Different geographic name for a unit and its parts.**—In changing the rank of a unit, the same name may not be applied both to the unit as a whole and to a part of it. For example, the Astoria Group should not contain an Astoria Sandstone, nor the Washington Formation, a Washington Sandstone Member.

(g) **Undesirable restriction.**—When a unit is divided into two or more of the same rank as the original, the original name should not be used for any of the divisions. Retention of the old name for one of the units precludes use of the name in a term of higher rank. Furthermore, in order to understand an author's meaning, a later reader would have to know about the modification and its date, and whether the author is following the original or the modified usage. For these reasons, the normal practice is to raise the rank of an established unit when units of the same rank are recognized and mapped within it.

Article 20.—**Abandonment.** An improperly defined or obsolete stratigraphic, lithodemic, or temporal unit may be formally abandoned, provided that (a) sufficient justification is presented to demonstrate a concern for nomenclatural stability, and (b) recommendations are made for the classification and nomenclature to be used in its place.

Remarks. (a) **Reasons for abandonment.**—A formally defined unit may be abandoned by the demonstration of synonymy or homonymy, of assignment to an improper category (for example, definition of a lithostratigraphic unit in a chronostratigraphic sense), or of other direct violations of a stratigraphic code or procedures

prevailing at the time of the original definition. Disuse, or the lack of need or useful purpose for a unit, may be a basis for abandonment; so, too, may widespread misuse in diverse ways which compound confusion. A unit also may be abandoned if it proves impracticable, neither recognizable nor mappable elsewhere.

(b) **Abandoned names.**—A name for a lithostratigraphic or lithodemic unit, once applied and then abandoned, is available for some other unit only if the name was introduced casually, or if it has been published only once in the last several decades and is not in current usage, and if its reintroduction will cause no confusion. An explanation of the history of the name and of the new usage should be a part of the designation.

(c) **Obsolete names.**—Authors may refer to national and provincial records of stratigraphic names to determine whether a name is obsolete (see Article 7b).

(d) **Reference to abandoned names.**—When it is useful to refer to an obsolete or abandoned formal name, its status is made clear by some such term as "abandoned" or "obsolete," and by using a phrase such as "La Plata Sandstone of Cross (1898)". (The same phrase also is used to convey that a named unit has not yet been adopted for usage by the organization involved.)

(e) **Reinstatement.**—A name abandoned for reasons that seem valid at the time, but which subsequently are found to be erroneous, may be reinstated. Example: the Washakie Formation, defined in 1869, was abandoned in 1918 and reinstated in 1973.

CODE AMENDMENT

Article 21.—**Procedure for Amendment.** Additions to, or changes of, this Code may be proposed in writing to the Commission by any geoscientist at anytime. If accepted for consideration by a majority vote of the Commission, they may be adopted by a two-thirds vote of the Commission at an annual meeting not less than a year after publication of the proposal.

FORMAL UNITS DISTINGUISHED BY CONTENT, PROPERTIES, OR PHYSICAL LIMITS

LITHOSTRATIGRAPHIC UNITS

Nature and Boundaries

Article 22.—**Nature of Lithostratigraphic Units.** A lithostratigraphic unit is a defined body of sedimentary, extrusive igneous, metasedimentary, or metavolcanic strata which is distinguished and delimited on the basis of lithic characteristics and stratigraphic position. A lithostratigraphic unit generally conforms to the Law of Superposition and commonly is stratified and tabular in form.

Remarks. (a) **Basic units.**—Lithostratigraphic units are the basic units of general geologic work and serve as the foundation for delineating strata, local and regional structure, economic resources, and geologic history in regions of stratified rocks. They are recognized and defined by observable rock characteristics; boundaries may be placed at clearly distinguished contacts or drawn arbitrarily within a zone of gradation. Lithification or cementation is not a necessary property; clay, gravel, till, and other unconsolidated deposits may constitute valid lithostratigraphic units.

(b) **Type section and locality.**—The definition of a lithostratigraphic unit should be based, if possible, on a stratotype consisting of readily accessible rocks in place, e.g., in outcrops, excavations, and mines, or of rocks accessible only to remote sampling devices, such as those in drill holes and underwater. Even where remote methods are used, definitions must be based on lithic criteria and not on the geophysical characteristics of the rocks, nor the implied age of their contained fossils. Definitions must be based on descriptions of actual rock material. Regional validity must be demonstrated for all such units. In regions where the stratigraphy has been established through studies of surface exposures, the naming of new units in the subsurface is justified only where the subsurface section differs materially from the surface section, or where there is doubt as to the equivalence of a sub-

surface and a surface unit. The establishment of subsurface reference sections for units originally defined in outcrop is encouraged.

(c) **Type section never changed.**—The definition and name of a lithostratigraphic unit are established at a type section (or locality) that, once specified, must not be changed. If the type section is poorly designated or delimited, it may be redefined subsequently. If the originally specified stratotype is incomplete, poorly exposed, structurally complicated, or unrepresentative of the unit, a principal reference section or several reference sections may be designated to supplement, but not to supplant, the type section (Article 8e).

(d) **Independence from inferred geologic history.**—Inferred geologic history, depositional environment, and biological sequence have no place in the definition of a lithostratigraphic unit, which must be based on composition and other lithic characteristics; nevertheless, considerations of well-documented geologic history properly may influence the choice of vertical and lateral boundaries of a new unit. Fossils may be valuable during mapping in distinguishing between two lithologically similar, non-contiguous lithostratigraphic units. The fossil content of a lithostratigraphic unit is a legitimate lithic characteristic; for example, oyster-rich sandstone, coquina, coral reef, or graptolitic shale. Moreover, otherwise similar units, such as the Formación Mendez and Formación Velasco mudstones, may be distinguished on the basis of coarseness of contained fossils (foraminifera).

(e) **Independence from time concepts.**—The boundaries of most lithostratigraphic units may transgress time horizons, but some may be approximately synchronous. Inferred time-spans, however measured, play no part in differentiating or determining the boundaries of any lithostratigraphic unit. Either relatively short or relatively long intervals of time may be represented by a single unit. The accumulation of material assigned to a particular unit may have begun or ended earlier in some localities than in others; also, removal of rock by erosion, either within the time-span of deposition of the unit or later, may reduce the time-span represented by the unit locally. The body in some places may be entirely younger than in other places. On the other hand, the establishment of formal units that straddle known, identifiable, regional disconformities is to be avoided, if at all possible. Although concepts of time or age play no part in defining lithostratigraphic units nor in determining their boundaries, evidence of age may aid recognition of similar lithostratigraphic units at localities far removed from the type sections or areas.

(f) **Surface form.**—Erosional morphology or secondary surface form may be a factor in the recognition of a lithostratigraphic unit, but properly should play a minor part at most in the definition of such units. Because the surface expression of lithostratigraphic units is an important aid in mapping, it is commonly advisable, where other factors do not countervail, to define lithostratigraphic boundaries so as to coincide with lithic changes that are expressed in topography.

(g) **Economically exploited units.**—Aquifers, oil sands, coal beds, and quarry layers are, in general, informal units even though named. Some such units, however, may be recognized formally as beds, members, or formations because they are important in the elucidation of regional stratigraphy.

(h) **Instrumentally defined units.**—In subsurface investigations, certain bodies of rock and their boundaries are widely recognized on borehole geophysical logs showing their electrical resistivity, radioactivity, density, or other physical properties. Such bodies and their boundaries may or may not correspond to formal lithostratigraphic units and their boundaries. Where other considerations do not countervail, the boundaries of subsurface units should be defined so as to correspond to useful geophysical markers; nevertheless, units defined exclusively on the basis of remotely sensed physical properties, although commonly useful in stratigraphic analysis, stand completely apart from the hierarchy of formal lithostratigraphic units and are considered informal.

(i) **Zone.**—As applied to the designation of lithostratigraphic units, the term "zone" is informal. Examples are "producing," "mineralized zone," "metamorphic zone," and "heavy-mineral zone." A zone may include all or parts of a bed, a member, a formation, or even a group.

(j) **Cyclothems.**—Cyclic or rhythmic sequences of sedimentary rocks, whose repetitive divisions have been named cyclothems, have been recognized in sedimentary basins around the world. Some cyclothems have been identified by geographic names, but such names are considered informal. A clear distinction must be maintained between the division of a stratigraphic column into cyclothems and its

division into groups, formations, and members. Where a cyclothem is identified by a geographic name, the word *cyclothem* should be part of the name, and the geographic term should not be the same as that of any formal unit embraced by the cyclothem.

(k) **Soils and paleosols.**—Soils and paleosols are layers composed of the in-situ products of weathering of older rocks which may be of diverse composition and age. Soils and paleosols differ in several respects from lithostratigraphic units, and should not be treated as such (see "Pedostratigraphic Units," Articles 55 et seq).

(l) **Depositional facies.**—Depositional facies are informal units, whether objective (conglomeratic, black shale, graptolitic) or genetic and environmental (platform, turbiditic, fluvial), even when a geographic term has been applied, e.g., Lantz Mills facies. Descriptive designations convey more information than geographic terms and are preferable.

Article 23.—Boundaries. Boundaries of lithostratigraphic units are placed at positions of lithic change. Boundaries are placed at distinct contacts or may be fixed arbitrarily within zones of gradation (Fig. 2a). Both vertical and lateral boundaries are based on the lithic criteria that provide the greatest unity and utility.

Remarks. (a) **Boundary in a vertically gradational sequence.**— A named lithostratigraphic unit is preferably bounded by a single lower and a single upper surface so that the name does not recur in a normal stratigraphic succession (see Remark b). Where a rock unit passes vertically into another by intergrading or interfingering of two or more kinds of rock, unless the gradational strata are sufficiently thick to warrant designation of a third, independent unit, the boundary is necessarily arbitrary and should be selected on the basis of practicality (Fig. 2b). For example, where a shale unit overlies a unit of interbedded limestone and shale, the boundary commonly is placed at the top of the highest readily traceable limestone bed. Where a sandstone unit grades upward into shale, the boundary may be so gradational as to be difficult to place even arbitrarily; ideally it should be drawn at the level where the rock is composed of one-half of each component. Because of creep in outcrops and caving in boreholes, it is generally best to define such arbitrary boundaries by the highest occurrence of a particular rock type, rather than the lowest.

(b) **Boundaries in lateral lithologic change.**—Where a unit changes laterally through abrupt gradation into, or intertongues with, a markedly different kind of rock, a new unit should be proposed for the different rock type. An arbitrary lateral boundary may be placed between the two equivalent units. Where the area of lateral intergradation or intertonguing is sufficiently extensive, a transitional interval of interbedded rocks may constitute a third independent unit (Fig. 2c). Where tongues (Article 25b) of formations are mapped separately or otherwise set apart without being formally named, the unmodified formation name should not be repeated in a normal stratigraphic sequence, although the modified name may be repeated in such phrases as "lower tongue of Mancos Shale" and "upper tongue of Mancos Shale." To show the order of superposition on maps and cross sections, the unnamed tongues may be distinguished informally (Fig. 2d) by number, letter, or other means. Such relationships may also be dealt with informally through the recognition of depositional facies (Article 22–1).

(c) **Key beds used for boundaries.**—Key beds (Article 26b) may be used as boundaries for a formal lithostratigraphic unit where the internal lithic characteristics of the unit remain relatively constant. Even though bounding key beds may be traceable beyond the area of the diagnostic overall rock type, geographic extension of the lithostratigraphic unit bounded thereby is not necessarily justified. Where the rock between key beds becomes drastically different from that of the type locality, a new name should be applied (Fig. 2e), even though the key beds are continuous (Article 26b). Stratigraphic and sedimentologic studies of stratigraphic units (usually informal) bounded by key beds may be very informative and useful, especially in subsurface work where the key beds may be recognized by their geophysical signatures. Such units, however, may be a kind of chronostratigraphic, rather than lithostratigraphic, unit (Article 75, 75c), although others are diachronous because one, or both, of the key beds are also diachronous.

(d) **Unconformities as boundaries.**—Unconformities, where recognizable objectively on lithic criteria, are ideal boundaries for lithostratigraphic units. However, a sequence of similar rocks may include an obscure unconformity so that separation into two units may be desirable but impracticable. If no lithic distinction adequate to define a widely recognizable boundary can be made, only one unit should be recognized, even though it may include rock that accumulated in different epochs, periods, or eras.

(e) **Correspondence with genetic units.**—The boundaries of lithostratigraphic units should be chosen on the basis of lithic changes and, where feasible, to correspond with the boundaries of genetic units, so that subsequent studies of genesis will not have to deal with units that straddle formal boundaries.

Ranks of Lithostratigraphic Units

Article 24.—Formation. The formation is the fundamental unit in lithostratigraphic classification. A formation is a body of rock identified by lithic characteristics and stratigraphic position; it is prevailingly but not necessarily tabular and is mappable at the Earth's surface or traceable in the subsurface.

Remarks. (a) **Fundamental unit.**—Formations are the basic lithostratigraphic units used in describing and interpreting the geology of a region. The limits of a formation normally are those surfaces of lithic change that give it the greatest practicable unity of constitution. A formation may represent a long or short time interval, may be composed of materials from one or several sources, and may include breaks in deposition (see Article 23d).

(b) **Content.**—A formation should possess some degree of internal lithic homogeneity or distinctive lithic features. It may contain between its upper and lower limits (i) rock of one lithic type, (ii) repetitions of two or more lithic types, or (iii) extreme lithic heterogeneity which in itself may constitute a form of unity when compared to the adjacent rock units.

(c) **Lithic characteristics.**—Distinctive lithic characteristics include chemical and mineralogical composition, texture, and such supplementary features as color, primary sedimentary or volcanic structures, fossils (viewed as rock-forming particles), or other organic content (coal, oil-shale). A unit distinguishable only by the taxonomy of its fossils is not a lithostratigraphic but a biostratigraphic unit (Article 48). Rock type may be distinctively represented by electrical, radioactive, seismic, or other properties (Article 22h), but these properties by themselves do not describe adequately the lithic character of the unit.

(d) **Mappability and thickness.**—The proposal of a new formation must be based on tested mappability. Well-established formations commonly are divisible into several widely recognizable lithostratigraphic units; where formal recognition of these smaller units serves a useful purpose, they may be established as members and beds, for which the requirement of mappability is not mandatory. A unit formally recognized as a formation in one area may be treated elsewhere as a group, or as a member of another formation, without change of name. Example: the Niobrara is mapped at different places as a member of the Mancos Shale, of the Cody Shale, or of the Colorado Shale, and also as the Niobrara Formation, as the Niobrara Limestone, and as the Niobrara Shale.

Thickness is not a determining parameter in dividing a rock succession into formations; the thickness of a formation may range from a feather edge at its depositional or erosional limit to thousands of meters elsewhere. No formation is considered valid that cannot be delineated at the scale of geologic mapping practiced in the region when the formation is proposed. Although representation of a formation on maps and cross sections by a labeled line may be justified, proliferation of such exceptionally thin units is undesirable. The methods of subsurface mapping permit delineation of units much thinner than those usually practicable for surface studies; before such thin units are formalized, consideration should be given to the effect on subsequent surface and subsurface studies.

(e) **Organic reefs and carbonate mounds.**—Organic reefs and carbonate mounds ("buildups") may be distinguished formally, if desirable, as formations distinct from their surrounding, thinner, temporal equivalents. For the requirements of formalization, see Article 30f.

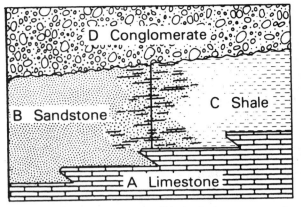

A.--Boundaries at sharp lithologic contacts and in laterally gradational sequence.

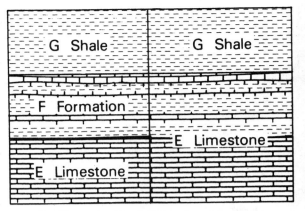

B.--Alternative boundaries in a vertically gradational or interlayered sequence.

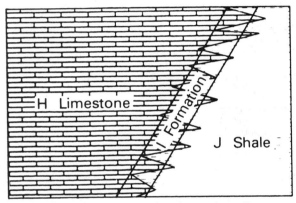

C.--Possible boundaries for a laterally intertonguing sequence.

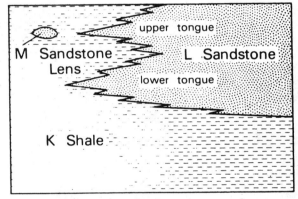

D.--Possible classification of parts of an intertonguing sequence.

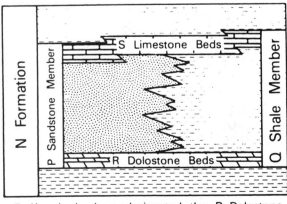

E.--Key beds, here designated the R Dolostone Beds and the S Limestone Beds, are used as boundaries to distinguish the Q Shale Member from the other parts of the N Formation. A lateral change in composition between the key beds requires that another name, P Sandstone Member, be applied. The key beds are part of each member.

EXPLANATION

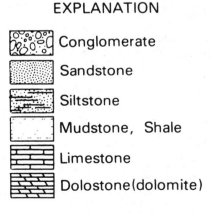

FIG. 2. Diagrammatic examples of lithostratigraphic boundaries and classfication.

(f) **Interbedded volcanic and sedimentary rock.**—Sedimentary rock and volcanic rock that are interbedded may be assembled into a formation under one name which should indicate the predominant or distinguishing lithology, such as Mindego Basalt.

(g) **Volcanic rock.**—Mappable distinguishable sequences of stratified volcanic rock should be treated as formations or lithostratigraphic units of higher or lower rank. A small intrusive component of a dominantly stratiform volcanic assemblage may be treated informally.

(h) **Metamorphic rock.**—Formations composed of low-grade metamorphic rock (defined for this purpose as rock in which primary structures are clearly recognizable) are, like sedimentary formations, distinguished mainly by lithic characteristics. The mineral facies may differ from place to place, but these variations do not require definition of a new formation. High-grade metamorphic rocks whose relation to established formations is uncertain are treated as lithodemic units (see Articles 31 et seq).

Article 25.—**Member.** A member is the formal lithostratigraphic unit next in rank below a formation and is always a part of some formation. It is recognized as a named entity within a formation because it possesses characteristics distinguishing it from adjacent parts of the formation. A formation need not be divided into members unless a useful purpose is served by doing so. Some formations may be divided completely into members; others may have only certain parts designated as members; still others may have no members. A member may extend laterally from one formation to another.

Remarks. (a) **Mapping of members.**—A member is established when it is advantageous to recognize a particular part of a heterogeneous formation. A member, whether formally or informally designated, need not be mappable at the scale required for formations. Even if all members of a formation are locally mappable, it does not follow that they should be raised to formational rank, because proliferation of formation names may obscure rather than clarify relations with other areas.

(b) **Lens and tongue.**—A geographically restricted member that terminates on all sides within a formation may be called a lens (lentil). A wedging member that extends outward beyond a formation or wedges ("pinches") out within another formation may be called a tongue.

(c) **Organic reefs and carbonate mounds.**—Organic reefs and carbonate mounds may be distinguished formally, if desirable, as members within a formation. For the requirements of formalization, see Article 30f.

(d) **Division of members.**—A formally or informally recognized division of a member is called a bed or beds, except for volcanic flow-rocks, for which the smallest formal unit is a flow. Members may contain beds or flows, but may never contain other members.

(e) **Laterally equivalent members.**—Although members normally are in vertical sequence, laterally equivalent parts of a formation that differ recognizably may also be considered members.

Article 26.—**Bed(s).** A bed, or beds, is the smallest formal lithostratigraphic unit of sedimentary rocks.

Remarks. **(a) Limitations.**—The designation of a bed or a unit of beds as a formally named lithostratigraphic unit generally should be limited to certain distinctive beds whose recognition is particularly useful. Coal beds, oil sands, and other beds of economic importance commonly are named, but such units and their names usually are not a part of formal stratigraphic nomenclature (Articles 22g and 30g).

(b) **Key or marker beds.**—A key or marker bed is a thin bed of distinctive rock that is widely distributed. Such beds may be named, but usually are considered informal units. Individual key beds may be traced beyond the lateral limits of a particular formal unit (Article 23c).

Article 27.—**Flow.** A flow is the smallest formal lithostratigraphic unit of volcanic flow rocks. A flow is a discrete, extrusive, volcanic body distinguishable by texture, composition, order of superposition, paleomagnetism, or other objective criteria. It is part of a member and thus is equivalent in rank to a bed or beds of sedimentary-rock classification. Many flows are informal units. The designation and naming of flows as formal rock-stratigraphic units should be limited to those that are distinctive and widespread.

Article 28.—**Group.** A group is the lithostratigraphic unit next higher in rank to formation; a group may consist entirely of named formations, or alternatively, need not be composed entirely of named formations.

Remarks. (a) **Use and content.**—Groups are defined to express the natural relationships of associated formations. They are useful in small-scale mapping and regional stratigraphic analysis. In some reconnaissance work, the term "group" has been applied to lithostratigraphic units that appear to be divisible into formations, but have not yet been so divided. In such cases, formations may be erected subsequently for one or all of the practical divisions of the group.

(b) **Change in component formations.**—The formations making up a group need not necessarily be everywhere the same. The Rundle Group, for example, is widespread in western Canada and undergoes several changes in formational content. In southwestern Alberta, it comprises the Livingstone, Mount Head, and Etherington Formations in the Front Ranges, whereas in the foothills and subsurface of the adjacent plains, it comprises the Pekisko, Shunda, Turner Valley, and Mount Head Formations. However, a formation or its parts may not be assigned to two vertically adjacent groups.

(c) **Change in rank.**—The wedge-out of a component formation or formations may justify the reduction of a group to formation rank, retaining the same name. When a group is extended laterally beyond where it is divided into formations, it becomes in effect a formation, even if it is still called a group. When a previously established formation is divided into two or more component units that are given formal formation rank, the old formation, with its old geographic name, should be raised to group status. Raising the rank of the unit is preferable to restricting the old name to a part of its former content, because a change in rank leaves the sense of a well-established unit unchanged (Articles 19b, 19g).

Article 29.—**Supergroup.** A supergroup is a formal assemblage of related or superposed groups, or of groups and formations. Such units have proved useful in regional and provincial syntheses. Supergroups should be named only where their recognition serves a clear purpose.

Remark. (a) **Misuse of "series" for group or supergroup.**—Although "series" is a useful general term, it is applied formally only to a chronostratigraphic unit and should not be used for a lithostratigraphic unit. The term "series" should no longer be employed for an assemblage of formations or an assemblage of formations and groups, as it has been, especially in studies of the Precambrian. These assemblages are groups or supergroups.

Lithostratigraphic Nomenclature

Article 30.—**Compound Character.** The formal name of a lithostratigraphic unit is compound. It consists of a geographic name combined with a descriptive lithic term or with the appropriate rank term, or both. Initial letters of all words used in forming the names of formal rock-stratigraphic units are capitalized.

Remarks. (a) **Omission of part of a name.**—Where frequent repetition would be cumbersome, the geographic name, the lithic term, or the rank term may be used alone, once the full name has been introduced; as "the Burlington," "the limestone," or "the formation," for the Burlington Limestone.

(b) **Use of simple lithic terms.**—The lithic part of the name should indicate the predominant or diagnostic lithology, even if subordinate lithologies are included. Where a lithic term is used in the name of a lithostratigraphic unit, the simplest generally

acceptable term is recommended (for example, limestone, sandstone, shale, tuff, quartzite). Compound terms (for example, clay shale) and terms that are not in common usage (for example, calcirudite, orthoquartzite) should be avoided. Combined terms, such as "sand and clay," should not be used for the lithic part of the names of lithostratigraphic units, nor should an adjective be used between the geographic and the lithic terms, as "Chattanooga Black Shale" and "Biwabik Iron-Bearing Formation."

(c) **Group names.**—A group name combines a geographic name with the term "group," and no lithic designation is included; for example, San Rafael Group.

(d) **Formation names.**—A formation name consists of a geographic name followed by a lithic designation or by the word "formation." Examples: Dakota Sandstone, Mitchell Mesa Rhyolite, Monmouth Formation, Halton Till.

(e) **Member names.**—All member names include a geographic term and the word "member;" some have an intervening lithic designation, if useful; for example, Wedington Sandstone Member of the Fayetteville Shale. Members designated solely by lithic character (for example, siliceous shale member), by position (upper, lower), or by letter or number, are informal.

(f) **Names of reefs.**—Organic reefs identified as formations or members are formal units only where the name combines a geographic name with the appropriate rank term, e.g., Leduc Formation (a name applied to the several reefs enveloped by the Ireton Formation), Rainbow Reef Member.

(g) **Bed and flow names.**—The names of beds or flows combine a geographic term, a lithic term, and the term "bed" or "flow;" for example, Knee Hills Tuff Bed, Ardmore Bentonite Beds, Negus Variolitic Flows.

(h) **Informal units.**—When geographic names are applied to such informal units as oil sands, coal beds, mineralized zones, and informal members (see Articles 22g and 26a), the unit term should not be capitalized. A name is not necessarily formal because it is capitalized, nor does failure to capitalize a name render it informal. Geographic names should be combined with the terms "formation" or "group" only in formal nomenclature.

(i) **Informal usage of identical geographic names.**—The application of identical geographic names to several minor units in one vertical sequence is considered informal nomenclature (lower Mount Savage coal, Mount Savage fireclay, upper Mount Savage coal, Mount Savage rider coal, and Mount Savage sandstone). The application of identical geographic names to the several lithologic units constituting a cyclothem likewise is considered informal.

(j) **Metamorphic rock.**—Metamorphic rock recognized as a normal stratified sequence, commonly low-grade metavolcanic or metasedimentary rocks, should be assigned to named groups, formations, and members, such as the Deception Rhyolite, a formation of the Ash Creek Group, or the Bonner Quartzite, a formation of the Missoula Group. High-grade metamorphic and metasomatic rocks are treated as lithodemes and suites (see Articles 31, 33, 35).

(k) **Misuse of well-known name.**—A name that suggests some well-known locality, region, or political division should not be applied to a unit typically developed in another less well-known locality of the same name. For example, it would be inadvisable to use the name "Chicago Formation" for a unit in California.

LITHODEMIC UNITS

Nature and Boundaries

Article 31.—**Nature of Lithodemic Units.** A lithodemic[6] unit is a defined body of predominantly intrusive, highly deformed, and/or highly metamorphosed rock, distinguished and delimited on the basis of rock characteristics. In contrast to lithostratigraphic units, a lithodemic unit generally does not conform to the Law of Superposition. Its contacts with other rock units may be sedimentary, extrusive, intrusive, tectonic, or metamorphic (Fig. 3).

Remarks. (a) **Recognition and definition.**—Lithodemic units are defined and recognized by observable rock characteristics. They are the practical units of general geological work in terranes in which rocks generally lack primary stratification; in such terranes they serve as the foundation for studying, describing, and delineating lithology, local and regional structure, economic resources, and geologic history.

(b) **Type and reference localities.**—The definition of a lithodemic unit should be based on as full a knowledge as possible of its lateral and vertical variations and its contact relationships. For purposes of nomenclatural stability, a type locality and, wherever appropriate, reference localities should be designated.

(c) **Independence from inferred geologic history.**—Concepts based on inferred geologic history properly play no part in the definition of a lithodemic unit. Nevertheless, where two rock masses are lithically similar but display objective structural relations that preclude the possibility of their being even broadly of the same age, they should be assigned to different lithodemic units.

(d) **Use of "zone ."**—As applied to the designation of lithodemic units, the term "zone" is informal. Examples are: "mineralized zone," "contact zone," and "pegmatitic zone."

Article 32.—**Boundaries.** Boundaries of lithodemic units are placed at positions of lithic change. They may be placed at clearly distinguished contacts or within zones of gradation. Boundaries, both vertical and lateral, are based on the lithic criteria that provide the greatest unity and practical utility. Contacts with other lithodemic and lithostratigraphic units may be depositional, intrusive, metamorphic, or tectonic.

Remark. (a) **Boundaries within gradational zones.**—Where a lithodemic unit changes through gradation into, or intertongues with, a rock-mass with markedly different characteristics, it is usually desirable to propose a new unit. It may be necessary to draw an arbitrary boundary within the zone of gradation. Where the area of intergradation or inter-tonguing is sufficiently extensive, the rocks of mixed character may constitute a third unit.

Ranks of Lithodemic Units

Article 33.—**Lithodeme.** The lithodeme is the fundamental unit in lithodemic classification. A lithodeme is a body of intrusive, pervasively deformed, or highly metamorphosed rock, generally non-tabular and lacking primary depositional structures, and characterized by lithic homogeneity. It is mappable at the Earth's surface and traceable in the subsurface. For cartographic and hierarchical purposes, it is comparable to a formation (see Table 2).

Remarks. (a) **Content.**—A lithodeme should possess distinctive lithic features and some degree of internal lithic homogeneity. It may consist of (i) rock of one type, (ii) a mixture of rocks of two or more types, or (iii) extreme heterogeneity of composition, which may constitute in itself a form of unity when compared to adjoining rock-masses (see also "complex," Article 37).

(b) **Lithic characteristics.**—Distinctive lithic characteristics may include mineralogy, textural features such as grain size, and structural features such as schistose or gneissic structure. A unit distinguishable from its neighbors only by means of chemical analysis is informal.

(c) **Mappability.**—Practicability of surface or subsurface mapping is an essential characteristic of a lithodeme (see Article 24d).

Article 34.—**Division of Lithodemes.** Units below the rank of lithodeme are informal.

Article 35.—**Suite.** A *suite* (metamorphic suite, intrusive suite, plutonic suite) is the lithodemic unit next higher in rank to lithodeme. It comprises two or more associated lithodemes of the same class (e.g., plutonic, metamorphic). For cartographic and hierarchical purposes, suite is comparable to group (see Table 2).

---

[6]From the Greek *demas, -os:* "living body, frame".

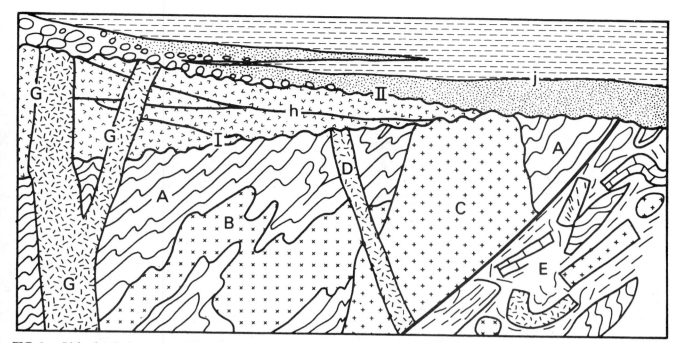

FIG. 3.—Lithodemic (upper case) and lithostratigraphic (lower case) units. A *lithodeme* of *gneiss* (**A**) contains an *intrusion* of diorite (**B**) that was deformed with the gneiss. A and B may be treated jointly as a *complex*. A younger *granite* (**C**) is cut by a dike of *syenite* (**D**), that is cut in turn by unconformity I. All the foregoing are in fault contact with a *structural complex* (**E**). A *volcanic complex* (**G**) is built upon unconformity I, and its feeder dikes cut the unconformity. Laterally equivalent volcanic strata in orderly, mappable succession (**h**) are treated as lithostratigraphic units. A *gabbro* feeder (**G'**), to the volcanic complex, where surrounded by gneiss is readily distinguished as a separate lithodeme and named as a *gabbro* or an *intrusion*. All the foregoing are overlain, at unconformity II, by sedimentary rocks (**j**) divided into formation and members.

Remarks. (a) **Purpose.**—Suites are recognized for the purpose of expressing the natural relations of associated lithodemes having significant lithic features in common, and of depicting geology at compilation scales too small to allow delineation of individual lithodemes. Ideally, a suite consists entirely of named lithodemes, but may contain both named and unnamed units.

(b) **Change in component units.**—The named and unnamed units constituting a suite may change from place to place, so long as the original sense of natural relations and of common lithic features is not violated.

(c) **Change in rank.**—Traced laterally, a suite may lose all of its formally named divisions but remain a recognizable, mappable entity. Under such circumstances, it may be treated as a lithodeme but retain the same name. Conversely, when a previously established lithodeme is divided into two or more mappable divisions, it may be desirable to raise its rank to suite, retaining the original geographic component of the name. To avoid confusion, the original name should not be retained for one of the divisions of the original unit (see Article 19g).

Article 36.—**Supersuite.** A supersuite is the unit next higher in rank to a suite. It comprises two or more suites or complexes having a degree of natural relationship to one another, either in the vertical or the lateral sense. For cartographic and hierarchical purposes, supersuite is similar in rank to supergroup.

Article 37.—**Complex.** An assemblage or mixture of rocks of *two or more genetic classes*, i.e., igneous, sedimentary, or metamorphic, with or without highly complicated structure, may be named a *complex*. The term "complex" takes the place of the lithic or rank term (for example, Boil Mountain Complex, Franciscan Complex) and, although unranked, commonly is comparable to suite or supersuite and is named in the same manner (Articles 41, 42).

Remarks (a) **Use of "complex."**—Identification of an assemblage of diverse rocks as a complex is useful where the mapping of each separate lithic component is impractical at ordinary mapping scales. "Complex" is unranked but commonly comparable to suite or supersuite; therefore, the term may be retained if subsequent, detailed mapping distinguishes some or all of the component lithodemes or lithostratigraphic units.

(b) **Volcanic complex.**—Sites of persistent volcanic activity commonly are characterized by a diverse assemblage of extrusive volcanic rocks, related intrusions, and their weathering products. Such an assemblage may be designated a *volcanic complex*.

(c) **Structural complex.**—In some terranes, tectonic processes (e.g., shearing, faulting) have produced heterogeneous mixtures or disrupted bodies of rock in which some individual components are too small to be mapped. *Where there is no doubt that the mixing or disruption is due to tectonic processes*, such a mixture may be designated as a structural complex, whether it consists of two or more classes of rock, or a single class only. A simpler solution for some mapping purposes is to indicate intense deformation by an overprinted pattern.

(d) **Misuse of "complex".**—Where the rock assemblage to be united under a single, formal name consists of diverse types of a *single class* of rock, as in many terranes that expose a variety of either intrusive igneous or high-grade metamorphic rocks, the term "intrusive suite," "plutonic suite," or "metamorphic suite" should be used, rather than the unmodified term "complex." Exceptions to this rule are the terms *structural complex* and *volcanic complex* (see Remarks c and b, above).

Article 38.—**Misuse of "Series" for Suite, Complex, or Supersuite.** The term "series" has been employed for an assemblage of lithodemes or an assemblage of lithodemes and suites, especially in studies of the Precambrian. This practice now is regarded as improper; these assemblages are suites, complexes, or supersuites. The term "series" also has been applied to a sequence of rocks resulting from a

succession of eruptions or intrusions. In these cases a different term should be used; "group" should replace "series" for volcanic and low-grade metamorphic rocks, and "intrusive suite" or "plutonic suite" should replace "series" for intrusive rocks of group rank.

Lithodemic Nomenclature

Article 39.—**General Provisions.** The formal name of a lithodemic unit is compound. It consists of a geographic name combined with a descriptive or appropriate rank term. The principles for the selection of the geographic term, concerning suitability, availability, priority, etc, follow those established in Article 7, where the rules for capitalization are also specified.

Article 40.—**Lithodeme Names.** The name of a lithodeme combines a geographic term with a lithic or descriptive term, e.g., Killarney Granite, Adamant Pluton, Manhattan Schist, Skaergaard Intrusion, Duluth Gabbro. The term *formation* should not be used.

Remarks. (a) **Lithic term.**—The lithic term should be a common and familiar term, such as schist, gneiss, gabbro. Specialized terms and terms not widely used, such as websterite and jacupirangite, and compound terms, such as graphitic schist and augen gneiss, should be avoided.

(b) **Intrusive and plutonic rocks.**—Because many bodies of intrusive rock range in composition from place to place and are difficult to characterize with a single lithic term, and because many bodies of plutonic rock are considered not to be intrusions, latitude is allowed in the choice of a lithic or descriptive term. Thus, the descriptive term should preferably be compositional (e.g., gabbro, granodiorite), but may, if necessary, denote form (e.g., dike, sill), or be neutral (e.g., intrusion, pluton[7]). In any event, specialized compositional terms not widely used are to be avoided, as are form terms that are not widely used, such as bysmalith and chonolith. Terms implying genesis should be avoided as much as possible, because interpretations of genesis may change.

Article 41.—**Suite Names.** The name of a suite combines a geographic term, the term "suite," and an adjective denoting the fundamental character of the suite; for example, Idaho Springs Metamorphic Suite, Tuolumne Intrusive Suite, Cassiar Plutonic Suite. The geographic name of a suite may not be the same as that of a component lithodeme (see Article 19f). Intrusive assemblages, however, may share the same geographic name if an intrusive lithodeme is representative of the suite.

Article 42.—**Supersuite Names.** The name of a supersuite combines a geographic term with the term "supersuite."

MAGNETOSTRATIGRAPHIC UNITS

Nature and Boundaries

Article 43.—**Nature of Magnetostratigraphic Units.** A magnetostratigraphic unit is a body of rock unified by specified remanent-magnetic properties and is distinct from underlying and overlying magnetostratigraphic units having different magnetic properties.

Remarks. (a) **Definition.**—Magnetostratigraphy is defined here as all aspects of stratigraphy based on remanent magnetism (paleomagnetic signatures). Four basic paleomagnetic phenomena can be determined or inferred from remanent magnetism: polarity, dipole-fieldpole position (including apparent polar wander), the non-dipole component (secular variation), and field intensity.

(b) **Contemporaneity of rock and remanent magnetism.**—Many paleomagnetic signatures reflect earth magnetism at the time the rock formed. Nevertheless, some rocks have been subjected subsequently to physical and/or chemical processes which altered the magnetic properties. For example, a body of rock may be heated above the blocking temperature or Curie point for one or more minerals, or a ferromagnetic mineral may be produced by low-temperature alteration long after the enclosing rock formed, thus acquiring a component of remanent magnetism reflecting the field at the time of alteration, rather than the time of original rock deposition or crystallization.

(c) **Designations and scope.**—The prefix *magneto* is used with an appropriate term to designate the aspect of remanent magnetism used to define a unit. The terms "magnetointensity" or "magnetosecularvariation" are possible examples. This Code considers only polarity reversals, which now are recognized widely as a stratigraphic tool. However, apparent-polar-wander paths offer increasing promise for correlations within Precambrian rocks.

Article 44.—**Definition of Magnetopolarity Unit.** A magnetopolarity unit is a body of rock unified by its remanent magnetic polarity and distinguished from adjacent rock that has different polarity.

Remarks. (a) **Nature.**—Magnetopolarity is the record in rocks of the polarity history of the Earth's magnetic-dipole field. Frequent past reversals of the polarity of the Earth's magnetic field provide a basis for magnetopolarity stratigraphy.

(b) **Stratotype.**—A stratotype for a magnetopolarity unit should be designated and the boundaries defined in terms of recognized lithostratigraphic and/or biostratigraphic units in the stratotype. The formal definition of a magnetopolarity unit should meet the applicable specific requirements of Articles 3 to 16.

(c) **Independence from inferred history.**—Definition of a magneto-polarity unit does not require knowledge of the time at which the unit acquired its remanent magnetism; its magnetism may be primary or secondary. Nevertheless, the unit's present polarity is a property that may be ascertained and confirmed by others.

(d) **Relation to lithostratigraphic and biostratigraphic units.**—Magnetopolarity units resemble lithostratigraphic and biostratigraphic units in that they are defined on the basis of an objective recognizable property, but differ fundamentally in that most magnetopolarity unit boundaries are thought not to be time transgressive. Their boundaries may coincide with those of lithostratigraphic or biostratigraphic units, or be parallel to but displaced from those of such units, or be crossed by them.

(e) **Relation of magnetopolarity units to chronostratigraphic units.**—Although transitions between polarity reversals are of global extent, a magnetopolarity unit does not contain within itself evidence that the polarity is primary, or criteria that permit its unequivocal recognition in chronocorrelative strata of other areas. Other criteria, such as paleontologic or numerical age, are required for both correlation and dating. Although polarity reversals are useful in recognizing chronostratigraphic units, magnetopolarity alone is insufficient for their definition.

Article 45.—**Boundaries.** The upper and lower limits of a magnetopolarity unit are defined by boundaries marking a change of polarity. Such boundaries may represent either a depositional discontinuity or a magnetic-field transition. The boundaries are either polarity-reversal horizons or polarity transition-zones, respectively.

Remark. (a) **Polarity-reversal horizons and transition-zones.**—A polarity-reversal horizon is either a single, clearly definable surface or a thin body of strata constituting a transitional interval across which a change in magnetic polarity is recorded. Polarity-reversal horizons describe transitional intervals of 1 m or less; where the change in polarity takes place over a stratigraphic interval greater than 1 m, the term "polarity transition-zone" should be used. Polarity-reversal horizons and polarity transition-zones provide the boundaries for polarity zones, although they may also be contained within a polarity zone where they mark an internal change subsidiary in rank to those at its boundaries.

---

[7]Pluton—a mappable body of plutonic rock.

Ranks of Magnetopolarity Units

Article 46.—**Fundamental Unit.** A polarity zone is the fundamental unit of magnetopolarity classification. A polarity zone is a unit of rock characterized by the polarity of its magnetic signature. Magnetopolarity zone, rather than polarity zone, should be used where there is risk of confusion with other kinds of polarity.

Remarks. (a) **Content.**—A polarity zone should possess some degree of internal homogeneity. It may contain rocks of (1) entirely or predominantly one polarity, or (2) mixed polarity.

(b) **Thickness and duration.**—The thickness of rock of a polarity zone or the amount of time represented should play no part in the definition of the zone. The polarity signature is the essential property for definition.

(c) **Ranks.**—When continued work at the stratotype for a polarity zone, or new work in correlative rocks elsewhere, reveals smaller polarity zones, these may be recognized formally as polarity subzones. If it should prove necessary or desirable to group polarity zones, these should be termed polarity superzones. The rank of a polarity unit may be changed when deemed appropriate.

Magnetopolarity Nomenclature

Article 47.—**Compound Name.** The formal name of a magnetopolarity zone should consist of a geographic name and the term *Polarity Zone*. The term may be modified by *Normal, Reversed,* or *Mixed* (example: Deer Park Reversed Polarity Zone). In naming or revising magnetopolarity units, appropriate parts of Articles 7 and 19 apply. The use of informal designations, e.g., numbers or letters, is not precluded.

BIOSTRATIGRAPHIC UNITS

Nature and Boundaries

Article 48.—**Nature of Biostratigraphic Units.** A biostratigraphic unit is a body of rock defined or characterized by its fossil content. The basic unit in biostratigraphic classification is the biozone, of which there are several kinds.

Remarks. (a) **Enclosing strata.**—Fossils that define or characterize a biostratigraphic unit commonly are contemporaneous with the body of rock that contains them. Some biostratigraphic units, however, may be represented only by their fossils, preserved in normal stratigraphic succession (e.g., on hardgrounds, in lag deposits, in certain types of remanié accumulations), which alone represent the rock of the biostratigraphic unit. In addition, some strata contain fossils derived from older or younger rocks or from essentially coeval materials of different facies; such fossils should not be used to define a biostratigraphic unit.

(b) **Independence from lithostratigraphic units.**—Biostratigraphic units are based on criteria which differ fundamentally from those for lithostratigraphic units. Their boundaries may or may not coincide with the boundaries of lithostratigraphic units, but they bear no inherent relation to them.

(c) **Independence from chronostratigraphic units.**—The boundaries of most biostratigraphic units, unlike the boundaries of chronostratigraphic units, are both characteristically and conceptually diachronous. An exception is an abundance biozone boundary that reflects a mass-mortality event. The vertical and lateral limits of the rock body that constitutes the biostratigraphic unit represent the limits in distribution of the defining biotic elements. The lateral limits never represent, and the vertical limits rarely represent, regionally synchronous events. Nevertheless, biostratigraphic units are effective for interpreting chronostratigraphic relations.

Article 49.—**Kinds of Biostratigraphic Units.** Three principal kinds of biostratigraphic units are recognized: *interval, assemblage,* and *abundance* biozones.

Remark: (a) **Boundary definitions.**—Boundaries of interval zones are defined by lowest and/or highest occurrences of single taxa; boundaries of some kinds of assemblage zones (Oppel or concurrent range zones) are defined by lowest and/or highest occurrences of more than one taxon; and boundaries of abundance zones are defined by marked changes in relative abundances of preserved taxa.

Article 50.—**Definition of Interval Zone.** An interval zone (or subzone) is the body of strata between two specified, documented lowest and/or highest occurrences of single taxa.

Remarks. (a) **Interval zone types.**—Three basic types of interval zones are recognized (Fig. 4). These include the range zones and interval zones of the International Stratigraphic Guide (ISSC, 1976, p. 53, 60) and are:

1. The interval between the documented lowest and highest occurrences of a single taxon (Fig. 4A). This is the *taxon range zone* of ISSC (1976, p. 53).

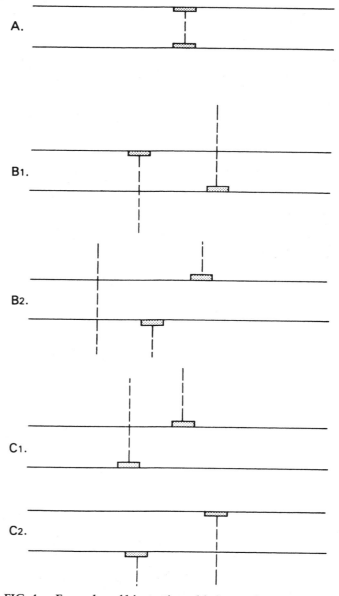

FIG. 4.—Examples of biostratigraphic interval zones. Vertical broken lines indicate ranges of taxa; bars indicate lowest or highest documented occurences.

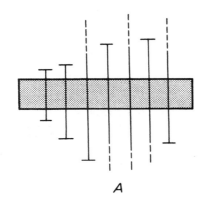

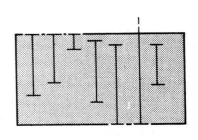

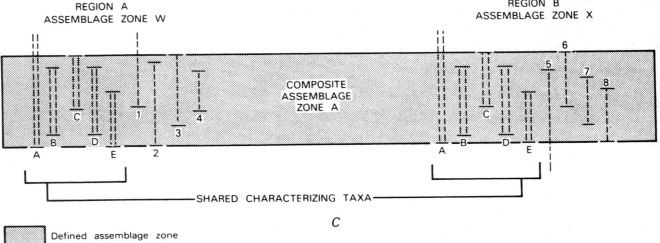

REGION A
ASSEMBLAGE ZONE W

REGION B
ASSEMBLAGE ZONE X

COMPOSITE
ASSEMBLAGE
ZONE A

SHARED CHARACTERIZING TAXA

C

☐ Defined assemblage zone

FIG. 5.—Examples of assemblage zone concepts.

2. The interval included between the documented lowest occurrence of one taxon and the documented highest occurrence of another taxon (Fig. 4B). When such occurrences result in stratigraphic overlap of the taxa (Fig. 4B-1), the interval zone is the *concurrent range zone* of ISSC (1976, p. 55), that involves only two taxa. When such occurrences do not result in stratigraphic overlap (Fig. 4B-2), but are used to partition the range of a third taxon, the interval is the *partial range zone* of George and others (1969).

3. The interval between documented successive lowest occurrences or successive highest occurrences of two taxa (Fig. 4C). When the interval is between successive documented lowest occurrences within an evolutionary lineage (Fig. 4C-1), it is the *lineage zone* of ISSC (1976, p. 58). When the interval is between successive lowest occurrences of unrelated taxa or between successive highest occurrences of either related or unrelated taxa (Fig. 4C-2), it is a kind of *interval zone* of ISSC (1976, p. 60).

(b) **Unfossiliferous intervals.**—Unfossiliferous intervals between or within biozones are the *barren interzones* and *intrazones* of ISSC (1976, p. 49).

Article 51.—**Definition of Assemblage Zone.** An assemblage zone is a biozone characterized by the association of three or more taxa. It may be based on all kinds of fossils present, or restricted to only certain kinds of fossils.

Remarks. (a) **Assemblage zone contents.**—An assemblage zone may consist of a geographically or stratigraphically restricted assemblage, or may incorporate two or more contemporaneous assemblages with shared characterizing taxa (*composite assemblage zones* of Kauffman, 1969) (Fig. 5c).

(b) **Assemblage zone types.**—In practice, two assemblage zone concepts are used:

1. The *assemblage zone* (or cenozone) of ISSC (1976, p. 50), which is characterized by taxa without regard to their range limits (Fig. 5a). Recognition of this type of assemblage zone can be aided by using techniques of multivariate analysis. Careful designation of the characterizing taxa is especially important.

2. The *Oppel zone*, or the *concurrent range zone* of ISSC (1976, p. 55, 57), a type of zone characterized by more than two taxa and having boundaries based on two or more documented first and/or last occurrences of the included characterizing taxa (Fig. 5b).

Article 52.—**Definition of Abundance Zone.** An abundance zone is a biozone characterized by quantitatively distinctive maxima of relative abundance of one or more taxa. This is the *acme zone* of ISSC (1976, p. 59).

Remark. (a) **Ecologic controls.**—The distribution of biotic assemblages used to characterize some assemblage and abundance biozones may reflect strong local ecological control. Biozones based on such assemblages are included within the concept of ecozones (Vella, 1964), and are informal.

Ranks of Biostratigraphic Units

Article 53.—**Fundamental Unit.** The fundamental unit of biostratigraphic classification is a biozone.

Remarks. (a) **Scope.**—A single body of rock may be divided into various kinds and scales of biozones or subzones, as discussed in

PEDOSTRATIGRAPHIC UNIT

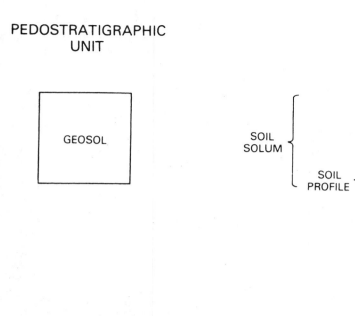

GEOSOL

PEDOLOGIC PROFILE OF A SOIL
(Ruhe, 1965; Pawluk, 1978)

SOIL SOLUM

SOIL PROFILE

O HORIZON	ORGANIC DEBRIS ON THE SOIL
A HORIZON	ORGANIC-MINERAL HORIZON
B HORIZON	HORIZON OF ILLUVIAL ACCUMULATION AND (OR) RESIDUAL CONCENTRATION
C HORIZON (WITH INDEFINITE LOWER BOUNDARY)	WEATHERED GEOLOGIC MATERIALS
R HORIZON OR BEDROCK	UNWEATHERED GEOLOGIC MATERIALS

FIG. 6.—Relationship between pedostratigraphic units and pedologic profiles. The base of a geosol is the lower clearly defined physical boundary of a pedologic horizon in a buried soil profile. In this example it is the lowest boundary of the B horizon because the base of the C horizon is not a clearly defined physical boundary. In other profiles the base may be the lower boundary of a C horizon.

the International Stratigraphic Guide (ISSC, 1976, p. 62). Such usage is recommended if it will promote clarity, but only the unmodified term *biozone* is accorded formal status.

(b) **Divisions.**—A biozone may be completely or partly divided into formally designated sub-biozones (subzones), if such divisions serve a useful purpose.

Biostratigraphic Nomenclature

Article 54.—**Establishing Formal Units.** Formal establishment of a biozone or subzone must meet the requirements of Article 3 and requires a unique name, a description of its content and its boundaries, reference to a stratigraphic sequence in which the zone is characteristically developed, and a discussion of its spatial extent.

Remarks. (a) **Name.**—The name, which is compound and designates the kind of biozone, may be based on:

1. One or two characteristic and common taxa that are restricted to the biozone, reach peak relative abundance within the biozone, or have their total stratigraphic overlap within the biozone. These names most commonly are those of genera or subgenera, binomial designations of species, or trinomial designations of subspecies. If names of the nominate taxa change, names of the zones should be changed accordingly. Generic or subgeneric names may be abbreviated. Trivial species or subspecies names should not be used alone because they may not be unique.

2. Combinations of letters derived from taxa which characterize the biozone. However, alpha-numeric code designations (e.g., N1, N2, N3 . . .) are informal and not recommended because they do not lend themselves readily to subsequent insertions, combinations, or eliminations. Biozonal systems based *only* on simple progressions of letters or numbers (e.g., A, B, C, or 1,2,3) are also not recommended.

(b) **Revision.**—Biozones and subzones are established empirically and may be modified on the basis of new evidence. Positions of established biozone or subzone boundaries may be stratigraphically refined, new characterizing taxa may be recognized, or original characterizing taxa may be superseded. If the concept of a particular biozone or subzone is substantially modified, a new unique designation is required to avoid ambiguity in subsequent citations.

(c) **Specifying kind of zone.**—Initial designation of a formally proposed biozone or subzone as an abundance zone, or as one of the types of interval zones, or assemblage zones (Articles 49–52), is strongly recommended. Once the type of biozone is clearly identified, the designation may be dropped in the remainder of a text (e.g., *Exus albus* taxon range zone to *Exus albus* biozone).

(d) **Defining taxa.**—Initial description or subsequent emendation of a biozone or subzone requires designation of the defining and characteristic taxa, and/or the documented first and last occurrences which mark the biozone or subzone boundaries.

(e) **Stratotypes.**—The geographic and stratigraphic position and boundaries of a formally proposed biozone or subzone should be defined precisely or characterized in one or more designated reference sections. Designation of a stratotype for each new biostratigraphic unit and of reference sections for emended biostratigraphic units is required.

PEDOSTRATIGRAPHIC UNITS

Nature and Boundaries

Article 55.—**Nature of Pedostratigraphic Units.** A pedostratigraphic unit is a body of rock that consists of one or more pedologic horizons developed in one or more lithostratigraphic, allostratigraphic, or lithodemic units (Fig. 6) and is overlain by one or more formally defined lithostratigraphic or allostratigraphic units.

Remarks. (a) **Definition.**—A pedostratigraphic[8] unit is a buried, traceable, three-dimensional body of rock that consists of one or more differentiated pedologic horizons.

(b) **Recognition.**—The distinguishing property of a pedostratigraphic unit is the presence of one or more distinct, differentiated, pedologic horizons. Pedologic horizons are products of soil development (pedogenesis) which occurred subsequent to formation of the lithostratigraphic, allostratigraphic, or lithodemic unit or units on

---

[8]Terminology related to pedostratigraphic classification is summarized on page 850.

which the buried soil was formed; these units are the parent materials in which pedogenesis occurred. Pedologic horizons are recognized in the field by diagnostic features such as color, soil structure, organic-matter accumulation, texture, clay coatings, stains, or concretions. Micromorphology, particle size, clay mineralogy, and other properties determined in the laboratory also may be used to identify and distinguish pedostratigraphic units.

(c) **Boundaries and stratigraphic position.**—The upper boundary of a pedostratigraphic unit is the top of the uppermost pedologic horizon formed by pedogenesis in a buried soil profile. The lower boundary of a pedostratigraphic unit is the lowest *definite* physical boundary of a pedologic horizon within a buried soil profile. The stratigraphic position of a pedostratigraphic unit is determined by its relation to overlying and underlying stratigraphic units (see Remark d).

(d) **Traceability.**—Practicability of subsurface tracing of the upper boundary of a buried soil is essential in establishing a pedostratigraphic unit because (1) few buried soils are exposed continuously for great distances, (2) the physical and chemical properties of a specific pedostratigraphic unit may vary greatly, both vertically and laterally, from place to place, and (3) pedostratigraphic units of different stratigraphic significance in the same region generally do not have unique identifying physical and chemical characteristics. Consequently, extension of a pedostratigraphic unit is accomplished by lateral tracing of the contact between a buried soil and an overlying, formally defined lithostratigraphic or allostratigraphic unit, or between a soil and two or more demonstrably correlative stratigraphic units.

(e) **Distinction from pedologic soils.**—Pedologic soils may include organic deposits (e.g., litter zones, peat deposits, or swamp deposits) that overlie or grade laterally into differentiated buried soils. The organic deposits are not products of pedogenesis, and O horizons are not included in a pedostratigraphic unit (Fig. 6); they may be classified as biostratigraphic or lithostratigraphic units. Pedologic soils also include the entire C horizon of a soil. The C horizon in pedology is not rigidly defined; it is merely the part of a soil profile that underlies the B horizon. The base of the C horizon in many soil profiles is gradational or unidentifiable; commonly it is placed arbitrarily. The need for clearly defined and easily recognized physical boundaries for a stratigraphic unit requires that the lower boundary of a pedostratigraphic unit be defined as the lowest *definite* physical boundary of a pedologic horizon in a buried soil profile, and part or all of the C horizon may be excluded from a pedostratigraphic unit.

(f) **Relation to saprolite and other weathered materials.**—A material derived by in situ weathering of lithostratigraphic, allostratigraphic, and(or) lithodemic units (e.g., saprolite, bauxite, residuum) may be the parent material in which pedologic horizons form, but is not a pedologic soil. A pedostratigraphic unit may be based on the pedologic horizons of a buried soil developed in the product of in-situ weathering, such as saprolite. The parents of such a pedostratigraphic unit are both the saprolite and, indirectly, the rock from which it formed.

(g) **Distinction from other stratigraphic units.**—A pedostratigraphic unit differs from other stratigraphic units in that (1) it is a product of surface alteration of one or more older material units by specific processes (pedogenesis), (2) its lithology and other properties differ markedly from those of the parent material(s), and (3) a single pedostratigraphic unit may be formed in situ in parent material units of diverse compositions and ages.

(h) **Independence from time concepts.**—The boundaries of a pedostratigraphic unit are time-transgressive. Concepts of time spans, however measured, play no part in defining the boundaries of a pedostratigraphic unit. Nonetheless, evidence of age, whether based on fossils, numerical ages, or geometrical or other relationships, may play an important role in distinguishing and identifying non-contiguous pedostratigraphic units at localities away from the type areas. The name of a pedostratigraphic unit should be chosen from a geographic feature in the type area, and not from a time span.

Pedostratigraphic Nomenclature and Unit

Article 56.—**Fundamental Unit.** The fundamental and only unit in pedostratigraphic classification is a geosol.

Article 57.—**Nomenclature.** The formal name of a pedostratigraphic unit consists of a geographic name combined with the term "geosol." Capitalization of the initial letter in each word serves to identify formal usage. The geographic name should be selected in accordance with recommendations in Article 7 and should not duplicate the name of another formal geologic unit. Names based on subjacent and superjacent rock units, for example the super-Wilcox–sub-Claiborne soil, are informal, as are those with time connotations (post-Wilcox–pre-Claiborne soil).

Remarks. (a) **Composite geosols.**—Where the horizons of two or more merged or "welded" buried soils can be distinguished, formal names of pedostratigraphic units based on the horizon boundaries can be retained. Where the horizon boundaries of the respective merged or "welded" soils cannot be distinguished, formal pedostratigraphic classification is abandoned and a combined name such as Hallettville-Jamesville geosol may be used informally.

(b) **Characterization.**—The physical and chemical properties of a pedostratigraphic unit commonly vary vertically and laterally throughout the geographic extent of the unit. A pedostratigraphic unit is characterized by the *range* of physical and chemical properties of the unit in the type area, rather than by "typical" properties exhibited in a type section. Consequently, a pedostratigraphic unit is characterized on the basis of a composite stratotype (Article 8d).

(c) **Procedures for establishing formal pedostratigraphic units.**—A formal pedostratigraphic unit may be established in accordance with the applicable requirements of Article 3, and additionally by describing major soil horizons in each soil facies.

ALLOSTRATIGRAPHIC UNITS

Nature and Boundaries

Article 58.—**Nature of Allostratigraphic Units.** An allostratigraphic[9] unit is a mappable stratiform body of sedimentary rock that is defined and identified on the basis of its bounding discontinuities.

Remarks. (a) **Purpose.**—Formal allostratigraphic units may be defined to distinguish between different (1) superposed discontinuity-bounded deposits of similar lithology (Figs. 7, 9), (2) contiguous discontinuity-bounded deposits of similar lithology (Fig. 8), or (3) geographically separated discontinuity-bounded units of similar lithology (Fig. 9), or to distinguish as single units discontinuity-bounded deposits characterized by lithic heterogeneity (Fig. 8).

(b) **Internal characteristics.**—Internal characteristics (physical, chemical, and paleontological) may vary laterally and vertically throughout the unit.

(c) **Boundaries.**—Boundaries of allostratigraphic units are laterally traceable discontinuities (Figs. 7, 8, and 9).

(d) **Mappability.**—A formal allostratigraphic unit must be mappable at the scale practiced in the region where the unit is defined.

(e) **Type locality and extent.**—A type locality and type area must be designated; a composite stratotype or a type section and several reference sections are desirable. An allostratigraphic unit may be laterally contiguous with a formally defined lithostratigraphic unit; a vertical cut-off between such units is placed where the units meet.

(f) **Relation to genesis.**—Genetic interpretation is an inappropriate basis for defining an allostratigraphic unit. However, genetic interpretation may influence the choice of its boundaries.

(g) **Relation to geomorphic surfaces.**—A geomorphic surface may be used as a boundary of an allostratigraphic unit, but the unit should not be given the geographic name of the surface.

(h) **Relation to soils and paleosols.**—Soils and paleosols are composed of products of weathering and pedogenesis and differ in many respects from allostratigraphic units, which are depositional units (see "Pedostratigraphic Units," Article 55). The upper boundary of a surface or buried soil may be used as a boundary of an allostratigraphic unit.

(i) **Relation to inferred geologic history.**—Inferred geologic history is not used to define an allostratigraphic unit. However, well-

[9]From the Greek *allo:* "other, different."

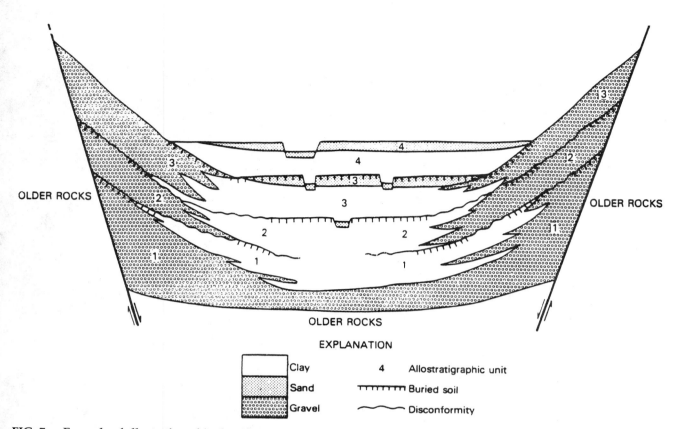

EXPLANATION

Clay		4	Allostratigraphic unit
Sand		⊤⊤⊤⊤⊤⊤	Buried soil
Gravel		⌒⌒⌒⌒	Disconformity

FIG. 7.—Example of allostratigraphic classification of alluvial and lacustrine deposits in a graben. The alluvial and lacustrine deposits may be included in a single formation, or may be separated laterally into formations distinguished on the basis of contrasting texture (gravel, clay). Textural changes are abrupt and sharp, both vertically and laterally. The gravel deposits and clay deposits, respectively, are lithologically similar and thus cannot be distinguished as members of a formation. Four allostratigraphic units, each including two or three textural facies, may be defined on the basis of laterally traceable discontinuities (buried soils and disconformities).

documented geologic history may influence the choice of the unit's boundaries.

(j) **Relation to time concepts.**—Inferred time spans, however measured, are not used to define an allostratigraphic unit. However, age relationships may influence the choice of the unit's boundaries.

(k) **Extension of allostratigraphic units.**—An allostratigraphic unit is extended from its type area by tracing the boundary discontinuities or by tracing or matching the deposits between the discontinuities.

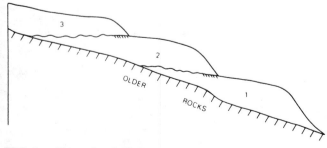

FIG. 8.—Example of allostratigraphic classification of contiguous deposits of similar lithology. Allostratigraphic units 1, 2, and 3 are physical records of three glaciations. They are lithologically similar, reflecting derivation from the same bedrock, and constitute a single lithostratigraphic unit.

Ranks of Allostratigraphic Units

Article 59.—**Hierarchy.** The hierarchy of allostratigraphic units, in order of decreasing rank, is allogroup, alloformation, and allomember.

Remarks. (a) **Alloformation.**—The alloformation is the fundamental unit in allostratigraphic classification. An alloformation may be completely or only partly divided into allomembers, if some useful purpose is served, or it may have no allomembers.

(b) **Allomember.**—An allomember is the formal allostratigraphic unit next in rank below an alloformation.

(c) **Allogroup.**—An allogroup is the allostratigraphic unit next in rank above an alloformation. An allogroup is established only if a unit of that rank is essential to elucidation of geologic history. An allogroup may consist entirely of named alloformations or, alternatively, may contain one or more named alloformations which jointly do not comprise the entire allogroup.

(d) **Changes in rank.**—The principles and procedures for elevation and reduction in rank of formal allostratigraphic units are the same as those in Articles 19b, 19g, and 28.

Allostratigraphic Nomenclature

Article 60.—**Nomenclature.** The principles and procedures for naming allostratigraphic units are the same as those for naming of lithostratigraphic units (see Articles 7, 30).

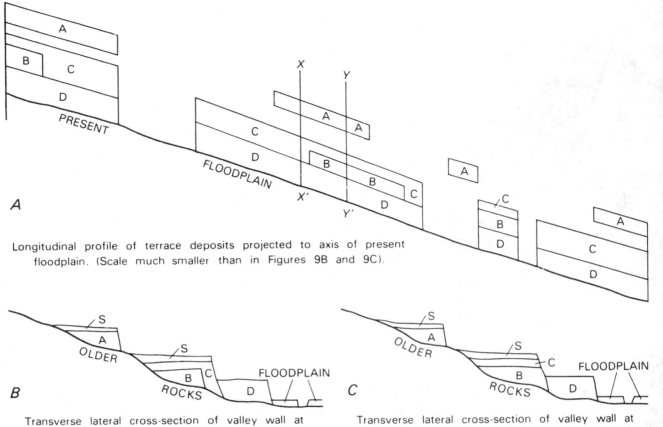

Longitudinal profile of terrace deposits projected to axis of present floodplain. (Scale much smaller than in Figures 9B and 9C).

Transverse lateral cross-section of valley wall at *X-X'* in Figure 9A.

Transverse lateral cross-section of valley wall at *Y-Y'* in Figure 9A.

FIG. 9.—Example of allostratigraphic classification of lithologically similar, discontinuous terrace deposits. A, B, C, and D are terrace gravel units of similar lithology at different topographic positions on a valley wall. The deposits may be defined as separate formal allostratigraphic units if such units are useful and if bounding discontinuities can be traced laterally. Terrace gravels of the same age commonly are separated geographically by exposures of older rocks. Where the bounding discontinuities cannot be traced continuously, they may be extended geographically on the basis of objective correlation of internal properties of the deposits other than lithology (e.g., fossil content, included tephras), topographic position, numerical ages, or relative-age criteria (e.g., soils or other weathering phenomena). The criteria for such extension should be documented. Slope deposits and eolian deposits (S) that mantle terrace surfaces may be of diverse ages and are not included in a terrace-gravel allostratigraphic unit. A single terrace surface may be underlain by more than one allostratigraphic unit (units B and C in sections b and c).

Remark. (a) **Revision.**—Allostratigraphic units may be revised or otherwise modified in accordance with the recommendations in Articles 17 to 20.

FORMAL UNITS DISTINGUISHED BY AGE

GEOLOGIC-TIME UNITS

Nature and Types

Article 61.—**Types.** Geologic-time units are conceptual, rather than material, in nature. Two types are recognized: those based on material standards or referents (specific rock sequences or bodies), and those independent of material referents (Fig. 1).

Units Based on Material Referents

Article 62.—**Types Based on Referents.** Two types of formal geologic-time units based on material referents are recognized: they are isochronous and diachronous units.

Article 63.—**Isochronous Categories.** Isochronous time units and the material bodies from which they are derived are twofold: geochronologic units (Article 80), which are based on corresponding material chronostratigraphic units (Article 66), and polarity-geochronologic units (Article 88), based on corresponding material polarity-chronostratigraphic units (Article 83).

Remark. (a) **Extent.**—Isochronous units are applicable worldwide; they may be referred to even in areas lacking a material record of the named span of time. The duration of the time may be represented by a unit-stratotype referent. The beginning and end of the time are represented by point-boundary-stratotypes either in a single stratigraphic sequence or in separate stratotype sections (Articles 8b, 10b).

Article 64.—**Diachronous Categories.** Diachronic units (Article 91) are time units corresponding to diachronous material allostratigraphic units (Article 58), pedostratigraphic units (Article 55), and most lithostratigraphic (Article 22) and biostratigraphic (Article 48) units.

Remarks. (a) **Diachroneity.**—Some lithostratigraphic and bio-stratigraphic units are clearly diachronous, whereas others have boundaries which are not demonstrably diachronous within the resolving power of available dating methods. The latter commonly are treated as isochronous and are used for purposes of chronocorrelation (see biochronozone, Article 75). However, the assumption of isochroneity must be tested continually.

(b) **Extent.**—Diachronic units are coextensive with the diachronous material stratigraphic units on which they are based and are not used beyond the extent of their material referents.

Units Independent of Material Referents

Article 65.—**Numerical Divisions of Time.** Isochronous geologic-time units based on numerical divisions of time in years are geochronometric units (Article 96) and have no material referents.

CHRONOSTRATIGRAPHIC UNITS

Nature and Boundaries

Article 66.—**Definition.** A chronostratigraphic unit is a body of rock established to serve as the material reference for all rocks formed during the same span of time. Each of its boundaries is synchtonous. The body also serves as the basis for defining the specific interval of time, or geochronologic unit (Article 80), represented by the referent.

Remarks. (a) **Purposes.**—Chronostratigraphic classification provides a means of establishing the temporally sequential order of rock bodies. Principal purposes are to provide a framework for (I) temporal correlation of the rocks in one area with those in another, (2) placing the rocks of the Earth's crust in a systematic sequence and indicating their relative position and age with respect to earth history as a whole, and (3) constructing an internationally recognized Standard Global Chronostratigraphic Scale.

(b) **Nature.**—A chronostratigraphic unit is a material unit and consists of a body of strata formed during a specific time span. Such a unit represents all rocks, and only those rocks, formed during that time span.

(c) **Content.**—A chronostratigraphic unit may be based upon the time span of a biostratigraphic unit, a lithic unit, a magnetopolarity unit, or any other feature of the rock record that has a time range. Or it may be any arbitrary but specified sequence of rocks, provided it has properties allowing chronocorrelation with rock sequences elsewhere.

Article 67.—**Boundaries.** Boundaries of chronostratigraphic units should be defined in a designated stratotype on the basis of observable paleontological or physical features of the rocks.

Remark. (a) **Emphasis on lower boundaries of chronostratigraphic units.**—Designation of point boundaries for both base and top of chronostratigraphic units is not recommended, because subsequent information on relations between successive units may identify overlaps or gaps. One means of minimizing or eliminating problems of duplication or gaps in chronostratigraphic successions is to define formally as a point-boundary stratotype only the base of the unit. Thus, a chronostratigraphic unit with its base defined at one locality, will have its top defined by the base of an overlying unit at the same, but more commonly another, locality (Article 8b).

Article 68.—**Correlation.** Demonstration of time equivalence is required for geographic extension of a chronostratigraphic unit from its type section or area. Boundaries of chronostratigraphic units can be extended only within the limits of resolution of available means of chronocorrelation, which currently include paleontology, numerical dating, remanent magnetism, thermoluminescence, relative-age crite-ria (examples are superposition and cross-cutting relations), and such indirect and inferential physical criteria as climatic changes, degree of weathering, and relations to unconformities. ideally, the boundaries of chronostratigraphic units are independent of lithology, fossil content, or other material bases of stratigraphic division, but, in practice, the correlation or geographic extension of these boundaries relies at least in part on such features. Boundaries of chronostratigraphic units commonly are intersected by boundaries of most other kinds of material units.

Ranks of Chronostratigraphic Units

Article 69.—**Hierarchy.** The hierarchy of chronostratigraphic units, in order of decreasing rank, is eonothem, erathem, system, series, and stage. Of these, system is the primary unit of worldwide major rank; its primacy derives from the history of development of stratigraphic classification. All systems and units of higher rank are divided completely into units of the next lower rank. Chronozones are non-hierarchical and commonly lower-rank chronostratigraphic units. Stages and chronozones in sum do not necessarily equal the units of next higher rank and need not be contiguous. The rank and magnitude of chronostratigraphic units are related to the time interval represented by the units, rather than to the thickness or areal extent of the rocks on which the units are based.

Article 70.—**Eonothem.** The unit highest in rank is eonothem. The Phanerozoic Eonothem encompasses the Paleozoic, Mesozoic, and Cenozoic Erathems. Although older rocks have been assigned heretofore to the Precambrian Eonothem, they also have been assigned recently to other (Archean and Proterozoic) eonothems by the IUGS Precambrian Subcommission. The span of time corresponding to an eonothem is an *eon*.

Article 71.—**Erathem.** An erathem is the formal chronostratigraphic unit of rank next lower to eonothem and consists of several adjacent systems. The span of time corresponding to an erathem is an *era*.

Remark. (a) **Names.**—Names given to traditional Phanerozoic erathems were based upon major stages in the development of life on Earth: Paleozoic (old), Mesozoic (intermediate), and Cenozoic (recent) life. Although somewhat comparable terms have been applied to Precambrian units, the names and ranks of Precambrian divisions are not yet universally agreed upon and are under consideration by the IUGS Subcommission on Precambrian Stratigraphy.

Article 72.—**System.** The unit of rank next lower to erathem is the system. Rocks encompassed by a system represent a time-span and an episode of Earth history sufficiently great to serve as a worldwide chronostratigraphic reference unit. The temporal equivalent of a system is a *period*.

Remark. (a) **Subsystem and supersystem.**—Some systems initially established in Europe later were divided or grouped elsewhere into units ranked as systems. *Subsystems* (Mississippian Subsystem of the Carboniferous System) and *supersystems* (Karoo Supersystem) are more appropriate.

Article 73.—**Series.** Series is a conventional chronostratigraphic unit that ranks below a system and always is a division of a system. A series commonly constitutes a major unit of chronostratigraphic correlation within a province, between provinces, or between continents. Although many

European series are being adopted increasingly for dividing systems on other continents, provincial series of regional scope continue to be useful. The temporal equivalent of a series is an *epoch*.

Article 74.—**Stage.** A stage is a chronostratigraphic unit of smaller scope and rank than a series. It is most commonly of greatest use in intra-continental classification and correlation, although it has the potential for worldwide recognition. The geochronologic equivalent of stage is *age*.

Remark. (a) **Substage.**—Stages may be, but need not be, divided completely into substages.

Article 75.—**Chronozone.** A chronozone is a non-hierarchical, but commonly small, formal chronostratigraphic unit, and its boundaries may be independent of those of ranked units. Although a chronozone is an isochronous unit, it may be based on a biostratigraphic unit (example: *Cardioceras cordatum* Biochronozone), a lithostratigraphic unit (Woodbend Lithochronozone), or a magnetopolarity unit (Gilbert Reversed-Polarity Chronozone). Modifiers (litho-, bio-, polarity) used in formal names of the units need not be repeated in general discussions where the meaning is evident from the context, e.g., *Exus albus* Chronozone.

Remarks. (a) **Boundaries of chronozones.**—The base and top of a *chronozone* correspond in the unit's stratotype to the observed, defining, physical and paleontological features, but they are extended to other areas by any means available for recognition of synchroneity. The temporal equivalent of a chronozone is a chron.
(b) **Scope.**—The scope of the non-hierarchical chronozone may range markedly, depending upon the purpose for which it is defined either formally or informally. The informal "biochronozone of the ammonites," for example, represents a duration of time which is enormous and exceeds that of a system. In contrast, a biochronozone defined by a species of limited range, such as the *Exus albus* Chronozone, may represent a duration equal to or briefer than that of a stage.
(c) **Practical utility.**—Chronozones, especially thin and informal biochronozones and lithochronozones bounded by key beds or other "markers," are the units used most commonly in industry investigations of selected parts of the stratigraphy of economically favorable basins. Such units are useful to define geographic distributions of lithofacies or biofacies, which provide a basis for genetic interpretations and the selection of targets to drill.

Chronostratigraphic Nomenclature

Article 76.—**Requirements.** Requirements for establishing a formal chronostratigraphic unit include: (i) statement of intention to designate such a unit; (ii) selection of name; (iii) statement of kind and rank of unit; (iv) statement of general concept of unit including historical background, synonymy, previous treatment, and reasons for proposed establishment; (v) description of characterizing physical and/or biological features; (vi) designation and description of boundary type sections, stratotypes, or other kinds of units on which it is based; (vii) correlation and age relations; and (viii) publication in a recognized scientific medium as specified in Article 4.

Article 77.—**Nomenclature.** A formal chronostratigraphic unit is given a compound name, and the initial letter of all words, except for trivial taxonomic terms, is capitalized. Except for chronozones (Article 75), names proposed for new chronostratigraphic units should not duplicate those for other stratigraphic units. For example, naming a new chronostratigraphic unit simply by adding "-an" or "-ian" to the name of a lithostratigraphic unit is improper.

Remarks. (a) **Systems and units of higher rank.**—Names that are generally accepted for systems and units of higher rank have diverse origins, and they also have different kinds of endings (Paleozoic, Cambrian, Cretaceous, Jurassic, Quaternary).
(b) **Series and units of lower rank.**—Series and units of lower rank are commonly known either by geographic names (Virgilian Series, Ochoan Series) or by names of their encompassing units modified by the capitalized adjectives Upper, Middle, and Lower (Lower Ordovician). Names of chronozones are derived from the unit on which they are based (Article 75). For series and stage, a geographic name is preferable because it may be related to a type area. For geographic names, the adjectival endings -an or -ian are recommended (Cincinnatian Series), but it is permissible to use the geographic name without any special ending, if more euphonious. Many series and stage names already in use have been based on lithic units (groups, formations, and members) and bear the names of these units (Wolfcampian Series, Claibornian Stage). Nevertheless, a stage preferably should have a geographic name not previously used in stratigraphic nomenclature. Use of internationally accepted (mainly European) stag names is preferable to the proliferation of others.

Article 78.—**Stratotypes.** An ideal stratotype for a chronostratigraphic unit is a completely exposed unbroken and continuous sequence of fossiliferous stratified rocks extending from a well defined lower boundary to the base of the next higher unit. Unfortunately, few available sequences are sufficiently complete to define stages and units of higher rank, which therefore are best defined by boundary-stratotypes (Article 8b).
Boundary-stratotypes for major chronostratigraphic units ideally should be based on complete sequences of either fossiliferous monofacial marine strata or rocks with other criteria for chronocorrelation to permit widespread tracing of synchronous horizons. Extension of synchronous surfaces should be based on as many indicators of age as possible.

Article 79.—**Revision of units.** Revision of a chronostratigraphic unit without changing its name is allowable but requires as much justification as the establishment of a new unit (Articles 17, 19, and 76). Revision or redefinition of a unit of system or higher rank requires international agreement. If the definition of a chronostratigraphic unit is inadequate, it may be clarified by establishment of boundary stratotypes in a principal reference section.

GEOCHRONOLOGIC UNITS

Nature and Boundaries

Article 80.—**Definition and Basis.** Geochronologic units are divisions of time traditionally distinguished on the basis of the rock record as expressed by chronostratigraphic units. A geochronologic unit is not a stratigraphic unit (i.e., it is not a material unit), but it corresponds to the time span of an established chronostratigraphic unit (Articles 65 and 66), and its beginning and ending corresponds to the base and top of the referent.

Ranks and Nomenclature of Geochronologic Units

Article 81.—**Hierarchy.** The hierarchy of geochronologic units in order of decreasing rank is *eon, era, period, epoch,* and *age*. Chron is a non-hierarchical, but commonly brief, geochronologic unit. Ages in sum do not necessarily equal epochs and need not form a continuum. An eon is the time represented by the rocks constituting an eonothem; era by an erathem; period by a system; epoch by a series; age by a stage; and chron by a chronozone.

Article 82.—**Nomenclature.** Names for periods and units of lower rank are identical with those of the corresponding chronostratigraphic units; the names of some eras and eons are independently formed. Rules of capitalization for chronostratigraphic units (Article 77) apply to geochronologic units. The adjectives Early, Middle, and Late are used for the geochronologic epochs equivalent to the corresponding chronostratigraphic Lower, Middle, and Upper series, where these are formally established.

POLARITY-CHRONOSTRATIGRAPHIC UNITS

Nature and Boundaries

Article 83.—**Definition.** A polarity-chronostratigraphic unit is a body of rock that contains the primary magnetic-polarity record imposed when the rock was deposited, or crystallized, during a specific interval of geologic time.

Remarks. (a) **Nature.**—Polarity-chronostratigraphic units depend fundamentally for definition on actual sections or sequences, or measurements on individual rock units, and without these standards they are meaningless. They are based on material units, the polarity zones of magnetopolarity classification. Each polarity-chronostratigraphic unit is the record of the time during which the rock formed and the Earth's magnetic field had a designated polarity. Care should be taken to define polarity-chronologic units in terms of polarity-chronostratigraphic units, and not vice versa.

(b) **Principal purposes.**—Two principal purposes are served by polarity-chronostratigraphic classification: (1) correlation of rocks at one place with those of the same age and polarity at other places; and (2) delineation of the polarity history of the Earth's magnetic field.

(c) **Recognition.**—A polarity-chronostratigraphic unit may be extended geographically from its type locality only with the support of physical and/or paleontologic criteria used to confirm its age.

Article 84.—**Boundaries.** The boundaries of a polarity chronozone are placed at polarity-reversal horizons or polarity transition-zones (see Article 45).

Ranks and Nomenclature of Polarity-Chronostratigraphic Units

Article 85.—**Fundamental Unit.** The polarity chronozone consists of rocks of a specified primary polarity and is the fundamental unit of worldwide polarity-chronostratigraphic classification.

Remarks. (a) **Meaning of term.**—A polarity chronozone is the worldwide body of rock strata that is collectively defined as a polarity-chronostratigraphic unit.

(b) **Scope.**—Individual polarity zones are the basic building blocks of polarity chronozones. Recognition and definition of polarity chronozones may thus involve step-by-step assembly of carefully dated or correlated individual polarity zones, especially in work with rocks older than the oldest ocean-floor magnetic anomalies. This procedure is the method by which the Brunhes, Matuyama, Gauss, and Gilbert Chronozones were recognized (Cox, Doell, and Dalrymple, 1963) and defined originally (Cox, Doell, and Dalrymple, 1964).

(c) **Ranks.**—Divisions of polarity chronozones are designated polarity subchronozones. Assemblages of polarity chronozones may be termed polarity superchronozones.

Article 86.—**Establishing Formal Units.** Requirements for establishing a polarity-chronostratigraphic unit include those specified in Articles 3 and 4, and also (1) definition of boundaries of the unit, with specific references to designated sections and data; (2) distinguishing polarity characteristics, lithologic descriptions, and included fossils; and (3) correlation and age relations.

Article 87.—**Name.** A formal polarity-chronostratigraphic unit is given a compound name beginning with that for a named geographic feature; the second component indicates the normal, reversed, or mixed polarity of the unit, and the third component is *chronozone*. The initial letter of each term is capitalized. If the same geographic name is used for both a magnetopolarity zone and a polarity-chronostratigraphic unit, the latter should be distinguished by an -an or -ian ending. Example: Tetonian Reversed-Polarity Chronozone.

Remarks: (a) **Preservation of established name.**—A particularly well-established name should not be displaced, either on the basis of priority, as described in Article 7c, or because it was not taken from a geographic feature. Continued use of Brunhes, Matuyama, Gauss, and Gilbert, for example, is endorsed so long as they remain valid units.

(b) **Expression of doubt.**—Doubt in the assignment of polarity zones to polarity-chronostratigraphic units should be made explicit if criteria of time equivalence are inconclusive.

POLARITY-CHRONOLOGIC UNITS

Nature and Boundaries

Article 88.—**Definition.** Polarity-chronologic units are divisions of geologic time distinguished on the basis of the record of magnetopolarity as embodied in polarity-chronostratigraphic units. No special kind of magnetic time is implied; the designations used are meant to convey the parts of geologic time during which the Earth's magnetic field had a characteristic polarity or sequence of polarities. These units correspond to the time spans represented by polarity chronozones, e.g., Gauss Normal Polarity Chronozone. They are not material units.

Ranks and Nomenclature of Polarity-Chronologic Units

Article 89.—**Fundamental Unit.** The polarity chron is the fundamental unit of geologic time designating the time span of a polarity chronozone.

Remark. (a) **Hierarchy.**—Polarity-chronologic units of decreasing hierarchical ranks are polarity superchron, polarity chron, and polarity subchron.

Article 90.—**Nomenclature.** Names for polarity chronologic units are identical with those of corresponding polarity-chronostratigraphic units, except that the term chron (or superchron, etc) is substituted for chronozone (or superchronozone, etc).

DIACHRONIC UNITS

Nature and Boundaries

Article 91.—**Definition.** A diachronic unit comprises the unequal spans of time represented either by a specific lithostratigraphic, allostratigraphic, biostratigraphic, or pedostratigraphic unit, or by an assemblage of such units.

Remarks. (a) **Purposes.**—Diachronic classification provides (1) a means of comparing the spans of time represented by stratigraphic units with diachronous boundaries at different localities, (2) a basis for broadly establishing in time the beginning and ending of deposition of diachronous stratigraphic units at different sites, (3) a basis for inferring the rate of change in areal extent of depositional processes, (4) a means of determining and comparing rates and durations of deposition at different localities, and (5) a means of comparing temporal and spatial relations of diachronous stratigraphic units (Watson and Wright, 1980).

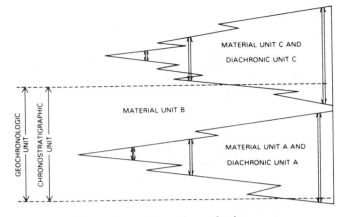
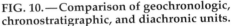

FIG. 10.—Comparison of geochronologic, chronostratigraphic, and diachronic units.

(b) **Scope.**—The scope of a diachronic unit is related to (1) the relative magnitude of the transgressive division of time represented by the stratigraphic unit or units on which it is based and (2) the areal extent of those units. A diachronic unit is not extended beyond the geographic limits of the stratigraphic unit or units on which it is based.

(c) **Basis.**—The basis for a diachronic unit is the diachronous referent.

(d) **Duration.**—A diachronic unit may be of equal duration at different places despite differences in the times at which it began and ended at those places.

Article 92.—**Boundaries.** The boundaries of a diachronic unit are the times recorded by the beginning and end of deposition of the material referent at the point under consideration (Figs. 10, 11).

Remark. (a) **Temporal relations.**—One or both of the boundaries of a diachronic unit are demonstrably time-transgressive. The varying time significance of the boundaries is defined by a series of boundary reference sections (Article 8b, 8e). The duration and age of a diachronic unit differ from place to place (Figs. 10, 11).

Ranks and Nomenclature of Diachronic Units

Article 93.—**Ranks.** A diachron is the fundamental and non-hierarchical diachronic unit. If a hierarchy of diachronic units is needed, the terms episode, phase, span, and dine, in order of decreasing rank, are recommended. The rank of a hierarchical unit is determined by the scope of the unit (Article 91 b), and not by the time span represented by the unit at a particular place.

Remarks. (a) **Diachron.**—Diachrons may differ greatly in magnitude because they are the spans of time represented by individual or grouped lithostratigraphic, allostratigraphic, biostratigraphic, and(or) pedostratigraphic units.

(b) **Hierarchical ordering permissible.**—A hierarchy of diachronic units may be defined if the resolution of spatial and temporal relations of diachronous stratigraphic units is sufficiently precise to make the hierarchy useful (Watson and Wright, 1980). Although all hierarchical units of rank lower than episode are part of a unit next higher in rank, not all parts of an episode, phase, or span need be represented by a unit of lower rank.

(c) **Episode.**—An episode is the unit of highest rank and greatest scope in hierarchical classification. If the "Wisconsinan Age" were to be redefined as a diachronic unit, it would have the rank of episode.

Article 94.—**Name.** The name for a diachronic unit should be compound, consisting of a geographic name followed by the term diachron or a hierarchical rank term. Both parts of the compound name are capitalized to indicate formal status. If the diachronic unit is defined by a single stratigraphic unit, the geographic name of the unit may be applied to the diachronic unit. Otherwise, the geographic name of the diachronic unit should not duplicate that of another formal stratigraphic unit. Genetic terms (e.g., alluvial, marine) or climatic terms (e.g., glacial, interglacial) are not included in the names of diachronic units.

Remarks. (a) **Formal designation of units.**—Diachronic units should be formally defined and named only if such definition is useful.

(b) **Inter-regional extension of geographic names.**—The geographic name of a diachronic unit may be extended from one region to another if the stratigraphic units on which the diachronic unit is based extend across the regions. If different diachronic units

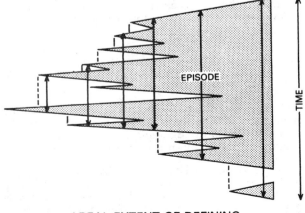

AREAL EXTENT OF DEFINING
MATERIAL UNIT OR UNITS

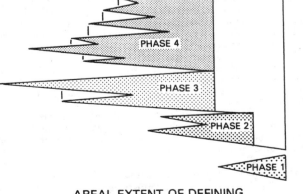

AREAL EXTENT OF DEFINING
MATERIAL UNIT OR UNITS

FIG. 11.—Schematic relation of phases to an episode. Parts of a phase similarly may be divided into spans, and spans into clines. Formal definition of spans and clines is unnecessary in most diachronic unit hierarchies.

in contiguous regions eventually prove to be based on laterally continuous stratigraphic units, one name should be applied to the unit in both regions. If two names have been applied, one name should be abandoned and the other formally extended. Rules of priority (Article 7d) apply. Priority in publication is to be respected, but priority alone does not justify displacing a well-established name by one not well-known or commonly used.

(c) **Change from geochronologic to diachronic classification.**— Lithostratigraphic units have served as the material basis for widely accepted chronostratigraphic and geochronologic classifications of Quaternary nonmarine deposits, such as the classifications of Frye et al (1968), Willman and Frye (1970), and Dreimanis and Karrow (1972). In practice, time-parallel horizons have been extended from the stratotypes on the basis of markedly time-transgressive lithostratigraphic and pedostratigraphic unit boundaries. The time ("geochronologic") units, defined on the basis of the stratotype sections but extended on the basis of diachronous stratigraphic boundaries, are diachronic units. Geographic names established for such "geochronologic" units may be used in diachronic classification if (1) the chronostratigraphic and geochronologic classifications are formally abandoned and diachronic classifications are proposed to replace the former "geochronologic" classifications, and (2) the units are redefined as formal diachronic units. Preservation of well-established names in these specific circumstances retains the intent and purpose of the names and the units, retains the practical significance of the units, enhances communication, and avoids proliferation of nomenclature.

Article 95.—**Establishing Formal Units.** Requirements for establishing a formal diachronic unit, in addition to those in Article 3, include (1) specification of the nature, stratigraphic relations, and geographic or areal relations of the stratigraphic unit or units that serve as a basis for definition of the unit, and (2) specific designation and description of multiple reference sections that illustrate the temporal and spatial relations of the defining stratigraphic unit or units and the boundaries of the unit or units.

Remark. (a) **Revision or abandonment.**—Revision or abandonment of the stratigraphic unit or units that serve as the material basis for definition of a diachronic unit may require revision or abandonment of the diachronic unit. Procedure for revision must follow the requirements for establishing a new diachronic unit.

GEOCHRONOMETRIC UNITS

Nature and Boundaries

Article 96.—**Definition.** Geochronometric units are units established through the direct division of geologic time, expressed in years. Like geochronologic units (Article 80), geochronometric units are abstractions, i.e., they are not material units. Unlike geochronologic units, geochronometric units are not based on the time span of designated chronostratigraphic units (stratotypes), but are simply time divisions of convenient magnitude for the purpose for which they are established, such as the development of a time scale for the Precambrian. Their boundaries are arbitrarily chosen or agreed-upon ages in years.

Ranks and Nomenclature of Geochronometric Units

Article 97.—**Nomenclature.** Geochronologic rank terms (eon, era, period, epoch, age, and chron) may be used for geochronometric units when such terms are formalized. For example, Archean Eon and Proterozoic Eon, as recognized by the IUGS Subcommission on Precambrian Stratigraphy, are formal geochronometric units in the sense of Article 96, distinguished on the basis of an arbitrarily chosen boundary at 2.5 Ga. Geochronometric units are not defined by, but may have, corresponding chronostratigraphic units (eonothem, erathem, system, series, stage, and chronozone).

PART III: ADDENDA

REFERENCES[10]

American Commission on Stratigraphic Nomenclature, 1947, Note 1– Organization and objectives of the Stratigraphic Commission: American Association of Petroleum Geologists Bulletin, v. 31, no. 3, p. 513–518.

—— , 1961, Code of Stratigraphic Nomenclature: American Association of Petroleum Geologists Bulletin, v. 45, no. 5, p. 645–665.

—— , 1970, Code of Stratigraphic Nomenclature (2d ed.): American Association of Petroleum Geologists, Tulsa, Okla., 45 p.

—— , 1976, Note 44—Application for addition to code concerning magnetostratigraphic units: American Association of Petroleum Geologists Bulletin, v. 60, no. 2, p. 273–277.

Caster, K. E., 1934, The stratigraphy and paleontology of northwestern Pennsylvania, Part 1, Stratigraphy: Bulletins of American Paleontology, v. 21, 185 p.

Chang, K. H., 1975, Unconformity-bounded stratigraphic units: Geological Society of America Bulletin, v. 86, no. 11, p. 1544–1552.

Committee on Stratigraphic Nomenclature, 1933, Classification and nomenclature of rock units: Geological Society of America Bulletin, v. 44, no. 2, p. 423–459, and American Association of Petroleum Geologists Bulletin, v. 17, no. 7, p. 843–868.

Cox, A. V., R. R. Doll, and G. B. Dalrymple, 1963, Geomagnetic polarity epochs and Pleistocene geochronometry: Nature, v. 198, p. 1049–1051.

—— , 1964, Reversals of the Earth's magnetic field: Science, v. 144, no. 3626, p. 1537–1543.

Cross, C. W., 1898, Geology of the Telluride area: U.S. Geological Survey 18th Annual Report, pt. 3, p. 759.

Cumming, A. D., J. G. C. M. Fuller, and J. W. Porter, 1959, Separation of strata: Paleozoic limestones of the Williston basin: American Journal of Science, v. 257, no. 10, p. 722–733.

Dreimanis, Aleksis, and P. F. Karrow, 1972, Glacial history of the Great Lakes–St. Lawrence region, the classification of the Wisconsin(an) Stage, and its correlatives: International Geologic Congress, 24th Session, Montreal, 1972, Section 12, Quaternary Geology, p. 5–15.

Dunbar, C. O., and John Rodgers, 1957, Principles of stratigraphy: Wiley, New York, 356 p.

Forgotson, J. M., Jr., 1957, Nature, usage and definition of marker-defined vertically segregated rock units: American Association of Petroleum Geologists Bulletin, v.41, no.9, p. 2108–2113.

Frye, J. C., H. B. Willman, Meyer Rubin, and R. F. Black, 1968, Definition of Wisconsinan Stage: U.S. Geological Survey Bulletin 1274-E, 22 p.

George, T. N., and others, 1969, Recommendations on stratigraphical usage: Geological Society of London, Proceedings no. 1656, p. 139–166.

Harland W. B., 1977, Essay review [of] International Stratigraphic Guide, 1976: Geology Magazine, v. 114, no. 3, p. 229–235.

—— , 1978, Geochronologic scales, in G. V. Cohee et al, eds., Contributions to the Geologic Time Scale: American Association of Petroleum Geologists, Studies in Geology, no. 6, p. 9–32.

Harrison, J. E., and Z. E. Peterman, 1980, North American Commission on Stratigraphic Nomenclature Note 52—A preliminary proposal for a chronometric time scale for the Precambrian of the United States and Mexico: Geological Society of America Bulletin, v. 91, no. 6, p. 377–380.

Henbest, L. G., 1952, Significance of evolutionary explosions for diastrophic division of Earth history: Journal of Paleontology, v. 26, p. 299–318.

Henderson, J. B., W. G. E. Caldwell, and J. E. Harrison, 1980, North American Commission on Stratigraphic Nomenclature, Report 8—Amendment of code concerning terminology for igneous and high-grade metamorphic rocks: Geological Society of America Bulletin, v. 91, no. 6, p. 374–376.

[10]Readers are reminded of the extensive and noteworthy bibliography of contributions to stratigraphic principles, classification, and terminology cited by the International Stratigraphic Guide (ISSC, 1976, p. 111–187).

Holland, C. H., and others, 1978, A guide to stratigraphical procedure: Geological Society of London, Special Report 10, p. 1–18.

Huxley, T. H., 1862, The anniversary address: Geological Society of London, Quarterly Journal, v. 18, p. xl–liv.

International Commission on Zoological Nomenclature, 1964: International Code of Zoological Nomenclature adopted by the XV International Congress of Zoology: International Trust for Zoological Nomenclature, London, 176 p.

International Subcommission on Stratigraphic Classification (ISSC), 1976, International Stratigraphic Guide (H. D. Hedberg, ed.): John Wiley and Sons, New York, 200 p.

International Subcommission on Stratigraphic Classification, 1979, Magnetostratigraphy polarity units—a supplementary chapter of the ISSC International Stratigraphic Guide: Geology, v. 7, p. 578–583.

Izett, G. A., and R. E. Wilcox, 1981, Map showing the distribution of the Huckleberry Ridge, Mesa Falls, and Lava Creek volcanic ash beds (Pearlette family ash beds) of Pliocene and Pleistocene age in the western United States and southern Canada: U. S. Geological Survey Miscellaneous Geological Investigations Map I-1325.

Kauffman, E. G., 1969, Cretaceous marine cycles of the Western Interior: Mountain Geologist: Rocky Mountain Association of Geologists, v. 6, no. 4, p. 227–245.

Matthews, R. K., 1974, Dynamic stratigraphy—an introduction to sedimentation and stratigraphy: Prentice-Hall, New Jersey, 370 p.

McDougall, Ian, 1977, The present status of the geomagnetic polarity time scale: Research School of Earth Sciences, Australian National University, Publication no. 1288, 34 p.

McElhinny, M. W., 1978, The magnetic polarity time scale; prospects and possibilities in magnetostratigraphy, in G. V. Cohee et al, eds., Contributions to the Geologic Time Scale, American Association of Petroleum Geologists, Studies in Geology, no. 6, p. 57–65.

McIver, N. L., 1972, Cenozoic and Mesozoic stratigraphy of the Nova Scotia shelf: Canadian Journal of Earth Science, v. 9, p. 54–70.

McLaren, D. J., 1977, The Silurian-Devonian Boundary Committee. A final report, in A. Martinsson, ed., The Silurian-Devonian boundary: IUGS Series A, no. 5, p. 1–34.

Morrison, R. B., 1967, Principles of Quaternary soil stratigraphy, in R. B. Morrison and H. E. Wright, Jr., eds., Quaternary soils: Reno, Nevada, Center for Water Resources Research, Desert Research Institute, Univ. Nevada, p. 1–69.

North American Commission on Stratigraphic Nomenclature, 1981, Draft North American Stratigraphic Code: Canadian Society of Petroleum Geologists, Calgary, 63 p.

Palmer, A. R., 1965, Biomere-a new kind of biostratigraphic unit: Journal of Paleontology, v. 39, no. l, p. 149–153.

Parsons, R. B., 1981, Proposed soil-stratigraphic guide, in International Union for Quaternary Research and International Society of Soil Science: INQUA Commission 6 and ISSS Commission 5 Working Group, Pedology, Report, p. 6–12.

Pawluk, S., 1978, The pedogenic profile in the stratigraphic section, in W C. Mahaney, ed., Quaternary soils: Norwich, England, GeoAbstracts, Ltd., p. 61–75.

Ruhe, R. V., 1965, Quaternary paleopedology, in H. E. Wright, Jr., and D. G. Frey, eds., The Quaternary of the United States: Princeton, N.J., Princeton University Press, p. 755–764.

Schultz, E. H., 1982, The chronosome and supersome–terms proposed for low-rank chronostratigraphic units: Canadian Petroleum Geology, v. 30, no. 1, p. 29–33.

Shaw, A. B., 1964, Time in stratigraphy: McGraw-Hill, New York, 365 p.

Sims, P. K., 1979, Precambrian subdivided: Geotimes, v. 24, no. 12, p. 15.

Sloss, L. L., 1963, Sequences in the cratonic interior of North America: Geological Society of America Bulletin, v. 74, no. 2, p. 94–114.

Tracey, J. I., Jr., and others, 1971, Initial reports of the Deep Sea Drilling Project, v. 8: U.S. Government Printing Office, Washington, 1037 p.

Valentine, K. W. G., and J. B. Dalrymple, 1976, Quaternary buried paleosols: A critical review: Quaternary Research, v. 6, p. 209–222.

Vella, P., 1964, Biostratigraphic units: New Zealand Journal of Geology and Geophysics, v. 7, no. 3, p. 615–625.

Watson, R. A., and H. E. Wright, Jr., 1980, The end of the Pleistocene: A general critique of chronostratigraphic classification: Boreas, v. 9, p. 153–163.

Weiss, M. P., 1979a, Comments and suggestions invited for revision of American Stratigraphic Code: Geological Society of America, News and Information, v. 1, no. 7, p. 97–99.

——, 1979b, Stratigraphic Commission Note 50–Proposal to change name of Commission: American Association of Petroleum Geologists Bulletin, v. 63, no. 10, p. 1986.

Weller, J. M., 1960, Stratigraphic principles and practice: Harper and Brothers, New York, 725 p.

Willman, H. B., and J. C. Frye, 1970, Pleistocene stratigraphy of Illinois: Illinois State Geological Survey Bulletin 94, 204 p.

APPENDIX I: PARTICIPANTS AND CONFEREES IN CODE REVISION

Code Committee

Steven S. Oriel (U.S. Geological Survey), chairman, Hubert Gabrielse (Geological Survey of Canada), William W. Hay (Joint Oceanographic Institutions), Frank E. Kottlowski (New Mexico Bureau of Mines), John B. Patton (Indiana Geological Survey).

Lithostratigraphic Subcommittee

James D. Aitken (Geological Survey of Canada), chairman, Monti Lerand (Gulf Canada Resources, Ltd.), Mitchell W. Reynolds (U.S. Geological Survey), Robert J. Weimer (Colorado School of Mines), Malcolm P. Weiss (Northern Illinois University).

Biostratigraphic Subcommittee

Allison R. (Pete) Palmer (Geological Society of America), chairman, Ismael Ferrusquia (University of Mexico), Joseph E. Hazel (U.S. Geological Survey), Erle G. Kauffman (University of Colorado), Colin McGregor (Geological Survey of Canada), Michael A. Murphy (University of California, Riverside), Walter C. Sweet (Ohio State University).

Chronostratigraphic Subcommittee

Zell E. Peterman (U.S. Geological Survey), chairman, Zoltan de Cserna (Sociedad Geológica Mexicana), Edward H. Schultz (Suncor, Inc., Calgary), Norman F. Sohl (U.S. Geological Survey), John A. Van Couvering (American Museum of Natural History).

Plutonic-Metamorphic Advisory Group

Jack E. Harrison (U.S. Geological Survey), chairman, John B. Henderson (Geological Survey of Canada), Harold L. James (retired), Leon T. Silver (California Institute of Technology), Paul C. Bateman (U.S. Geological Survey).

Magnetostratigraphic Advisory Group

Roger W. Macqueen (University of Waterloo), chairman, G. Brent Dalrymple (U.S. Geological Survey), Walter F. Fahrig (Geological Survey of Canada), J. M. Hall (Dalhousie University).

Volcanic Advisory Group

Richard V. Fisher (University of California, Santa Barbara), chairman, Thomas A. Steven (U.S. Geological Survey), Donald A. Swanson (U.S. Geological Survey).

Tectonostratigraphic Advisory Group

Darrel S. Cowan (University of Washington), chairman, Thomas W. Donnelly (State University of New York at Binghamton), Michael W. Higgins and David L. Jones (U.S. Geological Survey), Harold Williams (Memorial University, Newfoundland).

Quaternary Advisory Group

Norman P. Lasca (University of Wisconsin-Milwaukee), chairman, Mark M. Fenton (Alberta Research Council), David S. Fullerton (U.S. Geological Survey), Robert J. Fulton (Geological Survey of Canada), W. Hilton Johnson (University of Illinois), Paul F. Karrow (University of Waterloo), Gerald M. Richmond (U.S. Geological Survey).

Conferees

W. G. E. Caldwell (University of Saskatchewan), Lucy E. Edwards (U.S. Geological Survey), Henry H. Gray (Indiana Geological Survey), Hollis D. Hedberg (Princeton University), Lewis H. King (Geological Survey of Canada), Rudolph W. Kopf (U.S. Geological Survey), Jerry A. Lineback (Robertson Research U.S.), Marjorie E. MacLachlan (U.S. Geological Survey), Amos Salvador (University of Texas, Austin), Brian R. Shaw (Samson Resources, Inc.), Ogden Tweto (U.S. Geological Survey).

APPENDIX II: 1977–1982 COMPOSITION OF THE NORTH AMERICAN COMMISSION ON STRATIGRAPHIC NOMENCLATURE

Each Commissioner is appointed, with few exceptions, to serve a 3-year term (shown by such numerals as 80–82 for 1980–1982) and a few are reappointed.

American Association of Petroleum Geologists

Timothy A. Anderson (Gulf Oil Co.) 77–83, Orlo E. Childs (Texas Tech University) 76–79, Kenneth J. Englund (U.S. Geological Survey) 74–77, Susan Longacre (Getty Oil Co.) 78–84, Donald E. Owen (Cities Service Co.) 79–82, Grant Steele (Gulf Oil Co.) 75–78.

Association of American State Geologists

Larry D. Fellows (Arizona Bureau of Geology) 81–82, Lee C. Gerhard (North Dakota Geological Survey) 79–81, Donald C. Haney (Kentucky Geological Survey) 80–83, Wallace B. Howe (Missouri Division of Geology) 74–77, Robert R. Jordan (Delaware Geological Survey) 78–84, vice-chairman, Frank E. Kottlowski (New Mexico Bureau of Mines) 76–79, Meredith E. Ostrom (Wisconsin Geological Survey) 77–80, John B. Patton (Indiana Geological Survey) 75–78.

Geological Society of America

Clarence A. Hall, Jr. (University of California, Los Angeles) 78–81, Jack E. Harrison (U.S. Geological Survey) 74–77, William W. Hay (University of Miami) 75–78, Robert S. Houston (University of Wyoming) 77–80, Michael A. Murphy (University of California, Riverside) 81–84, Allison R. Palmer (Geological Society of America) 80–83, Malcolm P. Weiss (Northern Illinois University) 76–82, chairman.

United States Geological Survey

Earl E. Brabb (Menlo Park) 78–82, David S. Fullerton (Denver) 78–84, E. Dale Jackson (Menlo Park) 76–78, Kenneth L. Pierce (Denver) 75–78, Norman F. Sohl (Washington) 74–83.

Geological Survey of Canada

James D. Aitken (Calgary) 75–78, Kenneth D. Card (Kanata) 80–83, Donald G. Cook (Calgary) 78–81, Robert J. Fulton (Ottawa) 81–84, John B. Henderson (Ottawa) 74–77, Lewis H. King (Dartmouth) 79–82, Maurice B. Lambert (Ottawa) 77–80, Christopher J. Yorath (Sydney) 76–79.

Canadian Society of Petroleum Geologists

Roland F. deCaen (Union Oil Co. of Canada) 79–82, J. Ross McWhae (Petro Canada Exploration) 77–80, Edward H. Schultz (Suncor, Inc.) 74–77, 80–83, Ulrich Wissner (Union Oil Co. of Canada) 76–79.

Geological Association of Canada

W. G. E. Caldwell (University of Saskatchewan) 76–79, R. K. Jull (University of Windsor) 78–79, Paul S. Karrow (University of Waterloo) 81–84, Alfred C. Lenz (University of Western Ontario) 79–81, David E. Pearson (British Columbia Mines and Petroleum Resources) 79–81, Paul E. Schenk (Dalhousie University) 75–78.

Asociación Mexicana de Geólogos Petróleros

Jose Carillo Bravo (Petróleos Mexicanos) 78–81, Baldomerro Carrasco V., 75–78.

Sociedad Geólogica Mexicana

Zoltan de Cserna (Universidad Nacional Autónoma de México) 76–82.

Instituto de Geologia de la Universidad Nacional Autónoma de México

Ismael Ferrusquia Villafranca (Universidad Nacional Autónoma de México) 76–81, Fernando Ortega Gutiérrez (Universidad Nacional Autónoma de México) 81–84.

APPENDIX III: REPORTS AND NOTES OF THE AMERICAN COMMISSION ON STRATIGRAPHIC NOMENCLATURE

Reports (formal declarations, opinions, and recommendations)
1. Moore, Raymond C., Declaration on naming of subsurface stratigraphic units: AAPG Bulletin, v. 33, no. 7, p. 1280–1282, 1949.
2. Hedberg, Hollis D., Nature, usage, and nomenclature of time-stratigraphic and geologic-time units: AAPG Bulletin, v. 36, no. 8, p. 1627–1638, 1952.
3. Harrison, J. M., Nature, usage, and nomenclature of time-stratigraphic and geologic-time units as applied to the Precambrian: AAPG Bulletin, v. 39, no. 9, p. 1859–1861, 1955.
4. Cohee, George V., and others, Nature, usage, and nomenclature of rock-stratigraphic units: AAPG Bulletin, v. 40, no. 8, p. 2003–2014, 1956.
5. McKee, Edwin D., Nature, usage and nomenclature of biostratigraphic units: AAPG Bulletin, v. 41, no. 8, p. 1877–1889, 1957.
6. Richmond, Gerald M., Application of stratigraphic classification and nomenclature to the Quaternary: AAPG Bulletin, v. 43, no. 3, pt. I, p. 663–675, 1959.
7. Lohman, Kenneth E., Function and jurisdictional scope of the American Commission on Stratigraphic Nomenclature: AAPG Bulletin, v. 47, no. 5, p. 853–855, 1963.
8. Henderson, John B., W. G. E. Caldwell, and Jack E. Harrison, Amendment of code concerning terminology for igneous and high-grade metamorphic rocks: GSA Bulletin, pt. I, v. 91, no. 6, p. 374–376, 1980.
9. Harrison, Jack E., and Zell E. Peterman, Adoption of geochronometric units for divisions of Precambrian time: AAPG Bulletin, v. 66, no. 6, p. 801–802, 1982.

Notes (informal statements, discussions, and outlines of problems)

1. Organization and objectives of the Stratigraphic Commission: AAPG Bulletin, v. 31, no. 3, p. 513–518, 1947.
2. Nature and classes of stratigraphic units: AAPG Bulletin, v. 31, no. 3, p. 519–528, 1947.

3. Moore, Raymond C., Rules of geologic nomenclature of the Geological Survey of Canada: AAPG Bulletin, v. 32, no. 3, p. 366–367, 1948.

4. Jones, Wayne V., and Raymond C. Moore, Naming of subsurface stratigraphic units: AAPG Bulletin, v. 32, no. 3, p. 367–371, 1948.

5. Flint, Richard Foster, and Raymond C. Moore, Definition and adoption of the terms stage and age: AAPG Bulletin, v. 32, no. 3, p. 372–376, 1948.

6. Moore, Raymond C., Discussion of nature and classes of stratigraphic units: AAPG Bulletin, v. 21, no. 3, p. 376–381, 1948.

7. Records of the Stratigraphic Commission for 1947–1948: AAPG Bulletin, v. 33, no. 7, p. 1271–1273, 1949.

8. Australian Code of Stratigraphical Nomenclature: AAPG Bulletin, v. 33, no. 7, p. 1273–1276, 1949.

9. The Pliocene-Pleistocene boundary: AAPG Bulletin, v. 33, no. 7, p. 1276–1280, 1949.

10. Moore, Raymond C., Should additional categories of stratigraphic units be recognized?: AAPG Bulletin, v. 34, no. 12, p. 2360–2361, 1950.

11. Moore, Raymond C., Records of the Stratigraphic Commission for 1949–1950: AAPG Bulletin, v. 35, no. 5, p. 1074–1076, 1951.

12. Moore, Raymond C., Divisions of rocks and time: AAPG Bulletin, v. 35, no. 5, p. 1076, 1951.

13. Williams, James Steele, and Aureal T. Cross, Third Congress of Carboniferous Stratigraphy and Geology: AAPG Bulletin, v. 36, no. 1, p. 169–172, 1952.

14. Official report of round table conference on stratigraphic nomenclature at Third Congress of Carboniferous Stratigraphy and Geology, Heerlen, Netherlands, June 26–28, 1951: AAPG Bulletin, v. 36, no. 10, p. 2044–2048, 1952.

15. Records of the Stratigraphic Commission for 1951–1952: AAPG Bulletin, v. 37, no. 5, p. 1078–1080, 1953.

16. Records of the Stratigraphic Commission for 1953–1954: AAPG Bulletin, v. 39, no.9, p. 1861–1863, 1955.

17. Suppression of homonymous and obsolete stratigraphic names: AAPG Bulletin, v. 40, no. 12, p. 2953–2954, 1956.

18. Gilluly, James, Records of the Stratigraphic Commission for 1955–1956: AAPG Bulletin, v. 41, no. 1, p. 130–133, 1957.

19. Richmond, Gerald M., and John C. Frye, Status of soils in stratigraphic nomenclature: AAPG Bulletin, v. 31, no. 4, p. 758–763, 1957.

20. Frye, John C., and Gerald M. Richmond, Problems in applying standard stratigraphic practice in nonmarine Quaternary deposits: AAPG Bulletin, v. 42, no. 8, p. 1979–1983, 1958.

21. Frye, John C., Preparation of new stratigraphic code by American Commission on Stratigraphic Nomenclature: AAPG Bulletin, v. 42, no. 8, p. 1984–1986, 1958.

22. Records of the Stratigraphic Commission for 1957–1958: AAPG Bulletin, v. 43, no. 8, p. 1967–1971, 1959.

23. Rodgers, John, and Richard B. McConnell, Need for rock-stratigraphic units larger than group: AAPG Bulletin, v. 43, no. 8, p. 1971–1975, 1959.

24. Wheeler, Harry E., Unconformity-bounded units in stratigraphy: AAPG Bulletin, v. 43, no. 8, p. 1975–1977, 1959.

25. Bell, W. Charles, and others, Geochronologic and chronostratigraphic units: AAPG Bulletin, v. 45, no. 5, p. 666–670, 1961.

26. Records of the Stratigraphic Commission for 1959–1960: AAPG Bulletin, v. 45, no. 5, p. 670–673, 1961.

27. Frye, John C., and H. B. Willman, Morphostratigraphic units in Pleistocene stratigraphy: AAPG Bulletin, v. 46, no. 1, p. 112–113, 1962.

28. Shaver, Robert H., Application to American Commission on Stratigraphic Nomenclature for an amendment of Article 4f of the Code of Stratigraphic Nomenclature on informal status of named aquifers, oil sands, coal beds, and quarry layers: AAPG Bulletin, v. 46, no. 10, p. 1935, 1962.

29. Patton, John B., Records of the Stratigraphic Commission for 1961–1962: AAPG Bulletin, v. 47, no. 11, p. 1987–1991, 1963.

30. Richmond, Gerald M., and John G. Fyles, Application to American Commission on Stratigraphic Nomenclature for an amendment of Article 31, Remark (b) of the Code of Stratigraphic Nomenclature on misuse of the term "stage": AAPG Bulletin, v. 48, no. 5, p. 710–711, 1964.

31. Cohee, George V., Records of the Stratigraphic Commission for 1963–1964: AAPG Bulletin, v. 49, no. 3, pt. I of II, p. 296–300, 1965.

32. International Subcommission on Stratigraphic Terminology, Hollis D. Hedberg, ed., Definition of geologic systems: AAPG Bulletin, v. 49, no. 10, p. 1694–1703, 1965.

33. Hedberg, Hollis D., Application to American Commission on Stratigraphic Nomenclature for amendments to Articles 29, 31, and 37 to provide for recognition of erathem, substage, and chronozone as time-stratigraphic terms in the Code of Stratigraphic Nomenclature: AAPG Bulletin, v. 50, no. 3, p. 560–561, 1966.

34. Harker, Peter, Records of the Stratigraphic Commission for 1964–1966: AAPG Bulletin, v. 51, no. 9, p. 1862–1869, 1967.

35. DeFord, Ronald K., John A. Wilson, and Frederick M. Swain, Application to American Commission on Stratigraphic Nomenclature for an amendment of Article 3 and Article 13, Remarks (c) and (e), of the Code of Stratigraphic Nomenclature to disallow recognition of new stratigraphic names that appear only in abstracts, guidebooks, microfilms, newspapers, or in commercial or trade journals: AAPG Bulletin, v. 51, no. 9, p. 1868–1869, 1967.

36. Cohee, George V., Ronald K. DeFord, and H. B. Willman, Amendment of Article 5, Remarks (a) and (e) of the Code of Stratigraphic Nomenclature for treatment of geologic names in a gradational or interfingering relationship of rock-stratigraphic units: AAPG Bulletin, v. 53, no. 9, p. 2005–2006, 1969.

37. Kottlowski, Frank E., Records of the Stratigraphic Commission for 1966–1968: AAPG Bulletin, v. 53, no. 10, p. 2179–2186, 1969.

38. Andrews, J., and K. Jinghwa Hsü, A recommendation to the American Commission on Stratigraphic Nomenclature concerning nomenclatural problems of submarine formations: AAPG Bulletin, v. 54, no. 9, p. 1746–1747, 1970.

39. Wilson, John Andrew, Records of the Stratigraphic Commission for 1968–1970: AAPG Bulletin, v. 55, no. 10, p. 1866–1872, 1971.

40. James, Harold L., Subdivision of Precambrian: An interim scheme to be used by U.S. Geological Survey: AAPG Bulletin, v. 56, no. 6, p. 1128–1133, 1972.

41. Oriel, Steven S., Application for amendment of Article 8 of code, concerning smallest formal rock-stratigraphic unit: AAPG Bulletin, v. 59, no. 1, p. 134–135, 1975.

42. Oriel, Steven S., Records of Stratigraphic Commission for 1970–1972: AAPG Bulletin, v. 59, no. 1, p. 135–139, 1975.

43. Oriel, Steven S., and Virgil E. Barnes, Records of Stratigraphic Commission for 1972–1974: AAPG Bulletin, v. 59, no. 10, p. 2031–2036, 1975.

44. Oriel, Steven S., Roger W. Macqueen, John A. Wilson, and G. Brent Dalrymple, Application for addition to code concerning magnetostratigraphic units: AAPG Bulletin, v. 60, no. 2, p. 273–277, 1976.

45. Sohl, Norman F., Application for amendment concerning terminology for igneous and high-grade metamorphic rocks: AAPG Bulletin, v. 61, no. 2, p. 248–251, 1977.

46. Sohl, Norman F., Application for amendment of Articles 8 and 10 of code, concerning smallest formal rock-stratigraphic unit: AAPG Bulletin, v. 61, no. 2, p. 252, 1977.

47. Macqueen, Roger W., and Steven S. Oriel, Application for amendment of Articles 27 and 34 of stratigraphic code to introduce point-boundary stratotype concept: AAPG Bulletin, v. 61, no. 7, p. 1083–1085, 1977.

48. Sohl, Norman F., Application for amendment of Code of Stratigraphic Nomenclature to provide guidelines concerning formal terminology for oceanic rocks: AAPG Bulletin, v. 62, no. 7, p. 1185–1186, 1978.

49. Caldwell, W. G. E., and N. F. Sohl, Records of Stratigraphic Commission for 1974–1976: AAPG Bulletin, v. 62, no. 7, p. 1187–1192, 1978.

50. Weiss, Malcolm P., Proposal to change name of commission: AAPG Bulletin, v. 63, no. 10, p. 1986, 1979.

51. Weiss, Malcolm P., and James D. Aitken, Records of Stratigraphic Commission, 1976–1978: AAPG Bulletin, v. 64, no. 1, p. 136–137, 1980.

52. Harrison, Jack E., and Zell E. Peterman, A preliminary proposal for a chronometric time scale for the Precambrian of the United States and Mexico: GSA Bulletin, pt. I, v. 91, no. 6, p. 377–380, 1980.

Appendix B

Geologic Time Scales

B1

Cenozoic Time Scale (After Haq et al., 1987)

	Series	Stage	Relative Change of Coastal Onlap	Eustatic Curves (meters)	Magnetostratigraphy		
			Landward 1.0 — 0.5 — Seaward 0	250 200 150 100 50 0M	Polarity Epoch	Polarity Chronozone	Polarity[b]

Age (Ma)	Series	Stage	Polarity Epoch	Polarity Chronozone
0	Holocene			
	Pleistocene	Milazzian	1	C1
		Sicilian		
		Emilian	2	C2
	Pliocene U	Calabrian 1.65		
		Piacenzian 3.5	3	C2A
5	Pliocene L	Zanclean	4	C3
	Miocene Upper	Messinian 5.2	5	C3A
			6	
		Tortonian	7	C4
			8	
			9	C4A
10		6.3	10	
	Miocene Middle	Serravillian	11	C5
			12	
			13	
			14	C5A
15		Langhian 15.2 / 16.2	15	C5B
	Miocene Lower	Burdigalian	16	C5C
		20	17	C5D / C5E
20			18	C6
		Aquitanian	19	C6A
			20	
			21	C6B
25		25.2	22	C6C
	Oligocene[a] Upper	Chattan		C7 / C7A
				C8
				C9
30		30		C10
	Oligocene[a] Lower	Rupelian		C11
				C12
35		36		C13
	Eocene[a] Upper	Priabonian		C15 / C16
		39.4		C17
40		Bartonian 42		C18
				C19
	Eocene[a] Middle	Lutetian		C20
45				C21
		49		C22
50				C23
	Eocene[a] Lower	Ypresian		C24
55		54		C25
	Paleocene Upper	Thanetian		C26
60		60.2		C27
	Paleocene Lower	Danian		C28
65				C29
	Cretaceous Upper	66.5		C30
70		Maastrichtian		C31

[a] As noted in Chapter 18, the Eocene/Oligocene portion of the time scale is still undergoing revision.

[b] ▓, normal polarity; □, reversed polarity.

Mesozoic Time Scale (After Harland et al., 1982)

Age (Ma)	Eon	Era	Subera	Period	Epoch		Age	Best Age Estimate (Ma)	Relative Change of Coastal Onlap	Magnetostratigraphy Polarity[a]
55	Phanerozoic	Cenozoic	Tertiary	Paleogene	Paleocene	L	Thanetian	54.9	1.0 Landward — 0.5 — Seaward — 0	24r / 25 / 25r / 26 / 26r / 27 / 27r / 28 / 28r
60								60.2		
				Pg		E	Danian			
66.5		Mesozoic		Cretaceous	Late	Senonian	Maastrichtian	66.5		30 / 31 / 31r / 32 / 32r / 33
							Campanian	73		33r
							Santonian	83		34
							Coniacian	87.5		
							Turonian	88.5		
							Cenomanian	91		
100							Albian	97.5		
					Early		Aptian	113		M0 / M1n / M1 / M3 / M5
							Barremian	119		M9 / M10Nn / M11
						Neocomiam	Hauterivian	125		M12 / M14 / M16n
							Valanginian	131		M17 / M19n
				K			Berriasian	138		M20 / M22n / M24 / M25An / M26n
150				Jurassic	Late	Malm	Tithonian	144		M29
							Kimmeridgian	150		
							Oxfordian	156		
					Middle	Dogger	Callovian	163		
							Bathonian	169		
							Bajocian	175		
							Aalenian	181		
					Early	Lias	Toarcian	188		
200							Pliensbachian	194		
							Sinemurian	200		
				J			Hettangian	206		
				Triassic	Late		Rhaetian	213		
							Norian	219		
							Carnian	225		
					Middle		Ladinian	231		
							Anisian	238		
248				Tr	Early		Scythian	243		
250		Paleozoic		Permian	Late		Tatarian	248		
							Kazanian	253		
							Ufimian	258		
							Kungurian	263		
					Early		Artinskian	268		
							Sakmarian			
				P			Asselian	286		
300										

[a] ▓, normal polarity; ☐, reversed polarity.

Paleozoic Time Scale (After Harland et al., 1982)

Age (Ma)	Eon	Era	Period	Epoch	Age	Best Age Estimate (Ma)	Relative Change of Coastal Onlap
					Rhaetian	219	
				Late	Norian	225	
		Mesozoic	Triassic		Carnian	231	
					Ladinian		
				Middle	Anisian	238	
					Scythian	243	
248			Tr	Early	Tatarian	248	
250				Late	Kazanian	253	
					Ufimian	258	
			Permian		Kungurian	263	
				Early	Artinskian	268	
			P	Gzelian	Sakmarian	286	
				Kasimovian	Asselian	296	
300				Moscovian			
				Bashkirian		315	
			Carboniferous	Serpukhovian		320	
	Phanerozoic	Paleozoic				333	
				Visean			
350			C	Tournaisian		352	
						360	
				Late	Famennian	367	
					Frasnian	374	
			Devonian	Middle	Givetian	380	
					Eifelian	387	
					Emsian	394	
400				Early	Siegenian	401	
			D		Gedinnian	408	
				Pridoli		414	
				Ludlow		421	
			Silurian	Wenlock		428	
			S	Llandovery		438	
				Ashgill		448	
450				Caradoc		458	
				Llandeilo		468	
			Ordovician	Llanvirn		478	
				Arenig		488	
500			O	Tremadoc		505	
				Merioneth		525	
				St Davids		540	
			Cambrian				
550				Caerfai			
					— ? —	570	
			€		Tommotian	590	
590				Ediacaran		630	
600		Sinian	Vendian V	Varangian		670	
800			Sturtian U			800	
	?					900	

Relative Change of Coastal Onlap: Landward ← | 1.0 0.5 0 | → Seaward

Precambrian Time Scale (After Harland et al., 1982)

Age (Ma)	Eon	Era	Period	Epoch	Age	Best Age Estimate (Ma)
				Merioneth		525
				St Davids		540
550	Phanerozoic	Paleozoic	Cambrian	Caerfai		
					−?−	570
			€		Tommotian	590
590				Ediacaran		
600	Proterozoic	Sinian	Vendian V			630
			Sturtian U	Varangian		670
						800
						900
1000		Riphean	Yurmatin Y			1050
						1350
			Burzyan B			1650
2000			Huronian H		Precambrian subera and period names have no international status	2100
						2400
						2500
		Swazian	Randian Ran			2630
						2800
3000	Archean					3000
						3500
			Isuan I			3750
4000						3900
	Priscoan		Hadean			4000
			Hde			
5000						

Glossary

accommodation rate Rate of basin subsidence that allows stratigraphic accumulation.

accretionary wedge Prism of rocks formed as the downgoing plate scrapes off material on its way into the subduction zone. It includes oceanic sediments (cherts and shales), ophiolites, and blueschists, all pervasively sheared into mélange.

accuracy "Closeness to truth" of a measurement.

acoustic impedance Difference in densities (and therefore acoustic properties) of a rock.

agglomerate Very coarse-grained pyroclastic deposit composed mainly of bomb-sized (>64 mm diameter) tephra.

aggradation Vertical accumulation without either progradation or retrogradation.

albitization Diagenetic growth of sodium plagioclase (albite) in a sedimentary rock.

allochemical From the Greek *allos*, "elsewhere" or "from outside." Bits and pieces of calcium carbonate material such as fossil fragments, ooids, and intraclasts that were transported a long distance in solution but that prior to accumulation as components in carbonate rock have a history of entrainment, transport, and deposition as clastic components.

allochthonous From the Greek *allos*, "elsewhere" or "from outside," and *chthonos*, "place." Formed somewhere else and transported to its current location.

allostratigraphic unit Stratigraphic unit defined by its bounding unconformities.

alluvial fan Cone- or fan-shaped body of sand and gravel dumped at the mouth of a canyon.

alternating field (AF) demagnetization Removal of magnetic overprints by subjecting a sample to a higher and higher alternating field.

amorphous silica Noncrystalline silica (SiO_2), most of which is produced from the weathering of feldspar to clay mineral. Amorphous silica lacks a regular internal structure, is highly soluble (100 to 200 ppm) in most natural waters, and is the main source of silica for skeleton-building, silica-secreting marine organisms.

anaerobic Said of a depositional setting (or an organism that inhabits it) that lacks appreciable free oxygen (i.e., <0.1 mL of oxygen per liter of seawater).

angular unconformity Type of unconformity where the beds beneath the erosional surface are tilted and eroded.

anthracite coal Also called hard coal. The highest-rank coal, containing 90% to 100% carbon, with almost all the moisture and volatiles of the original organic matter having been removed by compaction and very low-grade metamorphism.

antidune Primary sedimentary structure, specifically, low, undulating large-scale ridge of sand-sized sediment that forms under rapid flow conditions and gradually migrates upstream.

arenite One of two major sandstone groups (the other is **wacke**). Arenites are texturally "cleaner" and consist of predominantly sand-sized framework clasts in tangential contact with one another. Interstitial pores are either empty, or filled or partly filled with chemical cement and smaller amounts of silt- and clay-sized matrix (clast diameter 0.03 mm or less). Compositional varieties are categorized on the basis of the sand-sized framework grains into quartz arenite, lithic (abundant clasts of quartz and rock fragments) arenite, and arkosic (abundant clasts of quartz and feldspar) arenite.

argillite Mudrock that has been subjected to low-grade metamorphism.

arkosic Sandstone (and in some cases, conglomerate and breccia) that contains significant amounts of detrital feldspar (typically 25% or more) as framework grains.

assemblage zone Biostratigraphic zone based on the association of three or more taxa.

aulacogen Failed rift valley, which fills with sediment as it opens and is then abandoned.

authigenic Refers to material that is formed or generated in place, for example minerals that grow in place within a rock, such as quartz and feldspar overgrowths that develop around transported grains after they are deposited.

autochthonous Formed in the same area where it is now found; not transported or relocated.

autoclastic breccia Breccia that is generated where a delicate, brittle crust developed on the surface of cooling lava in motion is broken into angular, pancakelike blocks.

avulsion Abrupt abandonment of a segment of a river channel.

bafflestone Type of boundstone in the Dunham classification consisting of organisms that act as baffles.

bajada Series of alluvial fans that have coalesced where they meet each other.

ball-and-pillow structure Secondary sedimentary structure that consists of spherical to elliptical masses of internally laminated sandstone (or, rarely, limestone) formed as the result of their foundering into underlying, partially liquefied muds.

banded iron formation (BIF) Centimeter- to multicentimeter-thick interlayered alternating bands of chert and iron-rich minerals interpreted to be direct chemical precipitates on the seafloor. Most banded iron formations are early Proterozoic in age (1800 to 2400 million years old), and their abundance is used to argue for an early terrestrial atmosphere that lacked appreciable free oxygen.

barred basin Embayment or estuary connected to the open sea by a shallow, generally narrow entrance that intermittently (during high tide) allows seawater of normal salinity to enter, but which (during low tide) becomes isolated and a potential site of bedded evaporite formation.

base level of erosion Level below which erosion cannot occur.

base level of aggradation Level above which sediments cannot accumulate permanently.

base surge Sediment gravity flow that forms when steam-saturated eruption columns collapse and fan out across the ground surface.

basin analysis Integrated study of a sedimentary basin, including the sources and provenance of the sediment, the subsidence history, and the tectonic causes of basin formation.

bauxite White to gray to brown to yellowish-brown rock composed of extensively weathered clay minerals that are so rich in aluminum that they can be mined.

beach rock Recent carbonate sediments on the beach cemented in very short periods of time.

bedded (primary) chert Chert that occurs as ribbonlike individual bands, layers, or laminae produced by the in-place accumulation of siliceous components, such as diatoms, radiolaria, and sponge spicules, and recrystallized at the depositional site; hence it is primary in origin.

bedform Primary sedimentary structures like ripple marks, dunes, and plane beds that appear as three-dimensional features on the top surface of bedding planes.

bentonite Clay-rich band of sediment produced by the reaction between seawater and volcanic ash beds deposited on the seafloor.

berm Raised ridge of sand on a beach above mean high tide, which marks the top of the foreshore.

Bernoulli's principle Principle of hydraulics that the sum of velocity and pressure on an object in a uniformly moving fluid is constant; each increase in velocity decreases pressure and vice versa. The ability of moving fluids to entrain resting particles from the base of a stream channel is a practical consequence.

bindstone Type of boundstone in the Dunham classification consisting of organisms that encrust and bind.

biochronology Sequence of fossil faunas independent of the stratigraphic units from which they come.

biofacies Biological or paleontological characteristics of a sedimentary rock body.

biogenic sedimentary rock Sedimentary rock, like reef coral, radiolarian chert, and bioclastic limestone, in which organic activity either directly produces most of the rock components or organic metabolism sufficiently modifies the geochemical environment so that specific varieties of sediment form. The former variety is referred to as organic sedimentary rock, the latter as biochemical sedimentary rock.

biogenic sedimentary structure Generally large-scale, commonly three-dimensional features of sedimentary rocks like ichnofossils and stromatolites that are generated by the activity of organisms such as burrowing and feeding.

bioherm Mound-shaped structure built up by carbonate-secreting organisms, such as a reef.

biomarker Diagnostic organic compound in petroleum that allow individual petroleum deposits to be specifically linked to their source rocks.

biostratigraphy Use of fossils to correlate and date rocks.

bioturbation Variety of processes by which organic activity deforms or alters pre-existing sedimentary rocks. Bioturbation structures include tracks, trail, and plant root marks.

biozone Total temporal and geographic range of an organism.

bituminous coal Also called soft coal. Intermediate-rank coal that contains between 80% and 90% carbon, with much of the original moisture and volatiles of the original organic matter removed by compaction.

bivariate Graphical means of comparing two separate and distinct variables of a population, for example, mean grain size versus sorting.

block Any large (> 64 mm), angular, sharp-edged fragment of lava that congealed prior to explosive volcanism but can subsequently become tephra.

bolide Any hypervelocity extraterrestrial object regardless of origin (comet, asteroid, or space invader) that collides with the Earth.

bomb Any large (> 64 mm), rounded, bloblike pyroclastically produced mass of lava that congealed in flight.

Bouma sequence Sequence of deposits of a single turbidity current, from the graded bed at the base to a plane bed, then decreasing ripple laminations, capped by fine muds.

boundary layer effect Process by which normally laminar flow becomes disrupted and transformed into turbulent flow (enhancing the ability to entrain and transport material) in a thin zone of that occurs at and near the contact of the moving fluid with a fixed surface.

boundstone Carbonate rock type in the Dunham classification that consists of biogenically precipitated organic components bound together from origin, for example, a reef rock or a biolithite in the Folk classification.

braided stream Stream or river characterized by numerous shallow sandy channels and bars that anastomose like the braids in hair.

breccia Also called sharpstone. Lithified rubble; a terrigenous (siliciclastic) sedimentary rock containing a significant component of angular clasts coarser than 2 mm.

caliche Limestone precipitated as surface or near-surface crusts and nodules by the evaporation of moisture in semi-arid climates.

caliper Well-logging device that measures the diameter of the well.

carbonate compensation depth (CCD) Depth below which calcium carbonate dissolves at a faster rate than it is deposited.

cataclastic Fragmental components in sediment or sedimentary rocks generated by crushing, breaking, and brittle responses of rocks along fault planes.

catagenesis Process by which kerogen is converted, at elevated temperatures, into petroleum.

cement Crystalline material of varying composition (silica, calcium carbonate, hematite) that is precipitated within the intergranular spaces of clastic sediment, binding the individual grains together to form a solid, coherent sedimentary rock.

central tendency Measure of the averageness of grain size diameter or compositional data, variously stated as mean, mode, or median.

chemical remanent magnetization (CRM) Magnetization caused by chemical precipitation of iron oxides.

chickenwire structure   Textural term used for evaporate deposits in which an interlocking mosaic of slightly elongate nodular anhydrite patches alternates with thin dark strings of carbonate or clay, resulting in a pattern reminiscent of chickenwire.

C-horizon   Bottom layer of a soil that lies beneath the uppermost A-horizon and middle B-horizon; consists of unconsolidated rock material separating better-developed soil horizons above from essentially unweathered parent rocks below.

chronostratigraphic unit   Time-rock unit, or a unit of rock that is bounded by time planes.

clast   From the Greek *klastos,* "broken." Any fragment or chunk of material, for example, bits and pieces of pre-existing rocks, minerals, and organic remains, regardless of size, shape, or roundness (angularity).

clay mineral   Large group of fine-grained phyllosilicate minerals that are produced by the chemical weathering (mainly hydrolysis) of feldspar and other silicate minerals. Clay minerals are important components of soil, mudrock, and some sandstones. Specific common varieties include kaolinite, gibbsite, and illite.

claystone   Type of very fine-grained siliciclastic sediment composed predominantly of clay-sized (< 0.0039 mm diameter) particles.

COCORP (Consortium on Continental Reflection Profiling)   Cooperative agreement between Cornell University and Conoco that led to breakthroughs in deep seismic profiling and our understanding of crustal structure.

collapse (founder) breccia   Any breccia that is produced as the result of the collapse of the ceiling rock that caps any underlying open space such as a cavern.

collophane   Sedimentary apatite (phosphate) mineral of uncertain origin.

compaction   Reduction in volume of loose sediments in response to pressure due to burial.

compositional maturity   Measure of the degree to which clastic sedimentary rocks, especially sandstone, have been repetitively cycled to eventually produce only hard, chemically resistant components like quartz and quartzite as opposed to less stable constituents like olivine and basalt. Levels of compositional maturity range from immature to mature and supermature.

concretion   Irregular shape formed by precipitation of groundwater cements within a sedimentary rock.

conglomerate   Also called roundstone or puddingstone, conglomerate is lithified gravel, a terrigenous (siliciclastic) sedimentary rock containing a significant component of rounded clasts coarser than 2 mm.

contact   Boundary between two rock units.

continental rise   Gently sloping wedge of sediments that builds up at the base of the continental slope through accumulation of turbidites.

continental shelf   Region of the seafloor that is a submarine continuation of the coastal plain. It is usually very shallow and low in relief, and it slopes gradually seaward to the shelf-slope break.

continental slope   More steeply sloping seafloor below the edge of the continental shelf.

contour current   Slow-moving, density-driven, bottom-hugging current that moves along the surface of the outer continental shelf, the continental slope, and the continental rise parallel with depth contours (bathymetric lines).

contourite   Deposit formed by a contour current.

convergent margin   Plate boundary where the plates are colliding and subducting; an active margin.

convolute bedding   Secondary sedimentary structure (most common in fine sand and silt) that consists of a series of tightly folded and/or faulted sedimentary beds or laminations. Probably generated by plastic deformation of water-rich sediment shortly after deposition, convolute bedding is commonly confined to single sedimentary units, with the immediately overlying and underlying beds remaining undeformed.

correlation   Correspondence or matching of beds.

craton   Stable core of a continent consisting of shield and platform areas.

crevasse splay   Sedimentary deposit (largely sand) that pours out of a breach in the natural levee and forms a small delta on a floodplain or interdistributary bay; a marsh.

cumulative frequency curve   Smooth curve fitted to a cumulative histogram that smooths out the distribution in order to more accurately characterize it.

cumulative histogram   Technical name for a cumulative bar diagram in which progressively finer size or compositional classes are systematically added to the preceding class so that the final bar totals 100% of the population.

Curie point   The temperature at which magnetization in an igneous rock is locked in.

cyclothem   Rhythmically repeating, vertically ordered sedimentary rock sequences in which individual lithologies supersede one another in a predictable fashion. Many coal seams occur as components within cyclothemic sequences.

debris flow   Type of sediment gravity flow that consists of a plastic slurry of poorly sorted clastic grains mixed with water.

degradational vacuity   Gap in the rock record formed by local erosion.

delta   Body of sediment formed when a river dumps its load into a lake or the sea.

density current (fluid-assisted gravity flow)   Types of sediment transport in which clastic grains are transported downslope principally by gravity rather than solely by the motion of the moving current.

deposition   Process by which material carried by various agents like wind or running water either as clastic detritus or components in solution are no longer transported and instead accumulate as sediment.

depositional setting   All aspects of the environment in which a sediment is deposited, including salinity, water depth, water temperature, current type and strength. The environment of deposition is most clearly summarized in terms of geomorphic setting, for example, a river delta, alluvial fan, or barrier island.

detrital remanent magnetization (DRM)   Magnetization caused by alignment of magnetic particles in a sedimentary rock.

detrital sedimentary rock   Sedimentary rock composed principally of bits and pieces of pre-existing rocks and

minerals produced by weathering and subsequent transport from the weathering site to the depositional area.

diachronic unit Stratigraphic unit that is explicitly time-transgressive.

diagenesis All changes (short of metamorphism) in the texture, composition, and other physical aspects of a sediment or sedimentary rock after it is deposited until the time it is examined and analyzed. Diagenetic processes include compaction, recrystallization, selective solution, and cementation.

diamictite Very poorly sorted paraconglomerate in which pebble-sized and coarser clasts are widely dispersed and float in a sandy or muddy matrix (like tillite and tilloid). Also called **diamixtite.**

diastem Subtle (possibly invisible) unconformity.

diatoms Microscopic algae that secrete a pair of shells made of silica.

dipmeter Well-logging device that measures the dip of the beds within a well.

directional structure Primary sedimentary structures whose internal characteristics can be used to infer the direction currents were flowing when the sediment containing these structures was deposited.

disconformity Type of unconformity where the beds above and below the erosional surface are parallel, but the erosional surface is still visible.

dismicrite Carbonate rock type that consists almost entirely of fine-grained microcrystalline calcite grains that have been subjected to extensive organic alteration (bioturbation).

dispersal Directional pattern that characterizes the pattern by which a sediment is distributed from source area to its area of deposition.

distal Toward the far end of a fan.

distributary mouth bar Sandbar that forms across the mouth of a delta distributary channel, usually where the freshwater flow of the river slows down as it meets the sea and dumps its load of sediment.

divergent margin Accreting (mid-ocean ridge) plate boundary where the plates are spreading or pulling apart.

dolomite Ordered carbonate mineral of composition $CaMg(CO_3)_2$.

dolostone Carbonate rock type composed substantially of the mineral dolomite.

dropstone Large (pebble-, cobble-, or boulder-sized) clasts floating in a (laminated) mudstone that are released from melting blocks of floating ice and drop into fine-grained, relatively deeper-water sediment.

dysaerobic Said of a depositional setting (or an organism that inhabits it) that lacks much free oxygen (between 0.1 and 1.0 ml of oxygen per liter of seawater).

Eh Short for redox potential, a measure of the potential for either oxidation (loss of electrons) or reduction (gain of electrons) of an ion in a given depositional setting. The value of Eh includes both a magnitude (in volts) and a sign (positive or negative). Eh depends largely on the availability of free atmospheric oxygen.

eluviation Downward movement of both solid and dissolved material in a soil from the A-horizon to the B-horizon.

entrain Process by which fragments of pre-existing rocks and minerals are picked up for transport by a moving current of water, air, ice, or mass movement.

epeiric sea Epicontinental sea, formed when high sea levels drowned the continents.

episodic sedimentation Sedimentation in irregular, episodic intervals.

erosion Processes by which pre-existing rocks and minerals exposed at or near the Earth's surface are first weathered (physically disintegrated and chemically decomposed) into soil, loose material, and dissolved components and then transported away or stripped from the weathering site.

eugeosyncline Rapidly subsiding deepwater trough on the outer edge of a continent, often full of volcaniclastic debris from the seaward direction. It is now thought to be formed by the continental rise turbidites of the passive margin.

event stratigraphy Correlation by events (such as the peak of transgression or regression in each rock unit) rather than correlation by the rock unit itself.

exotic (allochthonous) terrane Piece of crust that is not part of the continent where it is found but came from somewhere else.

extrabasinal Refers to weathered clastic sedimentary particles from a source area outside or adjacent to the sedimentary basin in which they accumulate.

extraformational Said of detrital components that are derived from the weathering of rock units below or adjacent to the sedimentary rock of which they are a part (for example, granite pebbles in a conglomerate).

facies map Map showing areal variation in facies of a given stratigraphic unit.

facies Appearance or aspect of a sedimentary rock body as contrasted with another part of the body.

fan delta Alluvial fan that pours immediately into a deep lake or the ocean so that coarse fan deposits build out into deep water.

fanglomerate Conglomerate formed in an alluvial fan.

fenestral porosity Pores in limestone formed by gas bubbles and shrinkage.

first-order cycle (supercycle) Global sedimentary cycle that lasts 200 to 400 million years.

flame structure Secondary sedimentary structure, typically due to loading, that appears as wavelike or flamelike tongues of mud that extend upward into overlying, generally coarser clastic material.

flaser bedding Variety of ripple bedding characterized by thin, discontinuous streaks of mud intermittently interbedded within cross-laminated sand and silt layers.

floatstone Carbonate rock in the Dunham classification that is supported by calcareous silt- and clay-sized matrix; less than 10% of the allochemical grains exceed 2 mm in diameter; allochems are not in tangential contact.

flocculate Process by which very small, clay- and silt-sized particles aggregate (due to electrostatic attraction) and as a consequence settle out of a moving or stationary fluid.

fluidity index Ratio of sand-sized framework grains to interstitial silt- and clay-sized (< 0.03 mm diameter) matrix in

a sandstone. The high ratio of wackes reflects their deposition from fluids of high density and viscosity; the low ratio of arenites represents deposition by running water, longshore currents, or wind.

fluidized sediment flow Type of density current in which concentrated masses of clasts supported by pore water between grains move downslope propelled by gravity.

flysch Preorogenic deepwater turbidites, shales, and cherts.

formation Rock unit that is lithologically distinctive and mappable at some scale.

fossil fuels Various solid, liquid, and gaseous materials, such as coal, petroleum (oil and natural gas) asphalt, tar sands, and oil shale, that are made up in substantial part of undecayed organic tissue that can be burned to produce thermal energy for transportation, heating, electricity, and other energy uses.

fourth-order cycle Sedimentary cycle that lasts 200,000 to 500,000 years.

framework Typically coarser component of clastic sediment or sedimentary rock, for example, pebble and sand-sized clasts, that collectively form a rigid arrangement of self-supporting larger grains that surround intergranular spaces that may be either empty or filled with chemical cement or finer-grained clastic components (matrix).

frequency curve Smooth curve fitted to a histogram that smooths out the distribution in order to more accurately characterize it.

geochronologic unit Unit of geologic time.

geode Hollow, globular body, often with either an empty center, or with crystals that have grown inward to fill the center.

geopetal structure Primary sedimentary structure such as cross-bedding and graded bedding that can be used to infer stratigraphic direction, i.e., which way is up in a sedimentary rock sequence, that is; the overall relation of the top and bottom of a given layer.

geophone Small seismograph used to measure reflected seismic waves for seismic stratigraphy.

glacial flour Clay- and silt-sized particles formed by rocks ground up in a glacier.

graded bedding Primary sedimentary structure in clastic sedimentary rocks in which clast diameter decreases upward from the base of a bed to the top.

grain flow Type of rapid sediment gravity flow that consists of cohesionless sediment dispersed in air and maintained in transport by the dispersive pressure generated by grain-to-grain collisions.

grainstone Carbonate rock type in the Dunham classification that is grain supported. Allochems are in tangential contact, fewer than 10% exceed 2 mm in diameter, and the interstitial pores contain no lime mud.

graphic correlation Shaw's method of correlating stratigraphic sections by placing them as axes on a bivariate graph.

Greenhouse Earth Time intervals in terrestrial history characterized by abnormally high global temperatures that on average (due to excess CO_2 in the atmosphere) lie above 30° to 40°C with hot, very humid climate, high sea levels, and no polar ice caps.

greensand Glauconite-rich shallow marine sand (or sandstone).

growth fault Syndepositional fault formed when dense, overloaded sediments on the front of a delta slope slump downward.

GSSP (Global Standard Stratotype Section and Point) Officially designated type section and standard point where a particular system or series is defined.

guano Fecal matter of birds and bats that can be diagenetically altered to phosphorite.

half-life Time it takes for half of a radioactive parent isotope to decay to a daughter isotope.

hardground Solid, crusty planar surface found in some horizons in carbonate rocks. Hardgrounds are produced in shallow-water depositional areas as a consequence of carbonate precipitation and in many instances mark depositional hiatuses.

herringbone cross-bedding Interbedded cross-bedded sedimentary layers in which vertically juxtaposed crossbeds are oriented in drastically different directions, most typically directly opposite one another, as is the case with tidal current generated cross-beds.

heterochrony Time-transgressive biostratigraphic occurrences.

hiatus Period of non-deposition.

histogram (or **bar diagram**) Technical name for a standard bar diagram in which the height of each bar is proportional to the abundance of the sample in that size or compositional class.

homotaxis Similarity in order of appearance of fossils.

hummocky cross-stratification Low mounds and hollows on the seafloor that form lenticular structures in cross-section. They are believed to be formed by storms.

humus Generally dark, chemically stable organically rich component of soil consisting of organic materials and partially decomposed as well as undecomposed rock and mineral matter.

hydraulic equivalency Term used to describe two clasts that, regardless of whether the controlling factors are density, shape, roundness, or grain size diameter, settle out of a fluid with the same velocity.

hydraulic jump Stationary surface wave of water generated where rapid turbulent flow abruptly changes to tranquil flow, producing a sudden increase in fluid depth.

hydraulics Science of the nature, controls, and manner of the flow of water and other fluids.

hylaloclastic breccia Breccia generated by the rapid cooling and contraction that occur where still-moving masses of hot lava come into direct contact with water, for example, on the rind of pillow lavas.

Icehouse Earth Time intervals in terrestrial history characterized by abnormally low global temperatures that on average lie near 0°C, resulting in a world in which ice rather than water is the dominant form of H_2O.

ichnofacies Assemblage or grouping of diverse ichnofossils (trace fossils, especially tracks, trails, and burrows) that can be used to infer a particular depositional environment.

ichnofossil (also called *Lebenspuren* and trace fossil) Trace fossils as opposed to body fossils, i.e., tracks, trails, borings, and other structures that are the product of organic activity but are not actual bits and pieces of organic remains.

ignimbrite Deposit produced by a gravity-driven cloud of ground-hugging tephra and gas that moves downslope at velocities up to 200 km/hr.

iluviation Accumulation of both solid and dissolved materiel in the B-horizon of a soil.

imbrication Preferred fabric orientation in which the long axes of detrital clasts are systematically parallel with one another (in fluvial gravels, commonly parallel to current vector and downstream).

impact (fallback) breccia Breccia composed of bits and pieces of material fragmented by the collision of Earth with an extraterrestrial object and almost immediately fell back to the impact site.

index fossil Fossil that is highly useful in biostratigraphy.

inheritance Especially in clastic grain size analysis, textural attributes of a particular sedimentary deposit may largely mirror the textural characteristics of the source materials from which that deposit was derived.

interdistributary bays and marshes Regions between deltaic distributaries that are flooded and marshy and trap mostly muddy sediments.

intertonguing Where two sedimentary bodies interfinger in a complex zigzag pattern.

interval zone Biostratigraphic zone defined by the interval between two consecutive first or last occurrences.

intrabasinal Refers to clastic sedimentary particles that are derived from the physical disintegration of materials within the sedimentary basin in which they now occur.

intraclast "Intrabasinal" allochemical grains of carbonate derived directly from sediment within which the intraclast-bearing carbonate rock is deposited (as opposed to "extrabasinal" limestone clasts).

intraformational Said of detrital clasts that are derived directly from the sedimentary rock of which they are a part (for example, rip-up shale pebbles and limestone intraclasts).

ironstone Conventional siliciclastic sedimentary rock (typically siltstone or mudrock) that contains iron in excess of 15%. Most ironstones are Phanerozoic in age and represent transported lateritic soils that accumulate where rivers reach the sea.

isopach Line on a map through points of equal stratigraphic thickness.

Jacob's staff Rod of a fixed length attached to a clinometer, used for measuring stratigraphic section in dipping rocks.

kerogen General term used to refer to extensively decomposed and altered insoluble organic material that is the parent matter from which oil and natural gas are derived by catagenesis.

kurtosis Statistical measure of the peakedness or flatness of a population. For grain size distribution, a measure of the degree of sorting in the central 50% of the population versus the two tails. Grain size distributions may be excessively peaked (leptokurtic), normally peaked (mesokurtic), or deficiently peaked (platykurtic).

lacuna Gap in the stratigraphic record formed by a hiatus and degradational vacuity.

lacustrine Pertaining to lakes.

ladderback Interference ripples that produce a ladder-like pattern.

lahar Mudflow that consists of water-saturated volcanic material.

laminar flow Fluid flow in which individual particles of matter (masses of water and air) move uniformly as sub-parallel sheets or filaments of material. In other words, flow lines (streamlines) are generally parallel with one another and the base of the fluid.

laminated pebbly (or **cobbly** or **bouldery**) **mudrock** Predominantly fine-grained mudrock in which scattered ice-rafted or pyroclastically produced coarser clasts float. Laminations in the mudrock alternately bend down abruptly beneath the large clasts and drape over them.

lamination Fine-scale (typically thinner than 1 cm) internal banding or layering present in some sediment and sedimentary rocks. Laminations can parallel bedding planes as well as cross-bedding surfaces.

lapilli Textural term applied to tephra ranging from 2 to 64 mm in diameter.

lateral accretion surfaces Depositional surfaces within a point bar, which form as the point bar grows and migrates laterally.

lateral sedimentation Tendency of some sedimentary rock bodies (such as prograding deltas or migrating point bar sequences) to accumulate by lateral growth so that they are internally time-transgressive.

laterite Orange to bright red clay-rich soils rich in iron oxide that are typically formed under warm, humid climatic conditions.

Lebenspuren (also called trace fossil) See **ichnofossil**.

LeChatelier's principle Chemical principle that changing any system initially at equilibrium will trigger additional changes that tend to restore that equilibrium; in practical terms, adding to the reactants side of a reaction will produce a corresponding increase in products.

Liesegangen **bands** Bands of color caused by precipitation of minerals (such as iron oxides), which resemble bedding but may or may not actually parallel any primary sedimentary structures.

lignite coal Lowest rank of coal, also called brown coal, containing roughly 70% carbon and significant amounts of moisture and volatiles.

lineage zone Biostratigraphic zone defined by the first appearance of successive taxa within a single lineage.

liquification Process by which water-saturated sediment flows as a fluid, generally as a consequence of increased pore pressure (as is the case with quicksand).

lithic Refers to sandstone (and in some cases, conglomerate and breccia) that contains significant amounts (25% or more) of rock fragments.

lithification Processes by which loose, unconsolidated sediment is altered into a cohesive sedimentary rock.

lithodemic unit Stratigraphic designation of an igneous or metamorphic rock unit.

lithofacies Lithologic appearance or aspect of a sedimentary rock body.

lithostratigraphic unit   Rock unit defined by its lithologic features only.

lodgment till   Till deposited by the melting of a glacier in a basin.

longitudinal bar   Sandbar in a braided stream that is parallel to the direction of flow.

luminescence   Emission of light (other than incandescent glowing) from a substance in response to its exposure to irradiation with light of a different wave length. For example, small differences in species of quartz irradiated with a beam of electrons may help decipher the provenance of different grains.

lysocline   Depth below which calcium carbonate reaches undersaturation.

magnetic polarity time scale   Global change of the Earth's magnetic polarity, used to correlate rocks around the world.

magnetometer   Instrument that measures magnetic direction and intensity.

magnetostratigraphy   Correlation by the changes in magnetic polarity recorded in rocks as they formed.

mass spectrometer   Instrument that measures the amounts of isotopes of different masses within a sample.

matrix   Finer-grained clastic components that enclose or fill the intergranular spaces that occur between coarser framework grains in clastic sediment or sedimentary rock.

maturation   Series of processes by which undecayed organic matter (mainly sapropel, fine-grained organic matter that accumulates subaqueously in anaerobic bodies of water) is converted first to kerogen and subsequently to petroleum.

measurement error   "Plus or minus" precision of a measurement.

median   Value of the middle item in a set of data, for example, median grain size.

mélange   Mixture of rocks found in the accretionary prism, which are pervasively sheared by the stresses of subduction.

metamorphic resetting   Tendency of high-grade metamorphic rocks to lose some parent or daughter isotope so that the ratio is a measure of when the rock cooled enough to lock in the isotopes in the crystals.

meteoritic   Fragmental components in sediment or sedimentary rocks generated by the impact of extraterrestrial, high-velocity bodies that hit the terrestrial surface fragmenting the target rocks into bits and pieces.

micrite   Fine-grained particles (< 4 microns diameter) of microcrystalline calcite mud matrix that can either fill the pore spaces between allochemical grains or constitute an entire carbonate rock.

migration   Process by which petroleum moves under pressure and as a consequence of buoyancy from source rocks where it forms to reservoir rocks from which it can be recovered. The expulsion of mature petroleum compounds from source rocks is primary migration; the subsequent rise of petroleum into overlying reservoir rocks because of its buoyancy is secondary migration.

Milankovitch cycles   Cycles of the Earth's orbital variation (eccentricity of the elliptical orbit; tilt of the Earth's axis; precession of the Earth's axis) that control global temperatures and the timing of ice ages.

miogeosyncline   Gradually subsiding shallow marine basins on the flank of the craton, now thought to be equivalent to the continental shelf.

modal size and class   Value of the most frequently occurring item in a set of data; grain size diameter class that is most abundant and the specific most frequently occurring grain size diameter.

molasse   Post-orogenic fluvial and deltaic deposits eroded from recently uplifted mountain ranges.

mudflow   Slurrylike mass of liquefied mud that moves downslope under the force of gravity.

mudrock   Collective term that refers to all siliciclastic sedimentary rocks composed predominantly of silt-sized and clay-sized particles.

mudstone   (1) In the Dunham classification of limestone, a limestone that consists principally of very fine-grained microcrystalline mud (grains < 0.03 mm) with less than 10% allochemical components (grains > 03 mm). (2) Lithified mud, a variety of terrigenous sedimentary rock that consists of a mixture of clay-sized (< 0.0039 mm) and silt-sized (0.0625 to 0.0039 mm) clasts.

mylonite   Fine-grained, powdery microbreccia generated by the extreme granulation and shearing along a fault zone.

natural levee   Wall of sand and mud that forms a high bank on each side of a river, formed when a flood recedes.

natural remanent magnetization (NRM)   Total magnetization of a rock before demagnetization of its components.

neomorphism   Typically in tandem diagenetic processes affecting carbonate rocks that can involve both inversion (transformation of aragonite to calcite) and overall recrystallization (to a coarser or finer grain size).

nodular (secondary) chert   Subspherical to ellipsoidal centimeter to tens-of-centimeter diameter bodies of chert most commonly found as anastamosing masses roughly coincidental with the bedding planes in shallow marine shelf carbonates. Nodular cherts are of secondary (replacement) origin and represent post-depositional migration of dissolved silica and its reprecipitation probably by bacterial action.

nodule   Small, somewhat rounded object formed by cementation, typically around a core, within a body of sedimentary rock.

nonconformity   Type of unconformity where sedimentary beds overlie an erosional surface carved into igneous or metamorphic rocks.

numerical age   Age of a rock unit in number of years before the present ("absolute age").

oligomict   Variety of breccia or conglomerate in which most of the coarse clasts consist of a very few, generally resistant rock or mineral types such as quartz and chert.

olistholith   Large exotic submarine slump block.

olisthostrome   Chaotic assemblage of brecciated blocks formed by a submarine landslide.

oncolite   Ovoid ball of sediment, usually several centimeters in diameter, formed by a rolled-up algal mat.

ooids   Also called oolites and ooliths. General term applied to coated allochemical carbonate grains (typically with a diameter of < 2 mm) that contain some sort of nucleus around

which occur a series of thin, concentrically laminated layers of fine-grain calcite or aragonite. Carbonate rocks composed almost entirely of ooids are also referred to as oolites or oolitic limestone.

ophiolite Piece of oceanic crust sliced off, uplifted, and plastered onto a continent. It consists of pillow lavas from the mid-ocean rift, sheeted dikes, and layered gabbros.

Oppel zone Assemblage zone as used by Albert Oppel.

orthochemical From the Greek *orthos*, "straight" or "true." Untraveled calcium carbonate constituents, specifically microcrystalline calcite matrix (micrite) or microcrystalline sparry cement (spar) that occur within carbonate rocks at the precise point where they were initially formed.

orthoconglomerate (or **orthobreccia**) Very coarse clastic sedimentary rock that consists mainly of gravel-sized (> 2 mm diameter) clasts that are in tangential contact supporting an intact framework but have a finer-grained (sand-sized and finer) matrix that amounts to no more than 15%.

overlapping (concurrent) range zones Biostratigraphic principle of defining zones by the overlap of two stratigraphic ranges.

oxbow lake Crescent-shaped lake filled with mud, which forms when meanders are abandoned.

packstone Carbonate rock type in the Dunham classification that is grain-supported. Allochems are in tangential contact but less than 10% exceed 2 mm in diameter; the interstitital mud consists of silt- and clay-sized mud of any composition (but is generally calcareous).

paleogeography Areal distribution during some point in time of both local topographic features and landforms (such as mountain fronts, alluvial fans, floodplains, and deltas) as well as broad regional features such as continental blocks, oceanic trenches, and volcanic island arcs.

paleopedology Study of ancient soils (paleosols), especially for the purpose of inferring ancient climates and past patterns of weathering.

paleosol Buried and ancient soils that predate the modern soils found in a region. Analysis of paleosols allows earlier climatic conditions to be inferred.

palimpsest effect Overprinting of modern processes of the continental shelf on ancient features formed when the shelf was subaerially emergent and affected by high-energy river processes. A palimpsest in classical archeology is a parchment or papyrus that has been written over, partially obscuring the original text with a later text.

paraconformity (obscure unconformity) Type of unconformity where the beds above and below the erosional surface are parallel, and there are no visible erosional effects; the gap is apparent only due to time missing, as shown by biostratigraphy.

paraconglomerate (or **parabreccia**) Very coarse clastic sedimentary rock with an unstable, nonintact framework. It consists of at least 15% sand-sized and finer-grained matrix (which can commonly exceed 50%) in which coarser grains float.

parasequence Relatively conformable successions of genetically related beds bounded by surfaces that represent abrupt change in water depth.

peat Wide variety of incompletely decayed and decomposed plant remains that accumulates in bogs, marshes, and swamps where free oxygen is not abundant.

ped Any naturally formed discrete component of soil such as a block or crumb.

pedalfer Typical soil type best developed in temperate conditions, typified by well-defined soil zones, including an organic-rich A-horizon.

pedocal Typical soil type best developed in desert and semiarid conditions; typically thin and rich in calcium carbonate.

pelagic Formed or deposited in the open ocean.

petromict Variety of breccia or conglomerate in which the coarse clasts are varied in composition and resistivity, for example, quartz mixed with shale mixed with volcanic rock fragments.

pH Conventional measure of the acidity (excess of H^+) or basicity (OH^-) of a solution, expressed as $-\log_{10}[H^+]$, with neutral at 7, acid from 1 to < 7, and base from > 7 to 14.

phosphorite Any sedimentary phosphate deposit, regardless of origin, that contains abnormally high (> 20%) concentrations of P_2O_5 occurring in a variety of minerals such as fluorapatite.

phreatic zone Zone of saturation that extends from the water table to unporous, impervious bedrock.

phyllosilicate Silicate family in which the basic structure includes a series of silicate tetrahedral sheets together with other structural and compositional components. Micas and clay minerals are phyllosilicates and exhibit characteristic flaky cleavage.

pinch-out Lateral thinning out of sedimentary rock units until they vanish.

pisoid Spherical to subspherical coated grains of calcium carbonate similar in shape and internal organization to ooids, but coarser (typically with grain diameter > 2.0 mm).

plane bed Sediment bed that is a broad, flat, essentially featureless planar surface produced in certain circumstances under both low-velocity and high-velocity conditions.

point-bar sequence Sequence of sedimentary structures formed in the inside bend, or point bar, of a meandering river.

precision Reproducibility of a measurement.

pressure solution Dissolution of solid framework grains by the pressure exerted between them as they compact against one another.

primary directional structure Generally large-scale, commonly three-dimensional features of sedimentary rocks (best studied in outcrop) such as cross-bedding, ripple marks, and graded bedding that form at the same time that the sediment containing them is deposited, usually in response to the flow conditions that carry and deposit the sediment.

primary dolomite Dolomite crystals that form directly from solution with no intervening stage as a calcium carbonate mineral such as calcite or aragonite.

primary porosity Pore space that exists in a sediment as it is deposited.

probability graph Standard method for displaying grain size distribution data. Cumulative percentages for individual grain size diameter classes are plotted on probability paper. The vertical (ordinate) axis is a log probability scale; the horizontal (abscissa) axis is an arithmetic scale. Actual

grain size distributions can be readily compared with a normal bell-shaped distribution, which on probability paper is a straight line.

prodelta Basal slope of a delta.

prograde Building out of land into a lake or the sea by excess sediment supply.

provenance All characteristics of a sediment or sedimentary rock source area, including source area composition, source area location (distance and direction), and source area relief. From the French *provenir*, "to come forth."

proximal Toward the head of a fan.

pseudomatrix Apparent matrix of clay minerals between framework grains that were formed by the breakdown of clay-rich framework grains (such as shales or schists), rather than by original deposition of clay between the framework grains.

pyroclastic airfall deposit Pyroclastic deposit that consists of volcanic ejecta thrown into the air and subsequently falling directly back to the terrestrial surface where it accumulates.

pyroclastic surge deposit Volcaniclastic deposit produced by the rapid, episodic-to-discontinuous downslope movement of pyroclastic material, gas, and in some cases, water.

pyroclastic Said of rocks or fragmental components in sediment or sedimentary rocks generated by explosive volcanism rather than physical (mechanical) weathering (epiclastic).

QFL plot Triangular diagram that plots the percentages of quartz and quartzite, feldspars, and lithic fragments in a sandstone.

QmFLt plot Triangular diagram that plots the percentages of monocrystalline quartz, feldspars, and total lithic fragments (including quartzite).

radiolaria Amoebalike protistans that secrete a shell of opaline silica.

recrystallization Diagenetic process in which the original fine-grained mineral has formed into larger crystals, while the chemistry of the mineral stays the same.

regression Relative fall of sea level off the land.

relative age Age of one rock unit with respect to another (younger or older).

remanence Amount of magnetism that remains in a rock.

replacement Diagenetic process in which one mineral has taken the place of another.

reservoir rocks Predominantly porous sedimentary rock types like sandstone, limestone, and dolomite that contain significant quantities of petroleum.

residence time Average time interval in years during which a component remains dissolved in seawater until it is chemically or biogenically extracted. Residence time is easily calculated by dividing total mass of an ion in seawater by its mean annual flux (the amount of that ion entering and leaving the sea yearly).

resolution Ability to discriminate between events closely spaced in time.

retrogradation Net transgression.

reverse (inverse) grading Primary sedimentary structure in clastic sedimentary rocks in which clast diameter increases upward from the base of a bed to the top, in contrast to normal graded bedding.

rose diagram Two-dimensional circular graph that displays statistical variations in directional sedimentary structure orientation, so called because vectors are grouped into classes that appear as petal-like fans (with petal length proportional to abundance).

rudstone Type of carbonate rock in the Dunham classification that is grain supported. Allochems are in tangential contact and more than 10% exceed 2 cm in diameter; interstitial pores are filled with silt- and clay-sized limy mud.

sabkha From the Arabic *sebkha*, "salt flat." Coastal marine supratidal mudflats, particularly in the arid and semi-arid regions of the Middle East.

salinity Total amount of solid material dissolved in water, expressed as both percent and parts per million (ppm). The salinity of normal seawater is 3.5% or 35,000 ppm.

saltation Process of bedload clastic sediment transport in which particles are transported by temporary suspension, hopping, skipping, and jumping downstream in an irregular, discontinuous fashion.

sand ribbon (tidal ridge) Linear sandbar up to 40 m high, 200 m wide, and 15 km long that forms parallel to the tidal flow direction on shallow shelves.

sapropel Inclusive term for any fine-grained organic material (most consists of the soft, organic tissue of plankton) that accumulates subaqueously in anaerobic bodies of water.

second-order cycle (synthem) Global sedimentary cycle that lasts 10 to 100 million years.

secondary dolomite Replacement dolomite: crystals of the mineral dolomite that were converted from pre-existing crystals of calcite or aragonite.

secondary porosity Pore space that occurs in a sediment or sedimentary rock as a consequence of post-depositional processes such as recrystallization and solution.

secondary sedimentary structures Generally large-scale, commonly three-dimensional features of sedimentary rocks (best studied in outcrop) such as sandstone dikes, load structures, convolute bedding, concretions, and stylolites that form shortly after deposition and before final lithification.

sediment Solid bits and pieces of material (fragments of rocks and minerals) produced by weathering, transported by various agents like wind, ice, running water, and mass movement, and either deposited or precipitated in layers on, at, or near the Earth's surface normally as loose, unconsolidated material.

sedimentary masking effect Kind of sediment accumulating at any point in time and space as a consequence of the rate at which sediment of different types are supplied; e.g., if carbonate sediment accumulates at only a modestly high rate compared to the rate at which detrital silt and sand is supplied, the resultant sediment will be classified as mudrock (albeit quite calcareous) rather than simply carbonate.

sedimentary petrology Branch of geology dealing with the origin, features, classification, nomenclature, and history of sedimentary rocks and the Earth's stratigraphic record.

sedimentary tectonics All aspects of the dynamic setting that prevails during the deposition of sedimentary rocks,

both the local nature of deformation (compressional fold belts, down faulted tensional grabens) and the regional context, for example, proximity to active or passive plate boundaries.

sedimentation unit Layer or stratum representing an interval of time of unspecified duration in which the sediment deposited and the conditions of deposition are relatively constant.

sedimentology Study of sedimentary rocks, their classification, origin, and interpretation, and the processes by which sediment is produced, transported, and deposited.

seismic sequence Unconformity-bounded package of rock on a seismic profile.

seismic stratigraphy Method of recording seismic reflections from subsurface layers and interpreting their meaning.

sequence Relatively conformable succession of related strata bounded by unconformities and their correlative conformities.

sericite Fine-grained muscovite and clays, usually formed by the alteration of feldspars.

shale Any mudrock that exhibits lamination or fissility (the ability to break in thin, papery sheets) or both.

shelf-slope break Abrupt change in gradient from the continental shelf to the continental slope.

silcrete Silica precipitated as surface or near-surface crusts and nodules by the evaporation of moisture in semi-arid climates.

siliciclastic Terrigenous material that consists almost entirely of bits and pieces of the principal rock-forming minerals like quartz and feldspar and clays.

siltstone Type of fine-grained siliciclastic sediment composed predominantly of silt-sized (.0625 to .0039 mm diameter) particles.

skewness Statistical measure of the asymmetry of a population, specifically the lack of coincidence of the mean, median, and mode of a distribution. For grain size distribution, a measure of the difference in degree of sorting between the coarser and finer halves of the population.

soil profile Vertical sequence of zones or bands that occurs within the band of soil weathered from exposure at the surface of the Earth where it is in contact with the atmosphere and elements.

soil Mass of material found in place (untransported) as a consequence of the physical breakdown (mechanical weathering) and chemical alteration of pre-existing rocks and minerals.

sole mark Primary structures (flute casts, groove casts) or secondary structures (load casts) that appear as three-dimensional topographic features visible on the bottom (underside) of sedimentary beds.

solution breccia Type of very coarse breccia produced as a consequence of the collapse of an unsupported cavern ceiling into the underlying space generated by dissolution of soluble components like limestone and evaporite.

sonde Torpedolike device pulled up through the mud of a recently drilled well that is used for well-logging.

sorting Variation in the grain size of clasts in sediment or sedimentary rocks and the dynamic process by which the variation in particle size diameter from the mean or average is achieved.

source rocks Moderately deep-seated organically rich porous and impermeable sedimentary rocks in which kerogen is altered to petroleum under moderately elevated temperatures and from which petroleum is expelled and migrates to reservoir rocks.

spar Moderately coarse interlocking crystalline grains (.02 to 0.10 mm in diameter) of typically clear calcium carbonate that commonly fills the pore space separating allochemical components, cementing them together to form a cohesive carbonate rock.

sphericity Shape or form of clastic sedimentary particles, typically as a consequence of their sphericity and roundness. Shape is commonly defined in terms of the ratio among long, short, and intermediate axes of a clast.

Stokes' law Mathematical expression that summarizes the various factors (namely, particle diameter, shape, and surface roughness, fluid viscosity, and gravitational force) that control the velocity with which a clast settles through a fluid.

stratification Banding or layering visible at some scale in all sedimentary rocks that reflects variations over time in the nature of the material that accumulated.

stratigraphic cutoff Arbitrary termination of a formation caused by disappearance or pinch-out of another formation which is part of its definition.

stratigraphy Subdiscipline of geology that examines and explains the complex distribution of sedimentary rocks in time and space in order to understand Earth's history.

stratotype Officially designated type section of a lithostratigraphic unit (such as a formation).

streamline Lines of fluid flow readily visible when colored dye is injected into a moving stream of water or air.

stromatolite Biogenic sedimentary structures that occur as domelike hemispherical laminations (mainly in carbonate rocks) produced by the trapping and binding activities during the successive upward growth of (blue-green) cyanobacterial algal mats, most often in the shallow intertidal and supratidal zone of the ocean.

structure contour map Map that shows the contour of the top of a bed in the subsurface.

stylolite Sedimentary structure generated by selective pressure solution in carbonate rocks. Stylolites appear as thin, jagged, toothlike seams in which insoluble silicate minerals like clay and quartz are concentrated.

subcrop map Geologic map of the units occurring below an erosional surface.

submarine fan Fan-shaped deposit of turbidites and pelagic muds formed at the mouth of a submarine canyon, which builds up the continental rise.

supercrop map ("worm's eye view" map) Map showing the units at the base of a sequence above an unconformity, as seen from below ("worm's eye view").

suspension Process of clastic sediment transport in which particles are moved downstream while remaining more or less continually within the body of moving fluid.

syntaxial overgrowth Framework grain (for example a fossil fragment) that has been overgrown with calcite that has precipitated with the same optical orientation, producing a composite grain that exhibits non-undulatory extinction.

taxon range zone Biostratigraphic zone based on the first and last occurrence of a single taxon.

tectonic hinge Boundary between the craton and the passive margin or between the shelf and the basin where subsidence rates increase rapidly.

teilzone Partial (or local) biostratigraphic range zone.

tektite Generally small (silt-, sand-, granule-, or pebble-sized) rounded spherical silicate glass bodies that are believed to be of extraterrestrial origin, namely recongealed droplets of material produced by impacts of objects with the Moon and Earth.

tephra Clastic material of any size or composition generated by pyroclastic activity (explosive volcanism).

tephrochronology Dating by volcanic ash layers.

tephrostratigraphy Correlation by volcanic ash layers.

terrigenous ("from the Earth") **material** Material principally composed of bits and pieces of pre-existing rocks generated by physical and chemical weathering.

textural maturity Measure of the degree to which clastic sedimentary rocks, especially sandstone, have been sorted and cleaned up during transport. Textural maturity can be measured by the degree to which clay-sized matrix has been removed and how well sorted and rounded the framework grains are. Levels of textural maturity range from immature through submature and mature to supermature.

thermal demagnetization Removal of magnetic overprints by heating and cooling a sample in field-free space at higher and higher temperatures.

thermal remanent demagnetization (TRM) Magnetization caused by cooling a rock below its Curie point.

third-order cycle (mesothem) Sedimentary cycle that lasts 1 to 10 million years.

tidal bedding Alternation of thin laminae of sand and mud ("heterolithic bedding") formed by the alternating energies of ebb and flood tides.

tidal flat Region of sand and mud in the intertidal zone that is exposed during low tide.

tidal sand waves. Submarine sand ridges 3 to 15 m tall and 150 to 500 m in wavelength, formed by tidal currents on shallow shelves.

till Type of glacial drift that consists exclusively of clastic material deposited directly by the ice.

tillite Glacial till that has been lithified by compaction and cementation.

tilloid Very coarse-grained paraconglomerates and parabreccias that are deposited by both dry and wet gravity-driven mass movements like debris flows and turbidity currents.

tool mark Any sedimentary structure (groove casts, skip marks, prod marks) produced by the impact of solid objects that were carried in a bottom-hugging current and were bouncing or dragged along the underlying upper surface of muddy sediment.

trace fossil (also called *Lebensspuren*) See **ichnofossil**.

traction Process of bedload clastic sediment transport in which particles are dragged and/or rolled along the bed of moving fluid.

transform margin Plate boundary where the plates are neither diverging nor converging, but sliding past one another on strike-slip faults.

transgression Relative rise of sea level onto the land.

transportation Process by which weathered material is removed from the weathering site and carried to its area of deposition.

transporting agent Mechanism by which a given sediment is transported from the weathering site to its area of deposition. Common depositional agents include running water, wind, glaciers, and mass movement.

transverse (linguoid) bar Sandbar in a braided stream that forms across the direction of flow as the channel migrates laterally.

triangle facies map Three facies (typically sandstone, shale, and limestone) are represented by different patterns on a triangular diagram and mapped with those patterns.

turbidite Sedimentary bed or layer that represents deposition from a turbidity current.

turbidity current Type of gravity-driven density current in which clasts remain in suspension supported by upwelling fluid turbulence.

turbulent flow Fluid flow in which individual particles of matter (masses of water and air) move in random, haphazard fashion with eddies that descend and rise; flow lines (streamlines) generally crisscross one another in an intertwining fashion.

Udden-Wentworth scale Grain diameter millimeter scale used to separate clasts into standard classes or categories. The scale has a fixed ratio between each successive size class: each size class is precisely half as large or twice as large as its neighboring classes (for example, 4 to 2 mm, 2 to 1 mm, 1 to 0.5 mm, and so on).

unconformity A substantial gap or break in the rock record.

vadose zone Zone of aeration that lies between the land surface and the water table.

volcaniclastic Any clastic sedimentary rock that contains abundant fragments of volcanic rock and glass, regardless of clast size or origin of the clasts.

volcaniclastic-flow deposit Deposit produced by the remobilization and downslope transport of tephra.

volcanogenic Any sedimentary rock that forms as a direct consequence of active volcanism.

wacke One of two major sandstone groups (the other is **arenite**). Wackes are texturally immature and consist of predominantly sand-sized framework clasts that float in a matrix of finer-grained (clast diameter 0.03 mm or less) silt- and clay-sized particles (amounting to at least 15%) that fills or partly fills the interstitial pore spaces. Quartz-rich, feldspar-rich, and rock fragment-rich categories are differentiated on the basis of their preponderance as framework grains.

wackestone Carbonate rock type in the Dunham classification that consists of 10% or more of allochems floating in silt- and clay-size mud that may be of any composition but is most generally calcareous.

Walther's law of correlation of facies Facies that occur in conformable vertical succession of strata also occur in laterally adjacent sedimentary environments.

washover deposit   Deposit of sand washed into the back-barrier lagoon by storms that breach the dune field at the crest of the barrier.

wave-ripple cross-bedding   Sedimentary structure with irregular bundles of cross-bedding forming bundled sets.

weathering   Various processes by which pre-existing rocks and minerals exposed at or near the surface of the Earth to organic activity and the elements (rain, ice, atmospheric gases) physically disintegrate and chemically decompose into soil, ions in solution, and transported bits and pieces.

well-logging   Measurement of the properties (electrical, chemical, seismic, and so on) of a well.

whole-rock analysis   Dating by measuring the entire rock (phenocrysts plus groundmass) rather than individual crystals.

Wilson cycle   Cycle by which mountain systems are produced by closing and opening of ocean basins.

Zeolite   Class of hydrous alumino-silicate minerals that are familiar diagenetic alteration products in sedimentary rocks.

zone of accumulation   Middle, second, or B-horizon of a typical soil zone; also known as the zone of illuviation. The B-horizon occurs below the A-horizon and is characterized by the accumulation of silicate clay minerals and oxidized iron.

zone of leaching   Uppermost or A-horizon of a typical soil zone in which large amounts of humic material accumulate and from which most of the soluble components have been selectively removed mainly by dissolution.

ZTR index   Percentage of highly stable heavy trace minerals such as zircon, tourmaline, and rutile, an indicator of maturity.

Bibliography

Adams, A. E., W. S. Mackenzie, and G. Guildford, 1984. *Atlas of Sedimentary Rocks under the Microscope.* John Wiley and Sons, New York.

Ager, D. V., 1964. "The British Mesozoic Committee." *Nature,* 203: 1059.

Ager, D. V., 1973. *The Nature of the Stratigraphical Record,* 1st ed. Macmillan, London.

Ager, D. V., 1981. *The Nature of the Stratigraphical Record,* 2d ed. John Wiley, New York.

Aitken, J. D., 1978. "Revised Model for Depositional Grand Cycles, Cambrian of the Southern Rocky Mountains, Canada." *Bull. Can. Petrol. Geol.,* 26: 515–542.

Allen, J. R. L., 1970. "Sediments of the Modern Niger Delta: A Summary and Review." *SEPM Spec. Publ.,* 15: 138–151.

Allen, J. R. L., 1985. *Principles of Physical Sedimentology.* George Allen and Unwin, London.

Allen, P. A., and J. R. Allen, 1990. *Basin Analysis: Principles and Applications.* Blackwell Scientific Publications, Oxford.

Alvarez, L. W., W. Alvarez, F. Asaro, and H. V. Michel, 1980. "Extraterrestrial Cause for the Cretaceous-Tertiary Extinction." *Science,* 208: 1095–1108.

Alvarez, W., M. A. Arthur, A. G. Fischer, W. Lowrie, G. Napoleone, I. Premoli-Silva, and W. R. Roggenthen, 1977. "Upper Cretaceous-Paleocene Magnetic Stratigraphy at Gubbio, Italy: V. Type Section for the Late Cretaceous-Paleocene Geomagnetic Reversal Time Scale." *Geol. Soc. Amer. Bull.,* 88: 383–389.

Alvaro, M., R. Capote, and R. Vegas, 1979. "Un Modelo de Evolución Geotectónica para la Cadena Celtibérica." *Acta Geol. Hispanica,* 14: 172–177.

Anstey, N. A., 1982. *Simple Seismics.* International Human Resources Development Corp., Boston.

Arthur, M. A., and B. B. Sageman, 1994. "Marine Black Shales: Depositional Mechanisms and Environments of Ancient Deposits." *Ann. Rev. Earth Planet. Sci.,* 22: 499–551.

Arthur, M. A., W. E. Dean, R. M. Pollastro, G. E. Claypool, and P. A. Scholle, 1985. "Comparative Geochemical and Mineralogical Studies of Two Cyclic Transgressive Pelagic Limestone Units, Cretaceous Western Interior Basin, U.S." In L. M. Pratt, E. G. Kauffman, and F. B. Zelt (eds.), *Fine-Grained Deposits and Biofacies of the Cretaceous Western Interior Seaway: Evidence of Cyclic Sedimentary Processes,* 16–27. SEPM Field Trip Guidebook 4.

Ashley, G. H., et al., 1933. "Classification and Nomenclature of Rock Units." *Geol. Soc. Amer. Bull.,* 44: 423–459.

Asquith, G. B., 1982. *Basic Well Log Analysis for Geologists.* Amer. Assoc. Petrol. Geol. Methods in Exploration Series, Tulsa, Okla.

Aubry, M.-P., 1991. "Sequence Stratigraphy: Eustasy or Tectonic Imprint?" *J. Geophys. Res.,* 96: 6641–6679.

Aubry, M.-P., W. A. Berggren, J. A. Couvering, and F. Steininger, 1999. "Problems in Chronostratigraphy: Stages, Series, Unit and Boundary Stratotypes, Global Stratotype Section and Point, and Tarnished Golden Spikes." *Earth Sci. Rev.,* 46: 99–148.

Aubry, M.-P., W. A. Berggren, D. V. Kent, J. J. Flynn, K. D. Klitgord, J. D. Obradovich, and D. R. Prothero, 1988. "Paleogene Geochronology: A Critique and Response." *Paleoceanography,* 3: 707–742.

Austin, G. S., 1974. "Multiple Overgrowths of Detrital Quartz Sand Grains in the Shakopee Formation (Lower Ordovician) of Minnesota." *J. Sed. Petrol.,* 44: 358–362.

Baars, D. L., 1962. "Permian System of the Colorado Plateau." *Amer. Assoc. Petrol. Geol. Bull.,* 46: 149–218.

Bachhuber, F. W., S. Rowland, and P. W. Huntoon, 1987. "Geology of the Lower Grand Canyon and Upper Lake Mead by Boat—An Overview." *Ariz. Bur. Geol. Min. Tech., Geol. Surv. Branch Spec. Paper,* 5: 39–51.

Baganz, B. P., J. C. Home, and J. C. Ferm, 1975, "Carboniferous and Recent Mississippi Lower Delta Plains: A Comparison." *Trans. Gulf Coast Assoc. Geol. Socs.,* 25: 183–191.

Balasubramaniam, D. S., et al. (eds.), 1989. *Weathering: Its Products and Deposits,* vol. 1, *Processes;* vol. 2, *Deposits.* Theophrastus Publications, Athens, Greece.

Bally, A. W., 1980. "Basins and Subsidence—A Summary." In A. W. Bally, P. L. Bender, T. R. McGetchin, and R. I. Walcott (eds.), *Dynamics of Plate Interiors,* 5–20. Amer. Geophys. Union, Geodynamics Series 1.

Bard, E., B. Hamelin, R. G. Fairbanks, and A. Zindler, 1990. "Calibration of the ^{14}C Timescale over the Past 30,000 Years Using Mass Spectometric U-Th Ages from Barbados Corals." *Nature,* 345: 405–410.

Barrande, J., 1846. *Notice préliminaire sur le système silurien et les trilobites de la Bohême.* Leipzig.

Barrell, J., 1917. "Rhythms and the Measurement of Geologic Time." *Geol. Soc. Amer. Bull.,* 28: 745–904.

Barron, E. J., and W. M. Washington, 1982. "Cretaceous Climate: A Comparison of Atmospheric Simulations with the Geologic Record." *Palaeogeogr. Palaeoclimat., Palaeoecol.,* 40: 103–133.

Barron, E. J., and W. M. Washington, 1985. "Warm Cretaceous Climates: High Atmospheric CO_2 As a Plausible Mechanism." *Geophys. Monogr. Series,* 32: 546–553.

Barron, E. J., M. A. Arthur, and E. G. Kauffman, 1985. "Cretaceous Rhythmic Bedding Sequences: A Plausible Link between Orbital Variations and Climate." *Earth Planet. Sci. Lett.,* 73: 327–340.

Barwis, J. H., and M. O. Hayes, 1979. "Regional Patterns of Barrier Island and Tidal-Inlet Deposition Applied to Hydrocarbon Exploration." In R. S. Saxena (ed.), *Stratigraphic Concepts in Hydrocarbon Exploration,* 113–158. Short Course Notes, 78th Gulf Coast Geol. Soc.–SEPM Meeting.

Bassett, M. G., 1985. "Towards a Common Language in Stratigraphy." *Episodes,* 8(2): 87–92.

Basu, A., S. W. Young, L. J. Suttner, W. C. James, and G. H. Mack, 1975. "Reevaluation of the Use of Undulatory Extinction and Polycrystallinity in Detrial Quartz for Provenance Interpretation." *J. Sed. Petrol.,* 45: 873–882.

Bateman, P. C., 1965. *Geology of Tungsten Mineralization of the Bishop District, California.* U.S. Geol. Surv. Prof. Paper 470.

Bathurst, R. G. C., 1975. *Carbonate Sediments and Their Disgenesis,* 2d ed. Developments in Sedimentology 12. Elsevier, New York.

Bathurst, R. G. C., and L. S. Land, 1986. "Carbonate Depositional Environments. Part 5: Diagenesis." *Quart. J. Colo. Sch. Mines,* 81: 1–41.

Baum, G. R., and P. R. Vail, 1988. "Sequence Stratigraphic Concepts Applied to Paleogene Outcrops, Gulf and Atlantic Basins." *SEPM Spec. Publ.,* 42: 309–328.

Bentor, Y. K., (ed.), 1980. *Marine Phosphorites.* SEPM Spec. Publ. 29.

Berger, W. H., 1974. "Deep-Sea Sedimentation." In C. A. Burk, and C. L. Drake (eds.), *The Geology of Continental Margins,* 213–241. Springer-Verlag, New York.

Berger, W. H., A. W. H. Bé, and W. V. Sliter, 1975. *Dissolution of Deep-Sea Carbonates: An Introduction.* Spec. Publ. Cushman Foundation 13.

Berggren, W. A., 1971. "Tertiary Boundaries and Correlations." In B. M. Funnell, and W. R. Riedel (eds.), *The Micropaleontology of Oceans,* 693–809. Cambridge University Press, Cambridge.

Berggren, W. A., 1986. "Geochronology of the Eocene-Oligocene Boundary." In C. Pomerol, and I. Premoli-Silva (eds.), *Terminal Eocene Events,* 349–356. Elsevier, Amsterdam.

Berggren, W. A., and J. A. Van Couvering, 1978. "Biochronology." *Amer. Assoc. Petrol. Geol. Mem.,* 6: 39–55.

Berggren, W. A., and M.-P. Aubry, 1983. "Rb-Sr Isochron of the Eocene Castle Hayne Limestone, North Carolina— Further Discussion." *Geol. Soc. Amer. Bull.,* 94: 364–370.

Berggren, W. A., D. V. Kent, and J. J. Flynn, 1985. "Paleogene Geochronology and Chronostratigraphy." In N. J. Snelling (ed.), *The Chronology of the Geological Record,* 141–195. Geol. Soc. Lond. Mem. 10.

Berggren, W. A., D. V. Kent, and J. Hardenbol (eds.), 1995. *Geochronology, Time Scales, and Stratigraphic Correlation.* SEPM Spec. Publ. 54.

Berggren, W. A., D. V. Kent, J. D. Obradovich, and C. C. Swisher III, 1992. "Toward a Revised Paleogene Geochronology." In D. R. Prothero, and W. A. Berggren (eds.), *Eocene-Oligocene Climatic and Biotic Evolution,* 29–45. Princeton University Press, Princeton, N.J.

Berggren, W. A., M. C. McKenna, J. Hardenbol, and J. D. Obradovich, 1978. "Revised Paleogene Polarity Time-Scale." *J. Geol.,* 86: 67–81.

Bergman, K. M., and R. G. Walker, 1987. "The Importance of Sea-Level Fluctuations in the Formation of Linear Conglomerate Bodies: Carrot Creek Member of the Cardium Formation, Cretaceous Western Interior Seaway, Alberta, Canada." *J. Sed. Petrol.,* 57(4): 651–665.

Berner, R. A., 1971. *Principles of Chemical Sedimentology.* McGraw-Hill, New York.

Berry, W. B. N., 1987. *Growth of a Prehistoric Time Scale Based on Organic Evolution,* 2d ed. Blackwell Scientific Publications, Palo Alto, Calif.

Bertrand-Sarfati, J., and M. R. Walter, 1981. "Stromatolite Biostratigraphy." *Precamb. Res.,* 15: 353–371.

Beukes, N. J., and C. Klein, 1993. "Models for Iron Formation Deposition." In J. W. Schopf, and C. Klein (eds.), *The Proterozoic Biosphere,* 147–151. Cambridge University Press, Cambridge.

Beyrich, H. E. von, 1854. "Über die Stellung die hessischen Tertiar-bildungen." *K. Preuss. Akad. Wiss. Berlin Monats. Heft.,* Nov. 1854: 664–666.

Biscaye, P. E., 1965. "Mineralogy of Sedimentation of Recent Deep-Clay in the Atlantic Ocean and Adjacent Seas and Oceans." *Geol. Soc. Amer. Bull.,* 76: 803–832.

Blatt, H., 1982. *Sedimentary Petrology.* W. H. Freeman and Co., San Francisco.

Blatt, H., G. V. Middleton, and R. C. Murray, 1980. *Origin of Sedimentary Rocks,* 2d ed. Prentice-Hall, Englewood Cliffs, N. J.

Blum, M. D., 1993. "Genesis and Architecture of Incised Valley Fill Sequences: A Late Quaternary Example from the Colorado River, Gulf Coastal Plain of Texas." *Amer. Assoc. Petrol. Geol. Mem.,* 58: 259–284.

Boggs, S., Jr., 1987. *Principles of Sedimentation and Stratigraphy,* 1st ed. Merrill Publishing, Columbus, Ohio.

Boggs, S., Jr., 1992. *Petrology of Sedimentary Rocks.* Macmillan Publishing Co., New York.

Boggs, S., Jr., 1995. *Principles of Sedimentation and Stratigraphy,* 2d ed. Merrill Publishing, Columbus, Ohio.

Blatt, H., 1992. *Sedimentary Petrology,* 2d ed. W. H. Freeman and Co., New York.

Bouma, A. H., 1962. *Sedimentology of Some Flysch Deposits—A Graphic Approach to Facies Interpretation.* Elsevier, New York.

Bouma, A. H., W. R. Normark, and N. E. Barnes (eds.), 1985. *Submarine Fans and Related Turbidite Systems.* Springer-Verlag, New York.

Bourgeois, J., 1980. "A Transgressive Shelf Sequence Exhibiting Hummocky Cross-Stratification: The Cape Sebastian Sandstone (Upper Cretaceous), Southwest Oregon." *J. Sed. Petrol.,* 50: 681–702.

Bowles, F. A., R. N. Jack, and I. S. E. Carmichael, 1973. "Investigations of Deep-Sea Volcanic Ash Layers from Equatorial Pacific Cores." *Geol. Soc. Amer. Bull.,* 84: 237–238.

Brenchley, P. J., J. D. Marshall, G. A. F. Carden, D. B. R. Robertson, D. G. F. Long, T. Meidla, L. Hints, and T. F. Anderson, 1994. "Bathymetric and Isotopic Evidence for a Short-Lived Late Ordovician Glaciation in a Greenhouse Period." *Geology,* 22: 295–298.

Brenner, R. L., and J. A. McHargue, 1988. *Integrative Stratigraphy.* Prentice-Hall, Englewood Cliffs, N.J.

Brett, C. E., and G. C. Baird, 1986. "Comparative Taphonomy: A Key to Paleoenvironmental Interpretations Based on Fossil Preservation." *Palaios,* 1: 207–227.

Bridge, J. S., 2003. *Rivers and Floodplains: Forms, Processes, and Sedimentary Record.* Blackwell Science, New York.

Brock, A., and G. L. Isaac, 1974. "Paleomagnetic Stratigraphy and Chronology of Hominid-Bearing Sediments East of Lake Rudolf, Kenya." *Nature,* 247: 344–348.

Broecker, W. S., 1972. *Chemical Oceanography.* Harcourt Brace Jovanovich, New York.

Bromley, R. G., 1990. *Trace Fossils, Biology and Taphonomy.* Special Topics in Paleontology. Unwin and Hyman, London.

Bromley, R. G., and A. A. Eckdale, 1986. "Composite Ichnofacies and Tiering of Burrows." *Geol. Mag.,* 123: 59–65.

Bronger, A., and J. A. Catt (eds.), 1989. *Paleopedology: Nature and Application of Paleosols.* Catena Verlag, Destedt, Germany.

Brooks, J., (ed.), 1990. *Classic Petroleum Provinces.* Geol. Soc. Lond. Spec. Publ. 50.

Brooks, J., and A. J. Fleet (eds.), 1987. *Marine Petroleum Source Rocks.* Geol. Soc. Amer. Spec. Publ. 26.

Brunhes, B., 1906. "Recherches sur le direction d'aimantation des roches volcaniques." *J. Physique,* 5: 705–724.

Buchheim, H. P., and R. Biaggi, 1988. "Laminae Counts within a Synchronous Oil Shale Unit: A Challenge to the Varve Concept." *Geol. Soc. Amer. Abstr. Prog.,* 20(7): A217.

Bull, W. B., 1972. "Recognition of Alluvial Fan Deposits in the Stratigraphic Record." In J. K. Rigby and W. K. Hamblin (eds.), *Recognition of Ancient Sedimentary Environments,* 63–83. SEPM Spec. Publ. 16.

Bull, W. B., 1977. "The Alluvial Fan Environment." *Prog. Phys. Geog.,* 1: 222–270.

Burk, C. A., and C. L. Drake (eds.), 1974. *The Geology of Continental Margins.* Springer-Verlag, New York.

Burke, K. C. A., and J. F. Dewey, 1973. "Plume-Generated Triple Junctions: Key Indicators in Applying Plate Tectonics to Old Rocks." *J. Geol.,* 81: 406–433.

Burton, R., C. G. St. C. Kendall, and I. Lerche, 1987. "Out of Our Depth: On the Impossibility of Fathoming Eustasy from the Stratigraphic Record." *Earth Sci. Rev.,* 24: 237–277.

Busby, C. J., and R. V. Ingersoll, 1995. *Tectonics of Sedimentary Basins.* Blackwell, Cambridge, Mass.

Busch, R. M., and H. T. Rollins, 1984. "Correlation of Carboniferous Strata Using a Hierarchy of Transgressive-Regressive Units." *Geology,* 12: 471–474.

Busch, R. M., and R. R. West, 1987. "Hierarchical Genetic Stratigraphy: A Framework For Paleoceanography." *Paleoceanography,* 2: 141–164.

Byers, C. A., 1982. "Stratigraphy—The Fall of Continuity." *J. Geol. Educ.,* 30: 215–221.

Campion, K. M., J. M. Lohmar, and M. D. Sullivan. 1994. "Paleogene Sequence Stratigraphy, Western Transverse Ranges, California." Pacific Sec. SEPM Field Guide.

Carter, R. M., 1988. "Plate Boundary Tectonics, Global Sea-Level Changes, and the Development of the Eastern South Island Continental Margin, New Zealand, Southwest Pacific." *Marine and Petrol. Geol.,* 4: 1–80.

Cas, R. A. F., and J. V. Wright, 1987. *Volcanic Successions: Modern and Ancient.* George Allen and Unwin, London.

Catt, J. A., 1986. *Soils and Quaternary Geology: A Handbook for Field Scientists.* Clarendon Press, Oxford.

Cavelier, C., J. -J. Chateauneuf, C. Pomerol, D. Rabussier, M. Renard, and C. Vergnaud-Grazzini, 1981. "The Geological Events at the Eocene-Oligocene Boundary." *Palaeogeogr., Palaeoclimat., Palaeoecol.,* 36: 223–248.

Cerling, T. E., and F. H. Brown, 1982. "Tuffaceous Marker Horizons in the Koobi Fora Region and the Lower Omo Valley." *Nature,* 299: 216–222.

Cerling, T. E., J. M. Harris, B. J. MacFadden, M. G. Leakey, J. Quade, V. Eisenmann, and J. R. Ehleringer, 1997. "Global Vegetation Change through the Miocene/Pliocene Boundary." *Nature,* 389: 153–158.

Chamley, H., 1989. *Clay Sedimentology.* Springer-Verlag, Berlin.

Chamley, H., 1990. *Sedimentology.* Springer-Verlag, New York.

Chang, K. H., 1975. "Unconformity-Bounded Stratigraphic Units." *Geol. Soc. Amer. Bull.,* 86: 1544–1552.

Chilingarian, G. V., and T. F. Yen, 1978. *Bitumens, Asphalts, and Tar Sands.* Elsevier, New York.

Chlupác, I., H. Jaeger, and J. Zikmundova, 1972. "The Silurian-Devonian Boundary in the Barrandian." *Bull. Can. Petrol. Geol.,* 20: 104–174.

Choquette, P. W., and N. P. James, 1987. "Diagenesis #12. Limestones—3. The Deep Burial Environment." *Geoscience Canada,* 14: 3–35.

Christie-Blick, N., 1991. "Onlap, Offlap, and the Origin of Unconformity-Bounded Depositional Sequences." *Marine Geol.,* 97: 35–36.

Christie-Blick, N., and N. W. Driscoll, 1995. "Sequence Stratigraphy." *Ann. Rev. Earth Planet. Sci.,* 23: 451–478.

Christie-Blick, N., G. S. Mountain, and K. G. Miller, 1990. "Seismic Stratigraphic Record of Sea-Level Change." In *Studies in Geophysics: Sea-Level Change,* 116–140. National Research Council, National Academy Press, Washington, D.C.

Clark, D. N., and L. Tallboda, 1980. "The Zechstein Deposits of Southern Denmark." *Contr. Sediment.,* 9: 205–231.

Clarke, F., 1924. "The Data of Geochemistry." *U.S. Geol. Surv. Bull.,* 770: 841.

Clemmey, H., and N. Badham, 1982. "Oxygen in the Precambrian Atmosphere: An Evaluation of the Geological Evidence." *Geology,* 10: 141–146.

Cloetingh, S., 1988. "Intraplate Stresses—New Element in Basin Analyses." In K. Kleinspehn and C. Paola (eds.), *New Perspectives in Basin Analysis,* 205–223. Springer-Verlag, New York.

Cloud, P., 1973. "Paleoecological Significance of Banded Iron Formation." *Econ. Geol.,* 68: 1135–1143.

Cluff, R. M., M. L. Reinbold, and J. A. Lineback, 1981. "The New Albany Shale Group of Illinois." *Ill. State Geol. Surv. Circ.,* 518: 1–83.

Cohee, G. V., M. F. Glaessner, and H. D. Hedberg (eds.), 1978. *Contributions to the Geologic Time Scale.* Amer. Assoc. Petrol. Geol. Stud. Geol. 6.

Colbert, E. H., 1935. "Siwalik Mammals in the American Museum of Natural History." *Trans. Amer. Phil. Soc.,* 26: 1–401.

Coleman, J. M., 1976. *Deltas: Processes of Deposition and Models for Exploration.* Continuing Education Publishing Co., Champaign, Ill.

Collinson, J. D., and D. B. Thompson, 1982. *Sedimentary Structures.* George Allen and Unwin, London.

Compton, R. R., 1962. *Manual of Field Geology.* John Wiley, New York.

Condie, K. C., 1989. *Plate Tectonics and Crustal Evolution.* Pergamon Press, Oxford.

Conkin, J. E., B. F. Conkin, and L. Z. Lipschutz, 1980. "Devonian Black Shales in the Eastern United States." *Univ. Louisville Stud. Paleont. Strat.,* 12: 1–65.

Conybeare, C. E. B., 1979. *Lithostratigraphic Analysis of Sedimentary Basins.* Academic Press, New York.

Cook, F. A., L. D. Brown, and J. E. Oliver. 1980. "The Southern Appalachians and the Growth of Continents." *Sci. Amer.,* 243(4): 156–168.

Cook, P. J., 1976. "Sedimentary Phosphate Deposits." In K. H. Wolf (ed.), *Handbook of Stratabound and Stratiform Ore Deposits,* 503–506. Elsevier, New York.

Cooke, B., 1976. "Suidae from the Plio-Pleistocene Strata of the Rudolf Basin." In Y. Coppens, F. C. Howell, G. L. Isaac, and R. E. F. Leakey (eds.), *Earliest Man and Environments in the Lake Rudolf Basin,* 251–263. University of Chicago Press, Chicago.

Cowie, J. A., and M. R. W. Johnson, 1985. "Late Precambrian and Cambrian Geological Time-Scale." In N. J. Snelling (ed.), *The Chronology of the Geological Record,* 47–64. Geol. Soc. Lond. Mem. 10.

Cowie, J. A., W. Ziegler, and J. Remane, 1989. "Stratigraphic Commission Accelerates Progress, 1984 to 1989." *Episodes,* 12(2): 79–83.

Cox, A. V., 1982. "Magnetostratigraphic Time Scale." In W. B. Harland (ed.), *A Geologic Time Scale,* 63–84. Cambridge University Press, Cambridge.

Crimes, T. P., 1975. "The Stratigraphical Significance of Trace Fossils." In R. W. Frey (ed.), *The Study of Trace Fossils,* 109–130. Springer-Verlag, New York.

Crowell, J. C., 1978. "Gondwana Glaciation, Cyclothems, Continental Positioning, and Climate Change." *Amer. J. Sci.,* 278: 1345–1372.

Crowell, J. C., 1982. "The Tectonics of the Ridge Basin, Southern California." In J. C. Crowell, and M. H. Link (eds.), *Geologic History of Ridge Basin, Southern California,* 25–42. Pacific Sec. SEPM Spec. Publ. 22.

Crowell, J. C., and L. A. Frakes, 1970. "Phanerozoic Glaciation and the Causes of Ice Ages." *Amer. J. Sci.,* 268: 193–224.

Crowell, J. C., and M. H. Link (eds.), 1982. *Geologic History of the Ridge Basin, Southern California.* Pacific Sec. SEPM, Spec. Publ. 22.

Crowley, T. J., and S. K. Baum, 1991. Towards Reconciliation of Late Ordovician (440 Ma) Glaciation with very high CO_2 levels. *J. Geophys. Res.,* 96: 22,597–22,610.

Cubitt, J. M., and R. A. Reyment (eds.), 1982. *Quantitative Stratigraphic Correlation.* John Wiley, New York.

Cummins, W. A., 1962. "The Greywacke Problem: Liverpool and Manchester." *Geol. Jour.,* 3: 51–72.

Curry, D., 1985. "Oceanic Magnetic Lineaments and the Calibration of the Late Mesozoic-Cenozoic Time-Scale." In N. J. Snelling (ed.), *The Chronology of the Geological Record,* 269–272. Geol. Soc. Lond. Mem. 10.

Curry, D., and G. S. Odin, 1982. "Dating of the Palaeogene." In G. S. Odin (ed.), *Numerical Dating in Stratigraphy,* 607–630. John Wiley, New York.

Curtis, G. H., R. E. Drake, T. Cerling, and J. Hampel, 1975. "Age of the KBS Tuff in Koobi Fora Formation, East Rudolf, Kenya." *Nature,* 258: 395–398.

Dalrymple, G. B., 1972. "Potassium-Argon Dating of Geomagnetic Reversals and North American Glaciations." In W. W. Bishop, and J. A. Miller (eds.), *Calibration of Hominid Evolution,* 107–134. Scottish Academic Press, Edinburgh.

Dalrymple, G. B., 1979, "Critical Tables for Conversion of K-Ar Ages from Old to New Constants." *Geology,* 7: 558–560.

Dalrymple, G. B., 1991. *The Age of the Earth.* Stanford University Press, Palo Alto, Calif.

Dalrymple, G. B., A. Cox, and R. R. Doell, 1965. "Potassium-Argon Age and Paleomagnetism of the Bishop Tuff, California." *Geol. Soc. Amer. Bull.,* 75: 665–674.

Dalrymple, G. B., and M. A. Lanphere, 1969. *Potassium-Argon Dating.* W. H. Freeman and Co., San Francisco.

Daly, R. A., 1907. "The Limeless Ocean of Precambrian Time." *Amer. J. Sci,.* 23: 93–115.

Davis, R. A., Jr., 1992. *Depositional Systems,* 2d ed. Prentice-Hall, New York.

Dana, J. D., 1873. "On Some Results of the Earth's Contraction from Cooling, Including a Discussion of the Origin of Mountains, and the Nature of the Earth's Interior." *Amer. J. Sci.,* 5: 423–443.

Daubrée, A., 1879. *Etudes Synthetiques de Geologie Experimentale.* Dunod, Paris.

Davaud, E., and J. Guex, 1978. "Traitement analytique 'manuel' et algorithmique de problèmes complexes de corrélations biochronologiques." *Eclogae Geol. Helv.,* 71: 581–610.

Davies, G. R., and S. D. Ludlam, 1973. "Origin of Laminated and Graded Sediments, Middle Devonian of Western Canada." *Geol. Soc. Amer. Bull.,* 84: 3527–3546.

Davis, R. A., Jr. (ed.), 1978. *Coastal Sedimentary Environments.* Springer-Verlag, New York.

Davis, R. A., Jr., 1983. *Depositional Systems—A Genetic Approach to Sedimentary Geology.* Prentice-Hall, New York.

Davis, R. A., Jr., and R. L. Ethington (eds.), 1976. *Beach and Nearshore Sedimentation.* SEPM Spec. Publ. 24.

Davis, R. A., Jr., and D. Fitzgerald, 2003. *Beaches and Coasts.* Blackwell Science, New York.

Dean, W. E., and R. P. Anderson, 1982. "Theoretical versus Observed Successions from Evaporites." In C. R. Handford, R. G. Loucks, and G. R. Davies (eds.), *Depositional and Diagenetic Spectra of Evaporites,* 74–85. SEPM Core Workshop No. 3.

Dean, W. E., and B. C. Schreiber (eds.), 1978. *Notes on a Short Course on Marine Evaporites.* SEPM Short Course No. 4.

Deffeyes, K. S., F. J. Lucia, and P. K. Weyl, 1965. "Dolomitization of Recent and Plio-Pleistocene Sediments by Marine Evaporite Waters in Bonaire, Netherlands Antilles." *SEPM Spec. Publ.,* 13: 71–88.

de Graciansky, P.-C., J. Hardenbol, T. Jacquin, and P. R. Vail (eds.), 1998. *Mesozoic and Cenozoic Sequence Stratigraphy of European Basins.* SEPM Spec. Publ. 60.

Denton, G. H., and T. J. Hughes, 1981. *The Last Great Ice Sheets.* John Wiley, New York.

DePaolo, D. J., and B. L. Ingram, 1985. "High-Resolution Stratigraphy with Strontium Isotopes." *Science,* 227: 938–941.

Desborough, G. A., 1978. "A Biogenic Chemical Stratified Lake Model for the Origin of Oil Shale of the Green River Formation: An Alternative to the Playa-Lake Model." *Geol. Soc. Amer. Bull.,* 89: 961–971.

Devoy, R. J., (ed.), 1987. *Sea Surface Studies.* Croon Helm, London.

Dewey, J. F., and J. M. Bird, 1970. "Mountain Belts and the New Global Tectonics." *J. Geophys. Res.,* 75: 2625–2647.

Dewey, J. F., and K. C. A. Burke, 1974. "Hotspots and Continental Breakup: Implications for Collisional Orogeny." *Geology,* 2: 57–60.

Dickinson, W. R., (ed.), 1974. *Tectonics and Sedimentation.* SEPM Spec. Publ. 22.

Dickinson, W. R., 1982. "Composition of Sandstones in Circum-Pacific Subduction Complexes and Fore-Arc Basins." *Amer. Assoc. Petrol. Geol. Bull.,* 66: 121–137.

Dickinson, W. R., 1993. "Exxon Global Cycle Chart: An Event for Every Occasion? Comment." *Geology,* 21: 282–283.

Dickinson, W. R., and C. A. Suczek, 1979. "Plate Tectonics and Sandstone Composition." *Amer. Assoc. Petrol. Geol. Bull.,* 63: 2164–2182.

Dickinson, W. R., and H. Yarborough, 1976. "Plate Tectonics and Hydrocarbon Accumulations." *Amer. Assoc. Petrol. Geol. Educ. Course Note Series,* 1: 1–55.

Dickinson, W. R., and R. Valloni, 1980. "Plate Settings and Provenance of Sands in Modern Ocean Basins." *Geology,* 8: 82–86.

Dickinson, W. R., L. S. Beard, G. R. Brakenridge, J. L. Exjavec, R. C. Ferguson, K. F. Inman, R. A. Knepp, F. A. Lindberg, and P. T. Ryberg, 1983. "Provenance of North American Phanerozoic Sandstones in Relation to Tectonic Setting." *Geol. Soc. Amer. Bull.,* 94: 222–235.

Dickinson, W. R., T. F. Lawton, and K. F. Inman, 1986. "Sandstone Detrital Modes, Central Utah Foreland Region: Stratigraphic Record of Cretaceous-Paleogene Tectonic Evolution." *J. Sed. Petrol.,* 56: 276–293.

Dietrich, R. V., and B. J. Skinner, 1979. *Rocks and Minerals.* John Wiley, New York.

Dietz, R. S., 1963. "Collapsing Ccontinental Rises, an Actualistic Concept of Geosynclines and Mountain Building." *J. Geol.,* 71: 314–333.

Dietz, R. S., and J. C. Holden, 1974. "Collapsing Continental Rises: Actualistic Concept of Geosynclines—A Review." *SEPM Spec. Publ.,* 19: 14–25.

Dineley, D. L., 1984. *Aspects of a Stratigraphic System: The Devonian.* John Wiley, New York.

Dobrin, M. B., 1976. *Introduction to Geophysical Prospecting,* 3d ed. McGraw-Hill, New York.

Donnelly, A. T., 1976. "The Refugian Stage of the California Tertiary: Foraminifera, Zonation, Geologic History, and Correlations with the Pacific Northwest." University Calif. Berkeley, unpubl. doct. dissert.

Donovan, D. T., and E. J. W. Jones, 1979. "Causes of Worldwide Changes in Sea Level." *J. Geol. Soc. Lond.,* 136: 187–192.

Donovan, S. K., 1994. *The Paleobiology of Trace Fossils.* Johns Hopkins University Press, Baltimore.

Dott, R. H., Jr., 1964. "Wacke, Graywacke and Matrix—What Approach to Immature Sandstone Classification?" *J. Sed. Petrol.,* 34: 625–632.

Dott, R. H., Jr. 1983. "Episodic Sedimentation—How Normal Is Average? How Rare Is Rare? Does It Matter?" *J. Sed. Petrol.,* 53: 5–23.

Dott, R. H., Jr., and R. H. Shaver (eds.), 1974. *Modern and Ancient Geosynclinal Sedimentation.* SEPM Spec. Publ. 19.

Doyle, L. J., and O. H. Pilkey (eds.), 1979. *Geology of Continental Slopes,* SEPM Spec. Publ. 27.

Drake, R. E., G. H. Curtis, T. E. Cerling, B. W. Cerling, and J. Hampel, 1980. "KBS Tuff Dating and Geochronology of Tuffaceous Sediments in the Koobi Fora and Shungura Formations, East Africa." *Nature,* 283: 368–372.

Dunbar, C. O., and J. Rodgers, 1957. *Principles of Stratigraphy.* John Wiley, New York.

Dunham, R. J., 1962. "Classification of Carbonate Rocks According to Depositional Texture." *Amer. Assoc. Petrol. Geol. Mem.,* 1: 108–121.

Dunham, R. J., 1970. "Stratigraphic Reef Versus Ecologic Reefs." *Amer. Assoc. Petrol. Geol. Bull.,* 54: 1931–1932.

Dutton, C. E., 1882. "Tertiary History of the Grand Cañon District." *U.S. Geol. Surv. Monogr.,* 2: 1264.

Dzulynski, S., and E. K. Walton, 1965. *Sedimentary Features of Flysch and Graywackes.* Developments in Sedimentology—7. Elsevier, New York.

Edwards, L. E., 1982a. "Quantitative Biostratigraphy: The Methods Should Suit the Data." In J. M. Cubitt and R. A. Rayment (eds.), *Quantitative Stratigraphic Correlation,* 45–60. John Wiley, New York.

Edwards, L. E., 1982b. "Numerical and Semi-Objective Biostratigraphy: Review and Predictions." *Proc. 3d N. Amer. Paleont. Conv.,* 1: 47–52.

Ehlers, E. G., and H. Blatt, 1982. *Petrology—Igneous, Metamorphic, and Sedimentary.* W. H. Freeman and Co., San Francisco.

Ehrlich, R., and B. Weinberg, 1970. "An Exact Method for Characterization of Grain Shape." *J. Sed. Petrol.,* 40: 205–212.

Ehrlich, R., S. J. Crabtree, S. K. Kennedy, and R. L. Cannon, 1984. "Petrographic Image Analysis, I. Analysis of Reservoir Pore Complexes." *J. Sed. Petrol.,* 54: 1365–1378.

Eicher, D. L., 1976. *Geologic Time,* 2d ed. Prentice-Hall, Englewood Cliffs, N.J.

Einsele, G., 1985. "Response of Sediments to Sea Level Change in Differing Subsiding Storm-Dominated Marginal Epeiric Basins." In U. Bayer, and A. Seilacher (eds.), *Sedimentary and Evolutionary Cycles,* 68–97. Springer-Verlag, Berlin.

Einsele, G., W. Ricken, and A. Seilacher (eds.), 1991. *Cycles and Events in Stratigraphy.* Springer-Verlag, Berlin.

Einsele, G., and U. Bayer, 1991. "Asymmetry in Transgressive-Regressive Cycles in Shallow Seas and Passive Continental Margin Settings." In G. Einsele, W. Ricken, and A. Seilacher (eds.), *Cycles and Events in Stratigraphy.* 660–681. Springer-Verlag, Berlin.

Ekdale, A. A., R. G. Bromley, and S. G. Pemberton (eds.), 1984. *Ichnology: The Use of Trace Fossils in Sedimentology and Stratigraphy.* SEPM Short Course Notes 15.

Elder, W. P., 1988. "Geometry of Upper Cretaceous Bentonite Beds: Implications about Volcanic Source Areas and Paleowind Patterns, Western Interior, United States." *Geology,* 16: 835–838.

Elderfield, H., 1986. "Strontium Isotope Stratigraphy." *Palaeogeogr., Palaeoclimat., Palaeoecol.,* 57: 71–90.

Elderfield, H., J. M. Gieskes, P. A. Baker, R. K. Oldfield, C. J. Hawkesworth, and R. Miller, 1982. "$^{87}Sr/Sr^{86}$ and $^{18}O/^{16}O$ Ratios, Interstitial Water Chemistry, and Diagenesis in Deep-Sea Carbonate Sediments of the Ontong Java Plateau." *Geochim. Cosmochim. Acta,* 45: 2201–2212.

Eldredge, N., and S. J. Gould, 1972. "Punctuated Equilibria: An Alternative to Phyletic Gradualism." In T. J. M. Schopf (ed.), *Models in Paleobiology,* 82–115. Freeman, Cooper, San Francisco.

Eldredge, N., and S. J. Gould, 1977. "Evolutionary Models and Biostratigraphic Strategies." In E. G. Kauffman, and J. E. Hazel (eds.), *Concepts and Methods of Biostratigraphy,* 25–40. Dowden, Hutchinson, and Ross, Stroudsburg, Penn.

Embry, A. F., and J. E. Klovan, 1971. "A Late Devonian Reef Tract on the Northeastern Banks Island, N. W. T." *Bull. Can. Petrol. Geol.,* 19: 730–781.

Embry, A. F., and J. E. Klovan, 1972. "Absolute Water Depth Limit of Late Devonian Paleoecological Zones." *Geol. Rundschau.,* 61: 672–686.

Emery, K. O., 1968. "Relict Sediments on Continental Shelves of the World." *Amer. Assoc. Petrol. Geol. Bull.,* 52: 445–464.

Emery, K. O., and H. Meyers, 1996. "Relict Sediments of the Continental Shelves of the World." *Amer. Assoc. Petrol. Geol. Bull.,* 52: 445–464.

Emiliani, C., 1966. "Isotopic Paleotemperatures." *Science,* 154(3751): 851–857.

Emiliani, C., 1955. "Pleistocene Temperatures." *J. Geol.,* 63: 538–578.

Emry, R. J., 1973. "Stratigraphy and Preliminary Biostratigraphy of the Flagstaff Rim Area, Natrona County, Wyoming." *Smithsonian Contrib. Paleobiol.,* 25: 1–20.

Ensley, R. A., and K. L. Verosub, 1982. "Biostratigraphy and Magnetostratigraphy of Southern Ridge Basin, Central Transverse Ranges, California." In J. C. Crowell, and M. H. Link (eds.), *Geologic History of Ridge Basin, Southern California,* 13–24. Pacific Sec. SEPM Spec. Publ. 22.

Epstein, A. G., J. P. Epstein, and L. D. Harris, 1977. "Conodont Color Alteration—An Index to Organic Metamorphism." *U.S. Geol. Surv. Prof. Paper,* 995.

Eskola, P., 1915. "Om sambandet mellan demisk och mineralogisk sammansättning hos Orijärvitraktens metamorfa bergarter." *Bull. Comm. Géol. Finlande,* 44: 1–145.

Ethridge, F. G., and R. M. Flores (eds.), 1981. *Recent and Ancient Nonmarine Depositional Environments: Models for Exploration.* SEPM Spec. Publ. 31.

Ethridge, F. G., T. J. Jackson, and A. D. Youngberg, 1981. "Floodbasin Sequences of a Fine-Grained Meander Belt Subsystem: The Coal-Bearing Lower Wasatch and Upper Fort Union Formations, Southern Powder River Basin, Wyoming." *SEPM Spec. Publ.,* 31: 191–212.

Eugster, H. P., 1970. "Chemistry and Origin of the Brines of Lake Magadi, Kenya." *Mineral. Soc. Amer. Spec. Publ.,* 3: 215–235.

Eugster, H. P., and L. A. Hardie, 1975. "Sedimentation in an Ancient Playa-Lake Complex: The Wilkins Peak Member of the Green River Formation of Wyoming." *Geol. Soc. Amer. Bull.,* 86: 319–334.

Evernden, J. F., and G. H. Curtis, 1965. "The Present Status of Potassium-Argon Dating of Tertiary and Quaternary Rocks." *Int. Assoc. Quat. Res., 6th Cong., Warsaw,* 1: 643–651.

Evernden, J. F., D. E. Savage, G. H. Curtis, and G. T. James, 1964. "Potassium-Argon Dates and the Cenozoic Mammalian Chronology of North America." *Amer. J. Sci.,* 262: 145–198.

Eyles, C. H., N. Eyles, and A. D. Miall, 1985. "Models of Glaciomarine Sedimentation and Their Application to the Interpretation of Ancient Glacial Sequences." *Palaeogeogr., Palaeoclimat., Palaeoecol.,* 51: 15–84.

Eyles, N., and C. H. Eyles, 1992. "Glacial Depositional Systems." In R. G. Walker and N. P. James (eds.), *Facies Models: Response to Sea Level Change,* 73–100. Geoscience Canada Reprint Series 1. Geological Association of Canada, Toronto.

Faure, G., 1986. *Principles of Isotope Geology,* 2d ed. John Wiley, New York.

Fenwick, I., 1985. "Paleosols, Problems of Recognition and Interpretation." In J. Boardman (ed.), *Soils and Quaternary Evolution,* 3–21. Wiley Interscience, New York.

Ferm, J. C., 1974. "Carboniferous Environmental Models in Eastern United States and Their Significance." *Geol. Soc. Amer. Spec. Paper,* 148: 79–95.

Findlater, I. C., 1978. "Isochronous Surfaces within the Plio-Pleistocene Sediments East of Lake Turkana." In W. W. Bishop (ed.), *Geological Background to Fossil Man,* 415–420. Scottish Academic Press, Edinburgh.

Fischer, A. G., 1961. "Stratigraphic Record of Transgressing Seas in Light of Sedimentation on the Atlantic Coast of New Jersey." *Amer. Assoc. Petrol. Geol. Bull.,* 45: 1656–1666.

Fischer, A. G., 1964. "The Lofer Cyclothems of the Alpine Triassic." *Kans. State Geol. Surv. Bull.,* 169: 107–149.

Fischer, A. G., 1981. "Climatic Oscillations in the Biosphere." In M. H. Nitecki (ed.), *Biotic Crises in Ecological and Evolutionary Time,* 102–131. Academic Press, New York.

Fischer, A. G., 1982. "Long-Term Climatic Oscillations Recorded in Stratigraphy." In W. H. Berger, and J. C. Crowell (eds.), *Climate in Earth History,* 97–104. National Academy Press, Washington, D.C.

Fischer, A. G., 1984. "The Two Phanerozoic Supercycles." In W. A. Berggren, and J. A. Van Couvering (eds.), *Catastrophes in Earth History,* 129–150. Princeton University Press, Princeton, N.J.

Fischer, A. G., 1986. "Climatic Rhythms Recorded in Strata." *Ann. Rev. Earth Planet. Sci.,* 14: 351–376.

Fischer, A. G., T. Herbert, and I. Premoli-Silva, 1985. "Carbonate Bedding Cycles in Cretaceous Pelagic and Hemipelagic Sequences." In L. M. Pratt, E. G. Kauffman, and F. B. Zelt (eds.), *Fine-Grained Deposits and Biofacies of the Cretaceous Western Interior Seaway: Evidence of Cyclic Sedimentary Processes,* 1–10. SEPM Field Trip Guidebook 4.

Fisher, R. V., and J. M. Rensberger, 1972. "Physical Stratigraphy of the John Day Formation." *Univ. Calif. Publ. Geol. Sci.,* 101: 1–45.

Fisher, R. V., and H. U. Schmincke, 1984. *Pyroclastic Rocks.* Springer-Verlag, Berlin.

Fisher, R. V., and H. U. Schmincke, 1994. "Volcaniclastic Sediment Transport and Deposition." In K. Pye (ed.), *Sediment Transport and Depositional Processes.* Blackwell Scientific Publications, Boston.

Fitch, F. J., and J. A. Miller, 1970. "Radioisotopic Age Determinations of Lake Rudolf Artefact Site." *Nature,* 226: 226–228.

Fitch, F. J., P. J. Hooker, and J. A. Miller, 1976. "⁴⁰Ar/³⁹Ar Dating of the KBS Tuff in Koobi Fora Formation, East Rudolf, Kenya." *Nature,* 263: 740–744.

Flesch, G. A., and M. D. Wilson, 1974. "Petrography of the Morrison Formation (Jurassic), Ojito Spring Quadrangle, Sandoval, New Mexico." *New Mexico Geol. Soc. Guidebook, 25th Field Conference,* 185–195.

Flint, R. F., J. E. Sanders, and J. Rodgers, 1960. "Diamictite, a Substitute Term for Symmictite." *Geol. Soc. Amer. Bull.,* 71: 1809–1810.

Flügel, E., 1982. *Microfacies Analysis of Limestones.* Springer-Verlag, Berlin.

Flynn, J. J., 1986. "Correlation and Geochronology of Middle Eocene Strata from the Western United States." *Palaeogeogr., Palaeoclimat., Palaeoecol.,* 55: 335–406.

Flynn, J. J., B. J. MacFadden, and M. C. McKenna, 1984. "Land Mammal Ages, Faunal Heterochrony, and Temporal Resolution in Cenozoic Terrestrial Sequences." *J. Geol.,* 92: 687–705.

Folk, R. F., 1951. "Stages of Textural Maturity." *J. Sed. Petrol.,* 21: 127–130.

Folk, R. F., 1966. "A Review of Grain-Size Parameters." *Sedimentology,* 6: 73–93.

Folk, R. L., 1959. "Practical Petrographic Classification of Limestones." *Amer. Assoc. Petrol. Geol. Bull.,* 43: 1–38.

Folk, R. L., 1962, "Special Subdivision of Limestone Types." *Amer. Assoc. Petrol. Geol. Mem.,* 1: 62–84.

Folk, R. L., 1974. *Petrology of Sedimentary Rocks.* Hemphill's, Austin, Texas.

Folk, R. L., and L. S. Land, 1975. "Mg/Ca Ratio and Salinity: Two Controls over Crystallization of Dolomite." *Amer. Assoc. Petrol. Bull.,* 59: 60–68.

Franca, A. B., and P. E. Potter, 1991. "Stratigraphy and Reservoir Potential of Glacial Deposits of the Itararé Group (Carboniferous to Permian), Paraná Basin, Brazil." *Amer. Assoc. Pet. Geol. Bull.,* 75: 62–85.

Frankel, C., 1999. *The End of the Dinosaurs.* Cambridge University Press, Cambridge.

Frey, R. W., 1972. "Trace Fossils of the Fort Hays Limestone Member of Niobrara Chalk (Upper Cretaceous), West-Central Kansas." *Univ. Kans. Paleont. Contrib.,* 58: 1–72.

Frey, R. W., and S. G. Pemberton, 1984. "Trace Fossil Facies Models." In R. G. Walker (ed.), *Facies Models,* 189–207. Geological Association of Canada, Toronto.

Frey, R. W., and S. G. Pemberton, 1985. "Biogenic Structures in Outcrops and Cores. I. Approaches to Ichnology." *Bull. Can. Petrol. Geo.,* 33: 72–115.

Frey, R. W., and A. Seilacher, 1980. "Uniformity in Marine Invertebrate Ichnology." *Lethaia,* 13: 183–207.

Friedman, G. M., 1962. "Comparison of Moment Measures for Sieving and Thin Section Data in Sedimentary Petrological Studies." *J. Sed. Petrol.,* 32: 15–29.

Friedman, G. M., 1967. "By Name Processes and Statistical Parameters Compared for Size Frequency Distribution of Beach and River Sands." *J. Sed. Petrol.,* 37: 327–354.

Friedman, G. M., (ed.), 1969. *Depositional Environments in Carbonate Rocks.* SEPM Spec. Publ. 14.

Friedman, G. M., and J. E. Sanders, 1978. *Principles of Sedimentology.* John Wiley, New York.

Friedman, G. M., J. E. Sanders, and D. C. Kopaska-Merkel, 1992. *Principles of Sedimentary Deposits.* Macmillan, New York.

Froelich, P. N., M. L. Bender, N. A. Luedtke, G. R. Heath, and T. DeVries, 1982. "The Marine Phosphorus Cycle." *Amer. J. Sci.,* 282: 474–511.

Frost, S. H., M. P. Weiss, and J. B. Saunders (eds.), 1977. *Reefs and Related Carbonates–Ecology and Sedimentology.* Amer. Assoc. Petrol. Geol. Stud. Geol. 4.

Fryberger, S. G., T. S. Ahlbrandt, and S. Andrews, 1979. "Origin, Sedimentary Features, and Significance of Low-Angle Eolian 'Sand Sheet' Deposits, Great Sand Dunes National Monument and Vicinity, Colorado." *J. Sed. Petrol.,* 49: 733–746.

Fulthorpe, C. S., 1991. "Geological Controls on Seismic Sequence Resolution." *Geology,* 19: 61–65.

Gale, N. H., 1985. "Numerical Calibration of the Paleozoic Timescale: Ordovician, Silurian, and Devonian Periods." In N. J. Snelling (ed.), *The Chronology of the Geological Record,* 81–88. Geol. Soc. Lond. Mem. 10.

Galloway, W. A., 1978. *Exploration for Stratigraphic Traps in Terrigenous Clastic Depositional Systems.* Amer. Assoc. Petrol. Geol. Short Course Notes 3.

Galloway, W. A., and D. K. Hobday, 1996. *Terrigenous Clastic Depositional System,* 2d ed. Springer-Verlag, New York.

Garrels, R. M., 1987. "A Model for the Deposition of the Microbanded Precambrian Iron Formation." *Amer. J. Sci.,* 287: 81–106.

Garrels, R. M., and F. T. Mackenzie, 1971. *Evolution of Sedimentary Rocks.* W. W. Norton, New York.

Garrels, R. M., E. A. Perry, and F. T. Mackenzie, 1973. "Genesis of Precambrian Iron-Formations and the Development of Atmospheric Oxygen." *Ecol. Geol.,* 68: 1173–1179.

Gast, P. W., 1962. "The Isotopic Composition of Strontium and the Age of Stone Meteorites, I." *Geochim. Cosmochim. Acta,* 26: 927–943.

Gilbert, C. M., 1938. "Welded Tuff in Eastern California." *Geol. Soc. Amer. Bull.,* 49: 1829–1862.

Gilbert, G. K., 1914. "Interpretation of Anomalies of Gravity." *U.S. Geol. Surv. Prof. Paper 85:* 29–37.

Gilluly, J., J. C. Reed, Jr., and W. Cady, 1970. "Sedimentary Volumes and Their Significance." *Geol. Soc. Amer. Bull.,* 81: 353–376.

Gilreath, J. A., J. S. Healy, and J. N. Yelverton, 1969. "Depositional Environments Defined by Dipmeter Interpretation." *Gulf Coast Assoc. Geol. Soc. Trans.,* 19: 101–109.

Ginsburg, R. N., (ed.), 1975. *Tidal Deposits, A Casebook of Recent Examples and Fossil Counterparts.* Springer-Verlag, New York.

Given, R. K., and B. H. Wilkinson, 1987. "Dolomite Abundance and Stratigraphic Age: Constraints on Rates and Mechanisms of Phanerozoic Dolostone Formation." *J. Sed. Petrol.,* 57: 1068–1078.

Glasby, G. P., (ed.), 1977. *Marine Manganese Deposits.* Elsevier, Amsterdam.

Glass, B. P., and J. R. Crosbie, 1982. "Age of Eocene/Oligocene Boundary Based on Extrapolation from North American Microtektite Layer." *Amer. Assoc. Petrol. Geol. Bull.,* 66: 471–476.

Glass, B. P., C. M. Hall, and D. York, 1986. "^{40}Ar/^{39}Ar Laser-Probe Dating of North American Tektite Fragments from Barbados and the Age of the Eocene-Oligocene Boundary." *Chem Geol.,* 69: 181–186.

Glass, B. P., M. B. Swincki, and P. A. Swort, 1979. "Australasian, Ivory Coast, and North American Tektite Strewn Fields: Size, Mass, and Correlation, with Geomagneetic Reversal and other Earth Events." *Proc. Lunar Planet Sci. Conf.,* 10: 2535–2545.

Gleadow, A. J. W., 1980. "Fission Track Age of the KBS Tuff and Associated Hominid Remains in Northern Kenya." *Nature,* 284: 228–230.

Gloppen, T. G., and R. J. Steel, 1981. "The Deposits, Internal Structure and Geometry in Six Alluvial Fan-Fan Delta Bodies (Devonian-Norway)—Study in the Significance of Bedding Sequence in Conglomerates." *SEPM Spec. Publ.,* 31: 49–69.

Goddard, E. N., et al., 1948. *Rock Color Chart.* Geol. Soc. Amer., Boulder, Colo.

Gold, T., 1999. *The Deep Hot Biosphere.* Springer-Verlag, New York.

Goldich, S. S., 1938. "A Study in Rock Weathering." *J. Geol.,* 46: 17–58.

Goodwin, P. W., and E. J. Anderson, 1980. "Punctuated Aggradational Cycles: A General Hypothesis of Stratigraphic Accumulation." *Geol. Soc. Amer. Abstr. Prog.,* 12(7): 435.

Goodwin, P. W., and E. J. Anderson, 1985. "Punctuated Aggradational Cycles: A General Hypothesis of Episodic Stratigraphic Accumulation." *J. Geol.,* 93: 515–533.

Gould, H. R., 1970. "The Mississippi Delta Complex." *SEPM Spec. Publ.,* 15: 3–30.

Gould, S. J., 1987. *Time's Arrow, Time's Cycle.* Harvard University Press, Cambridge, Mass.

Gould, S. J., and N. Eldredge, 1977. "Punctuated Equilibria: The Tempo and Mode of Evolution Reconsidered." *Paleobiology,* 3: 115–151.

Gradstein, F. M., F. P. Agterberg, J. C. Brower, and W. J. Schwarzacher (eds.), 1985. *Quantitative Stratigraphy.* D. Reidel Publishing Co., Dordrecht, Netherlands.

Gradstein, F. M., F. P. Agterberg, M.-P. Aubry, W. A. Berggren, J. J. Flynn, R. Hewitt, D. V. Kent, K. D. Klitgord, K. G. Miller, J. D. Obradovich, J. G. Ogg, D. R. Prothero, and G. E. C. Westermann, 1988. "Sea Level History." *Science,* 241: 599–601.

Graton, L. C., and H. J. Fraser, 1935. "Systematic Packing of Spheres—With Particular Relation to Porosity and Permeability." *J. Geol.,* 43: 785–909.

Gregg, J. M., and D. F. Sibley, 1984. "Epigenetic Dolomitization and the Origin of Xenotopic Dolomitic Textures." *J. Sed. Petrol.,* 54: 908–931.

Gregor, C. B., 1968. "The Rate of Denudation in Post-Algonkian Time." *Proc. K. Ned. Akad. Wet. Ser. 8 Phys. Sci.,* 71: 22–30.

Gregor, C. B., R. M. Garrels, F. T. Mackenzie, and J. B. Maynard (eds.), 1988. *Chemical Cycles in the Evolution of the Earth.* John Wiley and Sons, New York.

Gressly, A., 1838. "Observations géologiques dur le Jura Soleurois." *Neue Denkschr. Alls. Schweizerische Gesellsch. ges. Naturo.,* 2: 1–112.

Griffin, J. J., H. Windom, and E. D. Goldberg, 1968. "The Distribution of Clay Minerals in the World Ocean." *Deep-Sea Research* 15(4): 433–459.

Griffiths, J. C., 1967. *Scientific Methods in the Analysis of Sediments.* McGraw-Hill, New York.

Griffiths, C., and F. Gradstein, 1997. *Essentials of Quantitative Stratigraphy.* Chapman and Hall, London.

Grim, R. E., 1968. *Clay Mineralogy,* 2d ed. McGraw-Hill, New York.

Guex, J., 1977. "Une nouvelle méthode d'analyse biochronologique, note préliminaire." *Bull. Soc. Vaud. Sci. Nat.,* 351: 73.

Hall, J., Jr., 1859. *Paleontology,* vol. 3. Geological Survey of New York. Van Benthuysen, Albany.

Hallam, A., 1963. "Major Epeirogenic and Eustatic Changes since the Cretaceous, and Their Possible Relationship to Crustal Structure." *Amer. J. Sci.,* 261: 397–423.

Hallam, A., 1981. *Facies Interpretation and the Stratigraphic Record.* W. H. Freeman and Co., San Francisco.

Hallam, A., 1988. "A Reevaluation of Jurassic Eustasy in the Light of New Data and the Revised Exxon Curve." *SEPM Spec. Publ.,* 42: 261–274.

Hallam, A., 1992. *Phanerozoic Sea-Level Changes.* Columbia University Press, New York.

Hamblin, W. K., 1965. "Internal Structures of Homogenous Sandstones." *Kansas Geol. Surv. Bull.,* 175(1): 1–37.

Hambrey, M. J., and W. B. Harland (eds.), 1981. *Earth's Pre-Pleistocene Glacial Record.* Cambridge University Press, Cambridge.

Hancock, J. M., 1977. "The Historic Development of Biostratigraphic Correlation." In E. G. Kauffman, and J. E. Hazel (eds.), *Concepts and Methods of Biostratigraphy,* 3–22. Dowden, Hutchinson, and Ross, Stroudsburg, Penn.

Haq, B. U., J. Hardenbol, and P. R. Vail, 1988. "Mesozoic and Cenozoic Chronostratigraphy and Cycles of Sea-Level Change." *SEPM Spec. Publ.,* 42: 71–108.

Haq, B. U., J. Hardenbol, and P. R. Vail, 1987. "Chronology of Fluctuating Sea-Levels Since the Triassic." *Science,* 235: 1156–1167.

Haq, B. U., T. R. Worsley, L. H. Burckle, K. G. Douglas, L. D. Keigwin, Jr., N. D. Opdyke, S. M. Savin, M. H. Sommer, E. Vincent, and F. Woodruff, 1980. "Late Miocene Marine Carbon-Isotope Shift and Synchroneity of Some Phytoplanktonic Biostratigraphic events." *Geology,* 8: 427–431.

Hardenbol, J., and W. A. Berggren, 1978. "A New Paleogene Numerical Time Scale." *Amer. Assoc. Petrol. Geol. Mem.,* 6: 213–234.

Hardie, L. A., 1984. "Evaporites: Marine or Non-Marine?" *Amer. J. Sci.,* 284: 193–240.

Hardie, L. A., 1986. "Carbonate Tidal Flat Deposition." *Quart. J. Colo. Sch. Mines,* 81: 3–74.

Hardie, L. A., 1987. "Dolomitization: A Critical View of Some Current Views." *J. Sed. Petrol.,* 57: 166–183.

Hardie, L. A., 1991. "On the Significance of Evaporites." *Ann. Rev. Earth Planet. Sci.*, 19: 131–168.

Hardie, L. A., 1996. "Secular Variation in Seawater Chemistry: An Explanation for the Coupled Secular Variation in the Mineralogies of Marine Limestones and Potash Evaporites over the Past 600 m.y." *Geology*, 24: 279–283.

Hardie, L. A., and H. P. Eugster, 1970. "The Evolution of Closed-Basin Brines." *Mineral. Soc. Amer. Spec. Publ.*, 3: 273–290.

Hardie, L. A., and H. P. Eugster, 1971. "The Depositional Environment of Marine Evaporites: A Case for Shallow, Clastic Accumulation." *Sedimentology*, 16: 187–220.

Hardie, L. A., and E. A Shinn, 1986. "Carbonate Depositional Environments, Modern and Ancient. Part 3: Tidal Flats." *Quart. J. Colo. Sch. Mines*, 81: 1–74.

Hardie, L. A., J. P. Smoot, and H. P. Eugster, 1978. "Saline Lakes and Their Deposits: A Sedimentological Approach." *Int. Assoc. Sediment. Spec. Publ.*, 2: 7–41.

Hardy, R., and M. Tucker, 1988. "X-Ray Powder Diffraction of Sediments." In M. Tucker (ed.), *Techniques in Sedimentology*, 191–228. Blackwell Scientific Publications, Oxford.

Harland, W. B., A. V. Cox, P. G. Llewellyn, C. A. G. Pickton, A. G. Smith, and R. Walters, 1982. *A Geologic Time Scale.* Cambridge University Press, Cambridge.

Harms, J. C., 1979. "Primary Sedimentary Structures." *Ann. Rev. Earth Planet. Sci.*, 7: 227–248.

Harris, A. G., 1979. "Conodont Color Alteration, an Organo-Mineral Metamorphic Index, and Its Application to Appalachian Basin Geology." *SEPM Spec. Publ.*, 26: 3–16.

Harris, P. M., C. H. Moore, and J. L. Wilson, 1985. "Carbonate Depositional Environments, Modern and Ancient. Part 2: Carbonate Platforms." *Quart. J. Colo. Sch. Mines*, 80: 1–60.

Harris, W. B., and P. D. Fullagar, 1989. "Comparison of Rb-Sr and K-Ar Dates of Middle Eocene Bentonite and Glauconite, Southeastern Atlantic Coastal Plain." *Geol. Soc. Amer. Bull.*, 101: 573–577.

Harris, W. B., and V. A. Zullo, 1980. "Rb-Sr Glauconite Isochron of the Eocene Castle Hayne Limestone, North Carolina." *Geol. Soc. Amer. Bull.*, 93: 587–592.

Harris, W. B., P. D. Fullagar, and J. A. Winters, 1984. "Rb-Sr Glauconite Ages, Sabinian, Claibornian and Jacksonian Units, Southeastern Atlantic Coastal Plain, U.S.A." *Palaeogeogr., Palaeoclimat., Palaeoecol.*, 47: 53–76.

Harrison, C. G. A., I. McDougall, and N. D. Watkins, 1979. "A Geomagnetic Field Reversal Time Scale Back to 13.0 Million Years before Present." *Earth Planet. Sci. Lett.*, 42: 143–152.

Harshbarger, J. W., C. A. Repenning, and J. H. Irwin, 1957. "Stratigraphy of the Uppermost Triassic and the Jurassic Rocks of the Navajo Country (Colorado Plateau)." *U.S. Geol. Surv. Prof. Paper*, 291: 1–74.

Hattin, D. E., 1975. "Stratigraphic Study of the Carlile-Niobrara (Upper Cretaceous) Unconformity in Kansas and Northeastern Nebraska." *Geol. Assoc. Can. Spec. Paper*, 13: 31–54.

Hattin, D. E., 1982. "Stratigraphy and Depositional Environments of the Smoky Hill Chalk Member, Niobrar Chalk (Upper Cretaceous) of the Type Area, Western Kansas." *Kans. Geol. Surv. Bull.*, 255: 1108.

Hattin, D. E., 1985. "Distribution and Significance of Widespread, Time-Parallel Pelagic Limestone Beds in Greenhorn Limestone (Upper Cretaceous) of the Central Great Plains and Southern Rocky Mountains." *SEPM Field Trip Guidebook*, 4: 28–37.

Haug, E., 1907. *Traité de Géologie.*, 1: 1–536. Libr. Armand Cohn, Paris.

Hay, W. W., 1972. "Probabilistic Stratigraphy." *Eclogae Geol. Helv.*, 65: 255–266.

Hay, R. L., 1975. *Geology of Olduvai Gorge.* University of California Press, Berkeley, Calif.

Hay, W. W., and J. R. Southam, 1978. "Quantifying Biostratigraphic Correlation." *Ann. Rev. Earth Planet. Sci.*, 6: 353–375.

Hays, J. D., and N. D. Opdyke, 1967. "Antarctic Radiolaria, Magnetic Reversals, and Climatic Change." *Science*, 158: 1001–1011.

Hays, J. D., and W. L. Pitman, 1973. "Lithospheric Plate Motion, Sea-Level Changes and Climatic and Ecological Consequences." *Nature*, 246: 18–21.

Hays, J. D., and N. J. Shackleton, 1976. "Globally Synchronous Extinction of the Radiolarian *Stylatractus universus.*" *Geology*, 4: 649–652.

Hays, J. D., J. Imbrie, and N. J. Shackleton, 1976. "Variations in the Earth's Orbit—Pacemaker of the Ice Ages." *Science*, 194: 1121–1132.

Hazel, J. E., 1977. "Use of Certain Multivariate and Other Techniques in Assemblage Zonal Biostratigraphy: Examples Utilizing Cambrian, Cretaceous, and Tertiary Benthic Invertebrates." In E. G. Kauffman, and J. Hazel (eds.), *Concepts and Methods in Biostratigraphy*, 187–212. Dowden, Hutchinson, and Ross, Stroudsburg, Penn.

Heckel, P. H., 1974. "Carbonate Buildups in the Geological Record: A Review." *SEPM Spec. Publ.*, 18: 90–154.

Heckel, P. H., 1977. "Origin of Phosphatic Black Shale Facies in Pennsylvanian Cyclothems of Midcontinent North America." *Amer. Assoc. Petrol. Geol. Bull.*, 61: 1045–1068.

Heckel, P. H., 1980. "Paleogeography of Eustatic Model for Deposition of Midcontinent Upper Pennsylvanian Cyclothems." In T. D. Fouch, and E. R. Magathan (eds.), *Paleozoic Paleogeography of the West-Central United States*, 197–215. Rocky Mtn. Sec. SEPM, Paleogeography Symposium 1.

Heckel, P. H., 1986. "Sea-Level Curves for Pennsylvanian Eustatic Marine Transgressive-Regressive Depositional Cycles along the Midcontinent Outcrop Belt, North America." *Geology*, 14: 330–334.

Heckel, P. H., 1995. "Glacial-Eustatic Base-Level-Climatic Model for Late Middle to Late Pennsylvanian Coal-Bed Formation in the Appalachian Basin." *J. Sed. Res.*, B65: 348–357.

Hedberg, H. D., 1976. *International Stratigraphic Guide. A Guide to Stratigraphical Classification, Terminology, and Procedure.* John Wiley, New York.

Heezen, B. C., and C. D. Hollister, 1971. *The Face of the Deep.* Oxford University Press, New York.

Heirtzler, J. R., G. O. Dickson, E. M. Heron, W. C. Pittman, III, and X. Le Pichon, 1968. "Marine Magnetic Anomalies, Geomagnetic Field Reversals, and Motions of the Ocean Floor and Continents." *J. Geophys. Res.*, 73: 2119–2136.

Heller, P. L., P. D. Komar, and D. R. Pevear, 1980. "Transport Processes in Ooid Grains." *J. Sed. Petrol.*, 50: 943–952.

Hess, J., M. L. Bender, and J. G. Schilling, 1986. "Seawater $^{87}Sr/^{86}Sr$ Evolution from Cretaceous to Present—Applications to Palaeoceanography." *Science*, 231: 979–984.

Hesse, R., 1988. "Diagenesis #13. Origin of Chert: Diagenesis of Biogenic Siliceous Sediments." *Geoscience Canada*, 45: 1717–1792.

Hesse, R., 1989. "Silica Diagenesis: Origin of Inorganic and Replacement Cherts." *Earth Sci. Rev.*, 26: 253–284.

Hewlett, J. S., and D. W. Jordan, 1993. "Stratigraphic and Combination Traps Within a Seismic Sequence Framework, Miocene Stevens Turbidites, Bakersfield Arch, California." *Amer. Assoc. Petrol. Geol. Mem.*, 58: 135–162.

Hillhouse, J. W., J. W. M. Ndombi, A. Cox, and A. Brock, 1977. "Additional Results on Paleomagnetic Stratigraphy of the Koobi Fora Formation, East of Lake Turkana (Lake Rudolf), Kenya." *Nature*, 256: 411–415.

Hillhouse, J. W., T. E. Cerling, and F. H. Brown, 1986. "Magnetostratigraphy of the Koobi Fora Formation, Lake Turkana, Kenya." *J. Geophys. Res.*, 9(B11): 11581–11595.

Hobday, D. K., and A. J. Tankard, 1978. "Transgressive-Barrier and Shallow-Shelf Interpretation of the Lower Paleozoic Peninsula Formation, South Africa." *Geol. Soc. Amer. Bull.*, 89: 1733–1744.

Hohn, M. E., 1978. "Stratigraphic Correlation by Principal Components: Effect of Missing Data." *J. Geol.*, 86: 524–532.

Hohn, M. E., 1982. "Properties of Composite Sections Constructed by Least Squares." In J. M. Cubitt, and R. A. Reyment (eds.), *Quantitative Stratigraphic Correlation*, 107–117. John Wiley, New York.

Holland, H. D., 1973. "The Oceans As a Possible Source of Iron in Iron-Formations." *Econ. Geol.*, 68: 1169–1172.

Holland, H. D., 1984. *The Chemical Evolution of the Atmosphere and Oceans.* Princeton University Press, Princeton, N.J.

Holmes, A., 1911. "The Association of Lead with Uranium in Rock Minerals, and Its Application to the Measurement of Geologic Time." *Proc. R. Soc. London A*, 85: 248–256.

Holmes, A., 1913. *The Age of the Earth.* Harper, New York.

Hook, R. W., and J. C. Ferm, 1985. "A Depositional Model for the Linton Tetrapod Assemblage (Westphalian D, Upper Carboniferous) and Its Palaeo-Environmental Significance," *Phil. Trans. R. Soc. London B*, 311(1148): 101–109.

Hooke, R. L., 1967. "Processes on Arid-Region Alluvial Fans." *J. Geol.*, 75: 438–460.

Horne, J. C., and J. C. Ferm, 1976. *Carboniferous Depositional Environments in the Pocahontas Basin, Eastern Kentucky and Southern West Virginia: A Field Guide.* Department of Geology, University of South Carolina.

Horne, J. C., J. C. Ferm, F. T. Caruccio, and B. P. Baganz, 1978. "Depositional Models in Coal Exploration and Mine Planning in the Appalachian Region." *Amer. Assoc. Petrol. Geol. Bull.*, 62: 2379–2411.

Horowitz, A. S., and P. E. Potter, 1971. *Introductory Petrography of Fossils.* Springer-Verlag, Berlin.

House, M. R., 1985. "The Ammonoid Time-Scale and the Ammonoid Evolution." In N. J. Snelling (ed.), *The Chronology of the Geological Record*, 273–283. Geol. Soc. Lond. Mem. 10.

Howell, D. G., 1990. *Tectonics of Suspect Terranes.* Chapman and Hall, New York.

Howell, D. J., and J. C. Ferm, 1980. "Exploration Model for Pennsylvanian Upper Delta Plain Coals, Southwest West Virginia." *Amer. Assoc. Petrol. Geol. Bull.*, 64: 938–941.

Hsü, K. J., 1972. "Origin of Saline Giants: A Critical Review after the Discovery of the Mediterranean Evaporite." *Earth Sci. Rev.*, 8: 371–396.

Hsü, K. J., 1974. "Melanges and Their Distinction from Olistostromes." *SEPM Spec. Publ.*, 19: 321–333.

Hsü, K. J., 1983. *The Mediterranean Was a Desert.* Princeton University Press, Princeton, N.J.

Hsü, K. J., 1989. *Physical Principles of Sedimentology: A Readable Text for Beginners and Experts.* Springer-Verlag, Berlin.

Hubbard, R. J., 1988. "Age and Significance of Sequence Boundaries on Jurassic and Early Cretaceous Rifted Continental Margins." *Amer. Assoc. Petrol. Geol. Bull.*, 72: 49–72.

Hughes, J. D., 1984. "Geology and Depositional Setting of the Late Cretaceous Upper Bearpaw and Lower Horseshoe Canyon Formation in the Dodds-Round Hill Coalfield of Central Alberta—A Computer-Based Study of Closely Spaced Exploration Data." *Geol. Surv. Can. Bull.*, 361: 1–81.

Hunt, C. B., and D. R. Mabey, 1966. *General Geology of Death Valley: Stratigraphy and Structure.* U.S. Geol. Surv. Prof. Paper 494-A.

Huntoon, P. W., 1977. "Cambrian Stratigraphic Nomenclature and Ground-Water Prospecting Failures on the Hualapai Plateau, Arizona." *Ground Water*, 15: 426–433.

Hurford, A. J., A. J. W. Gleadow, and C. W. Naeser, 1976. "Fission-Track Dating of Pumice from the KBS Tuff, East Rudolf, Kenya." *Nature*, 263: 738–740.

Hutchinson, R. W., and G. G. Engels, 1970. "Tectonic Significance of Regional Geology and Evaporite Lithofacies in Northeastern Ethiopia." *Phil. Trans. R. Soc. A*, 267: 313–329.

Huxley, T. H., 1862. "The Anniversary Address." *Quar., J. Geol. Soc. Lond.*, 18: xl–liv.

Hyne, N. J., 1984. *Geology for Petroleum Exploration, Drilling and Production.* McGraw-Hill, New York.

Illing, L. V., 1954. "Bahamian Calcareous Sands." *Amer. Assoc. Petrol. Geol. Bull.*, 38: 1–95.

Imbrie, J., and K. P. Imbrie, 1979. *Ice Ages: Solving the Mystery.* Enslow Publishing, Short Hills, N.J.

Ingersoll, R. V., 1988. "Tectonics of Sedimentary Basins." *Geol. Soc. Amer. Bull.*, 100: 1704–1719.

Israelsky, M. C., 1949. "Oscillation Chart." *Amer. Assoc. Petrol. Geol. Bull.*, 33: 92–98.

Izett, G. A., 1981. "Volcanic Ash Beds: Records of Upper Cenozoic Silicic Pyroclastic Volcanism in the Western United States." *J. Geophys. Res.*, 86(B11): 10, 200–10, 222.

Izett, G. A., and C. W. Naeser, 1976. "Age of the Bishop Tuff of Eastern California As Determined by Fission-Track Method." *Geology*, 4: 587–590.

Izett, G. A., R. E. Wilcox, H. A. Powers, and G. A. Desborough, 1970. "The Bishop Ash Bed, a Pleistocene

Marker Bed in the Western United States." *Quat. Res.,* 1: 121–132.

Jaeger, J.-J., and J.-L. Hartenberger, 1975. "Pour utilization systématique de niveaux-repréres en biochronologie mammalienne." *3e Réunion Annuelle Sci. Terre,* 201.

James, N. P., 1984. "Introduction to Carbonate Facies Models: Shallowing-Upward Sequences in Carbonates; Reefs." In R. G. Walker (ed.), *Facies Models,* 2d ed., 209–244. Geoscience Canada Reprint Series 1. Geological Association of Canada, Toronto.

James, N. P., and P. W. Choquette, 1983. "Diagenesis #6. Limestones, the Sea Floor Diagenetic Environment." *Geoscience Canada,* 10: 162–179.

James, N. P., and P. W. Choquette, 1984. "Diagenesis #9. Limestones, the Meteoric Diagenetic Environment." *Geoscience Canada,* 11: 161–194.

James, N. P., and I. G. MacIntyre, 1985. "Carbonate Depositional Environments, Modern and Ancient. Part 1: Reefs." *Quart. J. Colo. Sch. Mines,* 80: 1–70.

Johanson, D., and M. Edey, 1981. *Lucy, the Beginnings of Humankind.* Simon and Schuster, New York.

Johnson, J. G., 1987. "Unconformity-Bounded Stratigraphic Units: Discussion." *Geol. Soc. Amer. Bull.,* 93: 443.

Johnson, N. M., N. D. Opdyke, G. D. Johnson, E. H. Lindsay, and R. A. K. Tahirkeli, 1982. "Magnetic Polarity Stratigraphy and Ages of Siwalik Grouop Rocks of the Potwar Plateau, Pakistan." *Palaeogeogr., Palaeoclimat., Palaeoecol.,* 37: 17–42.

Jones, B. C., P. F. Carr, and A. J. Wright, 1981. "Silurian and Early Devonian Geochronology—A Reappraisal, with New Evidence from the Bungonian Limestone." *Alcheringa,* 5: 197–207.

Jones, H. A., 1965. "Ferruginous Oolites and Pisoids." *J. Sed. Petrol.,* 35: 838–845.

Jopling, A. V., 1967. "Origin of Laminae Deposited by the Movement of Ripples along a Stream Bed: Laboratory Study." *J. Geol.,* 75: 287–305.

Jordan, T. E., and D. B. Flemings, 1990. "A Theoretical Evaluation of Sequence Stratigraphy in a Clastic Foreland Basin." *Geol. Soc. Amer. Abstr. Prog.,* 22(2): 26–27.

Jordan, T. E., and P. B. Flemings, 1991. "Large-Scale Stratigraphic Architecture, Eustatic Variation, and Unsteady Tectonism: A Theoretical Evaluation." *J. Geophys. Res.,* 96: 6681–6699.

Karig, D. E., and G. F. Sharman, III, 1975. "Subduction and Accretion in Trenches." *Geol. Soc. Amer. Bull.,* 86: 377–389.

Katz, A., Y. Kolodny, and A. Nissenbaum, 1977. "The Geochemical Evolution of the Pleistocene Lake Lisan-Dead Sea System." *Geochim. Cosmochim. Acta,* 41: 1609–1626.

Kauffman, E. G., 1969. "Cretaceous Marine Cycles of the Western Interior." *Mountain Geol.,* 6: 227–245.

Kauffman, E. G., 1984. "Paleobiogeographic and Evolutionary Response Dynamics in the Cretaceous Western Interior Seaway of North America." *Geol. Assoc. Can. Spec. Paper,* 27: 273–306.

Kauffman, E. G., 1985. "Cretaceous Evolution of the Western Interior Basin of the United States." In L. M. Pratt, E. G. Kauffman, and F. B. Zelt (eds.), *Fine-Grained Deposits and Biofacies of the Cretaceous Western Interior Seaway; Evidence of Cyclic Sedimentary Processes,* iv–xiii. SEPM Field Trip Guidebook 4.

Kauffman, E. G., and J. E. Hazel (eds.), 1977. *Concepts and Methods of Biostratigraphy.* Dowden, Hutchinson, and Ross, Stroudsburg, Penn.

Kauffman, E. G., and L. M. Pratt, 1985. "Field Reference Section." In L. M. Pratt, E. G. Kauffman, and F. B. Zelt (eds.), *Fine-Grained Deposits and Biofacies of the Cretaceous Western Interior Seaway—Evidence of Cyclic Sedimentary Processes,* 1–26. SEPM Field Trip Guidebook 4.

Kay, M., 1951. *North American Geosynclines.* Geol. Soc. Amer. Mem. 48.

Keller, J., W. B. F. Ryan, and D. Ninkovitch, 1978. "Explosive Volcanic Activity in the Mediterranean over the Past 200,000 Years As Recorded in Deep-Sea Sediments." *Geol. Soc. Amer. Bull.,* 84: 591–604.

Kendall, A. C., 1984. "Evaporites." In R. G. Walker (ed.), *Facies Models,* 2d ed., 259–286. Geoscience Canada Reprint Series 1. Geological Association of Canada, Toronto.

Kendall, C. G. St. C., and I. Lerche, 1988. "The Rise and Fall of Eustasy." *SEPM Spec. Publ.,* 42: 3–17.

Kendall, C. G. St. C., and P. A. d'E. Skipwith, 1969. "Holocene and Shallow-Water Carbonate and Evaporite Sediments of khor al Bazam, Abu Dhabi, Southwest Persian Gulf." *Amer. Assoc. Petrol. Geol. Bull.,* 53: 841–869.

Kendall, C. G. St. C., P. Moore, G. Whittle, and R. Cannon, 1992. "A Challenge: Is It Possible to Determine Eustasy, and Does It Matter?" *Geol. Soc. Amer. Mem.,* 180: 93–107.

Kennett, J. P., (ed.), 1980. *Magnetic Stratigraphy of Sediments.* Benchmark Papers in Geology 54. Dowden, Hutchinson, and Ross, Stroudsburg, Penn.

Kennett, J. P., 1982. *Marine Geology.* Prentice-Hall, Englewood Cliffs, N.J.

Kimberley, M. M., 1979. "Origin of Oolitic Iron Formations." *J. Sed. Petrol.,* 49: 111–132.

King, P. B., 1977. *The Evolution of North America,* 2d ed. Princeton University Press, Princeton, N.J.

Kirschvink, J. L., R. L. Ripperdan, and D. A. Evans, 1997. "Evidence for a Large-Scale Early Cambrian Reorganization of Continental Masses by Inertial Interchange Polar Wander." *Science,* 227: 541–545.

Klein, C., and Beukes, N. J., 1990. "Geochemistry and Sedimentology of a Facies Transition to Iron-Formation Deposition in the Early Proterozoic Transvaal Supergroup, South Africa." *Econ. Geol.,* 84: 1733–1774.

Klein, G. D., 1970. "Depositional and Dispersal Dynamics of Intertidal Sand Bars." *J. Sed. Petrol.,* 40: 1095–1127.

Klein, G. D., 1972a. "Determination of Paleotidal Range in Clastic Sedimentary Rocks." *24th Int. Geol. Cong., Comptes Rendus,* 6: 397–405.

Klein, G. D., 1972b. "A Sedimentary Model for Determining Paleotidal Range: Reply." *Geol. Soc. Am. Bull.,* 83: 539–546.

Klein, G. D., 1977. *Clastic Tidal Facies.* Continuing Education Publishing Co., Champaign, Ill.

Kleinspehn, K. L., and C. Paola (eds.), 1988. *New Perspectives in Basin Analysis.* Springer-Verlag, New York.

Koch, P. L., J. C. Zachos, and P. D. Gingerich, 1992. "Correlation between Isotopic Records in Marine and Continental Carbon Reservoirs near the Paleocene/Eocene Boundary." *Nature,* 358: 319–322.

Kocurek, G., and R. H. Dott, Jr., 1983. "Jurassic Paleogeography and Paleoclimate of the Central and South-

ern Rocky Mountains Region." In M. W. Reynolds, and E. D. Dolly (eds.), *Mesozoic Paleogeography of the West-Central United States*, 101–116. Rocky Mtn. Sec. SEPM.

Kocurek, G., and K. G. Havholm, 1993. "Eolian Sequence Stratigraphy—A Conceptual Framework." *Amer. Assoc. Petrol. Geol. Mem.,* 58: 393–410.

Koepnick, R. B., R. E. Denison, and D. A. Dahl, 1988. "The Cenozoic Seawater $^{87}Sr/^{86}Sr$ Curve: Data Review and Implications for Correlation of Marine Strata." *Paleoceanography,* 3: 743–756.

Koss, J. E., F. G. Ethridge, and S. A. Schumm, 1994. "An Experimental Study on the Effects of Base-Level Change on Fluvial, Coastal Plain, and Shelf Systems." *J. Sed. Res.,* B64: 90–98.

Koster, E. H., and R. J. Steel (eds.), 1984. *Sedimentology of Gravels and Conglomerates.* Can. Soc. Petrol. Geol. Mem.10.

Kottlowski, F. E., 1965. *Measuring Stratigraphic Sections.* Holt, Rinehart and Winston, New York.

Kraus, M. J., 1984. "Sedimentology and Tectonic Setting of Early Tertiary Quartzite Conglomerates, Northwest Wyoming." In E. H. Koster, and R. J. Steel (eds.), *Sedimentology of Gravels and Conglomerates,* 203–216. Can. Soc. Petrol. Geol. Mem. 19.

Krauskopf, K. B., 1959. "The Geochemistry of Silica in Sedimentary Environments." *SEPM Spec. Publ.,* 7: 4–19.

Krauskopf, K. B., 1967. *Introduction to Geochemistry.* McGraw-Hill, New York.

Krinsley, D., and J. Koornkamp, 1973. *Atlas of Quartz Sand Surface Textures.* Cambridge University Press, Cambridge.

Krumbein, W. C., 1934. "Size Frequency Distribution of Sediments." *J. Sed. Petrol.,* 4: 65–77.

Krumbein, W. C., and L. L. Sloss, 1963. *Stratigraphy and Sedimentation.* W. H. Freeman and Co., San Francisco.

Kuenen, P. H., and C. I. Migliorini, 1950. "Turbidity Currents As a Cause of Graded Bedding." *J. Geol.,* 58: 91–127.

Kumar, N., 1973. "Modern and Ancient Barrier Sediments: New Interpretation Based on Stratal Sequence in Inlet-Filling Sands and on Recognition of Nearshore Storm Deposits." *Ann. N.Y. Acad. Sci.,* 220(5): 245–340.

La Fon, N. A., 1981. "Offshore Bar Deposits of Semilla Sandstone Member of Mancos Shale (Upper Cretaceous), San Juan Basin, New Mexico." *Amer. Assoc. Petrol. Geol. Bull.,* 65: 706–721.

LaBrecque, J. L., et al., 1983. "Contributions to the Paleogene Stratigraphy in Nomenclature, Chronology, and Sedimentation Rates." *Palaeogeogr., Palaeoclimat., Palaeoecol.,* 42: 91–125.

LaBrecque, J. L., S. C. Cande, and R. D. Jarrard, 1985. "Intermediate Wavelength Magnetic Anomaly Field of the North Pacific and Possible Distributions." *J. Geophys. Res., B,* 90(3): 2549–2564.

LaBrecque, J. L., D. V. Kent, and S. C. Cande, 1977. "Revised Magnetic Polarity Time Scale for Late Cretaceous and Cenozoic Time." *Geology,* 5: 330–335.

Laffitte, R., W. B. Harland, H. K. Erben, W. H. Blow, W. Haas, N. F. Hughes, W. H. C. Ramsbottom, P. Rat, H. Tintant, and W. Ziegler, 1972. "Some International Agreement on Essentials of Stratigraphy." *Geol. Mag.,* 109: 1–15.

Lajoie, J., and J. Stix, 1992. "Volcaniclastic Rocks." In R. G. Walker, and N. P. James (eds.), *Facies Models:*

Response to Sea Level Change, 3d ed. Geoscience Canada Reprint Series 1. Geological Association of Canada, Toronto.

Laming, D. J. C., 1966. "Imbrication, Paleocurrents, and Other Sedimentary Features in the Lower New Red Sandstone, Devonshire, England." *J. Sed. Petrol.,* 36: 940–959.

Land, L. S., 1985. "The Origin of Massive Dolomite." *J. Geol. Educ.,* 33: 112–125.

Land, L. S, and T. F. Goreau, 1967. "Submarine Lithification of Jamaican Reefs." *Amer. Assoc. Pet. Geol. Bull.,* 53: 457–462.

Land, L. S., F. T. Mackenzie, and S. J. Gould, 1967. "Pleistocene History of Bermuda." *Geol. Soc. Am. Bull.,* 78: 993–1006.

Langford, R., and M. A. Chan, 1988. "Flood Surfaces and Deflation Surfaces within the Cutler Formation and Cedar Mesa Sandstone (Permian), Southeastern Utah." *Geol. Soc. Amer. Bull.,* 100: 1541–1549.

Langstaff, C. S., and D. Morrill, 1981. *Geologic Cross Sections.* International Human Resources Development Corporation, Boston.

Laporte, L. F., (ed.), 1974. *Reefs in Time and Space.* SEPM Spec. Publ. 18.

Laporte, L. F., 1967. "Carbonate Deposition Near Mean Sea-Level and Resultant Facies Mosaic: Manlius Formation (Lower Devonian) of New York State." *Amer. Assoc. Petrol. Geol. Bull.,* 51: 73–101.

Laporte, L. F., 1969. "Recognition of a Transgressive Carbonate Sequence within an Epeiric Sea: Helderberg Group (Lower Devonian) of New York State." *SEPM Spec. Publ.,* 14: 98–119.

Laporte, L. F., 1975. "Carbonate Tidal Deposits of the Early Devonian Manlius Formation of New York State." In R. N. Ginsburg (ed.), *Tidal Deposits: A Casebook of Recent Examples and Fossil Counterparts,* 243–250. Springer-Verlag, Berlin.

Lapworth, C., 1879. "On the Tripartite Classification of the Lower Palaeozoic Rocks." *Geol. Mag.,* 6: 1–15.

Lavoisier, A., 1789. "Observations générales sur les couches horizontales, qui ont été deposées par la mer, et sur les consequences, qu'on peut tirer de leurs dispositions, relativement à l'anciennée du globe terrestre." *Acad. Sci. Mem.,* 351–371.

Lazarus, D. B., and D. R. Prothero, 1984. "The Role of Stratigraphic and Morphologic Data in Phylogeny Reconstruction." *J. Paleont.,* 58: 163–172.

Leakey, L. S. B., J. F. Evernden, and G. H. Curtis, 1961. "Age of Bed 1, Olduvai." *Nature,* 191: 478–479.

Leakey, R. E., 1978. *Origins.* E. P. Dutton: New York.

Leeder, M. R., 1982. *Sedimentology, Process and Product.* George Allen and Unwin, London.

LeFournier, J., and G. M. Friedman, 1974. "Rate of Lateral Migration of Adjoining Sea-Marginal Sedimentary Environments Shown by Historical Records, Authie Bay, France." *Geology,* 2: 497–498.

Lerman, A., and M. Meybeck (eds.), 1988. *Physical and Chemical Weathering in Geochemical Cycles.* Kluwer Academic, Dordrecht, Netherlands.

LeRoy, L. W., and J. W. Low, 1954. *Graphic Problems in Petroleum Geology.* Harper and Brothers, New York.

Levell, B. K., J. Braakman, and K. W. Rutten, 1988. "Oil-Bearing Sediments of Gondwana Glaciations in Oman." *Amer. Assoc. Petrol. Geol. Bull.,* 72: 775–796.

Levorsen, A. I., 1967. *Geology of Petroleum.* W. H. Freeman and Co., San Francisco.

Lewin, R., 1987. *Bones of Contention: Controversies in the Search for Human Origins.* Simon and Schuster, New York.

Lewis, D. W., 1984. *Practical Sedimentology.* Hutchison Ross Publications, Stroudsburg, Penn.

Lindholm, R. C., 1987. *A Practical Approach to Sedimentology.* George Allen and Unwin, London.

Lindsey, D. A., (ed.), 1972. *Sedimentary Petrology and Paleocurrents of the Harebell Formation, Pinyon Conglomerate, and Associated Coarse Clastic Deposits, Northwestern Wyoming.* U.S. Geol. Surv. Prof. Paper 734B.

Lineback, J. A., 1968. "Turbidites and Other Sandstone Bodies of the Borden Siltstone (Mississipian) in Illinois." *Ill. Geol. Surv. Circ.,* 425: 1–29.

Lineback, J. A., 1970. "Stratigraphy of the New Albany Shale in Indiana." *Ind. Geol. Surv. Bull.,* 44: 1–73.

Liro, L. M., 1993. "Sequence Stratigraphy of a Lacustrine System: Upper Fort Union Formation (Paleocene), Wind River Basin, Wyoming, U.S.A." *Amer. Assoc. Petrol. Geol. Mem.,* 58: 317–334.

Lisitzin, A. P., 1972. *Sedimentation in the World Ocean.* SEPM Spec. Publ. 17.

Lofton, C. L., and W. M. Adams, 1971. "Possible Future Petroleum Provinces of the Eocene and Paleocene, Western Gulf Basin." *Amer. Assoc. Petrol. Geol. Mem.,* 15: 855–886.

Loope, D. B., 1985. "Episodic Deposition and Preservation of Eolian Sands, a Late Paleozoic Example from Southeastern Utah." *Geology,* 13: 73–76.

Loucks, R. G., D. G. Bebout, and W. E. Galloway, 1977. "Relationship of Porosity Formation and Preservation to Sandstone Consolidation History, Gulf Coast Lower Tertiary Frio Formation." *Gulf Coast Assoc. Geol. Soc. Trans.,* 27: 109–120.

Loucks, R. G., and J. F. Sarg (eds.), 1993. *Carbonate Sequence Stratigraphy.* Amer. Assoc. Petrol. Geol. Mem. 57.

Lowe, D. R., 1982. "Sediment Gravity Flows: II. Depositional Models with Special Reference to the Deposits of High-Density Turbidity Currents." *J. Sed. Petrol.,* 52: 279–297.

Lowrie, W., and W. Alvarez, 1981. "One Hundred Million Years of Geomagnetic Polarity History." *Geology,* 9: 392–397.

Lowrie, W., W. Alvarez, and I. Premoli-Silva, 1980. "Lower Cretaceous Magnetic Stratigraphy in Umbrian Pelagic Carbonate Rocks." *Geophys. J. R. Astron. Soc.,* 60: 263–281.

Lowrie, W., W. Alvarez, G. Napoleone, K. Perch-Nielsen, I. Premoli-Silva, and M. Toumarkine, 1982. "Paleogene Magnetic Stratigraphy in Umbrian Pelagic Carbonate Rocks. The Contessa Sections, Gubbio." *Geol. Soc. Amer. Bull.,* 93: 414–432.

Lumsden, D. N., 1985. "Secular Variations in Dolomite Abundance in Deep-Marine Sediments." *Geology,* 13: 766–769.

Lumsden, D. N., 1988. "Characteristics of Deep-Marine Dolomite." *J. Sed. Petrol.,* 58: 1023–1051.

Lunine, J. I., 1999. *Earth: Evolution of a Habitable World.* Cambridge University Press, New York.

Lupe, R., and T. S. Ahlbrandt, 1979. "Sediments of the Ancient Eolian Environments—Reservoir Inhomogeneity." *U.S. Geol. Surv. Prof. Paper,* 1052: 241–252.

Lyell, Sir Charles, 1830–3. *Principles of Geology,* 3 vols. John Murray, London.

Maglio, V. J., 1972. "Vertebrate Faunas and Chronology of Hominid-Bearing Sediments East of Lake Rudolf, Kenya." *Nature,* 239: 379–385.

Maglio, V. J., and H. B. S. Cooke (eds.), 1972. *Evolution of African Mammals.* Harvard University Press, Cambridge, Mass.

Maliva, R. G., and R. Siever, 1988. "Pre-Cenozoic Nodular Cherts: Evidence for Opal-CT Precursors and Direct Quart Replacement." *Amer. J. Sci.,* 288: 798–809.

Maliva, R. G., and R. Siever, 1989. "Nodular Chert Formation in Carbonate Rocks." *J. Geol.,* 97: 421–433.

Mallory, V. S., 1959. *Lower Tertiary Biostratigraphy of the California Coast Ranges.* Amer. Assoc. Petrol. Geol., Tulsa, Okla.

Mankinen, E. A., and G. B. Dalrymple, 1979. "Revised Geomagnetic Polarity Time Scale for the Interval 0–5 m.y. B.P." *J. Geophys. Res.,* 84: 615–626.

Mann, K. O., and H. R. Lane (eds.), 1995. *Graphic Correlation.* SEPM Spec. Publ. 53.

Maples, C. G., and R. P. West (eds.), 1992. *Trace Fossils.* Paleontological Society, Knoxville, Tenn.

Markevich, V. P., 1960. "The Concept of Facies." *Int. Geol. Rev.,* 2: 376–379, 498–507, 582–604.

Mason, B., 1966. *Principles of Geochemistry.* John Wiley and Sons, New York.

Matthews, R. K., 1984. *Dynamic Stratigraphy,* 2d ed. Prentice-Hall, Englewood Cliffs, N.J.

McArthur, J. M., 1994. "Recent Trends in Strontium Isotope Stratigraphy." *Terra Nova,* 6: 331–358.

McArthur, J. M., R. J. Howarth, and T. R. Bailey, 2001. "Strontium Isotope Stratigraphy: LOWESS Version 3: Best Fit to the Marine Sr-Isotope Curve for 0–509 Ma and Accompanying Look-Up Tables for Deriving Numerical Age." *J. Geol.,* 109: 155–170.

McBride, E. F., 1962. "Flysch and Associated Beds of the Martinsburg Formation (Ordovician), Central Appalachians." *J. Sed. Petrol.,* 32: 39–91.

McBride, E. F., (ed.), 1979. *Silica in Sediments: Nodular and Bedded Chert.* SEPM Reprint Series 8.

McDonald, D. A., and R. C. Surdam (eds.), 1984. *Clastic Diagenesis.* Amer. Assoc. Petrol. Geol. Mem. 37.

McDougall, I., 1985. "K-Ar and ^{40}Ar/^{39}Ar Dating of the Hominid-Bearing Pliocene-Pleistocene Sequence at Koobi Fora, Lake Turkana, Northern Kenya." *Geol. Soc. Amer. Bull.,* 96: 159–175.

McDougall, I., and T. M. Harrison, 1988. *Geochronology and Thermochronology by the ^{40}Ar/^{39}Ar Method.* Oxford University Press, New York.

McDougall, I., R. Maiaer, P. Sutherland-Hawkes, and A. J. W. Gleadow, 1980. "K-Ar Age Estimates for the KBS Tuff, East Turkana, Kenya." *Nature,* 284: 230–234.

McDougall, K., 1980. "Paleoecological Evaluation of Late Eocene Biostratigraphic Zonations of the Pacific Coast of North America." *J. Paleont.,* 54(4): 1–75 (supplement).

McElhinney, M. W., 1973. *Palaeomagnetism and Plate Tectonics.* Cambridge University Press, Cambridge.

McKee, E. D., 1982. *The Supai Group of the Grand Canyon.* U.S. Geol. Surv. Prof. Paper 1173.

McKee, E. D., and C. E. Resser, 1945. "Cambrian History of the Grand Canyon Region." *Carnegie Inst. Wash. Publ.,* 563: 1–168.

McKee, E. D., and R. C. Gutschick, 1969. "History of Redwall Limestone of Northern Arizona." *Geol. Soc. Amer. Mem.,* 114: 1–726.

McKelvey, V. E., J. S. Williams, R. P. Sheldon, E. R. Cressman, T. M. Cheney, and R. S. Swanson, 1959. "The Phosphoria, Park City, and Shedhorn Formations in the Western Phosphate Field." *U.S. Geol. Surv. Prof. Paper,* 313–A: 1–47.

McKerrow, W. S., R. St. J. Lambert, and L. R. M. Cocks, 1985. "The Ordovician, Silurian, and Devonian periods." In N. J. Snelling (ed.), *The Chronology of the Geological Record,* 73–80. Geol. Soc. Lond. Mem. 10.

McKerrow, W. S., R. S. St. J. Lambert, and V. E. Chamberlain, 1980. "The Ordovician, Silurian, and Devonian Time Scales." *Earth Planet. Sci. Lett.,* 51: 1–8.

McLaren, D. J., 1973. "The Silurian-Devonian Boundary." *Geol. Mag.,* 110: 302–303.

McLaren, D. J., 1977. "The Silurian-Devonian Boundary Committee: A Final Report." In *Silurian-Devonian Boundary,* 1–34. IUGS Series A 5.

McManus, J., 1988. "Grain Size Determination and Interpretation." In M. Tucker (ed.), *Techniques in Sedimentology,* 63–85. Blackwell Scientific Publications, Oxford.

McPhee, J., 1980. *Basin and Range.* Farrar, Straus and Giroux, New York.

McPhee, J., 1989. *The Control of Nature.* Farrar, Straus and Giroux, New York.

Meckel, L. D., 1967. "Origin of Pottsville Conglomerates (Pennsylvanian) in the Central Appalachians." *Geol. Soc. Amer. Bull.,* 78: 223–258.

Miall, A. D., 1986. "Eustatic Sea-Level Changes Interpreted from Seismic Stratigraphy: A Critique of the Methodology with Particular Reference to the North Sea Jurassic Record." *Amer. Assoc. Petrol. Geol. Bull.,* 70: 131–137.

Miall, A. D., 1991. "Stratigraphic Sequences and their Chronostratigraphic Correlation." *J. Sed. Petrol.,* 61: 497–505.

Miall, A. D., 1992. "Exxon Global Cycle Chart: An Event for Every Occasion?" *Geology,* 20: 787–790.

Miall, A. D., 1997. *The Geology of Stratigraphic Sequences.* Springer-Verlag, New York.

Miall, A. D., 1999. *Principles of Sedimentary Basin Analysis.* 3d ed. Springer-Verlag, New York.

Miall, A. D., and C. E. Miall, 2001. "Sequence Stratigraphy As a Scientific Enterprise: The Evolution and Persistence of Conflicting Paradigms." *Earth Sci. Rev.,* 54: 321–348.

Middleton, G. V., and A. H. Bouma (eds.), 1973. *Turbidites and Deep Water Sedimentation.* Pacific Sec. SEPM Short Course, Anaheim, Calif.

Middleton, G. V., and M. A. Hampton, 1976. "Subaqueous Sediment Transport and Deposition by Sediment Gravity Flows." In D. J. Stanley, and D. J. P. Swift (eds.), *Marine Sediment Transport and Environmental Management,* 197–218. John Wiley and Sons, New York.

Middleton, L. T., and R. C. Blakey, 1983. "Processes and Controls on the Intertonguing of the Kayenta and Navajo Formations, Northern Arizona: Eolian-Fluvial Interactions." In M. E. Brookfield and T. S. Ahlbrandt (eds.), *Eolian Sediments and Processes,* 613–624. Elsevier, Amsterdam.

Miller, F. X., 1977. "The Graphic Correlation Method in Biostratigraphy." In E. G. Kauffman, and J. Hazel (eds.), *Concepts and Methods in Biostratigraphy,* 165–186. Dowden, Hutchinson, and Ross, Stroudsburg, Penn.

Miller, J., 1988. "Cathodoluminescence Microscopy." In M. Tucker (ed.), *Techniques in Sedimentology,* 174–190. Blackwell Scientific Publications, Oxford.

Milliken, K. L., E. F. McBride, and L. S. Land, 1989. "Subsurface Dissolution of Heavy Minerals, Frio Formation Sandstones of the Ancestral Rio Grande Province, South Teas." *Sedim. Geol.,* 68: 187–199.

Milliman, J. D., 1974. *Marine Carbonates.* Springer-Verlag, Berlin.

Millot, G., 1987. *Geology of Clays: Weathering, Sedimentology, Geochemistry.* Springer-Verlag, New York.

Mitchum, R. M., Jr., 1977. "Seismic Stratigraphy and Global Changes in Sea Level. Part 1: Glossary of Terms used in Seismic Stratigraphy." *Amer. Assoc. Petrol. Geol. Mem.,* 26: 205–212.

Mitchum, R. M., Jr., P. R. Vail, and J. B. Sangree, 1977. "Stratigraphic Interpretation of Seismic Reflection Patterns in Depositional Sequences." *Amer. Assoc. Petrol. Geol. Mem.,* 26: 135–143.

Molostovsky, E. A., M. A. Peuzner, D. M. Petchersky, V. D. Rodionov, and A. N. Khramov, 1976. "Phanerozoic Magnetostratigraphic Scale and Geomagnetic Field Inversion Regime." *Geomagn. Res.,* 17: 45–52.

Montanari, A., A. Deino, R. Drake, B. D. Turrin, D. J. DePaolo, G. S. Odin, G. H. Curtis, W. Alvarez, and D. M. Bice, 1988. "Radioisotopic Dating of the Eocene-Oligocene Boundary in the Pelagic Sequence of the Northern Apennines." In I. Premoli-Silva, R. Coccioni, and A. Montanari (eds.), *The Eocene-Oligocene Boundary in the Marche-Umbria Basin (Italy).* IUGS Commission on Stratigraphy, Ancona, Italy.

Montanari, A., R. Drake, D. M. Bice, W. Alvarez, G. H. Curtis, B. Turrin, and D. J. DePaolo, 1985. "Radiometric Time Scale for the Upper Eocene and Oligocene Based on K/Ar and Rb/Sr Dating of Volcanic Biotites." *Geology,* 13: 596–599.

Moore, C. H., 1989. *Carbonate Diagenesis and Porosity.* Elsevier, New York.

Moore, C. H., Jr., E. H. Graham, and L. S. Land, 1976. "Sediment Transport and Dispersal across the Deep Fore-Reef and Island Slope (–55 m to –305 m), Discovery Bay, Jamaica." *J. Sed. Petrol.,* 46: 174–187.

Moore, D. M., and R. C. Reynolds, Jr., 1989. *X-Ray Diffraction and the Identification and Analysis of Clay Minerals.* Oxford University Press, Oxford.

Moore, G. F., and D. E. Karig, 1976. "Development of Sedimentary Basins on the Lower Trench Slope." *Geology,* 4: 693–697.

Moore, R. C., 1949. "Meaning of Facies." *Geol. Soc. Amer. Mem.,* 39: 1–34.

Moore, R. C., 1955. "Invertebrates and the Geological Time Scale." *Geol. Soc. Amer. Spec. Paper,* 62: 547–573.

Moore, R. C., et al., 1944. "Correlation of Pennsylvanian Formations of North America." *Geol. Soc. Amer. Bull.,* 55: 657–706.

Morgan, J. P., and R. H. Shaver (eds.), 1970. *Deltaic Sedimentation: Modern and Ancient.* SEPM Spec. Publ. 15.

Mörner, N.-A., 1981. "Revolution in Cretaceous Sea-Level Analysis." *Geology,* 9: 344–346.

Mörner, N.-A., 1987a. "Models of Global Sea-Level Changes." In M. J. Tooley, and I. Shennon (eds.), *Sea-Level Changes,* 332–355. Basil Blackwell, Oxford.

Mörner, N.-A., 1987b. "Pre-Quaternary Long-Term Changes in Sea Level." In R. V. N. Devoy (ed.), *Sea-Surface Studies, A Global View,* 242–263. Croon Helm, London.

Morse, J. W., and F. W. Mackenzie, 1990. *Geochemistry of Sedimentary Carbonates.* Elsevier, New York.

Mound, J. E., and J. X. Mitrovica, 1998. "True Polar Wander As a Mechanism for Long-Term Sea-Level Variation. *Ann. Geophys.,* 16: 57.

Mullins, H. T., 1986. "Carbonate Depositional Environments, Modern and Ancient. Part 4: Periplatform Carbonates." *Quart. J. Colo. Sch. Mines,* 81: 1–63.

Murphy, M. A., 1977. "On Chronostratigraphic Units." *J. Paleont.,* 52: 123–219.

Murray, J., and A. F. Renard, 1891. *Report on Deep-Sea Deposits Based on Specimens Collected during the Voyage of H.M.S. Challenger in the Years 1873–1876.* H.M.S.O., Edinburgh.

Mutti, E., 1985. "Turbidite Systems and Their Relations to Depositional Sequences." *NATO Adv. Study Inst. Ser. C,* 148: 65–73.

Mutti, E., and F. Ricci Lucchi, 1972. "Turbidites of the Northern Apennines: Introduction to Facies Analysis." *Geology Rev.,* 20(2): 125–166, and translated 1978 by T. H. Nilsen; reprinted by the American Geological Institute, Falls Church, Va.

Naeser, C., 1979. "Thermal History of Sedimentary Basins: Fission-Track Dating of Subsurface Rocks." *SEPM Spec. Publ.,* 26: 109–112.

Nahon, D. B., 1991. *Introduction to the Petrology of Soils and Chemical Weathering.* John Wiley and Sons, New York.

Nealson, K. H., and C. R. Myers, 1990. "Iron Reduction by Bacteria: A Potential Role in the Genesis of Banded Iron Formations." *Amer. J. Sci.,* 290-A: 35–45.

Nelson, C. H., and T. H. Nilsen (eds.), 1984. *Modern and Ancient Deep-Sea Fan Sedimentation.* SEPM Short Course No. 14.

Ness, G., S. Levi, and G. Couch, 1980. "Marine Magnetic Anomaly Timescales for the Cenozoic and Late Cretaceous: A Précis, Critique, and Synthesis." *Rev. Geophys. Space Phys.,* 18: 753–770.

Nier, A. O., 1940. "A Mass Spectrometer for Routine Isotope Abundance Measurements." *Rev. Sci. Instrum.,* 11: 212–216.

Ninkovitch, D., and N. J. Shackleton, 1975. "Distribution, Stratigraphic Position, and Age of Ash Layer 'L' in the Panama Basin Region." *Earth Planet Sci. Lett.,* 27: 20–34.

Normark, W. R., 1978. "Fan Valleys, Channels, and Depositional Lobes on Modern Submarine Fans: Characters for Recognition of Sandy Turbidite Environments." *Amer. Assoc. Petrol. Geol. Bull.,* 62: 912–931.

North American Commission on Stratigraphic Nomenclature, 1983. "North American Stratigraphic Code." *Amer. Assoc. Petrol. Geol. Bull.,* 67: 841–875.

North, F. K., 1985. *Petroleum Geology,* 1st ed. Allen and Unwin, Boston.

North, F. K., 1990. *Petroleum Geology,* 2d ed. Unwin-Hyman, London.

Nriagu, J. O., and P. B. Moore (eds.), 1984. *Phosphate Minerals.* Springer-Verlag, New York.

Obradovich, J. D., 1988. "A Different Perspective on Glauconite As a Chronometer for Geologic Time Scale Studies." *Paleoceanography,* 3: 757–770.

Obradovich, J. D., and W. A. Cobban, 1975. "A Time-Scale for the Late Cretaceous of the Western Interior of North America." *Geol. Assoc. Can. Spec. Paper,* 13: 31–54.

Odin, G. S., 1978. "Isotopic Dates for the Paleogene Time Scale." *Amer. Assoc. Petrol. Geol. Stud. Geol.,* 6: 247–257.

Odin, G. S., (ed.), 1982. *Numerical Dating in Stratigraphy,* 2 vols. John Wiley, New York.

Odin, G. S., 1985. "Remarks on the Numerical Scale of Ordovician to Devonian Time." In N. J. Snelling (ed.), *The Chronology of the Geological Record,* 93–98. Geol. Soc. Lond. Mem. 10.

Odin, G. S., and D. Curry, 1981. "L'échelle numerique des temps paléogénes en 1981." *C. R. Acad. Sci. Paris,* 293(II): 1003–1006.

Odin, G. S., and D. Curry, 1985. "The Palaeogene Time-Scale: Radiometric Dating Versus Magnetostratigraphic Approach." *J. Geol. Soc. Lond.,* 142: 1179–1188.

Odin, G. S. and A. Matter, 1981. "De glauconarium origine." *Sedimentology,* 28(5) 611–641.

Odom, I. E., T. W. Doe, and R. H. Dott, Jr., 1976. "Nature of Feldspar–Grain Size Relations in Some Quartz-Rich Sandstones." *J. Sed. Petrol.,* 45: 862–870.

Olsen, P. E., 1984. "Periodicities of Lake-Level Cycles in the Late Triassic Lockatong Formation of the Newark Basin (Newark Supergroup), New Jersey and Pennsylvania." In A. Berger, J. Imbrie, J. Hays, G. Kukla, and B. Saltzman (eds.), *Milankovitch and Climate,* Part 1, 129–146. The Hague, Netherlands.

Olsson, R. K., 1988. "Foraminiferal Modeling of Sea-Level Change in the Late Cretaceous of New Jersey." *SEPM Spec. Publ.,* 42: 289–298.

Opdyke, N. D., B. Glass, J. D. Hays, and J. Foster, 1966. "Paleomagnetic Study of Antarctic Deep-Sea Cores." *Science,* 154: 349–357.

Opdyke, N. D., L. H. Burckle, and A. Todd, 1974. "The Extension of the Magnetic Time Scale in Sediments of the Central Pacific Ocean." *Earth Planet. Sci. Lett.,* 22: 300–306.

Oppel, A., 1856–58. *Die Juraformation Englands, Frankreichs, und des südwestlichen Deutschlands.* Stuttgart.

Orbigny, A. D. d', 1842. *Paléontologie francaise, terraines jurasiques.* Pt. 1, *Cephalopodes.* Masson, Paris.

Osborn, H. F., and W. D. Matthew, 1909. "Cenozoic Mammal Horizons of Western North America." *U.S. Geol. Surv. Bull.,* 361: 1–138.

Owen, D. E., 1987. "Commentary: Usage of Stratigraphic Terminology in Papers, Illustrations, and Talks." *J. Sed. Petrol.,* 57: 363–372.

Palmer, M. R., and H. Elderfield, 1985. "Sr Isotope Composition of Sea Water over the Past 75 Myr." *Nature,* 314: 526–528.

Park, R. G., 1988. *Foundations of Structural Geology.* Blackie and Son, Ltd., Glasgow.

Parkinson, N., and C. Summerhayes, 1985. "Synchronous Global Sequence Boundaries." *Amer. Assoc. Petrol. Geol. Bull.,* 69: 685–687.

Parrish, R., and J. C. Roddick, 1985. *Geochronology and Isotope Geology for the Geologist and Explorationist.* Geol. Assoc. Canada, Cordilleran Section, Short Course No. 4.

Passega, R., 1964. "Grain Size Representation by CM Patterns As a Geological Tool." *J. Sed. Petrol.,* 34: 830–847.

Payton, C. E., (ed.), 1977. *Seismic Stratigraphy—Applications to Hydrocarbon Exploration.* Amer. Assoc. Petrol. Geol. Mem. 26.

Pemberton, S. G., J. A. MacEachern, and R. W. Frey, 1992. "Trace fossil facies models: environmental and allostratigraphic significance." In R. G. Walker, and N. P. James (eds.), *Facies Models: Response to Sea Level Change,* 3d ed. Geoscience Canada Reprint Series 1. Geological Association of Canada, Toronto.

Petrakis, L., and D. W. Gromaly, 1980. "Coal Analysis, Characterization and Petrography." *J. Chem. Educ.,* 57: 689–694.

Pettijohn, F. G., 1957. *Sedimentary Rocks,* 2d ed. Harper, New York.

Pettijohn, F. J., 1975. *Sedimentary Rocks,* 3d ed. Harper, New York.

Pettijohn, F. J., and P. E. Potter, 1964. *Atlas and Glossary of Primary Sedimentary Structures.* Springer-Verlag, New York.

Pettijohn, F. J., P. E. Potter, and R. Siever, 1972. *Sand and Sandstone,* 1st ed. Springer-Verlag, New York.

Pettijohn, F. J., P. E. Potter, and R. Siever, 1987. *Sand and Sandstone,* 2d ed. Springer-Verlag, New York.

Pitman, W. C., III., 1978. "Relationship between Eustacy and Stratigraphic Sequences of Passive Margins." *Geol. Soc. Amer. Bull.,* 89: 1389–1403.

Playford, P. E., 1980. "Devonian 'Great Barrier Reef' of Canning Basin, Western Australia." *Amer. Assoc. Petrol. Geol. Bull.,* 64: 814–840.

Playford, P. E., 1981. *Devonian Reef Complexes of the Canning Basin, Western Australia.* Geol. Soc. Aust. 5th Aust. Geol. Conv. Field Excursion Guidebook.

Playford, P. E., 1984. "Platform-Margin and Marginal-Slope Relationships in Devonian Reef Complexes of the Canning Basin." In P. G. Purcell (ed.), *The Canning Basin,* 189–214. Proc. Geol. Soc. Aust., Petrol. Explor. Soc. Aust. Symposium, Perth.

Playford, P. E., and D. C. Lowry, 1966. "Devonian Reef Complexes of the Canning Basin, Western Australia." *Geol. Surv. West. Aust. Bull.,* 118: 1–150.

Plumley, W. J., 1948. "Black Hills Terrace Gravels: A Study in Sediment Transport." *J. Geol.,* 56: 526–577.

Poag, C. W., 1977. "Biostratigraphy in Gulf Coast Petroleum Exploration." In E. G. Kauffman, and J. Hazel (eds.), *Concepts and Methods in Biostratigraphy,* 213–233. Dowden, Hutchinson, and Ross, Stroudsburg, Penn.

Poag, C. W., 1999. *Chesapeake Invader.* Princeton University Press, Princeton, N.J.

Poag, C. W., and J. S. Schlee, 1984. "Depositional Sequences and Stratigraphic Gaps on Submerged United States Atlantic Margin." *Amer. Assoc. Petrol. Geol. Mem.,* 36: 165–182.

Poldervaart, A., 1955. "Chemistry of the Earth's Crust." In A. Poldervaart, *Crust of the Earth,* 119–144. Geol. Soc. Am. Spec. Paper 62.

Poore, R. Z., L. Tauxe, S. F. Percival, Jr., and J. L. LaBrecque, 1982. "Late Eocene-Oligocene Magnetostratigraphy and Biostratigraphy at South Atlantic DSDP Site 527." *Geology,* 10: 508–511.

Porter, J. W., R. A. Price, and R. G. McCrossan, 1982. "The Western Canada Sedimentary Basin." *Phil. Trans. R. Soc. A,* 305(1489): 169–173.

Posamentier, H. W., and D. P. James, 1993. "An Overview of Sequence-Stratigraphic Concepts: Uses and Abuses." *Spec. Publ. Int. Assoc. Sedimentol.,* 18: 3–18.

Posamentier, H. W., and P. R. Vail, 1988. "Eustatic Controls on Clastic Deposition II—Sequence and Systems Tract Models." *SEPM Spec. Publ.,* 42: 125–154.

Posamentier, H. W., M. T. Jervey, and P. R. Vail, 1988. "Eustatic Controls on Clastic Deposition I—Conceptual Framework." *SEPM Spec. Publ.,* 42: 109–124.

Posamentier, H. W., C. P. Summerhayes, B. U. Haq, and G. P. Allen (eds.), 1993. *Sequence Stratigraphy and Facies Associations.* Int. Assoc. Sedimentol. Spec. Publ. 18.

Potter, P. E., 1978. "Petrology and Chemistry of Modern Big River Sands." *J. Geol.,* 86: 423–449.

Potter, P. E., and F. J. Pettijohn, 1977. *Paleocurrents and Basin Analysis,* 2d ed. Springer-Verlag, New York.

Potter, P. E., J. B. Maynard, and W. A. Pryor, 1980. *Sedimentology of Shale.* Springer-Verlag, New York.

Powell, J. W., 1875. *Exploration of the Colorado River of the West and Its Tributaries.* Government Printing Office, Washington, D.C.

Powers, M. C., 1953. "A New Roundness Scale for Sedimentary Particles." *J. Sed. Petrol.,* 23: 117–119.

Powers, M. C., 1982. "Comparison Chart for Estimating Roundness and Sphericity." *Amer. Geol. Inst.,* Data Sheet 18.1.

Pratt, L. M., 1985. "Isotopic Studies of Organic Matter and Carbonate in Rocks of the Greenhorn Marine Cycle." In L. M. Pratt, E. G. Kauffman, and F. B. Zelt (eds.), *Fine-Grained Deposits and Biofacies of the Cretaceous Western Interior Seaway: Evidence of Cyclic Sedimentary Processes,* 38–48. SEPM Field Trip Guidebook 4.

Pratt, L. M., E. G. Kauffman, and F. B. Zelt (eds.), 1985. *Fine-Grained Deposits and Biofacies of the Cretaceous Western Interior Seaway: Evidence of Cyclic Sedimentary Processes.* SEPM Field Trip Guidebook 4.

Pray, L. C., and R. C. Murray, 1965. *Dolomitization and Limestone Diagenesis.* SEPM Spec. Publ. 13.

Press, F., and Siever, R., 1986. *Earth,* 4th ed. W. H. Freeman and Co., New York.

Prothero, D. R., 1982. "How Isochronous Are Mammalian Biostratigraphic Events?" *Proc. 3d N. Amer. Paleont. Conv.,* 2: 405–409.

Prothero, D. R., 1985a. "Chadronian (Early Oligocene) Magnetostratigraphy of Eastern Wyoming: Implications for the Eocene-Oligocene Boundary." *J. Geol.,* 93: 555–565.

Prothero, D. R., 1985b. "North American Mammalian Diversity and Eocene-Oligocene Extinctions." *Paleobiology,* 11: 389–405.

Prothero, D. R., 1990. *Interpreting the Stratigraphic Record.* W. H. Freeman and Co., New York.

Prothero, D. R., 1994a. *The Eocene-Oligocene Transition: Paradise Lost.* Columbia University Press, New York.

Prothero, D. R., 1994b. "The Late Eocene-Oligocene Extinctions." *Ann. Rev. Earth Plant. Sci.,* 22: 145–165.

Prothero, D. R., 1999. "Does Climatic Change Drive Mammalian Evolution?" *GSA Today,* 9(9): 1–5.

Prothero, D. R., 2001a. "Chronostratigraphic Calibration of the Pacific Coast Cenozoic: A Summary." *Pacific Section SEPM Book,* 91: 377–394.

Prothero, D. R., 2001b. "Magnetostratigraphic Tests of Sequence Stratigraphic Correlations from the Southern California Paleogene." *J. Sed. Res. B,* 71: 525–535.

Prothero, D. R., and J. M. Armentrout, 1985. "Magnetostratigraphic Correlation of the Lincoln Creek Formation, Washington: Implications for the Age of the Eocene-Oligocene Boundary." *Geology,* 13: 208–211.

Prothero, D. R., C. R. Denham, and H. G. Farmer, 1982. "Oligocene Calibration of the Magnetic Polarity Timescale." *Geology,* 10: 650–653.

Prothero, D. R., C. R. Denham, and H. G. Farmer, 1983. "Magnetostratigraphy of the White River Group, and Its Implications for Oligocene Geochronology." *Palaeogeogr., Palaeoclimat., Palaeocol.,* 42: 151–166.

Prothero, D. R., and R. H. Dott, Jr., 2003. *Evolution of the Earth,* 7th ed. McGraw-Hill, New York.

Prothero, D. R., and C. C. Swisher III, 1992. "Magnetostratigraphy and Geochronology of the Terrestrial Eocene-Oligocene Transition in North America." In D. R. Prothero, and W. A. Berggren (eds.), *Eocene-Oligocene Climatic and Biotic Evolution,* 46–74. Princeton University Press, Princeton, N.J.

Prothero, D. R., and E. H. Vance, Jr., 1996. "Magnetostratigraphy of the Upper Middle Eocene Coldwater Sandstone, Central Ventura County, California." In D. R. Prothero, and R. J. Emry (eds.), *The Terrestrial Eocene-Oligocene Transition in North America,* 140–156. Cambridge University Press, Cambridge.

Prothero, D. R., and K. E. Whittlesey, 1998. "Magnetostratigraphy and Biostratigraphy of the Chadronian, Orellan, and Whitnian Land Mammal 'Ages' in the White River Group." *Geol. Soc. Amer. Spec. Paper,* 325: 39–61.

Pye, K., (ed.), 1994. *Sedimentary Transport and Depositional Processes.* Blackwell Scientific Publ., Cambridge, Mass.

Quade, J., T. E. Cerling, and J. R. Bowman, 1989. "Development of the Asian Monsoon Revealed by Marked Ecological Shift during the Latest Miocene in Northern Pakistan." *Nature,* 342: 163–166.

Quenstedt, F. A., 1856–58. *Der Jura.* H. Laupp, Tübingen.

Raaf, J. F. M. de, J. R. Boersma, and A. van Gelder, 1977. "Wave Generated Structures and Sequences from a Shallow Marine Succession, Lower Carboniferous, County Cork, Ireland." *Sedimentology,* 4: 1–52.

Rahmani, R. A., and R. M. Flores (eds.), 1985. *Sedimentology of Coal and Coal-Bearing Sequences.* Intl. Assoc. Sed. Spec. Publ. 7.

Ralph, E. K., 1971. "Carbon-14 Dating." In H. N. Michael, and E. K. Ralph (eds.), *Dating Techniques for the Archeologist,* 1–48, M.I.T. Press, Cambridge, Mass.

Ramos, A., A. Sopeña, and M. Perez-Arlucea, 1986. "Evolution of Buntsandstein Fluvial Sedimentation in the Northwest Iberian Ranges (Central Spain)." *J. Sed. Petrol.,* 56: 862–875.

Ramos, A., and A. Sopeña, 1983. "Gravel Bars in Low Sinuosity Streams (Permian and Triassic, Central Spain)." In J. D. Collinson, and J. Lewin (eds.), *Modern and Ancient Fluvial Systems,* 301–302. Spec. Publ., Int. Ass. Sediment. 6.

Ramsbottom, W. H. C., 1979. "Rates of Transgression and Regression in the Carboniferous of Northwest Europe." *J. Geol. Soc. Lond.,* 136: 147–153.

Rankin, D. W., T. W. Stern, J. C. Reed, Jr., and M. F. Newell, 1969. "Zircon Age of Felsic Volcanic Rocks in the Upper Precambrian of the Blue Ridge, Appalachian Mountains." *Science,* 166: 741–744.

Rau, W. W., 1958. "Stratigraphy and Foraminiferal Zonation in Some Tertiary Rocks of Southwestern Washington." *U.S. Geol. Surv. Oil and Gas Invest.,* Chart OC 57.

Rau, W. W., 1966. "Stratigraphy and Foraminifera of the Satsop River Area, Southern Olympic Peninsula, Washington." *Wash. Div. Mines and Geol. Bull.,* 53: 1–66.

Raup, D. M., 1985. *The Nemesis Affair: A Story of the Death of Dinosaurs and the Ways of Science.* W. W. Norton, New York.

Raup, D. M., and Stanley, S. M., 1976. *Principles of Paleontology,* 2d ed. W. H. Freeman and Co., San Francisco.

Reading, H. G., (ed.), 1996. *Sedimentary Environments and Facies,* 3d ed. Blackwell Scientific Publications, Oxford.

Reineck, H.-E., and I. B. Singh, 1973. *Depositional Sedimentary Environments.* Springer-Verlag, New York.

Reineck, H.-E., and F. Wunderlich, 1968a. "Classification and Origin of Flaser and Lenticular Bedding." *Sedimentology,* 11: 99–104.

Reineck, H.-E., and F. Wunderlich, 1968b. "Zeitmessungen und Bezeitenschichten." *Natur und Museum,* 97: 193–197.

Reinhardt, J., and W. R. Sigleo (eds.), 1988. *Paleosols and Weathering through Geologic Time: Principles and Applications.* Geol. Soc. Amer. Spec. Paper 216.

Repenning, C. A., 1967. "Palearctic-Nearctic Mammalian Dispersal in the Late Cenozoic." In D. M. Hopkins (ed.), *The Bering Land Bridge,* 288–311. Stanford University Press, Stanford, Calif.

Retallack, G. J., 1983. *Late Eocene and Oligocene Paleosols from Badlands National Park, South Dakota.* Geol. Soc. Amer. Spec. Paper 193.

Retallack, G. J., 1985. "An Excursion Guide to Fossil Soils of the Mid-Tertiary Sequence in Badlands National Park, South Dakota." In J. E. Martin (ed.), "Fossiliferous Cenozoic Deposits of Western South Dakota and Northwestern Nebraska." *Dakoterra* 2(2): 277–301.

Retallack, G. J., 1988. "Field Recognition of Paleosols. In J. Reinhardt, and W. R. Sigleo (eds.), *Paleosols and Weathering through Geologic Time: Principles and Applications*, 1–20. Geol. Soc. Amer. Spec. Paper 216.

Retallack, G. J., 1997. "Neogene Expansion of the North American Prairie." *Palaios,* 12: 380–390.

Retallack, G. J., 2001. *Soils of the Past.* Blackwell Science, London.

Retallack, G. J., and C. R. Feakes, 1987. "Trace Fossil Evidence for Late Ordovician Animals on Land." *Science,* 235: 61–63.

Reynolds, D. J., M. S. Steckler, and B. J. Coakley, 1990. "The Role of Sediment Load in Sequence Stratigraphy: The Influence of Flexural Isostasy and Compaction." *J. Geophys. Res.,* 96: 6931–6949.

Ricci Lucchi, F., and E. Valmori, 1980. "Basin-Wide Turbidites in a Miocene Oversupplied Deep-Sea Plain." *Sedimentology,* 27: 241–270.

Rickard, L. V., 1962. "Late Cayugan (Upper Silurian) and Helderbergian (Lower Devonian) Stratigraphy in New York." *N.Y. State Mus. Sci. Serv. Bull.,* 386: 1–157.

Rigby, J. K., and W. K. Hamblin (eds.), 1972. *Recognition of Ancient Sedimentary Environments.* SEPM Spec. Publ. 16.

Ricci Lucchi, F., 1995. *Sedimentographica: A Photograph Atlas of Sedimentary Structures,* 2d ed. Columbia University Press, New York.

Riley, J. P., and R. L. Chester (eds.), 1976. *Chemical Oceanography,* vol. 5, 2d ed. Academic Press, New York.

Ronov, A. B., 1964. "Common Tendencies in the Chemical Evolution of the Earth's Crust, Ocean, and Atmosphere." *Geochem. Int.,* 1: 713–737.

Ronov, A. B., 1968. "Probable Changes in the Composition of Seawater during the Course of Geological Time." *Sedimentology,* 10: 25–43.

Ronov, A. B., 1972. "Evolution of Rock Composition and Geochemical Processes in the Sedimentary Shell of the Earth." *Sedimentology,* 19: 157–172.

Ronov, A. B., V. E. Khain, A. N. Balukhovsky, and K. B. Seslavinsky, 1980. "Quantitative Analysis of Phanerozoic Sedimentation." *Sed. Geol.,* 25: 311–325.

Ronov, A. B., and A. A. Yaroshevsky, 1969. "Chemical Composition of the Earth's Crust." In P. J. Hart (ed.), *The Earth's Crust and Upper Mantle,* 37–57. Amer. Geophys. Union, Geophys. Monograph 13.

Ross, C. A., and J. R. P. Ross, 1985. "Late Paleozoic Depositional Sequences Are Synchronous and Worldwide." *Geology,* 13: 194–197.

Ross, D. A., 1982. *Introduction to Oceanography.* John Wiley, New York.

Rudwick, M. J. S., 1978. "Charles Lyell's Dreams of a Statistical Palaeontology." *Palaeontology,* 21: 225–244.

Rust, I. C., 1973. "The Evolution of the Paleozoic Cape Basin, Southern Margin of Africa." In A. E. M. Nairn, and F. H. Stehli (eds.), *The Ocean Basins and Margins,* 247–276. Plenum, New York.

Rust, I. C., 1977. "Evidence of Shallow Marine and Tidal Sedimentation in the Ordovician Graafwater Formation, Cape Province, South Africa." *Sed. Geol.,* 18: 123–133.

Ryder, R. T., T. D. Fouch, and J. H. Elison, 1976. "Early Tertiary Sedimentation in the Western Uinta Basin, Utah." *Geol. Soc. Amer. Bull.,* 87: 496–512.

Ryer, T. A., 1977. "Patterns of Cretaceous Shallow Marine Sedimentation, Coalville and Rockport Areas, Utah." *Geol. Soc. Amer. Bull.,* 88: 177–188.

Salvador, A., 1994. *International Stratigraphic Guide.* Geological Society of America, Boulder, Colo.

Sandberg, P. A., 1975. "New Interpretation of Great Salt Lake Ooids and of Ancient Nonskeletal Carbonate Mineralogy." *Sedimentology* 22: 497–537.

Sandberg, P. A., 1983. "An Oscillating Trend in Phanerozoic Nonskeletal Carbonate Mineralogy." *Nature,* 305: 19–22.

Sandberg, P. A., 1985. "Nonskeletal Aragonite and CO_2 in the Phanerozoic and Proterozoic." In E. T. Sundquist, and W. S. Broecker (eds.), *The Carbon Cycle and Atmospheric CO_2: Natural Variations, Archean to Present.* American Geophysical Union, Washington D.C.

Sanders, J. E., 1970. "Coastal-Zone Geology and Its Relationship to Water Pollution Problems." In A. A. Johnson (ed.), *Water Pollution in the Greater New York Area,* 23–25. Gordon and Breach, New York.

Sarna-Wojcicki, A. M., S. D. Morrison, C. E. Meyer, and J. W. Hillhouse, 1987. "Correlation of Upper Cenozoic Tephra Layers between Sediments of the Western United States and the Eastern Pacific Ocean, and Comparison with Biostratigraphic and Magnetostratigraphic Age Data." *Geol. Soc. Amer. Bull.,* 98: 207–223.

Scheltema, R. S., 1977. "Dispersal of Marine Invertebrate Organisms: Paleobiogeographic and Biostratigraphic Implications." In E. G. Kauffman, and J. E. Hazel (eds.), *Concepts and Methods of Biostratigraphy,* 73–108. Dowden, Hutchinson, and Ross, Stroudsburg, Penn.

Schenck, H. G., and R. M. Kleinpellm, 1936. "Refugian Stage of the Pacific Coast Tertiary." *Amer. Assoc. Petrol. Geol. Bull.,* 20: 215–255.

Schenck, H. G., and S. W. Muller, 1941. "Stratigraphic Terminology." *Geol. Soc. Amer. Bull.,* 52: 1419–1426.

Schermerhorn, L. J. G., 1974. "Late Precambrian Mictites: Glacial and/or Nonglacial?" *Amer. J. Sci.,* 274: 673–824.

Schieber, J., 1989. "Facies and Origin of Shales from the Mid-Proterozoic Newland Formation, Belt Basin, Montana, U.S.A." *Sedimentology,* 36: 209–219.

Schieber, J., 1993. "Evidence for High-Energy Events and Shallow-Water Deposition in the Chattanooga Shale, Devonian, Central Tennessee, U.S.A." *Sed. Geol.,* 93: 193–208.

Schlee, J. S., and L. F. Jansa, 1981. "The Paleoenvironment and Development of the Eastern North American Continental Margin." *Colloque C3: Géologie des marges continentales. 26th IGC, Oceanologica Acta,* 4: 71–80.

Schmaltz, R. F., 1969. "Deep-Water Evaporite Deposition in a Genetic Model." *Amer. Assoc. Petrol. Geol. Bull.,* 10: 798–823.

Schneidermann, N., and P. M. Harris, 1985. *Carbonate Cements.* SEPM Spec. Publ. 36.

Schoch, R. M., 1989. *Stratigraphy: Principles and Methods.* Van Nostrand Reinhold, New York.

Scholl, D. W., and M. S. Marlow, 1974. "Sedimentary Sequence in Modern Pacific Trenches and the Deformed Circum-Pacific Eugeosyncline." *SEPM Spec. Publ.,* 19: 193–211.

Scholle, P. A., 1978. *A Color Illustrated Guide to Carbonate Rock Constituents, Textures, Cements, and Porosities.* Amer. Assoc. Petrol. Geol. Mem. 27.

Scholle, P. A., 1979. *A Color Illustrated Guide to Constituents, Textures, Cements, and Porosities of Sandstones and Associated Rocks.* Amer. Assoc. Petrol. Geol. Mem. 28.

Scholle, P. A., and P. R. Schluger (eds.), 1979. *Aspects of Diagenesis.* SEPM Spec. Publ. 26.

Scholle, P. A., and D. Spearing (eds.), 1982. *Sandstone Depositional Environments.* Amer. Assoc. Petrol. Geol. Mem. 31.

Scholle, P. A., D. G. Bebout, and C. H. Moore (eds.), 1983. *Carbonate Depositional Environments,* Amer. Assoc. Petrol. Geol. Mem. 33.

Schreiber, B. C., (ed.), 1988. *Evaporites and Hydrocarbons.* Columbia University Press, New York.

Schultz, R. F., 1969. "Deep Water Evaporite Deposition: A Genetic Model." *Amer. Assoc. Petrol. Geol. Bull.,* 53: 798–823.

Schumm, S. A., 1993. "River Response to Baselevel Change: Implications for Sequence Stratigraphy." *J. Geol.,* 101: 279–294.

Schwartz, R. K., 1975. "Nature and Genesis of Some Storm Washover Deposits." *U.S. Army Corps Eng. Coastal Eng. Res. Center Tech. Mem.,* 61: 69.

Scoffin, T. P., 1987. *An Introduction to Carbonate Sediments and Rocks.* Blackie and Son, Ltd., London.

Scott, A. C., (ed.), 1987. *Coal and Coal-Bearing Strata: Recent Advances.* Geol. Soc. Lond. Spec. Publ. 32.

Secord, J. A., 1986. *Controversy in Victorian Geology: The Cambrian-Silurian Dispute.* Princeton University Press, Princeton, N.J.

Selley, R. C., 1978. *Ancient Sedimentary Environments,* 2d ed. Cornell University Press, Ithaca, N.Y.

Selley, R. C., 1982. *An Introduction to Sedimentology.* Academic Press, London.

Selley, R. C., 2000. *Applied Sedimentology,* 2d ed. Academic Press, San Diego.

Sellwood, B. W., 1975. ""Lower Jurassic Tidal Flat Deposits, Bornholm, Denmark." In R. N. Ginsburg (ed.), *Tidal Deposits,* 93–101. Springer-Verlag, New York.

Semikhatov, M. A., 1980. "On the Upper Precambrian Stromatolite Standard of Northern Eurasia." *Precamb. Res.,* 16: 235–247.

Sepkoski, J. J., and A. H. Knoll, 1983. "Precambrian-Cambrian Boundary: The Spike is Driven and the Monolith Crumbles." *Paleobiology,* 9: 199–206.

Shackleton, N. J., and N. D. Opdyke, 1973. "Oxygen Isotope and Paleomagnetic Stratigraphy of Equatorial Pacific Core V28–238: Oxygen Isotope Temperatures and Ice Volumes on a 10^5 Year and 10^6 Year Scale." *Quat. Res.,* 3: 39–55.

Shackleton, N. J., and N. D. Opdyke, 1976. "Oxygen Isotope and Paleomagnetic Stratigraphy of Equatorial Pacific Core V28–239, Late Pliocene to Latest Pleistocene." *Geol. Soc. Amer. Mem.,* 145: 449–464.

Shanmugam, G., and R. J. Moiola, 1982. "Eustatic Control of Turbidites and Winnowed Turbidites." *Geology,* 10: 131–135.

Shanmugam, G., and R. J. Moiola, 1985. "Is the Turbidite Facies Association Scheme Valid for Interpreting Ancient Submarine Fan Environments?" *Geology,* 13: 234–237.

Shanmugam, G., R. J. Moiola, and J. E. Damuth, 1985. "Eustatic Control of Submarine Fan Development." In A. H. Bouma, W. R. Normark, and N. E. Barnes (eds.), *Submarine Fans and Related Turbidite Systems,* 23–28. Springer-Verlag, New York.

Shapiro, A. H., 1961. *Shape and Flow: The Fluid Dynamics of Drag.* Doubleday, New York.

Sharpton, V. L., and P. D. Ward (eds.), 1990. *Global Catastrophes in Earth History: An Interdisciplinary Conference on Impacts, Volcanism, and Mass Mortalities.* Geol. Soc. Amer. Spec. Paper 247.

Shaw, A. B., 1964. *Time in Stratigraphy.* McGraw-Hill, New York.

Sheldon, R. P., 1963. *Physical Stratigraphy and Mineral Resources of Permian Rocks in Western Wyoming.* U.S. Geol. Surv. Prof. Paper 313-B.

Shelton, J. S., 1966. *Geology Illustrated.* W. H. Freeman and Co., San Francisco.

Shepard, F. P., F. B. Phleger, and T. J. van Andel (eds.), 1960. *Recent Sediments of the Northwest Gulf of Mexico.* Amer. Assoc. Petrol. Geol., Tulsa, Okla.

Sheridan, R. E., 1974. "Atlantic Continental Margin of North America." In C. A. Burk, and C. L. Drake (eds.), *The Geology of Continental Margins,* 391–407. Springer-Verlag, New York.

Sheridan, R. E., 1983. "Phenomenon of Pulsation Tectonics Related to the Breakup of the Eastern North American Continental Margin." *Init. Rept. Deep-Sea Drill. Proj.,* 76: 897–909.

Sheridan, R. E., 1987a. "Pulsation Tectonics As the Control of Continental Breakup." *Tectonophys.,* 143: 59–73.

Sheridan, R. E., 1987b. "Pulsation Tectonics As the Control of Long-Term Stratigraphic Cycles." *Paleoceanography,* 2: 97–118.

Sheriff, R. E., 1978. *A First Course in Geophysical Exploration and Interpretation.* International Human Resource Development Corp., Boston.

Sheriff, R. E., 1980. *Seismic Stratigraphy.* International Human Resource Development Corp., Boston.

Shields, A., 1936. *Anbuendung der Ahnlich Keitsmechanik und der Turbulenz Forschung auf die Geschiebebewegung.* Mitl. Preuss. Vers. Anst. Wasserb. Schiffb. Berlin, Heft 26.

Shimer, H. W., and R. R. Shrock, 1944. *Index Fossils of North America.* M.I.T. Press, Cambridge, Mass.

Shinn, E. A., and R. N. Ginsburg, 1964. "Formation of Recent Dolomite in Florida and the Bahamas." *Amer. Assoc. Petrol. Geol. Bull.,* 48: 547 (abstract).

Shinn, E. A., R. N. Ginsburg, and R. M. Lloyd, 1965. "Recent Supratidal Dolomite from Andros Island, Bahamas." *SEPM Spec. Publ.,* 13: 112–123.

Shinn, E. A., R. M. Lloyd, and R. N. Ginsburg, 1969. "Anatomy of a Modern Carbonate Tidal-Flat, Andros Island, Bahamas." *J. Sed. Petrol.,* 39: 1202–1228.

Sibley, D. F., and H. Blatt, 1976. "Intergranular Pressure Solution and Cementation of the Tuscarora Orthoquartzite." *J. Sed. Petrol.,* 45: 881–896.

Sieveking, G., and M. B. Hart (eds.), 1986. *The Scientific Study of Flint and Chert.* Cambridge University Press, Cambridge.

Siever, R., 1988. *Sand.* W. H. Freeman and Co., New York.

Silver, L. T., and P. H. Schultz (eds.), 1982. *Geological Implications of Impacts of Large Asteroids and Comets on the Earth.* Geol. Soc. Amer. Spec. Paper 190.

Sloss, L. L., 1963. "Sequences in the Cratonic Interior of North America." *Geol. Soc. Amer. Bull.,* 74: 93–114.

Sloss, L. L., 1972. "Synchrony of Phanerozoic Sedimentary-Tectonic Events of the North American Craton and the Russian Platform." *24th Int. Geol. Congr.,* 6: 24–32.

Sloss, L. L., 1979. "Global Sea-Level Change—A View from the Craton." *Amer. Assoc. Petrol. Geol. Mem.,* 29: 461–467.

Sloss, L. L., 1982. "The Midcontinent Province: United States." *Geol. Soc. Amer., DNAG Spec. Publ.,* 1: 27–39.

Sloss, L. L., 1984. "Comparative Anatomy of Cratonic Unconformities." *Amer. Assoc. Petrol. Geol. Mem.,* 36: 1–6.

Sloss, L. L., 1991. "The Tectonic Factor in Sea Level Change: A Countervailing View." *J. Geophys. Res.,* 96(B4): 6609–6617.

Sloss, L. L., 1993. "Tectonic Episodes of Cratons: Conflicting North American Concepts." *Terra Nova,* 4: 320–328.

Sloss, L. L., W. C. Krumbein, and E. C. Dapples, 1949. "Integrated Facies Analysis." *Geol. Soc. Amer. Mem.,* 39: 94–124.

Smith, D. G., 1987. "Meandering Riverpoint for Lithofacies Models: Modern and Ancient Examples." *SEPM Spec. Publ.,* 19: 83–91.

Smosna, R., 1989. "Compaction Law for Cretaceous Sandstones of Alaska's North Slope." *J. Sed. Petrol.,* 59: 572–584.

Snelling, N. J., 1985. "An Interim Time-Scale." In N. J. Snelling (ed.), *The Chronology of the Geological Record,* 261–265. Geol. Soc. Lond. Mem. 10.

Snyder, W. S., and C. Spinosa, 1993. "Exxon Global Cycle Chart: An Event for Every Occasion? Comment." *Geology,* 21: 283–284.

Soares, P. C., P. M. B. Landim, and W. S. Fulfair, 1978. "Tectonic Cycles and Sedimentary Sequences in the Brazilian Intracratonic Basins." *Geol. Soc. Amer. Bull.,* 89: 181–191.

Sonnenfeld, P., 1984. *Brines and Evaporites.* Academic Press, New York.

Spearing, D., 1974. *Summary Sheets of Sedimentary Deposits.* Geol. Soc. Amer. Maps and Charts Series MC-8.

Srinivasan, M. S., and J. P. Kennett, 1981. "A Review of Neogene Planktic Foraminiferal Biostratigraphy: Applications in the Equatorial and South Pacific." *SEPM Spec. Publ.,* 32: 395–432.

Stach, E., 1975. *Handbook of Coal Petrology,* 2d ed. Gebrüder Borutraeger, Berlin.

Stanley, D. J., and D. J. P. Swift (eds.), 1976. *Marine Sediment Transport and Environmental Management.* John Wiley, New York.

Stanley, S. M., 1989. *Earth and Life through Time,* 2d ed. W. H. Freeman and Co., New York.

Stanley, S. M., and L. A. Hardie, 1998. "Secular Oscillation in the Carbonate Mineralogy of Reef-Building and Sediment-Producing Organisms Driven by Tectonically Forced Shifts in Seawater Chemistry." *Paleogeogr. Paleoclimat., Paleoecol.,* 144: 3–19.

Stanley, S. M., and L. A. Hardie, 1999. "Hypercalcification: Paleontology Links Plate Tectonics and Geochemistry to Sedimentology." *GSA Today,* 9(2): 1–7.

Stanton, R. J., 1960. "Paleoecology of the Upper Miocene Castaic Formation, Los Angeles County, California." California Institute of Technology, unpubl. doct. dissert.

Stanton, R. J., 1966. "Megafauna of the Upper Miocene Castaic Formation, Los Angeles County, California." *J. Paleo.,* 40: 21–40.

Steel, R. J., 1988. "Coarsening-Upward and Coarse-Skewed Fan Bodies." In W. Nemec, and R. J. Steel (eds.), *Fan Deltas,* 75–83. Blackie and Sons, Glasgow, Scotland.

Steel, R. J., and T. G. Gloppen, 1980. "Late Caledonian (Devonian) Basin Formation, Western Norway—Signs of Strike-Slip Tectonics During Infilling." *Spec. Publ., Int. Assoc. Sedimentol.,* 4: 79–103.

Steel, R. J., A. Spinnangr, S. Mäehle, H. Nilsen, and S. L. Roe, 1977. "Coarsening-Upward Cycles in the Alluvium of Hornelen Basin (Devonian), Norway: Sedimentary Response to Tectonic Events." *Geol. Soc. Amer. Bull.,* 88: 1124–1134.

Steiger, R. H., and E. Jäger, 1977. "Subcommission on Geochronology: Convention on the Use of Decay Constants in Geo- and Cosmochronology." *Earth Planet. Sci. Lett.,* 36: 359–362.

Stille, H., 1936. *Wege und Ergebnisse der geologischtektonischen Forschung.* Kaiser Wilhelm Ges., Berlin.

Stonecipher, S. A., 2000. *Applied Sandstone Diagenesis: Practical Petrographic Solutions for a Variety of Common Exploration, Development, and Production Problems.* SEPM Short Course Notes 50.

Stowe, K. S., 1979. *Ocean Science.* John Wiley, New York.

Sugden, P. E., and B. S. John, 1976. *Glaciers and Landscape: A Geomorphological Approach.* Edward Arnold, London.

Summerhayes, C. P., 1986. "Sea Level Curves Based on Seismic Stratigraphy: Their Chronostratigraphic Significance." *Palaeogeogr., Paleoclimat., Palaeoecol.,* 57: 27–42.

Surdam, R. C., and C. A. Wolfbauer, 1975. "Green River Formation, Wyoming: A Playa-Lake Complex." *Geol. Soc. Amer. Bull.,* 86: 335–345.

Suttner, L. J., A. Basu, and G. H. Mack, 1981. "Climate and Origin of Quartz Arenites." *J. Sed. Petrol.,* 51: 1235–1246.

Swift, D. J. P., 1968. "Coastal Erosion and Transgressive Stratigraphy." *J. Geol.,* 76: 444–456.

Swift, D. J. P., 1974. "Continental Shelf Sedimentation." In C. A. Burk and C. L. Drake (eds.), *The Geology of Continental Margins,* 117–135. Springer-Verlag, Berlin.

Swift, D. J. P., D. B. Duane, and T. F. McKinney, 1973. "Ridge and Swale Topography of the Middle Atlantic Bight, North America: Secular Response to the Holocene Hydraulic Regime." *Marine Geol.,* 15: 227–247.

Swift, D. J. P., D. B. Duane, and O. H. Pilkey (eds.), 1972. *Shelf Sediment Transport: Process and Product.* Dowden, Hutchinson, and Ross, Stroudsburg, Penn.

Swift, D. J. P., D. J. Stanley, and J. R. Curray, 1971. "Relict Sediments on Continental Shelves: A Reconsideration." *J. Geol.,* 79: 322–346.

Swisher, C. C., III, and D. R. Prothero, 1990. "Single-Crystal ^{40}Ar/^{39}Ar Dating of the Eocene-Oligocene Transition in North America." *Science,* 249: 760–762.

Tankard, A. B., and D. K. Hobday, 1977. "Tide-Dominated Back-Barrier Sedimentation, Early Ordovician Cape Basin, Cape Peninsula, South Africa." *Sed. Geol.,* 18: 135–159.

Tarling, D. H., 1983. *Palaeomagnetism: Principles and Applications in Geology, Geophysics and Archaeology.* Chapman and Hall, London.

Tauxe, L., R. Butler, and J. C. Herguera, 1987. "Magnetostratigraphy: In Pursuit of Missing Links." *Rev. Geophys.,* 25: 939–950.

Taylor, J. C. M., 1978. "Sandstone Diagenesis." *J. Geol. Soc. Lond.,* 135: 1–133.

Taylor, S. R., and S. M. McClennan, 1985. *The Continental Crust: Its Composition and Evolution.* Blackwell Scientific Publications, Oxford.

Tazieff, H., 1970. "The Afar Triangle." *Sci. Amer.,* 222(2): 32–51.

Tedford, R. H., 1970. "Principles and Practices of Mammalian Geochronology in North America." *Proc. North Amer. Paleont. Conv.,* F: 666–703.

Teichert, C., 1958a. "Cold- and Deep-Water Coral Banks." *Amer. Assoc. Petrol. Geol. Bull.,* 42: 1064–1082

Teichert, C., 1958b. "Concepts of Facies." *Amer. Assoc. Petrol. Geol. Bull.,* 42: 2718–2744.

Thaler, L., 1972. "Datation, zonation, et mammifères." *Bur. Rech. Géol. Minières Mém.,* 77: 411–424.

Thompson, G. R., and J. Hower, 1973. "An Explanation for Low Radiometric Ages from Glauconite." *Geochim. Cosmochim. Acta,* 37: 1473–1491.

Thorne, J., and A. B. Watts, 1984. "Seismic Reflectors and Unconformities at Passive Continental Margins." *Nature,* 311: 365–368.

Tillman, R. W., and C. W. Siemars (eds.), 1984. *Siliciclastic Shelf Sediments.* SEPM Spec. Publ. 34.

Tillman, R. W., D. J. P. Swift, and R. G. Walker (eds.), 1985. *Shelf Sands and Sandstone Reservoirs.* SEPM Short Course No. 13.

Tissot, B. P., and D. H. Welte, 1984. *Petroleum Formation and Occurrence,* 2d ed. Springer-Verlag, Berlin.

Trendall, A. F., and R. C. Morris (eds.), 1983. *Iron-Formation, Facts and Problems.* Elsevier, Amsterdam.

Truswell, J. F., 1972. "Sandstone Sheets and Intrusions from Coffee Bay, Transkei, South Africa." *J. Sed. Petrol.,* 42: 578–583.

Tucker, M. E., 1981. *Sedimentary Petrology, An Introduction.* Blackwell Scientific Publications, Oxford.

Tucker, M. E., 1982. *The Field Description of Sedimentary Rocks.* Open University Press, Milton Keynes, England.

Tucker, M. E., 1991. *Sedimentary Petrology: An Introduction to the Origin of Sedimentary Rocks.* Blackwell Scientific Publications, Oxford.

Tucker, M. E., 1992, "Limestones through Time." In G. C. Brown, C. J. Hawkesworth, and R. C. L. Wilson (eds.), *Understanding the Earth,* 347–363. Cambridge University Press, Cambridge.

Tucker, M. E., 2001. *Sedimentary Petrology: An Introduction to the Origin of Sedimentary Rocks,* 3d ed. Blackwell Science, Oxford.

Tucker, M. E., and V. P. Wright, 1990. *Carbonate Sedimentology.* Blackwell Scientific Publications, Oxford.

Tye, R. S., J. S. Hewlett, P. R. Thompson, and D. K. Goodman, 1993. "Integrated Stratigraphic and Depositional-Facies Analysis of Parasequences in a Transgressive Systems Tract, San Joaquin Basin, California." *Amer. Assoc. Petrol. Geol. Mem.,* 58: 99–134.

Ulrich, E. O., 1916. "Correlation by Displacement of the Strandline and the Function and Proper Use of Fossils in Correlation." *Geol. Soc. Amer. Bull.,* 27: 451–490.

Vail, P. R., 1987. "Seismic Stratigraphic Interpretation Using Sequence Stratigraphy." In. A. W. Bally (ed.), *Atlas of Seismic Stratigraphy,* vol. 1. *Amer. Assoc. Petrol. Geol. Stud. Geol.,* 27:11–14.

Vail, P. R., F. Audemard, S.A. Bowna, D. N. Eisner, and G. Perez-Cruz, 1989. "Chronostratigraphy from Sequence Stratigraphy." *Réunion Annuelle Sci. Terre,* 13: 137.

Vail, P. R., J. Hardenbrol, and R. G. Todd, 1984. "Jurassic Unconformities, Chronostratigraphy and Sea Level Changes from Seismic Stratigraphy and Biostratigraphy." *Amer. Assoc. Petrol. Geol. Mem.,* 36: 129–144.

Vail, P. R., R. M. Mitchum, Jr., and S. Thompson, III, 1977a. "Global Cycles of Relative Changes in Sea Level." *Amer. Assoc. Petrol. Geol. Mem.,* 26: 83–98.

Vail, P. R., R. G. Todd, and J. B Sangree, 1977b. "Chronostratigraphic Significance of Seismic Reflections." *Amer. Assoc. Petrol. Geol. Mem.,* 26: 99–116.

Van Andel, T. H., G. R. Heath, and T. C. Moore, 1975. "Cenozoic History and Paleoceanography of the Central Equatorial Pacific." *Geol. Soc. Amer. Mem.,* 143: 1–134.

Van Houten, F. B., 1982. "Phanerozoic Oolitic Ironstones — Geologic Record and Facies Model." *Ann. Rev. Earth Planet. Sci.,* 10: 441–457.

Van Houten, F. B., and M. A. Arthur, 1989. "Temporal Patterns among Phanerozoic Oolitic Ironstones and Ocean Anoxia." In T. P. Young, and W. E. G. Taylor (eds.), *Phanerozoic Ironstones,* 33–49. Geol. Soc. Lond. Spec. Publ. 46.

Van Wagoner, J. C., R. M. Mitchum Jr., K. M. Campion, and V. D. Rahmanian, 1990. *Siliclastic Sequence Stratigraphy in Well Logs, Cores, and Outcrops: Concepts for High-Resolution Correlation of Time and Facies.* Amer. Assoc. Petrol. Geol. Methods in Exploration Series 7.

Van Wagoner, J. C., H. W. Posamentier, R. M. Mitchum, Jr., P. R. Vail, J. F. Sarg, T. S. Loutit, and J. Hardenbol, 1988. "An Overview of the Fundamentals of Sequence Stratigraphy and Key Definitions." *SEPM Spec. Publ.,* 42: 39–46.

Vidal, G., and V. Zoubek (eds.), 1981. "Biostratigraphic Schemes." *Precamb. Res.,* 15: 95–96.

Visher, G. S., 1969. "Grain Size Distribution and Depositional Processes." *J. Sed. Petrol.,* 39: 1074–1106.

Vogt, P. R., 1975. "Changes in Geomagnetic Reversal Frequency at Time of Tectonic Change: Evidence for Coupling between Core and Upper Mantle Processes." *Earth Planet. Sci. Lett.,* 25: 313–321.

Von Engelen, O. D., 1930. "Type Form of Faceted and Striated Glacial Pebbles." *Amer. J. Sci.,* 19: 9–16.

Walker, R. G., (ed.), 1984. *Facies Models,* 2d ed. Geoscience Canada Reprint Series 1. Geological Association of Canada, Toronto.

Walker, R. G., 1990. "Facies Modeling and Sequence Stratigraphy." *J. Sed. Petrol.,* 60: 777–786.

Walker, R. G., and F. J. Pettijohn, 1971. "Archean Sedimentation: An Analysis of the Minnetak Basin, Northwestern Ontario." *Geol. Soc. Am. Bull.,* 82: 2099–2130.

Walker, R. G., and N. P. James (eds.), 1992. *Facies Model: Response to Sea Level Change,* 3d ed. Geoscience Canada Reprint Series 1. Geological Association of Canada, Toronto.

Walker, T. R., and J. C. Harms, 1972. "Eolian Origin of Flagstone Beds, Lyons Sandstone (Permian), Type Area, Boulder County, Colorado." *Mountain Geol.,* 9: 279–288.

Walter, M., (ed.), 1976. *Stromatolites.* Elsevier, New York.

Walther, J., 1894. *Einleitung in die Geologie als historische Wissenschaft, Bd. 3, Lithogenesis der Gegenwart,* 534–1055. G. Fischer, Jena.

Wanless, H. R., and J. M. Weller, 1932. "Correlation and Extent of Pennsylvanian Cyclothems." *Geol. Soc. Amer. Bull.,* 43: 1003–1016.

Ward, C. R., (ed.), 1984. *Coal Geology and Coal Technology.* Blackwell Scientific Publications, Oxford.

Warme, J. E., R. G. Douglas, and E. L. Winterer (eds.), 1981. *The Deep-Sea Drilling Project: A Decade of Progress.* SEPM Spec. Publ. 32.

Warren, J., 1991. "Sulfate-Dominated Sea-Marginal and Platform Evaporative Settings." In J. L. Melvin (ed.), *Evaporites, Petroleum and Mineral Resource.* Blackwell Scientific Publications, Oxford.

Warren, J. K., 1989. *Evaporite Sedimentology.* Prentice-Hall, Englewood Cliffs, N.J.

Warren, J. K., 2000. "Dolomite: Occurrence, Evolution and Economically Important Associations." *Earth Sci. Rev.,* 52: 1–81.

Warren, J. K., and C. G. St. C. Kendall, 1985. "Comparison of Sequences Formed in Marine Sabkha (Subaerial) and Saline (Subaqueous) Settings—Modern and Ancient." *Amer. Assoc. Petrol. Geol. Bull.,* 69: 1013–1023.

Watts, A. B., 1982. "Tectonic Subsidence, Flexure, and Global Changes in Sea Level." *Nature,* 297: 469–474.

Watts, A. B., and J. Thorne, 1984. "Tectonics, Global Changes in Sea Level and Their Relationship to Stratigraphical Sequences in the U.S. Atlantic Continental Margin." *Mar. Petrol. Geol.,* 1: 319–339.

Weaver, C. E., 1989. *Clays, Muds, and Shales.* Developments in Sedimentology 44. Elsevier, Amsterdam.

Wedepohl, K. H., 1971. *Geochemistry.* Holt, Rinehart, and Winston, New York.

Weimer, P., and H. W. Posamentier, 1993. "Recent Developments and Applications in Siliciclastic Sequence Stratigraphy." *Amer. Assoc. Petrol. Geol. Mem.,* 58: 3–12.

Weimer, R. J., 1992. "Developments in Sequence Stratigraphy: Foreland and Cratonic Basins." *Amer. Assoc. Petrol. Geol. Bull.,* 76: 965–982.

Weller, J. M., 1947. "Relations of the Invertebrate Paleontologist to Geology." *J. Paleont.,* 21: 570–575.

Weller, J. M., 1958. "Stratigraphic Facies Differentiation and Nomenclature." *Amer. Assoc. Petrol. Geol. Bull.,* 42: 609–639.

Weller, J. M., 1960. *Stratigraphic Principles and Practice.* Harper and Brothers, New York.

Wetherill, G. W., 1971. Of Time and the Moon." *Science,* 173: 383–392.

Wheeler, H. E., 1964. "Baselevel, Lithospheric Surface, and Time-Stratigraphy." *Geol. Soc. Amer. Bull.,* 75: 599–610.

White, T. D., and J. M. Harris, 1977. "Suid Evolution and Correlation of African Hominid Localities." *Science,* 198(4312): 13–21.

Wilgus, C. K., B. S. Hastings, C. G. St. C. Kendall, H. W. Posamentier, C. A. Ross, and J. C. Van Wagoner (eds.), 1988. *Sea-Level Changes: An Integrated Approach.* SEPM Spec. Publ. 42.

Wilkinson, B. H., 1982. "Cyclic Cratonic Carbonates and Phanerozoic Calcite Seas." *J. Geol. Educ.,* 30: 189–203.

Wilkinson, B. H., J. M. Budai, and R. K. Give, 1984. "Episodic Accumulation and the Origin of Formation Boundaries in the Helderberg Group of New York State: Comment." *Geology,* 12: 572–573.

Williams, D. F., 1988. "Evidence for and against Sea-Level Changes from the Stable Isotope Record of the Cenozoic." *SEPM Spec. Publ.,* 42: 31–37.

Williams, E. C., J. C. Ferm, A. L. Guber, and R. E. Bergenback, 1964. *Cyclic Sedimentation in the Carboniferous of Western Pennsylvania.* Guidebook 29th Ann. Field Conf. Penn. Geol., Penn. State University, University Park, Penn.

Williams, H. S., 1901. "The Discrimination of Time Values in Geology." *J. Geol.,* 9: 570–585.

Wilson, J. C., and E. F. McBride, 1978. "Composition and Porosity Evolution of Pliocene Sandstones, Ventura Basin, California." *Amer. Assoc. Petrol. Geol. Bull.,* 72: 664–681.

Wilson, J. L., 1975. *Carbonate Facies in Geologic History.* Springer-Verlag, New York.

Wood, H. E., II, R. W. Chaney, J. Clark, E. H. Colbert, G. L. Jepsen, J. B. Reeside, and C. Stock, 1941. "Nomenclature and Correlation of the North American Continental Tertiary." *Geol. Soc. Amer. Bull.,* 52: 1–48.

Woodburne, M. O., 1977. "Definition and Characterization in Mammalian Chronostratigraphy." *J. Paleont.,* 51: 220–234.

Woodburne, M. O., (ed.), 1987. *Cenozoic Mammals of North America: Geochronology and Biostratigraphy.* University of California Press, Berkeley.

Woodburne, M. O., 1989. "Hipparion Horses: A Pattern of Worldwide Dispersal and Endemic Evolution." In D. R. Prothero, and R. M. Schoch (eds.), *The Evolution of Perissodactyls,* 197–233. Oxford University Press, Oxford.

Woodburne, M. O., (ed.), 2003. *Late Cretaceous and Cenozoic Mammals of North America.* Columbia University Press, New York.

Worsley, T. R., D. Nance, and J. B. Moody, 1984. "Global Tectonics and Eustasy for the Past 2 Billion Years." *Marine Geol.,* 58: 373–400.

Wright, V. P., 1986. *Paleosols: Their Recognition and Interpretation.* Princeton University Press, Princeton, N.J.

Wunderlich, F., 1970. "Genesis and Environment of the Nellenköpfchenschichten (Lower Emsian, Rheinian Devon) at Locus Typicus in Comparison with Modern Coastal Environment of the German Bay." *J. Sed. Petrol.,* 40: 102–130.

Yang, C. T., 1996. *Sediment Transport: Theory and Practice.* McGraw-Hill. New York.

Yang, C.-S., and S.-D. Nio, 1993. "Application of High-Resolution Sequence Stratigraphy to the Upper Rotliegend in the Netherlands Offshore." *Amer. Assoc. Petrol. Geol. Mem.,* 58: 285–316.

Yen, T. F., and G. V. Chilingarian (eds.), 1976. *Oil Shale.* Elsevier, New York.

York, D., and R. M. Farquhar, 1972. *The Earth's Age and Geochronology.* Pergamon Press, Oxford.

Young, T. P., and W. E. G. Taylor (eds.), 1989. *Phanerozoic Ironstones.* Geol. Soc. Lond. Spec. Publ. 46.

Zeller, E. J., 1964. "Cycles and Psychology." *Geol. Surv. Kansas Bull.,* 169: 631–636.

Zenger, D. H., J. B. Dunham, and R. L. Ethington (eds.), 1980. *Concepts and Models of Dolomitization.* SEPM Spec. Publ. 28.

Zingg, T., 1935. "Beiträge zur Schotteranalyse." *Schweiz. Mineralog. Petrog. Mitt.,* 15: 39–140.

Index

Note: Page numbers followed by f, t, and b indicate figures, tables, and boxed material.